International Scientific Conference Interstroymeh (ISM 2019)

IOP Conference Series: Materials Science and Engineering
Volume 786

Kazan, Russia
12 - 13 September 2019

ISBN: 978-1-7138-4717-5
ISSN: 1757-8981

Printed from e-media with permission by:

Curran Associates, Inc.
57 Morehouse Lane
Red Hook, NY 12571

CURRAN ASSOCIATES INC.
proceedings
.com

Some format issues inherent in the e-media version may also appear in this print version.

Institute of Physics
Dirac House, Temple Back
Bristol BS1 6BE UK

Phone: 44 1 17 929 7481
Fax: 44 1 17 920 0979

techtracking@iop.org

Additional copies of this publication are available from:

Curran Associates, Inc.
57 Morehouse Lane
Red Hook, NY 12571 USA
Phone: 845-758-0400
Fax: 845-758-2633
Email: curran@proceedings.com
Web: www.proceedings.com

TABLE OF CONTENTS

Editorial Preface

The annual International Science and Technology Conference Interstroymeh-2019 was held at Kazan State University of Architecture and Engineering in Kazan, Russia, September 12-13, 2019. The host university is one of the country's leading universities conducting construction research.

The conference is dedicated to discussing contemporary achievements of scientists, opportunities and challenges to implementing developments aimed to improve energy efficiency, reliability of different-purpose machines, as well as smart urban mobility: prospects for using modern information and transport technologies to improve the safety and quality of transport services.

More than one hundred people from ten countries took part in the conference.

We express our gratitude to the conference participants and wish them new insights and discoveries.

Conference Chairs.

Peer review statement

All papers published in this volume of *IOP Conference Series: Materials Science and Engineering* have been peer reviewed through processes administered by the proceedings Editors. Reviews were conducted by expert referees to the professional and scientific standards expected of a proceedings journal published by IOP Publishing.

Optimization of the internal damping rubber elements of the small diameter roller wheel of tracked system

S. A. Korostelev[1] and A. F. Verbilov[2,*]

[1]Department of land transport-technological systems, Polzunov Altai State Technical University, Barnaul, Altai Region, Russia

[2]Firearms and Technical Training Department, Barnaul Law Institute of the Ministry of Internal Affairs of Russia, Barnaul, Altai Region, Russia

* E-mail: korsan73@mail.ru

Abstract. The article presents the formulation and solution of the problem of optimization of the internal damping rubber elements of the small diameter roller wheel of tracked system. The search for the optimal design is based on the results of calculating the stress-strain state of the rubber element caused by the assembly of the roller and loads during subsequent operation. As an objective function, the specific energy of deformation of the rubber was chosen. The geometric parameters of the rubber element of the roller wheel in the initial state and after installation in the structure are taken as variable parameters. Explicit and functional constraints in the form of inequations are imposed on the variable parameters. The results of solving the optimal design problem are presented.

1. Introduction

One of the heavily loaded elements of the tracked system is the roller wheel (or basic running roller), which is subjected to significant dynamic loads during its rolls along the contact surface of the track chain [1–3]. Dynamic loads cause increased dynamic stresses in the contact between the bandage of the roller wheel and the track treadmill, as well as high volumetric stresses that act in the links and pins of the track chain [1, 4 - 6]. High contact stresses contribute to intensive wear of the metal bandage of the roller wheel and the link treadmill. High volumetric dynamic stresses reduce the fatigue strength of links and pins and, together with wear, are the main factors that limit the life of a tracked system.

To reduce dynamic loads, power internal rubber and rubber-metal shock absorbers are installed in the roller wheels of the tracked system [7]. Such elements with a relatively small specific gravity allow providing a reduction in dynamic loads and increasing the durability of the elements of the caterpillar mover.

When designing a caterpillar mover with power rubber-metal elements, it is necessary to take into account the conflicting requirements [6 - 9] imposed on its structure. On the one hand, high requirements are made on the strength and stiffness of the structural elements of the tracked mover, providing a margin in connection with increased wear. On the other hand, it is necessary to create a structure with a minimum mass. When solving such problems, it is necessary to apply optimal design methods.

In this paper, the formulation and solution of the problem of optimization of the internal damping rubber elements of the roller wheel of tracked systems are presented. Also proposals aimed at improving the structures to increase their durability were made.

2. Features of the use of rubber elements in the constructions of roller wheels

The variety in the geometry of the rubber elements of the roller wheels determines the difference of their stiffness and damping characteristics. The degree of pressing, radial and axial stiffness during secondary loading of rubber elements of the same overall size can differ from each other by several times. This, in turn, determines a wide range of possibilities for their application, allowing for each specific structure of the roller wheel to use the most rational geometric shape of the rubber elements. An optimal combination of rubber and metal elements and a reasonable choice of their structural parameters allow getting the stiffness characteristics which are necessary to reduce dynamic loads.

During operation, rubber elements are subjected to the following types of deformation: deformations associated with the installation and assembly of rubber elements in a roller wheel (rubber elements may be subjected to compression along the end surface), pressing along the inner radius, pressing on the outside radius [7]. Rubber elements can be made in the form of ordinary rubber elements or in the form of rubber-metal, i.e. having metal fittings. When moving along the treadmill of the track, the roller wheel is subjected to shock and dynamic effects in the radial and axial directions associated with the discreteness of the chain design, with vertical vibrations of the tracked vehicle body, with hook load, with speed and uneven road surface.

When assembling, rubber elements undergo large deformations, while the initial shape and geometry of the element changes significantly. As a result, significant stress and strain concentrators appear. As theoretical and experimental studies show, the decisive role in fatigue failure is played by concentrators arising from deformation during secondary loading of rubber elements. The location and the size of the concentrator depend on the initial geometry of the rubber element, the assembly conditions and the loads arising during operation.

The combination of constant and variable mechanical loads as a result of assembly and subsequent radial and axial loading increases cyclic strain and stress in the concentration aria and causes fatigue failure of rubber. Moreover, the imposition of secondary inelastic deformations of radial and axial loading on the equilibrium deformations of the assembly significantly changes both the amplitude and average components of local stresses and deformations.

The cyclic nature of the modes of operation of the internal damping elements, viscoelastic properties of rubber cause heat generation in the rubber elements. The existing uneven distribution of the temperature field along the length and thickness of the rubber rings is a source of temperature stresses with different gradients within the considered section.

3. Example of the stress-strain state assessment of the internal damping rubber element of the roller wheel with a small diameter

As an example, the design of the roller wheel with internal damping ring (Figure 1), the prototype of which was subjected to field tests, is presented.

IOP Conf. Series: Materials Science and Engineering **786** (2020) 012001 doi:10.1088/1757-899X/786/1/012001

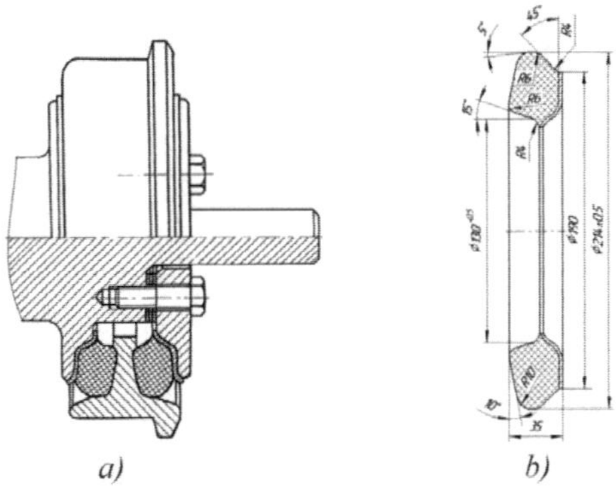

Figure 1. The roller wheel with a small diameter (a) and the design of the rubber element of the internal damping. (b)

The calculation of the stress-strain state of the rubber structural elements shown in Figure 1 included two stages. At the first stage of the calculation, the stress-strain state of the rubber element is due to the assembly of the support roller, that is, under pressing into the outer metal band and subsequent compression in the axial direction. At the second stage, the calculation was performed taking into account the loading by radial force.

For the calculation, a software package was used, the feature of which is the ability to automatically take into account changes in the boundary conditions in the process of deformation. As a result of solving the task of assembly and secondary loading with radial force, the following calculated data were obtained: patterns of displacement of nodes of the finite element model, distribution of stress tensor components over the rubber elements section, distribution of the specific potential strain energy over the section (Figure 2). The figures show a cross section of a rubber element located in the plane of action of the radial force in the compression aria.

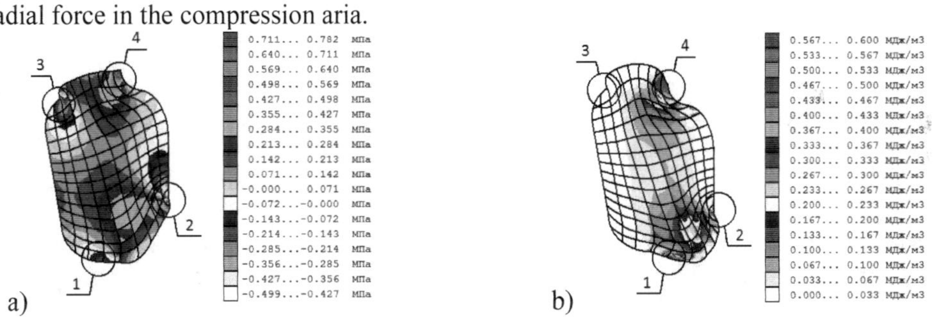

Figure 2. Tangential stresses after assembly (a), specific strain energy (b) caused by radial force (axial compression 3 mm, radial displacement 6 mm).

For the considered construction of the roller wheel, the concentration of tangential stresses and the specific strain energy during assembly (axial compression) is observed in zones 1 - 4. In those zones the rubber elements either come into contact with the sharp edges of the metal parts or are the boundary contact areas between rubber and metal. During secondary loading by radial force the concentration of the specific strain energy is observed in areas 1, 2, and 4. Area 3 is unloaded. Under the action of radial force, the rubber is squeezed out between the metal parts of the rim (zone 1) and a flange rigidly connected to the hub (zone 2).

Field tests of the considered experimental support roller showed that the durability of rubber elements is 2000 hours. The first fatigue cracks were observed in the areas of concentration of specific strain energy (Figure 2, b). Subsequent operation was accompanied by delamination of the rubber from the metal base in zones 2 and 4, an increase in temperature, and thermomechanical destruction of the rubber elements along the line connecting zones 1 and 4.

The considered internal damping rubber elements of have an irrational design, since during assembly and secondary loading with radial force there is an uneven distribution of stresses and specific strain energy over the cross section of the rubber element. The distribution of these parameters has distinct concentration areas.

4. Proposal approach of optimal design of rubber elements of roller wheels to increase their durability

A reserve for increasing the operability of the construction of small diameter roller wheels with internal rubber damping element is the use of deformation limiters.

When designing the elements of the internal damping of the roller wheel of the caterpillar mover it is necessary [6, 7]:

-to provide the possibility of placing rubber elements in the construction of the roller wheel;

-to ensure the absence of an influx of rubber on the metal elements of the roller wheel;

-to eliminate the possibility of penetration of the abrasive between the metal and the rubber element;

-to ensure that the tangential stresses caused by the assembly do not exceed the permissible values;

-to provide characteristics of radial and axial stiffness of the rubber elements of the roller wheel, allowing to reduce dynamic loads;

-the temperature of the rubber elements must not exceed the permissible values;

-to provide for the design of the roller wheel to limit the radial deformation of the rubber elements;

-to minimize the mass of the unsprung part of the roller wheel, i.e. weight of metal bandage.

To assess the fatigue endurance of a rubber element under a complex mechanical loading regime (simultaneous action of dynamic and static stress), equations based on the energy criterion are used [10, 11]. The equation for describing the dependence of the fatigue endurance of rubber on the strain energy has the following form [10]:

$$N = \left(W_P k_T / W\right)^{n_W}, \qquad (1)$$

where N is the number of loading cycles before the crack; n_W - the coefficient of fatigue endurance of rubber; W_p is the fracture work under single loading; W is the strain energy under cyclic loading; k_T - the coefficient of the change of strength from temperature.

In the case of a complex stress-strain state, the strain energy is determined using the Treolar potential [10], which is a function of the invariants (or components) of the tensor of the measure of finite strains and depends on the coordinates. Estimation of the highest value of the specific strain energy is associated with the determination of finite strains, which are a solution to the boundary value problem with a system of nonlinear partial differential equations. To solve this problem, an algorithm based on a step-by-step procedure is used. The numerical implementation of the algorithm is performed using the finite element method [12]. In addition to assessing the stress-strain state, the results obtained are used to determine the stiffness characteristics of rubber elements.

Thus, in the problems of optimal design of the internal damping rubber elements of the roller wheels of caterpillar systems, the objective function and functional constraints imposed on the variational parameters are not set analytically, but in the form of an algorithm. To solve such problems, optimization methods related to direct search methods are used [13-15]. In this paper, to solve optimization problems, the complex Box method is used [13, 14]. This approach is a modification of the Nelder – Mead simplex method [15], which allows one to take into account the constraints on various parameters, both explicitly and functional limitations.

The formulation of the problem of optimal design of the internal damping rubber element on the example of the small diameter roller wheel (Figure 3) will be considered. As the objective function, the maximum value of the specific strain energy during secondary loading with radial force will be chosen.

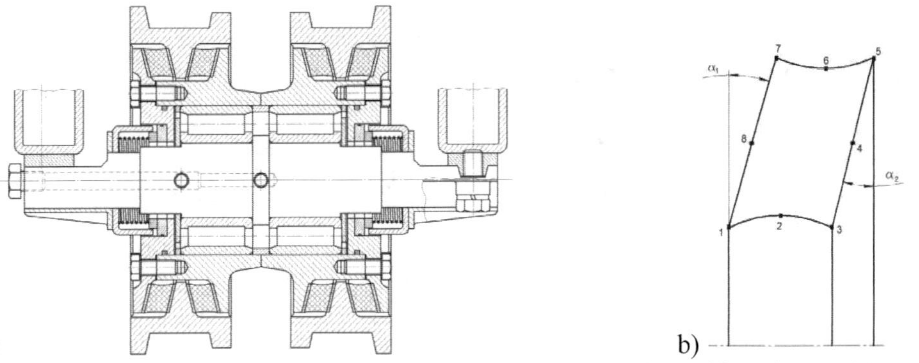

a) b)

Figure 3. Small diameter roller wheel with an internal damping rubber element (a), cross section of a rubber element with locations of variational parameters (b).

Variational parameters x_j are: the coordinates of the nodal points of the cross section, which determine the shape of the rubber element and geometric dimensions prior to assembly in the initial undeformed state (Figure 3, b); the value of compression acting on the rubber elements in the axial direction after assembling the roller.

The mathematical formulation of the optimization problem has the form: to find a vector $\vec{x}$ that minimizes the objective function

$$Z = \min_{\Gamma}\left[\ \max_{\Omega}(W)\ \right], \tag{2}$$

where Ω is the cross section zone of the of the element with a variational boundary Γ.

Constraints on variational parameters (geometric) are set by the system of inequations:

$$x_j^- \le x_j \le x_j^+,\ (\text{j=1, 2,…,m}), \tag{3}$$

where x_j^-, x_j^+ are the limits of variation of the components of the vector $\vec{x}$; m – the number of independent parameters.

Functional constraints are set by the system of inequations:

$$F_z - [F_z] > 0;$$

$$\max_{\Omega}\left(|\tau_{rz}|\right) - [\tau_{rz}] < 0;$$

$$K_{r\min} \le K_r \le K_{r\max};$$

$$K_{Z\min} \le K_Z \le K_{Z\max}, \tag{4}$$

where F_z - the axial force providing the necessary friction force; τ_{rz} - the tangential stresses due to assembly; $K_{r\min}$, $K_{r\max}$ - the lower and upper limits of radial stiffness; $K_{Z\min}$, $K_{Z\max}$ - the lower and upper limits of axial stiffness.

5. Calculation results

As a result of solving the optimal design problem, a cross-sectional shape of the rubber element of the internal depreciation of the track roller is obtained (Figure 4, a). The tangential stresses caused by the deformation of the rubber elements in the axial direction during assembly of the roller (Figure 4, b) are less than the permissible values. The objective function - the specific strain energy caused by the secondary loading of rubber elements by radial force (Figure 4, c) - decreases at least four times in comparison with the previously considered design (Figures 1, 2) with a radial displacement (6 mm) and 12 times with an additional limiter of the radial deformation of rubber elements due to structural measures.

IOP Conf. Series: Materials Science and Engineering **786** (2020) 012001 doi:10.1088/1757-899X/786/1/012001

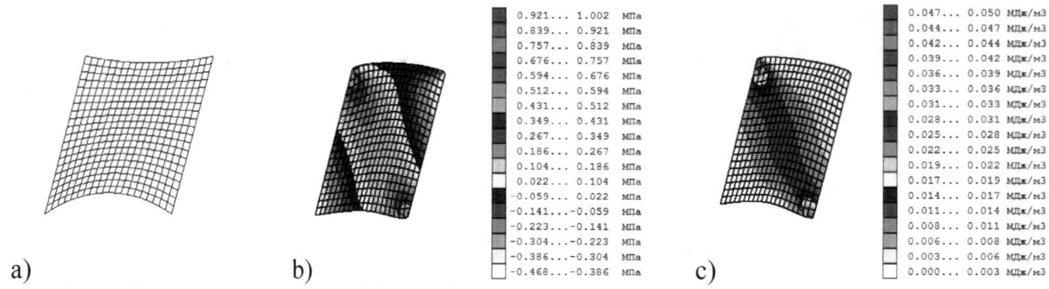

Figure 4. The cross section of the rubber element (a), tangential stresses after assembly (b), specific strain energy caused by radial force (c).

6. Conclusions

As a result of solving the optimal design problem, the shape and design parameters of the internal damping rubber element of the roller wheel of the tracked system were calculated. which made it possible to reduce the specific strain energy under radial loading to 0.05 MJ/m^3. This value is 4 times less than the same parameter of the existing roller when exposed to radial forces of the same magnitude. In this case, the radial stiffness of the rubber elements of the roller wheels of the optimal and existing design is the same.

References

[1] Campanelli M, Shabana A A and Choi J H 1998 Chain Vibration and Dynamic Stress in Three-Dimensional Multibody Tracked *Vehicles Multibody System Dynamics* **2(3)** 277-316

[2] Wong J Y 1997 Dynamics of Tracked Vehicles *Vehicle System Dynamics* **28(2-3)** 197-219

[3] Shabana A A 2010 Computational Dynamics (New York: John Wiley & Sons Ltd) 528

[4] Choi J H, Lee H C and Shabana A A 1998 *Spatial Dynamics of Multibody Tracked Vehicles Part I*: Spatial Equations of Motion Vehicle System Dynamics **29(1)** 27-49

[5] Lee H C, Choi J H and Shabana A A 1998 *Spatial Dynamics of Multibody Tracked Vehicles Part II*: Contact Forces and Simulation Results Vehicle System Dynamics **29(2)**113-7

[6] Korostelev S A, Verbilov A F and Kovalev V V 2013 Influence of elastic characteristics of the road wheels of tracked system on their dynamic *loading Izvestia of the Samara Research Center of the Russian Academy of Sciences* **15(2)** 515-8

[7] Korostelev S A 2015 Definition the stress-strain state of rubber elements of internal shock absorbers of the road wheels of caterpillar tracks *Izvestia of the Samara Research Center of the Russian Academy of Sciences* **17(4)** 793-8

[8] Korostelev S A, Verbilov A F and Kovalev V V 2012 Theoretical study of dynamic load of rubber-metal joints of the tracked systems with radial deformation limiters *Izvestia of the Samara Research Center of the Russian Academy of Sciences* **14(2)** 381-3

[9] Korostelev S A 2010 Assessment of the tense-deformed state of the rubber element of the rubber-metal joints of the tracked systems in assembly and torsion Tractors and agricultural machines 11 26-9

[10] Hromov M K 1983 On pattern of variations of the in fatigue endurance of rubber Kauchuk I Rezina 6 29-38

[11] Zhang J, Xue F, Wang Y, Zhang X, Han S 2018 Strain energy-based rubber fatigue life prediction under the influence of temperature *Royal Society Open Science* **5(10)** 180951

[12] Zhao J, Li Q, Shen X 2008 Finite Element Analysis and Structure Optimization for Improving the Fatigue Life of Rubber Mounts Journal of Macromolecular Science Part A: Pure and Applied Chemistry 6 479-84

[13] Bunday B 1984 Basic Optimization Methods (London: Edward Arnold , Bedford Square)

[14] Box M J 1965 A New Method of Constrained Optimization and a Comparision with Other Methods Computer Journal 8 42-52
[15] Nelder J A, Mead R 1965 A Simplex Method for Function Minimization Computer Journal (7): 308-313

Influence of friction forces in limiter-eyelet contact on twisting angles of the rubber elements of tracked system hinges

S A Korostelev[1], A V Gorbachev[1], A F Verbilov[2] and V V Kovalev[2]

[1] Department of land transport-technological systems, Polzunov Altai State Technical University, Barnaul, Altai Region, Russia

[2] Firearms and Technical Training Department, Barnaul Law Institute of the Ministry of Internal Affairs of Russia, Barnaul, Altai Region, Russia

Abstract. The article presents a study of the influence of the radial deformation limiters of rubber-metal compounds of tracked systems on the loading of rubber elements. Description of the mathematical model of the dynamic behaviour of a tracked propulsion unit with rubber-metal hinges is presented. Dynamic calculations using the software package based on the equations were performed. The analysis of the results of the computer calculation revealed the influence of friction forces and moments in the contact zone of the limiter-eyelet.

1. Introduction

The tracked system is widely used in construction and road vehicles. Elements of the tracked system are exposed to high dynamic loads affecting strength, reliability and durability. Reducing dynamic loads of the tracked systems is an important engineering task. One of the effective ways in this issue is the use of rubber elements in the mechanism of tracked system. Power rubber elements are applied in the constructions of the return and idler rollers, road wheels, suspension elements as well as in the tracked chain.

During the designing of the tracked systems it is very important to assure high requirements on metal intensity, durability, stiffness and strength of all the elements. One of the elements limiting the resource of the caterpillar engine is the connection of adjacent links. Structurally connecting links can be performed in a variety of ways, but most often used connections in the form of hinges. There are open and closed hinges, which can include rubber elements that act as an elastic bond or just seal the connection. Used rubber metal joint (RMJ) with force rubber elements have a high enough carrying capacity and good compensatory properties, high malleability, increased damping. The main drawback of the structures of the high-speed vehicles with the RMJ is that the rubber elements fully perceive the forces and moments acting on the hinge, which leads to large deformations of rubber and rapid destruction of rubber elements.

2. The rubber-metal hinge compound with rigid limiters of radial deformations

To increase the durability of the rubber elements of the connections of tracked chain the construction of rubber metal joint with a rigid limiter of radial deformation of rubber elements are used. When designing a sequentially rubber-metal hinge (Figure 1) it is necessary to ensure equal angular stiffness of rubber elements, located in the eyelets of the connected links. This condition ensures an equal twisting angle for all rubber elements and the same shearing stresses.

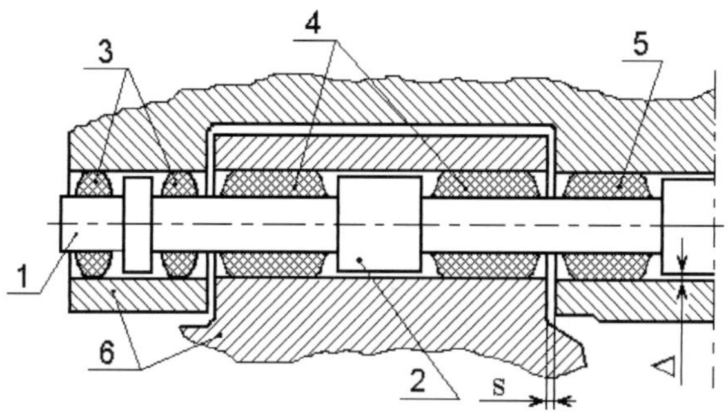

Figure 1. Rubber-metal hinge compound of track chain links.

1 - metal pin fitting; 2 - limiter of radial deformation; 3 - rubber elements of the lateral triple eyelets; 4 - rubber elements of double eyelets; 5 - rubber elements of the central eyelet; 6 - adjacent links.

When under the influence of a stretching force in the chain the magnitude of the radial deformation of rubber elements becomes equal to the magnitude of the radial gap between the limiter and the eyelet, the limiter comes into contact with the surface of the eyelet, with a mutual angular displacement of the link and pin. At this moment the friction force is caused, which is proportional to the force of normal pressure in the contact area (as shown in figure 2).

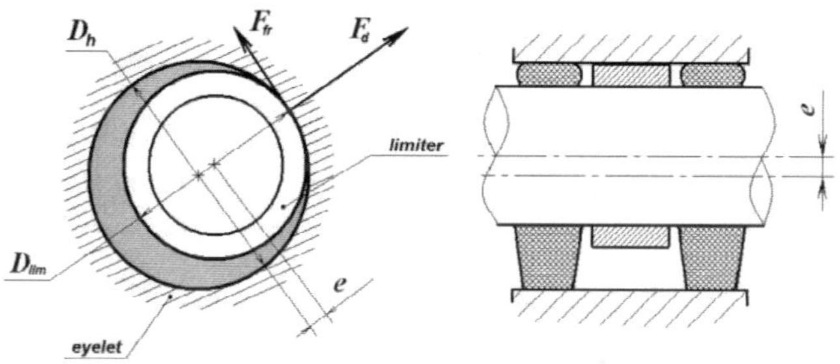

Figure 2. Interaction between the track and the radial deformation limiter.

Therefore, additional force factors are the friction forces and their moments, which additionally load adjacent elements [3]. The total radial and angular stiffness of the rubber elements of the adjacent links of the hinge are always differ. For example, for existing RMJ with 5 eyelets to ensure equal total angular stiffness of elements located in double and triple eyelets, the total radial rigidity of rubber elements of double eyelets will always be higher than triples [2]. Therefore, limiters of triple eyelets will always come into contact earlier, and the friction force and friction torque will also always be higher at the contact of the limiters of triple eyelets. Therefore, one of the tasks that must be solved when determining the loads acting on the rubber elements of the hinge connection is to evaluate the influence of the friction moment in the contact of the limiter-eyelet on the twisting angle of the rubber elements of the adjacent links [3].

3. The mathematical model of the dynamic behavior of a tracked propulsion unit with rubber-metal hinges

To assess the influence of friction forces in the contact of the limiter-eyelet, the methods to study the dynamics of solid systems are used [4 - 7].

The exact mathematical description of the phenomena occurring in a tracked vehicle is associated with great difficulties. The complication of mathematical models for describing the mechanical behavior of a tracked system requires a considerable amount of time when solving on a computer [8-10]. In this paper, a number of assumptions are made that simplify the mathematical model of the dynamic behavior of the elements of a tracked propulsor [10 - 12].

-a mechanical system consisting of a body of the tracked vehicle, a sprocket wheel, a guide wheel and elements of the depreciation-tensioning device, suspension arms, road rollers, tracks makes a flat movement;

-links, road rollers, levers, wheels are absolutely rigid, non-deformable elements;

-connections between the elements of the tracked system are realized in the form of elastic, viscoelastic connections or absolutely rigid contact;

-dissipative forces in frictional bonds are dismissively small;

-elastic-plastic connections between the links of the tracked system and the ground are realized.

The motion of a mechanical system with holonomic constraints is described by the Lagrange equations, which are second-order differential equations for generalized coordinates

$$\frac{d}{dt}\left(\frac{\partial T}{\partial \dot{q}_i}\right)-\frac{\partial T}{\partial q_i}=Q_i+\sum \lambda_k \frac{\partial \Phi_k}{\partial q_i}. \quad (1)$$

The kinetic energy of an individual element is determined by the equation

$$T=\frac{1}{2}m_i\left(\dot{x}_i^2+\dot{y}_i^2\right)+\frac{1}{2}\cdot J_i\dot{\varphi}_i^2 \quad (2)$$

The generalized forces acting on the ***i***-element caused by the force connection with the ***j***-element

$$Q_i^x=F_{ij}^x;$$

$$Q_i^y=F_{ij}^y;$$

$$Q_i^\varphi=F_{ij}^x\left(-\xi_{ij}\sin\varphi_i-\eta_{ij}\cos\varphi_i\right)+F_{ij}^y\left(\xi_{ij}\cos\varphi_i-\eta_{ij}\sin\varphi_i\right). \quad (3)$$

The result is a system of differential algebraic equations connecting the vector of unknown displacements of the mechanical system

$$\begin{aligned}&[M]\{\ddot{q}\}+[C]\{\dot{q}\}+[K]\{q\}+[\Phi_q]^T\{\lambda\}=\{P(t)\}\\&\{\Phi(q,t)\}=\{0\}\end{aligned}, \quad (4)$$

where $[M]$ is the global mass matrix of the system; $[K]$ is the global stiffness matrix of the system; $[C]$ is the global damping matrix of the system; $\{q\}$ is the vector of generalized coordinates; $\{\dot{q}\}$is the vector of generalized velocities; $\{\ddot{q}\}$is the vector of generalized accelerations; $\{\lambda\}$ - vector of Lagrange multipliers; $\{P(t)\}$ is the vector of external forces;

$$[\Phi_q]=\begin{bmatrix}\frac{\partial \Phi_1}{\partial q_1} & \frac{\partial \Phi_1}{\partial q_2} & \cdots & \frac{\partial \Phi_1}{\partial q_n}\\ \frac{\partial \Phi_2}{\partial q_1} & \frac{\partial \Phi_2}{\partial q_2} & \cdots & \frac{\partial \Phi_2}{\partial q_n}\\ \vdots & \vdots & \ddots & \vdots\\ \frac{\partial \Phi_m}{\partial q_1} & \frac{\partial \Phi_m}{\partial q_2} & \cdots & \frac{\partial \Phi_m}{\partial q_n}\end{bmatrix}.$$

The system of nonlinear differential-algebraic equations should be supplemented with initial and boundary conditions [4, 5]. That system of equations (4) includes *3N* nonlinear differential equations of the second order (*N* is the number of elements of the mechanical system) and m kinematic connection equations, which are nonlinear algebraic equations. System (4) has a differentiation index of three [14]. To solve the system it is necessary to downgrade the index. By double-differentiation of the equations of the system (4) it is possible to transform it to the following form [5-7,10,11]

$$\begin{bmatrix} [M] & [\Phi_q]^T \\ [\Phi_q] & 0 \end{bmatrix} \begin{Bmatrix} \{\ddot{q}\} \\ \{\lambda\} \end{Bmatrix} = \begin{Bmatrix} \{P\} \\ \{\gamma\} \end{Bmatrix}, \tag{5}$$

where

$$\{P\} = \{P(t)\} - [C]\{\dot{q}\} - [K]\{q\};$$

$$\{\gamma\} = -\{\Phi_{tt}\} - \left([\Phi_q]\{\dot{q}\}\right)_q \{\dot{q}\} - 2[\Phi_{qt}]\{\dot{q}\}.$$

In expression (5) the lower indexes denote the derivative of the function of the kinematic constraints by the generalized coordinates and time.

To solve the system of differential-algebraic equations (5) numerical methods are used [15, 16]. In order to stabilize the numerical algorithm when solving the system of equations (5) the vector $\{\gamma\}$ is replaced by the expression [4-6]

$$\{\gamma^*\} = \{\gamma\} - 2\alpha[\Phi_q]\{\dot{q}\} - (\beta)^2[\Phi_q]\{q\}, \tag{6}$$

where $\alpha > \beta$ and $\beta \neq 0$.

Large displacements of the elements of tracked systems from the initial state, the presence of kinematic equations of links, nonlinear characteristics of stiffness and damping of viscoelastic bonds, which are nonlinear functions of deformation and deformation rate, complicate the solution of the system and impose restrictions on the numerical methods used to solve it.

In the present work, as one of the stages of increasing the stability of the numerical solution of system (5), an algorithm of the linearization of the equations of motion and expressions describing holonomic kinematic constraints is used.

After linearization, the system of differential algebraic equations (5) takes the form as

$$\begin{bmatrix} [M] & [\overline{\Phi}_{\Delta q}]^T \\ [\overline{\Phi}_{\Delta q}] & 0 \end{bmatrix} \begin{Bmatrix} \{\Delta\ddot{q}\} \\ \{\Delta\lambda\} \end{Bmatrix} = \begin{Bmatrix} \{\overline{P}\} \\ \{\overline{\gamma}^*\} \end{Bmatrix}, \tag{7}$$

where

$$\{\overline{P}\} = \{\overline{P}(t)\} - [\overline{C}]\{\Delta\dot{q}\} - [\overline{K}]\{\Delta q\}.$$

From system (7), a following expression for determining the increment of the Lagrange multipliers at each step can be obtained [17]

$$\{\Delta\lambda\} = \left([\overline{\Phi}_{\Delta q}][M]^{-1}[\overline{\Phi}_{\Delta q}]^T\right)^{-1}\left([\overline{\Phi}_{\Delta q}][M]^{-1}\{\overline{P}\} - \{\overline{\gamma}^*\}\right). \tag{8}$$

The resulting expression is substituted into the first block of equations of system (7)

$$[M]\{\Delta\ddot{q}\} = \{\overline{P}\} - [\overline{\Phi}_{\Delta q}]^T\left([\overline{\Phi}_{\Delta q}][M]^{-1}[\overline{\Phi}_{\Delta q}]^T\right)^{-1}\left([\overline{\Phi}_{\Delta q}][M]^{-1}\{\overline{P}\} - \{\overline{\gamma}^*\}\right). \tag{9}$$

To solve the system of differential-algebraic equations, finite-difference formulas of the Newmark β-scheme are used [17]

$$\{\Delta\ddot{q}_{t+\Delta t}\}=\frac{1}{\beta\Delta t^2}(\{\Delta q_{t+\Delta t}\}-\{\Delta q_t\})-\frac{1}{\beta\Delta t}\{\Delta\dot{q}_t\}-\left(\frac{1}{2\beta}-1\right)\{\Delta\ddot{q}_t\};$$

$$\{\Delta\dot{q}_{t+\Delta t}\}=\frac{\alpha}{\beta\Delta t}(\{\Delta q_{t+\Delta t}\}-\{\Delta q_t\})-\left(\frac{\alpha}{\beta}-1\right)\{\Delta\dot{q}_t\}-\Delta t\left(\frac{\alpha}{2\beta}-1\right)\{\Delta\ddot{q}_t\}, \quad (10)$$

where α, β are the parameters of the method; Δt - the length of the time interval.

Substituting expressions (10) into the system of linearized differential-algebraic equations (9), the initial system is transformed into a system of linear algebraic equations, for the solution of which direct methods are used. Using the values of displacements, as a result of solving this system, we determine the speed and acceleration using expressions (10). Then, using the expression (8), it is possible to determine the increments of the Lagrange multipliers.

Sequential step-by-step solution of the system of equations (7) allows to determine the position of the elements of a tracked propulsor at a specific point in time as well as displacements, speeds and accelerations.

4. The simulation analysis

The above equations are the basis of the software package for calculating the dynamic loads acting on the rubber elements of the track-type propulsion unit. The software package allows determining the displacements, speeds and accelerations of the elements of the tracked systems, which are used to determine the forces acting on the rubber elements.

In Figure 3 changes in the twisting angle of the rubber element when moving the hinge around the tracked propulsor's perimeter are presented. The effect of the friction torque caused by the contact of the limiter and the eyelet is not taken into account. The results were obtained for a tracked tractor of class 3 when driving in first gear with a hook load. It can be seen from the figure that the rubber elements of the double eyelet both on the driving wheel and on the guides are twisted at a greater angle. In this case, the maximum value of the twisting angle does not exceed 6.0°.

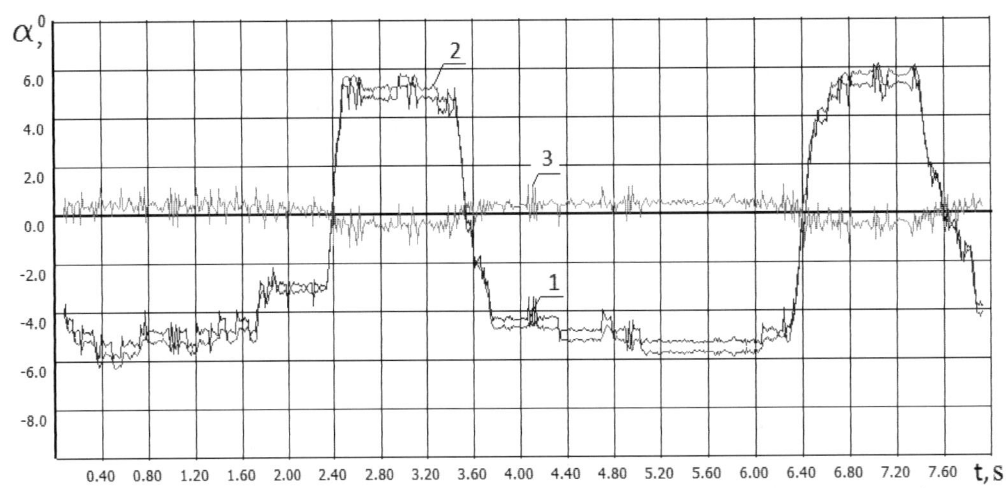

IOP Conf. Series: Materials Science and Engineering **786** (2020) 012002 doi:10.1088/1757-899X/786/1/012002

Figure 3. Twisting angle of the rubber elements of the RMJ (the tractor's speed of movement is 0.93 m/s, without the influence of the limiter): 1 - triple eyelets; 2 - double eyelets; 3 - the difference of the twist angles of the rubber elements of the triple and double eyelets.

In Figure 4 the change in the twist angle of rubber elements, taking into account the influence of friction in the contact limiter-eyelet is presented. The twisting angle of the rubber elements of the double eyelets when twisting the hinge after the engagement is 1.0° more than the rubber elements of the triple eyelets. It should be noted that in this area rubber elements are subject to maximum radial deformation, the value of which is limited by the radial clearance. The combined effect of radial deformation and twisting leads to greater damage to the rubber elements of the double eyelets in comparison with triple rubber elements [1, 2].

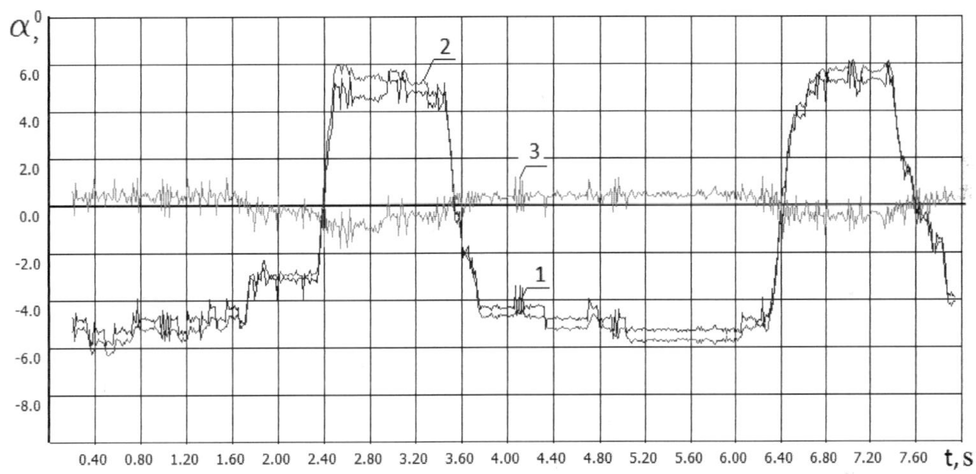

Figure 4. Twisting angle of the rubber elements of the RMJ (the tractor's speed of movement is 0.93 m/s, taking into account the the influence of the limiter): 1 - triple eyelets; 2 - double eyelets; 3 - the difference of the twist angles of the rubber elements of the triple and double eyelets

5. Conclusion

Friction forces in the contact area of the limiter-eyelet increase the twisting angle of the rubber elements of double eyelets on the leading section by 0.5°. For the RMJ with radial deformation limiters considered here rubber elements of the double eyelets on the leading section are twisted by 1.0° more than the rubber elements of triple eyelets. To create a uniform structure with an equal strength and increase the durability of the hinged connection of the links of a tracked propulsion units, the angular stiffness of the rubber elements located in the double eyelets must be greater than the angular stiffness of the rubber elements of the triple eyelets.

References

[1] Korostelev S A 2010 Assessment of the tense-deformed state of the rubber element of the rubber-metal joints of the tracked systems in assembly and torsion Tractors and agricultural machines (11) 26-9

[2] Korostelev S A 2012 Effect of the radial and axial deformation of the rubber element of the rubber-metal joint of the tracked systems on its stressed-strained state Izvestia of the Samara Research *Center of the Russian Academy of Sciences* **14(2)** 378-80

[3] Korostelev S A, Verbilov A F and Kovalev V V 2012 Theoretical study of dynamic load of rubber-metal joints of the tracked systems with radial deformation limiters Izvestia of the Samara Research *Center of the Russian Academy of Sciences* **14(2)** 381-3

[4] Shabana A A 2005 Dynamics of Multibody Systems (Cambridge: Cambridge University Press) 374

[5] Shabana A A 2010 Computational Dynamics (New York: John Wiley & Sons Ltd) 528
[6] Flores P, Koshy C S, Lankarani H M et al 2011 Numerical and experimental investigation on multibody systems with revolute clearance joints *Nonlinear Dynamics* **65(4)** 383-98
[7] Flores P 2010 A parametric study on the dynamic response of planar multibody systems with multiple clearance joints *Nonlinear Dynamics* **61(4)** 633-53
[8] Wallin M, Aboubakr A K, Jayakumar P et al 2013 A comparative study of joint formulations: application to multibody system tracked vehicles *Nonlinear Dynamics* **74(3)** 783-800
[9] Choi J H, Lee H C and Shabana A A 1998 Spatial Dynamics of Multibody Tracked Vehicles Part I: *Spatial Equations of Motion Vehicle System Dynamics* **29(1)** 27-49.
[10] Lee H C, Choi J H and Shabana A A 1998 Spatial Dynamics of Multibody Tracked Vehicles Part II: *Contact Forces and Simulation Results Vehicle System Dynamics* **29(2)** 113-7
[11] Beck R R and Wehage R A 1979 The Modeling and Simulation of Two Coupled M-113 Armored Personnel Carriers Proceedings of the Tenth Annual Pittsburgh Conference on *Modeling and Simulation* **10(2)** 353-9
[12] Haug E J, Wehage R A, Barman N C 1981 Design Sensitivity Analysis of Planar Mechanism and Machine Dynamics Journal of Mechanical Design (103) 561-70
[13] Choi J H, Campanelli M, Shabana A A and Wehage R A 1998 Approximation Methods in the Nonlinear Analysis of Planar Tracked Vehicles *Vehicle System Dynamics* **29(3)** 181-211
[14] Harrier E 1999 Solution of ordinary differential equations (Moscow: Mir) 685
[15] Haug E J, Negrut D and Iancu M 1997 A State-Space-Based Implicit Integration Algorithm for Differential-Algebraic Equations of Multibody Dynamics Mechanics Based Design of Structures and Machines **25(3)** 311-34
[16] Negrut D, Haug E J and German H C 2003 An Implicit Runge-Kutta Method for Integration of Differential Algebraic Equations of Multibody *Dynamics Multibody System Dynamics* **9(2)** 121-42
[17] Yudakov A A and Boykov V G 2013 Numerical methods of integrating motion equations of multi-component mechanical systems, based on methods of direct integration of equations of the dynamics of the finite elements method Bulletin of the Udmurt University. Mathematics.Mechanics. Computer science (1):131-44

Fatigue crack path and propagation rate at weld root in cruciform joint

A Y Shlepetinskiy [1,*], K P Manzhula [2] and A G Saveliev [3]

[1] Engineering Centre JSC «REP Holding», Saint Petersburg, Russia

[2] Peter the Great St.Petersburg Polytechnic University, Russia, 195251, Saint Petersburg, Polytechnicheskaya, 29

[3]Moscow Automobile and Road Construction State Technical University, 125319, Moscow, Leningradsky Prospect, 64.

* E-mail: A.Shlepetinskiy@reph.ru

Abstract. The results of experimental research of the path and propagation rate of fatigue cracks in cruciform welded joint with a lack of fusion are presented. It is shown that the cracks are initiated at and propagates from the all non-fused weld root faces. Experimental results are compared with the data of computer simulation in Ansys, and with numerical calculations of fatigue life using the received equations of stress intensity factors at the weld root.

1. Introduction

In the construction of machines and structures with complex configurations, there are constructional-technological lacks of fusion. Constructional lack of fusion is often caused by the impossibility of welding the root of the weld due to the constraints of the geometric dimensions. Technological lack of fusion is usually the result of a violation of welding technology. Being stress concentrators, lack of fusion under cyclic loading causes fatigue cracks to appear. Welded joints with lack of fusion by the degree of danger of fatigue failure are assigned to the most dangerous classes FAT36 [1], K4 [2], W [3, 4], C [5]. In [6–8], it was shown that the weld root (tip lack of fusion) is characterized by different degrees of sharpness and therefore the feature of weld root can be considered as almost ready macrocrack or concentrator with radii at the tip.

In this paper, sharp crack-like lack of fusion is considered. To assess the fatigue life of a joint with sharp crack-like lack of fusion, it is necessary to know the stress intensity factors (SIF) at the tip, the crack path, the crack propagation rate, and the loading cycle parameters.

Numerical researches of SIF at the weld root were carried out in [8–12], in which the SIF dependences on the influence of the joint geometrical parameters and the crack size were received.

2. Experimental and numerical studies

In this work, the crack propagation path and rate from the weld root in a cruciform welded joint were determined experimentally. The geometrical dimensions of the studied samples are shown in Figure 1. The plates thickness were 40 mm. The sample thickness was 16 mm. To reduce the probability of fatigue cracks appearance at the weld toe, they were additionally mechanically processed with the radii shown in Figure 1.a. Macrophotography of the samples showed that the gap between the non-fused weld root faces was averages 0.4 mm.

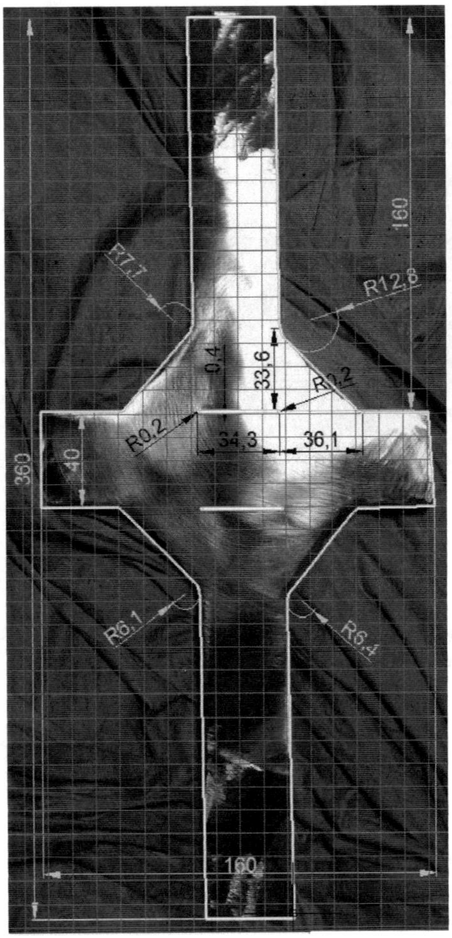

a)

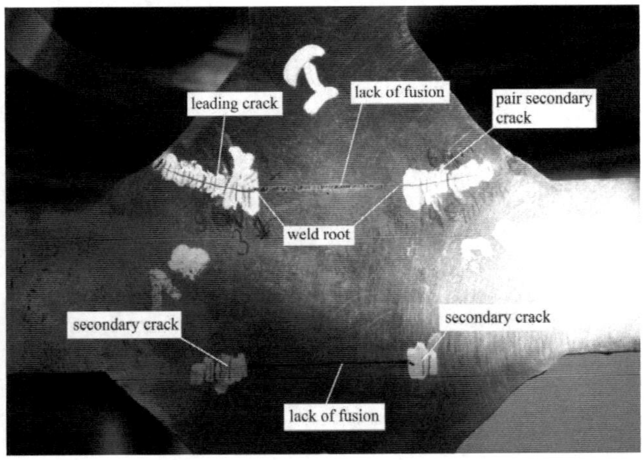

b)

Figure 1. Experimental sample with dimensions –a); leading and secondary fatigue cracks in experimental sample – b)

The joint was made of rolled low alloy structural steel for welding construct steel 09G2S. The chemical composition and mechanical properties of rolled steel 09G2S are presented in tables 1 and 2.

Table 1. Chemical composition low alloy structural steel for welding construct steel 09G2S of experimental sample (%).

C	Si	Mn	S	P	Cr	Ni	Cu	N
0,1	0,7	1,49	0,008	0,018	0,04	0,04	0,08	0,005

Table 2. Mechanical properties low alloy structural steel for welding construct steel 09G2S of experimental sample.

Steel grade and category	Tensile strength, MPa	Yield strength, MPa	Relative extension, %	Impact strength after mechanical aging,t+20, J/cm^2	Impact strength KCU t-40, J/cm^2
09G2S-12	510	360	30	85	117

Welding was carried out by a semiautomatic device in a protective gas environment (80% argon, 20% carbon dioxide) with welding wire 08G2S without welding the root in such a way that non-welded between base plate and attachment plates.

The tests were carried out on an INSTRON 8806 servohydraulic machine. A Dynacell model strain gauge force sensor is mounted on a fixed traverse, which, together with a switching and recording measurement system, is a force-measuring device of the system. The Dynacell sensor has a built-in accelerometer, which makes it possible to compensate for the error in force measurement resulting from the action of inertial forces of moving masses (grippers, devices, etc.).

The loading frequency in all experiments was the same and was 5 Hz, i.e. 18000 cycles took place in one hour of continuous loading. The loading cycle was a slightly shifted to the side of the tensile cycle (R=0). The minimum value of the load in the cycle was 1500 N. The parameters of the load, the frequency of loading, the restrictions on displacements and the number of cycles passed were controlled using a computer. The crack existence and position was determined in a visual way using a Brinell loupe. The crack tip position at a certain moment of the passed number of cycles was recorded by a mark, with an error of no more than 0.5 mm, and a microphotograph of the crack path was taken. Since the weld roots have insignificant roundings, a certain period of macrocrack formation was observed in the experiments. Researches have shown that fatigue cracks develop from all weld roots, but at different rates. There is always a leading crack, the formation period of which is the shortest and growing faster than the others, which can be called secondary (Figure 1, b).

The crack front, especially in the initial period, corresponds to ¼ of the ellipse, which is associated with the formation of a crack from the free border of the sample and the possible slight eccentricity of the installation of the sample. Figure 2 shows the lines of the crack front propagation on the left side and the right side of the sample, corresponding to the marked number of loading cycles.

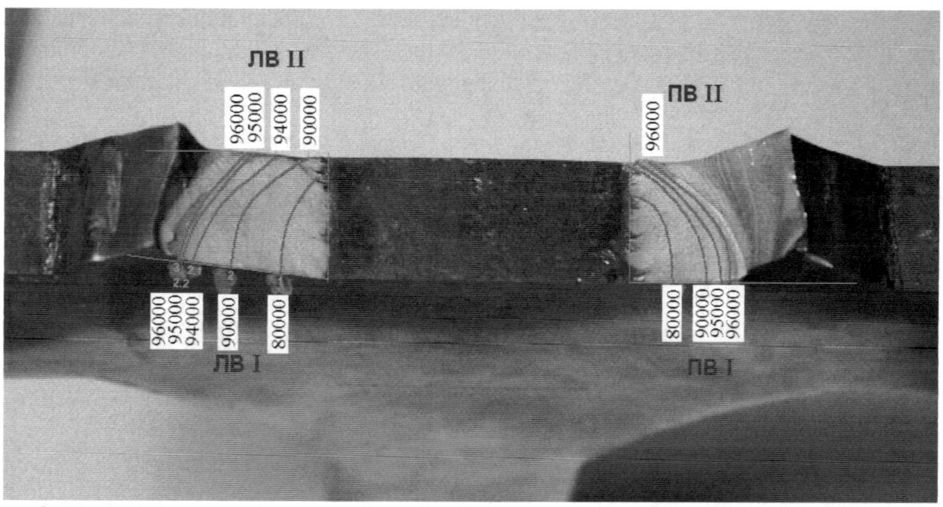

Figure 2. Marked the crack front on a destroyed sample No. 2 with a fixed number of loading cycles.

The position of the crack tip was determined visually using a Brinell magnifier. After a certain number of cycles, the tip of the crack was fixed with a mark with an error of no more than 0.5 mm and the trajectory was photographed. On the basis of the obtained images, the crack length was determined by the graphical method.

In [8] the equation for SIF in the weld root in cruciform welded joint with axial longitudinal tensile load and using the method of forming the experiment plan [13] was obtained

$$K_I = b_0 - 2.33(b_1 + b_2 + b_5) - 4b_3 - 1.67b_4 + 3.33b_1\frac{K_b}{t_1} + 3.33b_2\frac{a}{t_1} + 5b_3\frac{K_g}{K_b} + 0.067b_4t_2 +$$
$$+3.33b_5\frac{t_1}{t_2} + b_{12}\left(3.33\frac{K_b}{t_1} - 2.33\right)\left(3.33\frac{a}{t_1} - 2.33\right) + b_{14}\left(3.33\frac{K_b}{t_1} - 2.33\right)(0.067t_2 - 1.67) +$$
$$+b_{15}\left(3.33\frac{K_b}{t_1} - 2.33\right)\left(3.33\frac{t_1}{t_2} - 2.33\right) + b_{24}\left(3.33\frac{a}{t_1} - 2.33\right)(0.067t_2 - 1.67) +$$
$$+b_{25}\left(3.33\frac{a}{t_1} - 2.33\right)\left(3.33\frac{t_1}{t_2} - 2.33\right) + b_{45}(0.067t_2 - 1.67)\left(3.33\frac{t_1}{t_2} - 2.33\right) +$$
$$+b_{124}\left(3.33\frac{K_b}{t_1} - 2.33\right)\left(3.33\frac{a}{t_1} - 2.33\right)(0.067t_2 - 1.67) +$$
$$+b_{125}\left(3.33\frac{K_b}{t_1} - 2.33\right)\left(3.33\frac{a}{t_1} - 2.33\right)\left(3.33\frac{t_1}{t_2} - 2.33\right) +$$
$$+b_{245}\left(3.33\frac{a}{t_1} - 2.33\right)(0.067t_2 - 1.67)\left(3.33\frac{t_1}{t_2} - 2.33\right), \quad (1)$$

The coefficients b_p are polynomials of the relative size of the crack increment and, in general, can be written as

$$b_p = \beta_{0p} + \beta_{1p}\cdot\left(\frac{a'}{w}\right) + \beta_{2p}\cdot\left(\frac{a'}{w}\right)^2 + \beta_{3p}\cdot\left(\frac{a'}{w}\right)^3 + \beta_{4p}\cdot\left(\frac{a'}{w}\right)^4, \quad (2)$$

where β_{qp} *(q=0,1,…,4)* is the coefficient before the corresponding degree of the multiplier *(a'/w)*, defined by table 3, *p=0,1,2,…,5,12,…,245* - index indicating the corresponding parameter or group of parameters.

Table 3. The values of the equation (2) coefficients, obtained as a result of numerical experiment.

	β0p	β1p	β2p	β3p	β4p
b0	10,89	11,06	54,03	-136,15	155,96
b1	-2,56	-0,04	-9,67	17,76	-25,59
b2	3,40	-0,98	8,15	-15,28	18,22
b3	-0,33	1,79	6,76	-24,03	37,66
b4	3,62	4,91	8,22	-21,15	34,29
b5	2,69	1,89	25,74	-64,90	56,18
b12	-1,37	-1,46	2,91	-6,26	1,97
b14	-0,86	-0,12	-3,90	8,41	-12,95
b15	-0,63	1,27	-14,28	38,71	-31,89
b24	1,13	-0,26	1,36	0,61	-2,91
b25	0,67	0,59	-3,28	7,70	0,32
b45	0,90	2,05	-4,90	15,58	-14,66
b124	-0,45	-1,41	7,71	-19,44	16,34
b125	-0,31	-0,18	0,01	0,05	-5,75
b245	0,516	0,628	3,246	-8,634	9,554

The equation (1) gives more accurate results compared to equations [11,14]. When $K_b = K_g = K$ and $t_1 = t_2 = t$, the parameters like in experimental samples, the dependence (1) can be represented as

$$K_I = 1.87 + \frac{K}{t}\left(1.254 - 10.077\frac{a}{t} - 0.311a\right) + \frac{a}{t}(12.309 + 0.522t) + 0.079t + 0.031K +$$
$$+\left(\frac{a'}{w}\right)\cdot\left(-12.1 + \frac{K}{t}\left(6.21 - 4.811\frac{a}{t} - 0.636a\right) + \frac{a}{t}(10.877 + 0.271t) + 0.134t + 0.377K\right) +$$
$$+\left(\frac{a'}{w}\right)^2\cdot\left(48.073 + \frac{K}{t}\left(-13.556 - 3.833\frac{a}{t} - 0.237a\right) + \frac{a}{t}(0.673 + 0.769t) + 0.607t - 0.37K\right) \quad (3)$$

In the finite element modeling (FEM) program Ansys, a crack growth was simulated [8]. In the crack tip surrounding, there were two layers of singular elements with a radius of the first layer of 0.01 mm. The root crack growth path was determined by the action of stress intensity factors K_I and K_{II}, which were determined at each step of crack propagation, and the angle in the direction of which the crack will grow, from the condition of the normal to maximum normal tensile stress [15]. The values of the coefficient K_{II} was less than 0.2 K_I throughout the crack path.

According to the measurement results, trajectories and the dependence of the crack length on the number of loading cycles were described (Figure 3). The stress range of the tensile cycle was σ=77 MPa in sample 1, σ=88 MPa in sample 2, σ=96 MPa in sample 3.

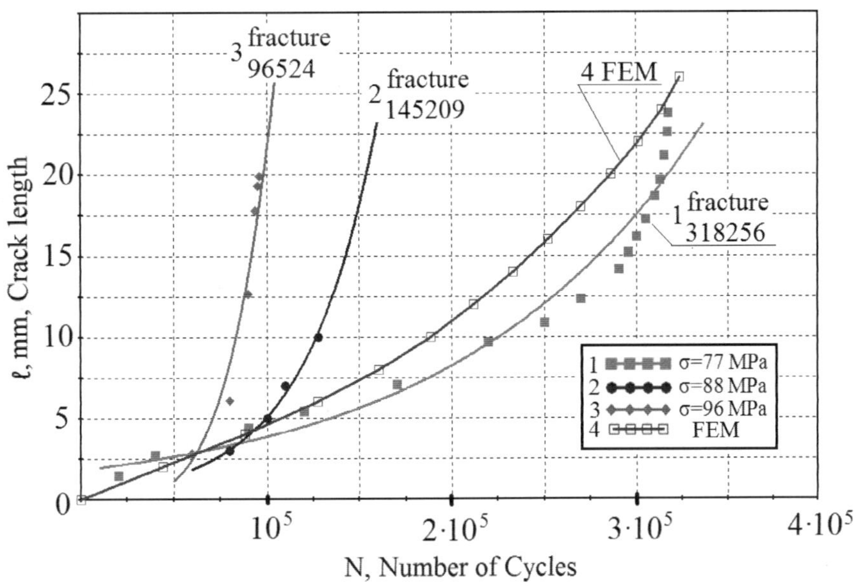

Figure 3. The leading cracks propagation in the experimental samples - lines 1, 2, 3 (calculated points and approximate curves) and crack propagation by the FEM calculating at σ = 70 MPa, C = 4.75 · 10^{-12} MPa · $m^{0.5}$, n = 3 - line 4.

The number of cycles was calculated simply by integration using the trapezoid method $N_i = N_{i-1} + \frac{1}{0.5(v_i + v_{i-1})(K_i - K_{i-1})}$, where v is the crack propagation rate. The parameters of the Paris equation were C=4,75·10^{-12}MPa·$m^{0,5}$, n=3. In Figure 3 points 1 correspond to the experimentally obtained in sample 1, and the line to approximate curves. The curve 4 is calculated by the FEM.

In Figure 4 for sample 1 shows a comparison of the experimental real crack paths with the FEM calculated crack paths.

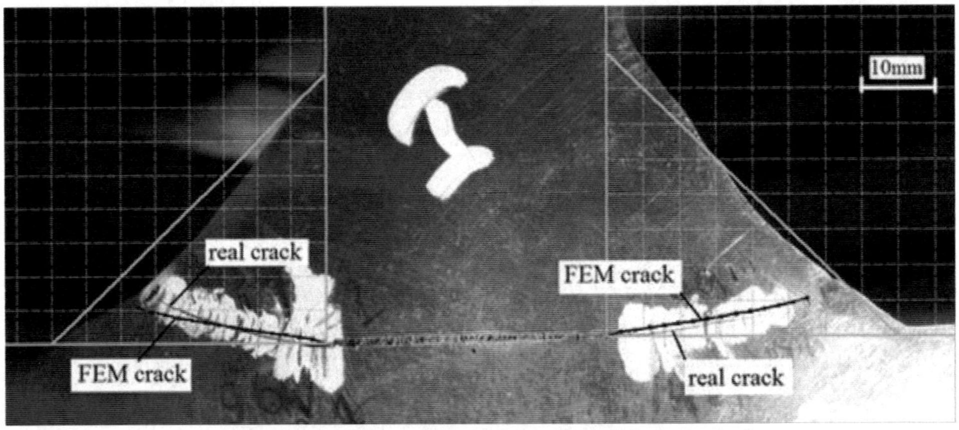

Figure 4. Comparison the crack path in sample 1.

The shift to the right along the number of cycles of the experimental points and curve 1 relatively to curve 4, obtained by FEM simulation, is explained by the fact that when calculating the FEM, a complete crack was considered, which reached a threshold value at the tip ΔK_{th} == 9.3 MPa·m$^{0.5}$ (for 09G2C, [16]). In the experimental samples, concentrators with radii were observed at the weld root, so there was an incubation period for a macrocrack formation.

3. Conclusion

The experimental research of cracks initiation and propagation from the weld root showed that fatigue cracks grow from all non-fused weld root faces. The fatigue cracks propagation rate is not the same, there is one crack that grows faster than the others. The FEM simulation of the crack path corresponds well enough to the experimentally observed trajectories of fatigue cracks from weld roots. Calculations of crack growth periods to critical sizes using the Paris equation and the dependences K_I on the geometry parameters of a welded cruciform joint obtained in the work agree well with the experimental results and can be used for practical applications. In this case, it is necessary to take into account the correction for the period of formation of a macrocrack at the lack of fusion tip.

References

[1] Hobbacher A Recommendations for Fatigue Design of Welded Joints and International Institute of Welding doc. XIII-2151r4-07 / XV-1254r4-07. - Paris, 2008. – 149
[2] DIN 15018 Part 1. Cranes. Steel structures Verification and analyses, Berlin, 1984. – 38
[3] EN 1993-1-9 Eurocode 3: Design of steel structures. Part 1.9: Fatigue // European Committee for Standardisation. - Brussels, 2004 31
[4] DNV-RP-C203. Fatigue Design of Offshore Steel Structures: Recommended practice // Det Norske Veritas. – Norway, 2011. – 142
[5] ANSI AWS D1.1/D1.1M:2004 Structural Welding Code – Steel.- American Welding Society.- Florida, 2004.- 541
[6] Manzhula K P and Shlepetinski A Yu 2016 Stress and Strain Concentration in Weld-Joint Flaws/ Russian Engineering Research, **36(9),** 722–6 © Allerton Press, Inc., 2016.

IOP Conf. Series: Materials Science and Engineering **786** (2020) 012003 doi:10.1088/1757-899X/786/1/012003

[7] Shlepetinski A Yu and Manzhula K P 2019 Size of a Zone Dangerous by Damage at the Root of Cruci-form Weld Joint/ Lecture Notes in Mechanical Engineering. - Springer Nature Switzerland AG, 181-189

[8] Shlepetinskiy A J and Manzhula K P Stress 2010 intensity factors for a crack from a lack of fusion in a fillet weld of criform joint // *Nauchno-tekhnicheskie vedomosti SPbGPU* (St. Petersburg State Polytechnical University Journal).**4** 137-146

[9] Shlepetinskiy A Yu and Manzhula K P 2011 Stress intensity factors for weld root crack in t-joint loaded shear force // *Nauchno-tekhnicheskie vedomosti SPbGPU* (St. Petersburg State Polytechnical University Journal) **3** 122 - 9

[10] Murakami Y 1990 (Ed) Spravochnik po koefficientam intensivnosti napriazhenii (Stress Intensity Factors Handbook): in 2 books. - Moscow. Mir, 1016

[11] Frank, K H 1979 The fatigue strength of fillet welded connections: Ph.D. Thesis / K.H.Frank. - Lehigh University,

[12] Noblett, J E 1996 A stress intensity factor solution for root defects in fillet and partialpenetration welds: TWI Research Report 575/1996 /– UK, Cambridge: TWI,

[13] Adler Yu P, Markova E V and Granovskiy Yu V 1976 Planirovaniye eksperimenta pri poiske optimal'nykh usloviy (Planning an experiment when searching for optimal conditions). Moscow, Nauka Publishers 279

[14] PD 6493. Guidance on Methods for Assessing the Acceptability of Flaws in Fusion Welded Structures. Appendix J:29. – London: British Standard Institution, 1997

[15] Pestrikov V M and Morozov E M 2002 Mechanika razrushenia tverdih tel (Fracture mechanics of solids: a course of lectures) – St.Petersburg, Profeccion Publishers, 320

[16] Panasuk V V 1988 Mechanika razrushenia i prochnost metallov(Fracture mechanics and strength of metals. Reference Guide), Ed., Kiev. 90

Mathematical model of increase in durability of road machines taking into account operating conditions

Vladimir Zorin[1,*] and Le Tchong Tuan[2]

[1] Moscow automobile and road state technical university, Russia, Moscow

[2] Moscow automobile and road state technical university, Russia, Moscow

[*] Email: madi-dm@list.ru

Abstract. In this article results of researches and development of a mathematical model of extension of a resource and increase in durability of the road machines functioning in tropical conditions of Vietnam taking into account operating conditions are considered. By means of the software of Matlab the original rated program of determination of optimum parameters of frequency of maintenance and repair is developed; the number of repairs and categories of maintenance which need to be carried out throughout all life cycle of the road machine for ensuring best values of indicators of durability of machines taking into account operating conditions.

1. Introduction

It is known that the organization of effective use and quality of functioning of road machines (are defined by the level of their reliability) play an important role in ensuring quality of construction of roads and minimization of damage to the national economy [1, 2, 13-15].

Increase in a resource and increase in durability of road machines requires carrying out the complex of actions covering spheres of design, production and operation. The efficiency of actions for increase in durability of road machines is estimated by the sum of costs of production and operation of the machine throughout all life cycle taking into account operating conditions [7, 9, 10, 12, 13].

2. Methods

As criterion of optimization at a solution of an objective it is selected a minimum of specific total costs of $C(t)$ of technical maintenance, repair of the machine and compensation of losses because of decline in production and increase in a consumption of fuels and lubricants owing to deterioration in technical condition of the machine.

Vasilyev V.O. and Dombrovsky K.P. were the first who conducted researches of models of optimization of resources of machines. In work [12] Vasilyev V.O. offered a method of definition of optimum operating time of engines with use of criterion of the smallest average cost value of maintenance of machines. Dombrovsky K.P. [7] suggested to set best value of a resource of operation of the machine equal to number of years through which increase in average repair cost becomes comparable or exceeds the value of annual norm of depreciation.

The method of definition of an operation life of cars offered Tokareva E.G. deserves attention [11] and Selivanov A.I. [8]. In these works, optimum is understood as service life which provides the minimum value of costs of run unit for the entire period of operation.

In the field of operation of the road-building equipment the greatest distribution was gained by the optimization models described in works [9, 10]. Target function is the cornerstone of these models:

$$C(t) = \frac{n_c(t)}{M}\frac{C_{dep}+C_r+C_{los}}{t_{r1}} + \sum_i \frac{C_{tmi}}{t_{tmi}} \rightarrow min. \quad (1)$$

where

$n_c(t)$- number of repair cycles for service life of the road machine before write-off;

C_{dep}- the average value of depreciation charges is over one repair cycle of the road machine;

C_r-costsof elimination of failures and faults of the road machine in the course of repairs for one repair cycle;

C_{los}- costs of compensation of losses owing to decline in production and increase in a consumption of lubricants at wear of details of the road machine during one repair cycle;

C_{tm}- costs of technical maintenance of one road machine;

t_{r1}- resource before the first major repair;

C_{tmi}- costs of technical maintenance of road machines;

t_{tmi}- frequency of technical maintenance;

M- coefficient, is defined as follows:$M = 1 + k_r[n_c(t) - 1]$.

Here k_r- the coefficient equal to the resource relation after major repair to a resource of the new road machine:$k_r = t'_{r1}/t_{r1}$;t'_{r1}- a resource after major repair.

According to provisions of Technical regulations of TPTC 010/2011 "About safety of machines and the equipment" [6], the machine which underwent major repair on the technical condition should conform to regulatory requirements.

Provided that machine resources after the major repairs first, second, etc. are equal among themselves, it is possible to define depreciation charges for one repair cycle as follows:

$$C_{dep} = \frac{[C_{ma}+C_{mr}(n_c(t)-1)]}{n_c(t)}. \quad (2)$$

where

C_{ma} - the cost of the road machine;

C_{mr} - the cost of major repair.

Meanwhile costs of compensation of lossesC_{los}declines in production and increases in a consumption of lubricants are function which depends on technical condition of the machine (I) and is reflected degree dependence (taking into account operating conditions):

$$C_{los} = c.I^\beta = K_s.c_0.I^\beta = K_s^{\beta+1}c_0 b^\beta t_{tm}^{(\alpha-1)\beta} t_{r1}^\beta. \quad (3)$$

where

β-exponent caused by construction and functional purpose of interface;

c - the coefficient depending on an operation mode and operating conditions of the road machine;

c_0 - the coefficient defined under operating conditions the road machine, which is not depending on climatic conditions;

K_s- the coefficient determined by technical rigidity of climate [16].

On the other hand, total costs of elimination of failures and faults in the course of repairs C_r are the function depending on operating time t_{r1} accounting of operating conditions it is possible to define as follows:

$$C_r = \int_0^{t_{r1}} e.t^\gamma = \frac{e}{\gamma+1}.t_{r1}^{\gamma+1}. \quad (4)$$

After the conversions made by means of expressions (2), (3) and (4), expression (1) will take a form:

$$C(t) = \frac{C_{ma}+C_{mr}(n_c(t)-1)+n_c(t)K_s c_0 I_{lim}^{\beta}}{t_r} + \frac{n_c(t)et_r^{\gamma}}{(\gamma+1)M^{\gamma+1}} + \sum_i \frac{C_{tmi}}{t_{tmi}} \rightarrow min. \quad (5)$$

Here t_r - a full average resource: $t_r = M.\ t_{r1}$;

I_{lim} – an indicator of a limit status: $I_{lim} = K_s . b . t_{r1}^{\alpha}$

3. Result

Let's take derivative C(t) on the variable $x = t_r$ and having equated it to zero. Then from expression (5) we will receive:

$$\frac{\partial C(t)}{\partial t_r} = -\frac{C_{ma}+C_{mr}(n_c(t)-1)+n_c(t)K_s c_0 I_{lim}^{\beta}}{t_r^2} + \frac{n_c(t)e\gamma t_r^{\gamma-1}}{(\gamma+1)M^{\gamma+1}} = 0. \quad (6)$$

or

$$(\gamma+1)M^{\gamma+1}\left[C_{ma} + C_{mr}(n_c(t)-1) + n_c(t)K_s c_0 I_{lim}^{\beta}\right] - n_c(t)e\gamma t_r^{\gamma+1} = 0. \quad (7)$$

From here, the optimal value of a resource of the road machine is defined with operating conditions by the following expression:

$$t_{r.opt} = \left\{\frac{M^{\gamma+1}(\gamma+1)\left[C_{ma}+C_{mr}(n_c(t)-1)+n_c(t)K_s c_0 I_{lim}^{\beta}\right]}{n_c(t)e\gamma}\right\}^{\frac{1}{\gamma+1}}. \quad (8)$$

If to accept that during operation of the machine before the first technical maintenance of expense for elimination of failures and faults of the machine are equal to zero ($C_r=0$), then the equation (5) corresponds in a look:

$$C(t) = \frac{K_s b\left[C_{ma}+C_{mr}(n_c(t)-1)+n_c(t)K_s c_0 I_{lim}^{\beta}\right]}{M.I_{lim}} . t_{tm}^{\alpha-1} + \frac{C_{tm}}{t_{tm}} \rightarrow min. \quad (9)$$

Having taken derivative *C(t)* on the variable $x = I_{lim}$ and having equated it to zero, from expression (9) we will receive:

$$\frac{\partial C(t)}{\partial I_{lim}} = \frac{K_s b t_{tm}^{\alpha-1}}{M}\left[\frac{(\beta-1)K_s c_0 n_c(t) I_{lim}^{\beta} - [C_{ma}+C_{mr}(n_c(t)-1)]}{I_{lim}^2}\right] = 0. \quad (10)$$

From here, best value of an indicator of a limit status of the machine:

$$I_{lim.opt} = \left[\frac{C_{ma}+C_{mr}(n_c(t)-1)}{(\beta-1)K_s c_0 n_c(t)}\right]^{\frac{1}{\beta}}. \quad (11)$$

Then, if best value of an indicator of a limit status is more than corresponding value set by criterion of impossibility of further operation ($I_{lim.opt} > I_{lim}$), in calculations we accept value I_{lim}.

Further, we will take derivative *C(t)* on the variable $x = t_{tm}$ and having equated it to zero, from expression (9) we will receive:

$$\frac{\partial C(t)}{\partial t_{tm}} = \frac{K_s b(\alpha-1)\left[C_{ma}+C_{mr}(n_c(t)-1)+n_c(t)K_s c_0 I_{lim}^{\beta}\right]}{M.I_{lim}} . t_{tm}^{\alpha-2} - \frac{C_{tm}}{t_{tm}^2} = 0. \quad (12)$$

From here, optimum maintenance rate taking into account operating conditions:

$$t_{tm.opt} = \left\{\frac{M.I_{lim}.C_{tm}}{K_s b(\alpha-1)\left[C_{ma}+C_{mr}(n_c(t)-1)+n_c(t)K_s c_0 I_{lim}^{\beta}\right]}\right\}^{\frac{1}{\alpha}}. \quad (13)$$

If set the number of services for optimal resource equal to N_{tm}:

$$N_{tm} = \frac{K_s(\alpha-1)\left[C_{ma}+C_{mr}(n_c(t)-1)+n_c(t)K_s c_0 I_{lim}^{\beta}\right]}{MC_{tm}}. \quad (14)$$

then expression (13) can be rewritten in a look:

$$t_{tm.opt} = \left(\frac{I_{lim}}{bN_{tm}}\right)^{\frac{1}{\alpha}}. \quad (15)$$

Therefore (14), (15) and (8) we can create a mathematical model for determination of optimum parameters of durability of the road machine taking into account operating conditions from the equations as follows:

IOP Conf. Series: Materials Science and Engineering **786** (2020) 012004 doi:10.1088/1757-899X/786/1/012004

$$\begin{cases} N_{tm} = \frac{K_s(\alpha-1)\left[C_{ma}+C_{mr}(n_c(t)-1)+n_c(t)K_s c_0 I_{lim}^{\beta}\right]}{MC_{tm}} \\ t_{tm.opt} = \left(\frac{I_{lim}}{bN_{tm}}\right)^{\frac{1}{\alpha}} \\ t_{r.opt} = \left\{\frac{M^{\gamma+1}(\gamma+1)\left[C_{ma}+C_{mr}(n_c(t)-1)+n_c(t)K_s c_0 I_{lim}^{\beta}\right]}{n_c(t)e\gamma}\right\}^{\frac{1}{\gamma+1}} \end{cases} . \quad (16)$$

If best value of an indicator of a limit status is less or to equally corresponding value set by criterion of impossibility of further operation ($I_{lim.opt} \leq I_{lim}$) when calculating the value $I_{lim.opt}$ is accepted. Having made corresponding changes to expression (11), we will receive a mathematical model (16) in the following look:

$$\begin{cases} N_{tm} = \frac{K_s(\alpha-1)\beta[C_{ma}+C_{mr}(n_c(t)-1)]}{MC_{tm}(\beta-1)} \\ t_{tm.opt} = \left(\frac{I_{lim.opt}}{bN_{tm}}\right)^{\frac{1}{\alpha}} \\ t_{r.opt} = \left\{\frac{M^{\gamma+1}(\gamma+1)\beta[C_{ma}+C_{mr}(n_c(t)-1)]}{n_c(t)e\gamma(\beta-1)}\right\}^{\frac{1}{\gamma+1}} \end{cases} . \quad (17)$$

On the basis of mathematical models (16) and (17) we build algorithms of determination of optimum parameters of durability of the road machines functioning in tropical conditions (on the example of Vietnam):

- Step 1: Define coefficient of $K_s$which characterizes severity of tropical climate of Vietnam in comparison with a temperate climate;
- Step 2: Determination of the optimum parameters connected with durability of the road machine.

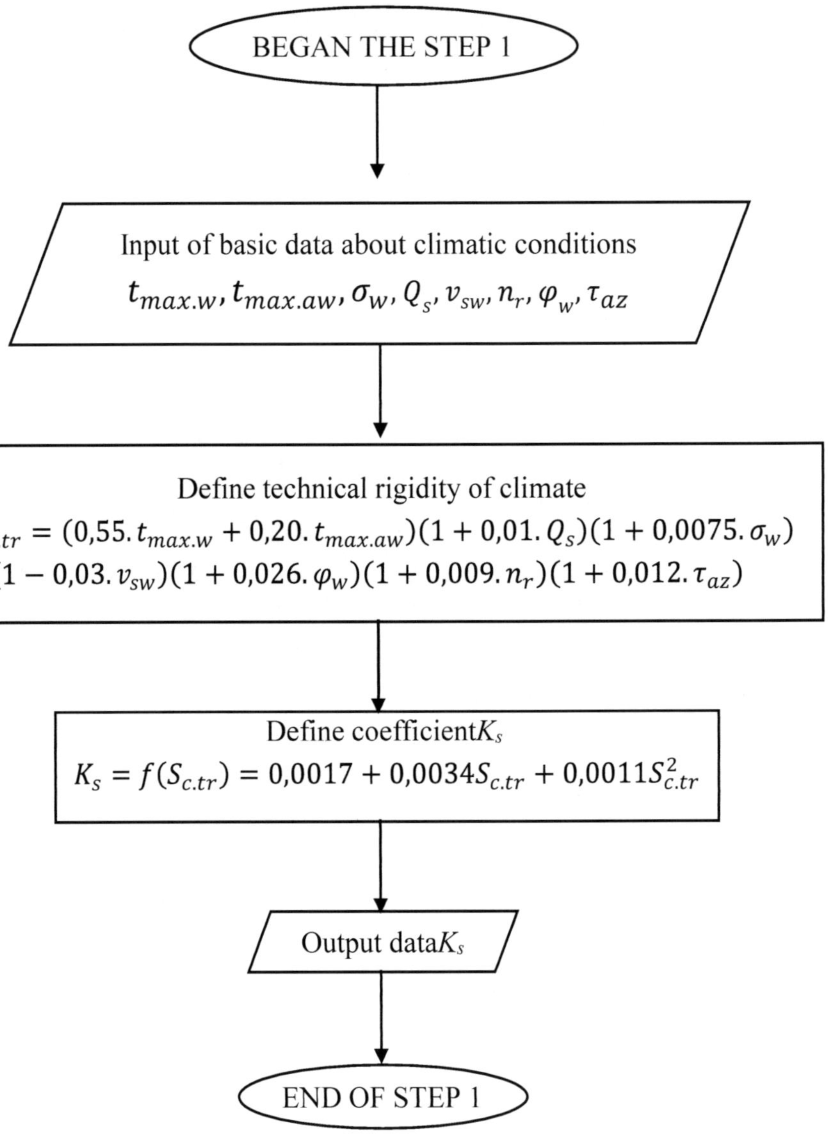
BEGAN THE STEP 1
Input of basic data about climatic conditions
$t_{max.w}, t_{max.aw}, \sigma_w, Q_s, v_{sw}, n_r, \varphi_w, \tau_{az}$
Define technical rigidity of climate
$S_{c.tr} = (0{,}55.t_{max.w} + 0{,}20.t_{max.aw})(1 + 0{,}01.Q_s)(1 + 0{,}0075.\sigma_w)$
$.(1 - 0{,}03.v_{sw})(1 + 0{,}026.\varphi_w)(1 + 0{,}009.n_r)(1 + 0{,}012.\tau_{az})$
Define coefficientK_s
$K_s = f(S_{c.tr}) = 0{,}0017 + 0{,}0034S_{c.tr} + 0{,}0011S_{c.tr}^2$
Output dataK_s
END OF STEP 1

IOP Conf. Series: Materials Science and Engineering **786** (2020) 012004 doi:10.1088/1757-899X/786/1/012004

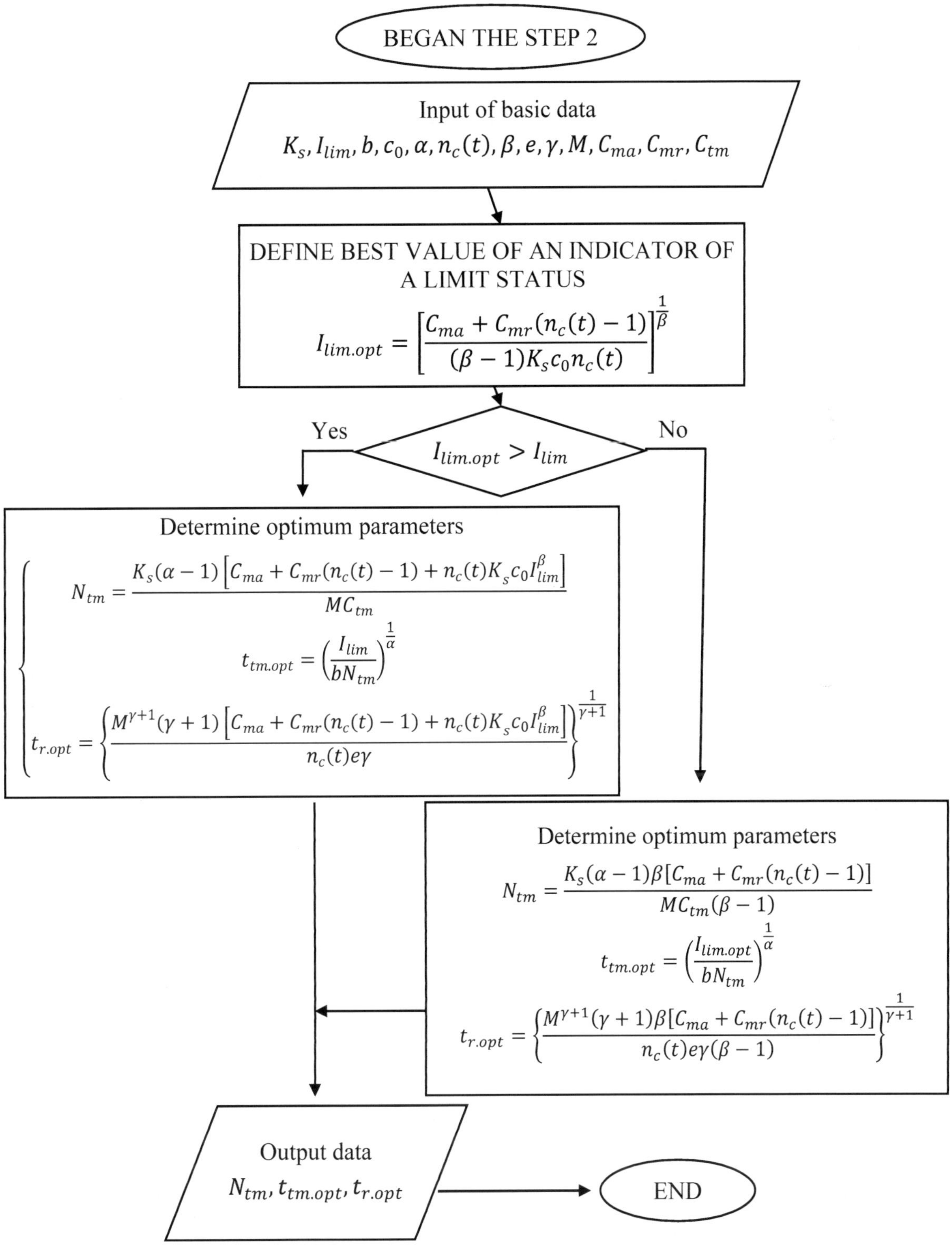

On the basis of an algorithm by means of the software of Matlab, we build the graphic program of calculation of parameters. In Figure 1 the example of start of the program is given.

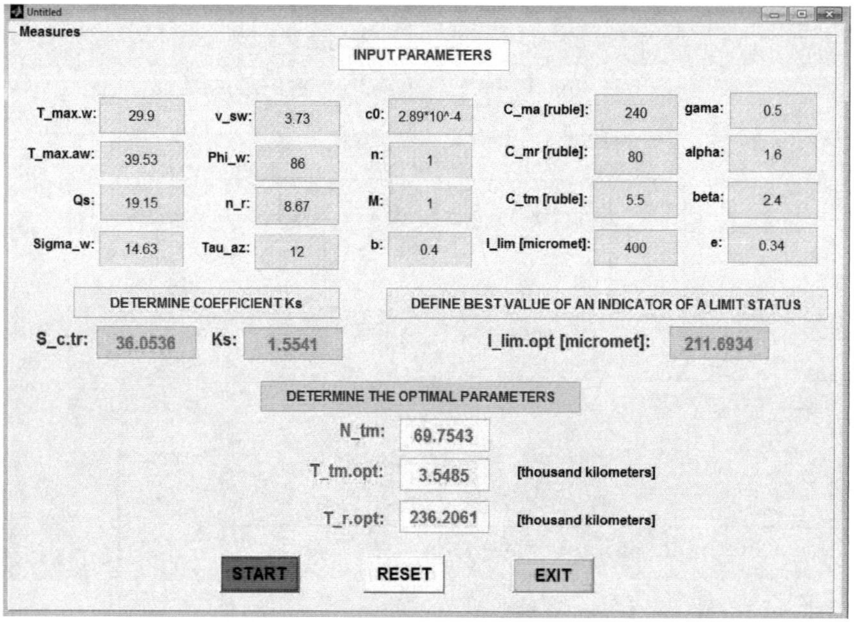

Figure 1. An example of calculation of optimum parameters of the excavators working in Hanoi

4. Conclusion

Taking note of operating conditions on reliability and efficiency of use will allow to prove the choice of models of road machines when forming parks of means of mechanization of construction, more precisely to plan terms of carrying out construction works, to make the reasonable diagrammaintenanceandrepair of machines during construction works, to calculate need of the park for spare parts and operational materials. The most important direction of reliability augmentation, extension of a resource and increase in durability of road machines in the conditions of aggressive influence of the corrosion environment is use of polymeric composition materials at technical maintenance and repair [2-5].

References

[1] Alferov A K and Petrov I V 1980 Ensuring operability of construction machines. Stroyizdat, Moscow, 136

[2] Baurova N I 2008 Diagnosing and repair of machines using polymeric materials. ISBN 978-5-94385-038-7, 280

[3] Baurova N I 2009 Polymeric repair materials of machines. UDC 629.7.07, 46

[4] Baurova N I and Zorin V A 2016 Use of polymeric composition materials by production and repair of machines. Textbook grant: UDC 621-049.32-036, Moscow Administrative Road Inspectorate of, 264

[5] Baurova N I and Zorin V A 2017 Use of polymeric composition materials in mechanical engineering. Manual of ISBN 978-5-16-104622-7 (online). the 2nd prod., reslave. and additional, 301

[6] Decision of the Commission of the Customs Union of October 18, 2011 N 823 "On the adoption of the technical regulations of the Customs Union" On the safety of machinery and equipment "", Moscow

[7] Dombrovsky K I 1944 Organization of repair of engines and optimum service life. *Equipment of the railroads* **5** 6-9

[8] Selivanov A I 1971 Bases of the theory of aging of machines. the 2nd prod, Mechanical engineering, 408
[9] Sheining A M , Filippov B I and Zorin V A , etc. 1992 Operation of road machines. Textbook of ISBN 5-277-00778-4, Transport,Moscow, 328
[10] Sheining A M and Sheining V A 1981 Algorithms and programs of a solution of optimum problems of reliability of machines, Moscow Administrative Road Inspectorate, 112
[11] Tokarev G G Rational service life cars, Avtotratransizdat,Moscow, 79
[12] Vasilyev V O 1962 Assignments and expenses on resuming of property in the railway enterprise. Materials on the issue of resuming of fixed capital of the railroads. Works of Economic bureau at NKPS, 18-26
[13] Zorin V A 2005 Bases of operability of technical systems. Textbook: UDC 629.7.07, ISBN 5-902048-51-6, Magistr publishing house, 536
[14] Zorin V A 1998 Increase in durability of road-building machines by improvement of a system of technical maintenance and repair. Thesis of the Doctor of Engineering: 05.05.04, 411
[15] Zorin V A 2015 Reliability of mechanical systems Textbook: ISBN 978-5-16-010252-8, M: INFRA-M, 380
[16] Zorin V A and Le Trong 2019 Tuan. Reliability assessment of the road machines working in tropical conditions. *The scientific and technical bulletin of the Bryansk state university* **1** 45-49

IOP Conf. Series: Materials Science and Engineering **786** (2020) 012005 doi:10.1088/1757-899X/786/1/012005

Indicators of the efficiency for using of road-building machines at various levels of management decisions

A I Dotsenko[1] and K P Mandrovskiy[1]

[1]Moscow Automobile and Road Construction State Technical University (MADI), 125319, Moscow, Leningradsky prospect, 64, RUSSIAN FEDERATION

Abstract. Methods of technical and economic evaluation of road-construction machines are highly developed but their operation requires a wide list of initial data which is very difficult to obtain without monitoring systems. In modern conditions, economic factors determine the possibility of machines' operation and the technical excellence of the machine is viewed by considering the cost of ownership and operation. This determines the relevance of the application of technical and economic assessments. In order to ensure the feasibility of applying of such technical and economic indicators towards the performance evaluation, they need to be adapted for use in monitoring systems. Modern monitoring systems allow you to monitor the technical and operational characteristics without a comparison with the indicators of the economic sphere. There is almost complete uncertainty in the methods of selection and evaluation of equipment that could be used in practice in modern conditions characterized by the presence in the market of equipment of the same purpose from different manufacturers. In such conditions, a system for collecting and processing the information with feedback would be convenient - a monitoring system for effectiveness evaluation. Feedback is necessary for creatingthe control actions in order to increase the efficiency or maintain it at satisfactory level - performance management. The development of methods of both technical and economic assessment which are suitable for using for road-construction machines and adapted for implementation in monitoring systems is a relevant topic of scientific research.

The concept of efficiency applies not only to machines but also to enterprises which determines the hierarchy of performance indicators for different decision-making levels. Indicators of this hierarchy can be applied to the evaluation of the technical and economic efficiency of a certainroad-construction machine [1].

Natural indicators (length, time, area, volume, mass, etc.) that occur at the driver's level, form more complex, intermediate indicators (metal consumption, productivity, energy intensity, etc.), which determine, in turn, indicators of the next level (for example, the full unit cost of production) [2; 3].

There may be the following hierarchy of indicators presented in real time by monitoring (actual), related to differentdecision-making levels:

- level of operator and/or on-board computer: performance indicators; indicators of the functional state of the aggregates; indicators of stock sufficiency of materials (technological, fuel and lubricants);

- level of the work supervisor at the facility: indicators of the integrated productivity of the unit (or group) of the machines; performance schedule, indicators (alarming) of the operability; indicators (alarming) of theposition or direction of movement of the self-propelled machine; indicators of the physiological state of the driver or operator, indicators (alarming) of the quality of work of the machines;
- level of the Head of the enterprise: integrated indicators of the implementation of the schedule; quality indicators (emergency) of operation; indicators (emergency) of machine operability; indicators (emergency) of the schedule;

level of regional and sectoral leader: indicators of risk to the life or health of the population of the region, indicators of the threat to the environmental situation; generalized indicators of the implementation of the schedule of works on the objects by contractors.

The efficiency of equipment exploitation may depend on decisions which the implementation is mandatory directly in the process of work as well as on strategic decisions that are appropriate for further using of equipment. Due to the fact that the collection and processing of data in automated systems may take some time, to simplify data processing it is advisable to separate the indicators obtained during the monitoring which need to be processed in real time, and the indicators that can be processed with a delay.

To simplify the separation of indicators into the two considered groups, it is advisable to distinguish such generalizing features as spatial position, state of internal systems, and product quality. For the spatial position, the following subgroups can be distinguished (real-time mode): maintaining a given trajectory of the working body in a local coordinate system tied to the object of impact; tracking the position and speed of movement of machines in the global coordinate system. The following subgroups can be distinguished for the state of the internal systems (real-time mode): registration of operator state indicators; registration of performance indicators of machines. For product quality, the following subgroups can be distinguished (real-time mode): measurement of machine performance indicators (deviation from design parameters of the surface, degree of compaction, etc.) moving in field with the global coordinate system; measurement of the performance of stationary machines in the local coordinate system.

A natural indicator of the effectiveness of machines must be available for instrumental control in order to enable automated accumulation of the indicator values in processing and decision making. Among the physiological parameters of the operator there are may be: blood pressure level; heart rate; body temperature.

Among the natural indicators of the efficiency of moving machines there are the following:

- indicators characterizing the work of machine units: independent timer; the crankshaft rotational speed, rotational speed of the driving element of the propulsion unit (sprocket or wheel); fuel consumption, frequency of pressure pulsations in the hydraulic system; consumption of technological materials (if provided by the type of machine); CO and NOx content in engine exhaust;
- indicators of the scope of work performed by the machine: the volume of load carried by the machine with the moldboard; the distance of material movement dragged by moldboard; the time of material movement; the volume of material moved by the machine with a bucket; the distance of material movement by bucket; material transfer time by bucket; the volume of the excavation (pit, trench, well, ditch, etc.) dug by the machine; the time to create a notch; the volume of material transported by the machine; material transportation distance; material transportation time; surface area processed by the machine; surface processing time;
- indicators characterizing the quality of work of machines: the discrepancy between the actual and designedmarks of surfaces; the discrepancy between the actual and designed degrees of compaction; the discrepancy between the actual and standard distribution densities of reagents; the discrepancy between the actual and regulatory conditions of the surface of the artificial coating.

Indicators of the work quality can be controlled by modern monitoring systems, but there are no systems that allow to obtain the integrated technical and economic indicators [4–12], i.e. this direction does not need a development.

When reaching to higher decision-making levels the natural indicators stop to be relevant but are included in (i.e. define) more complex calculated indicators.

The calculated indicators at the level of the facility manager allow an operational assessment of the effectiveness, allowing to determine the machines or groups of machines that need attention with the following formation of a control action. These indicators include:

- shift time utilization rate (determined by the duration of the engine with a nominal power or close to the last);
- productivity of machine (determined by the volume and range of the transferred material; the volume of laid material; the amount of laid material; theoretical speed; slipping; resistance to displacement; soil strength; size of the tool, working time in traction mode, etc.);
- resource capacityof production unit (for example, fuel consumption per unit volume of extracted, displaced or compacted soil, laid cement or asphalt concrete, distributed reagent, etc.).

As for the Head's level of the enterprise, in addition to indicators of the direct assignment of the created object, it is necessary to form an idea of the work efficiency at the facility, the efficiency of the enterprise and the industry. It is also necessary to have an idea about the possible environmental consequences of working activity (for example, transport construction). Thus, for this level there are technical, economic and sanitary-environmental indicators. Technical and economic indicators are important for all interested groups - both for the work producer (contractor), and for the customer, who may be presented as an administrative organizations (authorities) and private organizations. Sanitary and environmental indicators in the context of a well-established relationship with a relatively low responsibility can be considered as significant only for the customer whobelongs to the administrative structures.

These two groups of indicators essentially contradict with each other, since the process of any construction is accompanied by environmental degradation. Sanitary and environmental indicators, as a rule, are transferred to the category of restrictions consisting in establishing standards for harmful effects on the environment and humans. In case of going beyond the limits of the norm, legal proceedings are initiated against the responsible persons (physical, legal)andend in fines. Thus, restrictions on sanitary and environmental indicatorsare ultimately reduced to indicators of the economic sphere.

A significant indicator of the work's manufacturer (the contractor) is the profit $p_i$for the reporting period [13]:

$$p_i = \frac{D_\Sigma - Z_\Sigma}{T_1}, \tag{1}$$

where $D_\Sigma$is the sum of all (gross and other) incomes for the period T_1; Z_Σ- current expenses and losses for the periodT_1; T_1- the duration of the construction of the object of transport construction.

The value $D_\Sigma$may define the sum of the contract fixed during the tender won by the manufacturer (contractor). The value Z_Σ consists of the operating costs $Z_1$and lossesZ_2on eliminating of negative effects of the construction such as changingof hydrogeological conditions [13]:

$$Z_\Sigma = Z_1 + Z_2. \tag{2}$$

The value T_1isusually fixed by the contract and depends on the amount of work $Q_\Sigma$on the object stipulated by the contractand the productivity $P_1$of the machines provided by the contractor [13]:

$$T_1 = \frac{Q_\Sigma}{P_1}. \tag{3}$$

The contractor's profitp_i being a possible criterion of its effectivenessdepends on external factors (D_Σ, Q_Σ,Z_2) which the contractor cannot influenceand factors determined by the contractor (P_1,Z_1) [13]:

$$p_i = \frac{P_1(D_\Sigma - Z_1 - Z_2)}{Q_\Sigma}. \quad (4)$$

Thus, operating costs and efficiency can be attributed to the largest indicators of the contractor's self-control. If these indicators are combined,then the unit cost of production will be obtained.

As the main criterion for assessing the technical and economic efficiencyit is advisable to choose the unit cost of production including the cost of ownership (SEPV):

$$SEPV = \frac{C_1}{P_2} + \frac{1}{T_2 \cdot T_3 \cdot n_1 \cdot P_2} \cdot C_0, \quad (5)$$

where $C_1$is the cost of machine-shift with reduced costs of ownership (thousand rubles); P_2- operational replacement performance (production unit / shift); T_2- the life of the machine (years); T_3- the number of days in a year (days); n_1- the number of shifts in day; C_0- the cost of the car (thousand rubles).

At the regional and federal level, it is necessary to guide by higher level indicators in order to assess the effectiveness of the implementation of budgetary funds and to assess the sanitary and environmental safety of the region. To do this, you can use the following indicators: the timing of the construction of the object; the expected profit of other economic entities from the commissioning of the object; socio-economic consequences of the facility commissioning; budget expenditures oncontrol compliance with the rules of construction and eliminate its negative consequences.

1. Conclusion

Thus, by depending on the level of decision making there is a significant number of indicators. Some of these indicators are used for direct control on the machines, so they must be available in real time. The other part can be obtained with some delay. To assess the effectiveness of road-construction machines, the decision-making levels of the Head of the company and the work manager at the facility are important,whose operating costs and productivity can be attributed to the largest indicators of self-control, i.e. there is the owner's possibility to correct these indicators.

Reference

[1] Shestopalov K K and Mandrovskiy K P 2015 Overview and analysis of remote-control systems for road machines // *Road Power* **61** 34-7

[2] Golkina G E 2015 Possibilities of ERP-systems to control the cost of production// *Modern trends in the development of science and technology* **1-5** 44-7

[3] Tikhonov A A 2007 Production cost and financial performance: what comes first?// *Vestnik of the Financial University* **1** 132-41

[4] DotsenkoA I and Mandrovskiy K P 2018 Improving the quality of asphalt concrete pavement of roads by monitoring the parameters of road machines in the course of work [Electronic resource] // Interstroymeh-2018 Materials of the international scientific and technical conference - M : National Research Moscow State University of Civil Engineering, 223-7 - Access mode: http://mgsu ru/resources/izdatelskaya-deyatelnost/izdaniya/izdaniya-otkr-dostupa/2018/conf_inter_stroy_mex2018 pdf

[5] Concord-GPS Monitoring system of transport objects [Electronic resource] - Access mode: http://www.konkord-gps.ru/ (access date: 07.28.15).

[6] NPP "Transnavigation". Automated system for monitoring passenger traffic [Electronic resource]. - Access mode: http://www.transnavi.ru/projects/asmpp/about/about.php (access date: 07/01/17).

[7] Saarenketo T Monitoring the condition of roads with low traffic [Electronic resource]. - Access mode: http://www roadex.org/wp-content/uploads/2014/01/RussianTranslation4.pdf (access date: 06.30.17).

[8] Mandrovskiy K P 2016 Analysis of monitoring systems for road-building machines and the concept of a performance management system// *Vestnik of the Moscow Automobile and Road Construction State Technical University* (MADI) **1(44)** 26-33

[9] Mandrovskiy K P 2016 Possible prospects for the development of monitoring systems for road machines in performance management and technical audit // Mechanization of construction 77 - 10 47-52

[10] BankerR D, Charnes A and Cooper W W 1984 Some Models for Estimating Technical and Scale Inefficiencies in Data Envelopment Analysis // *Management Science* **30(9)** 1078-92

[11] Cooper W W, Seiford L M and Tone K 2006 Data envelopment analysis. A comprehensive text with models, applications, references and DEA-solver software.– 2-nd edition. New York: Springer, 528

[12] Sakhapov R, Makhmutov M M, Nikolaeva R and Gatiatullin M 2018 Asphalt granulate coating for roadsides International Scientific Conference on Energy, Environmental and Construction Engineering (EECE-2018) electronic edition. "MATEC Web of Conferences" C. 03014.

[13] Kustarev G V, Shestopalov K K, Karasev G N and Mandrovskiy K P 2010 Development of scientific foundations and electronic resources of the monitoring system for assessing the effectiveness of road-construction, airfield and utility vehicles: a report on research and development. - M.: Moscow Automobile and Road Construction State Technical University (MADI) 50

Structural analysis of Braking Energy-Conservative Systems of Transport Technological System

V S Isakov[1,*], A V Erejskiy[2] and A A Kvetkin[3]

[1] Russia, Rostov region, Novocherkassk, Platov Prospect, 176

[2] Russia, Rostov region, Novocherkassk, Bakunin St., 14

[3] Russia, Rostov region, Rostov-on-Don, 40 years of Victory Prospect, 37 Apt. 42

* E-mail: vs.isakov@mail.ru

Abstract. The possibilities of application of the ways and means allowing beneficial use of kinetic energy of moving masses of various transport and transport-technological machines in the course of braking are considered. The concepts of development of inertial and energy-storage braking means for different types of actuating units are presented. The generalized block schematic diagram of a multimass object with the inertial energy-storage braking system and morphological matrix of possible technical solutions are offered. Various offers on realization of inertial and energy-storage braking systems in the designs of technological machines are shown and the results of the researches of energy-storage braking are presented.

1. Introduction

The modern mechanical engineering assigns tasks to engineers to create the multipurpose machines and the equipment, universal technical objects, adaptive and indifferent devices capable to change the functions, structures or parameters in the automatic mode depending on external and (or) internal factors. The considerable group of such machines is made of transport and technological ones, providing all variety of construction, materials-handling, roadway, underground mining processes and works[1].

The basis for such combination is the unity of structures, uniformity of working and executive bodies, transmissions, blocks unification, control devices and other modules. However, not all the blocks and modules are adequately researched and covered in literature. Among them a specific place is held by the brake and traction and brake modules capable to provide efficient use of the braked objects kinetic energy through the creation of braking power or brake energy recuperation and its use in further work.

Unfortunately, the majority of transport-technological machines is equipped with the brake means transforming kinetic or potential energy of the moving objects to thermal energy with its dispersion into the environment. Energy is spent for creation of brake power (torque) in such means in addition.

2. Energy-Storage Braking Conceptual Baseline

The first mentions in literature on efficient use when braking kinetic energy of the moving vehicle belong to the crew of "horse car" and the railway train. In the first case wheels kinetic energy when braking was transferred in to the vehicle body hoisting mechanism, i.e. it was transformed into potential energy. At start-off the crew fell and via the same transmission gear it rotated wheels, helping horses to pull away. In the second case, the American scientist Gibbs [6] offered the inertial brake using the efforts arising in train hitches when braking the engine.

The modern inertial brake energy-storage and energy-saving means allowing to carry out recovery of energy of the braked object in one guise or another are used on the motor transport (coasting brakes and the combined drive gears) [1.3], on the trains of underground locomotive haulage [8], on mine rise [9], on rotary platforms of excavators [9], bridge cranes [4], etc.

Actually, by recuperation [2] is meant the return of kinetic or potential energy which is saved up by an object to its source or the consumer. Predominantly, this refers to the transformation of kinetic energy of the moving vehicle to electric one with its return to contact system or to the accumulator unit. Such systems are rather widely used on locomotive transport and motor transport with the hybrid drive, on the lifting equipment (at load-lowering).

The energy transformation into other kind can be absent. In this case kinetic energy of an object is transferred to the special storage device – a flywheel, and if necessary it is restored to an object without transformation either [3.4]. The scheme of such drive is shown in Figure 1.

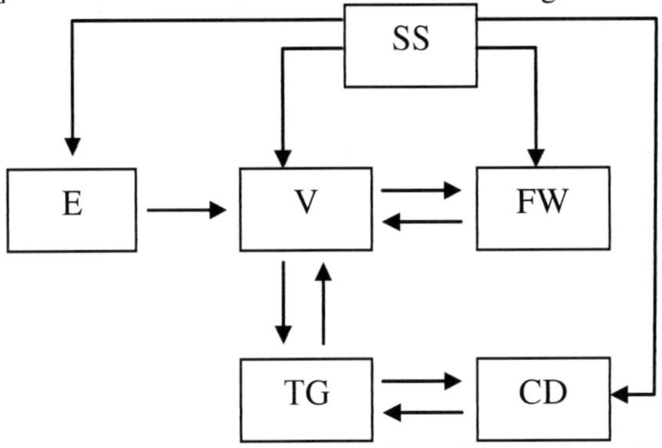

Figure 1. Actuator configuration with inertia braking energy-storage system.
SS – Supervision System; E – Engine; V – Variator; FW – Fly Wheel;
TG – Transmission Gear; CD – Controller Device.

The engine (E) through the variator (V) and the transmission gear (TG) transfers torque to the controller device (CD) in the mode of working motion. At the same time the variator allows the supervision system to slow down the flywheel shaft. Thus, in the mode of working motion the flywheel remains motionless.

In the mode of braking SS unblocks FW, switches (blocks) the engine. As a result kinetic energy of CD and TG through V is transferred to FW. At the maximum acceleration of FW or in case the speed of FW exceeds the speed of the output shaft of V, the supervision system disconnects the flywheel and blocks the output shaft of V. The fly wheel continues to rotate by inertia.

In the mode of starting of SS connects FW through V and TG to the controller device. E can be disconnected. In case the torque value of FW becomes less necessary, SS switches E.FW and E work simultaneously. At the stop of FW the supervision system blocks its output shaft. The drive works as in the mode of working motion.

Generally, the stores can be mechanical (flywheel-type, springing, gravity, etc.), chemical, thermal, electric, hydraulic (hydraulic-pneumatic). Each type of the energy store has the advantages and shortcomings which need to be considered at designing of energy-storage drives of certain motor-cars.

The general requirements to the stores should be considered the saved-up energy conservation with the time minimum losses, the smoothness of transition of their traction mode into braking one and back, a possibility of energy return regulation to the system at start-off, the steady motion and other purposes.

As it is seen from the description of the modes, the energy potential of CD and TG in the mode of braking is to change from the maximum value to zero. At the same time the kinetic energy of FW is to change from zero to the maximum value.

If we take the ratio of the ωoactuator speed and the ωmfly wheel speed for transfer ratio of the im transmission gear between the flywheel and the actuator, then it is seen from the diagram that the transmission gear should have variable gear ratio from ∞ at the beginning of the braking process up to 0 at a motor-car stop. At the same time the transfer ratio should be changed smoothly. To a greater extent, the application of a planetary differential reduction device as the transmission gear meets the case that is proved in [4.5].

The authors have conducted theoretical and experimental researches of the energy-storage braking process which showed operating capacity and efficiency of the braking system in which the fly wheel store is attached by means of the differential planetary reduction device to the actuator drive.

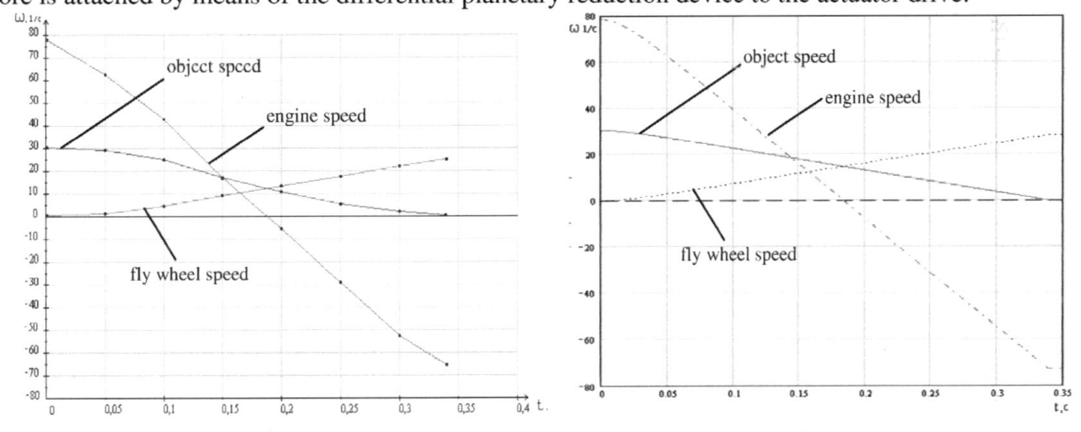

Figure 2. Energy-Storage Drive Gear Elements Speeds Change Diagrams.

In figure 2 the diagrams of the fly wheel braked object and the traction engine speeds change in the course of energy-storage braking on the test stand (Figure 2a), and as a result of the mathematical model equations solution are presented (Figure 2b). The researches have shown the possibility of the adjustable braking of the object to a full stop and the fly wheel acceleration during the entire period of braking. The volume of the kinetic energy which is saved up in the fly wheel store can reach more than 60% of initial energy of the braked object.

The differential reduction devices usage proposals in the energy-storage drive gears of the wheeled vehicles are shown the works [5,6].

3. Multimass Objects Inertia Braking Systems Synthesis Methodology Validation

As it is proved in the works [1.9] kinematic structure of any technical object, equipped with the inertial energy-storage braking system and storage, includes the closed kinematic circuit with the constant or variable value of voltage. For example, for the vehicles connected consistently the value of voltage is defined by a number of parameters:

$$H = f(m_1; m_2; \dots; m_n; \dot{x}_1; \dot{x}_2; \dots; \dot{x}_n; \ddot{x}_1; \ddot{x}_2; \dots; \ddot{x}_n; T_1; T_2; \dots; T_n; \psi);$$
$$\text{at} -|H_{max}| \le H_i \le +|H_{max}|; \qquad (1)$$

where $\dot{x}_1; \dot{x}_2; \dots; \dot{x}_n; \ddot{x}_1; \ddot{x}_2; \dots; \ddot{x}_n$ are corresponding speeds and acceleration;
$m_1; m_2; \dots; m_n$ are connected in series technical objects masses ;
$T_1; T_2; \dots; T_n$ are corresponding masses braking strains value and character;

ψ is tire-to-surface or rail friction coefficient.

The voltage value defines the drive capacity for energy-storage in collaboration with the variator constructional features.

Thus, varying the braking devices parameters, transformation and storage of energy, we can improve and synthesize new traction braking systems.

The general approaches to variable search of new technical solutions on the basis of functional and structural morphological tables are stated in works by V.V. Dloukhy, S. Nurakov, V.P. Bykov, A.V. Andreychikov, A.G. Voytov, etc. K. Rot [1] developed the system of catalogs and tables which allowed to systematize separate knots and devices, providing completeness of engineering development of products. D.P. Volkov, V.Ya. Krikun, K.K. Shestopalov, V.I. Balovnev, L.A. Chmara [1] and some other scientists were engaged in the development of the generalized structural-morphological schemes and systematization of functional modules in the field of construction and materials-handling machines. In the works of G.Sh. Hazanovich, Yu.M. Lyashenko, A.S. Nosenko and other authors [1,7,9] the questions of synthesis of new technical solutions of mine loading and transport modules of heading machines on the basis of constructive and functional structure are considered. The creation of structural-morphological classifications is constructed by V.I. Solod and G.I. Solod in relation to a number of mining machines taking into account their specifics and structural systematization.

The principle of consecutive synthesis of models of admissible options of creation of separate elements, blocks and system in general with the subsequent choice on synthesizable model of structure of a system of the best option of its realization and development is exploited by A.D. Tsvirkun, V.M. Odrin and others. A.N. Drovnikov, V.S. Isakov [1] offered the generalized model and the technique allowing to develop the functional mechanisms of materials-handling, construction and mining machines using the intense closed circuits.

Such approaches allow to systematize and unify systems, modules, blocks, and to choose the most rational and perspective constructive decisions, to systematize the general principles and to unify calculation procedures. Thus, in the works by L.A. Chmara, A.P. Holodov, V.F. Scherbakov, A.N. Drovnikov, G.M. Vodyanik, V.S. Isakov [1.9] the researches in the field of creation of recuperative systems of drives for construction mining machines with use of hydro-pneumatic accumulators and thermal accumulators are presented.

In the work [1] the attempt to create the classification of the tandem-coupled machines and their modules based on possible types of the connected masses movements is stated. The scheme depicted in Figure 1 does not consider the nature of the masses movement, the possibility of their mutual movement and change of the mechanism structure including "singular points".

Systematization of any technical systems having the intense closed kinematic circuits in the structure and a search technique on its basis of new technical means is sufficiently detailed in the works [1.9]. However, the technique does not consider the system of energy-storage, mobility the coupled elements mobility, etc.

In this regard, we can propose the generalized structural diagram (Figure 3) of a multimass object with the inertial energy-storage braking system and a morphological matrix of possible technical solutions based on the scheme depicted in Figure 1.

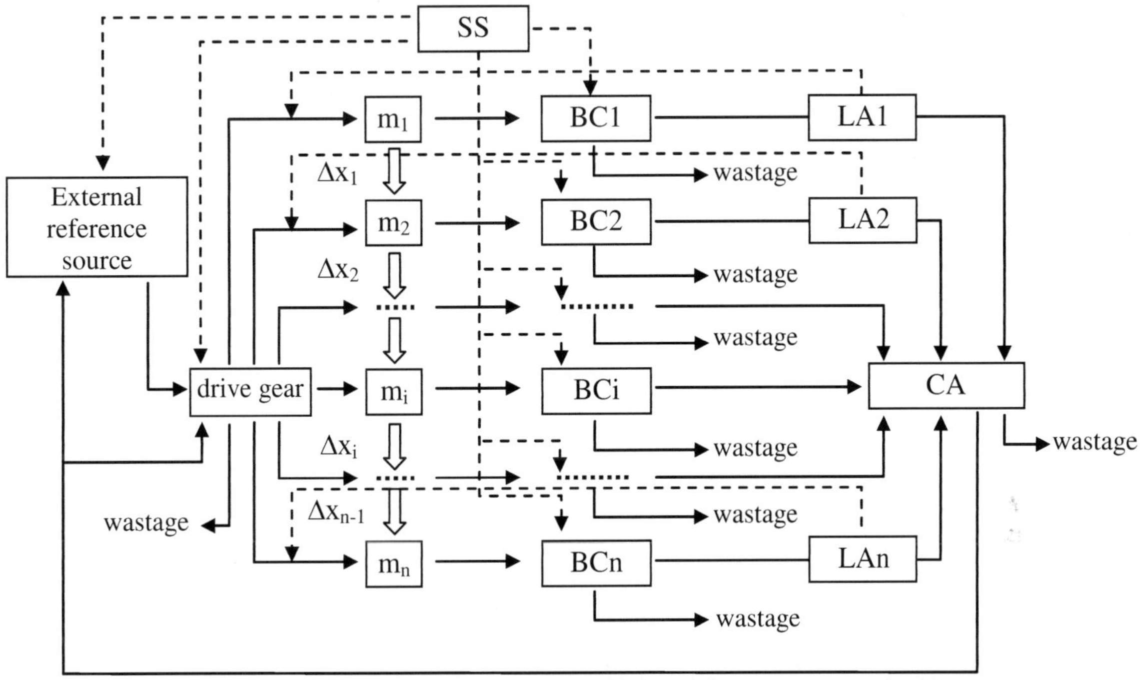

Figure3. – Multimass object structural scheme with inertial energy-storage braking system.

$m_1; m_2; ...; m_n$ are masses of the series-connected objects;

Δx1; Δx2;…; Δxi;… Δxn-1 are moving variable couplings permitting masses relative movement;

BC1; BC2;…;BCi;…BCn are brake-conventors making kinetic energy transfer of the moving object into another type and transmission to its accumulator;

LA1; LA2;…LAn are local accumulators;

CA is central accumulator.

SS is supervision system.

The inertial traction-braking system provides the set mode of the movement of the objects group $m_1; m_2; ...; m_n$ due to braking at least one of the masses. Other objects accomplish braking, using the efforts arising in the conjunctions Δx1; Δx2;…; Δxi;… Δxn-1 allowing relative motion of the masses.

Transformation of energy to another type can be absent. The wastage when braking, transforming and storage of energy are inevitable and natural. Depending on design and functional features of the brake converter BC1; BC2;…;BCi;…BCn wastage value will be various.

Thus, the possible solutions matrix will include:

$$M = \{\sum_{n-1}^{1} m_i x \sum_{n-1}^{0} \Delta x_i x \sum_{n}^{1} BC_i x \sum_{n}^{0} LA_i x \sum_{1}^{0} CA\} \quad (2)$$

Moving variable couplings can accept zero value, i.e. they are not interconnected. For example, movement and braking of the trolley with simultaneous load-lowering. Local accumulators in the presence of central can be absent, i.e. they possess value “0”. Transmission of energy is carried out to central accumulator directly. CA can possess values “0” or “1”. The matrix can include the additional couplings providing transmission of energy to the consumer i.e. coupling $Ln_i \rightarrow m_i$.

A variety of masses and the nature of their movement, and consequently the storage, it is possible to show on derived schemes from the matrix (Figure 4):

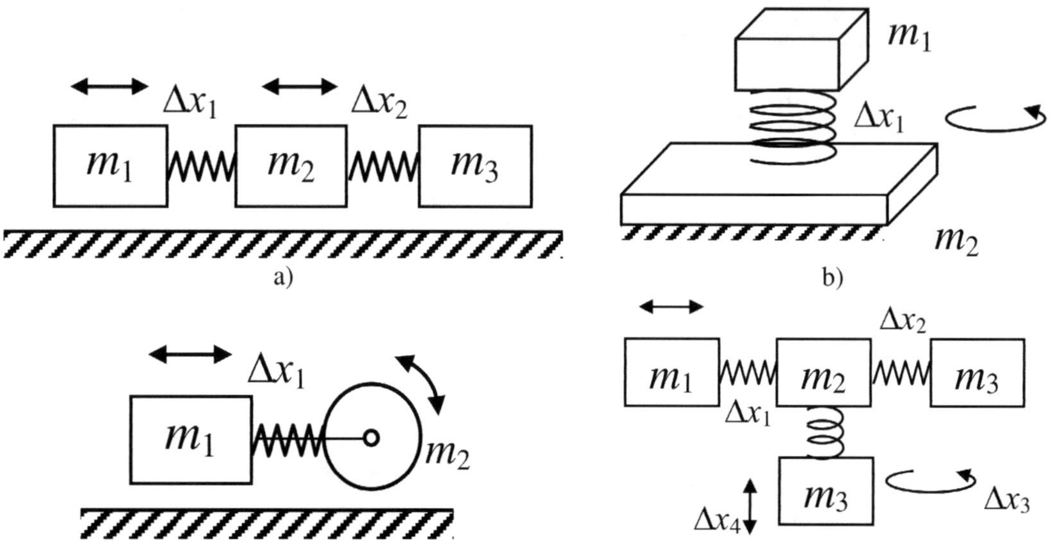

Figure 4. Derived elementary schemes.

In Figure 4a the scheme of the tandem-coupled objects m_1; m_2; $m_3$is shown, having moving constraints Δx_1; Δx_2.Such scheme corresponds, for example, to the underground locomotive haulage train. In Figure 4b the scheme of the single-mass object, having rotary motion about motionless (referred to the reference) masses, is depicted. The work of the excavator rotary platform can be an example: braking in motion for digging bucket unloading and return of the platform and a digging bucket to a drift. The scheme 4c presents the dual-mass system having the mass m_1 moving linearly and which is movably connected to the mass $m_2$and also the mass m_2 moving rotationally and having rather motionless referencem_1. For example, tractor-hauled roller having overrunning brake and at the same time a brake converter of the roll rotary motion relating to the tow-barattached to the tractive vehicle. Figure 4d displays movement of three masses simultaneously connected in tandem and one mass rotating concerning the mass 2 which is rather motionless reference for mass 4. As an example we can present the movement of the jib type crane crab with the boom swing, simultaneous load-lowering and movement of the very crane (theoretical case).

The researches [9] conducted by the authors showed that at traction-brake systems operation the maximum effect is reached by the first four-five consistently connected objects. In Figure 5 the braking process parameters of the train equipped with inertial brakes are presented.

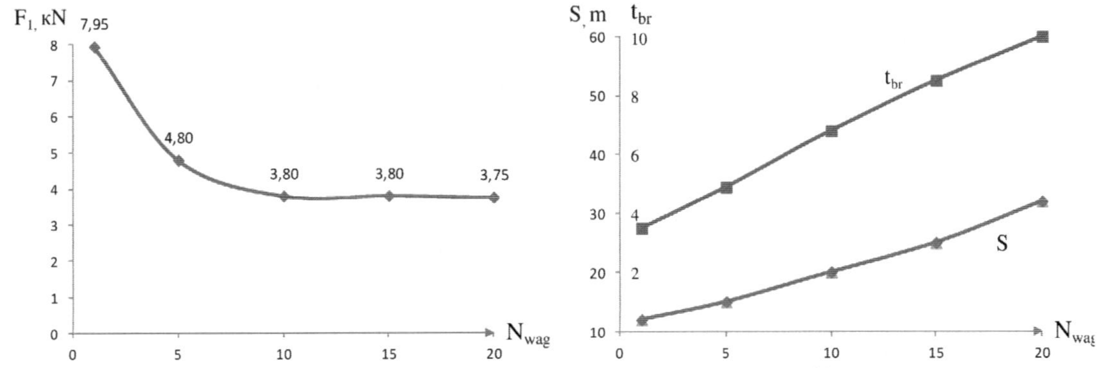

a) b)

Figure 5. a,b – The interdependencies of braking strain (a) and braking path (b) on number of in sequence connected wagonettes in operation.

The working capacity and efficiency of energy-storage braking application is also confirmed while investigating the dynamics of the hydraulic excavator rotary platform with the hydraulic impulse storage unit which is built in hydraulic system. In considering the studied system functioning dynamics the interrelation of all links (lift arms, arms, digging bucket) is taken into account.

In Figure 6 the changes of pressure in drain lines of hydraulic actuators in the course of the rotary platform braking are presented.

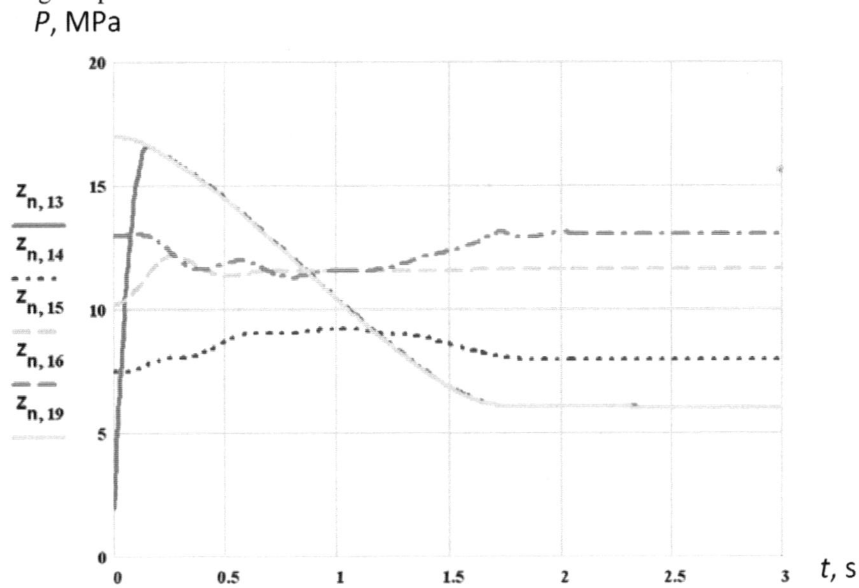

Figure 6. – Pressure change in drain lines of the hydraulic actuator $Z_{n,13}$, actuation cylinder of the lift arm $Z_{n,14}$, arm $Z_{n,15}$, digging bucket $Z_{n,16}$, accumulator $Z_{n,19}$.

The results of the conducted researches showed a possibility of accumulation in the hydraulic impulse storage unit to 90% of the kinetic energy of the rotary platform.

4. Conclusion

The conducted analysis of the existing energy-storage and energy-conserving brake devices (including developed and tested by the authors of the article) allowed to establish their operability and efficiency of functioning in various operating modes for the tandem-coupled multimass objects with the common support platform, to reveal the general structural regularities of such devices for construction, materials-handling, roadway and mining machines.

The multimass object principal structural model developed on this basis with the inertial energy-storage brake system can be used in the structural-morphological analysis and synthesis of new energy-storage systems and generic modules, development of consistent methodological approaches to justification and calculation of their rational parameters.

The developed methodology of the structural-morphological analysis allows to mark out standard structures taking into account hierarchical levels of specification: from the general functional and structural diagrams at the high hierarchical levels to the general basic and constructive – on the lowest, taking into account concrete characteristics of the designed technical object.

Theoretical and experimental studies confirmed a possibility of use of the central accumulator of multimass objects energy for reasonable redistribution of energy resources among potential consumers.

References

[1] Isakov V S 2005 Inertial transport unit on the basis of closed kinematic circuits. – Rostov-on-Don.: Magazine publ. house. «Higher educational institution publ. house North-Caucasian region», – 140

[2] Leonov I V , Leonov D I and Barbashov N N 2012 Animation model of materials-handling machine with energy storage device / Izvestiya of higher educational institutions. Machine engineering, **9** – 45-50

[3] Gulia N V and Serkh A G 1984 Fly wheel energy-storage usage possibilities // *Industrial transport*, **11** – 4-6

[4] Erejskiy V D , Erejskiy A V , Isakov V S , Krapivin D M , Gudikov G G and Kataev G N 2002 Transport vehicle energy-storage brake method / Patent of invention 2193706 by demand 2000129423

[5] Isakov V S, Erejskiy V D and Erejskiy A V 2001 Energy-storage brake application investigations results / The works of international scientific and technical conference Interstroymekh – 2001. Saint-Petersburg: Publ.house SPSTU 135-139

[6] Wilson M 1975 American scientists and inventors (translated from English) - Moscow.: Znanie, – 135

[7] Pachikov I S , Ostapenko V A and Gasyukov V S 1976 Mine car brake operating on pressing force grasping / *Ukranian Coal*, Kiev **9**

[8] Isakov V S, Erejskiy V D, Polezhaev V G and Ivanov O P 1979 Self-contained hydraulic brake unit / Coal machine engineering / *CSIEI Coal* **7** 3-4

[9] Vodyanik G M and Isakov V S 2006 Mathematical modeling of energy-storage and energy-conserving brake systems on the basis of closed kinematic circuits / SRSTU – Novocherkassk: Magazine staff «Higher education institutions publ. house. Electrical engineering»

[10] Jungmeister D A 2002 Mining machines units building-up on the basis of morphological analysis. – Saint-Petersburg. Mining institute. – Saint-Petersburg 152

Improving the efficiency of existing methods of diagnosing the hydraulic drive of road-building machines

U A Piramatov[1] and K G Pugin[1,2]

[1]Federal State-Funded Educational Institution of Higher Professional Education «Perm National Research Polytechnic University» Perm, Russia

[2]Federal State-Funded Educational Institution of Higher Professional Education «Perm State Agro-Techological University» Perm, Russia

Abstract. The article presents modern methods of diagnosing hydraulic drive. The drawbacks of modern diagnostic methods are revealed. The negative consequences of the effect of stagnation of the working fluid. The results of studies of working fluid samples are given. Suggested ways to solve the existing problem. A method for determining the nodes subject to the effect of stagnation of the working fluid is proposed. The effect of solving the problem of stagnation of the working fluid is determined.

More attention is paid to monitoring the condition of the hydraulic drive of road-building machines, this is due to the desire to increase the service life of the hydraulic drive, as well as to minimize the number of unexpected failures of the hydraulic drive elements. In road construction machines, 70-90% of hydraulic failures are associated with contamination of the working fluid. [1] The service life of the hydraulic drive and the number of failures of its elements directly affect the final economic indicators of road construction machines, this is especially important for the operation of machines in countries with seasonality of construction work. Economic efficiency is one of the key parameters determining the cost of work. Reducing the cost of work will increase the profits of road-building companies, and the possibility of increasing profitability makes it important to increase the reliability and reliability of the hydraulic drive and its elements.

Today, there are many methods for monitoring the condition of the hydraulic drive of road-building machines. These diagnostic methods are based on the analysis of the elements of the hydraulic drive or on the analysis of the state of the working fluid. A prominent representative of the first group is the statistical parameter method; this diagnostic method determines the condition of the hydraulic drive by comparing the parameters of leakage, flow, pressure and efficiency of the hydraulic drive elements. Diagnostic methods based on the analysis of the state of the working fluid are based on the study of the working fluid for its purity. The purity of the working fluid is regulated by GOST 17216-2001, this standard defines 19 classes of purity of the working fluid, according to this standard, the working fluid is considered clean if it contains less than 0.064% of pollutants. Also important is the parameter of the acidity of the working fluid, determined by the number of milligrams of potassium hydroxide necessary to neutralize all the acidic components contained in 1 gram of the working fluid, the maximum permissible value is 1.5 milligrams.

More attention is paid to monitoring the condition of the hydraulic drive of road-building machines, this is due to the desire to increase the service life of the hydraulic drive, as well as to minimize the number of unexpected failures of the hydraulic drive elements. In road construction machines, 70-90% of hydraulic failures are associated with contamination of the working fluid. [1] The service life of the hydraulic drive and the number of failures of its elements directly affect the final economic indicators of road construction machines, this is especially important for the operation of machines in countries with seasonality of construction work. Economic efficiency is one of the key parameters determining the cost of work. Reducing the cost of work will increase the profits of road-building companies, and the possibility of increasing profitability makes it important to increase the reliability and reliability of the hydraulic drive and its[] elements.

Statoparametric method is widely used due to high efficiency, but for the operation of this method it is necessary to determine the main parameters of the hydraulic drive of a fully functional machine, and then compare the parameters with them. [2] Another advantage is the ability to determine the state of individual elements of the hydraulic system. [3] The disadvantages of the system are the need for expensive research equipment, the research process is also very labor intensive, and the need to introduce special equipment to the hydraulic system requires relocating the equipment to a clean room to prevent contamination of the system, Felix Angie, Jennifer Harding and Jacqueline Glass describe in their article problems on the diagnosis of hydraulic, faced by enterprises operating equipment in the field, also noted that this problem complicates economic planning and increases the number of unanticipated failures. [4] Technological progress is not in place, new diagnostic methods are replacing the old, more advanced and less labor-intensive. Methods of analyzing the state of the hydraulic drive based on the state of the working fluid are becoming more common in recent years, the company Caterpillar Inc. In their cars, she implemented an additional option, by which payment the enterprise operating equipment gets the opportunity to send samples of the working fluid to specialized laboratories. The laboratories analyze the working fluid, on the basis of which the frequency of replacement of the working fluid is determined, and on the basis of the contamination intensity of the working fluid and on the type of pollutant, the enterprise operating the machine is provided with information on the presence of an element in the system with increased wear. [5] In addition, the Hitachi EX5500-6 excavator implements a system for monitoring the state of the working fluid in real time.

Over time, the hydraulic drive of the road construction machines became more and more complicated; at present, this has led to the possibility of the emergence in the hydraulic drive of zones in which the direct functions of the working fluid are not performed. If there are nodes in the system that are distant from the hydraulic distributor, a situation may arise in which the working fluid will not be updated due to the fact that the volume of fluid in the pipeline is higher than the volume of fluid in the cylinder cavity. In connection with the complication of the hydraulic drive of the road-building machines, the question of the division of the hydraulic drive into zones with complete, partial and obstructed circulation of the working fluid is becoming increasingly important. The division of the hydraulic drive into zones will improve modern methods of diagnosing the hydraulic drive.

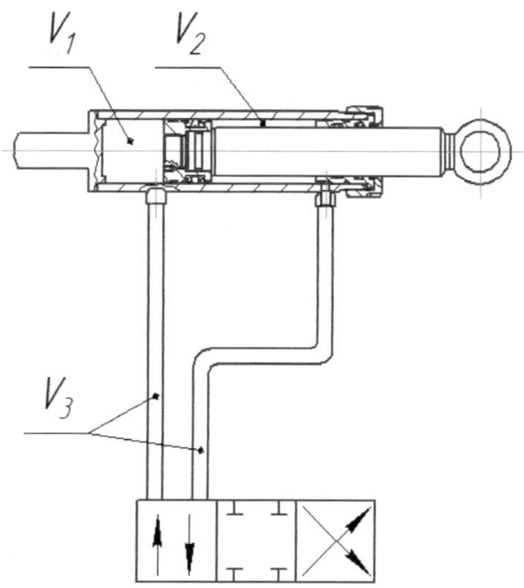

Figure 1. simplified hydraulic circuit.

Figure 1 shows a simplified diagram with a hydraulic cylinder and a hydraulic distributor, in that case, if the volume of working fluid in the piston cavity (V1) is less than the volume of working fluid in the pipeline (V3), the process of updating the working fluid will be difficult. The main functions of the working fluid are lubrication of rubbing surfaces, cleaning from wear elements, as well as heat dissipation. In case of problems with the circulation of the working fluid, respectively, there are problems with the lubricity, cleaning and cooling functions of the working fluid, this in turn can lead to a serious reduction in the resource of remote elements of the hydraulic drive.

Over time, not only diagnostic methods and hydraulic actuators are improved, but also hydraulic fluid. Modern hydraulic fluid incorporates many additives that have anti-corrosion, anti-foam, anti-oxidant properties. In the event of a stagnation effect of the working fluid, the additives will more quickly deplete their resources, as a result of which they will lose their properties, which in turn leads to a further reduction in the life of the hydraulic drive elements. [6] In addition, possible inclusions of inclusions designed to improve the properties of the working fluid in the sediment, respectively, increasing the degree of contamination of the working fluid.

To verify the existence of this effect, sampling of the working fluid from the HBM TG190TA-4 grader was made. The working fluid was replaced in April 2018, sampling was carried out in March 2019, the motor grader produced 3,000 machine hours, in the summer the motor grader was involved in road construction works, in the winter period road maintenance works were carried out, the dozer was accounted for about 5-10% of the work. Samples were taken from the piston cavity of the main blade and from the rod cavity of the bulldozer blade, this was done in order to bring the samples from the main cylinder of the main blade to the working fluid circulating in the system as close as possible, and to take a sample of the working fluid from the conditions of a difficult update.

Table 1. The results of laboratory studies of the working fluid.

№	Sample name	Total acid number, [mg] KOH for 1 g of oil according to GOST 5985	Mass fraction of mechanical impurities, [%] according to GOST 6370
1	Blade hydraulic cylinder (piston cavity)	0,58	0,92

2	Hydraulic cylinder of the remote blade (rod cavity)	0,27	0,68

As a result of the study of samples of the working fluid (Table 1), the value of acidity and the mass fraction of mechanical impurities were determined. In the samples there are serious deviations of the indicators, the values in the dozer dump are lower than in the main dump, this is due to the low workload of the dozer dump. Before sampling, the full working stroke of the hydraulic cylinders was performed, respectively, the difference in acidity values and the mass fraction of mechanical impurities indicates the difficulty of updating the working fluid. Studies have also been conducted to determine the types of mechanical impurities.

Table 2. Studies of the type and amount of metals in the samples

№	Sample	Mechanical impurities		
		Lead, [mg / kg]	Iron [mg / kg]	Chromium [mg / kg]
1	Hydraulic cylinder of the main dump	2,36	9,27	0,07
2	Cylinder of Dozer Blade	0,42	8,8	-

The study (table 2) revealed the presence of lead and iron in the samples, and chromium inclusions were also found in the hydraulic cylinder of the main blade. By the amount of impurities in the samples, it is also clear that updating the working fluid is difficult. It is especially worth noting the presence of chromium in the sample from the main blade and its absence in the hydraulic cylinder of the bulldozer blade, this gives us reason to assume that the degree of renewal of the working fluid in the hydraulic cylinder of the bulldozer blade is minimal.

It is worth noting that the process of replacing the working fluid in the hydraulic drive will not be a solution to the current problem. The process of replacing the working fluid consists in draining the working fluid from the hydraulic tank and then filling it with a new fluid. [7] In this case, the fluid from the pipeline practically does not drain, for units with difficulty updating the working fluid, this leads to a gradual accumulation of wear elements and, accordingly, to a decrease in the life of the hydraulic drive elements.

It is possible to solve the existing problem by introducing additional filters into the pipeline, introducing check valves in the hydraulic drive, or introducing an additional hydraulic distributor. In addition, a technique is needed to determine the nodes that have problems with the circulation of the working fluid.

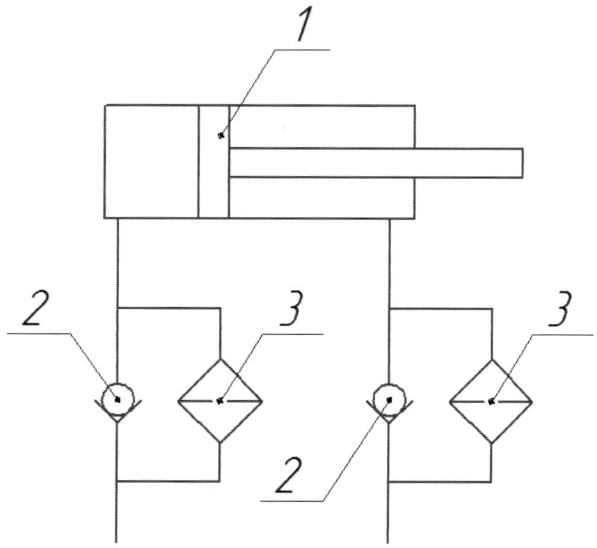

Figure 2. filter plant
1- hydraulic cylinder, 2- check valve, 3- filter.

Figure 2 shows the configuration with the introduction of additional filters into the system (3) that clean the working fluid from contaminants. In addition, check valves (2) are also provided in the layout, one for each line. Check valves are embedded in the filter assembly in order to force the working fluid to circulate through the filter. When the rod extends, the working fluid flowing into the piston cavity of the hydraulic cylinder (1) from the distributor passes through the check valve, and in the case of fluid movement in the opposite direction, the check valve forcibly passes the working fluid through the filter of the rod cavity. The advantage of this arrangement is its simplicity, which allows to reduce the contamination of the working fluid without the need for high costs. A serious disadvantage of this arrangement is the remaining negative effect of the developed additives, since the filtering unit ensures the cleaning of the working fluid exclusively from the elements of mechanical wear. The introduction of the filter unit will significantly reduce the wear of the rubbing surfaces of the hydraulic cylinder, but will retain the negative impact of the additives developed in the remote nodes.

IOP Conf. Series: Materials Science and Engineering **786** (2020) 012007 doi:10.1088/1757-899X/786/1/012007

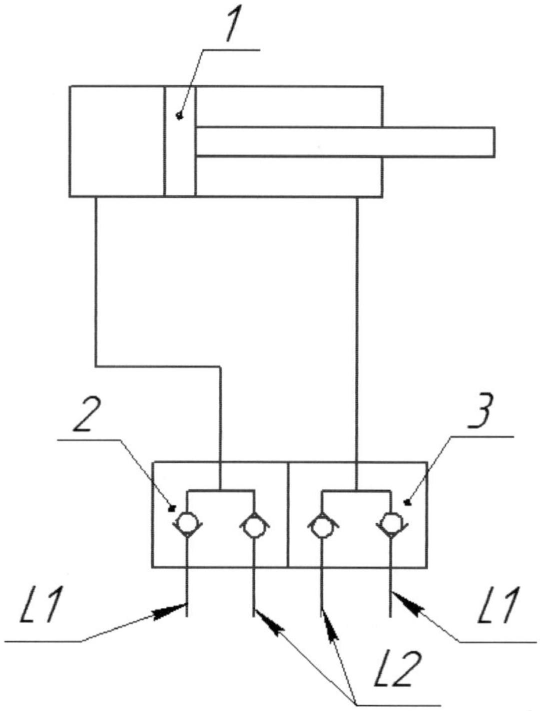

Figure 3. forced circulation unit
1- hydraulic cylinder, 2- block of reciprocating valves of the piston cavity,
3- block of check valves of the rod line.

The introduction of additional check valves (Figure 3) will improve the circulation of the working fluid in the nodes remote from the control valve, which will completely solve the problem of stagnation of the working fluid. The principle of operation is the introduction of two additional lines to the hydraulic tank (L2), which discharge the working fluid from the working cavities to the hydraulic tank, and the hydraulic cylinder (1) will receive liquid from the hydraulic tank through the existing highways (L1), which ultimately will ensure full circulation of the working tank. fluid at remote sites. When the hydraulic cylinder rod extends, the working fluid flows from the L1 line of the piston cavity check valve block (2), and from the rod cavity, the working fluid flows through the rod valve check valve block (3) through the floor of the L2 line. The introduction of additional check valves will reduce the amount of mechanical impurities in the working fluid involved in the work, as well as provide an update of the additives in the working area. A significant disadvantage of this arrangement is the need to introduce additional pipelines.

For the development of methods for identifying nodes with difficult circulation of working fluid, it is necessary to conduct full-scale experiments to determine the degree of updating of the working fluid due to mixing. In identifying the nodes in which the circulation of the working fluid is difficult, it is necessary to introduce elements into the hydraulic actuator that ensure the forced circulation of the working fluid or its cleaning. The introduction of an additional unit for cleaning the working fluid will reduce the effect of the effect of obstructed circulation of the working fluid. It is necessary to note the high load on the filters in the remote nodes and ensure their proper maintenance. Introduction to the zones remote from the hydraulic distributor of the components ensuring the circulation of the working fluid, in the form of check valves or additional hydraulic distributors.

The introduction of additional methods for determining nodes subject to stagnation of working fluid and additional elements designed to make working fluid in remote sites meet the regulatory requirements of purity, will increase the efficiency of methods for diagnosing hydraulic drive based on the analysis of working fluid, which will increase the service life of hydraulic drive construction and road machines and reduce the number unforeseen failures. This will ultimately lead to a decrease in machine downtime, an increase in its service life, a reduction in the cost of repairing the hydraulic drive elements, and an increase in efficiency.

References

[1] Logvinov L M , Kovalev M A and Khablo I I 2007 Improving the reliability of hydraulic systems of aircraft due to the analysis of the parameters of particles of pollution of the working fluid // *Bulletin of the Samara State Aerospace University* **1** 196-200
[2] Chilikin A A and Trushin N N 2014 Comparative analysis of modern methods for diagnosing the state of hydraulic systems. // *Bulletin of TSU. Technical science* **3** 117-127
[3] A Helwig and K Maier 2015 An Optoelectronic Monitoring System for Aviation Hydraulic Fluids// Procedia Engineering **120** 233-236
[4] Felix Ng , Jennifer A and Harding 2017 Jacqueline Glass. Improving hydraulic excavator performance through in line hydraulic oil contamination monitoring// *Mechanical Systems and Signal Processing* **83** 176–193
[5] Androsov V V 2017 Modern methods of diagnosing hydraulic systems of machines for environmental engineering // "The role of environmental management of rural areas in ensuring the sustainable development of the agricultural sector" Materials of the international scientific-practical conference Moscow
[6] Blum K H and Ellenrieder K 2017 Validated Hydraulic Fluids for Increased Hydraulic Life //*ATZoffhighway worldwide* **10(1)** 50-55
[7] Hydraulic fluids: Controlling contamination in hydraulic fluids// *Filtration & Separation* 2010 **47(3),** 28-30

Energy-saving technology and equipment for the production of composite binder using local raw materials

A A Romanovich[1,*], E Amini[2], D M Annenko[1] and E I Chekhovskoy[1]

[1]BSTU named after V.G. Shukhov, 46 Kostyukova Street, Belgorod, 308012, Russian Federation

[2]K. N. Toosi University of Technology, Tehran, Iran

*E-mail: AlexejRom@yandex.ru

Abstract. The article presents the technology and energy-saving equipment for producing a composite binder with a reduction in the energy cost reduction of the process up to 50%, due to the synergistic effect of mechanical activation of the raw material mixture with the replacement of the clinker component with a mineral hydroactive additive. The energy efficiency of the line is achieved by the joint work of the press-roll unit, and the rotor-vortex mill of superfine, which allow the grinding of raw materials to a dispersion with a specific surface area of 4000...4500 cm2/g. The resulting equation for calculating the effort of grinding materials, which is implemented between the conical rollers in the press-roll unit (PRU). It has been established that the power spent on grinding the material and overcoming the friction in the axles of the pivots rolls and proposed the equation for calculating its variable.

1. Introduction

The rational use of mineral resources in the construction industry is one of the main issues in solving the tasks of energy and resource saving. At the same time, the building materials industry can be a unique waste disposal of other industries to produce new composite binders and a wide range of high-performance materials based on them. Every year in the world about 26 billion tons of rock mass is extracted from the bowels of the planet, about 4.5 billion tons of various minerals are processed, of which only from 2 to 10% becomes a useful product, 2.7 billion tons pass into the category of industrial waste of the field of use, which, as a rule, are not determined [1-3]. Mining is accompanied by the accumulation of billions of cubic meters of waste rock, which occupy hundreds of thousands of hectares of land. As a result of the accumulation on the surface of our planet of powerful man-made formations, environmental pressure occurs. By the beginning of the XXI century mankind has accumulated such amount of man-made waste that they can be called new deposits. Therefore, the problem of their use with minimal processing costs is acute.

Today, the problem of waste reduction is closely related to the issues of sustainable economic development: in conditions of ever-growing demand for raw materials, while reducing the available reserves and the ultimate exhaustion of natural resources.

2. Main part.

The solution to the problem of industrial waste depends on a several factors, but the most effective solution is the acceptance of wasteless technology. With the integrated use of raw materials, industrial waste from some industries can be the raw materials of others. Rational integrated use of mineral natural resources, involvement in the production of man-made waste and secondary products of various industries for the production of building materials will significantly reduce the burden on the environment and lead to resource and energy saving. One of the locomotives of construction, ready for processing and use of man-made waste is the cement industry.

In our country, with account for production capacity of cement enterprises, there has been observed an acute shortage of binders. According to the forecasts of the Ministry of regional development, which developed a strategy for the development of the building materials industry until 2020, the need for the construction industry in cement in ten years will almost double (to 98 million tons). One way to quickly increase the volume of mineral binders is to obtain composite binders (OCB) using local raw materials and man-made waste. The use of OCB with the use of industrial waste will not only reduce the consumption of clinker component in the binder, but also to obtain high-performance building materials based on them. It is known that the reserves of waste in the form of slag of electric steel furnaces of the Oskolsky electrometallurgical works are more than 4 million tons, with an annual replenishment of about 500 thousand tons. Currently, there are a number of developments on the use of slag in the production of building materials, construction, repair and reconstruction of roads [4-7]. It has been established that the use of slag materials in the foundations of roads and components of asphaltic concrete road, as sand aggregate and mineral powder, reduces the cost of construction and improves the quality and reliability of the road surface.

Also known technology for producing mineral binders using slag waste from blast furnaces of metallurgical production, one of the main components of which is the fine grinding of its components, including slags.

However, the initial product, slag, is a fairly hard material with a grain strength of P = 300-320 kg/cm2, it has abrasive properties and has larger dimensions than the granule size of the mineral additive, which requires grinding and comminution. At this time, there is a wide variety of grinding units and systems used for grinding materials and industrial wastes, which have different strength characteristics. The analysis of technical and economic efficiency from the use of existing and developed grinding systems [9-12] made it possible to identify as the most effective those systems that implement the principle of stepwise grinding of materials with the removal of the coarse grinding stage into a separate unit that performs a more economical method of destruction charge (volume- shearing deformation), than, for example, impact and abrasion. Scientists of BSTU named after V.G. Shukhov developed industrial technology and energy-saving equipment to obtain a composite binder with a decrease in the energy intensity of the process up to 50% due to the synergistic effect during the mechanical activation of the raw mix with the replacement of the clinker component of the mineral hydroactive additive (Figure 1). The technological process (Figure 1) is based on the sequential input of components in dispersed phases into the raw material mixture in the grinding path and at the stage of product separation with certain dispersed characteristics.

The increase in the energy efficiency of the line is achieved by the joint operation of the press-roller unit, which is the development of scientists of the BSTU named after V.G. Shukhov, and the rotary vortex mill superfine comminution new design. In this case, a synergistic effect is achieved. The technology of obtaining a composite binder implies the grinding of raw materials to dispersion with a specific surface area of 4000 ... 4500 cm2 /g.

IOP Conf. Series: Materials Science and Engineering **786** (2020) 012008 doi:10.1088/1757-899X/786/1/012008

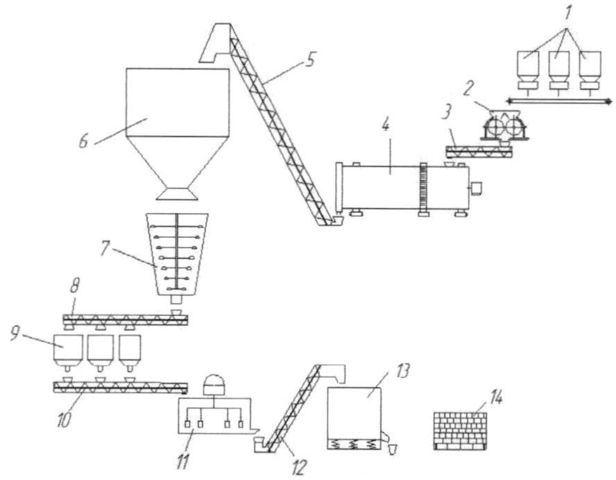

Figure 1. The process flowsheet of the production of composite materials: 1,2,3 – hoppers for raw materials; 2 – press roller unit; 3,5,8,10,12- feed screw; 4- dryer drum; 6 – storage bin; 9-additive bin; 11- mixing machine; 13- packing machine; 14-finished-products storage area.

Reducing the specific energy consumption is achieved, including through the use of a press-roll unit (PRU) at the stage of preliminary grinding with working bodies implementing volume - shearing deformation of materials (Figure 2).

a) b)

Figure 2. Press-roll unit: a) general form, b) view of the rolls in the unit [13].

At this technological line is set down the press-roll unit, having the following technical characteristics. (Table 1)

However, the application of energy-saving technology using various regional raw materials and industrial wastes requires scientifically-based recommendations for calculating and designing the energy parameters of the main grinding units of the line. One of the important parameters for the application of the press-roller unit is the power consumption of the engine, which directly depends on the grinding force of the materials realized in the cone-shaped rollers.

Table 1. The technical characteristics of press-roll unit.

№	Technical characteristics	Parameter value
1	Geometrical dimension of rolls	0.4×0.3 m
2	Dimension tapering of rolls	0.3
3	Roll opening	(3-8)×10^{-3} m
4	Circumferential speed roll rotation	0.8 m/sec
5	Generating capacity of drive rolls	2×7.5 kW
6	Maximum backward force	90×10^{4} N
7	Unit capacity	5-7 tons per hour

To determine the force arising between the conical rollers in the process of grinding materials, assume that from a certain angle γ_1, which corresponds to the coating thickness H_1, deformation of the crushed charge is proportional to the specific pressure P_1 (Figure 3). Then instantaneous value of angle γ_T corresponds to the specific pressure P_T. On the basis of the above it can be inferred that:

$$\gamma_T H_1 = \gamma_1 H_T \text{ or } P_T H_2 = P_2 H_T \text{ ,} \tag{1}$$

where P_2 - centerline pressure, MPa.

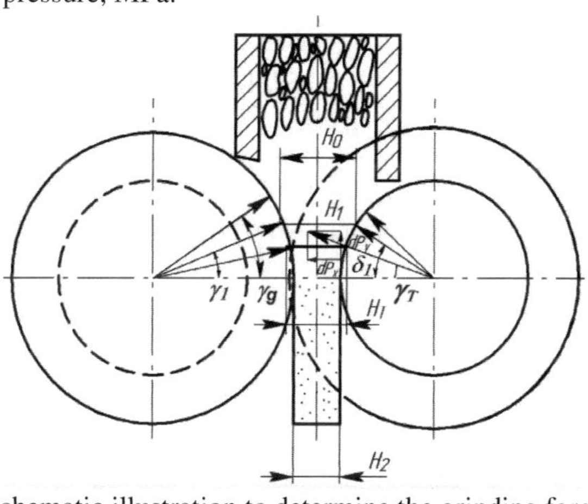

Figure 3. Schematic illustration to determine the grinding force of materials.

Considering the fact that $H_1 - H_2 = 2R_{av}(1-\cos\gamma_1)$, where R_{av} - midradius of rolls; H_2 thickness of the layer of material on the center line, m.

Using the ratio (1), deduce:

$$H_2 = \frac{2R_{av}(1-\cos\gamma_1)}{\gamma_1 - \gamma_2} . \tag{2}$$

From figure 3 exist, that the thickness of the current material layer is equal to:

$$H_T = 2R_{av}\cos\gamma_T - 2R_{av}\cos\gamma_1 . \tag{3}$$

Then variable expression (1) with regard for (2) and (3) will take the following form:

$$\frac{P_T}{P_2} = \frac{2R_{av}\cos\gamma_T - 2R_{av}\cos\gamma_1}{2R_{av}(1-\cos\gamma_1)/(\gamma_1-\gamma_2)} . \tag{4}$$

Perform shortcut in recognition that measure of the angle γ_2 on line centers equal to 0°, than formula (4) be written as:

$$P_T = P_2 \frac{\cos\gamma_T - \cos\gamma_1}{1-\cos\gamma_1} . \tag{5}$$

Let us introduce axis of reference «x» and «y», and project the elementary force dP on axis «x», its projection equals:

$$dP_x = dP_T \cos\gamma_T \cos\alpha , \tag{6}$$

where α - angle of inclination forming the working surface of the rolls.

On the entire width of the working surface of the roll, taking into account (5) having:

$$dP_x = \frac{P_2 B R_{av}}{1-\cos\gamma_1}(\cos\gamma_T - \cos\gamma_1)\cos\alpha\cos\gamma_T d\gamma_T , \tag{7}$$

or $$P_x = \frac{P_2 B R_{av}\cos\alpha}{1-\cos\gamma_1}\left(\int_0^{\gamma_1}\cos\gamma_T d\gamma - \cos\gamma_1\int_0^{\gamma_1} d\gamma\right)\int_0^{\gamma_1}\cos\gamma_T d\gamma .$$

1) Upon integrating:

$$P_x = \frac{P_2 B R_{av}\cos\alpha}{1-\cos\gamma_1}(\sin\gamma_T - \gamma_1\cos\gamma_1)\sin\gamma_1 . \tag{8}$$

For the sake of simplicity next computation makes use Maclaurin formula and distribute the sine and cosine in row, in which we restrict ourselves to the first two variables, and discard the rest as small values. Then the variable expression takes (8) the following form:

$$P_x = \frac{2P_2 B R_{av}\cos\alpha}{\gamma_1^2}\left[\gamma_1 - \frac{\gamma_1^3}{6} - \gamma_1\left(1-\left(\frac{\gamma_1}{2}\right)^2\right]\right. . \tag{9}$$

After the appropriate transformation, we obtain an equation for calculating the effort of grinding materials between the PRU tapered rollers.

$$P_x = P_2 B R_{av}\gamma_1\cos\alpha . \tag{10}$$

Determine the motor capacity, spent on grinding materials in PRU with conical rollers, according to the following equation:

$$N_{dv} = \frac{N_1 + N_2}{\eta} , \tag{11}$$

where N_1 and N_2 - supplied power for grinding material and overcoming friction in the roll axles, respectively , W ; η - coefficient drive efficiency.

The supplied power for grinding material in the tapered rolls can be determined by the equation:

$$N_1 = M_p\omega , \tag{12}$$

where M_p – resulting moment, arising from PRU in process a tapered roll profile when grinding material, Nwm; ω - angular velocity roll rotation, radian per second.

Determine the value resulting moment that is necessary apply to rolls to ensure their rotation during the grinding process. According to analytical model, presented in the figure 4, we find that:

IOP Conf. Series: Materials Science and Engineering **786** (2020) 012008 doi:10.1088/1757-899X/786/1/012008

$$M_p = 2\frac{F}{2}R_{av} = FR_{av} \, , \tag{13}$$

where $\frac{F}{2}$ - circumferential force, have an effect on the average radius of the roll.

The circumferential force can be determined by the following equation:

$$F = 2P_{\Sigma} \sin \gamma_H \, , \tag{14}$$

where P_{Σ} – summarized thrust force, N; γ_H - neutral angle degree.

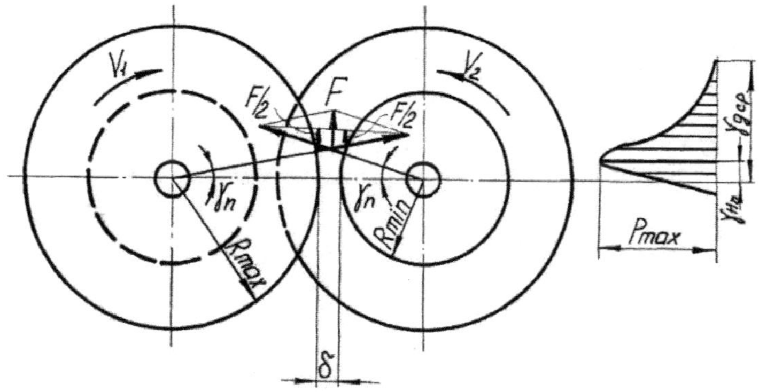

Figure 4. Scheme for determining the resulting moment.

$$P_x = 0{,}7P_2BR_{av}\gamma_1 \cos\alpha \, , \tag{15}$$

where P_2 – unit pressure of grinding, N/m^2, B - width of rolls, m.

Then make an assumption, that $\sin\gamma_H \approx \gamma_1$, since it is very small, we get:

$$N_1 = 1{,}4P_2\omega BR^2{}_{av}\gamma_i{}^2 \cos\alpha \, . \tag{16}$$

The supplied power on the overcome friction in roll axles, equal to:

$$N_2 = 2\pi d_c \theta_p f_k n \, , \tag{17}$$

where d_c – roll axle diameter, m; θ_p - resultant force from the total spacer effort and the force of gravity of the roll, is equal to:

$$\theta_p = \sqrt{P_{\Sigma}{}^2 + G_B{}^2} \, , \tag{18}$$

where f_k - coefficient of rolling friction.

As a case in point determine the intake power of drive PRU with the following source data:

D_{av} =0.4 m, B = 0.1 m, P_2=100 $\cdot 10^6$ N/m^2 , n =0.25 s^{-1}, δ – $4\cdot10^{-3}$m, γ_H =4^0, d_c = 0.01m, $f = 0.001; \eta = 0{,}85; G_v = 979.68$ N;

To construct the graphical dependence of power on the size of rolls taper and the specific grinding pressure. The calculated data are presented in table 2.[6-8]

Table 2. The effect of rolls taper on the supplied power on grinding at different pressures.

K Coefficient	The supplied power on grinding materials, W			
	P_2=100·10^6, N/m^2	P_2=150·10^6, N/m^2	P_2=200·10^6, N/m^2	P_2=250·10^6, N/m^2
0	4271.2	6406.8	8542.4	10678.0
0.15	4257.98	6386.97	8515.96	10645.0
0.3	4218.7	6328.1	8437.4	10547.0
0.5	4125.6	6188.4	8251.3	10314.1
1.25	3698.9	5548.4	7397.8	9247.3
2	3020.2	4530.3	6040.4	7550.6

The graphical dependence of the effect of the taper value on the supplied power on grinding the clinker at different specific pressures, obtained experimentally, is shown in Figure 5.

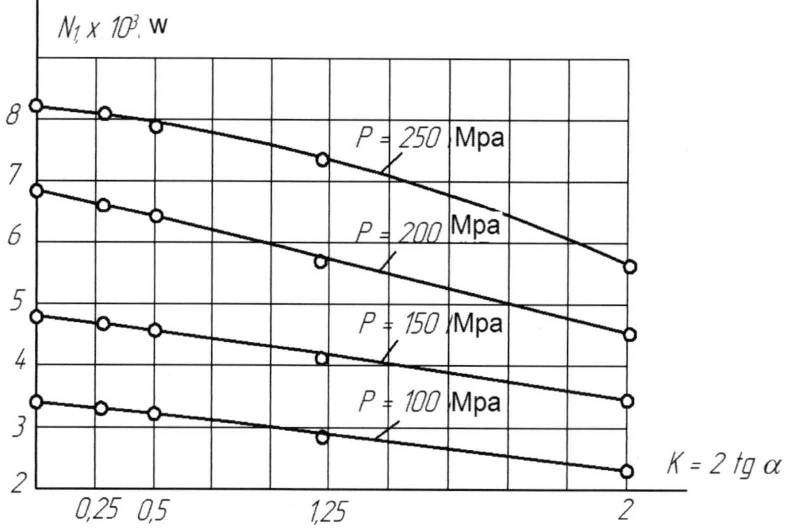

Figure 5. The dependence of the effect of the taper value on the supplied power on grinding the clinker at different specific pressures, obtained experimentally.

Thus, the obtained analytical expressions with sufficient adequately reflect the real process, the discrepancies between the calculated values at P_u = 100 MPa with rolls taper equal to 0.3 ($N_{dv} = 4.218 \cdot 10^3 w$) and experimentally obtained values ($N_{dv} = 3.755 w$) less than 10%, which fairly adequately reflects the actual process.

The use of these materials in a finely divided state will create conditions not only for the disposal of industrial waste, but also for the development of promising technologies and processes for obtaining cheaper building, thermal insulation and other materials.

3. Summary

It is hereby recommended that an industrial technology and energy-saving equipment for obtaining a composite binder with a reduction in the energy intensity of the process up to 50% due to the synergistic effect during the mechanical activation of the raw mix with the replacement of the clinker component of the mineral hydroactive additive is proposed. The energy efficiency of the line is achieved by the joint operation of the press-roller unit and the rotary-vortex mill of superfine grinding, which allow the grinding of raw materials to dispersion with specific surface area of 4000 ... 4500 cm2 / g.

IOP Conf. Series: Materials Science and Engineering **786** (2020) 012008 doi:10.1088/1757-899X/786/1/012008

We obtained the equation for calculating the effort of grinding materials, realized between the conical rolls of PRU, is obtained. It was established that the power expended on grinding the material and overcoming friction in the roll axles and derived an equation for calculating its value, which reflects the actual process with sufficient accuracy.

References

[1] G A Denisevich 2001 Mechanics of a granular medium. *Problems of mechanics* **3** 91-3

[2] M A Verdiyan 1977 The Process of grinding solids, results of science and technology, *Processes and devices of chemical technology* **5** 5-90

[3] A M Gridchin 2006 Improving the efficiency of road construction through the use of anisotropic raw materials, M.: Ed. *Association of construction universities* 2 486-788

[4] M Mersman 2005 Technology for the modernization of cement plants of the company KHD Humboldt Wedag GmbH. *Cement and Its application* **3** 40-3

[5] M Sucker 2006 Installations for the production of roller presses cement company, Cement, lime, gypsum. 2 60-4

[6] A A Romanovich, M A Romanovich, A I Belov and E I Chekhovskoy 2018 Energy-saving technology of obtaining composite binders using technogenic wastes, IOP Conference Series: Journal of Physics: Conf. Series. 1118 012035

[7] A A Romanovich, R S Kolesnikov and M A Romanovich 2018 Study of device for precompaction and uniform supply of materials to working bodies of aggregate, Materials Science and Engineering: Conf. Series. 042052

[8] A A Romanovich, E I Chekhovskoy, M A Romanovich and I V Apukhtina 2017 Calculation of the power of the unit drive for obtaining cube-shaped rubble, *Bulletin of BSTU* **7** 111-5

[9] P N Malov and N S Tarabanov 1968 On the effect of elastic consequences on the properties of pressed carbon materials, *Chemistry of fuel* **1** 132-3

[10] V L Valkevich, G I Fmedlider, T V Fortunatov 1970 Dispersion and elastic expansion of oxide granular masses during pressing, Refractories. 45-8

[11] R Ya Popilsky, F V Kondrashev 1988 Pressing ceramic powders. *Metallurg* 27-8

[12] A A Romanovich 2011 Study of the effect of the speed of rotation of the rolls on the output indicators of the grinding process and the development of recommendations for improving the wear resistance of their working surfaces 4 71-3

IOP Conf. Series: Materials Science and Engineering **786** (2020) 012009 doi:10.1088/1757-899X/786/1/012009

The definition of force parameters when grinding materials with anisotropic texture pressure in the press roller unit

A A Romanovich[1,*], E Amini[2], P V Trishin[1], E I Chekhovskoy[1] and A S Serebrennikov[1]

[1]BSTU named after V.G. Shukhov, 46 Kostyukova Street, Belgorod, 308012, Russian Federation

[2]K. N. Toosi University of Technology, Tehran, Iran

*E-mail: AlexejRom@yandex.ru

Abstract. The article deals with the issues of improving the efficiency of grinding materials with an anisotropic texture. The main specific features and conditions of the organization of the grinding process, including the conditions of loading materials with an anisotropic texture, the creating conditions of their directed motion and application of force load. The article explains the design of the press-roll unit (PRU) with a device for the directed supply of anisotropic materials is given, description its scheme and the process of work. The conditions of destruction of an anisotropic material between eccentrically mounted rolls are considered and an equation for the calculation of the specific force required for the destruction of anisotropic materials in eccentric rolls of PRU is obtained. The comparative data of values of specific values of crushing pressure at grinding of organogenic limestone obtained experimentally in PRU and by calculation are presented.

1. Introduction

Modern construction industry is based on the processing of great amount of materials with different mineralogical composition and physical and mechanical properties (clay, sand, limestone, granite, basalt, etc.). At the same time, the process of their grinding is accompanied by significant material and energy costs. It is known that about 10% of the world's electricity is spent on crushing and grinding processes [1].

The increasing needs for processing of natural raw materials to create new building materials and mixtures, as well as the need to dispose of man-made dumps and waste of various industries will probably require even greater energy costs for grinding.

At the same time, the resource development of ores and other minerals tens of billions of tons of rocks are annually sent to the dump for storage, which in their mineralogical composition can be used in the production of a wide range of building materials. By the way, the extracted rocks differ from the traditional raw materials of the construction industry in their geological origin, mineralogical composition, texture and physical and mechanical properties. A special position among them is occupied by anisotropic materials characterized by various physical and mechanical parameters of the medium (compressive resistance, tensile strength, bending strength, Young's modulus, shear strength, Poisson's ratio, dielectric, magnetic permeability, etc.) table 1.

Much research of scientists has been done [2-4] confirm the direct dependence of the anisotropy coefficient $\hat{E}_{an}$:

$$\hat{E}_{an} = \frac{\sigma_1}{\sigma_2}, \quad (1)$$

where σ_1, σ_2 - respectively, the tensile strength of an anisotropic material in directions perpendicular and parallel to the banding mineral structure.

These circumstances determine the great attention of scientists and experts to the development of their processing technology and the constructive improvement of crushing and grinding equipment for the purpose save energy when grinding materials with anisotropic texture. In this regard, a special relevance arises in the directional and uniform supply of aggregates uses for the destruction of materials with anisotropic texture, as it is possible to provide a directed force on the crushed material in the direction of its lowest strength, which will significantly reduce the energy consumption for grinding.

2. Main part

When crushing anisotropic rocks, the efficiency of the destruction process is determined by many factors: the strength characteristics of materials, their structural and textural characteristics and, most importantly, the method and direction of application of destructive forces. The analysis of scientific and technical researches of processes of destruction of isotropic and anisotropic materials in various crushing and grinding units shows that at the organization of process of crushing of anisotropic materials it is necessary to consider not only their specific features, but also conditions of the organization of the process:

-conditions of loading of materials with anisotropic structure and their ordered motion;

-the direction of force application;

-process flowsheet of organization the process of grinding materials at each stage of their processing.

Table 1. Ultimate stress limit of anisotropic materials depending on the force application.

Material short text	Compressive resistance, MPa		Coefficient of anisotropy
	Perpendlcularly slaty cleavage	Parallely slaty cleavage	
Limestones, organogenic	95	65	1.46
Metamorphic shists (land deposit Kursk Magnetic Anomaly)	130	59	2.2
Amphibolite (land deposit Kursk Magnetic Anomaly)	145	75	1.93
Quartzite sandstone banded	260	190	1.37

The existing range of crushing equipment both in Russia and abroad does not take into account the features of the crushed raw materials.

The analysis of the conditions of the movement of the material and its grinding compression in the press-roll units indicates the feasibility of the use of feed devices roll type, which provide a uniform distribution of the charge layer across the width of the working bodies of PRU, its effective pre-compaction and directed supply of the material layer in the inter-roll space. At the level of the invention, we have developed the construction design of a press-roll unit with a device for the directed supply of anisotropic materials. (Figure 1).

The press-roll unit consists of rolls 1, 2 eccentrically mounted on the frame, over which the roll device is located inside the bin, consisting of two movable cheeks 5, in contact with the rollers 3,4, connected to each other by guide leverage. The principle of operation of the unit. The material having an anisotropic texture, for example, organogenic limestone and others, is fed to receiving bin, where it is fed to the rolls 3 and 4 along the movable cheeks 5, captured by them, rotated in a forward-rearward location, compacted, evenly distributed across the width of the rolls and sent to the inter-roll space.

Further, the material is captured by eccentric rolls 1, 2, where its crush - shearing deformation is realized, due to the different circumferential speeds of the working surfaces of the rolls. The opening between the spring-actuated rolls 3,4, determined within the magnitude of the value $H = R_{av}(1-\cos\alpha) + \delta$ (where α - angle of the rolls, R_{av} - average radius of the rolls, δ - eccentric roll opening). (Figure 1)

a)

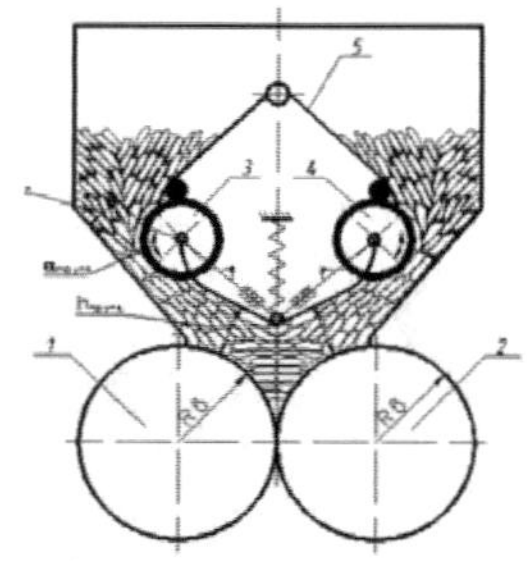

b)

Figure 1. Press-roll unit with a device for directed supply of anisotropic materials: a – photo of the experimental setup, b – scheme of work.

Field researches confirmed the performance capability, construction design of PRU and high efficiency from the use of materials with an anisotropic texture when grinding.

However, the complexity of the implementation of these units in production is largely determined by the lack of a method for calculating the magnitude of the effort required for the destruction of anisotropic materials. Press roller units with eccentrically mounted rollers allow Implementing effective conditions of crush and shear directional force on anisotropic material.

The force effect realized between eccentric rolls depends largely on size of the grinding force applied area, which is limited to certain angles, which have a significant impact not only on the amount of material they capture, but also on the output process parameters. Therefore, the determination of the magnitude of the angles in the eccentric rolls is the important task when calculating the force of grinding materials in the press-roller units.

It has been established that a layer of particles with an anisotropic texture during the process of grinding particles in PRU with eccentric rolls and device for directional feeding, first undergoes directional movement, width distribution and compaction by the device rollers [7-10] (Figure 2), and then compacted and oriented particles of it in a uniform layer along the thickness h_{com} and width B of the rolls are fed into the inter-roll space. Entered into the zone of capture of material α_{def}, material undergoes deformation between eccentrically mounted rolls, where they are subjected to crushing-shear deformation at variable radii and peripheral speeds of the working surfaces of the rolls.

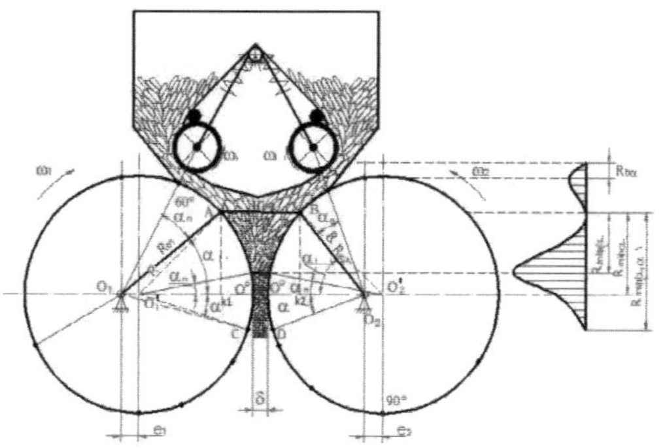

Figure 2. Schematic illustration parameters diagram of eccentrically mounted rolls.

Maximum grinding forces for grinding material is achieved in the zone of neutral angle α_n (Figure 2) and the elastic expansion of the compressed particles, facilitating the release of materials from the inter-roll space, is limited by the angle of elastic expansion α_{com}.

With equality when $R_1 = R_2 = R$ ($e = 0$) the width of the zone of capture of the material rolls equal to:

$$H_0 = 2R - 2R\cos\alpha_{defR} + \delta = 2R(1 - \cos\alpha_{defR}) + \delta . \quad (2)$$

Then the roll deformation angle is:

$$\alpha_{defR} = \arccos\left(1 - \frac{H_o - \delta}{2R}\right) = \arccos\left[1 - \frac{\delta(k_{com} - 1)}{2R}\right], \quad k_{com} = \frac{H_o}{\delta} . \quad (3)$$

Accordingly, the current value of the layer thickness in the particle deformation zone, its elastic expansion are equal to:

$$H_{com.\alpha} = 2R(\cos\alpha_{defR} - \cos\alpha_0) , \quad (4)$$

$$H_{com.\alpha} = 2R(\cos\alpha_{comR} - \cos\alpha_0) . \quad (5)$$

Determine the magnitude of the deformation angles for eccentrically mounted rolls, when $e_1 = e_2 = e$. From right-angled triangles, radius constrained R_1 = O$_1$A and R_1 = O$_2$B and verticals lowered to horizontal axes K_1 and K_2 when $R_1 = R_2 = R$, there is as result $AK_1 = B\hat{E}_2 = R\sin\alpha_{defR}$.

When the eccentric rolls are in a position where the eccentricity values are maximum and opposite from the triangles ΔO_1AK_1 and $\Delta O_2'BK_2$ arise from:

$$AK_1 = (R\cos\alpha_{defR} + e)\cdot tg\alpha_{def\,Re\,1} , \quad (6)$$

$$B\hat{E}_2 = (R\cos\alpha_{defR} - e)\cdot tg\alpha_{def\,Re\,2} . \quad (7)$$

Then measure of the angles of roll deformation equals:

$$tg\alpha_{def\,Re\,1} = \frac{AK}{R\cos\alpha_{defR} + e} = \frac{R\cos\alpha_{defR}}{R\cos\alpha_{defR} + e} ; \quad (8)$$

$$tg\alpha_{def\,Re\,2} = \frac{BN}{R\cos\alpha_{defR} - e} = \frac{R\cos\alpha_{defR}}{R\cos\alpha_{defR} - e}$$

or

$$\alpha_{def\,Re\,1} = arctg\frac{R\cos\alpha_{defR}}{R\cos\alpha_{defR} + e} ; \quad (9)$$

$$\alpha_{def\,Re\,2} = arctg\frac{R\cos\alpha_{defR}}{R\cos\alpha_{defR} - e}. \tag{10}$$

With the aim of measuring neutral angles, consider the forces acting in the zone of the neutral angle α_n (Figure 3). In the same zone, an elementary area is acted upon by a maximum pressure force. P_{max} and roll friction P_f, which can be written as:

$$dP_{max} = Đ_{max} \cdot BR_1 d\alpha_{n_1} = Đ_{max} \cdot BR_2 d\alpha_{n_2};$$
$$dP_f = f \cdot Đ_{max} \cdot BR_1 d\alpha_{n_1} = f \cdot Đ_{max} \cdot BR_2 d\alpha_{n_2}, \tag{11}$$

where B - roll width, m; f - coefficient of external friction.

The equilibrium equation of elementary forces acting in the zone of a neutral angle can be written as:

$$-\int_{\alpha_{n1}}^{\alpha_{def1}} f_{\alpha_{n1}} \cdot Đ_{def1} \cdot B_1 \cdot R_1 \cdot \cos\alpha_i \cdot d\alpha_i + \int_0^{\alpha_{n1}} f_{\alpha_{n1}} \cdot Đ_{def1} \cdot B_1 \cdot R_1 \cdot \cos\alpha_i \cdot d\alpha_i +$$
$$+\int_0^{\alpha_{def1}} Đ_{def1} \cdot R_1 \cdot \sin\alpha_i \cdot d\alpha_i = \int_{\alpha n2}^{\alpha def_{def2}} f_{\alpha_{i2}} \cdot Đ_{def2} \cdot B_2 \cdot R_2 \cdot \cos\alpha_i \cdot d\alpha_i - \tag{12}$$
$$-\int_0^{\alpha_{n2}} f_{\alpha_{n2}} \cdot Đ_{def2} \cdot B_2 \cdot R_2 \cdot \cos\alpha_i \cdot d\alpha_i - \int_0^{\alpha_{def2}} Đ_{def2} \cdot B_2 \cdot R_2 \cdot \sin\alpha_i \cdot d\alpha_i$$

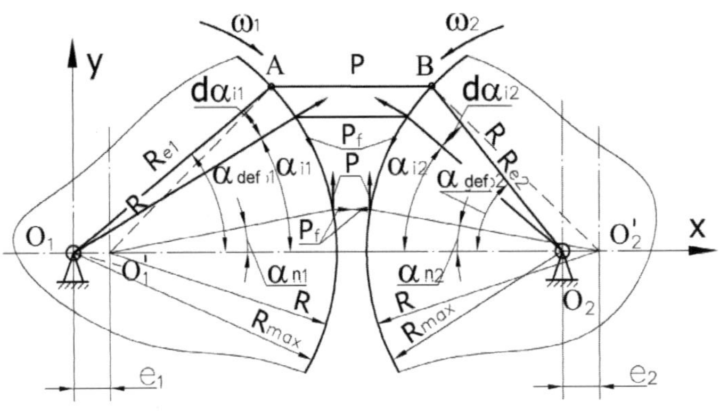

Figure 3. Schematic illustration for the calculation of neutral angles.

Considering that in the zone of neutral angle $Đ_{def1} = Đ_{def2}$, $B_1 = B_2$, which are constant, then after on rearrangement we get:

$$\int_0^{\alpha_{n1}} f_{\alpha_{n1}} \cdot \cos\alpha_i \cdot d\alpha_i + \int_0^{\alpha_{def1}} \sin\alpha_i \cdot d\alpha_i - \int_{\alpha_{n1}}^{\alpha_{def1}} f_{\alpha_{n1}} \cdot \cos\alpha_i \cdot d\alpha_i =$$
$$= \int_{\alpha_{n2}}^{\alpha_{def2}} f_{\alpha_{n2}} \cdot \cos\alpha_i \cdot d\alpha_i - \int_0^{\alpha_{n2}} f_{\alpha_{n2}} \cdot \cos\alpha_i \cdot d\alpha_i - \int_0^{\alpha_{def2}} \sin\alpha_i \cdot d\alpha_i \tag{13}$$

Integrating and tailored $\alpha_{def1} = \alpha_{defR1}$ (8) and $\alpha_{def_2} = \alpha_{defR_2}$ (9) equalization (12) can be written as:

$$\alpha_{n1} + \alpha_{n2} = \frac{1}{4f_{\alpha_n}} \left[\begin{array}{l} arctg \dfrac{R\cos\alpha_{defR}}{R\cos\alpha_{defR} + e} \left(2f_{\alpha_f} - arctg \dfrac{R\cos\alpha_{defR}}{R\cos\alpha_{defR} + e} \right) + \\ + arctg \dfrac{R\cos\alpha_{defR}}{R\cos\alpha_{defR} - e} \left(2f_{\alpha_f} - arctg \dfrac{R\cos\alpha_{defR}}{R\cos\alpha_{defR} - e} \right) \end{array} \right]. \quad (14)$$

After passing the zone of maximum effort $P_{\max}$ is the result an elastic expansion of the layer of crushed particles occurs in the zone of the angle of elastic expansion, the value of which can be determined by the radius $R = R_{av} = \frac{(R + \mathring{a}) + (R - \mathring{a})}{2}$ (where e – current eccentricity value) by analogy with (2):

$$\alpha_{comR} = \arccos\left[1 - \frac{\delta'(k_{com} - 1)}{2r}\right], \quad (15)$$

where δ' - the thickness of the layer of particles after its elastic expansion, m; k_{com} - coefficient of elastic expansion, $k_{com} = \frac{\delta'}{\delta}$.

To determine the specific grinding force, we consider the conditions for the destruction of anisotropic material between eccentrically mounted rolls, assuming that the normal stresses in the anisotropic material are uniformly distributed along the roll arc in the deformation zone, and the greatest grinding efforts are achieved along the line connecting the centers of the rolls to the neutral zone of action angle rolls.

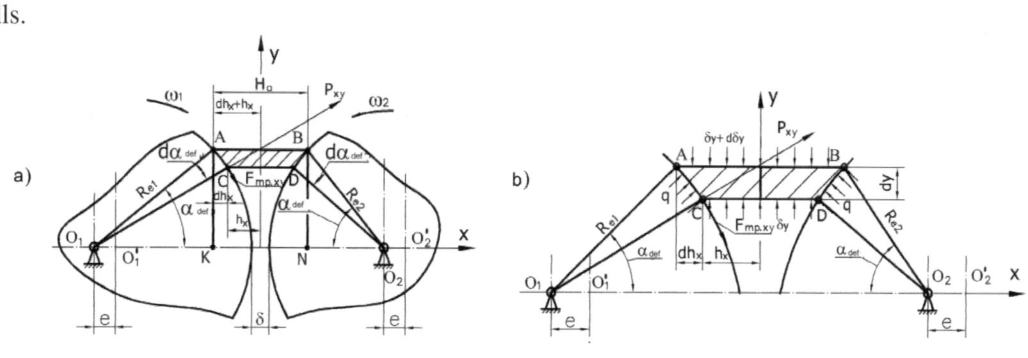

Figure 4. Schema forces for determining the forces grinding.

The sum of projections of forces, arising in the elementary layer on the coordinate axis "x" and "y" is equal to:

$$\Sigma F_x = 0, \; R_{e1} q d\alpha_{def} \cos\alpha_{def} + R_{e1} q d\alpha_{def} f \sin\alpha_{def} - \delta_x dy = 0; \quad (16)$$

$$\sum F_y = 0, \; \delta_y h_x - (\delta_y + d\delta_y)(h_x + dh_x) + R_{e1} q d\alpha_{def} \sin\alpha_{def} - f\, R_{e1} q d\alpha_{def} \cos\alpha_{def} = 0, \quad (17)$$

where R_{e1} - the radius of the roll, m; q - specific force, arising on the surface of the roll, N/m^2; α_{def} - the angle degree of deformation; f - coefficient of external friction; δ_x - direct stress, arising in the deformable layer, N/m^2.

After the corresponding transformations and differentiation, we obtain an equation for calculating the specific force required for the destruction of anisotropic materials in eccentric rolls of PRU.

$$Đ_i = \frac{0.71 fL\alpha_k\sigma_{sz}}{k_{an}}(tg\gamma - f_T)/e^{\frac{\xi fL\delta}{S\ \cos\alpha}}) \cdot \frac{H_0 \cdot tg\alpha_{def}}{\Delta h f_ò (f_ò + tg\alpha_{def})} \cdot \left[1+\frac{f_ò}{tg\alpha_{def}}\left(1-\frac{2(Í_0-\delta)}{H_o}\right)-\frac{(tg\alpha_{def}-f_ò)}{tg\alpha_{def}}\left(\frac{H_o}{h_x}\right)^{f_ò / tg\alpha_{def}}\right]. \quad (18)$$

where, G_{sz} - ultimate resistance anisotropic media under compression, N/m^2; $tg\alpha_{def}$ - angle of deformation of the batch layer; f, f_T - respectively, the coefficient of internal and external friction; S - the area of the crushing chamber, m^2; L – the perimeter crushing cavity, m; α_k; ξ - coefficient side thrust; $tg\gamma$ - the average value of the angle of inclination of the contact area of anisotropic particles to three mutually perpendicular coordinate axes; H_o- the thickness of the layer of particles at the beginning of deformation, m; Δh - deformation of the batch layer during its destruction, m; k_{an} – anisotropy factor of material; α - the angle of maximum effort; δ - gap of the roll opening, m.

From the graphical dependence (Figure 5) built on the equation (17) showed that the increase in the radius of the rolls entails an increase in the grinding force. This indicates that an increase in the radius of the rolls leads to an increase in the thickness of the layer of material captured by the rolls and the area of the maximum force zone, and, consequently, L and S, requires large grinding forces.

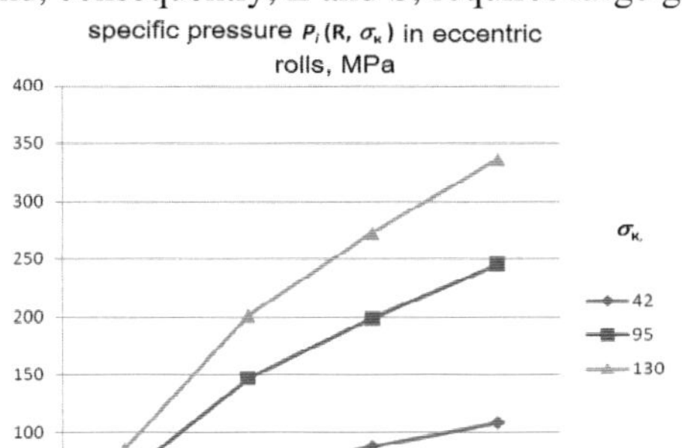

Figure 5. The influence of the inverted divisor of the radius of the rolls on the specific fracture pressure in the PRU of anisotropic materials with different compressive strength.

Grinding forces also increase with the growth of the initial strength of the crushed material, which makes it possible to make an important conclusion for practice that when grinding anisotropic materials it is necessary to create a force effect in the direction of their lowest strength. This requires the development of a device that carries out their directed supply to the rolls of PRU.

3. Summary

Thus, it was found that:

- the force effect that occurs between the eccentric rolls depends largely on the size of the site of application of grinding forces, which is limited to certain angles, which have a significant impact not only on the value of the captured piece of material, but also on the output indicators of the process.
- layer of particles with an anisotropic texture in the process of grinding it into PRU with eccentric rolls h_{com} and device for directed feeding B, is first subjected to directional movement, width distribution and compaction by the rollers of the device, then the compacted and oriented particles are fed into the inter-roll space with a uniform layer in thickness and width of the rolls.

The material received in the capture zone α_{def} is subjected to deformation between eccentrically mounted rolls, where they are crushed and shear deformation at variable radii and circumferential speeds of the working surfaces of the rolls.

4. Conclusion

As a result of analytical studies, the equations for calculating the value of the angular parameters are obtained:

- the force effect realized in eccentric rolls when grinding materials with an anisotropic texture depends not only on the construction design of the working bodies, but also on the force application.-
- in the article obtained an equation that allows us to determine the value of the current resistance of the crushing-shear deformation of the charge layer based on the geometric parameters of the press-roll grinder with eccentric rolls and physical and mechanical characteristics of the crushed anisotropic material. The comparison of the specific values of the grinding pressure in the grinding of organogenic limestone ($G_{sz} = 95$ N/m^2), available from experiments in PRU with R =0.5m, is 141 MPa and by means of a calculation is 148.6 MPa, showed that the discrepancy does not exceed 10%. Thus, the obtained equation (16) with adequate accuracy reflects the real process.

References

[1] G.A. Denisevich 2001 Mechanics of a granular medium. Problems of mechanics 3 91-93

[2] M. A. Verdiyan 1977 The Process of grinding solids, results of science and technology, Processes and devices of chemical technology 5 5-90

[3] A.M. Gridchin 2006 Improving the efficiency of road construction through the use of anisotropic raw materials, M.: Ed. Association of construction universities 2 486-788

[4] M. Mersman 2005 Technology for the modernization of cement plants of the company KHD Humboldt Wedag GmbH. Cement and Its application 3 40-43

[5] M. Sucker 2006 Installations for the production of roller presses cement company, Cement, lime, gypsum 2 60-64

[6] A.A. Romanovich, M.A. Romanovich and Ebrahim Amini 2018 Study of process of materials grinding in energy-saving complex, IOP Conf. Series: *Journal of Physics*: Conf. Series **1118** 012033

[7] A.A. Romanovich, R.S. Kolesnikov and M.A. Romanovich 2018 Study of device for precompaction and uniform supply of materials to working bodies of aggregate, Materials Science and Engineering: Conf. Series 042052

[8] A.A. Romanovich, E.I. Chekhovskoy, M.A. Romanovich and I.V. Apukhtina 2017 Calculation of the power of the unit drive for obtaining cube-shaped rubble, Bulletin of BSTU named after V.G. Shukhov 7 111-115

[9] V.A. Zhulai, Yu.F. Ustinov and M.A. Romanovich 2019 Adequacy assessment of the simulation of construction machines' dynamic transmission systems, Materials Science and Engineering: Conf. Series 560 012087

[10] L.G. Romanovich, S.A. Deryabina and M.A. Romanovich 2018 Engineering activities in Russian Universities, International Conference Economy in the Modern World 61 139-144

IOP Conf. Series: Materials Science and Engineering **786** (2020) 012010 doi:10.1088/1757-899X/786/1/012010

Selection of optimal vibration parameters molding reinforced concrete products

V G Vasiliev

Moscow State University of Civil Engineering (National Research University), Yaroslavskoeshosse, 26, Moscow, 129337, Russian Federation

Abstract. The efficiency of reinforced concrete products vibration moldingdepends largely on the rational modes of vibration on the compacted concrete mix and the selected parameters of the equipment used. The article summarizes the results of experimental studies on the compaction of concrete mixes in molds that are freely installed on the vibration platform through elastic pads. It is shown that the forming concrete products in non-fixed mold provides for effective compaction by selecting the parameters of the oscillating system "vibrating table - elastic element - a mold with concrete mix".The obtained results open a new prospect in the development and improvement of vibration moldingmachines with free set molds.

1. Introduction

It is impossible to improve the efficiency of vibration moldingof reinforced concrete products, especially of dry mixes, without choosing effective vibration modes and rational parameters of vibrating platforms providing high density and homogeneity of compacted concrete mixes [1-6]. When molding reinforced concrete products on vibration platforms under the production conditions, the mold may collide with the surface of the vibrating table due to it being unreliably attached to the vibrating table, the inconsistency of the static moments of the unbalances and other factors.

It is possible to improve the efficiency of vibration molding significantly when forming reinforced concrete products in molds freely installed (without fastening) on vibration platforms via elastic elements (pads), with an appropriate choice of parameters of the system "vibrating table - elastic element - mold with concrete mix", as shown in [7-11]. The theoretical prerequisites for taking into account the interaction of the elements of the oscillating system when compacting concrete mixes on vibrating plates are considered in [10–13].

The article summarizes the results of experimental studies on the vibration compaction of concrete mixtures in the molds freely installed on the vibrating platforms through elastic pads, while varying the parameters of the system "vibrating table - elastic element - mold with a mixture".

2. Research methods

Vibration platforms with harmonic oscillations were used in the studies. The selection of the mass ratio of the platform vibrating parts and the shape, elastic parameters and dimensions of the pads [3, 8, 11, 12] plays an important role in implementing the oncoming movement of the vibrating platform surface

IOP Conf. Series: Materials Science and Engineering **786** (2020) 012010 doi:10.1088/1757-899X/786/1/012010

and the shape defining the impact mode. Under certain parameters of the system, a mold with concrete mixture may come off the surface of the vibrating table and collide with it. When forming large-sized products, it is also necessary to take into account the peculiarities of the wave front propagation through the product thickness [14, 15, 16].

The most important parameter of the system "vibrating table - elastic element - a mold with concrete mix" is the coefficient of elasticity (rigidity) of the pad [10, 12, 17]. This coefficient (C) is determined by the shape, size and dynamic modulus of elasticity of the pad material. The manufacturers are recommended to use sheets of rubber or plastic as pads [6, 8, 18, 19].

The range of changing stiffness of the pads used for a vibrating platform with a capacity of 1.2 tons was determined by the equation:

$$C = \frac{(1,7...2,6)M_1M_2}{M_1+M_2}, \text{ N/m}, \quad (1)$$

where M_1 and M_2 are the mold weight with the concrete mixture and the weight of the vibration table, respectively.

In the studies, the ratio M_1/M_2 changed from 1:2 to 1:5. The required bearing area of the pads S was determined on the basis of the calculated stiffness C, the magnitude of the dynamic modulus of elasticity E of the pad material, and the selected thickness by the formula:

$$C = \frac{E \cdot S}{\delta}. \quad (2)$$

The choice of rubber pads depending on the mold weight and the type of rubber for different frequencies of the driving force was made according to the nomograms given in [12].

The process of the mold interaction with the mixture and the working surface of the vibration platform was studied with using molds of 10x10x10 cm, 20x20x20 cm, and 450x450x900 cm. Oscillation pulses patterns of the mold and vibrating table that occur during collisions were recorded using holographic interferometry [20]. Also, the study involved measuring magnitudes of the mold accelerations for various parameters of the oscillatory system.

The nature and magnitude of the impulses during the collisions of the mold with the surface of the vibrating platform depend primarily on the vibration modes of the table, which are determined by the static moment of the unbalances K, the stiffness of the elastic elements and by the ratio of vibrating masses M_1/M_2[11, 12, 17].

Depending on these factors, the impact occurs in each period, every second period, after 3 periods of oscillation of the operated device. In our tests, we studied modes with stable impacts, during which one impact occurred for one period of changed driving force.

3. Results

Studies of the influence of the pad stiffness on the oscillation frequency, at which the shock-vibration mode of oscillations was implemented, were carried out with a fixed weight of the mold with a concrete mix M_1 = 900 kg. Pads with a modulus of elasticity of 6.3 MPa were used in the tests. With different values of the pads bearing area S, their rigidity C was:

$S_1 = 0.20$ m^2 $C_1 = 3.5 \cdot 10^6$ N/m;

$S_2 = 0.48$ m^2 $C_2 = 8.6 \cdot 10^6$ N/m;

$S_2 = 0.75$ m^2 $C_2 = 13.6 \cdot 10^6$ N/m.

Obtained results are given in Figure 1.

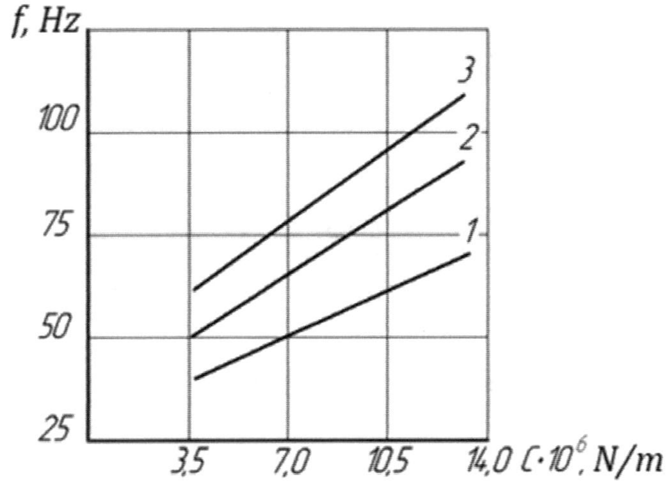

Figure 1. The influence of the pad stiffness C on the oscillation frequency f, at which the mold collides with the table. Static moments $K$of the unbalance weight: 1 – 3.52 10^3N·m; 2 – 6.18 10^3N·m; 3 – 13.4 10^3N·m.

As it can be seen from Figure 1, the frequency of stable collisions of the mold on the vibrating table increases significantly with an increase in the pad stiffness, which indicates the high dependence of a mode of the oscillating system "vibrating table - elastic element - mold with a mixture" on the stiffness of the elastic element. Thus, an increase in the pad stiffness from $3.52 \cdot 10^6$ to $13.4 \cdot 10^6$ N/m leads to an increase in the frequency of vibrations, at which the impact of the table with the mold occurs, by 1.75 times on average. It is also necessary to take into account that depending on the stiffness of the contact of the mold with the vibrating table, the amplitude-frequency spectrum of oscillations propagating in the concrete mix, which determines the intensity of the vibration compaction process, changes significantly [12, 17].

Consequently, by changing the stiffness of the pads and the static moment of the unbalance, it is possible to change the frequency of the mold impact with the vibrating platform table significantly, to form a spectrum of frequencies transmitted to the concrete mix, thus creating an optimal mode of forming concrete products.

As shown in [3, 4, 6, 9], one of the main criteria for the technical evaluation of the efficiency of a vibration machine is the maximum acceleration of the oscillation of a mold with a concrete mix at the time of the impact with the table of the vibrating platform.

Studies to determine the effect of M_1/M_2 ratio and the static moment K of the vibrating platform unbalances on the mold acceleration A_g, which it receives when interacting with the operated device through an elastic pad, were carried out with a fixed stiffness of the pad being $1.7 \cdot 10^6$ N/m. The parameters of the static moment of the weight unbalances took the following values: K – 3.52; 6.18; 9.1; $13.4 \cdot 10^3$N·m. The ratios M_1/M_2were chosen as 0.4; 1.0; 1.5; 2.0 (by changing the weight of the mold with a concrete mix). The test results are presented in Figure 2

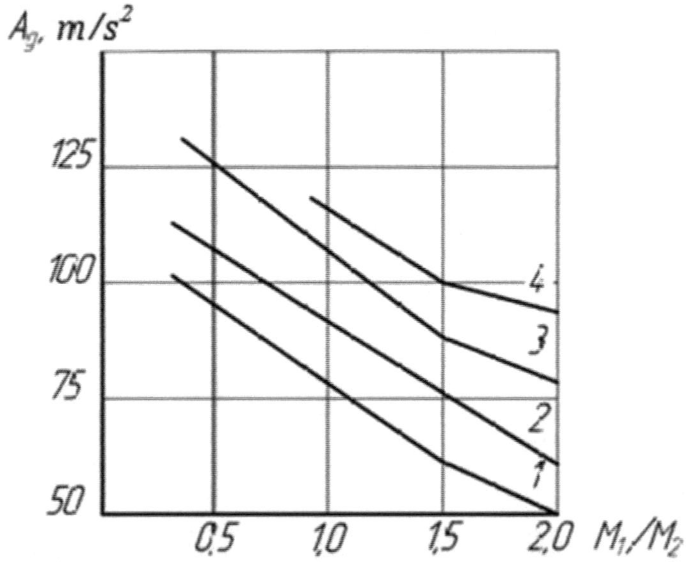

Figure 2. The dependence of the acceleration *Ag* of the mold with a concrete mixture when colliding withthe vibrating table on the weight ratio M_1/M_2. Static moments *K* of the unbalance weight:1 – 3. 52 10^3N·m; 2 – 6. 18 10^3N·m; 3 - K = 9.1·10^3N·m; 4 – 13. 4 10^3N·m.

As can be seen from Figure 2, the mold acceleration at the time of contact with the vibrating platform when changing the parameters of the system "vibrating table — elastic element — mold" varies over a wide range. With an increase in the weight of the mold with a concrete mixture, the mold acceleration regularly decreases at the moment of contact with the vibrating table. When the weight of the mold with the mixture is 2 times more than the weight of the vibrating table (M1/M2 = 2), the mold acceleration is within 4.5g ... 9g depending on the static moments of the unbalance weight.

If the weight of the mold with a concrete mixture and the weight of the vibrating table are equal ($M_1/M_2 = 1$), and make 900 kg, the mold acceleration at the moment of impact varies from 7g to 11g. In the case of lighter molds ($M_1/M_2 = 0.4$), the peak acceleration reaches 11g or more.

It should be noted that with an increase in the static moment of the weights of unbalances at any values of the weight ratio of the mold and the vibrating table, the mold acceleration increases. The study results show that if lighter molds are used, it is possible to obtain large accelerations of their movement in the process of forming concrete products.

4. Summary

Forming products from concrete and reinforced concrete in molds freely installed via elastic pads on vibrating platforms and creating conditions for the impact of the mold with the table surface through an elastic element allow changing the parameters of the oscillating system "vibrating table - elastic element - a mold with concrete mixture" to implement the optimal modes to consolidate concrete mixes.

The obtained results open a new prospect in the development and improvement of vibration molding machines with free set molds, since this eliminates the need for fasteners, which often do not work reliably. It also simplifies the design and reduces the metal capacity of the machine.

References

[1] O A Savinov and E V Lavrinovich 1986 Vibration technology of compaction and formation of concrete mixtures, M., Striyizdat

[2] S M Pulyaev, M A Stepanov, B A Kaitukov and others 2015 Mechanical equipment and technological complexes: manual, M , MGSU
[3] B V Gusev, A D Deminov, B I Kryukov and others, 1982 Shock-vibration technology of compaction of concrete mixes, M., Stroyizdat
[4] B V Gusev and V G Zazimko 1991 Vibration technology of concrete, Kiev, Budivelnik
[5] B V Gusev, V D Kudryavtseva 2012 Rational amplitude-frequency vibration compaction of concrete mix, *Bulletin of KIGIT*, **3** 9-11
[6] B V Gusev, E Z Axelrod, A I Zvezdov and others 1988 Manual on the technology of molding of reinforced concrete products, M , Stroyizdat
[7] P Rebu 1970 Vibrating the concrete, M , Stroyizdat,
[8] V N Shmigalsky 1968 Forming products on vibroplatforms, M , Stroyizdat
[9] V G Zazimko 1984 Technology of forming of reinforced concrete products with adjustable vibration parameters, Abstract of Doctor of Science thesis,
[10] B V Gusev and A S Faivusovich 2002 Technological mechanics of vibrating concrete mixtures, M , Scientific world
[11] V G Vasiliev 2015 The choice of optimal parameters of shock-vibrating tables, *J Mech of const* , **11** 30-3
[12] E G Zinoviev 1989 Forming of large-size reinforced concrete products with the use of controlled modes of vibration compaction, PhD work, M
[13] E A Sinyaeva 1982 Influence of the parameters of vertical-directional symmetric and asymmetric vibrations on the compaction of the concrete mix, Abstract dissert Cand tech ofsciences, NIIZHB, M
[14] S A Osmakov and F G Browde 1969 To the question of the dynamics of the vibrating column of concrete mix, in: Theory of concrete molding, M , NIIZHB 172-9
[15] V I Telichenko and V G Vasiliev 2017 The investigation of the spectrum of pulses in concrete under shock and vibration molding, *J Bulletin of BSTU*, **4** 80-5
[16] V G Vasiliev 2017 The choice of parameters of vibration effects in the molding of concrete products, *J Mech. of const.*, **11** 21-5
[17] R Sharapov and V Vasiliev 2017 Analysis of the spectrum distribution of oscillation amplitudes of the concrete mix at shock vibration molding, MATEC Web of Conferences, 117 21-5
[18] A A Borshchevsky and A S Ilyin 2009 Mechanical equipment of enterprises for the production of building materials and products, M , Alliance,
[19] V E Karamzin 1970 Some features of the behavior of the concrete mix at the easel vibration, in: Vibration technology of concrete products, M , Stroyizdat, 67-84
[20] E G Zinoviev, V G Vasiliev and E P Parfenov 1999 Development of methods for the study of the "machine - medium" in the implementation of various vibrations, M , Dep VINITI, *Households and buildings*, **745-B99**

SN-EVM unit for experimental studies of stability in circular cylindrical shells under combined loading

S V Cheremnykh* and P O Skudalov

Department of Structures and Facilities, Tver State Technical University, naberezhnaya Afanasiya Nikitina 22, Tver, 170026, RUSSIAN FEDERATION

*E-mail: cheremnykh_s.v@mail.ru

Abstract. The article summarizes results of experimental studies of stability in circular cylindrical shells under combined loading by compressive force, internal pressure and torsion at the SN-EVM unit, developed by the Department for Strength of Materials, Theory of Elasticity and Plasticity at Tver State Technical University. It gives description of the test facility and methodology of experiments for studying the elastoplastic properties of materials under combined loading. It cites the processed experimental results for comparison with the formulated theory of bifurcation for the thin-walled circular cylindrical shell.

1. Introduction

Cylindrical shells are widely used in various structures of machines and mechanisms. Thin-walled shells without stiffeners and most cost-efficient under certain conditions. Along with cost efficiency, they require very precise computational rationale most closely approximate to results of full-scale studies in order to enable selection of parameters and structural materials. A slightest imperfection, a structural defect in a shell nearly guarantees the loss of stability for a structure even in case of simple loading.

To date, we observe multiple formulated theories to calculate the circular cylindrical shells. The papers by [1-11] deal with tweaking the theory of elastoplastic stability for thin-walled circular cylindrical shells. These review the behavioral patterns of the shells under combined loading (compressive force, internal pressure and torsion) in view of the loading history at varying loading speeds.

Experimental studies for the stability loss process in thin-walled cylindrical shells under simple subcritical loading beyond the elastic limit have already been taken by various authors. However, they were mainly done for a restricted class of flat trajectories, i.e. in case of torsional compression or compression with an internal pressure on the shell.

The need for experimental testing of applicability limits for the suggested theory under combined loading required designing and constructing of a device that is able to solve the problems.

2. Description of SN-EVM test facility and its operating principle

State-of-the-art experimental systems to study the patterns of elastoplastic properties of materials under combined loading must meet a bunch of criteria: implementability of multiparameter loading while ensuring the reversal of loading; highly precise traceability of implemented processes in a widely varying range of loading speeds and specimen strain; process controllability, both in the stress, and in the strain space; clarity of visualizing the implemented processes; automated processing of test results. These criteria are fully met with the SN-EVM automated calculation and experimental facility, developed at the mechanical testing laboratory at the Department for Strength of Materials, Theory of Elasticity and Plasticity at Tver State Technical University.

The facility is comprised by a loading unit that implements a three-parameter effect on the specimen. The loads that act on the specimen are measured using a two-component axial load and torsion force gauge. An extensometer is applied to measure axial, circumferential and shear displacements of the shell. For online and high-quality monitoring of the experiment progress, the process data is graphically displayed on a screen. For general view of the experimental unit see Figure 1.

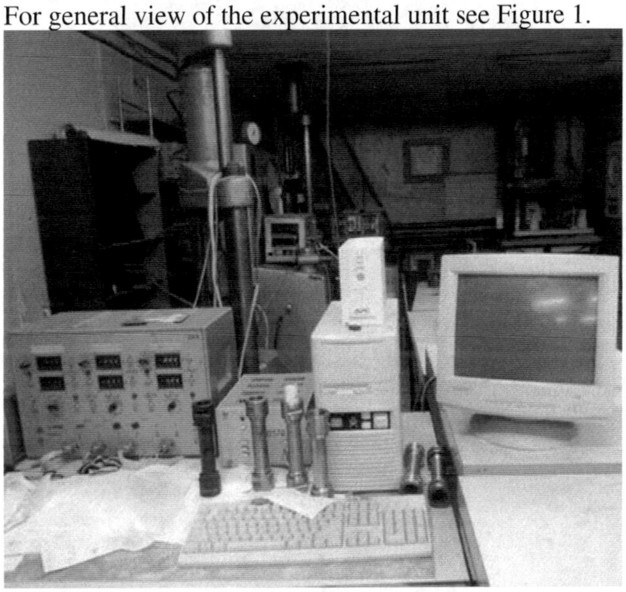

Figure 1. General view of the SN-EVM experimental unit.

Thin-walled circular cylindrical shells (Figure 2) serve as the specimens for testing with a facility. Steel specimen are fabricated by turning from a bar under the following technology: roughing of the external surface with the side allowance of 2 mm, drilling and boring of the internal surface with the side allowance of 0.1 mm, finishing of the internal surface by reaming, workpiece installation onto a mandrel and subsequent finishing of the specimen external surface by finish turning. The external and internal surfaces of specimens are machined to match precision class VII. The thin-walled tubular specimens have wall thickness $h = 1$ mm and $h = 0.5$ mm, middle surface radius $R = 15.5$ mm and $R = 15$ mm, respectively.

IOP Conf. Series: Materials Science and Engineering **786** (2020) 012011 doi:10.1088/1757-899X/786/1/012011

Figure 2. Stability testing specimen.

The axial straining of a specimen causes axial relative displacement of the upper and lower parts, which is recorded by elastic elements with strain gauges glued onto them. The specimen torsion results in the relative angular displacement of the upper part versus the lower one, which is also recorded by an elastic element with the glued strain gauges. Installing the Archimedean spiral on the top part of the extensometer improves the accuracy in measuring the angular strains by eliminating the nonlinearity in the output signal characteristic from the rotation angle.

Using the extensometer (Figure 3) the axial displacement of the specimen is measured with the error not more than ±6·10-3 mm, rotation angle ±7-10-3 rad. The displacement range is 2 mm axially, 0.25 rad angularly, and ± 0.8 mm radially. The measurement errors result in the linear approximation of the strain gauge's inherent characteristics.

Figure 3. Specimen assembly with extensometer.

Prior to mounting onto the specimen, the strain gauge is graded on a special device made from T-30 theodolite. The device enables to independently simulate the process of extension (compression) and strain gauge torsion with the accuracy of $1 \cdot 10^{-3}$ mm for axial strain and $1 \cdot 10^{-4}$ rad for rotation angle.

Clamps for the specimen used in the testing are comprised by two rings and a block-type fork, horizontally coupled in the reverse perpendicular direction and forming a cross-shaped Hooke's joint.

IOP Conf. Series: Materials Science and Engineering **786** (2020) 012011 doi:10.1088/1757-899X/786/1/012011

The block-type fork of the upper clamp is attached to the upper crosshead, and of the lower clamp – directly to the dual-component force gauge. The integrated unit of the lower clamp-force gauge is mounted to the lower moving crosshead. The clamps are easy to disassemble, which facilitates assembly and disassembly of the test module. The inner ring of clamps has two parallel through holes, fitted with rolls that have eccentric surfaces. The specimen heads have half-sectoral flats that are similar in shape. When the specimen is being mounted, the rolls fit into the relevant grooves of the flats after being rotated, thus achieving reliable transfer of axial forces and the torsional moment to the specimen. Due to availability of cross pieces, the clamps can be self-aligning under the action of the applied forces, which rules out any impact from misalignment of the axial force applied.

When the specimen is loaded with internal pressure, its ends are generally sealed with caps. This generates an additional tensile force, which complicates the experiment control and reduces the accuracy. To exclude this phenomenon, the SN-EVM unit uses a device resembling a solid cylindrical core with sealing elements on its ends. In the working part of the specimen length its diameter is by 2 to 5 mm below the specimen's inner diameter. The pressure force is locked to the hard core and does not influence the axial force control channel.

Executing of operations in the control computer starts from defining the input data, such as the experiment type attribute (control by strain or by loads), number of trajectory sections, loading speed. The following are defined locally for each section: trajectory type (a straight interval or a radial arc), proposed length of interval, interval slope angle, initial phase and arc curvature radius, number of turns, initial radius, spiral development direction. The input data are entered from the keyboard, and then scaled into the machine equivalents. Programmed conditions for transfer from one section to another could be definable interval lengths, achievement of a certain value by the measured parameter, fulfilment of a certain calculable condition, or mere button pressing on a control unit.

The experiments are done at constant straining or loading speed. To rule out the impact of relaxation effects on the test results, the straining speed is accepted equal to $\dot{S} = 10^{-6}$ s^{-1} in the area of elastoplastic strains and $\dot{S} = 5 \cdot 10^{-6}$ s^{-1} within the elasticity limits.

3. Results of experiments done at the SN-EVM unit

To verify the theory proposed in [12-17], we performed experiments at specimens from Steel 40X and Steel 45.

To verify the initial isotropy of the material, we compared the strain diagrams in simple loading processes. For the superimposed diagram see Figure 4 (numbers indicate the loads: 1 – torsion, 2 – compression, 3 – proportional compression with torsion at $\alpha_1 = 45°$, $\alpha_2 = 0°$). It enabled a conclusion that the material of the specimens is isotropic, since in the advanced plastic strains the spread in values of the stress vector modulus does not exceed 3%.

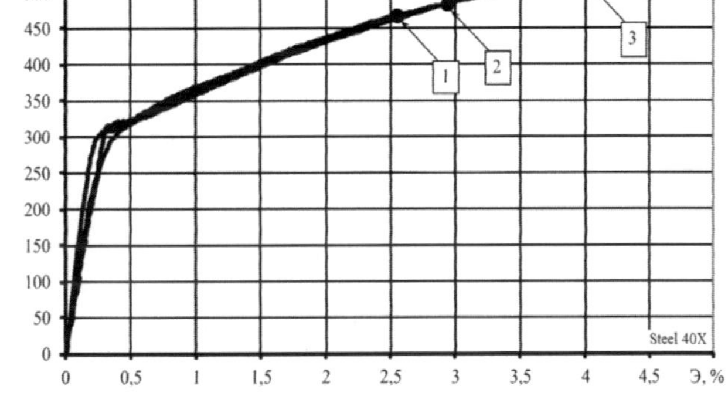

Figure 4. Superimposed diagram for strain of shells from steel 40X with wall thickness h ≈ $1 \cdot 10^3$ m.

The results obtained for complex loading processes are represented as a superimposed diagram in Figure 5 (numbers indicate the loads: 1 – torsion, 2 – compression, 3 – proportional compression with torsion and internal pressure at $\alpha_1 = 60°$, $\alpha_2 = 75°$). Comparison of strain diagrams enabled a conclusion that the material of specimens is conventionally isotropic, since in the advanced plastic strains the spread in values of the stress vector modulus does not exceed 14%.

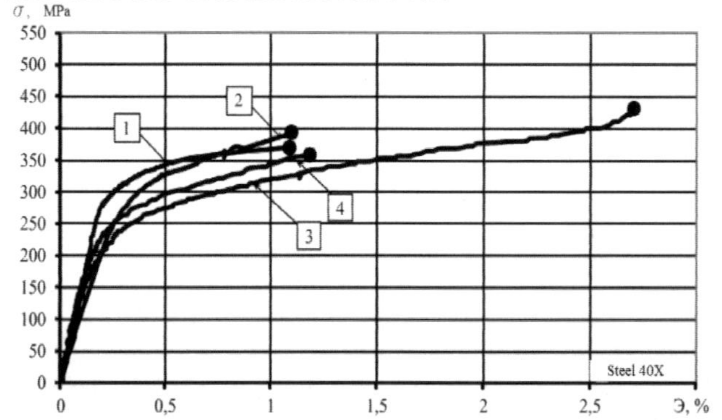

Figure 5. Superimposed diagram for strain of shells from steel 40X with wall thickness h ≈ $0.57 \cdot 10^3$ m.

4. Summary

Processing of results from experiments on torsion of shells from steel 40X with various material wall thickness also enabled a conclusion that in advanced plastic strains the lateral strain ratios for the material under review are close to the value of $\mu = 0.5$.

The constructed SN-EVM unit and the designed method of testing using the same enabled to obtain all the required experimental data to verify the proposed theory for stability of circular cylindrical shells and evaluate the degree of its compliance with the full-scale experiments.

References

[1] V G Zubchaninov, A A Alekseev, E G Alekseeva and V I Gultiaev 2007 Experimental verification of postulate of isotropy and mathematical modeling of elastoplastic deformation proc. following the complex angled nonanalytic trajectories, *Materials Physics and Mechanics* **32** 298–304

[2] V G Zubchaninov, A A Alekseev and V I Gultiaev 2017 Modeling of elastic-plastic deformation of work material along multielement piecewise zig-zag linear trajectories, Bulletin of the Perm National Research Polytechnic University 203–15

[3] V G Zubchaninov 2011 Isotropy postulate and the law of complex unloading of continua, *Mechanics of Solids* **46** 21–9

[4] V V Garanikov, V G Zubchaninov and N L Okhlopkov 1999 On deformation of metals along flat curvilinear trajectories of variable curvature Part 1. Vector and scalar properties, *Strength of Materials* **31** 331–5

[5] V G Zubchaninov and D Al'-Delemi Saadi 1993 Experimental investigation of the processes of complex loading of alloy amg6, *Strength of Materials* **25** 342–7

[6] A A Treschev, M V Spasskaya, M B Shereshevsky, A A Bobrishev and L N Shafigullin 2017 The influence of temperature differences for the analysis of thin orthotropic cylindrical shell, Scientific *Review Engineering and Environmental Sciences* **26** 528–37

[7] L N Shafigullin, A A Treschev, A V Romashina and V T Erofeev 2017 Concentration of stress on holes in a plate of different resistant material Astra Salvensis 213–25

[8] L N Shafigullin, A A Bobrishev, A N Shafigullina, V T Erofeev and A A Treshchev 2015 Development of the recommendations on selection of glass-fiber reinforced polyurethanes for vehicle parts, *International Journal of Applied Engineering Research* **10** 43758–62

[9] A A Treschev and M Y Delyagin 2013 Modeling of an isotropic shell of resistive graphite with the help of bulk finite elements taking into account the connectivity of stresses and temperatures *Materials Physics and Mechanics* **17** 59–70

[10] V S Bondar' and V V Danshin 2014 Mathematical simulation of cyclic thermoviscoplastic deformation and fracture of materials, *Strength of Materials* **46** 613–8

[11] V S Bondar', V V Danshin and D A Makarov 2014 Mathematical modeling of the processes of deformation and accumulation of damages during cyclic loadings, Bulletin of the Perm National Research Polytechnic University 125-52

[12] S V Cheremnykh, V G Zubchaninov and V I Gultyaev 2019 Deformation of cylindrical shells of steel 45 under complex loading, XXII Int Scientific Conf Construction the Formation of Living Environment 97 1–8

[13] V G Zubchaninov, A A Alekseev and V I Gultyaev 2014 On the construction of the yield surface of steel 45 and verification of the isotropy postulate on straight trajectories under multiple alternating loading, Bulletin of the Perm National Research Polytechnic University 71–88

[14] A A Treschev, V G Telichko and P Y Khodorovich 2014 Modeling the stress-strain state of thick cylindrical shells made of materials with complicated properties, *Materials Physics and Mechanics* **21** 38–50

[15] V V Garanikov, V G Zubchaninov and N L Okhlopkov 1999 Experimental study of vector and scalar properties of 9kh2 steel in deformation along flat curvilinear trajectories. Part 2. Astroid-shaped trajectories, *Strength of Materials* **31** 559–63

[16] V V Garanikov, V G Zubchaninov and N L Okhlopkov 1999 Experimental study of vector and scalar properties of 9kh2 steel in deformation along flat curvilinear trajectories Part 1. Logarithmic-spiral trajectories, *Strength of Materials* **31** 455–8

[17] V V Garanikov, V G Zubchaninov and N L Okhlopkov 2000 Complicated deformation of metals along a flat curved trajectory in the form of an asteroid, *International Applied Mechanics* **36** 130–6

IOP Conf. Series: Materials Science and Engineering **786** (2020) 012012 doi:10.1088/1757-899X/786/1/012012

Modeling the process of autoclaving treatment of cellular concrete products as control object

S Y Galitskov[1,*], K S Galitskov[1] and A S Bolkhovetsky[1]

[1] Samara State Technical University, Samara, Russian Federation

*E-mail: maes@samgtu.ru

Abstract. The authors consider the questions of modeling the process of autoclave treatment of cellular concrete products as a control object. The modeling method, performed in several stages, is suggested. At the first stage, a mathematical model of energy processes occurring in an autoclave, taking into account the internal heat generation during the formation of tobermorite in raw cellular concretes, is developed. The temperature dynamics of steam in an autoclave $T_n(t)$ has been determined. At the second stage, heat transfer in a cellular concrete is modeled for a given value of $T_n(t)$. As a result of computational experiments, the temperature dynamics $T_m(x, y, z, t)$ in the most characteristic parts of the cellular concrete was determined. The assessment of the adequacy of the developed model was made by comparing the results of modeling with the results of experimental studies obtained with the existing process equipment.

1. Introduction

The production of cellular concrete products with specified characteristics is largely determined by the algorithms controlling the main technological processes. The strength of cellular concrete products is determined mainly by the autoclaving treatment. In this regard, the development of a mathematical model of the technological process of autoclave treatment of cellular concrete products as a control object is highly relevant.

The description of the temperature field dynamics in raw cellular concretes during their autoclave treatment under conditions of nonstationarity of parameters, by analogy with [1,2], goes down to solving a system of differential equations with partial derivatives that take into account heat and mass transfer in the autoclave environment and raw cellular concretes.

The authors of this article, understanding the complexity and lack of knowledge of the process under consideration, taking into account the fact that the dynamics of the autoclaving treatment is characterized by a vector that includes a sufficiently large number of parameters, among which there are various internal unsteady connections, and focusing on creating a mathematical model of the process under study as a control object for which a certain degree of "roughness" is permissible, assumed it appropriate to consider the object in terms of the following assumptions and simplifications:

At the first stage it is proposed to consider raw concretes as thermally thin bodies. This allows us to create a fairly simple energy model of the dynamics of heat flows impacting the raw concretes. At the second stage, it is proposed to consider a single raw concrete example from the set loaded into the autoclave as an object with distributed parameters. The heat flux acting on the surface of this raw concrete will be determined at the first stage of modeling. As a result, at the second stage, the dynamics of thermal processes in the concrete is investigated, which allows to form an autoclave control algorithm taking into account the restrictions connected with the temperature distribution inside raw concrete. The third stage implies the use of an iterative approach to specify the dynamics of the energy model of the autoclaving process. This will allow to perform the next stage of modeling an object with distributed parameters, taking into account the refined nature of the heat flux impacting the raw concrete. This paper covers the first two stages.

The main purpose of autoclaving automation is to create conditions for a sustainable process when calcium hydrosilicates in the form of tobermorite are formed in raw concretes, ensuring the required strength of cellular concrete products [3].

The main technological requirements for this process are:

1) Absence of air in the pores of raw concrete;

2) Complete filling of raw concrete pores with saturated steam to ensure the chemical reactions;

Technological limitations apply to:

1) the rate of heating and cooling of raw concretes;

2) the pressure in the autoclave limit value;

The control object will be understood as the combination of heating performance, physical and chemical processes occurring during the autoclave treatment of the raw cellular concrete products. The state of the object will be characterized by **X** vector of the output coordinates, which include the temperature distribution over the product and the vapor pressure in the autoclave. We should take the steam flow rate g_1 supplied to the autoclave and the steam exhaust flow rate g_2 coming out of the autoclave as elements of **U** vector of control action. The main perturbations are the uncertainty (within the framework of known limitations) of the physical and mineralogical composition of the raw concretes, the variation in temperature values and steam pressure supplied to the autoclave from the heat source (boiler, steam accumulator), the unsteady internal heat generation processes and heat losses to the environment through the structural elements of the autoclave.

2. Modeling Goal

Development of an adequate mathematical model of the process of autoclaving cellular concrete products with focus on its use in combination of ACS for autoclaving treatment.

The authors consider the stages of pressure increase and curing in the autoclave before which the blowing and vacuuming, necessary to form well-defined initial conditions for the raw cellular concretes and the autoclave environment, are performed.

In the autoclave, with a radius of $r_{aut}=1.35\ m$ and a length of $l_{aut}=26.5\ m$ 12 raw cellular concrete blocks of *D400* density grade with the total mass $m_m{}^{12}=24\cdot10^3\ kg$ are placed. The mass of the elements of the autoclave internal structure is $m_{zvk}=22\cdot10^3\ kg$.

The following are accepted as the *initial conditions:*

1) The temperatures of the raw concretes placed in the autoclave and the elements of the internal structure of the autoclave are uniform across the volume and T_0 equals to $70°C$;

2) Air has been completely forced out of the pores of cellular concrete and the internal environment of the autoclave;

3) The initial steam pressure in the autoclave $P_0=56\cdot10^3$ Pa.

Assumptions:

1. We consider that each raw concrete is a plate uniformly treated with steam along all the surfaces. Taking into account that *2R* plate thickness is *0.64 m*, where *R=0.32 m*, we assume it a thin body in terms of heating;
2. We assume that, given the large inertia of the process, the temperature of the raw concrete T_m is equal to the temperature of the internal environment of the autoclave T_{aut};
3. We believe that the transfer of heat to the raw concretes, determined by the supply of saturated steam to the autoclave, is carried out by means of convective heat exchange and steam condensation on the surfaces of products and elements of the internal structure of the autoclave. Moreover, we take into account that, starting from a temperature *T* value of *165°C,* internal heat generation occurs in autoclaved products due to the chemical reaction of the formation of calcium hydrosilicates in the structure of the raw concrete;
4. We assume that the process of internal heat generation is determined by the formation of hydrosilicates only in the form of tobermorite, and its dynamics can be represented by a symmetric smooth curve [4];
5. We believe that we are aware of the range of values of the main characteristics describing the chemical and mineralogical composition of steamed raw cellular concretes of a given density grade;

3. Mathematical Description

As applied to the task, within the accepted assumptions, the heat exchange processes in the autoclave can be represented, in the first approximation, as an object with lumped parameters. In this case, the raw cellular concretes are presented in the form of thermally thin bodies (plates). The plate is heated by the total heat flow, for which the heat balance equation holds true:

$$F=F_m+F_{tb}-F_{ok}-F_0, \quad (1)$$

where F_m, F_{tb}, F_{ok}, F_0 are the heat flows generated in the autoclave by g_n flow rate of steam from the main line, heat generation in the raw cellular concretes during the formation of hydrosilicates, losses through the protective covering and steam consumption for filling the autoclave to a given pressure, respectively.

Heat transfer from steam to the raw concretes during their autoclave treatment is carried out both by means of convective heat exchange [5] and by heat transfer [6]. The resulting heat flow is represented by their sum:

$$F_m=F_n+F_k, \quad (2)$$

where F_n is the thermal capacity released during steam condensation, F_k is the residual heat of hot condensate. Convective heat transfer, which characterizes the transfer of heat from steam to the concretes during condensation, is described by the equation:

$$F_n=g_n r_n,$$

where g_n is the steam consumption, r_n is the specific heat of vaporization. Heat transfer from condensate to raw cellular concretes can be described as (Fourier thermal conductivity law):

$$F_k=a(T_k-T_m)S, \quad (3)$$

where a is the heat transfer coefficient, T_k and T_m are the vapor condensation temperature (a non-stationary characteristic described by $f(P_{abt})$ function of the pressure in the autoclave) and the temperature of the raw cellular concrete, respectively, S is the surface area involved in heat exchange for *12* considered raw concrete blocks with S=*300 m*^{2}.

Steam is supplied to the autoclave through an adjustable main valve, and is discharged by means of a steam battery valve and a relief valve together making up the system of controlled valves. Thus, steam flow rate, which determines the value F_n, is presented by:

$$g_n=g_m-g_{akk}-g_{nk}, \quad (4)$$

where the flow through the corresponding valve is determined by the ratio of pressure in a given area to its hydraulic resistance $g_m=P_m/R_{gm}$, $g_{akk}=P_{akk}/R_{gakk}$, and g_{nk} flow rate is constant when the condition $P_{abt}\geq P_{max}$ is met.

The heat flow of internal heat generation in raw cellular concretes during the formation of tobermorite F_{tb} seems to be a symmetrical smooth curve, with the accepted assumptions:

$$F_{tb}=Q_{tb}/(2t_1+t_2), \tag{5}$$

here Q is the amount of heat, t_1 and t_2 are the time intervals characterizing the intensity of heat release.

F_{ok} value of heat loss through protective covering in accordance with Fourier law of heat conduction is:

$$F_{ok}=\lambda S(T_{abt}-T_{nap})/l, \tag{6}$$

where λ, S, l are the coefficient of thermal conductivity, surface area and wall (lining) thickness of the autoclave, T_{abt}, T_{nap} are the temperatures inside the autoclave and the environment, respectively.

The amount of heat needed to create pressure in the autoclave:

$$F_0=g_0r_n, \tag{7}$$

where g_0 is steam consumption at the stage of pressure rise.

Thermal power $\triangleleft$ is used for heating raw cellular concrete F_b and steel elements of the internal structure of the autoclave F_{ct}:

$$F=F_b+F_{ct}, \tag{8}$$

Where $F_b=c_bm_b\frac{dT}{dt}$ and $F_{ct}=c_{ct}m_{ct}\frac{dT_{abt}}{dt}$, here the values of c_b, c_{ct}, m_b, m_{ct} are the heat capacity, the mass of autoclaved concrete and steel elements of the internal structure of the autoclave, respectively.

Taking into account the accepted assumptions, the heat balance equation can be written in the form of the well-known Fourier-Kirchhoff convective heat exchange equation:

$$c_bm_b\frac{dT_{abt}}{dt}+c_{ct}m_{ct}\frac{dT_{abt}}{dt}=g_nr_n+\alpha S(T_k-T_m)+\frac{Q_{tb}}{(2t_1-t_2)}-\frac{\lambda S(T_{abt}-T_{nap})}{l}-g_0r_n, \tag{9}$$

However, the process of autoclaving is controlled by using a regulator for a given programmed trajectory of P_{abt} pressure in the autoclave, whereas an observer of temperature P_{abt} is needed only to comply with the restrictions imposed on the pressure increase stage. In relation to the task, P_{abt} pressure in the autoclave can be obtained from the equation of state of an ideal gas.

Based on the obtained equations, the structure of the mathematical model of the autoclave treatment process as an object with lumped parameters was synthesized (Figure 1).

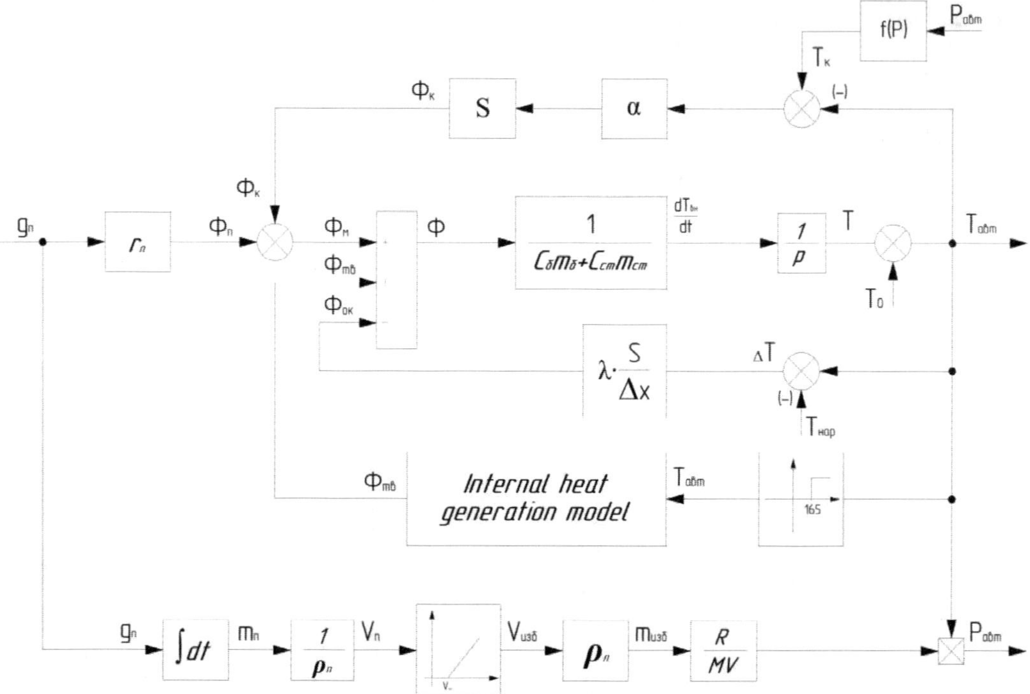

Figure 1. The structure of the mathematical model of the process of autoclave treatment as a control object.

Based on the proposed mathematical model, a computational model was developed using Matlab software (Figure 2).

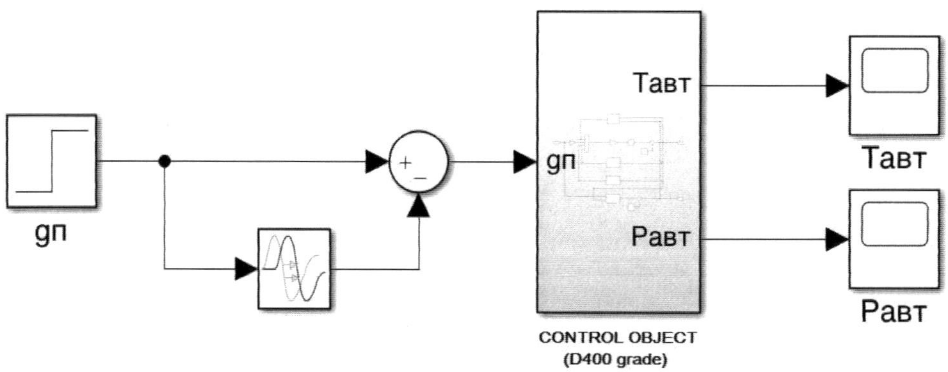

Figure 2. Computational model of the control object with lumped parameters.

The input of the system has $g_n=0.1$ *kg/s* flow rate of the steam line with stepped effect for $t_1=11700$ *s*, which corresponds to the duration of the pressure rise in the autoclave. Let us perform a series of experimental investigations of the control object with and without heat losses through the protective covers of the autoclave F_{ok} and the heat flow from the internal heat release F_{tb} as applied to D400 grade concrete. As a result, at the output of the control object, we obtain a family of pressure curves P_{abt} (Figure 3) and temperature in an autoclave T_{abt} (Figure 4) under given conditions.

IOP Conf. Series: Materials Science and Engineering **786** (2020) 012012 doi:10.1088/1757-899X/786/1/012012

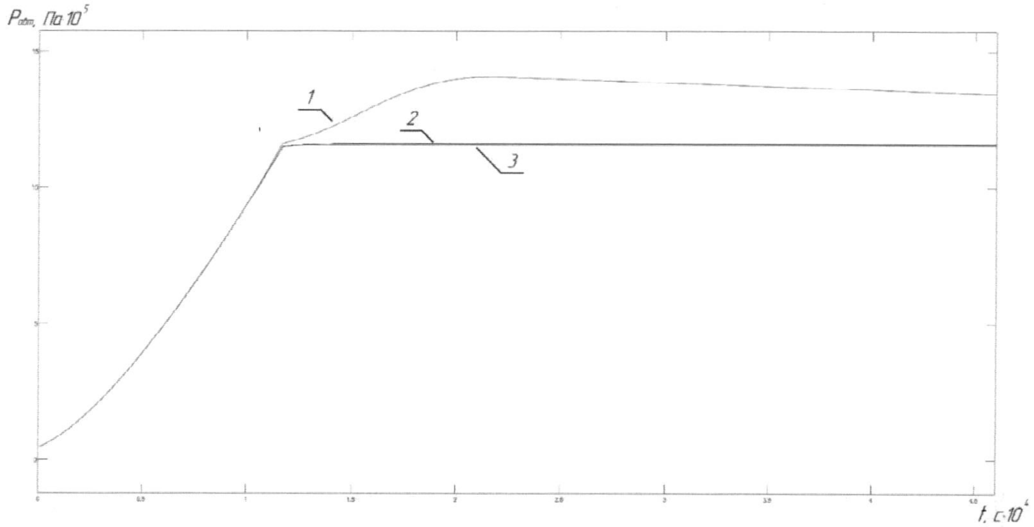

Figure 3. P_{abt} pressure curves in the autoclave: 1 – without loss of F_{ok} or F_{tb} heat release; 2 – with losses of F_{ok} without F_{tb} heat release; 3 – with losses of F_{ok} and F_{tb} heat release.

Similarly, T_{abt} temperature curves in an autoclave for D500 and D600 grade concrete are formed (Figure 5).

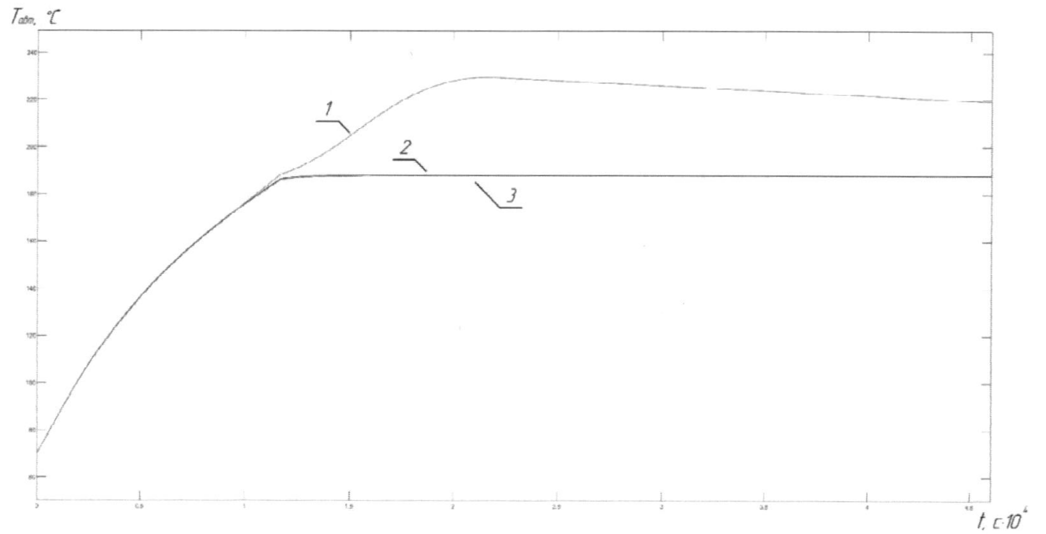

Figure 4. P_{abt} pressure curves in the autoclave. 1 – without loss of F_{ok} or F_{tb} heat release; 2 – with losses of F_{ok} without F_{tb} heat release; 3 – with losses of F_{ok} and F_{tb} heat release.

The curves obtained in the computational model quite accurately coincide with the production data given in [3], which confirms the adequacy of the developed model.

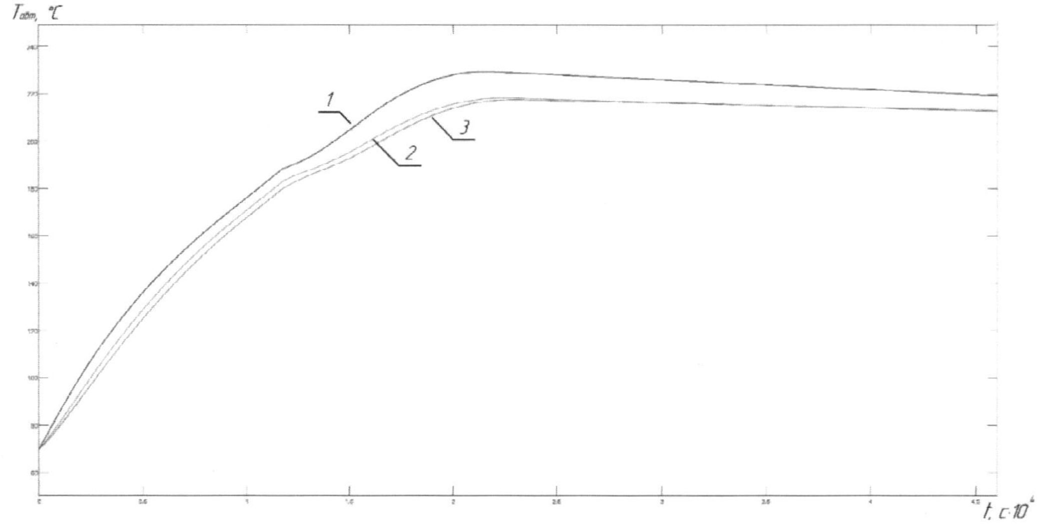

Figure 5. T_{abt} curves for temperature in an autoclave for autoclaved concretes by density grade: 1 – D400; 2 – D500; 3 – D600.

At the second stage of modeling, to find the temperature field $T_m(x,y,z,t)$ of the raw cellular concrete, at the stage of pressure increase and curing in the autoclave, we consider the D400 concrete, with measurements of $6x1.3x0.64$ m. The concrete is placed into the saturated steam environment, the temperature T_n of which varies as a function of time and is assumed to be equal to the temperature T_{abt} without taking into account the internal heat generation (Figure 4, 2) determined at the first stage of modeling. The temperature dynamics $T_n(t)$, the initial pressure P_n^0 of steam environment and the initial temperature distribution $T_m^0(x,y,z,0)$ over the volume of the concrete are set as initial conditions, with the initial pressure $P_n^0=0.056\ MPa$, and the initial temperature distribution over the volume of the concrete being uniform and having the value $T_m^0(x,y,z,0)=70°C$. The heat capacity and density of the raw concrete are determined by analogy with [7]. The thermal conductivity of raw concretes λ_m in the process of their autoclave treatment, obviously, differs from the grade value. Well-known experimental studies [8] present the data on the influence of well-defined factors on the heat transfer process in cellular concretes. The authors of the above-mentioned studies found that heating of autoclaved products largely depends on mass transfer, the intensity of which, in turn, depends on the presence of air in the pores of raw concretes, which impedes heat and mass transfer processes. The relationship of mass transfer coefficients and thermal conductivity can be determined by the expression [9]:

$$\beta = \frac{\alpha}{c_p \rho_m}, \qquad (10)$$

where c_p is the specific heat capacity of the concretes, ρ_m is the density of the concretes, α is the heat transfer coefficient, equal to the ratio of the concretes thermal conductivity λ_m to the thickness of the boundary layer δ:

$$\alpha = \frac{\lambda_m}{\delta}, \qquad (11)$$

The value β of the mass transfer coefficient for the autoclave processing of cellular concrete can be determined by jointly solving the differential equations of mass transfer [10], Navier-Stokes and continuity equations, or by conducting field experiments. Applied to this problem, an approximate value of β can be taken by analogy with [8]. Then, within the framework of the accepted assumptions, we suppose that λ_m value of thermal conductivity of concretes takes the value equivalent to λ_m[redacted], taking into account the mass exchange between steam and raw cellular concrete.

Internal heat generation is defined as a volume heat source, the dynamics of the specific volume heat generation of which is determined in accordance with [4], and its value:

$$q_{\mathrm{m}}^{\mathrm{tb}} = \frac{Q_{max}}{12}, \tag{12}$$

where Q_{max} is the maximum heat generation value for 12 concretes. It is known [3] that the internal heat generation in cellular concrete during its autoclave treatment is due to the exothermic reaction of tobermorite formation in its structure. The reaction starts when the raw cellular concrete reaches T_{tb} temperature. Let us set the value of 165°C for T_{tb} as a target set point, reaching which a volume heat source with the specified magnitude and volume specific heat dynamics previously described is generated within the raw concrete.

Let us conduct a number of computational experiments using SolidWorks *Flow simulation* software. The physical calculation time corresponds to the duration of the stages of pressure increase and curing, summing up to t_p=*36000 s*.

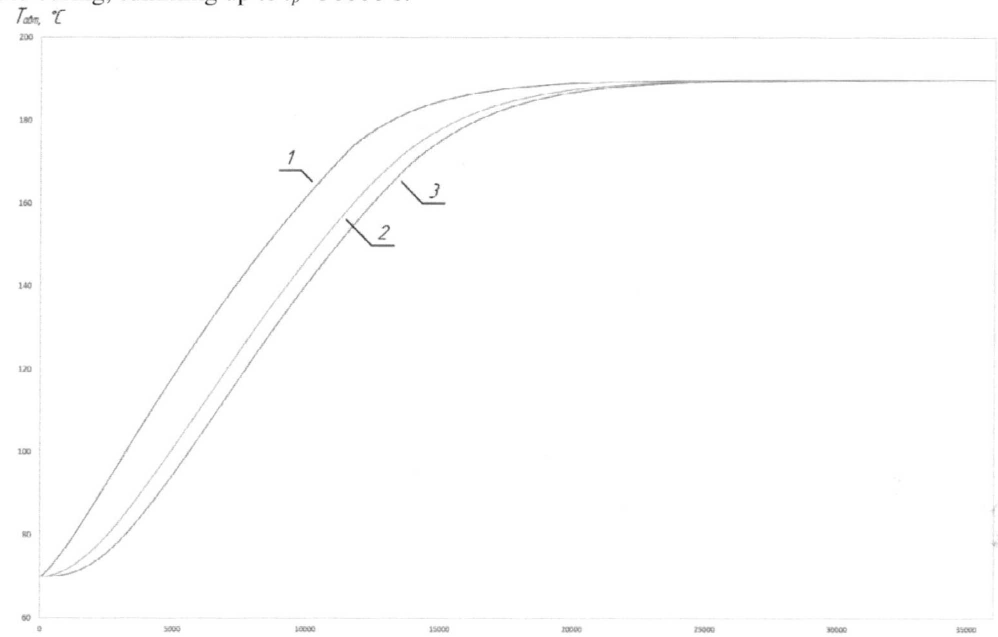

Figure 6. Dynamics of $T_m(x, y, z, t)$ temperature: 1 – on the surface of raw concrete; 2 – at ¼ thickness of the raw concrete; 3 – in the center of concrete

As a result of computational experiments, the temperature dynamics $T_m(x, y, z, t)$ in the most characteristic parts of the cellular concrete was determined on the model (Figure 6). The results of modeling thermal processes in raw cellular concrete with a sufficient degree of accuracy coincide with the data given in [3, 8], which proves the adequacy of the developed mathematical model.

4. Conclusion

The mathematical and computational model of the autoclave processing as a control object in the assumption of the concentration of parameters has been developed. The temperature dynamics of steam in an autoclave $T_n(t)$ has been determined. Computational experiments were carried out to determine $T_m(x, y, z, t)$ temperature dynamics at the most characteristic parts of the cellular concrete as an object with distributed parameters, taking into account the internal heat release during the formation of tobermorite. The assessment of the adequacy of the developed model was made by comparing the results of modeling with the results of experimental studies obtained with the existing process equipment.

IOP Conf. Series: Materials Science and Engineering **786** (2020) 012012 doi:10.1088/1757-899X/786/1/012012

References

[1] Makarov, B N 2005 Modeling and Control of Heat and Mass Transfer Processes During the Firing of Expanded Clay in Countercurrent Drum Unit Abstract of dissertation for the degree of candidate of technical sciences. Ivanovo 102

[2] Fadeev, A S , Galitskov, S Ya and Kivran, V K 2007 Mathematical Modeling of the Temperature Distribution on the Wall Surfaces of the Rotary Kiln for Expanded Clay. Computer technologies in science, practice and education. Proceedings of the 6th All-Russian Intercollegiate Scientific Practical Conference. SSTU – Samara 234

[3] Kaftaeva, M V and Rakhimbaev I 2013 Sh. Heat Release During the Synthesis of Hydrosilicates of Autoclaved Aerated Concrete. *International journal of applied and fundamental research scopus* **10** 373-6

[4] S Y Galitskov, V N Mihelkevich and A S Bolkhovetsky. Heat Generation Identification During Autoclaving of Cellular Concrete. Proceedings of the XVIII International Conference / Problems of control and modeling in complex systems. Samara, 2016

[5] Gorev, V A and Fomina, M V 2016 Simplified Calculation of Heat Transfer on Flat Surfaces. / Fire and Explosion Safety, **25(3)**

[6] A V Lykov 1967 Theory of Heat Conduction. Publishing house Higher School. M. 600

[7] Bolkhovetsky, A S 2016 Evaluation of the Influence of Non-Linearity of Raw Concrete Heat Capacity on the Heat Flow Dynamics // Mechanization and automation of construction Digest 18-21

[8] Belkin, Y M and Khaimsky, Z M 1973 Heat Transfer in Cellular Concrete and Effective Modes of Autoclave Treatment of Pproducts **2**

[9] Kutateladze, S S 1979 Fundamentals of the Theory of Heat Transfer - 5th Edition (revised and updated). - M.: Atomizdat 416

[10] M M Ishanhodjaeva 2012 Physical Cchemistry. Part I. Diffusion in Systems with a Solid Phase. SPSTUVP SP 35

Organization of recycling of construction scrap with the use of mobile crushing plants in Russia

D Yu Gustov

NRU MGSU, 129337, 26, Yaroslavskoye Shosse, Moscow, Russia

E-mail: GustovDU@mgsu.ru

Abstract. The issues of recycling construction scrap during the demolition of buildings, taking into account the specifics of the work on the demolition of buildings in Russia. The feasibility of the formation of "distribution cards" demolished buildings with an indication of the type and predicted amount of construction scrap. A typical extended set of construction and road machines has been proposed, which should be used when using mobile crushers. Suggestions are given on the mechanism for forming a set of machines according to the criterion of specific reduced costs. The calculated dependencies are given. It indicates which machines should be considered when choosing a kit, and which machines can be excluded from the analysis.

1. Introduction

Recycling issues are becoming more and more relevant given the complex economic and environmental approaches in the production of building materials. In many countries, the possibility of multiple use of resources in the chain of the life cycle is considered not as a lucky coincidence of circumstances, but as an obligatory requirement for organizations involved in the liquidation of previous facilities. Thus, in Russia, the Federal Law "On Production and Consumption Wastes" (clause 8 of Article 12) established [1]: "The dumping of waste, which includes useful components to be disposed of, is prohibited." Given this, some large construction companies do not they simply talk about the possibility of such work, but highlight this problem as a priority and position it as a separate type of their activity.

The issue of multiple processing of materials is particularly relevant for regions in which large-scale work is underway to reconstruct existing facilities, demolish other buildings and structures planned for recycling and construction in their place. It is in such places that it is possible to use construction scrap (or, as it is sometimes incorrectly called "construction waste", which is not the same thing) for its subsequent use as a filler for building materials in the future.

World practice shows that such questions are analyzed most actively in regions with active construction activities at the site of demolished old buildings. This is especially important for regions with dense buildings, among which are both European [2] and Asian countries [3], including those in the zone of possible ecological disasters of natural genesis [4].

Despite the fact that demolishing of civil buildings is extensively conducted in large Russian cities and has been described for a long time [5], almost no one solves the issues of assessing the life cycle of buildings and city blocks compared to demolition and renovation with repairs, as is customary in world practice [6]. Also, the issues of reducing mixed construction waste during demolition of buildings are not widely analyzed, which is relevant for dense building conditions [7].

A separate important problem is the fact that it is Russian modern scientific research that is necessary for the full implementation of the construction waste recycling scheme. This applies to the study of the bearing properties of composite masonry, built from waste secondary destruction of buildings, as well as the actual development of building materials from construction and demolition waste. Foreign work in this area [8, 9, 10] can only be used as a technique. It is impossible to use them fully due to the fact that buildings and structures demolished in Russian megalopolises have a different character and composition of construction scrap from those studied in the above-described works.

It should also be noted that some related issues [11] are not dealt with at all in Russia. There is no production of secondary reinforcing iron. Energy and carbon emissions during the processing of construction scrap and secondary products are not assessed.

Also, the specifics of using mobile crushers is a wide range of climatic temperatures, which has a greater impact on mobile equipment in comparison with stationary ones. This is due to the high level of hydrification of mobile equipment drives. This causes a change in the performance of technology in the transition from summer to winter temperatures.

Despite the above, today the situation is changing for the better, but the pace seems to be not high enough. But at the same time, it is impossible not to regret that the technology of crushing of construction scrap and the secondary production of materials directly at the construction site is not widely used. This is largely due to the fact that over the past years construction waste in Russia was simply dumped to landfills for disposal, despite the fact that this had a significant negative effect on the environment. The issues of preserving the territories used for landfills also seemed insignificant.

Today in the Russian market there are several manufacturers of crushing equipment in the mobile version. The most well-known suppliers in the Russian market are Rubble master, Resta, Sandvik. Also in our market there is a manufacturer "Drobmash", among whose products there are self-propelled crushing and screening machines, which are popular both in the Russian and foreign markets. At the same time, there is practically no scientific work on crushers of various degrees of mobility in Russia. And the analysis of their application is taken mostly from foreign literature [12, 13]. There are no books similar to [14] in Russian practice.

Also a significant deterrent is the fact that the practice of using sets of machines for this type of work is not accumulated, and there is also no clear understanding, in particular, expressed in the evaluation criteria, of when it is expedient to use stationary crushing equipment, both factory-made and mounted at the construction site, and when self-propelled crushing plants.

In this regard, we have begun work on the creation of sets of machines, as well as groups of criteria for the selection of such sets, which could be used in the demolition of buildings and structures of various sizes in terms of the volume of construction scrap formed during demolition.

2. Materials and methods

The solution of the task can be carried out in different ways, but in our opinion the solution is promising by drawing up a “distribution map” (area) of potential demolished (dismantled) objects. At the same time, it is necessary to classify buildings and structures by the type of material and the volume of construction scrap. It is also advisable to work out two possible "distribution cards":

- complete «distribution map» containing all objects potentially demolished in the near future;
- «distribution map2 of objects, the demolition of which is possible only with the formation of construction scrap.

The analogue of maps of the first type can be built on the basis of existing maps for large cities, which contain information about both demolished objects and those that are subject to reconstruction and restoration.

Maps oriented to the demolition of buildings are also being worked out by various organizations on the basis of various mapping software (Figure 1) [15] and can be used as a basis for development.

The second type map will allow to exclude buildings and structures whose demolition is possible with the help of lifting equipment without forming construction scrap (mainly such objects include dwellings of panel structures), which will allow optimizing the solution of logistic tasks when moving mobile crushing complexes. Analogues of the second type of cards currently do not exist, but for their basis you can also use the maps and software products described above, filling the information base of the escort with data on the nature of buildings and buildings, the type and predicted volume of waste generated during the demolition of a building.

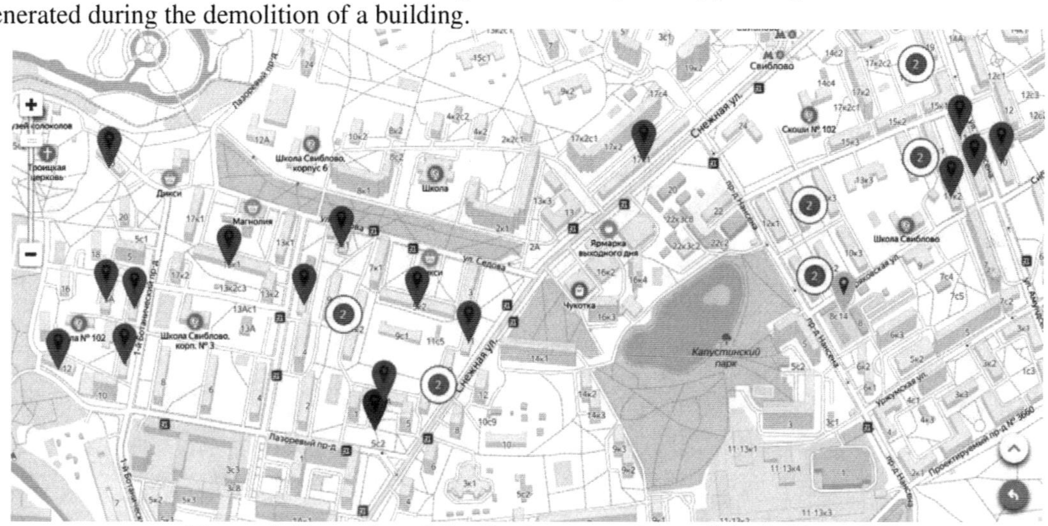

Figure 1. Examples of distribution map of demolished objects [15]

- Launch sites renovation 2020-2021, - Launch sites renovation after 2022

- Demolished five-story building.

The resulting "distribution map" allows us to predict the feasibility of using mobile crushing plants, the rational use of which is limited by a number of factors, including: the presence of stationary crushers near the demolished objects, the volume of work performed and the coordination of the performance of the mobile crushing complex with both the scope of work and performance set of machines that provide the demolition of the object. The latter will significantly improve the efficiency of the aggregate set of machines, by reducing downtime or crushing complex, or machines for demolition.

Previously, the scheme of forming a set of machines for processing construction scrap generated during the demolition of a building can be determined by technological operations with the chosen method of work production (Figure 2). At the same time, with an economically based application of a mobile crusher, the sequence of operations "demolition - preliminary crushing - final crushing - delivery to the consumer" turns out to be shorter by one operation compared to the crushing process at a stationary installation..

It is obvious that this efficiency will be realized only in the conditions of the demolition of several buildings localized in proximity to each other. At the same time, in case the transportation of the received crushed stone will be carried out for a short distance, then it will be expedient not even automobile transportation, but conveyor transport. This is confirmed by data obtained in related industries in the extraction of ore materials by the open method [16] when analyzing the transportation of different volumes of the product to different distances (Figure 3)

For the economically feasible formation of a set of machines used in the processing of construction scrap during the demolition of a building, methods were used to form optimal sets of machines based

IOP Conf. Series: Materials Science and Engineering **786** (2020) 012013 doi:10.1088/1757-899X/786/1/012013

on a single-bucket excavator [17]. Despite the fact that the processing of construction scrap is conducted by a mobile crushing plant, it is the excavator that is the "supporting" machine of sets of machines of different composition, using various specialized equipment.

The economic substantiation of the set of machines was carried out according to the criterion of the specific reduced costs of Y_{CP}, which are calculated as the ratio of the total costs C_{Σ} to the performance of the set of machines P_{Set}. The productivity of a set of machines was understood as work on the production of the final product, for example, secondary rubble.

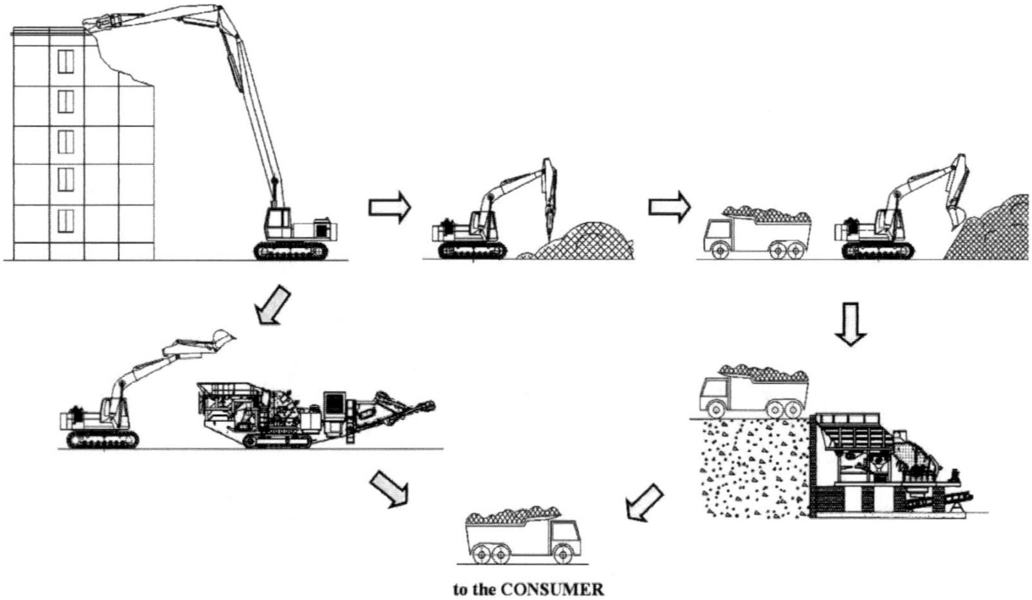

Figure 2. Comparison of technologies for processing construction scrap during the demolition of the building.

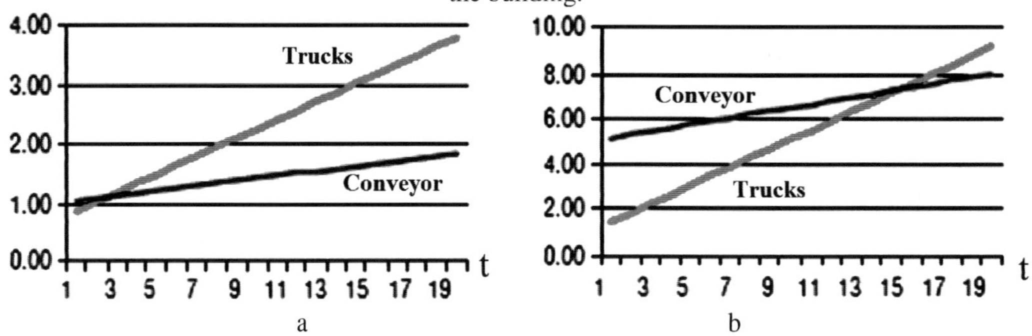

Figure 3. The cost of transporting products at a distance of 1.0 km (a) and 5.0 km (b) [16].

3. Research results

From Figure 3 it is clearly seen that increasing the transportation distance significantly reduces the efficiency of using the conveyor already at the initial stage. The economic effect with a transportation distance of 5.0 km is manifested in long-term use. However, at a distance of 1.0 km the effect can be traced from the initial stage.

Given that the demolition of the building and the use of recycled rubble are rarely in substantial proximity, the conveyor delivery of the finished product to the place of use can be recommended in a limited number of cases. Thus, with the exception of special situations, the main method of transportation will be the use of dump trucks.

Formed sets of machines for operations for schemes with stationary (located at the processing plant) and mobile (located at the construction site) crushers.

Typical equipment presented in the table.

Table 1. The distribution of sets of machines on operations.

<table>
<tr><th>Operations</th><th>Sets of machines</th><th>Set 1</th><th>Set 2</th></tr>
<tr><td>Building demolition equipment</td><td>Excavator with multi-processors (primary pulverizers)</td><td colspan="2">+</td></tr>
<tr><td>Equipment for the primary crushing of construction scrap for the crusher</td><td>Excavator with pulverizers (secondary pulverizers)
Excavators with hammers</td><td colspan="2">+</td></tr>
<tr><td>Equipment for loading crushed construction scrap into a dump truck</td><td rowspan="2">Excavators</td><td>+</td><td></td></tr>
<tr><td>Equipment for loading ground construction scrap into a mobile crusher</td><td></td><td>+</td></tr>
<tr><td>Equipment for the delivery of construction scrap to the crusher</td><td>Dump trucks</td><td>+</td><td></td></tr>
<tr><td>Equipment for the final crushing of construction scrap</td><td>Stationary crusher
Mobile crusher</td><td>+</td><td>+</td></tr>
<tr><td>Equipment for loading the final product in the dump truck</td><td>Loaders
Excavator Loaders</td><td colspan="2">+</td></tr>
<tr><td>Equipment for the delivery of the final product to the consumer</td><td>Dump trucks</td><td colspan="2">+</td></tr>
<tr><td>Equipment for the delivery of equipment to the construction site</td><td>Trailers
Trucks</td><td colspan="2">+</td></tr>
</table>

Set 1 – a set of machines with a stationary crusher (located at a processing plant)

Set 2 – a set of machines with a mobile crusher (located at the construction site)

As noted above, the economic rationale for a set of machines was conducted according to the criterion of the specific reduced costs Y_{CP}. In general, this value can be determined by the formula:

$$Y_{CP} = (\Sigma n_{Cr} \cdot C_{Cr} + \Sigma n_H \cdot C_H + \Sigma n_E \cdot C_E + \Sigma n_{TP} \cdot C_{TP} + \Sigma n_C \cdot C_C + \\ + \Sigma n_L \cdot C_L + \Sigma n_{TS} \cdot C_{TS} + \Sigma n_S \cdot C_S + \Sigma n_M \cdot C_M + \Sigma n_A \cdot C_A)/P_{Set} \quad (1)$$

n_{Cr} and C_{Cr} – the number of excavators with multi-processors (primary breakers) and reduced costs for their work;

n_H and C_H –number of excavators with hammers and with pulverizers (secondary pulverizers) and reduced costs for their work;

n_E and C_E – number of bucket excavators and reduced costs for their work;

n_{TP} and C_{TP} –number of dump trucks transporting construction scrap to the stationary crusher (primary dump trucks) and reduced costs for their work;

n_C and C_C –number of crushers and reduced costs for their work;

n_L and C_L –number of loaders and reduced costs for their work;

n_{TS} and C_{TS} –number of dump trucks transporting the final product to the consumer (secondary dump trucks) and reduced costs for their work;

n_S and C_S – number of machines that provide delivery of equipment to the construction site, and the costs of their work;

n_M and C_M – number of auxiliary workers and reduced costs for their work;

n_A and C_A – amount of auxiliary equipment and reduced costs for its work.

Auxiliary workers and equipment include personnel and equipment, the involvement of which in the technological process may be required taking into account the specific situation. In this case, any equipment that does not fit into the standard equipment (see table) can be used: lifting, mounting equipment, power tools, etc.

The analysis of the technology of work shows that part of the machinery and equipment performs the scope of work unrelated to the method of crushing of construction scrap (equipment for demolishing a building and; equipment for primary crushing of construction scrap for a crusher). Such machines can be excluded from the analysis of the effectiveness of the kit. At the same time, excluding operations that are similar in essence (equipment for loading the final product into a dump truck, equipment for delivering the final product to the consumer, etc.) should not be excluded because these operations will differ in cost, since such parameters as: transportation, the structure of access roads and, perhaps, even the model of the equipment used. Equipment for the delivery of equipment to the construction site can be partially excluded, since the delivery of tracked excavators will be present in both schemes of work, and the delivery of a mobile crushing plant - only in the second.

Considering the above, formula (1) will be converted for set 1:

$$Y_{CP1}= (\Sigma n_{E1} \cdot C_{E1} + \Sigma n_{TP} \cdot C_{TP} + \Sigma n_{CS} \cdot C_{CS} + \Sigma n_{L1} \cdot C_{L1} + \Sigma n_{TS1} \cdot C_{TS1} + \Sigma n_{M1} \cdot C_{M1} + \Sigma n_{A1} \cdot C_{Av})/P_{Set1} \quad (2)$$

for set 2:

$$Y_{CP2}= (\Sigma n_{E2} \cdot C_{E2} + \Sigma n_{CM} \cdot C_{CM} + \Sigma n_{L2} \cdot C_{L2} + \Sigma n_{TS2} \cdot C_{TS2} + \Sigma n_{SC} \cdot C_{SC} + \Sigma n_{M2} \cdot C_{M2} + \Sigma n_{A2} \cdot C_{A2})/P_{Set2} \quad (3)$$

: n_{CS} and C_{CS} – number of stationary crushers and reduced costs for their work;

n_{CM} and C_{CM} – number of mobile crushers and reduced costs for their work;

n_{SC} and C_{SC} – number of machines ensuring the delivery of mobile crushers to the construction site, and the reduced costs of their work.

In the case of the use of mobile screens at the construction site in order to divide the resulting crushed stone into fractions, the list of used machines should be expanded. In this case, in addition to taking into account the screeners themselves, it is necessary to take into account the method of their delivery to the construction site. This can be either an additional trailer or a trailer delivering a mobile crusher. During the calculations, this value can be taken into account as part of the crushing equipment or as ancillary equipment.

4. Conclusions:

1. The creation of "distribution maps" of demolished buildings with an indication of the type and predicted volume of construction scrap is effective for megacities.
2. Construction plans for buildings, roads, and other facilities should be consistent with the building demolition plan in order to minimize costs by using construction scrap directly from the dismantling facility without interim storage..
3. In the context of a comprehensive demolition plan for housing concentrated in one area, followed by building up the released areas, it is advisable to organize the production of secondary crushed stone and its subsequent use directly close to the demolished objects using mobile crushers.
4. In terms of the dismantling of a stand-alone building, the feasibility of using a mobile crusher should be confirmed by the economic justification in comparison with the classical scheme.
5. In the conditions of transportation of recycled rubble to the place of its further use at distances up to 1.0 km, it is advisable to use a conveyor. The efficiency of the conveyor will be the higher, the smaller the distance, and the greater the volume of the transported material.
6. Calculation of the economic efficiency of the use of a set of machines should be carried out without taking into account the cost of using machines that perform work unrelated directly to the method of crushing of construction scrap.

References

[1] Information on http://www.consultant.ru/document/cons_doc_LAW_19109/ (on Russian)
[2] P V Saez; M D Merino; C Porras-Amores 2012 "Estimation of construction and demolition waste volume generation in new residential buildings in Spain", *Waste management & research*, **30(2)** 137-46
[3] T Ding; JZ Xiao 2014 "Estimation of building-related construction and demolition waste in Shanghai", *Waste management*, **34(11)** 2327-34

[4] G Karunasena; D Amaratunga, D 2016 "Capacity building for post disaster construction and demolition waste management A case of Sri Lanka", *Disaster prevention and management*, **25(2)** 137-53

[5] M B Vasilyeva 2012 "Problem of Demolition First Mass Series in Complex Reconstruction of Residential Building", Conference: 3rd Forum of Young Researchers: In the Framework of International Forum on Education Quality, 288-291

[6] V, Weiler; H Harter; U Eicker 2017 "Life cycle assessment of buildings and city quarters comparing demolition and reconstruc tion with refurbishment", *Energy and buildings*, **134** 319-328

[7] BH Son; YC Kim; WH Hong 2015 "Measures to reduce mixed construction wastes during demolition of buildings", Asia Life Scieces, Supplement **12** 89-100

[8] C Jayasinghe 2016 WMCDJ Fonseka; YM Abeygunawardhene, "Load bearing properties of composite masonry constructed with recycled building demolition waste and cement stabilized rammed earth", *Construction and building materials*, **102** 471-477

[9] A Pathak; S Kumar 2014 VK "Development of Building Material from Geopolymerization of Construction and Demolition Waste (CDW)", *Transactions of the Indian ceramic society*, **73(2)** 133-7

[10] T Huang; F Shi; H Tanikawa; JL Fei; J Han 2013 "Materials demand and environmental impact of buildings construction and demolition in China based on dynamic material flow analysis", *Resources conservation and recycling*, **72** 91-101

[11] X W Cheng and H T Xi 2012 «Analysis Calculation on Single-Tooth Roll Mobile Crusher Mechanics Model», *Applied Mechanics and Materials*, **105-107** 2149-52

[12] X Y W and X G Yu 2013 «The Reformation Design of the Qidashan Semi-Mobile Crusher Station», *Applied Mechanics and Materials* **397-400** 924-7

[13] G W Cha, W H Hong, S W Shin 2013 «A Study on Energy Consumption and CO2 Emissions According to Manufacture of Recycled Rebar Using Wasted Steel from Building Demolition», *Applied Mechanics and Materials*, **357-360** 869-875

[14] F Pacheco-Torgal; V Tam; J Labrincha; Y Ding; J de Brito 2013 "Handbook of Recycled Concrete and Demolition Waste", Woodhead Publishing, 672

[15] Information on https://stroi mos ru/novaia-proghramma-rienovatsii-piatietazhiek (on Russian)

[16] J Kempas, A Ashadulin 2012 «Technologies and conditions for effective use of mobile large crushing plants and belt conveyors», *Mining Industry*, **4** 52-56(on Russian)

[17] Kudryavtsev E M , Gustov D Yu 2010 «Determining the performance of a single-bucket excavator with reverse shovel equipment», *Vestnik MGSU* [Proceedings of Moscow State University of Civil Engineering], **4(1)** 101-107(on Russian)

IOP Conf. Series: Materials Science and Engineering **786** (2020) 012014 doi:10.1088/1757-899X/786/1/012014

Formation of the network of spare elements warehouses

B G Kim[1,*], R A Nasrulloeva[1,2], Z N Shakir[1,3] and L V Zakrevskaja [1,4]

[1]Department of construction production, Vladimir State University, 87 Gorky Street Vladimir, Russia.

[2]Department of construction production, Vladimir State University,Tajikistan.

[3]Department of construction production, Vladimir State University,68 Karradaa Street Bagdad, Iraq.

[4]Department of construction production, Vladimir State University, 87 Gorky Street Vladimir, Russia.

*E-mail: kim_bg@mail.ru

Abstract: Discusses the redundant machinery construction enterprises. The method of calculation of a network of warehouses taking into account the possibility of different conditions of the reserve from the part to the backup machine. The technique allows you to determine the required number of warehouses of different levels, taking into account the time of delivery of the spare element from the warehouse on the other hand, it is necessary to take into account that the increase in inventory helps to slow down the turnover of reserve elements, increases the cost of the Insurance Fund.

1. Introduction

Reservation of the fleet of machines allows optimizing the cost of the formation of its size and storage capacity of spare parts. Russia's conditions are characterized by territorial disunity of facilities and weak road infrastructure. This is typical for such a mountainous country like Tajikistan.

However, rationalizing the warehousing network can dramatically reduce the cost of spare parts. As a rule, the researchers of this topic [1-7] simplify the solution of reviewing tasks by bringing to one level of warehouses and turning the hierarchical structure into a linear chain. In the works [8-12] the ways of solving the problem are outlined, but the final version is not given.

For the current state of modern construction is practicality characterized by of the objects being constructed separately. Insufficiently developed infrastructure, causing significant transport costs, long waiting period for the delivery of spare elements from manufacturing plants, development of technical centres and dealerships, raises the question of creating a spare parts supply network - peripheral regional network of warehouses.

Let us consider the functional hierarchy of spare elements in the system of servicing by the mechanisms and machines of building objects. Spare elements can be contingently classified into several groups:

- A spare technological set of machines and equipment for the production inherent in the organization of construction work in question.
- Backup construction machine.
- Reserve node, complete unit.
- Spare block.
- Detail, fixed connection.

It is obvious that a reserve technological set (if it is available or can be made up of reserve machines) should be located on the central base of a large or specialized organization such as mechanization management or in the place where the main construction objects are located. In case if one object has the priority in a building that is much higher than that of other elements; this group can be placed in the location of this separate object and used as a "hot" backup.

Approximately the same policy should be applied in respect of large building machines, placing reserve units on the central base of the unit or at the main support station.

A more difficult task is to form warehouses of deficient elements, the need for which is significant. Here we have two possible conditions:

a) The storage of reserve elements should be as close as possible to the consumer;

b) The increase in the number of warehouses reduces the efficiency of the reserve use.

The construction in the North, especially in the conditions of Siberia, is characterized by the territorial dissociation of building objects and poorly developed infrastructure. In this regard, the task of storing spare parts is of considerable relevance, since there are certain economic losses for any solution.

In general, the size of a separate warehouse must satisfy the following condition:

$$P_{nej} \leq P_{ye} \quad (1)$$

where:

P_{nej} - losses and costs associated with storing the e- element in the j- warehouse;

P_{ye} - losses and costs associated with the delivery of the e- element from the central warehouse.

However, entire losses across the warehouses will differ from the simple sum of loss amounts from each warehouse. This is due to the fact that the system of warehouses has the property of emergency.

The task of completing warehouses of the reserve equipment can be formulated as follows: a fleet of machines of the well-known nomenclature the revitalized structure is given in the directive order of each machine. Patterns of machine failure and the transition of the elements of the machine from the serviceable to the faulty are known. A set of warehouses with their organizational and economic characteristics is given. It is necessary to determine the required number of reserve elements and their placement in the system of warehouses subdivisions.

2. The solution to the problem

We carry out the classification of reserve elements based on their production needs. Let us determine the cost of the e- element at the facility supplied from the j- warehouse.

If all j- warehouses are denoted by N, then two options can occur: the e- element is requested at j- warehouse and the e- element is not requested at j- warehouse. Thus, we can make two sets of elements:

M^* is the set of e- elements that are requested through all warehouses, M^{**} is the set of e- elements requested only through separate warehouses.

Here it is necessary to determine the volume of reserve elements in each warehouse. Their number on a separate item in stock may vary in the range $0 \leq M_e^{**} \leq M^*$

In general, j warehouse contains I types of elements

$$I^n = \sum_{e=1}^{I} M_j \tag{2}$$

The volume of the warehouse J is determined by the expression

$$N_j = \sum_{r=1}^{R} \sum_{e=1}^{I} M_{ej} \tag{3}$$

Where: r {0,1,2,...,R} is a number of e- item in stock.

The total amount of spare parts in the contracting organization

$$N = \sum_{j=1}^{I} \sum_{r=1}^{R} \sum_{e=1}^{J}, M_{ej} = \sum_{j=1}^{I} N_j \tag{4}$$

The total cost in the spare elements equals:

$$\sum_{j=1}^{I} \sum_{r=1}^{R} \sum_{e=1}^{J}, M_{ej} = \sum_{j=1}^{I} \sum_{r=1}^{R} \sum_{e=1}^{J} S_{ejr} + Q_{ej} \tag{5}$$

Where: S_{ejr} Need of e- element requested by the j- warehouse;

Q_{ej} Reserve of the e- element in the j- warehouse in case of an abnormal supply and demand of this element.

Thus, the division of the number of warehouses on one side reduces the delivery time of the reserve element to the non-efficient machine, when on the other side, the amount of insurance reserve increases, which leads to a slowdown in the warehouse turnover.

The second important point in the formation of volumes and the nomenclature of warehouses is to determine the possibility of replacing the concrete element on the object.

Indeed, there is a need for one or another spare element (even the simplest detail) to be available at any object, but due to the difficulties of replacing it in the field, the machine must be sent to a central repair shop. Therefore, we form the new set M, which includes only those reserve elements, which, according to the technical principle, can be requested only from one or several warehouses.

So, in the network of warehouses, their significance and size should be classified. The first group of warehouses includes warehouses with a full or maximum set of reserve elements for a large construction organization.

This may be a central warehouse, a specialized repair warehouse or warehouses of mechanization organizations, a warehouse of a Stroymashservis type dealership company.

The second group includes warehouses of mechanization divisions bases (when the repair service is centralized), warehouses of priority facilities when servicing a significant amount of equipment at a small distance from each other and under the condition of good transport connection between facilities.

The third group includes warehouses of reference points, object-specific warehouses of large objects. The fourth group includes small warehouses, warehouses for distribution of tools with a minimum number and a narrow range of backup elements (hardware, filters, etc.)

This gradation allows us to systematize the distribution of spare elements and determine their rational volume. This technique can also be used by construction equipment manufacturers to supply their technical centres.

3. Conclusion

The given method provides justification of the hierarchical structure of warehouses and reserve sites of construction equipment and, first of all, is suitable for use by large holders of construction equipment, as well as for the construction of large objects (for example, Rogun HPP in Tajikistan). The technique provides the possibility of using spare sets of machines (individual machines) as a "hot" reserve.

References

[1] V S Lukinsky 2017 Inventory management in supply chains in 2 hours: textbook and workshop for undergraduate and graduate etc ; — Moscow: Trait, — 307
[2] Schreiber George 2016 Efficient inventory management. – M.: Alpina Publisher 304
[3] R Bley Effective store maintenance Techniques engineering plants and maintenance. One thousand nine hundred sixty-three.
[4] A Maybe 1968"Inventive control of technical characteristics of equipment and maintenance of the plant"
[5] G Nemes 1965 Theory and practice of PM at continuous technological plant "Technologies of mechanical engineering and maintenance of plant"
[6] A Sterlikova 2009 Inventory management in supply chains M. Infra – M 430
[7] L L Reshetnikov 2009 Conceptual provisions of the theory of management and operation of the fleet of machines. *Bulletin of Chelyabinsk State University* **1(139)** Economy 17 42-5
[8] B G Kim Ensuring the health and health parks building machines. Vladimir. Publishing house of VSU. 2004, 240
[9] R A Rudzki and Douglas A 2006 Smork Straight to the bottom line. J Ross Publishing. —304
[10] D J Baursaks and Kloss D J Logistics 2001 "Olympus-Business" 640
[11] BEA F K Dichtl, Schweitzer M 1999 Economy of the enterprise M , *Infra-M*, 928
[12] Elmaghraby W J 2000 supply contract competition and sourcing policy. Production Of Service Oper. Management **2(4)** 350-71

Risk factors affecting the implementation of construction projects in Iraq

B G Kim[1,*], Z N Shakir[1,2] and R A Nasrulloeva[1,3]

[1]Department of construction production, Vladimir State University, 87 Gorky Street Vladimir, Russia.

[2]Department of construction production, Vladimir State University,68 Karradaa Street Bagdad, Iraq.

[3]Department of construction production, Vladimir State University,Tajikistan.

[*]E-mail: kim_bg@mail.ru

Abstract. The construction industry of Iraq is widely associated with multiple influencing factors that due to their specific lead to increased uncertainty and multiple risks during the construction phase. The main objective of this article is to identify and evaluate key risk factors and their impact on the formation of car parks and the realisation construction projects in Baghdad (Iraq). In this study the risks associated with construction projects identified with the help of open questionnaire and interview with construction project specialists. In total, factors of risks were identified and categorized into five groups: financial, political, organizational, technical and legal. The study revealed that the top ten important risk factor are: inconsistency structure specifications(76.89%), changes in design(75.97%), late arrival of equipment and materials from the country of origin(75.74%), poor coordination between the client and the contractor(72.29%), lack of clarity in contractual obligations(71.17%), the occurrence of internal problems between the project parties(71.17%), dues payment delay not according to contract (68.09%) , insufficient financial allocations to complete the project(66.37%) , war (47.75%) and receipt of operational advances delay due to legal proceedings(46.58%).

1. Introduction

Construction in Iraq involves the issue of mechanization and the risks of reserving machines. Construction projects face multiple and severe risks throughout their lifecycle starting from decision-making stages to final project delivery. Risks are generally seen as events that affect project objectives, that leads to delayed delivery of the project or increase in cost and sometimes even affect the quality of work. As far as the construction industry is inherently risky, some of these risks are rather predictable or easily identifiable. Others may not be entirely foreseen. Over the past two decades, attention to the issue of risk has been widespread [5]. There have been many studies on risk management and the strategies of avoiding the risks, limiting their effects or transferring risks to another party. Construction companies in Iraq face complicated types of risks during the project implementation phase. The existence of these risks is time-wasting and results in resource depletion. It is necessary to identify the types of these risks and their impact on the project objectives to predict how they can be addressed and responded.

Risk is defined as an uncertain event potentially having a negative or positive effect [7-10] at least on one of the project objectives. The risk is also defined as the prospect of any future event that may affect plans of the project[12-14], including financial and temporal and therefore a deviation from what has been planned. There are several ways to classify the risks involved in the project. Risks can be classified internally or externally [11] or based on their primary sources [1-2,4] such as financial, legal, political and management. Internal risks are related to all types of projects, regardless of project size. External risks are directly linked to economic and political changes[15,17], changes in prevailing laws and unforeseen circumstances[16,18].

2. Methodology

The overall methodology of this study is largely based on a survey that took place based on the responses given on the questionnaire collected from construction companies of different sizes in Baghdad by mail and personal interviews with project managers and engineers employed in construction companies. The questionnaire covered two types of information. The first part of the questionnaire focused on obtaining general information about respondents such as company type, organization, experience, etc. The second part of the questions about construction risks was based on their primary sources. The risk factors for this study have been classified into five types of risk associated with construction projects: 1. Financial risks, 2. Political risks 3. Organizational risks 4. Technical risks 5.Legal risks .The questionnaire was submitted to 77 participants, 50 of whom fully responded to the questionnaire, all responses were analyzed afterwards.

Data processing and analysis were conducted with the use of Excel software, and the results showed that more than half of the sample size of respondents have more than 10 years of experience, which says for reliability of the information provided by respondents to portray the reality in the construction industry. The study adopted the following criterion for assessing the importance of risk according to risk categories: [6, 3]

$$\text{Frequency indicator (F.I.)} = \frac{\sum a\,(n/N)}{5}\ \% \tag{1}$$

$$\text{Severity index (S.I)} = \frac{\sum a\,(n/N)}{5}\ \% \tag{2}$$

Where (a) is the constant expressing weighting given to each response (range from 1 for very low to 5 for very high)

(n) is the frequency of the response

(N) is the total number of responses.

Combine these two criteria into the following equation to find the importance of each risk factor as the function:

$$\text{Importance Index (I.I) (\%)} = \frac{\text{F.I (\%)x S.I (\%)}}{100} \tag{3}$$

3. Results and discussion.

3.1 Financial risks.

As shown in the table (1), the financial risks are faced by all parties involved in the implementation of construction projects. The highest risk rating is in dues payment delay not according to contract 68.09 %. It sometimes causes lengthy negotiations between the contractor and the owner that results in bringing in design changes and may cause delays in payments by the owner to win the negotiations in his favor.

Table 1. Financial risks associated with construction projects.

Type of risk	F.I	S.I	Total risk I.I	Total risk Rink
Dues delay payment not according to contract	78.13	87.19	68.09	1
Insufficient financial allocations to complete the project	74.34	89.26	66.37	2
High prices of materials and reinforcing steel	48.44	62.06	30.08	3
Inflation and price volatility	45.28	40.82	18.48	4
Unexpected funding cuts	42	43.58	18.30	5

3.2 Political risks.

The table (2) shows that war is the most important factor affecting the political risks associated with construction projects ranked at 47.75 % because of its negative impact on the national economy in general.

Table 2. Political risks associated with construction projects.

Type of risk	F.I	S.I	Total risk I.I	Total risk Rink
War	68.08	69.83	47.75	1
Delayed arrival of equipment and materials due to deteriorating conditions	48.84	52.14	22.33	2
Project damage due to war conditions	50.12	41.66	20.87	3
Insecurity and thefts	37.60	41.58	15.63	4

3.3 Organizational risk.

Table (3) shows that the stable coordination between the client and the contractor is the most important factor affecting the organizational risks associated with construction projects ranked at 72.29 % due to the absence of the client or his representative during the implementation period.

IOP Conf. Series: Materials Science and Engineering **786** (2020) 012015 doi:10.1088/1757-899X/786/1/012015

Table 3. Organizational risk associated with construction projects.

Type of risk	F.I	S.I	Total risk	
			I.I	Rink
Poor coordination between the client and the contractor	86.57	85.82	72.29	1
Lack of clarity in contractual obligations	85.81	82.95	71.17	2
Delay in starting work	48.52	50.56	24.53	3
Events and holidays	51.10	47.13	24.08	4

3.4 Technical risks

Table (4) shows that inconsistency structure specifications are the most important factor affecting the technical risks associated with construction projects classified at 76.89 % due to the absence of the consultant for monitoring of the preparation of designs and at the construction stage. This results in a mismatch of approved designs with technical specifications.

Table 4. Technical risks associated with construction projects.

			Total risk	
Type of risk	F.I	S.I	I.I	Rink
Changes in design	87.13	87.20	75.97	2
Inconsistency structure specifications	85.74	89.68	76.89	1
Late arrival of equipment and materials	88.04	86.04	75.74	3
Inaccurate scheduling	75.10	75.33	56.57	4
Inaccurate surveys	75.11	74.97	56.30	5

3.5 Legal risks

Regarding the legal risks associated with construction projects, the study showed that legal disputes during the construction phase between the project parties topped the highest proportion of legal risk ranked at 46.58 % as showed in the table (5). It should be noted that these risks are of medium importance because all parties recognize the necessity of solving problems between the parties before reaching the judiciary, and the consequent access to time consuming justice for the high costs.

Table 5. Legal risks associated with construction projects.

Type of risk	F.I	S.I	Total risk	
			I.I	Rink
Receipt of operational advances delay due to legal proceedings	60.09	76.05	46.58	1
Legal disputes between the construction project parties	48.88	45.28	22.13	2
Difficulty with obtaining licenses and work permits	43.71	46.91	20.50	3

4. Conclusion.

The key point identified by this research is to explore key risk factors and identify those factors that can be encountered in construction projects in Iraq. The determinants for each type were shown in tables (1to 5). The study suggested that factors that have the highest impact on risk in construction sites are those that have a higher indication, technical risks, organizational risks, financial risks, political risks and legal risks. The ten most important factors are: inconsistency structure specifications, changes in design, late arrival of equipment and materials from the country of origin, poor coordination between the client and the contractor, lack of clarity in contractual obligations, the occurrence of internal problems between the project parties, payment delay in disagreement with a contract, receipt of operational advances delays due to legal proceedings, war and insufficient funds allocations.

References

[1] Al-Bahar G and Crandall K 1990 J Construction Engineering and Management 3 533-546
[2] Aleshin A 2001 *International Journal of Project Management* **19** 207-222
[3] Assaf S and Al-Hejj 2006 *International Journal of Project Management* **4** 349-357
[4] Baloi D and Priceh D 2003 *International Journal of Project Management* **21** 261–269
[5] Entisar K 2015 *Journal of Engineering* **4** 1-19
[6] Firas K 2015 *International Journal of Advanced Research in Engineering and Technology* **6** 35-48
[7] Kartam N and Kartam S 2001 *International Journal of Project Management* **19** 325-335
[8] Merna T and Al-Thani F 2005 Corporate Risk Management (England: John Wiley & Sons Ltd)163
[9] Muna H, Mohamed N and Omar A 2012 *Damascus University Journal of Engineering Sciences* **28** 129-150
[10] PMBOK 2004 A *Guide to the Project Management Body of Knowledge* (Pennsylvania: Project Management Institute) 187
[11] PMBOK 2013A *Guide to the Project Management Body of Knowledge*(Pennsylvania : Project Management Institute) 223
[12] kravchenko I and Kim B 2019 *Construction and road machines* **3** 45-49
[13] Kim B 1991 *Construction Mechanization* **11** 2-4
[14] Kim B and Kuznetsov K 1994 *Mechanization of construction* **7** 23-24

[15] Pankratov E and Pankratov O 2015 *Construction Economy* **3** 4-17
[16] Tuskaeva Z 2016 *Journal of Applied Engineering* **11** 4369-4376
[17] Asaul A 2004 *Construction Economics* **6** 16-25
[18] Asaul A 2005 *Construction Economics* **7** 14-25

IOP Conf. Series: Materials Science and Engineering **786** (2020) 012016 doi:10.1088/1757-899X/786/1/012016

Assessment of resources of road machines working in the conditions of tropical climate

Vladimir Zorin[1,*] and Le Tchong Tuan[2]

[1] Moscow automobile and road state technical university, Russia, Moscow

[2] Moscow automobile and road state technical university, Russia, Moscow

* Email: madi-dm@list.ru

Abstract. In this article results of researches are considered, formulas full gamma percent resource $T_{f.\gamma}$and a gamma percent resource before the first major repair of$T_{r.\gamma}$(is developed for definition at γ=80%) road-building machines, according to technical condition and operating conditions of the machine in the tropical climate of Vietnam. The corresponding values of resources of road-building machines in Vietnam were at the same time defined that formed a basis for exact planning terms of carrying out construction works, to make the reasonable diagram of maintenance and repair of machines for each climatic zone of Vietnam.

1. Introduction

Vietnam is developing country with the annual average speed of economic growth of 6-7% [0]. In it, infrastructure and a transport network in particular are upgraded and extend more and more. As a result the great demand of the equipment in service of the national economy, including road machines forms.

Characteristic of operating conditions of road machines in Vietnam is the hot, humid tropical climate of the country located near the sea [0]. It is necessary to evaluate fallback resources of operability of machines as a basis of planning of operation, technical maintenance and repair for the purpose of increase in efficiency of use of the road-building equipment in climatic conditions of Vietnam.

2. Methods

The key resource indicators of road machines are full gamma percent resource $T_{f.\gamma}$and a gamma percent resource before the first major repair of $T_{r.\gamma}$ (at γ=80%), determined by a formula [0, 0]:

$$\begin{cases} T_{f.\gamma} = k_\gamma . T_d . T_a \\ T_{r.\gamma} = T_{f.\gamma} / \sum_{i=1}^{n} C_{i-1} \end{cases} \quad (1)$$

where

T_d - depreciation service life, years;

T_a - annual operating time of the machine;

k_γ- the coefficient depending on a type of the distribution law of values of a resource, variation coefficient υ and probability values;

C - the coefficient considering reduction of a between-repairs cycle after major repair of the machine;

n - number of between-repairs cycles.

This formula is used for calculation of resource indicators of the new road machines working in normal conditions of operation (the temperate climate characteristic of the Moscow region). In severe climatic conditions of Vietnam generally import road machines having a considerable operating time therefore for calculation of a residual resource it is necessary to consider coefficient of technical rigidity of climate and also machine utilization coefficient on time are used. In this case the formula (1) will be transformed as follows:

$$\begin{cases} T_{f.\gamma} = k_{\gamma}.T_{d}.T_{a}.\frac{S_{c.t}}{S_{c.tr}}.k_{u} \\ T_{r.\gamma} = T_{f.\gamma}/\sum_{i=1}^{n} C_{i-1} \end{cases} \quad (2)$$

where k_u- machine utilization coefficient on time (for the majority of road machines in Vietnam is accepted$k_u \approx 0,8$);

$S_{c.t}$- coefficient of technical rigidity of a temperate climate (Moscow region);

$S_{c.tr}$- coefficient of technical rigidity of tropical climate in Vietnam.

Coefficients$S_{c.t}$and$S_{c.tr}$it is calculated by the following formulas[0]:

$$S_{c.t} = (0,75.t_{min.c} + 0,25.t_{min.ac})(1 + 0,015.\sigma_c)(1 + 0,07.v_{sc})(1 + 0,26.\varphi_c)(1 + 0,014.n_{f.c})(1 + 0,022.\tau_{bz}) \quad (3)$$

and

$$S_{c.tr} = (0,55.t_{max.w} + 0,20.t_{max.aw})(1 + 0,01.Q_s)(1 + 0,0075.\sigma_w)(1 - 0,03.v_{sw})(1 + 0,026.\varphi_w)(1 + 0,009.n_r)(1 + 0,012.\tau_{az}) \quad (4)$$

where

$t_{max.w}, t_{max.aw}, \sigma_w$- respectively mean value of maxima of air temperature, mean value of an absolute maximum of air temperature and average non-periodic amplitude of daily fluctuations of air temperature in three warmest months, °C;

Q_s, v_{sw}, n_r- respectively mean value of the monthly sum of solar radiation, average speed of wind and a median number of days with rains in three warmest months;

φ_w, φ_c- mean value of relative air humidity in three warmest months and in three coldest months in unit shares;

τ_{az} - action duration in months of average temperature of air is above zero.

$t_{min.c}, t_{min.ac}, \sigma_c$- respectively mean value of average minimum temperatures of air, mean value of an absolute minimum of air temperature and average non-periodic amplitude of daily fluctuations of air temperature in three coldest months, °C;

v_{sc}- average speed of wind in three coldest months, m/s;

$n_{f.c}$- mean value of number of days with fog and a blizzard in three coldest months, days;

τ_{bz}- action duration in months of average temperature of air is below zero.

3. Results

On the basis of the statistical data characterizing climatic conditions in a tropical zone (regions of Vietnam [0, 0]) and in a zone of a temperate climate (The Moscow region [0]), we can define coefficient of technical rigidity of climate, as shown in tab.1.

Table 1. Values of coefficient of technical rigidity of climate.

City / Parameter	Hanoi	Vinh	Da Nang	Gia Lai	Vung Tau	Moscow
$S_{c.tr}$ or $S_{c.t}$	36.05	36.58	35.73	28.92	33.10	28.48

It agrees [0], we obtain the data necessary for calculation of resource indicators of some typical road machines (tab.2.).

Table 2. Values of parameters for calculation of resource indicators.

Name machines	Annual operating time of machine T_a [h]	Depreciation service life of T_d, [years]	Number of between-repairs cycles n	Coeffitsent variations υ	Coefficient k_γ
Excavator SOLAR-140w	2400	10	2	0.30	0.74
Bulldozer T-140	2000	8	1	0.4	0.64
Grader DZ-122	1600	8	1	0.20	0.83
Wheel-loader WA350	1400	9	1	0.25	0.78

Substituting values of indicators from table 2 in a formula (2), we will receive values of a resource of road machines for operating conditions in tropical (Vietnam) and moderated (Moscow region) climatic zones (Figure 1).

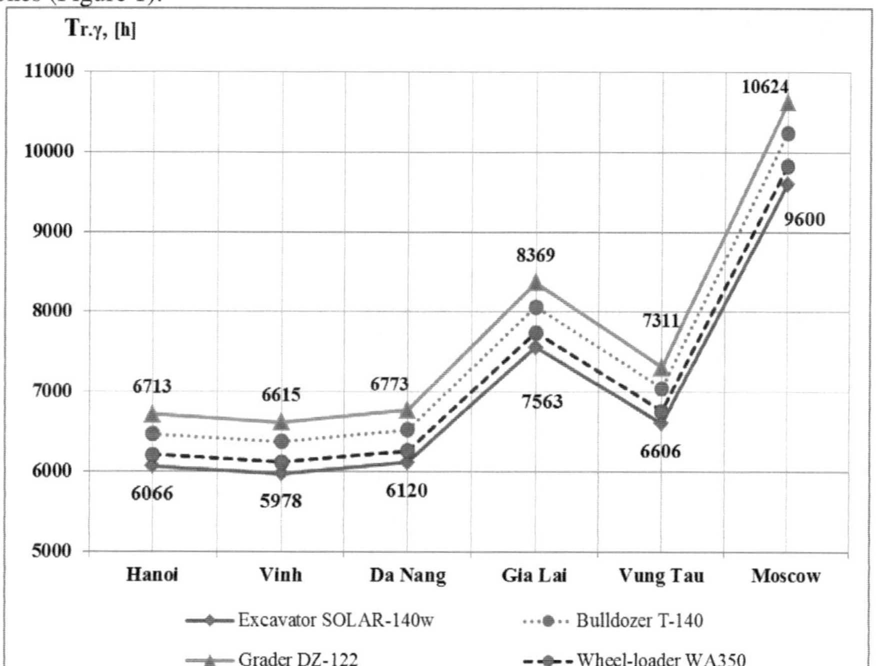

Figure 1. A gamma percent residual resource before the first major repair of Trγof the road machines working in tropical (Vietnam) and moderated (Moscow region) climatic zones.

From the diagram it is visible that value of a residual resource of road machines during the work in tropical climatic conditions of Vietnam, it is much less, than during the work in moderate climatic conditions Moscow area. Values of a residual resource of road machines during the work in the city of Vinh are the smallest (less than 1.61 times), and in the city of Gia Lai - the highest (less than 1.27 times), in comparison with conditions of a temperate climate (Moscow region). Thus, for operating conditions of Vietnam mean value of resources of road machines is 1.5 times less, than the corresponding value under normal conditions operation.

4. Conclusion

Taking note of operating conditions and technical condition at assessment reliability of road machines in general [0-0, 0, 0-0] and definition of resource indicators, in particular, will allow to plan precisely terms of carrying out construction works, to make the reasonable diagram of maintenance and repair of machines for each climatic zone of Vietnam.

References

[1] Alferov A K and Petrov I V 1980 Ensuring operability of construction machines Stroyizdat,Moscow, 136
[2] Balovnev V I 2014 Determination of the optimal parameters and the choice of road-building machines by analyzing the fourth coordinate. Textbook: UDC 625.8.002.5, 180
[3] Balovnev V I 2010 Determination of the optimal parameters and the choice of earthmoving machinery, depending on operating conditions. Textbook: UDC 625.76.08, 132
[4] Climate of Vietnam - https://en.wikipedia.org/wiki/Climate_of_Vietnam
[5] Intensity of solar radiation in regions of Vietnam - https://solare.vn/blogs/news/cuong-do-buc-xa-nang-luong-mat-troi-tung-khu-vuc-tai-viet-nam
[6] Karnaukhov N N, Merdanov Sh M, Schaefer V V and Ivanov A A 2005 Operation of machines in construction Manual of UDC 625-08 Tyum of GNGU, 393
[7] Koch P I 1981 Climate and reliability of machines, Mechanical engineering, 175
[8] Kutuzov V V 2012 Increase in efficiency of operation of construction and road machines taking into account change of their technical condition. Thesis of Candidate of Technical Sciences: 05.05.04, 255
[9] Maksimenko A N, Makatsariya D Yu and Kutuzov V V 2006 Organization of operation of construction and road machines taking into account their technical condition Bulletin of the Belarusian-Russian university **4(13)** 28-31
[10] Reliability of mechanical systems Textbook. M: INFRA-M, Moscow publishing house of 2015, 380
[11] Statistical data - General statistic office of Vietnam 2017 - https://www.gso.gov.vn/default.aspx?tabid=713
[12] Statistical data of social economic situation - General statistic office of Vietnamhttps://www.gso.gov.vn/default.aspx?tabid=621
[13] Statistical data: climatic Moscow - https://ru.wikipedia.org/wiki/Klimat_moskvy
[14] Zorin V A 1986 Bases of durability of construction and road machines, Mechanical engineering,Moscow, 245
[15] Zorin V A 2005 Bases of operability of technical systems Textbook: UDC 629 7 07, ISBN 5-902048-51-6, Magistr publishing house, 536

Modern pedagogical education requirements for students at construction and mechanical engineering faculties

P O Skudalov* and S V Cheremnykh

Department of Structures and Facilities, Tver State Technical University, naberezhnaya Afanasiya Nikitina 22,Tver, 170026, RUSSIAN FEDERATION

*E-mail: p.s@live.ru

Abstract. The professional activity of a fundamentalist teacher with classical education fully involved in the general educational and scientific process, and a teacher who successfully combines his professional activity with industrial one is considering. Professional characteristics of a teacher with a higher education and acquirements obtained at an industrial enterprise are compared to attainments of a teacher with fundamental pedagogical education to monitoring and eliminating gaps in students' knowledge.

1. Introduction

The statement that the system of higher education and the involvement of students in the production process is becoming global and is taken as an axiom by many researchers in this field [1]. In the Russian system of higher education, traditionally inclined to think within national boundaries, recently the fact of the globalization of the education system has also become generally accepted [2]. At the moment, the country's universities are facing an extremely ambitious task: by 2020, five of them should be in the top 100 universities in the world. Based on the tasks set, the system of higher education in Russia as a "test pilot" is one of the area's most subject to reform attempts.

2. The reasons for part-time teachers

Modern economic conditions require from higher education teachers unceasing professional, pedagogical and scientific activities [3-5]. According to Presidential Decree No. 597 of May 7, 2012, "On Measures to Implement State Social Policy," in the regions, by 2018, it was necessary to ensure an increase for teachers at higher professional education institutions and researchers' salaries to 200% of the average salary in the region. Of course, the overall wage level of university teachers and researchers has increased over the past six years, but their final increase does not measure up to the level set by presidential decree [6]. Therefore, like many other categories of working citizens, teachers have to combine their professional activities with different ways of earning, similar to the disciplines they teach.

Of course, there are a lot of ways of profession combinations and they may have more creative meanings. For example, as did the Czech humanist teacher Jan Amos Komensky, who compared a teacher with a gardener who lovingly grows plants in a garden, with an architect who carefully builds knowledge of every corner of a human being, with a sculptor carefully scaling and polishing the minds and souls of people, with the commander energetically leading offensive against barbarism and ignorance [7]. This phrase is absolute truth, which can be taken even as an ideal, but unfortunately, in the era of capitalism the familiar aphorism "Spiritual food cannot be full" immediately comes to mind. That is why the phrase of the head of the Government of the Russian Federation can rightly be considered fair, that "a modern teacher can not only receive a salary on a schedule, but also earn something else".

In addition to the financial component in the combination of professions, which is fundamental for the teacher [8], there are a number of advantages for students to be described in this article.

First of all, it is worth noting that, despite the significant differences between production and teaching, in general, their combination is appropriate. And the point is not only that as a result, students get the opportunity to keep abreast of the latest achievements not only of science, but also of technology, but also that this experience is very useful for the teacher himself [9]. After all, teaching implies the need to be aware of the problems of a particular field of knowledge as a whole [10], not allowing you to follow KozmaPrutkov's classic statement "A specialist is like a flux: its fullness is one-sided". Production practice is a reliable preventive measure against excessive "fixation" on a particular topic, which makes it difficult to see problems in a particular industry in a wider context and, ultimately, reduces the teacher's productivity. In addition, the process of production in a particular profession often stimulates the teacher to the emergence of new ideas, provides an opportunity to see the usual problems in new perspectives. Conversely, in the process of communicating with students, production ideas and concepts undergo a kind of "common sense test". By the degree of their clarity for the audience and by the way it reacts to different materials, what questions they ask, you can better understand the degree of validity of certain provisions, notice controversial and insufficiently elaborated points. Finally, the teaching experience lucidly expounds its ideas, gives a solid experience of public speaking, which is also important.

3. Differences in teaching approaches

Then you can try to identify the general and what may have its advantages or disadvantages when comparing a fundamentalist teacher with a classical education, fully immersed in the general educational and scientific process, and a teacher who successfully combines his professional activity with one or another industrial activity.

General:

- Teaching activity is a special type of social (professional) activity, which is aimed at the realization of the goals of education [11].

A teacher who graduated from an institution of higher education and then took advantage of the knowledge acquired in it at work, as well as a teacher who has a basic pedagogical education, provide the listener with the same curriculum and readable disciplines agreed upon by the Ministry of Science and Higher Education of the Russian Federation.

- Equally fulfilled the conditions of the employment contract for a particular position.
- Having the basic skills of a successful teacher, such as:

constructive activity in the selection of educational material and the construction of the pedagogical process [12];
organizational activities with the implementation of a system of actions aimed at the active involvement of students in various activities, as well as the organization of joint activities;
communicative activity and the establishment of pedagogically expedient relations of a teacher with pupils, other teachers, parents, members of the public [13];
control and evaluation component [14].

All these and many other components provide continuous learning activities in which the student masters a particular discipline and achieves the goals set by the teacher.

Skills and abilities of a teacher involved in production, distinguishing him from a fundamentalist teacher with a classical education:

- The adult generation of citizens is always striving to pass on to the younger generation the accumulated practical experience of mankind [15-17], so that society can be maintained and developed; this can be called culture. Therefore, a person who owns a culture is one of the main goals of society. To achieve this goal, a special activity is organized in society - pedagogical. Given the interconnectedness of concepts, it can be assumed that all pedagogical activity has its beginnings in practical, even more correct, experimental activities. But it is impossible to convey the whole culture through pedagogical activity, and for this, the necessary and necessary for mastering part of the practical experience that can be taught by the person who has it is selected.
- Pedagogical activity can be carried out by a socially educated person who has a pedagogical (professional) education. The subject of pedagogical activity is the whole part of the culture selected and intended to be transferred to the younger generation, the special processing of which makes it possible to achieve the set goal. But what to do with the ability to interest or give a student to relax and recover? It is always easier for production teachers to find interesting and not well-known examples from their many years of work.
- The means by which pedagogical activity is carried out are, first of all, natural properties and qualities belonging to a person: the ability to realize, understand, use knowledge and skill, as well as various kinds of tools, including books. But it is not uncommon when a teacher with production experience is ready to show modern, already completed work of his enterprise, capable of attracting trainees and interest them in their future profession.
- In the absence of pedagogical qualifications, a production teacher can master a number of courses or receive, for example, a diploma of "Higher School Teacher", where the future specialist will be presented with all the nuances of professional pedagogical activity.
- Of the minuses it can be noted that although the current legislation does not contain a ban on the combination of professions in combination for teachers and they can look for additional income without restrictions, but the teacher often does not have enough time and energy for this. Here it is time to remember the phrase by A.P. Chekhov, who later became a saying, "Chase two hares, you will not catch a single one".

4. Summary

It is worth noting that the presence in the faculty of a higher educational institution of a teacher involved in an adjacent profession is not at all obligatory, because the practical skills of students are promoted by practical training, which they carry out annually and by the results of which a report is made. But in the modern world this is becoming the norm, the pros and cons of which are displayed in this article.

References

[1] I Menter 2018 Defining teachers' professional knowledge: The interaction of global and national influences, *Education & Self Development* **13** 32-42

[2] M N Dudin, V V Bezbakh, E E Frolova and M V Galkina 2018 The Models of Higher Education in Russia and European Countries at the beginning of the XXIst century: the Main Directions of Development, *European Journal of Contemporary Education* **7(4)** 653-667

[3] E T Minasyan, V O Midova, O A Danko and D K Balakhanova 2017 Implementing New Performance Pay-Based Schemes in Higher Educational Institutions, *European Journal of Contemporary Education* **6(4)** 748-756

[4] A R Alaverdov 2019 Faculty Publication Activity and Resources for its Increase, *Higher Education in Russia* **28(2)** 23-36

[5] I E Nikulina 2018 Effective Contract at the University as a Driver of Work Quality Improvement of Scientific and Pedagogical Workers, *Higher Education in Russia* **27(5)** 9-19

[6] A S Elkina and E S Shevchenko 2016 Factors Affecting Demand in the Teacher Labour Market in Russia, *Higher Education in Russia* **4(200)** 81-90
[7] J A Komenský 1982 Selected pedagogical writings 1
[8] N I Arkhipova, E Y Koletvinova and O L Sedova 2012 The staffing problems of higher education systems in Russia, Bulletin of the RSUH series: Economics Control Right **10(90)** 11-25
[9] S D Reznik and O A Vdovina 2018 A Regional University Teacher: Evolution of Teaching Staff and Priority Activities, *European Journal of Contemporary Education* **7(4)** 790-803
[10] M Hattinger 2018 Researchers design conceptions of e-learning courses targeting industry practitioners' competence needs, International Journal of Continuing Engineering Education and Life-Long Learning **28(3/4)** 235-254
[11] V E Pisarev and T E Pisareva 2009 Theory of Pedagogy
[12] S A Molodyakov 2016 Problems and Professor Work at the University, *Higher Education in Russia* **3(199)** 91-98
[13] A S Robotova 2018 Optimistic Senses of Teaching at Higher School, *Higher Education in Russia* **2(220)** 66-77
[14] V V Fedorov,I V Dovgaleva, T R Barkaya and P O Skudalov 2016 The problems of applying modern technology of independent evaluation of quality in higher education, Bulletin of Tver state university Series: Pedagogy and psychology **2** 126-134
[15] A R Alaverdovand and T P Alaverdova 2016 Organization of the Process of Professional and Psychological Adaptation of Young University Professors, *Higher Education in Russia* **4(200)** 5-16
[16] Sakhapov R L and Absalyamova S G 2015 The use of telecommunication technologies in education network
Proceedings of 2015 12th International Conference on Remote Engineering and Virtual Instrumentation, **12** 14-17
[17] Sakhapov R L , Absalyamova S G 2018 Fourth industrial revolution and the paradigm change in engineering education International Scientific Conference on Energy, Environmental and Construction Engineering (EECE-2018) electronic edition "MATEC Web of Conferences" 12003

IOP Conf. Series: Materials Science and Engineering **786** (2020) 012018 doi:10.1088/1757-899X/786/1/012018

Model of infrastructure development of housing and utilities

J O Kustikova

Moscow State University of Civil Engineering, Yaroslavskoe Shosse, 26, Moscow, 129337, Russia

E-mail: yulia.kustikowa@yandex.ru

Abstract. Theoretical premises of sustainable development of the housing-and-municipal complex (HAMC) in a perspective of harmonious functioning of a triad of influence: economic-social-ecological (ESE) influencing factors are presented. The mechanism of implementation of technological innovations and creation of digital doubles, by object digitization of JKK with forming of a virtual system of vital maintenance is developed. The example of calculation with a different object combination of a system is presented to JKK showing complexity of obtaining adequate result without application of simulation models. It is proved that developing forecast model of the separate period of functioning it is necessary to represent risk scenario "a tree of events", with simulation of result of adaptive adaptation to the changing conditions and assessment of coherence of elements of a triad of ESE.

1. Introduction

Questions of the organization and management of construction are in details stated in the standard regulating documents which are updated taking into account the latest technologies, introduction of new materials and use of various means of technical maintenance [1.2].

Building of infrastructure in the most developed urbanized zones demands not so much improvement of quality of construction works how many modernization of the mechanism of the organization and management of the operational period of these objects, through its orientation to the consumer, his inquiries with optimization of business processes. Improvement of any system is based on experience of introduction innovative the practician, harmonizations of the previous achievements with future innovations.

Construction objects belong to objects of the superlong period of operation that demands deeper study of this period not so much for a single object, how many for complexes of objects, establishment of efficiency of sustainable development of the urbanized zones. Theoretical base of scientific research in the field of assessment of the competitive status of territories are developments of such scientists as M. Porteta, I. Ansoffa, A. Anuchinoj, O.N. Belenovoj, S.N. Glagoleva, A.G. Gradova, E. Korol, A. Rudycheva, E. Shchetinina and etc. [1-5].

2. Theoretical prerequisites of sustainable development of a housing-and-municipal complex in a perspective of harmonious functioning of a triad of influence of ESE

Structural personification of JKK is carried out step by step.

Structurally JKK has standard saturation the real estate objects of different function territorially stated taking into account zones of inclination and alienation, these are buildings of inhabited appointment, production and social structures, highways and the ave. Historically formation of these zones in terms of equilibrium development of triad of influence of ESE, happened taking into account the effective period of existence, so for components [12]:

- "economic" - is considered the vital period of existence of 50 and more years;
- "social" - the vital period of JKK in foreshortening of appeal to various age groups of the population, within 25 years of active (constant) stay in JKK borders;
- "ecological" - the vital period of change of ecoloading can be reduced to 5 years, and the variability is presented by the range of seasonal change of microclimate eco systems. In this analysis it is about complex indicator bio - and vegetable environment and firmness of buildings.

Obviously system contradiction demanding realization of preservation of comfortable living conditions for the person and natural identity of the environment.

The adopted programs of strategic development are already focused on smoothing of the revealed contradictions, i.e. intensive capital investments in the social sphere with vector on restoration of natural identity have begun, however the positive result is not possible without deep scientific and technical study of problem with representation of the adequate model of medium-term and long-term character considering characteristic risks.

It is offered to consider the available technological innovations and to realize the gear of creation of digital doubles, by object digitization of JKK and formation of virtual system of vital maintenance, the offered algorithm is presented on Figure.1

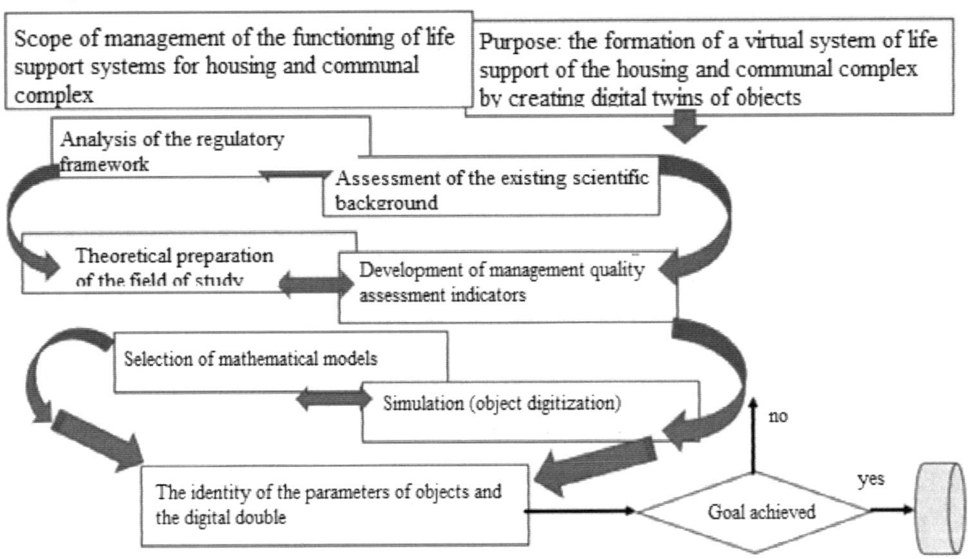

Figure 1. Mechanism of creation of digital doubles, object digitization of a housing-and-municipal complex and forming of a virtual system of vital maintenance.

Key select properties [1, 2,8]:

- integrity. It is indisputable what involves changes of one of parameters of a triad, violation of borders of "field" of harmonious functioning from which exit leads to a system imbalance. Given property the effect of a reverse is characteristic: the system status is estimated in a complex when changes of one variable cause change of all other variables. It is necessary to consider a diversity in change manifestation time, i.e. effect of the postponed influence. For example, for JKK, the social position of residents of the area will not allow to break to please a short-term economic benefit natural integrity by installation of additional buildings, laying is expensive or other, and virtual long-term forecasting of changes with the digital double will allow to find the alternative harmonized option of reconstruction of territories.

- differentsialization. The long-term period of functioning of JKK demands constant perfection, upgrade of outdated objects, use of the innovation technologies of life support. The functional triad of ESE for JKK is represented open system in the direction of a differentsialization and improvement. For JKK transition to autonomous systems of power supply, collecting garbage, its sorting and in the long term utilization in processing enterprises will allow to gain additional income from functioning of JKK, to socially rally inhabitants and to lower ecoload of nearby territories.

- balance. This property is executable in the long-term period, i.e. in residential quarters change of the owner is always an isolated fact, the period constantly living in residential areas often is measured by several generations that develops a certain stereotype of intra house life. For an ecosystem of JKK which is territorially in borders balance maintenance, consists in cultivation of plantings, their leaving and updating on the entire period of existence of JKK.

- recurrence of events. Objects of JKK have the different level of demand which cycles depend on season, of economic and social requests of inhabitants, in particular, the social demand of adjacent territories of a housing estate depends on season; frequency of garbage removal, maintenance rate of engineering systems of houses, recovery of a paving and other actions of the supporting character on the long-term period of existence of separate JKK is set.

Thus, theoretical assessment of harmonious existence of a triad of ESE for JKK allows to provide their effective functioning as on the medium-term, and long-term period.

Selecting in the system of influences key the external environment as the source of factors of production and information defining ability of conversion of these resources is implemented a possibility of object digitization with the subsequent by the personified management of efficiency of system functioning by means of digital doubles.

It is represented that achievement of equilibrium interaction is possible through flexibility property, ability to adapt to external and internal changes, on the long-term period that is also provided with electronic modules of management, for example multiplex, hardware of digital systems.

3. Mathematical modeling

Settlement mathematical the model of component coherence is presented in a dimensionless format that provides control of dynamics of change of indicators through influence infogram for various periods of operation of JKK with a vector of risk forecasts [14-16]. For the purpose of coverage of the maximum quantity of indicators of factorial influence the integrated indicator of complex assessment of a condition of a triad is offered:

$$I_{hcc} = f\left(k_{ec.bal.}; k_{soc.bal.}; k_{env.id.}\right) \tag{1}$$

where kec.bal.- indicators of level of economic balance;

ksoc.bal. - indicators of level of social balance;

kenv.id. – indicators of level of ecological identity.

Thus, the problem of quantitative assessment of indicators is solved that allows to create model of the digital double with territorial zoning in borders of influence of a single housing-and-municipal complex.

It is offered to accept basic values of key parameters as reference size. This assumption is proved by the fact that any construction objects before commissioning are estimated on compliance to standards, regarding quality, safety and the term of effective functioning in specific conditions.

The current state of an object, functioning parameters through the system of monitoring is transmitted to the logical module of control of change of a system in general, deviations of values are fixed in relative values on the module that allows to receive the size of an integrated indicator and to make expected model of its change, both on the medium-term period, and on long-term.

$$k_{ec.bal.;soc.bal.;env.id.} = \sum_{i=1}^{n} \left| \frac{X_i^f}{X_i^{ref}} - 1 \right| \quad (2)$$

A set of components of the expected models forming an integrated indicator when modeling can be minimized, in case of insignificant influence for the short-term analysis or to enter the maximum quantity to establish risks of the effect of decrease in an indicator postponed in time.

The presented mathematical model complemented with an algorithm of formation of basic indicators and introduction of additional parameters of assessment of sustainable development of objects of JKK allows, in real time to carry out monitoring of "field" of harmonious functioning of a triad of ESE for JKK. To establish limits of system balance, to estimate an economic condition as single and in a complex of objects of functioning, to consider the social inquiries changing in time, to fix ecoloading of territories.

Change of indicators and the description of factors of influence depends on depth of a research of external and internal environment of vital existence of JKK, i.e. the following stage of researches is stability assessment.

It is obvious that it is possible to provide stability of a system by creation of the mechanism of self-control, using basic provisions of the theory of risk and probability theory [14, 15]. Development risk scenarios "a tree of events" with a situational algorithm for the program module of the digital double is required, consistently setting imbalance parameters, receiving reactions of the JKK system components on change of indicators of economic expenses, social contradictions and ecological violations.

The following level of assessment of management efficiency is as a result formed: the expected model of change of parameters considering the speed of reaction to factorial influences, revealing critical and problem situations which model allows to accept the operating influences in reality for the purpose of their prevention, or decrease to a minimum of negative impact, in case of impossibility of full neutralization.

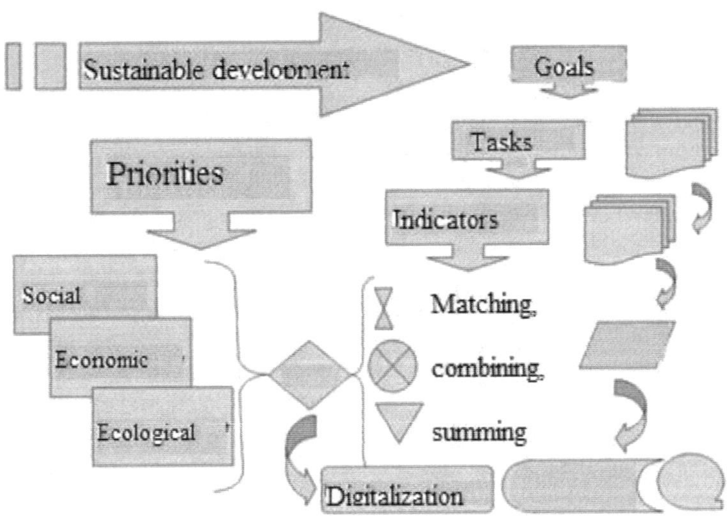

Figure 2. Technique of sustainable development of a system triad of ESE with factorial digitalization.

That is developing model of the separate period of functioning for the digital double it is necessary to offer risk scenario "a tree of events", with simulation of result of adaptive adaptation to the changing conditions and assessment of coherence of elements of a triad of ESE.

The mathematical model characterizing structure of the JKK system is represented in a format of matrixes of influence.

Let's assume that JKK is object presented to n of the joint complex structures of zones of inclination. Signs of structural association are the functional demand (ain) and territorial attractiveness (bin), i.e. the sign of a clustering is selected that of the JKK systems it is quite characteristic [8, 13].

Further we make matrixes of incidence [16]:

$$A_{Fd} = \begin{pmatrix} a_{11} & a_{12} & \dots & a_{1n} \\ a_{21} & a_{22} & \dots & a_{2n} \\ \dots & \dots & \dots & \dots \\ a_{n1} & a_{n2} & \dots & a_{nn} \end{pmatrix} \quad Б_{Ta} = \begin{pmatrix} b_{11} & b_{12} & \dots & b_{1n} \\ b_{21} & b_{22} & \dots & b_{2n} \\ \dots & \dots & \dots & \dots \\ b_{n1} & b_{n2} & \dots & b_{nn} \end{pmatrix} \tag{3}$$

For accounting of criteria deviations of size the settlement matrix is given to a rated look by representation of indicators in unit shares.

The sum of the normalized matrix forms the generalized indicator displaying the criteria "force of communication" of parameters.

$$\begin{pmatrix} a_{11} & a_{12} & \dots & a_{1n} \\ a_{21} & a_{22} & \dots & a_{2n} \\ \dots & \dots & \dots & \dots \\ a_{n1} & a_{n2} & \dots & a_{nn} \end{pmatrix} + \begin{pmatrix} b^H_{11} & b^H_{12} & \dots & b^H_{1n} \\ b^H_{21} & b^H_{22} & \dots & b^H_{2n} \\ \dots & \dots & \dots & \dots \\ b^H_{n1} & b^H_{n1} & \dots & b^H_{nn} \end{pmatrix} = \begin{pmatrix} c_{11} & c_{12} & \dots & c_{1n} \\ c_{21} & c_{22} & \dots & c_{2n} \\ \dots & \dots & \dots & \dots \\ c_{n1} & c_{n2} & \dots & c_{nn} \end{pmatrix} \tag{4}$$

The most connected are elements $c_{in} = c_{ni}$ having greatest values. For JKK objects it indicates harmony of a triad of ESE and is the basis of inclusion in uniform structure. The subsequent calculations will allow to carry out object ranging of the JKK elements and level of harmony of communications of a triad of ESE.

4. Scientific and practical support of the concept

The mechanism of collection of information is in real time presented in the form of the algorithm including the control system consisting of 4 levels (Figure3).

The module of collection of information, value of parameter is in real time provided by installation of monitor sensors, data transmissions in peripheral blocks.

The module of information processing includes a mathematical model with base of reference data on all controlled parameters. Information is compared, on the put algorithms the decision on the level of change of parameter and its influence on the general indicators is made. The control panel by a complex through submission of information on the monitor, its visualization is provided in formats of charts and indication.

All information has several storage locations, both an internal memory, and external resources, up to "Cloudy spaces".

Key result is decision-making by option of harmonious functioning, from standard, to the sentence alternative, in this case at the initial stage of implementation of model of management participation of the person in decision-making is supposed.

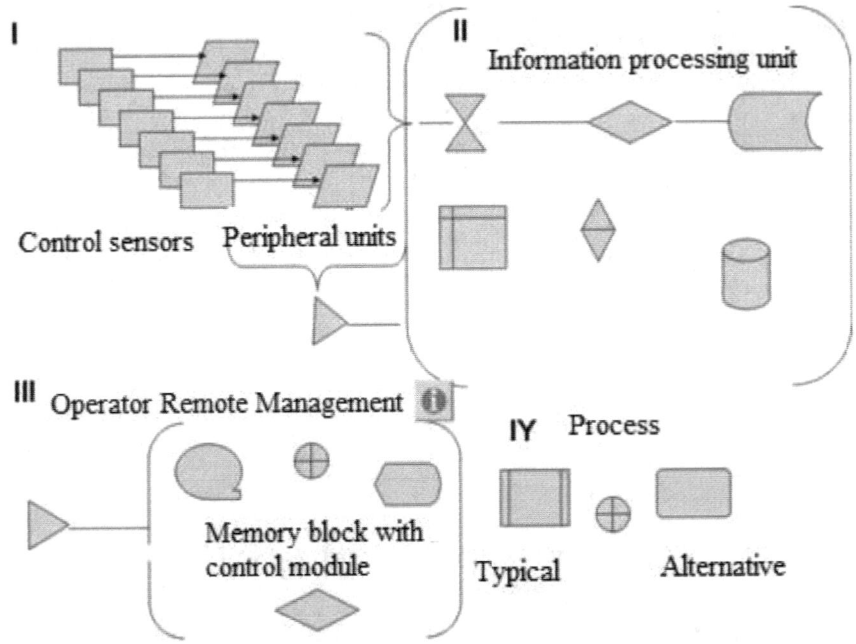

Figure 3. Algorithm management mechanism of gathering real-time information on the effective functioning of the triad ESE using hardware digital systems/

The estimated example is given for assessment of object saturation of JKK, the task of assessment of binomial coefficient by means of Pascal's triangle is set $C_n^k = \binom{n}{k}$. As basic data 10 elements (n) randomly combined (k) in the JKK [17] system are taken.

Design models with any object combination in the JKK system are received:

- for 6 object combinations:

$$y = 0{,}0021x^6 - 0{,}047x^5 + 0{,}4527x^4 - 2{,}1251x^3 + 5{,}1413x^2 - 5{,}0803x + 1{,}6667 \quad (5)$$

- for 5 object combinations:

$$y = 0{,}0096x^5 - 0{,}1789x^4 + 1{,}3365x^3 - 4{,}4726x^2 + 6{,}8811x - 3{,}6667 \quad (6)$$

- for 4 object combinations:

$$y = 0{,}0427x^4 - 0{,}6677x^3 + 3{,}8154x^2 - 8{,}3378x + 5{,}4167 \quad (7)$$

- for 3 object combinations:

$$y = 0{,}1286x^3 - 1{,}4172x^2 + 4{,}7269x - 4{,}1333 \quad (8)$$

Calculations show that at increase in quantity of the combined objects, and in the same measure it is possible to speak also about parameters of harmonious functioning, extent of influence changes much. It is at first sight possible to reject not meaning components as a little influencing a system in general, but this statement for the long-term period demands check that is implemented only in simulation models with the included settlement algorithm.

Visualization of this situation is presented on graphics (Figure4).

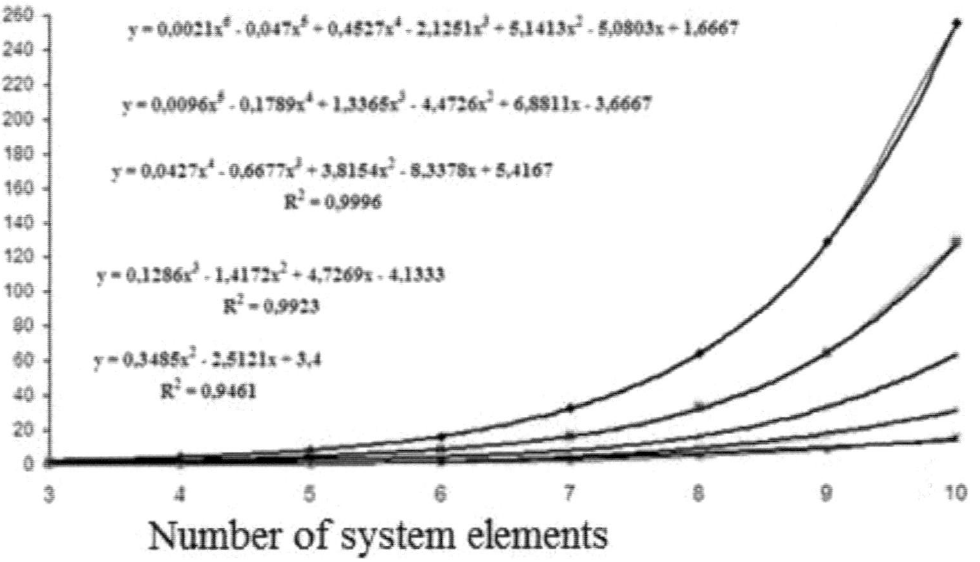

Figure 4. Assessment of object saturation with any combination in the JKK system.

The mechanism of the solution of a problem of quantitative assessment of indicators, by creation of model of the digital double with territorial zoning in borders of influence of a single housing-and-municipal complex using the modal analysis realizes various completing of objects of influence. Infogram shows value of number of the considered parameters of an object or system objects, the risk of a mistake increases in decision-making with increase in the combined combinations of parameters that once again confirms expediency of the translation of the initial analysis in a digital form with modeling of risk level and the subsequent transition to natural objects and the environment of influence.

5. Results and discussions

Numerous domestic and foreign researches show that to consider problems of sustainable development of the urbanized zones in a foreshortening of one branch component it is already not enough. The available technological and digital reserves move apart a framework of solvable tasks transferring a question to cross-disciplinary, for the guaranteed long-term harmonious functioning of a triad of influence of ESE. The prospects of modernization of the mechanism of management of objects of the construction sphere, in a perspective of development of technological innovations by improvement of the created digital doubles of a virtual system of vital maintenance are obvious [18-22].

6. Conclusion

It is analytically established that improvement of the mechanism of management and organization of process of operation of construction objects on the superlong period the practician demands accounting of the latest technologies and the innovation.

The mechanism of implementation of technological innovations and creation of digital doubles, by object digitization of JKK with forming of a virtual system of vital maintenance is developed.

The definition of the concept "field" of harmonious functioning of a triad of ESE for JKK having characteristic to it borders from which exit leads to a system imbalance is given.

The mathematical model of component coherence is presented in a dimensionless format that provides control of dynamics of change of indicators through influence infogram for the different temporary periods with a vector of risk forecasts.

The mechanism of a solution of a problem of quantitative assessment of indicators allowing to create model of the digital double with territorial zoning in borders of influence of a single housing-and-municipal complex is offered.

The example of calculation with a different object combination of a system for JKK showing complexity of obtaining adequate result without application of simulation models is presented.

It is proved that developing forecast model of the separate period of functioning it is necessary to represent risk scenario "a tree of events", with simulation of result of adaptive adaptation to the changing conditions and assessment of coherence of elements of a triad of ESE.

References

[1] Glagolev S N 2007 Methodological bases of formation of mechanism of adaptation of industrial enterprises in the face of changing market Wednesday: monograph -Belgorod: BSTU, – 163s

[2] Glagolev S N , Sevrjugin N S and Vlasova E A 2012 Concept elaboration of principles and criteria for the management of the quality management system for corporate services of motor transport enterprises J *Road transport enterprise* **7** 52-53

[3] Gusakova, E A 2004 Systems engineering organization of the life cycle of the construction object - M : New Millennium Foundation 256

[4] Korchagin V A , Lyapin S A and Sysoev D K 2007 Modeling the environmental-economic interaction of transport and the environment J *Science and technology of transport* **1** 47-51

[5] Korchagin V A and Sysoev D K 2007 Economic-mathematical model for assessing the effectiveness of industrial and environmental investments J Bulletin of higher educational institutions of the Black Soil Region **2** 86-89

[6] Porter M 2005 Competition: Trans from English / M , Porter - M : Williams Publishing House, - 608, Ill - Paral tit Angl

[7] Saaty T L and Vargas L G 1982 The Logic of Priorities: Applications in Business, Energy, Health and Transortation Boston, 299

[8] S N Glagolev, N G Gorshkova, N S Sevryugin and A A Konev 2013 General issues of the cluster approach in the formation of a regional roadside service system on the example of the Korochansky district of the Belgorod region *Science and technology of transport* **2** 63-66

[9] Kustikova Yu O and Podgorny S E 2016 Monitoring the technical condition of buildings and structures as an integral part of an integrated approach for the timely monitoring of urban environment J *Scientific Review* **11** 107-111

[10] Kustikova Yu O and Matushkina A S 2017 Techniques for reconstruction of the preserved housing stock J *Bulletin of MGSU* 12 **10(109)** 1090-1097

[11] Korol E A and Kagan P B 2015 Management systems of urban planning *International Journal of Applied Engineering Research* 10 **23** 43456-43460

[12] Sevryugina N and Kapurin P 2019 Smart house: model of resourse management and safety of vertical transport E3S Web of Conferences 97/ 010116 https://doi org/10.1051/e3sconf/20199701016

[13] Glagolev S N , Sevryugin N S , Kozlova N V and Konev A A 2014 On the acceptability of the territory for the formation of roadside service structures J Motor Transport Enterprise **8** 37-39

[14] Solozhentsev E D 2012 Risk management technologies (with logic and probabilistic models) Springer, 328

[15] Karasev V V 2014 Risk management technologies with logical-probabilistic models J Problems of risk analysis, **11(3)** 14-21

[16] V G Shukhov and Sevryugin N S 2013 Scientific and technical substantiation of the transport and logistics system of resource provision for the formed clusters and priority development zones of the Belgorod region [Text]: R & D report (Intermediate): A 21-12 / Belgorod State Technological University ; performance : A Konev [and etc] - Belgorod , - 138 - GR number 01201278487

[17] Glagolev S N , Sevryugin N S , Kozlova N V and Konev A A 2012 Formation of roadside service complexes as a component of the transport and logistics cluster in the general system of innovative development of the region J *Motor Transport Enterprise* **11** 42-44
[18] Pomorov S B 2011 Architectural and sociological study of roadside auto service complexes in the Altai Territory Polzunovskiy messenger - Barnaul **1** 299-302
[19] Sheshenya N L Mathematical models of forecast estimates of the geoecological state of urban areas (elibrary_25719970_45769784)
[20] Korol E , Kagan P , Barabanova T and Bunkina I 2016 Description of technological processes in construction using formal language International Journal of Applied Engineering Research **11(3)** 1691-1693
[21] Dontsov D G and Yushkova N G 2017 Principles of sustainable development of the territory and priorities of architectural and urban construction activity. AIP Conference Proceedings Proceedings of the III International Young Researchers Conference 050011
[22] Sakhapov R L and Absalyamova S G 2018 Fourth industrial revolution and the paradigm change in engineering education International Scientific Conference on Energy, Environmental and Construction Engineering (EECE-2018) electronic edition "MATEC Web of Conferences" 12003

IOP Conf. Series: Materials Science and Engineering **786** (2020) 012019 doi:10.1088/1757-899X/786/1/012019

Usage of discrete element method in the research of vibrating mills with circular vibrations of the grinding chamber

V S Bogdanov, S Y Lozovaya, Y M Fadin, N E Bogdanov*

Belgorod State Technological University, 46 Kostyukova street, Belgorod, Russian Federation

*E-mail: nikita.bogdanov.rus@gmail.com

Abstract. The paper reviewed the application of the discrete element method for researching the motion of the ball load in vibration mills with circular vibrations of the grinding chamber at different amplitude-frequency characteristics. Vibration mills are actively used in various industries, in particular for road construction in the production of bitumen or mineral components. The paper presents the model of the vibration mill used in numerical calculation, the main indicators of the load's kinematics providing the efficiency of the vibration mill, presents and analyzes the results of numerical experiments, determines the rational mode of operation of the machine.

1. Introduction

As the data used to determine the efficiency of grinding media movement based on the results of the numerical experiment, two main technological parameters of the material grinding process in the vibrating mill [1] were determined directly related to the kinematics load's [2,3,4]. Average speed of movement of the grinding bodies, U_{av}, m/s. This parameter is proportional to root values of the energy intensity – intensity motion of the particles loaded into vibrating mill [5]. The number of collisions of the grinding media determines the efficiency of movement of the load in the grinding chamber of the vibrating mill and the periodicity of exposure to the grinding material, Z.

The main determining technological parameter of the material grinding process in the vibration mill is the energy intensity. Energy intensity – the intensity of motion of the loading particles in the vibration mill, N_v. The energy intensity of the grinding process is proportional to the square of the grinding media velocity formula (1) [6]:

$$U_{av} \approx \sqrt{N_v} \tag{1}$$

Oscillatory motion of grinding media leads to their collisions. The number of collisions Z per unit time, per unit volume of the mixture of grinding media and material can be expressed in terms of determining values d_b, n_b and U using the method of dimensions [6]:

$$Z \approx d_b^2 n_b U \approx \sqrt{N_v} \tag{2}$$

where db – the size of grinding balls, m; nb – number of balls per unit loading volume.

From the analysis of formula (2) it follows that the number of collisions of grinding media, leading to the grinding of the material, riss with increasing energy intensity. We can say that almost all the technological characteristics of the vibration grinding are determined by the parameter of the energy intensity of the load. Studies show [7] that in vibration mills, the frequency of collisions of grinding media in the loading volume is several thousand times higher than the number of collisions occurring during grinding in ball drum mills.

2. Purpose of the EDEM software product

For research the parameters of the average speed of grinding media and their number of collisions, an electron-digital model of the vibration mill in the EDEM system was made and a numerical experiments were carried out. EDEM is one of the leading programs of DEM (discrete element method), used to simulate the behavior of bulk materials when they move and interact with each other and parts of the equipment [8].

The discrete element method used in EDEM can be considered as a generalization of the finite element method. While modeling the process by this method, the initial positions and velocities of various particles are set. Then, based on the initial data and given the physical laws of interaction of particles, the forces acting on each particle are calculated by solving mathematical equations. For each particle, the resulting force is calculated and the Cauchy problem is also solved at the selected time interval. The result is the initial data for the next calculation step. The calculations continue for the entire duration of the process of interest. It is worth noting that such calculations are very time-consuming and require a large amount of computing resources from the computer processor and often take a fairly long period of time.

EDEM allows quickly and easily create a model granular system of solid bodies with different parameters, EDEM combines the functionality of various mechanical, material and other physical properties in the process of modeling of the motion of solid particles. Finished models can be conveniently stored in the library and with changing the initial conditions, which simplifies and accelerates the process of re-calculations by the user. A large tool for analyzing ready-made models has extensive functionality for studying the output data and interpreting them in a visual form easy to understand (graphics, videos, tables, vector fields, screenshots, etc.).

Working with the EDEM software package can be divided into the following stages of creating a calculation model:

- set the global parameters of the calculations: the model of interaction of bodies, material properties of the bodies, friction and restriction coefficients, Shear's modulus and of the various other;
- creation of grinding body model, material selection;
- creation of grinding chamber geometry, material selection, setting of motion parameters (in particular vibrations), amplitude, frequency, angular velocity;
- the creation of a working environment and the input parameters of the calculation: the quantity or mass of grinding bodies, the generation rate of the grinding bodies fill factor etc.;
- calculation;
- analysis of calculation output.
-

3. Plan of research of vibration mill

To carry out the research motion of the load in the software product EDEM, an electron-digital model of a vibration mill (figure 1) with circular oscillations was created. Table 1 below lists the parameters of the vibration mill.

IOP Conf. Series: Materials Science and Engineering **786** (2020) 012019 doi:10.1088/1757-899X/786/1/012019

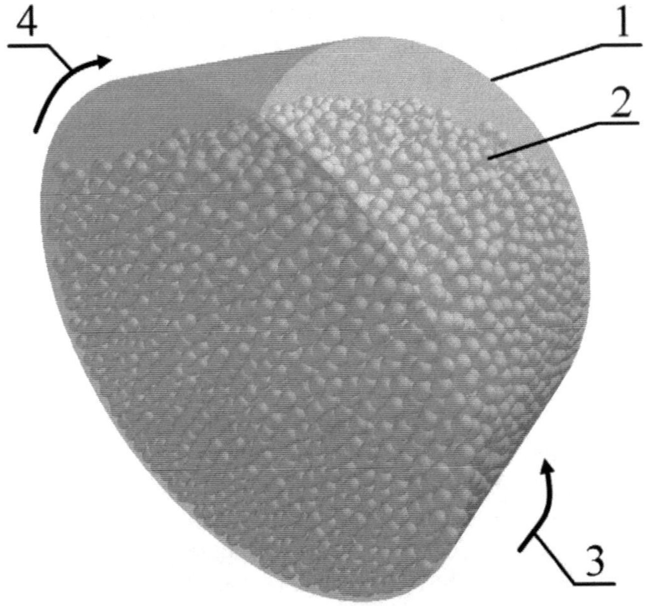

Figure 1. The model of the vibration mill
1 – grinding chamber, 2 – grinding bodies, 3 – direction of circulation of grinding bodies, 4 – direction of circular vibrations of the grinding chamber.

Table 1. The parameters of vibration mill model's.

Parameter	Value
Grinding chamber length, mm	250
The diameter of the grinding chamber, mm	219
The thickness of the wall of the grinding chamber, mm	10
Grinding chamber volume, m^3	0,00858
Number of grinding media	13500
The radius of the grinding body, mm	7
Load factor	0,85
The coefficient of restitution of the milling bodies	0,5
Coefficient of rolling friction of the grinding body	0,5
The coefficient of sliding friction of grinding media	0,01

Two series of different numerical experiments were carried out. The initial data of the first experiment included 10 experiments of circular vibrations of the grinding chamber with increasing amplitude from 2 mm to 9 mm, with the same frequency equal to 50 Hz. The initial data of the second one included 6 experiments, circular oscillations of the grinding chamber with increasing frequency from 25 to 70 Hz and the same amplitude equal to 3 mm.

The aim of these numerical experiments was to determine the effect of the amplitude and frequency of oscillations on the motion of grinding media in the grinding chamber of the vibration mill. Among the many parameters describe the output in the model EDEM Analyst to assess the milling process in vibration mill was selected:

1) the average speed of the grinding media, which is defined as the rate of change in the position of the particle along the axes X, Y and Z, and the addition of the values of the three vectors:

$$U_{av} = \left[\vec{U}_{av}\right] = \sqrt{U_x^2 + U_y^2 + U_z^2} \tag{3}$$

IOP Conf. Series: Materials Science and Engineering **786** (2020) 012019 doi:10.1088/1757-899X/786/1/012019

2) the number of collisions of balls, which is defined as the contact between the grinding media at the time of recording the control point of the simulation.
3) distribution of grinding media in zones depending on their speed.

4. Research result

As can be seen from the graph of the velocity dependence on the amplitude of the oscillations shown in Figure 2. the speed increases exponentially with the amplitude of the oscillations. The minimum average speed equal to 0.3076 m/s grinding bodies have an amplitude equal to 2 mm. the maximum average speed of grinding bodies equal to 1.499 m/s is observed at an amplitude of 9 mm. This is a consequence of the more active mechanical action of the grinding chamber on the grinding bodies, due to its greater amplitude of oscillations.

Based on the analysis of the graph of the number of collisions of grinding media of the mill from the amplitude shown in Figure 2, a gradual decrease in this value to the extreme point is noticeable. At an amplitude equal to 8 mm, the number of collisions of grinding media is 13288. Then there is an increase in this indicator to 16535 at an amplitude equal to 9 mm, which is slightly less than the maximum indicator in 17402 collisions recorded at an amplitude of 2 mm. The decrease in the number of collisions of grinding media in the range of vibration amplitude 2-8 mm, due to the fact that at these values, part of the load goes into a suspended state, and the free volume between the balls in the load increases, respectively, the number of contacts between the grinding media decreases.

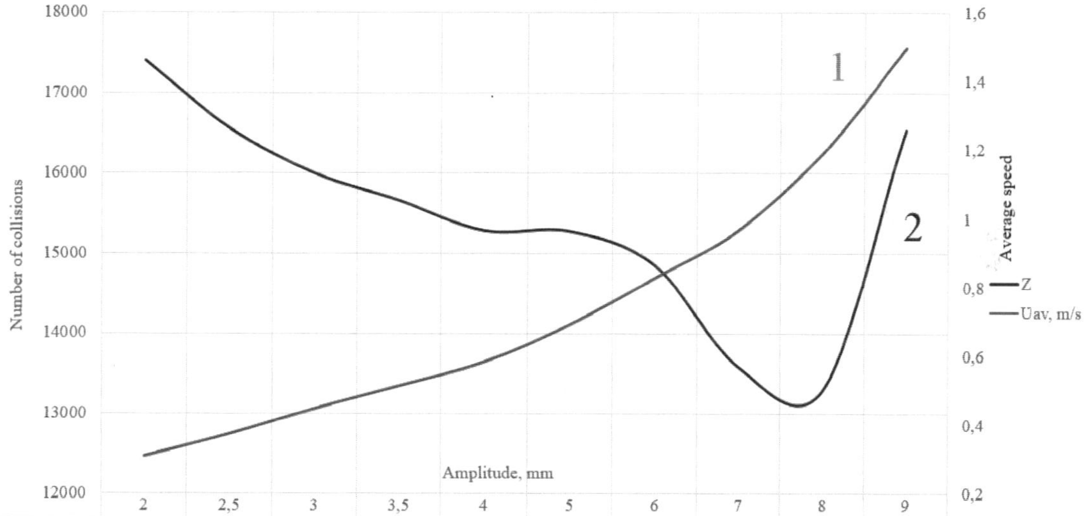

	2	2,5	3	3,5	4	5	6	7	8	9
Z	17402	16550	15992	15655	15280	15272	14844	13583	13288	16535
Uav, m/s	0,3076	0,373	0,447	0,5135	0,5832	0,692	0,8282	0,9675	1,192	1,499

Figure 2. Dependences of the average speed (1) and the number of collisions of grinding media (2) on the vibration amplitude of the grinding chamber of the vibration mill.

From the analysis of the graphs of the dependence of the average speeds of the grinding bodies and their number of collisions on the amplitude, it can be concluded that the most preferred mode of operation of the vibration mill with a body length equal to 250 mm, a diameter of 219 mm, a load factor equal to 0.85, loaded with grinding bodies of the same diameter of 14 mm. will be observed at an amplitude equal to 3 mm. With such parameters of circular oscillations, the number of collisions of grinding media at a time is still large enough, but at the same time, the growth of the amplitude to the extreme values shown in these experiments can lead to an increase in the power consumption of the vibration mill, since achieving such amplitude values will require the use of more powerful vibration exciters. The research notes that the optimal value for the vibration amplitude of the grinding chamber of the vibration mill is a range from 2 to 5 mm [6]. In addition, it is worth noting that with the growth of the amplitude, the vibration loads on the foundation will increase many times, and the wear of the grinding chamber and grinding bodies will accelerate.

After determining the optimal value of the amplitude equal to 3 mm, the effect of changes in the oscillation frequency on the loading motion in the vibration mill was researched by conducting a second series of calculations with increasing frequency. From the graph of the dependence of the average speed on the amplitude of the circular oscillations shown in Figure 3, it follows that the speed increases linearly with an increase in the oscillation frequency. The minimum average speed equal to 0.339 m/s grinding media have an amplitude equal to 25 Hz. The maximum average speed of grinding media equal to 0.543 m/s is observed in the experiment with a set frequency of 70 Hz.

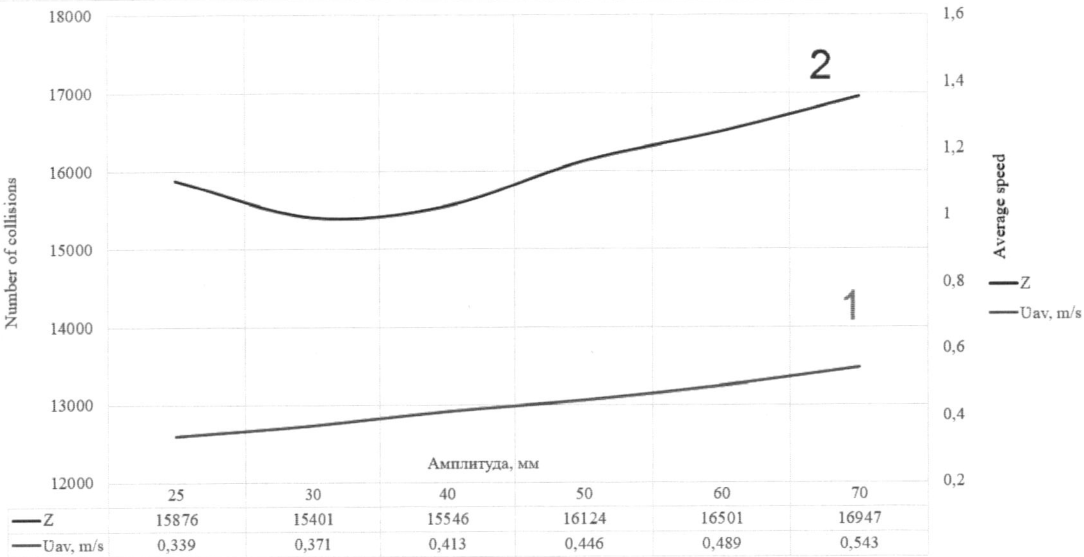

Figure 3. Graph of the dependence of the average speed (1) and the number of collisions (2) on the oscillation frequency of the grinding chamber of the vibration mill.

Based on the analysis of the graph of the dependence of the number of collisions of grinding media of the mill on the oscillation frequency shown in Figure 3, a decrease in this value to the minimum point at a frequency of 30 Hz is seen. At this frequency, the number of collisions of grinding media is equal to 15401, which is less than the index of the number of collisions of grinding media equal to 15876 at the lowest considered frequency of 25 Hz. This is followed by the growth of this parameter to 16947 at the maximum value of the investigated frequency equal to 70 Hz.

Further, a stagnant loading zone was researched in the model with a vibration amplitude of 3 mm and a frequency of 50 Hz, (which includes grinding bodies with an average speed of up to 0.4 m/c, shown in blue in figure 4), located in the Central part of the chamber and moves along a circular trajectory during grinding. From the end of the grinding chamber (Figure 4, (a) the stagnant zone is minimal in volume, this is due to the collision of grinding media with the end of the chamber and then Bouncing them off it. On a quarter (Figure 4, (b) and half (Figure 4, (c) the length of the grinding chamber, the stagnant zone is approximately one-third of the total volume of grinding bodies. Of the longitudinal section of the grinding chamber (Figure 4, (d) it follows that the stagnant zone extends to its entire length.

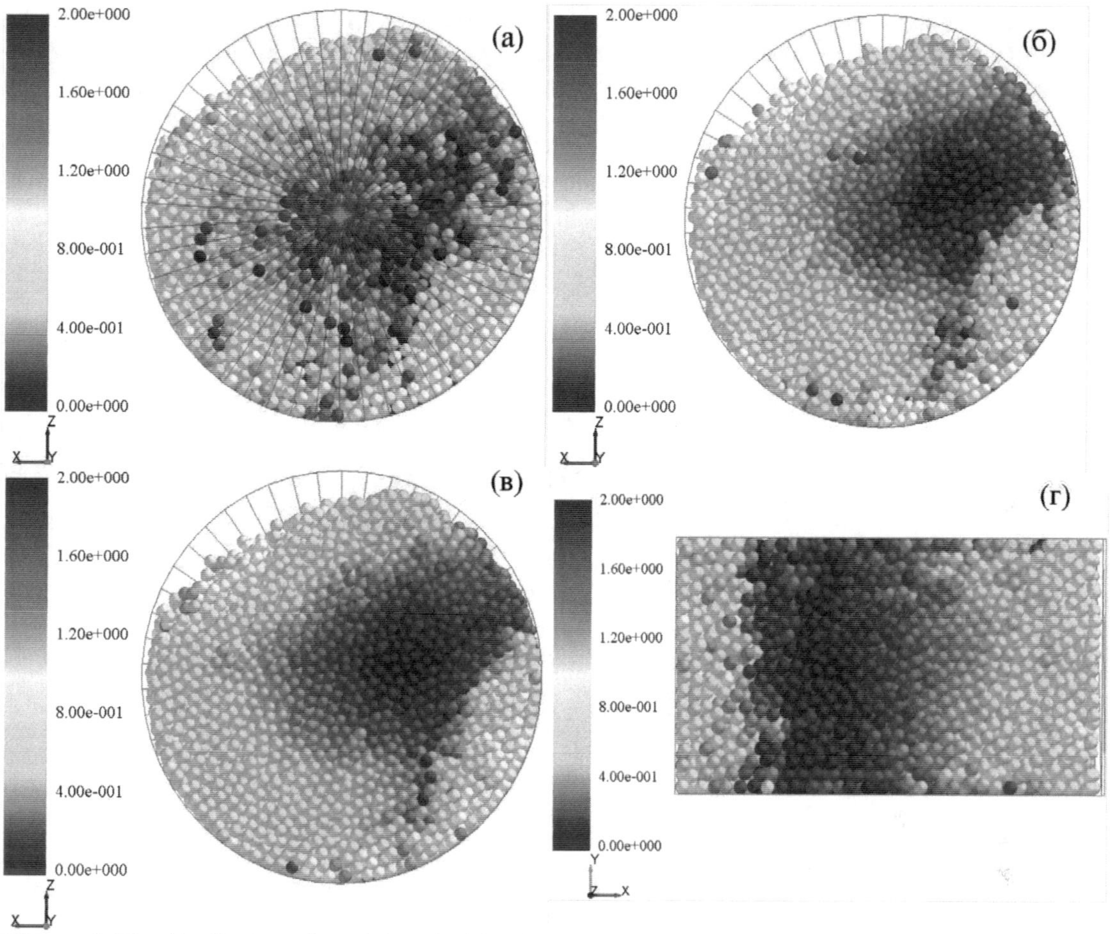

Figure 4. The distribution of particle velocities from the end of the vibration mill (a), a quarter of the length of the grinding chamber (b), half the length of the grinding chamber (b), in the longitudinal section (g).

5. Conclusions

In total, the most rational parameters of the operating mode for a laboratory sample with a body length equal to 250 mm, a diameter of 219 mm, a load factor equal to 0.85, loaded with grinding media of the same diameter of 14 mm will be a vibration amplitude equal to 3 mm and a vibration frequency equal to 50 Hz.

References

[1] Rose H E, Sullivan R M 1961 Vibration Mills and Vibration Milling (London: Constable) 195
[2] Bogdanov V S, Bogdanov N E, Gavrinov A V, Eltsov M Y and Karagodina K I 2017 Kinematics of material grinding in a vibration mill News of higher educational institutions. Construction **3** 76-83
[3] Gavrovon A V, Bogdanov N E, Karagodina K I and Shekhovcova Y A 2017 The influence of the grinding process mode in the vibration mill on the specific energy consumption The Bulletin of BSTU **5** 98-103
[4] Ryadinskaya L V, Lozovoi N M and Lozovaya S Y 2018 The kinematics of the load in equipment with the working chamber of different shapes (In the digest: Interstroymeh–2018 Digest of reports of the XXI International scientific and technical conference.) Ed. S. J. Galitskova (Kazan)169-172

[5] Pospelov A A, Miheev G G, Ushakov S G and Blinichev V N 1989 Experimental research of the influence of the type and size of grinding media on the speed of vibration grinding (Processes in granular media: Interuniversity scientific digest.) (Ivanovo) 8-11
[6] Lesin A D 1957 Vibration grinding of materials. Elements of the theory and method of calculation of the main parameters of vibrators. (Moscow: Promstroyizdat) 113
[7] Fedorov S M 2005 Improving the efficiency of the vibration mill for grinding of mineral raw materials (candidate dissertation) (Moscow: MGTU) 139
[8] EDEM 2.2 2009 (User guide. Revision 31) (DEM Solutions Ltd) 149

Ultimate loads, arising on the blade during exploitation, according to existed design positions for calculations

A G Saveliev[1,*], V A Mikhailovskaya[2]

[1]Russia, Moscow, Verescovaya St., home 16, apt. 27,

[2]Russia, Mytishi, Oktyabrskaya St., home 22, apt. 44

*E-Mail: prof.saveliev@yandex.ru

Abstract. In the paper a method for determining external loads on the working equipment of the bulldozeris presented. Using this method, it was revealed that the computations using the existing algorithm allow to obtain overestimated values of the loads, whatleads to a significant increase in the metal intensity of the machine design. Also in the article is shownaway for carrying outan experiment on the bulldozer stand, described the positions of bulldozer working equipment, according to which it will be tested and made the conclusions about the research.

1. Introduction

Currently, there are no clear recommendations on how to carry outdesign calculation and make a choice from different constructions of pushing arms. It is important becausepushing arms are one of the most important elements of the bulldozer structure, they should have minimum mass of the design, at the same time provide the necessary strength during exploitation. Therefore, the urgent task is to develop optimal designs of pushing bars for any kinematic schemes of bulldozer working equipment and exploitation conditions of the machine. To reach this goal, it is necessary to elaborate a way of calculating the bulldozer design, which would allow to get loads on the operating equipment close to real. Because the existing design calculation has some inaccuracies in determining of the applied working loads, for example, the longitudinal force consists of the static and dynamic component, but according to existed calculation, the maximum values of these components are takeninto consideration, regardless of what place this force is applied. Also, the method does not take into account the difference in the rates of increment and the nature of the dynamic and static efforts.

2. Goal

The purpose of the experimental research is to find the way how to get forces arising at the bladeduring the machine work and to test the strength of the elements of the bulldozer equipment at the existing design positions and according to the specified calculation. Moreover, to comparethe results of the specified calculation with the results of the existing method.

IOP Conf. Series: Materials Science and Engineering **786** (2020) 012020 doi:10.1088/1757-899X/786/1/012020

3. Analytical model

To determine the real loads on the blade of the bulldozer, they will be considered according to the existing method:

$$P = P_S + P_D, \tag{1}$$

P_S – static load;

P_D – dynamic load.

The maximum traction force occurswhen the machineis at the moment of track skidding [1]:

$$P_{ys} = G_d * \varphi_{max}, \tag{2}$$

whereP_{ys}– static force on the blade of the bulldozer, directed along its longitudinal axis;

G_B – bulldozer gravity;

φ_{max}– maximum friction coefficient.

It is possible to accept the meaning of the φ_{max} coefficient while skidding will be 20% [1].

Thus, the actual speed of the bulldozer will be reduced by 20% compared with the theoretical:

$$V_a = V_{th} * (1 - \delta), \tag{3}$$

V_a – actual speed of bulldozer;

V_{th} – theoretical speed of bulldozer.

The value of φ_{max}depends on the distance from the point of force application on the blade to the longitudinal axis of the bulldozer.

The ratio of $P_{ys}$from the distance l of the point of the force application on the blade was revealed by V. S. Berezin [2].

It should be noted that in determining the maximum loads, it is necessary to take into account the friction ellipse (that is anisotropy of friction of a tracked machines on the ground) [3]. This reflects the fact that during the slipping of the bulldozer tracks (i.e. with the adhesion coefficient in the longitudinal direction $\varphi = \varphi_{max}$)when the position of the load on the bladeis along the longitudinal axis of the bulldozer, we can disregard the lateral adhesion coefficient of the tracks ($\varphi' = \varphi'_{max}$) $\varphi = 0$.

To define the maximum possible static loads on the blade when the bulldozer run into a firm obstacle (look at the scheme on the Figure 1), we need to take into consideration friction ellipse (Figure 2).Maximum loads P_{ys} and P_{xs}become maximum together when their resultant is directed to the center of pressure (when the vertical force on the bladeis absent– into the projection of the center of the bulldozer gravity to the bearing surface. In this case, the blade does not slip over the obstacle. Using the friction ellipse (Figure 2), we can get the following equation.

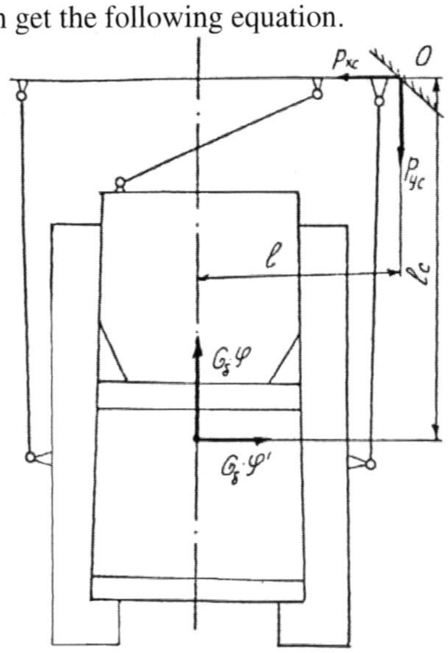

Figure 1. Scheme illustrated how bulldozed run into a firm obstacle.

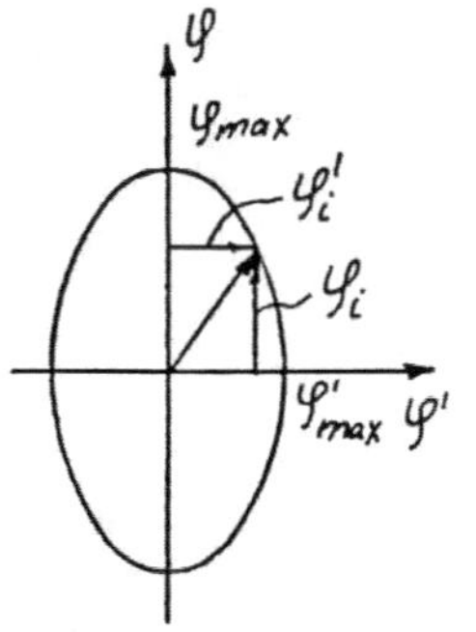

Figure 2. Ellipse of friction.

$$\frac{\varphi^2}{\varphi_{max}^2}+\frac{\varphi'^2}{\varphi_{max}'^2}=1$$

or

$$\varphi' = \varphi'_{max} * \sqrt{\varphi_{max}^2 - \varphi^2} \quad (4)$$

From the equation of the sum of the force moments in relation to the point O (Figure 1):

$$G_B * \varphi * l = G_B * \varphi' * l_c$$

or after simplification of the equation we can get the following line:

$$\varphi * l = \varphi' * l_c \quad (5)$$

Put (4) into (5), we can get:

$$\varphi = \frac{\varphi_{max} * \varphi'_{max} * l_c}{\sqrt{l^2 + \varphi_{max}^2 * l_c^{\ 2}}}$$

Static force on the blade in the horizontal direction:

$$P_s = G_B * \sqrt{\varphi^2 + \varphi'^2};\ P_{ys} = G_B * \varphi;\ P_{xs} = G_B * \varphi'. \quad (6)$$

Such forces on the blade occur when bulldozer workswith a significant speed and when the engine power is sufficient to achieve the maximum of the tractionforce.

When working at the high speeds, the tractionforce not always reach its maximum adhesion value.

The power balance equation for the earthmoving machine is [4]

$$N_{np} = N_e - N_{pt} - N_s - N_f \quad (7)$$

N_{np} – netpower, spent for the creating force on the blade;
N_e – the power of engine;
N_{pt} – power,spentduetoworking of thepowertrain;
N_s – power, spent during skidding.

$$N_s = \left(N_e - N_{pt}\right) * \delta \quad (8)$$

N_f – power, spenton the bulldozer moving

$$N_f = V_{pt} * G_B * f, \quad (9)$$

where$f$ – coefficient of rolling friction.

Power balance equation is:

$$P'_{ys} * V_{th} * (1-\delta) = \left(N_e - N_{pt}\right) * (1-\delta) + f * G_B * V_{th} * (1-\delta) \quad (10)$$

then

$$P'_{ys} = \frac{N_e - N_{pt}}{\omega_{sh} * i_{gr} * R_{sw}} - f * G_B \quad (11)$$

ω_{sh} – angular velocity of the mover shaft;
i_{gr} – power train gear-ratio;
R_{sw} – radius drive sprocket wheel.

Since $N_e$dependson ω_{sh}, i. e.$N_e = f(\omega_{sh})$, then P'_{ys}dependonthe speed of the bulldozer motion.Moreover, the greater the angular velocity of the mover shaft of the bulldozerω_{sh}, the smallerN_e and P'_{ys} [4].

To determine the actual speed, we need to determine the skidat the moment when the ground still holds force without disintegration:

$$P_{ys} = G_B * \varphi \text{or} \varphi = \frac{P_{ys}}{G_B}. \tag{12}$$

The value of φ depending on the slip δ is described by different equations [1]. Knowing δ, it is easy to find φ by using any of these equations. For example, for the equationdeveloped by A. P. Parfenov.

$$\varphi = \varphi_{max} - a * e^{-b\delta} \tag{13}$$

where a and b – coefficients, depended on the type of soil;

e – basement of natural logarithm.

Then$\delta = -\frac{1}{b} * \ln\left(\frac{\varphi_{max}-\varphi}{a}\right)$,

where$V_e = V_{th} * (1 - \delta)$.

Maximum of the lateral constituent of force on the blade we can find according following equation:

$$P_{xs} = G_B * \varphi = G_B * \varphi'_{max} * \sqrt{\varphi^2_{max} - \left(\frac{P_{ys}}{G_B}\right)^2} \tag{14}$$

The dynamic force will be found from the condition that the difference in kinematic energies before and after the hit transforms into energy of design deformation.

The kinetic energy of a bulldozer in a straight-line motion before hit:

$$\text{K}_1 = \frac{m * V^2_{bf}}{2} \tag{15}$$

wherem – massof bulldozer;

V_{bf} – bulldozer speed before a hit.

Kinetic energy after hit:

$$\text{K}_2 = \frac{m * V^2_{af}}{2} + \frac{J * \omega^2}{2} \tag{16}$$

whereV_{af} – bulldozer speed after a hit;

J – moment of inertia of the bulldozer around the axis of rotation;

ω – angular velocity of the axis of rotation of the earthmoving machine.

The rotation of the earthmoving machine after the impact occurs because the line of the dynamic force action does not pass through center of mass of the machine.

The force that appears on the blade when machine run into a rigid obstacle by its blade is determined from the thought that the change in kinetic energy during the impact is expended on elastic deformation.

$$P_d = \sqrt{2 * \text{C} * (\text{K}_1 - \text{K}_2)} \tag{17}$$

where C – stiffness.

SchemeoftheearthmovingmachinehitisshownintheFigure 3.

To determine the maximum possible values, we assume that when machine hits by their edges of the blade,the rotation occurs in planes I-I and II-II, but when it hits by the center of the blade – in plane III-III. All three planes are parallel to the direction of the machine motion and they pass through the center –point A and the points of hit O, O_1, O_2.

If we consider the rotation of earthmoving machine only in one of the planes I-I, II-II or III-III, then the task will be transformed from space into a flat one, since bulldozer does not have adegree of freedom to move down due to the limitations of the base surface.

From the theorem of impulses for plane motion:

$$m * \left(V_{bf} - V_{af}\right) = S \tag{18}$$
$$m * r_i^2 * \omega = S * l_i; \; l_i = A * O_i$$

where S – force impulse;

r_i – radius of inertia

$$V_{af} = \omega * l_i = -\varepsilon * V_{bf} \tag{19}$$

where ε – recovery coefficient;
V_{af} – speed after a hit;
V_{bf} – speed before a hit.
Then

$$\omega = \frac{l_i * V_{bf}}{l_i + r_i} * (1 + \varepsilon);\ V_{af} = \frac{l_i^2 - \varepsilon * r_i^2}{l_i^2 - r_i^2} * V_{bf} \tag{20}$$

Since $r_i^2 = \frac{a_i^2 + b_i^2}{12}$

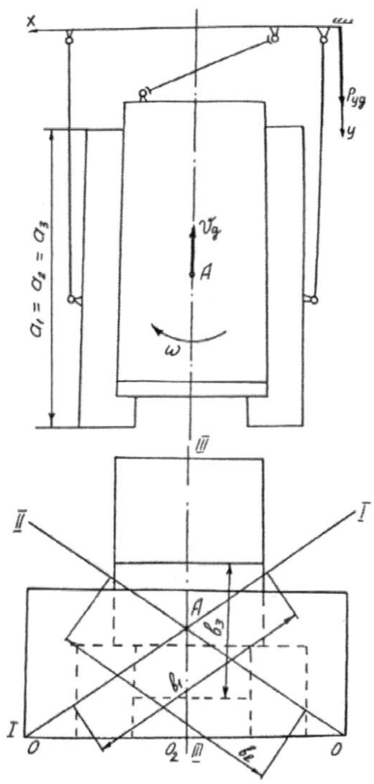

Figure 3. Scheme for the determination of a dynamic force on a blade.

From previous equations we have:

$$V_{af} = \frac{12 * l_i^2 - \varepsilon * r_i^2}{12 * l_i^2 + a_i^2 + b_i^2} * V_{bf} \tag{21}$$

$$\omega = \frac{12 * l_i * V_{\blacksquare} * (1 + \varepsilon)}{12 * l_i^2 + a_i^2 + b_i^2} \tag{22}$$

IntheplaneI-I$l_1 = OA$, in the planeII-II$l_2 = O_1A$, in the planeIII-III$l_3 = O_2A$.

The energy of deformation is maximum at ε = 0.

In the equation for determinationof the dynamic force, should be found the stiffness in the Y direction, since we will consider the dynamic force only in the Y direction because the force impulse is parallel to this axis. The stiffness of the bulldozer designwas sufficiently considered in the papers of A. M. Kholodov and N. F. Fedotov [5, 6]. In the horizontal plane, the stiffness of the earthmoving machine will influence on the stiffness of the pushing arms$C_{p.a.}$(Figure 4) in case when the mechanisms for compensating the tilt of the blade connect them to each other. In case when the compensation mechanism is made in the form of a diagonal rod, connecting the blade with the base tractor, the stiffness of the bulldozer design in all directions can be determined by the formula [7]:

$$C_1 = \alpha * \frac{G_{bt}}{m} \tag{23}$$

where G – gravity of the base tractor;
$\alpha = (0,9 \ldots 1,0)$ – coefficient of stiffness of the operating equipment per 1 kilogram of tractor mass.
Stiffness:

$$C = \frac{C_1 * C_2}{C_1 + C_2} \tag{24}$$

where C_2 – stiffness of the obstacle.

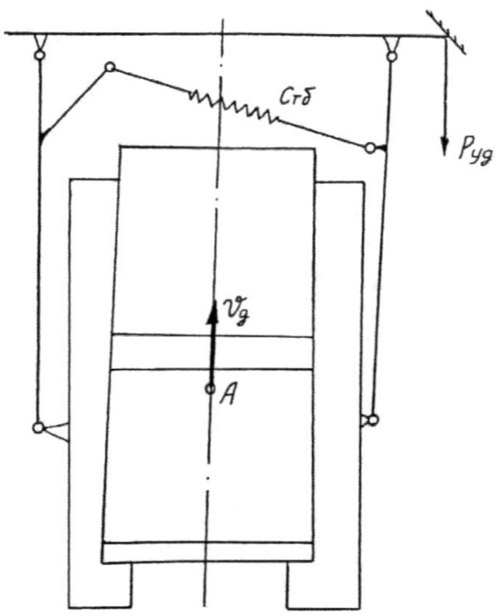

Figure 4. Scheme where is shown how to take into account stiffness of the pushing arms.

The equationsthat were found on the blade allow to get real values of the forces and let us to estimate the actual stress strain state of the metal elements of the working equipment earthmoving machineduring the exploitation. The values of the loads when they are determined by the existing method [7] are overestimated. The static load on the edge of the blade is less by following value:

$$\Delta P_{ys} = G_B * \varphi_{max} - G_B * \frac{\varphi_{max} * \varphi'_{max} * l_c}{\sqrt{l^2 + \varphi_{max}{}^2 * l_c{}^2}} \tag{25}$$

The difference of the found equation for the dynamic force in comparison with that found by the existing method is also significant when a tractor power sufficient to drive at the same speed on 1st gear

$$\Delta P = V_{th} * \sqrt{2 * C * m} - \sqrt{2 * C * (K_1 - K_2)} \tag{26}$$

The purpose of the experiment on the model is: verification of the design scheme of the dozer equipment with a non-rotary blade and without redundant constrains in the structure, revealing of the similarity criteria of the stressed strain state of the pushing arms. For this experiment the operating equipment of the DZ-110A and DZ-132 bulldozers were used.

DZ-110A bulldozer equipment is mounted on a T-130.T.G-1 tractor with a power of 118 kW. DZ-110A bulldozer is produced by the Chelyabinsk plant of road machines named after Kolushchenko and is one of the most massive on release. The bulldozer meets modern requirements, since it is not only allows significant change of the blade position when it is raised and lowered but also it lets the blade to skew to 12° (actually to 20°). Therefore, the study of attachments of this bulldozer is advisable.

4. Experiments and results

The program and procedureof thebulldozer stand testing was developed together withResearch and Production Association «VNIIStroydormash» and Moscow Automobile and Road construction State Technical University. A full-scale sample of the bulldozer equipment was made on the basis of a DET-250 tractor (bulldozer stand). The scheme of the bulldozer equipment corresponds to the scheme DZ-

132 [8]. The mass of the pushingarms of the bulldozer stand is less than the mass of the pushingarms of the DZ-132 bulldozer by 10%.

The purpose of the experimental research on the bulldozer stand is:

• Verification of the forces, arising on the blade at the existing design positions, for strength testing of the elements of the earthmovingmachine equipment.

• Determination of the stress-strain state of the push arms at additional design positions.

Foil stress gauges with a base of 15 mm were used to measure the voltages at the test points.

The stresses were measured at the following points of the push arms:

1. On the left arm along the axis at a distance of 1170 mm from the axis of the vertical pin of the hinge which attaches the arm to the blade, in front of the loopof the pitch rod (numbers of sensors15 and 16).

2. On the left arm from the top, along the axis, at a distance of 480 mm from the axis of the loops of the pitch rods at the rear (numbers of sensors 17 and 18).

3. On the right arm, at the rear of the pitch rod loop at the distance of 480 mm of the axis of this loop (numbers of sensors 19 and 20).

4. On the right arm from the top front of the loop at a distance of 1170 mm from the axis of the vertical pin of the hinge which attaches the arm to the blade (numbers of sensors 21 and 22, Figure 5).

5. On the right arm on the side from the inside at a distance of 1170 mm from the axis of the vertical pin of the hinge which attaches the arm to the blade (numbers of sensors 23 and 24).

The sensors were glued perpendicular to the axis of the bulldozer (odd sensors) and parallel (even sensors). The sensor 23 is glued parallel to the axis of the bulldozer, the sensor 24 is vertically.

The test points, where the stresses were determined, were chosen in order to determine the maximum stresses from the forces and stresses on the bladeaccording to additional design positions. The sensors were connected by a bridge circuit.

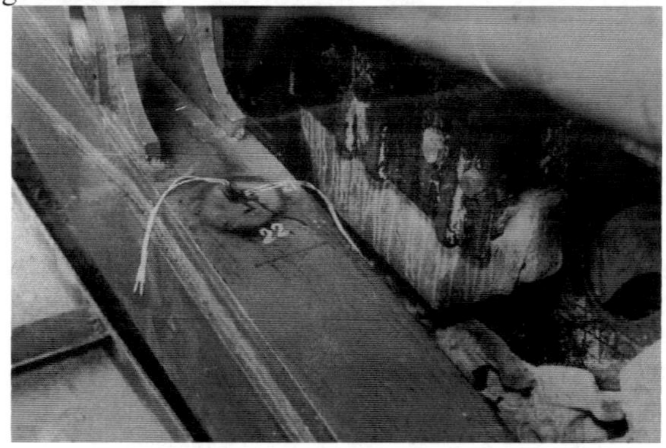

Figure 5. The fourth point on the design where stresses were measured.

Tesometric equipment was installed in the laboratory. The force on the blade was measured with the help of a taredstrainer up to 30 tons.

In Figure 12 shows on of the test moments when the bulldozer ready for testing, the load on the blade was transferred by a special hook.

Experimental studies included the following tests:

1. Hanging the bulldozer in the middle of the blade, on the left and right edge of the blade. The force is created by hydraulic cylinders for raising and lowering the blade. The blade rests on the hook, which is connected with the use of a strain-measuring load gageto the hook of an overhead crane (load capacity 15 tonnes). Strain-measuring load gage measures the vertical force on the blade, which is sufficient to overturn the bulldozer relative to the drive sprockets of the base tractor.

Figure 6. Bulldozer, prepared for the tests.

2. Test of the bulldozer traction: the force from the rigid load (8.2 tonnes × 5) on the strain-measuring load gage and hook was attached to the center of the blade, to the right side and the left side of the blade, parallel to the longitudinal axis of the bulldozer.

3. Determination of dynamic forces when bulldozer lowering their blade. Then the blade at the maximum lowering speed hit by its center, left and right edge of the blade (Figure 7) into a metal obstacle mounted on a hard concrete floor.

Figure 7. Example of the ninth test.

Thanks to the tests number 1 and 2, that were carried out, the maximum static horizontal, parallel to the longitudinal axis of the bulldozer and the vertical forces on the blade at the existing design positions were determined; test number 3 helped to reveal dynamic forces are when the blade hits a rigid obstacle while lowering the blade at maximum working speed; Preliminary tests showed that for the level of confidence probability $\gamma = 0.95$ the required number of experiments is $r = 5$. The calculation was carried out according to the well-known method [9-11].

Let us calculate the difference of the static force, according to the formula (25) of the existed and presented in this paper method. We will find how much the values that were gotten according to the existing method were overestimated.

$$\Delta P_{ys} = G_B * \varphi_{max} - G_B * \frac{\varphi_{max} * \varphi'_{max} * l_c}{\sqrt{l^2 + \varphi_{max}'^2 * l_c^2}} = 335000 * \left(0{,}71 - \frac{0{,}71 * 0{,}75 * 4468}{\sqrt{2155^2 + 0{,}75^2 * 4468^2}}\right) = 2{,}141 * 10^5\ N \quad (27)$$

5. Conclusions

1. Analysis of the correlations,receivedfor the determinationof the changingof the static force on the blade,depending on the coordinates of its application to the blade, shows that the maximum longitudinal static force generated by the bulldozer at the bottom corner of the blade is less than the longitudinal static force generated in the center of the blade cutting edge in 1.5 times.

2. The longitudinal static force at the lower corner of the blade, determined by the existing method, is in 1.5 times overestimated.

3. It was revealed experimentally that for the most common heavy bulldozers, for example the DZ-132 bulldozer, the longitudinal static force in the center of the cutting edge of the blade is in 1.5 times high the longitudinal static force in the lower corner of the blade.

4. The existing design positions, suggesting that the maximum longitudinal and transverse forces on the blade can act together, do not confirmby the real scheme of bulldozer loads. The adhesive coefficients of the tracks with the ground in the longitudinal and transverse directions are interconnected in such a way that the maximum value of one of them corresponds to the minimum value of the other.

5. It has been established experimentally that the stiffness of the metal structure of the bulldozer equipment in the longitudinal direction with the application of hit force to the lower corner of the blade and to the center of its cutting edge can be considered in the computations the same.

6. It has been established experimentally that for heavy bulldozers, for example bulldozer DZ-132, whenit pushes by a pushing arm into an rigid obstacle duringits turning, the stresses in the pushing armsbecome maximum.

References

[1] Magarillo B L , Pozin B M and Makarov P M 1973 The dependence of skidding on specific traction force of tractor / Problems of design and research of tractors and tractor engines Collectionofscientificpapers / ChPI – Chelyabinsk, 15-21

[2] Berezin V S 1973 Intensification of the bulldozer exploitation in difficult ground conditions by developingoperating equipment: PhD thesis – Moscow, – 146

[3] Opeiko F A 1948 Anisotropic friction 18-22

[4] ChudakovD A 1962 The basics of the theory of tractors and cars – Moscow: Selhozizdat, – 312

[5] Velednitskii U B and Yarkin A A 1970 Loading modes of the bulldozer on the tractorwith weight of 15 tonnes – Papers / VNIIstroydormash, –Release 47 – 27-35

[6] Ginsburg O V 1971 Features of traction dynamics and calculation of traction force of tracked tractor with hydromechanical transmission: PhD paper – Moscow: NATI, – 196

[7] Kononenko M A , Zelenskii N S and Gogadze B M 1981 Kinematics of bulldozer equipment when the blade tilts in the vertical plane / Intensification of work processes of road machines Collectionofscientificpapers / MADI – Moscow, – 51-3

[8] BalovnevV I , BerezinV S and SavelievA G 1984 Certificate of authorshipUSSR №1089203 The device for determining the initial loads acting on the bulldozer // Chelyabinsk plant of road machines named afterKolushchenko №**16**

[9] Zelenin A N ,Balovnev V I and Kerov I I 1975 Earth moving machines – Moscow : Mechanical Engineering, 424

[10] Saveliev A G 1984 Additional design positions of the elements of dozer equipment / Improving the efficiency of working bodies and units of road-building machines Collectionofscientificpapers/ MADI, 118-21

[11] Sakhapov R L , Nikolaeva R V , Gatiyatullin M H and Makhmutov M M 2017 Modeling of traction-coupling properties of wheel propulsor *Journal of Physics: Conference Series* 6 "6th International Conference on Mathematical Modelling in Physical Sciences, IC-MSQUARE 2017" **012033**

IOP Conf. Series: Materials Science and Engineering **786** (2020) 012021 doi:10.1088/1757-899X/786/1/012021

Influence of removable anti-skid device on the soil density

R L Sakhapov* and M M Makhmutov

Department «Road Building Machinery», Kazan State University of Architecture and Engineering, Russian Federation

*E-mail: rusakhapov@gmail.com

Abstract. The ways to improve the traction characteristics of the wheel MTA by the drive wheels loaders, the use of additional removable loads, the filling of the drive wheels chambers with liquid, the traction force of the tractor due to the increase in the coupling mass are studied. However, with the increase in the structural dimensions of the wheels and the use of dual tires improve the functional properties of the engines, however, significantly increase the cost of the engine, its dimensions, complicated handling of the machine.The use of removable anti-skid devices on the wheel drive increases the vertical load due to the weight of the anti-skid device. At the moment of interaction of the hook with the soil, the force required to introduce it into the soil, reduces the load on the wheel. A formula for determining the density of soils in the Wake of the mover is obtained. Depending on the magnitude of the force of introduction into the soil, two modes of interaction of the wheel with the soil are considered. To improve the traction properties of the engine, a new design of the device support was developed. After the introduction of the support to a dense layer of soil, it is opened, thereby increasing the contact area and sharply reducing the pressure of the wheel drive tires on the upper layers of the soil. Due to this process, the wheel (partially or completely) is not in contact with the topsoil. The equipment of propellers of wheeled vehicles with removable anti-skid devices, which allow to increase traction properties, reduce the compacting effect of the propulsor on the soil, seems to be a rational solution to the problem of increasing the traction characteristics of the driving wheels. The equipment of wheel machines movers with removable anti-skid devices allows to approximate the traction characteristics of wheel machines to the characteristics of crawler type movers due to the interaction of hooks with more compacted layers of the subsurface horizon soil.

1. Introduction

Depending on the location of the left and right driving wheel lugs arise when driving slopes and distortions of transport vehicles. Coupling properties are lost and determined by the traction ability only from removable lugs models (5) and (6).

In order to reduce these negative phenomena, it is necessary to develop devices to reduce vibration and distortions of the tractor (with a variable radius of the center of the wheel) and improve traction properties (forming a support in the subsurface layers of the soil). Development work in these areas is necessary. As an example, we can show the following development (patent for invention № 2349460

RF, IPC 60 in 15/26. A method of increasing the traction properties of the engine and the device for its implementation).

The anti-skid device performs the function of the bearing base and traction properties. The device consists of (Figure1) from the axis 1 fixed to the wheel rim disc by nuts 2, cheek 3 fixing the position and guide bracket 4. Inside the bracket moves the rack 5, fixed lock 6 with a spring 7. At the top of the rack there is a roller 8, swinging on the surface of the Cam 9, which is mounted on a fixed axis 10. In the lower part of a rack the support is fixed (Figure2) devices 11, the working surface of which is made in the form of a triangle, hinged on the axis 12.

The principle of operation of the device is as follows. When rolling the wheel on soft waterlogged soils, the rack 5, due to the shape of the Cam, introduces the support 12 in the closed state in the subsurface layer of the soil. When introducing the support 12, the soil pressing on its edges, opens the support of the device 11. Due to this process, the wheel (partially or completely) does not come into contact with the topsoil. The anti-skid device performs the function of the bearing base and traction properties. With further rolling of the wheel, the support 11, coming out from under the arable layer, loosens the soil and closes. In interaction with the surface of the rolling wheel enters the next device and the cycle is repeated.

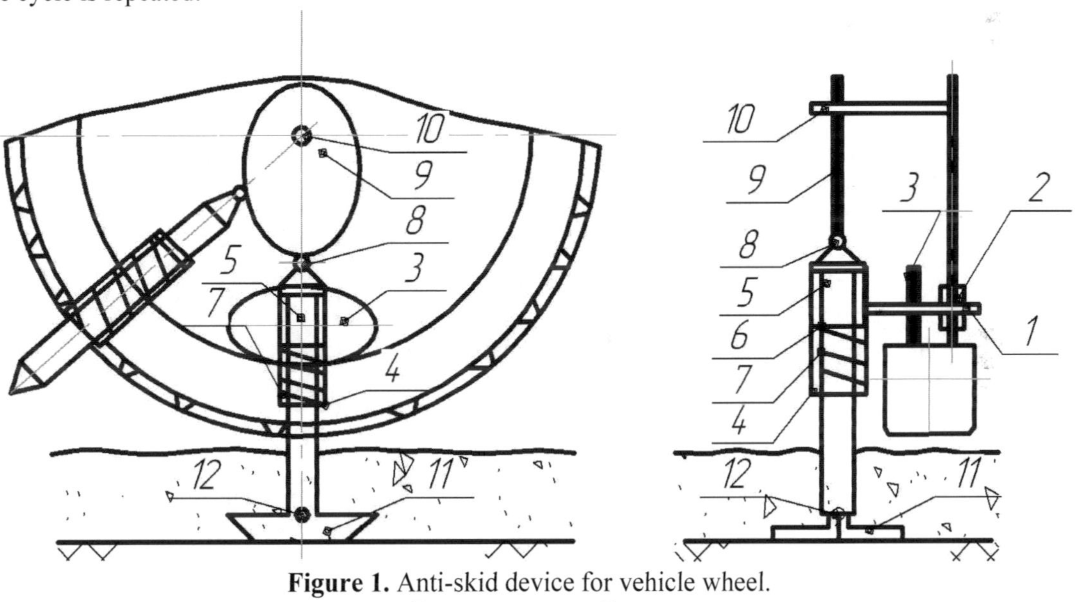

Figure 1. Anti-skid device for vehicle wheel.

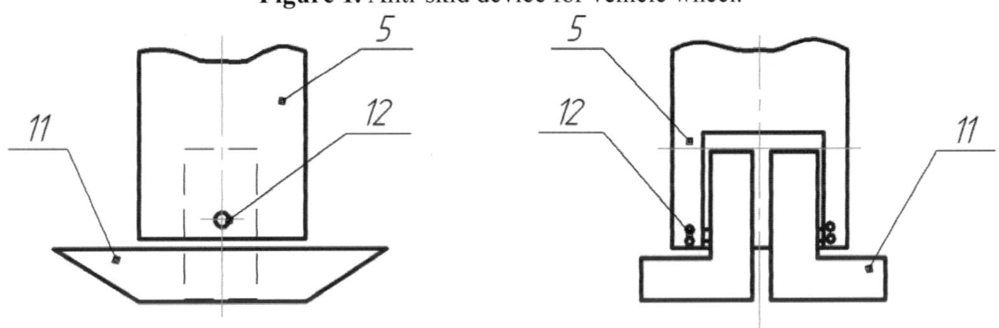

Figure 2. Anti-skid support.

The use of removable anti-skid devices on the wheel drive increases the vertical load due to the weight of the anti-skid device (Figure 3). However, at the moment of interaction of the hook with the soil, the force required to introduce it into the soil, reduces the load on the wheel. Therefore, at the moment, the density of the soil in the wake of the mover is defined as:

$$\gamma_{cl} = \gamma_0+\Delta \gamma_K+\Delta \gamma_Y -\Delta \gamma_B,, \tag{1}$$

where γ_0 - relative density of soil, respectively, the initial (before passage) and the track mover, kg/m^3; Δ y_k, $\Delta\gamma_Y$, $\Delta\gamma_B$ -the increment of the density of the soil, respectively, from the wheel loads, the device weight of snow and force the introduction of the hook in the soil, kg/m^3.

Expressing the density increment through the weight parameters, we obtain:

$$\gamma_{cl} = \gamma_0+k_G\cdot(G_k +Z_c\, m_y g -P_b), \tag{2}$$

where k_G - the coefficient of proportionality of the increment of the density of the soil per unit load, m^3, defined by the expression; g – gravitational acceleration, m/s^2.

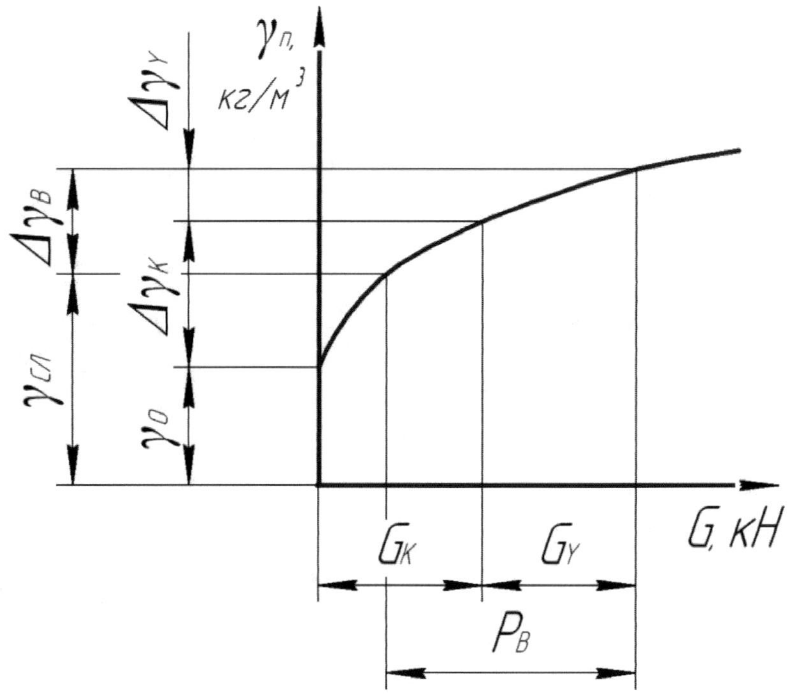

Figure 3. Effect of vertical load on soil density in the wake of the mover.

2. Results and discussion

The calculated values of the model (1) of certain intervals of variation of the studied factors (table. 2) presented in table.1. The analysis of the influence of the studied factors on the density of the soil in the Wake of the mover, at the Central level of variation shows them (table. 1) that with increasing height of the removable lugs on the wheel, regardless of their number, it decreases along a curve with a negative acceleration. For Z_C =2 PCs., 4 PCs., 6 PCs., conl respectively increases, in the range of h_C from 0 to 0.1 m, by 1.04 %, 1.02 % and 1.01 %. With an increase in h_C from 0.1 to 0.2 m, the usl increases, depending on the number of removable lugs, respectively, by 1.05; 1.03 and 1.02 %.

With the increase in the length of the removable lugs on the wheel, regardless of their number, the density of the soil in the track of the mover is reduced by a linear dependence of 5 kg/m3. The study of the hook load in these intervals of variation of the studied factors does not have a significant impact.

With the increase in the number of removable ground hooks on the wheel, regardless of their height, with a Central level of variation of factors, the density of the soil in the Wake of the mover increases along the curve with a positive acceleration. For h_C = 0 m, relative increases in the range Z_C from 2 to 4 pieces by 1.04 %, with h_C = 0.1 m – 1, 05 %, for h_C = 0.2 m – 1.06 %. By increasing the Z_C from 4 to 6 PCs., con increases, depending on the height of the removable lugs, respectively, 1.09; 1.10 and 1.12 %.

This increase in soil density is explained by the fact that with the increase in the number of hooks increases the load on the wheel.

With the increase in the length of removable lugs on the wheel, regardless of their height, the density of the soil in the Wake of the mover decreases proportionally. Moreover, with the increase in the height of the ground hooks, the intensity of the decrease in soil density increases, and for hC = 0.1 m it is 5 kg/m3, and for hC = 0.2 m – 10 kg/m^3. In order to determine the time of interaction with the soil, as a criterion for assessing the length of the less compacted part of the soil trace, which depends mainly on the strength of the introduction of the hook into the soil, its height, the number and slipping of the wheel.

Depending on the magnitude of the force of introduction into the soil, there are two modes of interaction of the wheel with the soil.

1. The implementation force of the hook is less than the load on the hook:

$$P_b \leq G_k + Z_c \cdot m_y g. \quad (3)$$

The density in the trace will be variable and will reach its minimum at the fully entered hook in the soil (Figure4). Therefore, taking into account the processes of slipping and the number of removable lugs on the wheel, we determine the area of the less compacted part of the track by the engine for the rotation of the wheel L_{cl}:

$$L_{cl} = (1-\delta_k) \cdot Z_c \cdot \sqrt{(R_K + h_C)^2 - (R_K - h_K - h_Z)^2} \;. \quad (4)$$

Table 1. Calculated values of soil density in the Wake of the mover γ_{cl} (kg/m^3).

Z_C, PCs.	h_C, m			l_C, m			P_{KP}, kN		
	0	0,1	0,2	0,05	0,10	0,15	4	5	6
2	1448	1432	1418	1438	1433	1428	1433	1433	1433
4	1463	1448	1433	1453	1448	1443	1448	1448	1448
6	1479	1464	1449	1469	1464	1459	1464	1464	1464

h_C, m	Z_C, PCs.			l_C, m			P_{KP}, kN		
	2	4	6	0,05	0,10	0,15	4	5	6
0	1448	1463	1479	1463	1463	1463	1463	1463	1463
0,1	1433	1448	1464	1453	1448	1443	1448	1448	1448
0,2	1418	1433	1449	1443	1433	1423	1333	1333	1333

l_C, m	Z_C, PCs.			h_C, m			P_{KP}, kN		
	2	4	6	0	0,1	0,2	4	5	6
0,05	1438	1453	1469	1463	1453	1443	1453	1453	1453
0,10	1433	1448	1464	1463	1448	1433	1448	1448	1448
0,15	1428	1443	1459	1463	1443	1423	1443	1443	1443

P_{KP}, *kN*	Z_C, PCs.			h_C, *m*			l_C, *m*		
	2	4	6	0	0,1	0,2	4	5	6
4	1433	1448	1464	1463	1448	1333	1453	1448	1443
5	1433	1448	1464	1463	1448	1333	1453	1448	1443
6	1433	1448	1464	1463	1448	1333	1453	1448	1443

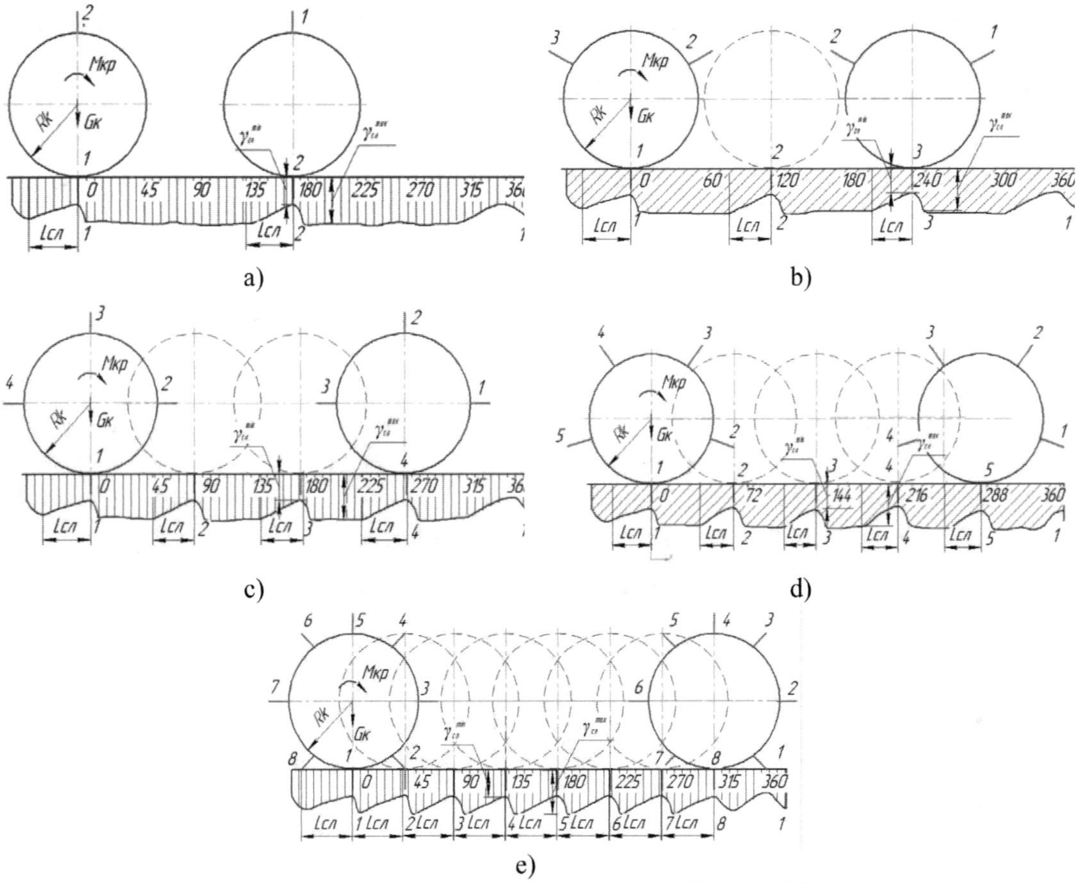

Figure 4. The plot of ground pressure wheels with claws for his one turn:
a-ZC =2 PCs, b-Zc =3 PCs, c-ZC =4 PCs, d-ZC =5 PCs, e – ZC =8 PCs.

The analysis of the influence of the studied factors on the zone of less compacted soil in the Wake of the mover shows (table. 2) that with increase in height of removable ground hooks on a wheel, irrespective of their quantity, it rises on a curve with negative acceleration at the Central level of variation of factors. With the increase in the length of the removable ground hooks on the wheel, regardless of their number, the zone of the less compacted area of the soil in the trace of the mover increases proportionally.

Table 2. Calculated values of the zones of the least compacted area of the soil in the wake of the propeller.

Z_C, Pcs	h_C, m			l_C, m			P_{KP}, kN		
	0	0,1	0,2	0,05	0,10	0,15	4	5	6
2	0	0,64	0,94	0,62	0,64	0,66	0,66	0,64	0,60
4	0	1,32	2,02	1,28	1,32	1,38	1,38	1,32	1,28
6	0	2,00	3,12	1,92	2,00	2,08	2,06	2,00	1,94

h_C, m	Z_C, Pcs			l_C, m			P_{KP}, kN		
	2	4	6	0,05	0,10	0,15	4	5	6
0	0	0	0	0	0	0	0	0	0
0,1	0,64	1,32	2,00	1,28	1,32	1,38	1,38	1,32	1,28
0,2	0,94	2,02	3,12	1,92	2,02	2,12	2,10	2,02	1,96

l_C, m	Z_C, Pcs			h_C, m			P_{KP}, kN		
	2	4	6	0	0,1	0,2	4	5	6
0,05	0,62	1,28	1,92	0	1,28	1,92	1,32	1,28	1,22
0,10	0,64	1,32	2,00	0	1,32	2,02	1,38	1,32	1,28
0,15	0,66	1,38	2,08	0	1,38	2,12	1,42	1,38	1,32

P_{KP}, kN	Z_C, Pcs			h_C, m			l_C, m		
	2	4	6	0	0,1	0,2	4	5	6
4	0,66	1,38	2,06	0	1,38	2,10	1,32	1,38	1,42
5	0,64	1,32	2,00	0	1,32	2,02	1,28	1,32	1,38
6	0,60	1,28	1,94	0	1,28	1,96	1,22	1,28	1,32

With an increase in the hook load, regardless of the number of removable lugs, the zone of less compacted soil in the Wake of the mover decreases along the curve with negative acceleration. With the increase in the number of removable ground hooks on the wheel, regardless of their height, the zone of less compacted soil in the Wake of the mover increases in proportion.

With an increase in the number of removable lugs on the wheel, regardless of the slip coefficient at the joint interaction of removable and tire lugs, the zone in the track of the engine increases along the curve with a positive acceleration at the Central level of variation of factors. For δ_{csh} =0,15, L_{cl} increases in the interval Z_c from 2 to 4 PCs of 9.6 %, with δ_{csh} =0,20 - 6,0 %, for δ_{csh} =0,25 - 4,2 %. With increasing Z_c from 4 to 6 PCs., L_{cl} increases, depending on the strength of the introduction of the hook into the soil, respectively, 9.6; 6.0 and 4.1 %.

$$P_b \geq G_k + Z_c \cdot m_y g. \quad (5)$$

In this mode of operation, the wheel is not in contact with the soil and the anti-skid device performs the function of the bearing base. The density in the trace is equal to the original. The negative side of this mode is that the contact of the wheel with the soil is lost, when moving, there are fluctuations in the unit with an amplitude equal to the height of the removable lugs and the frequency depending on the speed of movement and the number of lugs.

3. Conclusion

With an increase in the number of removable lugs on the wheel, regardless of the slip coefficient in the interaction of tire lugs, the zone in the track of the mover increases along the curve with a positive acceleration at the Central level of variation of factors. For δ_{csh} =0,20, L_{cl} increases in the interval Z_C from 2 to 4 PCs 10.6 %, with δ_{csh} =0,30 – 7.9 %, for δ_{csh} = 0,40 – 17.9 %. With increasing Z_c from 4 to 6 PCs, L_{cl} increases, depending on the strength of the introduction of the hook into the soil, respectively, 10.6; 6.0 and 6.8 %.

With an increase in the number of removable lugs on the wheel, regardless of the height of the removable lugs, the zone in the track of the mover increases along the curve with a negative acceleration at the Central level of variation of factors. For S = 0.1 m, L_{cl} increases in the interval Z_C from 2 to 4 PCs 7.7 %, with L_c = 0.2 m – 7.9 %, for s = 0.3 m – 6.9 %. By increasing the Z_c from 4 to 6 PCs, L_{cl} increases, depending on the height of the hook, respectively, 5,6; 6,4 and 5,9 %.

With the increase in the number of removable ground hooks on the wheel, regardless of the diameter of the wheel, the zone in the track of the engine increases in linear dependence at the Central level of

IOP Conf. Series: Materials Science and Engineering **786** (2020) 012021 doi:10.1088/1757-899X/786/1/012021

variation of factors. For D_k = 0.5 m, L_{cl} increases in the range of Z_C from 2 to 4 PCs. by 10.6 %, with D_k=1.0 m-by 7.9 %, for D_k = 1.5 m – by 5.0 %. With increasing Z_c from 4 to 6 PCs., L_{cl} increases, depending on the diameter of the wheel, respectively, 8.2; 6.4 and 5.0 %.

The main factors affecting the density in the track of the engine are the parameters of the wheels. For the purpose of effective choice of a set of tires parameters of optimization of a wheel on energy criterion and resistance to movement of a wheel with removable hooks shall be considered.

References

[1] Sakhapov, R.L., Absalyamova, S.G. The use of telecommunication technologies in education network. // Proceedings of 2015 12th International conference on Remote Engineering and Virtual Instrumentation. Bangkok. REV 2015. 25-27 Feabruary 2015. P. 14-17.

[2] Makhmutov M.M., Makhmutov M.M. An analytical model for volumetric collapse of the soil / Vestnik Kazanckogo gosudarstvennogo agrarnogo univerciteta. 2008. T. 3. № 4 (10). Pp. 156-157.

[3] Makhmutov, M. M. Optimization of parameters of taper of the strut device chains. Tractori i chelhozmachinye. Moscow. 2013. № 6. p. 31-33.

[4] Sakhapov R.L., Makhmutov M.M. Packing parameters of different types of soil // Izvestiya KGASU. 2015. № 2 (32). Pp. 294-299.

[5] Sakhapov R.L., Makhmutov M.M. The quality of compaction of subgrade // Izvestiya KGASU, № 2, 2015. – P. 289-294.

[6] Sakhapov R.L., Gabdullin T.R., Zagretdinov R.V. Modeling of control systems of road-building machinery // Izvestiya Samarskogo nauchnogo centra RAN. 2014. T.16. № 1 (2). Pp. 394-396.

[7] Patent RUS 86531 Device to reduce the harmful effects on the soil and improve the traction properties of the wheel drive // Autors: Makhmutov M.M., Makhmutov M.M., Fachutdinov M.H., Fachutdinov H.C., patentee Kazanskiy GAU, 19.03.2009.

[8] Nikolaeva, R. V., Sakhapov, R. L., Gabdullin T.R, Makhmutov, M. M. Motor Roads As a Factor of the Economic Potential of the Republic of Tatarstan, International Conference on Applied Economics, ICOAE 2015, T. 24, 2015. C.606-612.

[9] Sakhapov R. L., Nikolaeva R. V., Gatiyatullin M. H., Makhmutov M. M.Risk management model in road transport systems // Journal of Physics: Conference Series. 2016. T. 738. № 1. C. 012008.

[10] Sakhapov R.L., Nikolaeva R.V., Gatiyatullin M.H., Makhmutov M.M. Modeling the dynamics of the chassis of construction machines. Journal of Physics: Conference Series. 2016. T. 738. № 1. C. 012119.

[11] Sakhapov, R.L., Nikolaeva, R.V., Gatiyatullin, M.H., Makhmutov, M.M. Mathematical model of highways network optimization // Journal of Physics: Conference Series. 2017. T. 738. № 2. C. 012009.

[12] Sakhapov, R.L., Nikolaeva, R.V., Gatiyatullin, M.H., Makhmutov, M.M. Modeling of traction-coupling properties of wheel propulsor // Journal of Physics: Conference Series. 2017. T. 738. № 2. C. 012115.

Vehicle Counting at Road Intersections Using Video Data and Zone-Based Method with Additional Contour

O Ibryaeva[1,*], A Zhulev[1], V Shepelev[1], A Burzev[2]

[1]South Ural State University, Lenina avenue, 76, Chelyabinsk, Russia

[2]Kuzbass State Technical University named after T.F. Gorbachev, Ilyich st., 32a, Belovo, Russia

[*]Email: ibriaevaol@susu.ru

Abstract. The problem of counting vehicles and determining the directions of their movement at an intersection is considered. A new zone-based method for counting vehicles which uses an additional inner contour is proposed. The presence of an additional inner contour helps to partially expands zones of interest and increases our chances of catching a vehicle in time. YOLO and CSRT tracker are used for detection and tracking of objects. Full flowchart of the working process including detection, tracking and counting, is given. The algorithm for inner contour constructing is proposed and it is shown that this contour presence reduces the error in counting vehicles.

1. Introduction

Recently, due to the increase in the number of roads and vehicles, the task of controlling traffic with a smart transport system has become increasingly important. A large number of studies in this area are based on the analysis of video data by computer vision systems. The output values of this analysis are, for example, the class of the object, its speed, trajectory, number of objects, the presence of traffic jams and accidents, etc. The most challenging task is the real-time video analysis, taking into account the presence of shadows, bright sun, overlapping, rain, snow, fog. Despite numerous works and already existing solutions, we can say that the task of accurately calculating traffic, taking into account all these difficulties, is still far from complete.

Usually, in video analysis of traffic, the background subtraction (BS) method and its various modifications are used [0-0]. However, this method fails in the case of heavy traffic conditions. The presence of shadows and overlapping objects also prevent good object detection. There are various modifications of the BS method [0, 0], which try to solve these problems, but even they do not always cope. The BS method is even less suitable when, in addition to detecting objects, we also need to classify them, and we can do this either by comparing the relative size of the objects or their Histogram of Gradient. Truly, only deep convolutional networks can cope with the classification problem [0], however, they require a lot of time for their work. The YOLO network [0] perfectly solves this problem by detecting all objects of the frame "at a glance" (You Only Look Once), and providing a speed of 45 frames / seconds. In this paper, we have chosen YOLO method of detection and identification of objects.

YOLO detector is not enough for the task of counting vehicles. We have to track objects on each frame, to determine whether they are old or new in order not to count the same object several times. It is important not to lose or mix up previously found objects. The problem of tracking objects in a video sequence is extremely relevant and many works have been devoted to it, for example, [0-0]. This problem is also complicated by the possible overlap of objects, changes in illumination, fast speed of objects, similarities with the background, etc. In particular, trackers based on the discriminative correlation filter (DCF) method [0-0] have shown great performance in all standard benchmarks. Discriminative correlation methods learn a filter with a pre-defined response on the training image. The latter is obtained by slightly extending the region around the target to include background samples. However, the published DCF methods are not without drawbacks.

The Discriminative Correlation Filter with Channel and Spatial Reliability [0] overcomes both the problems of circular shift enabling an arbitrary search (and training) region size and the limitations related to the rectangular shape assumption. It shows state-of-the-art performance on standard benchmarks while running close to real-time on a single CPU and implemented in OpenCV as CSRT tracker. Our work combines the object detection unit with the CSRT tracker to count the traffic data.

There is a considerable amount of research in general purpose detection and tracking, but relatively little work on the application to large scale transportation tasks such as counting. Most works use counting systems based on a line crossing approach. Vehicles are counted every time their bounding boxes cross a user defined line [0]. In [0] a polygon is drawn in the scene and each side of the polygon corresponds to an entering or exiting virtual gate. In [0] counting is performed by determining the number of trajectories that are close to a source or a sink (entering and exiting points in the scene). Some works count vehicles when they are present in special zones [0].

The real challenge for video analysis is the case of an intersection - there are a lot of vehicles and one need to track the trajectory of each of them. At first glance an obvious approach is to keep track of which areas each car has visited and, based on this, make a conclusion about its direction of movement. However, due to the fast speed, the object may slip through the desired zone and be detected later. Another approach in through trajectories, which are then checked for intersections with a line or zone of interest is also not very successful, because due to the large number of objects, they may be confused, or also start too late due to the late detection of the object. In this paper, we use an approach based on tracking the entry of an object into certain zones. We propose to solve the problem of late object detection by introducing an additional inner contour.

The paper is organized as follows. In the next section, the algorithm for counting vehicles and determining the direction of their movement is described. The algorithm for constructing an additional contour is also given. Two algorithms (with and without an additional contour) are compared in Section 3. Finally, Section 4 contains the main conclusions of the work.

2. Proposed Method

2.1 Algorithm of Inner Contour construction

The direction of movement of the vehicle is determined by its presence in the zones of interest. The correctness of such determining depends on whether the object was detected in time. If the vehicle is going at high speed, YOLO may not be able to detect it in the desired zone and we cannot determine the arrival area of the object. The presence of an additional inner contour helps to partially expands zones of interest (Figure 1) and increases our chances of catching a vehicle in time.

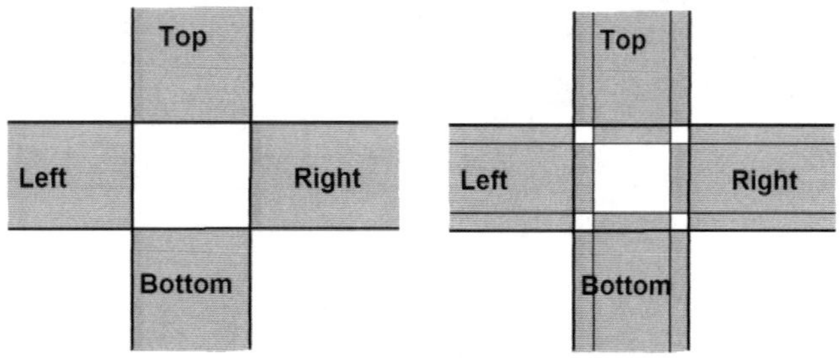

Figure 1. Zones of interest without and with additional inner contour.

Let us present the algorithm for constructing the inner contour. Given the four vertices $A(x_A, y_A), B(x_B, y_B), C(x_C, y_C), D(x_D, y_D)$ of the outer contour (Figure 2).

Algorithm for Inner Contour Construction

1. Find the equations of the lines $AC: \frac{x - x_A}{x_C - x_A} = \frac{y - y_A}{y_C - y_A}$ and $BD: \frac{x - x_B}{x_D - x_B} = \frac{y - y_B}{y_D - y_B}$.

2. Find the point $O(x_O, y_O)$ of intersection of two lines *AC* and *BD* from the system:

$$\begin{cases} x_O \cdot \frac{1}{x_C - x_A} - y_O \cdot \frac{1}{y_C - y_A} = \frac{x_A}{x_C - x_A} - \frac{y_A}{y_C - y_A} \\ x_O \cdot \frac{1}{x_D - x_B} - y_O \cdot \frac{1}{y_D - y_B} = \frac{x_B}{x_D - x_B} - \frac{y_B}{y_D - y_B} \end{cases} \quad (1)$$

We have $x_O = \frac{\Delta_1}{\Delta}, y_O = \frac{\Delta_2}{\Delta}$ where

$$\Delta = \begin{vmatrix} \frac{1}{x_C - x_A} & -\frac{1}{y_C - y_A} \\ \frac{1}{x_D - x_B} & -\frac{1}{y_D - y_B} \end{vmatrix}, \Delta_1 = \begin{vmatrix} \frac{x_A}{x_C - x_A} - \frac{y_A}{y_C - y_A} & -\frac{1}{y_C - y_A} \\ \frac{x_B}{x_D - x_B} - \frac{y_B}{y_D - y_B} & -\frac{1}{y_D - y_B} \end{vmatrix}, \Delta_2 = \begin{vmatrix} \frac{1}{x_C - x_A} & \frac{x_A}{x_C - x_A} - \frac{y_A}{y_C - y_A} \\ \frac{1}{x_D - x_B} & \frac{x_B}{x_D - x_B} - \frac{y_B}{y_D - y_B} \end{vmatrix} \quad (2)$$

3. Find the vectors $\overline{OA}$, $\overline{OB}, \overline{OC}, \overline{OD}$, and the vectors $\overline{AA'} = \overline{OA} \cdot k$, $\overline{BB'} = \overline{OB} \cdot k$, $\overline{CC'} = \overline{OC} \cdot k$, $\overline{DD'} = \overline{OD} \cdot k$.

4. Find the vertices *A'*, *B'*, *C'*, *D'* of the inner contour using the coordinates of the vertices *A*, *B*, *C*, *D* and vectors $\overline{AA'}$, $\overline{BB'}$, $\overline{CC'}$, $\overline{DD'}$.

The parameter *k* is responsible for how close the inner contour will be to the outer one and selected experimentally. The inner contour should not be too close in the outer, otherwise its meaning will disappear. At the same time, it should not be too far from the outer contour, since in this case again we will not get a good expansion of the zones of interest (because of the four large white rectangles in the Figure 2).

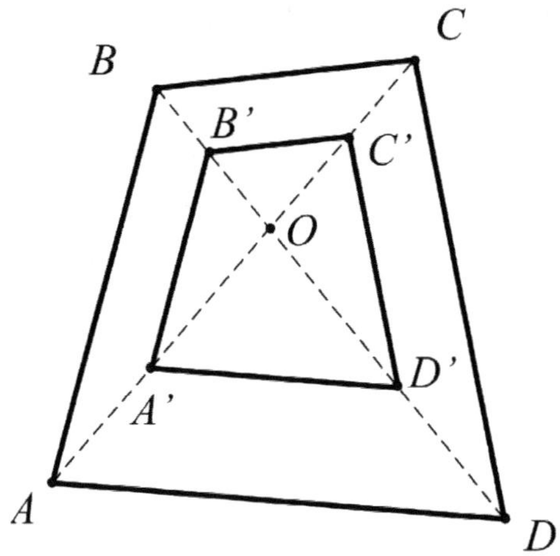

Figure 2. To algorithm of inner contour construction.

2.1 Zone-Based Method with Additional Inner Contour

The working process of the proposed method is depicted in Figs. 3, 4. We combine YOLO detector to recognize the vehicles in video data, followed by tracking using CSRT tracker from OpenCV. In YOLO working each frame is divided into MxM (ex. 13x13) grids, and each grid predicts B (ex.9) bounding boxes with the confidence scores. Each bounding box is associated with x, y, w, h, confidence score as predictions. Each grid cell predicts C (ex. 80) conditional class probabilities to indicate the likelihood of object in it.

We used YOLO trained on COCO dataset and it can predict 80 classes such as bicycle, car, truck, motorcycle, person etc. The confidence score of each bounding box and class predictions are combined to estimate the object YOLO is applied to the image every 15 frames in our 25 FPS video. The class "person" is deleted from the detection process. Each detected object is attributed by its ID, centroid and the bounding box. The detected objects are tracked across the video frames by CSRT tracker.

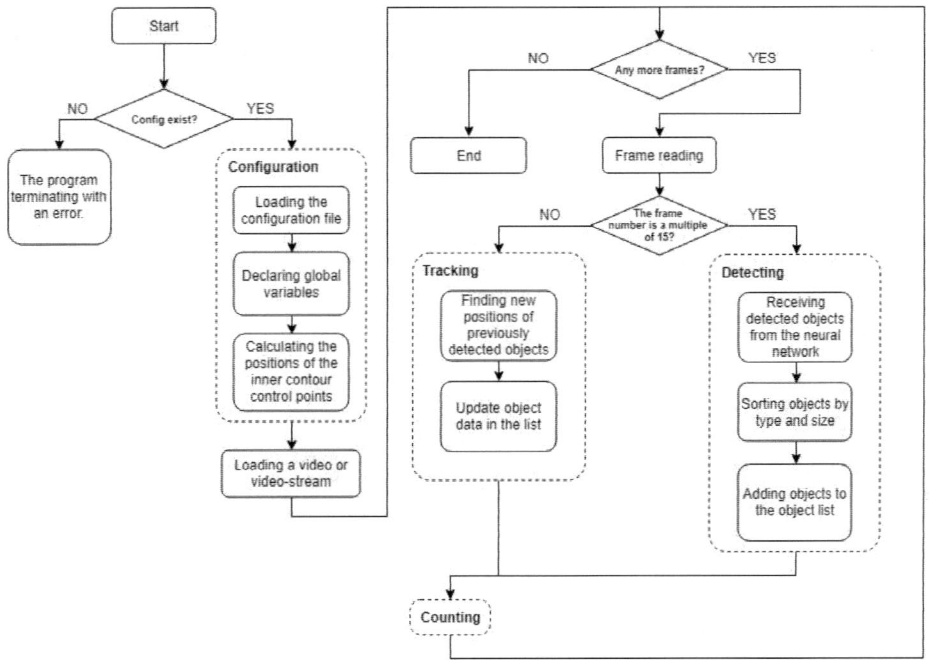

Figure 3. Flowchart of the working process.

The presented method uses both inner and outer contours. In order to investigate how much the inner contour improves the efficiency of the method, we compare it with the method which uses only one outer contour.

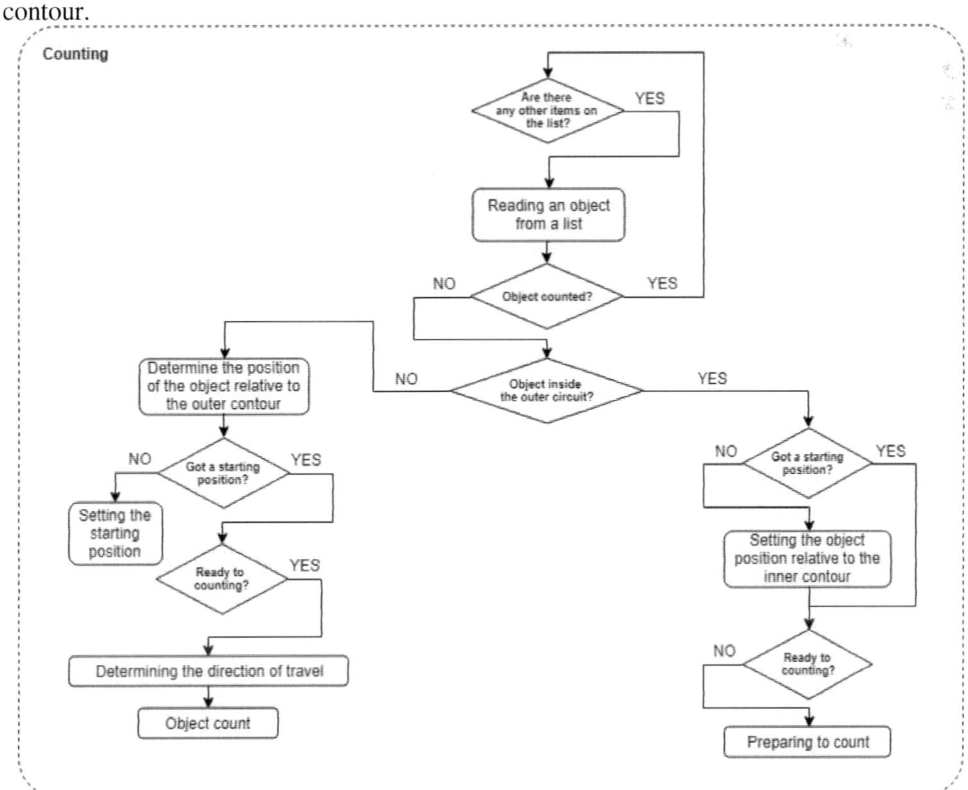

Figure 4. Flowchart of the proposed vehicle counting process.

The flowchart for this method is very similar and we did not give it. It does not contain calculating the position of the inner contour control points in block "Configuration" and setting the object position relative to the inner contour in the block "Counting". The next section compares these two methods.

3. Experimental analysis

To test the robustness of vehicle counting, 3 scenarios lasting 5 minutes, including day, night and rain are selected from the real video from the intersection of our city. To evaluate the quality of the counting result, the Absolute Percentage Error (*APE*)

$$APE = \frac{GT - N}{GT} \cdot 100\% \qquad (3)$$

was used. Here N and GT are respectively the number of vehicles counted automatically and the ground-truth number of vehicles.

The *APE* of the vehicle counting in our methods without and with the additional contour is listed in Table 1. All possible turning movements counts were considered.

The value $APE = 100\%$ is taken in those cases when no objects were found in this particular direction ($N = 0$), but in fact the cars passed there ($GT \neq 0$). There were cases when $GT=0$ in the reviewed video files. In these cases, we always had $N = 0$ and $APE = 0\%$ Of course, for the cases when $GT=N$ we also have $APE = 0\%$.

The last line shows the average *APE* for the video file. As one can see, the average *APE* decreases significantly in the case of the algorithm with additional inner contour.

Table 1. The Absolute Percentage Error values for Algorithms without/with Inner Contour.

Zones of interest	Algorithm without inner contour			Algorithm with inner contour		
	26.07.19 7:00- 7:05 Day	06.08.19 23:00-23:05 Night	02.08.19 13:50-13:55 Rain	26.07.19 7:00- 7:05 Day	06.08.19 23:00-23:05 Night	02.08.19 13:50-13:55 Rain
Top-Top	0	0	0	0	0	0
Top-Right	0	0	100	0	0	100
Top-Bottom	33.33	100	48.15	0	54.55	22.22
Top-Left	0	100	100	0	100	33.33
Right-Left	0	0	100	0	0	100
Right-Right	0	0	0	0	0	0
Right-Bottom	41.18	7.14	43.75	17.65	0	28.125
Right-Left	56.41	16.67	83.33	51.28	16.67	81.82
Bottom-Top	16.67	76.92	58.82	5.56	61.54	35.29
Bottom-Right	32	20	10.20	4	0	10.20
Bottom-Bottom	0	0	0	0	0	0
Bottom-Left	0	60	23.08	0	60	11.54
Left-Top	0	0	25	0	0	0
Left-Right	14.81	48.39	35	11.11	40.91	25
Left-Bottom	11.11	8.33	6.90	0	7.69	0
Left-Left	0	0	0	0	0	0
Average *APE*	12.84	27.34	39.64	5.60	21.34	27.97

4. Conclusions

In the paper, the zone-based method for counting vehicles is proposed. The method uses a new idea of expanding zones of interest due to the additional inner contour. The algorithm for the inner contour construction is also presented. The advantage of the proposed counting method is shown experimentally.

The absolute percentage error in vehicles counting when using the outer and inner contours is significantly less than when we use only the outer one. Note that the error values obtained in themselves are quite large, but this is due to insufficient network and tracker performance. Our goal was to demonstrate the advantage of using an inner contour and, as we can see, other things being equal, this method works better. As for improving the network and tracker, we will devote further work to this.

References

[1] C Pompanomchai, T Liamsanguan, V Vannakosit, 2008 Vehicle detection and counting from a video frame, Wavelet Analysis and Pattern Recognition **1** 356-361

[2] V Pham, B Lee 2015 Front-view car detection and counting with occlusion in dense traffic flow, International Journal of Control, *Automation and Sustems* **13(5)** 1150-1160

[3] H Jang, I Won, D Jeong 2014 Automatic Vehicle Detection and Counting Algorithm, International *Journal of Computer Science and Network Security* **14(9)** 99

[4] P Barcellos, C Bouvie, F Escouto 2015 A novel video based system for detecting and counting vehicles at user-defined virtual-loops, *Expert Systems with Applications* **42(4)** 1845-1856

[5] Y Xia, X Shi, G Song, Q Geng, Y Liu 2016 Towards improving quality of video-based vehicle counting method for traffic flow estimation, *Signal Processing* **120** 672-681

[6] G Salvi 2014 An automated nighttime vehicle counting and detection system for traffic surveillance, *Computational Science and Computational Intelligence* **1** 131-136

[7] C Ren, K He, R Girshick, D McAllester, Faster R 2015 CNN: Towards real-time object detection with region proposal network, *Advances in Neural Information Processing System* 91-99

[8] J Redmon, S Divvala, R Girshick, A Farhadi 2016 You only look once: Unified, real-time object detection, Proc of the IEEE Conference on Computer Vision and Pattern Recognition **7780460** 779-788

[9] Y Wu, J Lim, M Yang 2013 Online object tracking: A benchmark, Comp Vis Patt Recognition **6619156** 2411-2418

[10] M Kristan, J Matas, A Leonardis, T Vojir, R Pflugfelder, G Fernandez, G Nebehay, F Porikli, L Cehovin 2016 Novel performance evaluation methodology for single-target trackers, IEEE Trans Pattern Anal Mach Intel **38(11)** 2137-2155

[11] P Liang, E Blasch, H Ling 2015 Encoding color information for visual tracking: Algorithms and benchmark, *IEEE Trans Image Proc* **24(12)** 5630-5644

[12] M Mueller, N Smith, B Ghanem 2016 A benchmark and simulator for UAV tracking, *Lecture Notes in Computer Science* **9905** 445-461

[13] M Danelljan, G Hager, F S Khan, M Felsberg 2014 Accurate scale estimation for robust visual tracking, Proc *British Machine Vision Conference* 1-11

[14] M Danelljan, G Hager, F Shahbaz Khan, M Felsberg 2015 Learning spatially regularized correlation filters for visual tracking, Proc Int Conf Computer Vision 4310-4318

[15] J F Henriques, R Caseiro, P Martins, J Batista 2015 Highspeed tracking with kernelized correlation filters, IEEE Trans Pattern Anal Mach Intell **37(3)** 583-596

[16] Y Li, J Zhu, A 2014 scale adaptive kernel correlation filter tracker with feature integration, Proc European Conf Computer Vision 254-265

[17] A Lukezic, T Vojir, L Zajk, J Matas, M Kristan 2018 Discriminative correlation filter tracker with channel and spatial reliability, *International Journal of Computer Vision* **126(7)** 671-688

[18] K Jie, M Liu 2005 Computer vision based real-time information acquisition for trnsport traffic, Information Acquisition

[19] S Messelodi, C M Modena, M Zanin 2005 A computer vision system for the detection and classification of vehicles at urban road intersections, *Pattern Analysis and Applications* **8(1)** 17-31

[20] R Zhao, X Wang 2013 Counting vehicles from semantic regions, IEEE Transactions on Intelligent Transportation Systems **14(2)** 1016-1022

[21] J C Mendes, A G C Bianchi, A R P Junior 2015 Vehicle tracking and origin-destination counting system for urban environment, Proc of the 10th International Conference on Computer Vision Theory and Applications **3** 600-607

Innovative technology for monitoring pavements

R Sakhapov* and M Makhmutov

Kazan State University of Architecture and Engineering

E-mail: rusakhapov@gmail.com1

Abstract. The paper considers the advantages and nature of georadar monitoring of road surfaces comparing to traditional destructive methods of road exploration. The authors checked the thickness of the structural layers of pavement to determine appropriate repair work using the amplitude-frequency characteristics and the dielectric constant (the speed of signal transmission). The article examined the section of the highway and the examination data was recorded in a file, which allows further study and documentation of the material with the help of computer equipment.

1. Introduction

Theoretical basics of the application of radio waves for the solution of geological problems are laid down in the works of G. Lovi and G. Leimbach (1910-1912), A.A. Petrovsky (1925) and A.G. Tarkhov (1950). A great contribution to the development of georadiolocation and practical applications of the method was made by such scientists as M.L. Vladov, A.V. Starovoitov, S.V. Izyumov, S.V. Druchinin, A.S. Voznesensky, A.Y. Grinev, D. Daniels (Great Britain), M. Baneau (France), P. Annan (Canada), T. Saarenketo (Finland), J. Hughesman (Netherlands) and others. The principle of operation of radars is based on the emission of ultra-wideband electromagnetic pulses without a carrier into the underlying medium and recording their reflections from the interfaces of layers or objects.

The propagation velocities of longitudinal C_p and (more rarely) transverse C_s elastic waves are used as informative parameters of ultrasonic (US) control, as well as their attenuation coefficients a_P and a_s respectively. The frequency range used in measurements in an array is usually 30 to 150 kHz [4].

When determining the stresses in the contour rock massif, depending on the measurement scheme being implemented, the following methods are distinguished:

1) the method of ultrasonic transmitted waves;
2) ultrasonic logging method;
3) the method of reflected waves.

The method of ultrasonic passing waves (sounding) occupies a dominant position among ultrasonic methods of geocontrol. This is due, on the one hand, to the most perfect methodological and hardware support, and on the other hand to the possibility of using these methods in a wide range of controlled bases (from a few centimeters to several meters) both on samples and in an array rocks (variant of interwell sounding) [3].

One of the most frequently solved problems with the use of ultrasonic methods of geocontrolis to study the nature of the spatial distribution of stresses in the environments of the excavations and its dynamics over time.

The sequence of operations for the implementation of measurements by the sounding method is as follows. From the mine workings to the array, a set of parallel holes is drilled in a triangular or circular pattern, as shown in Figure1. Further, the borehole probes are placed in parallel boreholes. The acoustic transducer in one of the probes is used as a radiator for pulsed ultrasonic vibrations, and in the second - as a receiver. Then the probes move synchronously and discretely into the interior of the array. Measurement of the spatial distribution of informative parameters of ultrasonic testing, most often the value of C_p, reflects the spatial distribution of stresses.

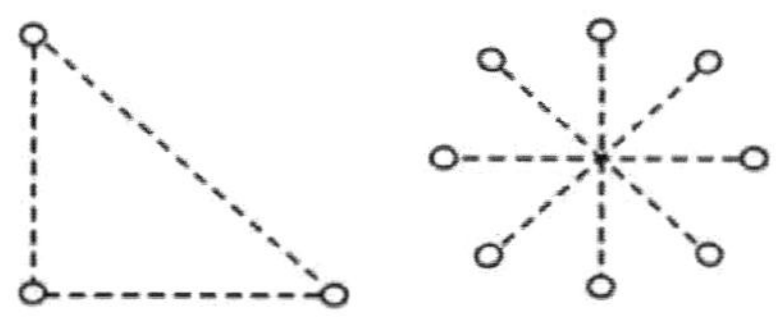

a b

Figure1. The location of the boreholes: a - in a triangular pattern; b - along a circular pattern.

US logging is a downhole method of geophysical studies, based on the study of the acoustic properties of rock formations intersected by the well and directly adjacent to it. The implementation of this method involves the use of a single borehole probe in which the radiator and one or several receivers of ultrasonic pulse signals are placed.

Such a probe moves along a linear profile located on the axis of the water-filled or forming a dry well. By the nature of the distribution of measured acoustic characteristics along the profile, the nature of the distribution of stresses in the contour array is judged.

The ultrasonic wave reflection method assumes radiation into the controlled region of the mass and subsequent recording of the ultrasonic signal reflected from the natural or artificial inhomogeneity in it. The method has a relatively limited application in the study of the geological environment, which can be explained by the significant loss of signal power on the doubled base of its propagation (from a combined acoustic transducer to reflective heterogeneity and vice versa), and also at the boundary between the host geomedium and heterogeneity, since the reflection coefficient from the latter can be much less than unity. In terms of amplitude-frequency characteristics, dielectric permeability (speed of signal transmission) in various soils, the soil type, its humidity and relative density are determined [7].

The purpose of the study is to increase the effectiveness of the application of georadar equipment for flaw detection of the pavement. Georadars are fixed on vehicles. During one session you can record profiles for several hundred kilometers long. Depending on the complexity of the geological conditions, one operator processes from 3 to 10 km of the recorded profile during a working day. During the road construction works, the compliance of the executed works with the approved project documentation, technical regulations and standards is estimated [5]. For example, when assessing the thickness of paved bituminous concrete, the common practice is core sampling, measurement of their thickness and laboratory tests for determination of density, compaction factor and water saturation.

The main characteristic describing the degree of conductivity of asphalt concrete is the volume specific resistance, defined by the formula:

$$P_v = \frac{R_V \cdot S}{h}, \quad (1)$$

where R_V–volume resistance of the sample, Ohm; S –electrode space, m^2;
h –sample'sthickness, m.

IOP Conf. Series: Materials Science and Engineering **786** (2020) 012023 doi:10.1088/1757-899X/786/1/012023

The electrical resistance depends on the mineral composition of the asphaltic concrete and varies widely. So, for gravel and rubble, which are conductors, Pv ~ 10^{-5} Ohm · m, and for some components characterized by dielectric properties, Pv ~ 10^{17} Ohm · m [11].

To solve the problems of determining the stress state of an array, it is important that the changes in the resistivity with increasing pressure in different materials can have a different character [6]. During the road surface observation, there was developed geological research equipment for flaw detection of the pavement (Figure 2.). For this purpose the device uses two geological prospecting sensors [8, 9]. The principle of the georadar is based entirely on radiolocation: radiation and fixation of reflected electromagnetic pulses. The impulse is produced by the device itself and, with the aid of a radiator (antenna), is sent to the studied environment.

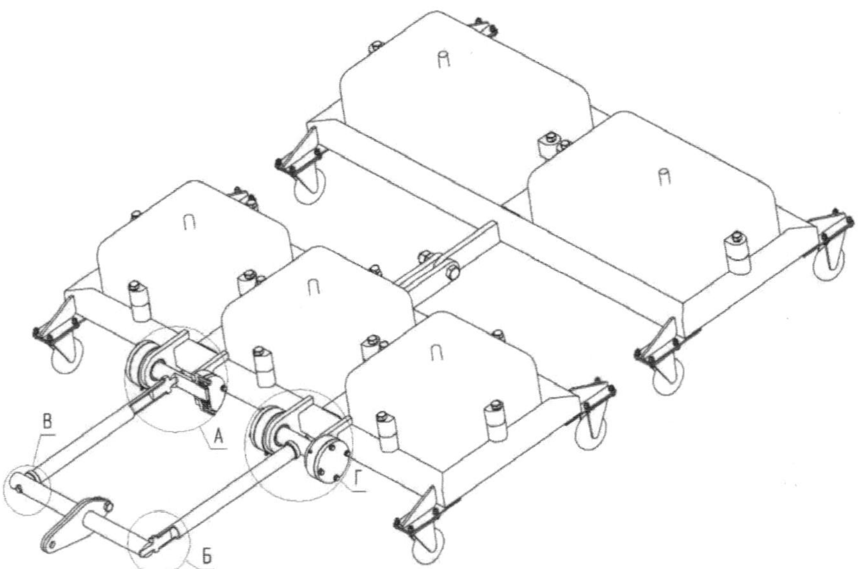

Figure2. General device of equipment for flaw detection.

The environment can be any material: concrete, soil, brick wall, etc. When scanning, the computer screen displays a radargram in the course of georadar movement, where in-phase lines are recorded in the form of black-and-white or colored bands, as well as noise [10]. The distance traveled by the georadar, the depth of sounding, the time of passage of the signal and special marks are indicated on a radargram, as a rule. Control drilling can be performed before scanning, during scanning and after scanning.

If the design information is unknown, then the test drilling can be performed before scanning. To do this, it is necessary to record the radargram on a short section, to allocate a homogeneous plot on it, mark it on the terrain and perform the control drilling. Based on the drilling results, adjustments should be made to the initial settings of the GPR.

Interpretation is reduced to the solution of the inverse physical and mathematical problem, which results in a formulation of an electrophysical model, and then - a section of the road structure. At this stage, taking into account the drilling data, the propagation velocities of electromagnetic waves in the layers and their thicknesses are being determined [1, 2]. The solution of the inverse problem is usually carried out in several stages, each of which contains the stages of processing and interpretation. The processing of field radargrams is performed in the case when interfering waves make it difficult to identify and trace useful waves. It should be borne in mind that as a result of processing, some of the information, including useful, is inevitably lost. The treatment procedure is divided into two stages: preliminary processing and information analysis. Interpretation of results is a process of constructing a section of pavement and consists of two main stages: kinematic and dynamic. The kinematic interpretation allows, according to the recorded times of useful waves, to restore the position of the boundaries and the distribution of the velocities of electromagnetic waves in the layers of the road structure.

At the same time, the economic effect in the reserach of 10 km of road structures using georadar technologies reaches at least 2.5 million rubles due to a reduction in the number of drilling operations and core sampling. Moreover, preliminary calculations show that the costs for subsequent maintenance work will decrease by at least 20 percent. Calculations also show that, by laying the track even 13% longer, but under the best soil and hydrogeological conditions, it is possible to save considerably on operating costs in the life cycle of the road, avoiding large expenses and protracted repair work.

FAO "ROSDORNII" developed the interstate standard GOST 32868-2014 "Roads of public use. Requirements for engineering and geological surveys", which regulates the application of the georadiolocation method in the exploration of roads for new construction, reconstruction and major repairs of highways, as well as in the process of road construction works.

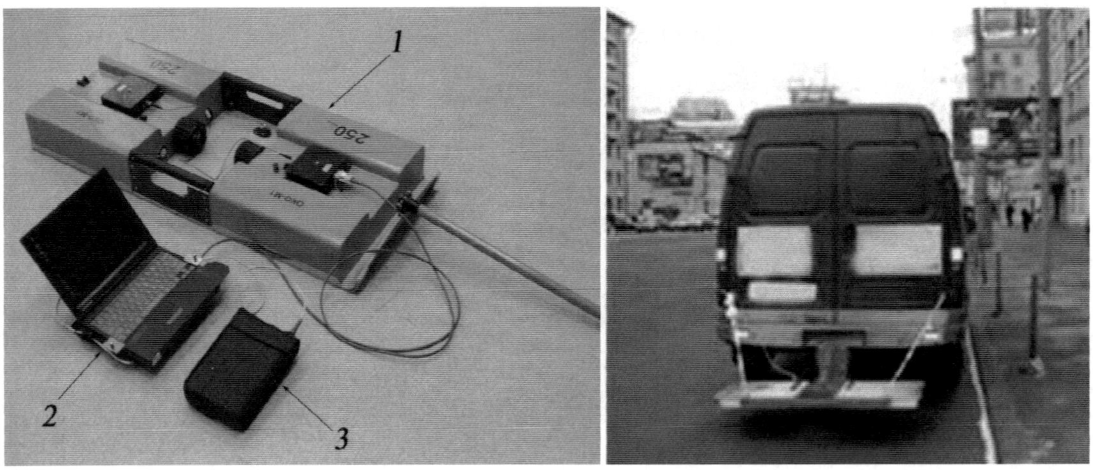

a b

Figure 3. WorkingsetofageoradarOKO-2 (a) andoperationona «Gazelle»car (b):1- an antenna unit including a transmitting and receiving antenna; 2 - device for collecting and processing information; 3 - external rechargeable battery.

There were developed two branch road methods: "Methodological recommendations for the application of georadars in the research of road structures" and ODM 218.2.037-2013 “Methodological recommendations for conducting survey work in the repair and repair of highways”. The functional view of the minimal components of the georadar OKO-2 is shown in Figure 3. As a rule, the package includes a base set and one or more antenna blocks: a control unit; antenna unit; optical converter; registration device (laptop and 2 mini-cameras); block of processing (control and processing unit); power supply unit with chargers; motion sensors; cables.

During the inspection a section of the highway in the Apastovsky District of the Republic of Tatarstan, the georadar worked using an antenna block with a frequency of 2500 MHz. The task of the study was to check the thickness of constructive layers of pavement for the purpose of assigning appropriate repair work, if necessary. The survey was conducted at an air temperature of +15 ° C, scanning at a speed of 30 to 35 km/h. The used method suggested continuous georadar scanning of the engineering geological section along the lines of profiles passing in the middle of each lane of the road. In the process of georadar movement, the laptop displayed a program with fixed lines of in-phase in the form of black-and-white or color bands and the noise. The radargram also showed the distance traveled by geo-radar, the depth of sounding, the time of the signal passage and special marks (Figure 4).

Analysis of radargrams in this section of the road in the longitudinal direction showed that the thickness of the asphalt concrete layer along the length of the road varies from 12 to 20 cm, and the thickness of the bed of crushed stone in the range of 12.0-18.0 cm. At the beginning of the section, the thickness of the asphalt concrete layer under the central lane of movement was 5 cm thicker than under the extreme bands and 10 cm thicker in the middle of the section.

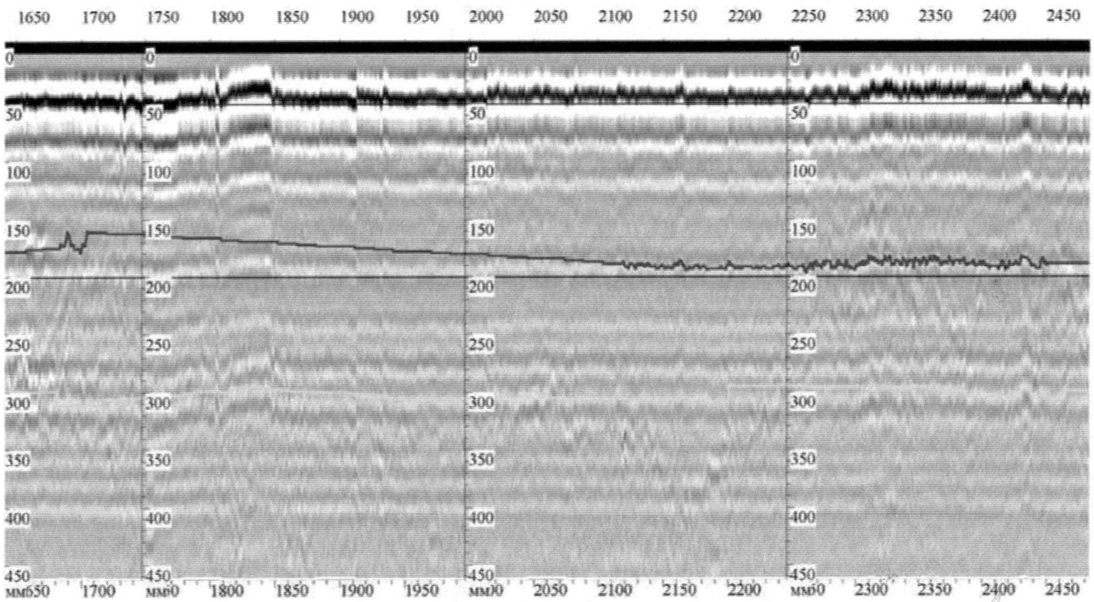

Figure 4. Radargram with fixed in-phase lines.

The method of georadiolocation is based on the study of the field of high-frequency electromagnetic waves [12]. The method relies on the difference in the rocks from the dielectric constant [13]. The radiated impulse, propagating in the environment or object under examination, is reflected from the boundaries with the change of the electrical characteristics - electrical conductivity and dielectric constant [14, 15]. The reflected signal is being received by the receiving antenna, and then amplified, digitized and stored. The frequency of the probing impulse of the georadar is in the range from 25 MHz to 3 GHz, the wavelengths in the investigated environmentd are in the range from several centimeters to several meters. The resolving power of georadar in depth research can be improved during the mathematical processing of the field material.

Monitoring of the road section with thegeoradar method allows us to draw the following conclusions:

- destruction of asphalt concrete pavement is associated with non-compliance with the thickness of the road structure;
- designation of the appropriate type of repair work will allow to extend the life of the pavement.

2. Conclusion

For the wide application of georadar technologies, we advise to develop a sectoral five-year scientific and technical program for the introduction of georadar technologies. The program envisages research on the development of multipurpose mobile georadar laboratories; improvement of software for automatic recognition of abnormal zones; determination of the physico-mechanical properties of road-building materials and soils by amplitude-frequency characteristics, etc.

Economic calculations have shown that the proposed version of the equipment for roadway defecting is more efficient and rational, the equipment is cheaper in value. Monitoring can identify the true causes of road damage and won't affect the environment. Based on the results of monitoring, it is possible to assign effective types of corresponding repair work that will ensure the operational reliability of highways for a longer period. Consequently, production and monitoring with the developed equipment is expedient and economically justified.

References

[1] Sakhapov R L and Absalyamova S G 2015 The use of telecommunication technologies in education network // Proceedings of 2015 12th International conference on Remote Engineering and Virtual Instrumentation Bangkok REV 2015 25-27 Feabruary 14-7
[2] Brekhman A I, Sakhapov R L and Absalyamova S G 2014 Innovative model of integration of education and business in the road construction industry *International Journal "Science & Engineering for Roads"* **3(69)** 39-41
[3] Vladov M L and Starovoitov A V 2002 GPR studies upper part of the section M Mosk University press
[4] Eremenko A V, Zverev E A, Klepikova S M and Osipenko M M 2004 GPR research in terms of reinforcement The report at the conference «Georadar-2004»
[5] Sakhapov R L, Makhmutov M M , Makhmutov M M and Zemdikhanov M M 2016 Theory of the functioning of mechanical energy in construction machines Mechanizationofconstruction T **77(8)** 17-21
[6] Makhmutov M M 2013 Optimization of parameters of taper of the strut device chains Tractoriichelhozmachinye **6** 31-3
[7] 2004 Methodical recommendations about application of GPR at inspection of road designs Moscow Rosavtodor 64-75
[8] Sakhapov R L and Makhmutov M M 2014 Influence of the studied factors on productivity of a road mill Izvestiya KGASU Kazan **4(30)** 497-502
[9] Sakhapov R L, Gabdullin T R and Zagretdinov R V 2014 the Modeling of control systems of road-building equipment IzvestiyaSamarskogonauchnogocentra RAN Samara 16 **1(2)** 394-396
[10] Sakhapov R L, Makhmutov M M , Zemdikhanov M M and Fazylov M A 2016 Calculation device for demeterova road marking Izvestiya KGASU Kazan **3** 250-254
[11] Silyanov V V and Sapegin L N 1973 Modeling algorithm for the study on EVM impact of the road conditions and means of regulation on the flow of cars Trudy MADI, **52** 139-148
[12] Sitnikov Y M, Devochkin O A and Silyanov V V 1975 Stage-by-stage improvement of transport and operational qualities of roads M Transport
[13] Semeykin N P and Pogosov V V 2004 Survey of roads with the georadar method M : Mir dorog;
[14] Sakhapov R L, Nikolaeva R V, Gatiyatullin M H and Makhmutov M M 2016 Risk management model in road transport systems // *Journal of Physics: Conference Series* **738(1)** 012008
[15] Sakhapov R L, Nikolaeva R V, Gatiyatullin M H and Makhmutov M M 2016 Modeling the dynamics of the chassis of construction machines *Journal of Physics: Conference Series* **738(1)** 012119

IOP Conf. Series: Materials Science and Engineering **786** (2020) 012024 doi:10.1088/1757-899X/786/1/012024

A Study of the Characteristics of Electric Fuel Pumps of Automobile Engines

A Gritsenko[1,*], V Shepelev[1], K Shubenkova[2], M M Zemdikhanov[3] and M H Gatiyatullin[3]

[1]South Ural State University, Lenina avenue, 76, Chelyabinsk, Russia

[2]Almaty Management University, Rozybakiyeva st., 227, Almaty, Kazakhstan

[3] Kazan State University of Architecture and Engineering, 420043, Kazan, Zelenaya str., 1, Russia

*E-mail: alexgrits13@mail.ru

Abstract. The article considers theoretical and experimental research of the fuel system of motor-and-tractor engines: the change in the current consumption of an electric fuel pump (EFP) from the degree of contamination ε of series elements in the system and leakage in the injection unit of the EFP. The environmental friendliness and efficiency during operation of mobile energy resources is put first. According to some studies, currently, there are 43.5 millions of passenger cars in Russia, 29.2% of which do not comply with EURO-2 toxicity standards. Among light commercial vehicles, 42.9% do not comply with EURO-2 toxicity standards. Among trucks, 62.9% of vehicles neither comply with EURO-2 toxicity standards. The electric fuel pump combines electronic and hydraulic elements. According to the statistics of failures, most of them (up to 35%) fall at the electric motor. The main factors affecting failures of EFPs are: increased dustiness and humidity, water ingress into fuel, violation of maintenance intervals, etc.

1. Introduction

Currently, the environmental friendliness during operation of mobile energy means (MEM) is put first [1]. According to some studies, currently, there are 43.5 millions of passenger cars in Russia, 29.2% of which do not comply with EURO-2 toxicity standards. At the same time, 16.4% of the car fleet meet EURO-5 and higher standards. As of the beginning of this year, there were 4.1 millions of cars in the segment of light commercial vehicles, 42.9% of which do not comply with EURO-2 toxicity standards, but only 3.1% of the car fleet meet EURO-5 standard [2]. There are 3.8 millions of registered trucks, 62.9% of which neither comply with EURO-2 toxicity standards, and only 4.4% of the fleet meet EURO-5 and higher standards.

Modern fuel supply systems for motor-and-tractor equipment are complex hydraulic, mechanical and electronic systems [3, 4, 5]. The electric fuel pump combines electronic and hydraulic elements [6]. According to the statistics of failures, most of them (up to 35%) fall at the electric motor [7, 8]. The main factors affecting failures of EFPs are: increased dustiness and humidity, water ingress into fuel, violation of maintenance intervals, etc. [9, 10]. The basis for maintaining the efficiency of the fuel supply system and the EFP in particular is test diagnostics. Taking this into consideration: the purpose of the research is to ensure the efficiency of electric fuel pumps for motor-and-tractor engines through the use of the technologies for test diagnostics of their technical condition.

2. Theoretical Research

The entire fuel system has various series and parallel hydraulic resistances, for example, the passage area changes on the coarse and fine filter depending on the degree of clogging of the filters [5], the fuel supply line, as well as leakages in the EFP injection unit [8]. It follows from the energy balance that the total pressure loss is equal to the sum of the pressure losses in all the sections connected in series [11]. The fuel supply process was divided into several sections (A, B, C) allowing us to study the operation of all the elements of the power supply system: 1. Section (A) – the fuel movement from the tank to the injection unit of the EFP; 2. Section (B) - the fuel movement from the injection unit of the EFP to the fuel filter; 3. Section (C) - the fuel movement from the fuel filter to the fuel rail. Taking this into consideration, we will write:

$$\Delta P_A + \Delta P_B + \Delta P_C = \sum_A^C P \tag{1}$$

where ΔP_A, ΔP_B, ΔP_C are the pressure drops in the sections A, B and C, Pa, respectively.

Any changes in the structural parameters (resistance, leakage in the fuel system) will lead to changes in the parameters of the electric fuel pump: current I, A, and supply voltage U, B. These parameters are indirect parameters of the structural ones: pressure in the fuel system P, kPa, and fuel supply Q, l/h [8, 12]. The dependence should be considered in the form of an equation of the energy balance of the EFP current source ETN (E_{del}) with the energy of the fluid movement in the fuel supply line (E_{trans}):

Let us consider the energy balance of EFP, J:

$$E_{del} = E_{trans} \tag{2}$$

where E_{del} is the energy delivered to the EFP, J; E_{trans} is the energy transmitted to the fuel supply line, J.

Let us present the left and the right sides in an expanded form:

$$I \cdot U \cdot t = \frac{U^2}{R} \cdot t + \frac{\Delta P}{\rho \cdot g} \cdot t + \mu \cdot S_{tot} \cdot \sqrt{\frac{2 \cdot \Delta P^3}{\rho}} \cdot t \tag{3}$$

where I is the current consumption of the EFP, A; U is the supply voltage of EFP, V; t is the pump operation time, sec; R is the resistance of the EFP, Ohm; ΔP is the pressure drop at the section, kPa; μ is the fuel consumption ratio; S_{tot} is the total passage area of the fuel supply line, which depends on the type of malfunction (clogging of the series elements or leakage in the injection unit of the EFP), m^2; ρ is the density, kg/m^3; g is the acceleration of gravity, m/s^2.

The total passage area of the fuel supply line when the series elements of the fuel system are contaminated is determined by the formula, m^2:

$$\mu \cdot S_{tot} = \mu \cdot S_1 + \mu \cdot S_2 \tag{4}$$

where S_1 is the passage area of the first clogged element, m^2; S_2 is the passage area of the second clogged element, m^2.

The total passage area of the fuel supply line in case of leakages in the EFP is determined by the formula, m^2:

$$\frac{1}{\mu \cdot S_{tot}} = \frac{1}{\mu \cdot S_1^{'}} + \frac{1}{\mu \cdot S_2^{'}} \tag{5}$$

where $S_1^{'}$ is the passage area of the fuel supply line from the pressure side, m^2; $S_2^{'}$ is the passage area of the leakages, m^2.

Let us equate the left and the right sides of equation (2); reducing by the operation time t in formula (3), we will express the current consumption of the EFP I from it, A:

$$I = \frac{\frac{U^2}{R} + \frac{\Delta P}{\rho \cdot g} + \mu \cdot S_{tot} \cdot \sqrt{\frac{2 \cdot \Delta P^3}{\rho}}}{U} \tag{6}$$

Thus, we established a theoretical dependence of the current consumption of the EFP on the electrical parameters of the drive motor (U, R) and the technical condition of the main EFP elements ($\mu \cdot S_{tot}$, ΔP).

Let us consider the equation of the dependence of the EFP power consumption for hydraulic losses on the electrical parameters of the drive motor and the technical condition of the main EFP elements [13, 14, 15]:

$$N_{hl} = I_{EFP} \cdot U_{EFP} - N_{EFP}^{эel.losses} - \mu \cdot S_C \cdot \sqrt{\frac{2 \cdot \Delta P_{FS}^3}{\rho_f}} \tag{7}$$

where I_{EFP} is the pump's current consumption, A; U_{EFP} is the pump's voltage consumption, V; $N_{EFP}^{el.losses}$ is the power of losses in the EFP electrical circuits, W; $N_{EFP}^{hyd.losses}$ is the power of hydraulic losses, W, ΔP_{FS} is the pressure drop in the fuel supply system, kPa; μ is the fuel consumption ratio, 0.8; S_c is the clearance area in the fuel supply system, or between the EFP housing and rollers, m^2; ρ_f is the fuel density, kg/m^3.

Substituting the values of the studied gasoline feed pump into equation (7), we obtained the dependencies of the power consumption of the EFP for hydraulic losses on the electrical parameters of the drive motor and the technical condition of the main EFP elements (Figure 1 (a, b) and Figure 2 (a, b))

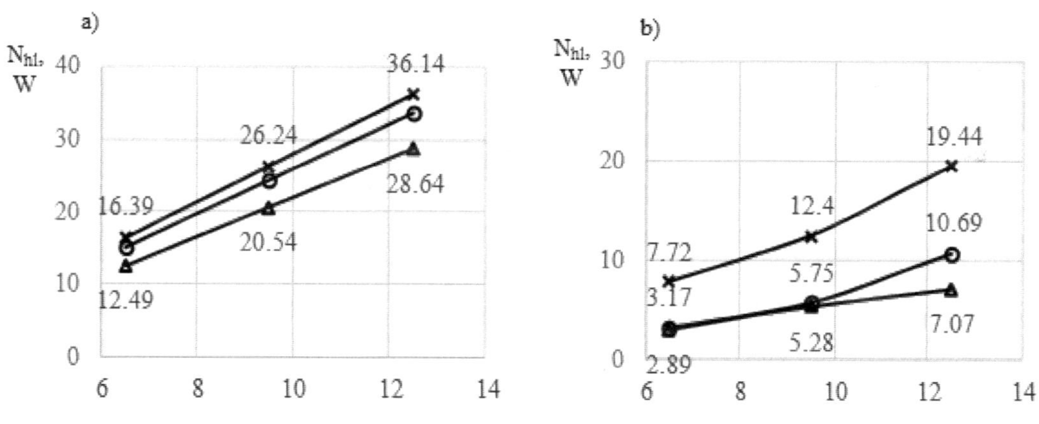

Figure 1. Power consumption for hydraulic losses due to an increase in the area of fuel leakages inside the gasoline pump through radial and end clearances:
a) S_C=0.00000001539 m^2; I=const; b) S_C=0.00000113 m^2; I=const.

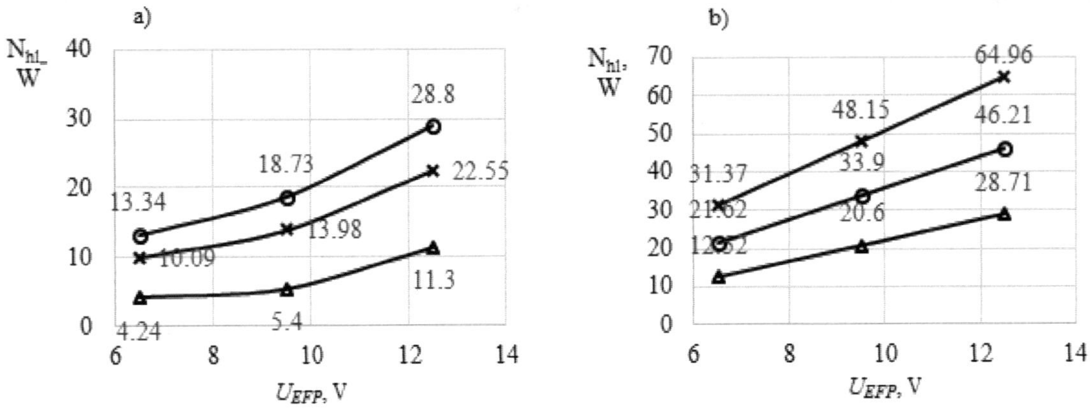

Figure 2. Power consumption for hydraulic losses when the clearance area of the flow passage of the fuel supply line is reduced:
a) S_C=0.00003847 m^2; *I*=const; b) S_C=0.00000001539 m^2; *I*=const.

Thus, with an increase in the clearance area between the EFP housing and rollers in Figure 1 (a, b) S_C=0.00000001539 m^2 to 0.00000113 m^2 the power consumption of the EFP is 2-4 times decreased, which is associated with hydraulic losses, namely N_{hl} =28.64-36.14 W to 7.07-19.44 W. Due to large clearances between the EFP rollers and housing, the pump body loses traction, the fuel is weakly injected into the line and returns to the inlet cavity [7].

Due to a decrease in the clearance area of the flow passage of the fuel supply line in Figure 2 (a, b) S_C=0.00003847 m^2 to 0.00000001539 m^2, the power consumption of the EFP is 2-2.5 times increased, which is associated with hydraulic losses, namely with N_{hl}=11.3-28.8 W to 28.71-64.96 W [18, 19]. Due to the increase in the fuel supply line resistance, which is equal to clogging of the fuel filters and the fuel wire, the fuel poorly enters the fuel line.

3. The Results of Experimental Research

This research enables to control the technical condition of the fuel supply system elements - clogging of the fuel filters and the fuel supply line and wear of the EFP injection unit. In case of an abnormal operation of the fuel pump, leakage in its injection unit, contamination of the series elements of the fuel system, the necessary fuel supply to the engine injectors is not ensured [9], therefore, the stability of its operation [13, 14] is violated. Besides, the operation of the EFP in these conditions is accompanied by overheating and, ultimately, seizure, as a result of a severe contamination on the fuel induction through the coarse filter.

We carried out experimental research and a subsequent data analysis simulating failures in the fuel system. The simulation results are presented in Figs. 3, 4, 5.

When analyzing the data in Figure 3, it has been established that with a decrease in the diameter of the fuel supply line section of elements contaminated in series (for example, clogging of the fine fuel filter, the fuel supply line), the current consumption level increases. The research was carried out at a decrease in the supply voltage of the EFP in the range from 12.5 to 6.5 V.

An increase in the current consumption of the EFP is connected with an increase in the hydraulic resistance in the fuel system, for this reason the EFP rotor rotates under a considerably larger load, which results in the absorption of more energy.

Moreover, it is clear from Figure 3 that a significant increase in the current rate is observed from an equivalent section of 1 mm. The larger the supply voltage of the EFP, the more the current flows in its windings.

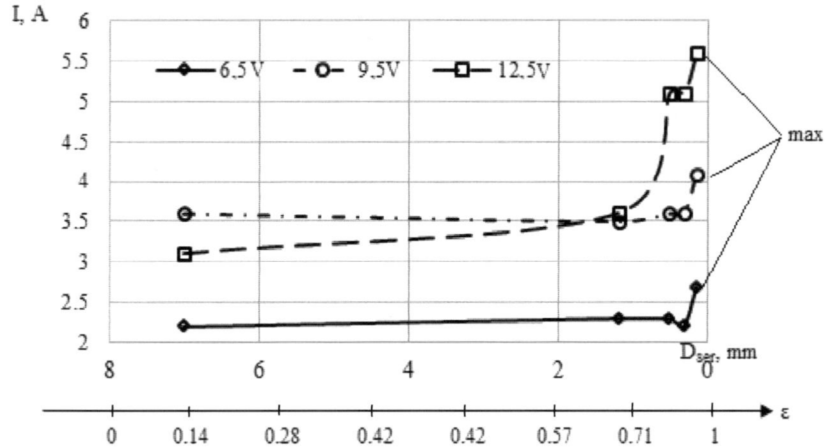

Figure 3. Regularities of changes in the current rate I, A when simulating clogging of fuel elements D_{ser}, mm (equivalent section diameter).

Let us analyze the data presented in Figure 4.

When analyzing the data in Figure 4, it has been established that with an increase in the diameter of leakages in the EFP pump body, the current consumption noticeably decreases. Due to the presence of clearances and leaks in the EFP injection unit, the fuel does not flow into the pumping cavity of the pump, as a result of which the rotor rotates freely, the consumed energy decreases with an increase in the equivalent leakage diameter.

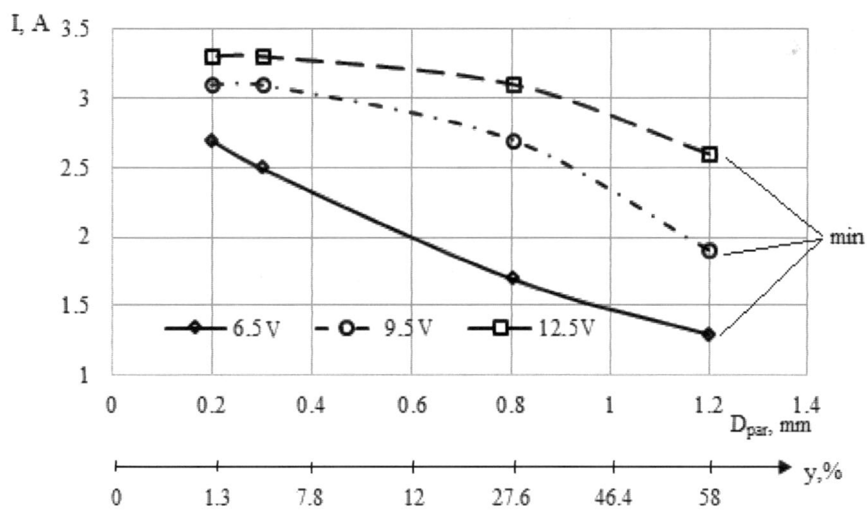

Figure 4. Regularities of changes in the current rate I, A when simulating leakages y, % of the EFP injection unit.

Figure 5 shows the dependence of the change in the current consumption of the EFP at simultaneous leakages (y, %) and contamination in the fuel system ε.

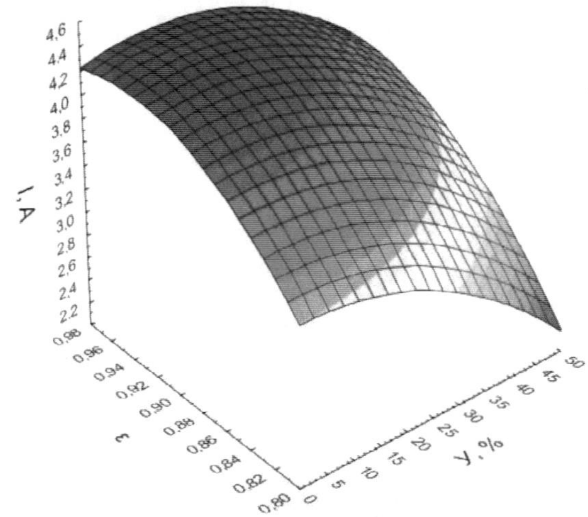

Figure 5. Regularities of changes in the current rate *I*, A at simultaneous leakages *y*, % in the EFP injection unit and contamination of series elements of the fuel system *ε*.

An analysis of Figure 5 shows that the current consumption increased to the maximum of 4.2 A with an increase in the contamination *ε* to 0.96, and decreased to 2.2 A with an increase in leakage to 50%.

4. Conclusions

In the experimental part, we studied the dependence of the current consumption of the EFP on the electrical parameters of the drive motor (*U, R*) and the technical condition of the main EFP elements ($\mu \cdot S_{tot}$, ΔP). Due to the research, we obtained graphical dependencies of the current rate on:

1) contamination of series elements of the fuel system *ε*, for example, fuel filters, fuel supply line;
2) fuel leakage through the EFP injection unit *y*, %;
3) simultaneous action of the contamination and leakage. The research has shown that with an increase in the contamination of the fuel system (Figure 3), the current consumption increases, moreover, it doubles to 6 A if the fuel supply line is almost completely blocked. In case of increasing leakages (Figure 4), the current rate decreased to the minimum value of 1.3 A at the maximum leakage of 58%. In case of a simultaneous action of the contamination and leakage (Figure 5), the indicators change as follows: the current consumption rises to the maximum value of 4.2 A with an increase in the contamination *ε* to 0.96, and decreases to 2.2 A with an increase in leakages to 50%.

References

[1] A Brown and D Kotori 2013 Dynamic vehicle powertrain model development and hardware-in-the-loop simulator for developing and measuring fuel efficient CO2 reducing technologies, SAE Technical Papers 2

[2] V N Kozhanov, A A Petelin, A V Gritsenk, V D Shepelev 2018 Reducing the toxicity of exhaust gases of a diesel engine by turning off part of its cylinders, Bulletin of SUSU (Series: Engineering) **18(2)** 34-44

[3] P Krause 2017 Periodical changes in fuel spray microstructure during injection for two fuel supply methods to standard fuel injectors in diesel engine, International Multidisciplinary Scientific GeoConference Surveying Geology and Mining Ecology Management **17(52)** 41-8

[4] M De Cesare, M Parotto, F Covassin, S Sgatti, Electric low pressure fuel pump control for fuel saving, SAE Technical Papers **2** (2013)

[5] F Stola, M De Cesare, L Lacchini, N Cavina, S Sohal 2015 Diesel exhaust fluid (DEF) supply system modelling for control and diagnosis applications, SAE International Journal of Engines **8(1)** 350-60
[6] F Stola, D Paolino, M Parotto, F Troina 2015 Electric feed pump simulation and control for fuel saving and system cost reduction, SAE Technical Papers
[7] A V Gritsenko, K V Glemba , D B Vlasov 2018 Detection of latent failures of electric fuel pumps of mobile energy in agriculture by the method of test diagnosis, Electric equipment: operation and repair **11** 50-7
[8] A M Plaksin, A V Gritsenko, M V Zapevalov, N V Kostyuchenkov 2018 Production Potential of Agriculture: Development Stages, State, Problems of Modernization, Chelyabinsk,
[9] P W Hou, T W Nichols, K T Miller, J J Bennett 2008 A bench test study of port fuel injection fouling, SAE Technical Papers
[10] F F Rovai, D K Tanaka, A Sinatora 2005 Wear and corrosion evaluation of electric fuel pumps with ethanol/gasoline blends, SAE Technical Papers
[11] R Ding, L Xiao, X Jin 2019 Robust Control for Electric Fuel Pump with Variant Nonlinear Loads Based on a New Combined Sliding Mode Surface, International Journal of Control, Automation and Systems **17(3)** 716-28
[12] S Som, A I Ramirez, D E Longman, S K Aggarwal 2011, Effect of nozzle orifice geometry on spray, combustion, and emission characteristics under diesel engine conditions, Fuel **90(3),** 1267-1276
[13] O Ivanov, R Kharak, O Kostenko, V Arendarenko, O Nazarenko, A Pushka, V Sarana 2019 Estimation model of the diesel engine fuel system with an electromechanical device to intensify fuel supply, Eastern-European Journal of Enterprise Technologies, **1** 50-9
[14] F Luo, C Wang, F Xue, B Ye, X Wu 2016 A study on the influence of fuel pipe on fuel injection characteristics of each nozzle hole in diesel injector, MATEC Web of Conferences 40

Features of formation of gas-thermal coatings with equal thickness on flat surfaces

U I Belenkiy, V I Kretinin, V A Sokolova, A V Teppoev, V A Markov and G K Parfenopulo

Saint Petersburg State Forest Technical University, 5, Institutskiy lane, Saint-Petersburg, 194021, Russia

Abstract. The article deals with the formation of uniform gas thermal coatings on flat surfaces. It is substantiated that gas-thermal coatings are formed due to the superposition of single wavy stripes. The technique of obtaining uniform thermal coatings depends on spraying modes. The dependences of the speed of movement and supply of a gas-flame burner on the spraying distance are theoretically substantiated and experimentally confirmed. The dependence of coating thickness on the speed of a spraying device was determined.

1. Introduction

One of the most effective ways to increase the durability of cutting working bodies of forestry machines interacting with the soil and tree-shrub vegetation is to use bimetallic working parts allowing for the effect of self-sharpening. The service life of these parts increases 2-3-fold [1, 2, 3, 4, 7, 9, 12].

The uneven coating thickness of the soil-cutting parts during the hardening of working bodies of forestry machines has an extremely negative effect on the stabilization and shaping of the blade profile which causes wear unevenness. When hardening soil-cutting parts by gas-flame spraying, it is necessary to evaluate the effect of changes in the technological modes of spraying on the equal thickness of coatings and determine their allowable tolerance field.

During the movement of a gas-flame burner on the part surface, the coating becomes wavy. The uneven thickness is due to the fact that the distribution of powder particles in the torch obeys the normal law, and the cross section of the deposited spot is described by the Gauss equation [5, 6, 11, 13]. As a result, the coating undulation is formed. When spraying flame on a flat surface, the coating is formed when the gun is moved along this surface with a speed υ. At the end of movement, the pistol moves to the supply S and moves in the opposite direction. The coating is formed by the superposition of one pass over another (Figure 1). The uniform thickness depends on the speed of movement of the gun and supply. [8, 14, 15]. Therefore, it is necessary to determine technological regimes of the spraying process in which the coating will be uniform in thickness, and the required thickness will be obtained for a minimum number of passes.

These modes are provided by setting appropriate values for flat surfaces: movement speed υ and spraying supply S.

The purpose is to identify and justify the algorithm for determining technological modes of flame spraying to form coatings of uniform thickness.

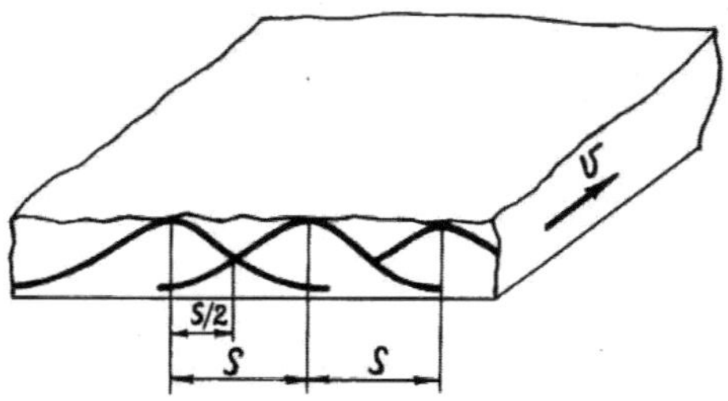

Figure 1. The formation of a coating on a flat surface.

2. Research method

With a straight movement of the sprayer relative to the substrate, a single coating band is formed on the surface. Its cross section thickness obeys the normal distribution law:

$$y_x = y_0 \exp(-\frac{x^2}{2\sigma^2}) \tag{1}$$

where y_0 and y_x – thickness of a single strip in the center of the spraying spot and at a distance *X* from the center; σ^2 – dispersion. It was recommended to use the term "dispersion radius" $\rho=\sigma$ – standard deviation.

The shape of a jet and the nature of distribution of sprayed particles over the spot cross section can be represented as Figure 2 with a dispersion field of random value y_x equal to 6 ρ.

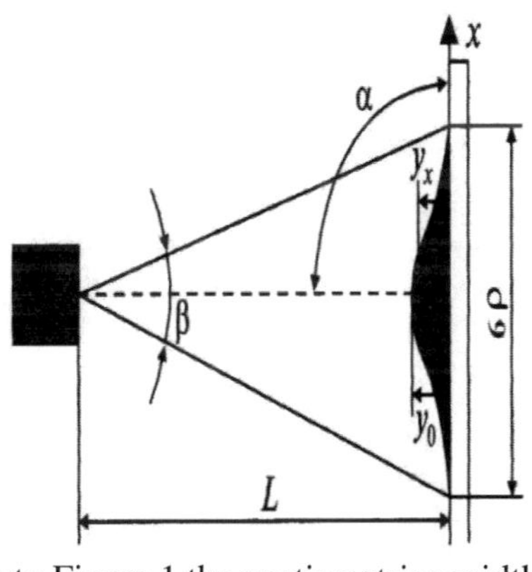

Figure 2. The shape of a jet and distribution of sprayed particles: β – jet opening angle, α –spraying angle, *L* – spraying distance, ρ – radius of dispersion.

According to Figure 1 the coating stripe width S (spraying supply) can be identified according to the Gauss law as a dispersion field of a random variable:

$$S=3\rho \tag{2}$$

where: ρ – dispersion radius.

The dispersion radius is determined by formula:

$$\rho = 1/3\, L\mathrm{tg}(\beta/2) \tag{3}$$

Then, from (2):

$$S=L\mathrm{tg}(\beta/2) \tag{4}$$

To prove expression (1), single strips of gas-flame coating were obtained at different spraying distances (Figure 3). A single strip was formed using an GN-2 gas burner. The material for spraying was a nickel-based PR-H70X17C4R4 self-fluxed alloy with a fraction of 100 μm.

Analysis of the spraying spot cross section (Figure 3), and measurements of the curve coordinates (Figure 4), (Table 1) confirm that the cross section of the spray spot is described by the Gauss equation:

$$y = y_0\, e^{-Kx^2} \tag{5}$$

where y_0 – coating thickness at $x = 0$, *K* – coefficient depending on the spraying distance.

IOP Conf. Series: Materials Science and Engineering **786** (2020) 012025 doi:10.1088/1757-899X/786/1/012025

Figure 3. Spray spot cross-section by the coating thickness: 1) L = 170 mm; 2) L = 200 mm.

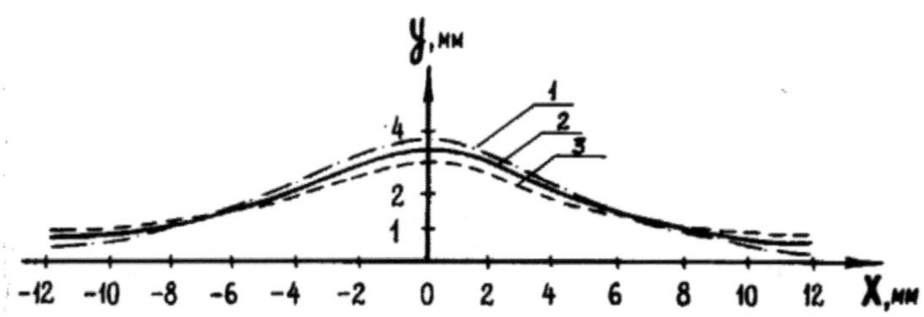

Figure 4. Spray spot cross section curves: Spraying duration t = 90 s; spraying distance L: 1) 140 mm, 2) 170 mm, 3) 200 mm.

Table 1. The results of measurement of the coordinates X and Y and the spraying spot cross section.

The distance from the axis of the spraying spot, X, mm	Spraying distance – L, mm								
	140			170			200		
	Coating thickness – Y, mm								
	1	2	3	1	2	3	1	2	3
-12	0.41	0.43	0.42	0.48	0.51	0.46	0.62	0.57	0.60
-10	0.69	0.68	0.73	0.82	0.84	0.77	0.87	0.93	0.91
-8	1.17	1.15	1.19	1.1	1.13	1.16	1.17	1.15	1.14
-6	1.69	1.62	1.65	1.58	1.55	1.53	1.48	1.46	1.5
-4	2.2	2.23	2.19	2.06	2.04	2.03	1.89	1.94	1.92
-2	2.99	2.97	3.01	2.73	2.72	2.73	2.51	2.49	2.53
0	3.43	3.37	3.39	3.13	3.18	3.15	2.93	2.87	2.89
2	3.02	3.05	2.97	2.74	2.79	2.75	2.5	2.52	2.48
4	2.24	2.21	2.17	2.06	2.02	2.05	1.93	1.90	1.88
6	1.63	1.68	1.65	1.56	1.58	1.61	1.47	1.49	1.51
8	1.18	1.16	1.20	1.19	1.16	1.18	1.14	1.11	1.16
10	0.67	0.7	0.71	0.81	0.85	0.78	0.92	0.94	0.86
12	0.45	0.39	0.40	0.48	0.53	0.51	0.59	0.61	0.63

Table 2. The measurement results for the radius of the t spraying spot and calculation data for the coefficient K depending on the spraying distance.

Spraying distance, L, mm	Spraying spot radius						Coefficient $K, {}^{1}/_{mm^2}$
	R_1	R_2	R_3	R_4	R_5	$\bar{R}$	
140	16	17	16	15	16	16,0	0,028
170	18	18	18	19	19	18,6	0,024
200	21	20	20	21	21	20,6	0,02
230	22	23	22	24	22	22,6	0,016

The y and x coordinates were measured using an MIM-7 instrumental microscope.

3. Research results

The study of the dependence of the spot radius (Table 2) on the spraying distance shows that with its increase, the radius increases, and the coating thickness at $x = 0$ decreases (Figure 4). Therefore, coefficient K which takes into account this dependence should be entered into equation (5). This coefficient characterizes the angle of jet opening β, and determines design features of spraying devices.

In general, the coating thickness at any point is determined by formula:

$$h = y_0 e^{-Kx^2} + y_0 e^{-K(x-S)^2} + \dots + y_0 e^{-K(x-(n-1)S)^2} \quad (6)$$

where S – spraying supply, K – coefficient of layering of single bands depending on the spraying distance.

The coating thickness at points with coordinates that are multiples of the spraying supply will be considered as a superposition of two passes (Figure 1).

Then between the passes, the thickness is a sum of two members at $x=S/2$:

$$h = 2y_0 e^{-\frac{KS^2}{4}} \quad (7)$$

On the other hand, the total thickness of the coating should be equal toy_0:

$$y_0 = 2y_0 e^{-\frac{KS^2}{4}} \quad (8)$$

Transforming equation (8), we have an expression to determine supply S:

$$S = 2\sqrt{\frac{ln2}{K}} = \frac{1,665}{\sqrt{K}} \quad (9)$$

Thus, calculation if supply S of the spraying gun is reduced to calculating coefficient K. Let us transform expression (5):

$$K = \frac{lny_0 - lny}{x^2} \quad (10)$$

where y, x current coordinates of the spraying spot, mm.

Using Table 1 and Table 2, the dependences of K (Figure 5) and S (Figure 6) on the spraying distance were constructed.

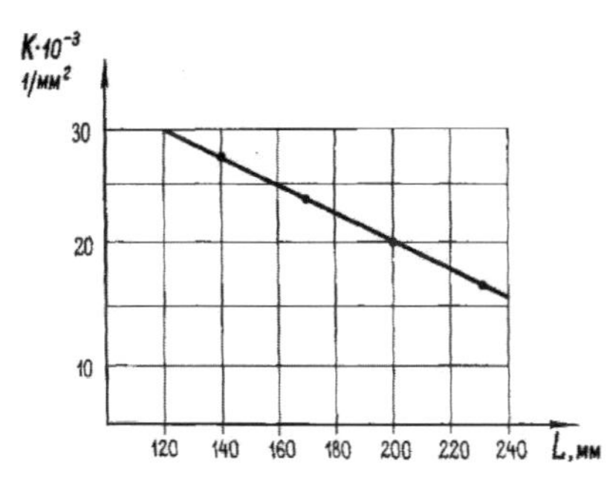

Figure 5. Dependence of coefficient K on the spraying distance.

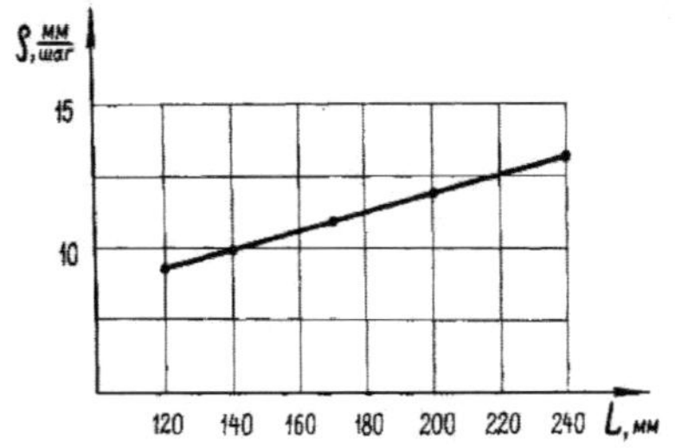

Figure 6. Dependence of supply S on the spraying distance.

To determine the speed of movement of the spray device, it is necessary to determine its relationship with the thickness of a single coating strip.

Knowing the performance of the spray device and the spraying time, let us determine the mass of the sprayed layer:

$$m = G_p \cdot K_{mu} \cdot t \quad (11)$$

where G_p – spraying performance, K_{mu} – material use coefficient, t – spraying time.

The mass of the sprayed layer can be expressed in terms of density of the layer material γ_l and its volume V_l:

$$m = \gamma_l \cdot V_l \quad (12)$$

The volume of a single strip of coverage is equal to:

$$V_l = F \cdot \upsilon \cdot t \quad (13)$$

where F – area under the curve of normal distribution, υ - speed of movement of the spray spot, t – spraying time:

$$F = y_0 \int_{-\infty}^{+\infty} \exp\left(-\frac{x}{S}\right) dx = y_0 \cdot S \cdot \sqrt{\pi} \quad (14)$$

From (11-14), the thickness of the single band y_0 can be determined

$$y_0 = \frac{G_p \cdot K_{mu}}{\upsilon \cdot \gamma_l \cdot S \cdot \sqrt{\pi}} \quad (15)$$

From (13), the speed of movement of the spraying device can be determined

$$\upsilon = \frac{G_p \cdot K_{mu}}{y_0 \cdot \gamma_l \cdot S \cdot \sqrt{\pi}} \quad (16)$$

If the capacity of the device is known, by setting the spraying distance and the coating thickness, it is possible to determine the speed of movement. The gas-flame spraying capacity is 0.00055 kg/s.

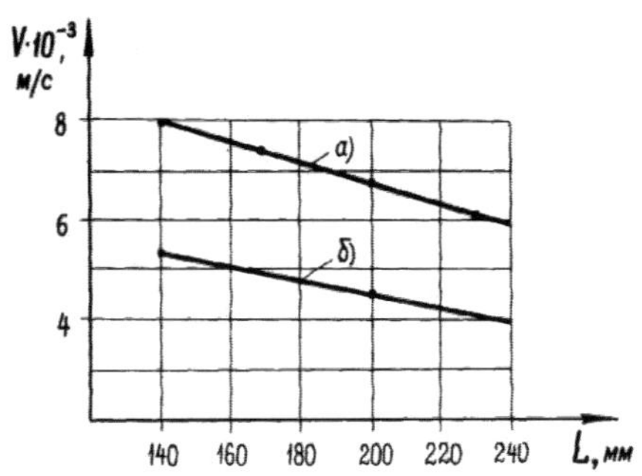

Figure 7. Dependence of the speed of movement of the gun on the spraying distance at the coating thickness: a) h = 1mm, b) h = 1.5 mm.

4. Conclusion

As can be seen from Figure 7, an increase in the spraying distance causes a decrease in the speed of movement of the gun due to a change in the maximum thickness of the spraying spot.

The modes of application of wear-resistant coatings from self-fluxing alloys depend on the design of the apparatus used, the type and granulation of the powder, and other factors. This creates serious obstacles for the current control of dimensions of sprayed coatings. The required thickness can be obtained by stabilizing or regulating technological parameters. Thus, the dependences help calculate nominal values and the control range for technological parameters of the shape control according to specified requirements for the average thickness, the coating waviness and the spraying mode.

References

[1] Turichin G A, Zemlyakov E V, Pozdeeva E Y, Tuominen J and Vuoristo P 2012 Technological possibilities of laser cladding with the help of powerful fiber lasers Metal Science and Heat Treatment **54(3-4)**:139-44 DOI: 10 1007/s11041-012-9470-y

[2] Kretinin V I , Shariga A D , Dyachok N G , Zakhartsov G S and Shentsov A I 2004 Improvement in technology for sorting the sintering product

[3] Kondrateva A S, Bespalova P G, Filatov L A, Tanklevskaya E M, Pavlov S I and Alexandrov S E Gas-sensitive properties of thin nickel oxide films *Russian Journal of Applied Chem-istry* 2017 Jun 1;**90(6)**, 846-52 DOI: 10 1134/S1070427217060039

[4] Vasilyev A A, Kolbasnikov N G, Rudskoy A I, Sokolov D F and Sokolov S F 2017 Kinetics of Structure Formation in the Heating of Cold-Rolled Automotive Steel Sheet Steel in Translation **47(12)** 830-8 DOI: 10 3103/S0967091217120130

[5] Sokolov D F, Vasil'ev A A, Kolbasnikov N G and Sokolov S F 2011 Influence of dynamic austenite decomposition on the microstructure and mechanical properties of pipe steel Steel in Translation **41(4)** 351-5 DOI: 10 3103/S096709121104022X

[6] Ogoltcov A, Sokolov D, Sokolov S and Vasilyev A 2017 Practical Use of Computer Model STAN 2000 for Improvement and Creation of Regimes for Hot Rolling of Steels on SEV-ERSTAL Mill 2000 InMaterials Science Forum **879**, 2543-8 Trans Tech Publications DOI: 10 4028/www scientific net/MSF 879 2543

[7] Zhitelev P, Vasilyev A, Sokolov S, Sokolov D and Paligin R 2016 Investigation and Modeling of Recrystallization of Cold Rolled Automotive Steels InIOP Conference Series: Materials Science and Engineering Apr (Vol **124(1)** 012112) IOP Publishing DOI: 10 1088/1757-899X/124/1/012112

[8] Budelovskiy D I, Petrovich S Y, Lipin V A and Andreeva, V D Mechanical 2016 alloying of powder alloy Al - Si - Ni with nanosized carbon additions Tsvetnye Metally **9(77-82)** DOI: 10 17580/tsm 2016 09 11

[9] Sokolova V A , Gryazkin A V , Beliaeva N V , Petrik V V and Smirnov, A P Bipolar 2018 electrokinetic dehydration of wood by electro-osmosis of various breeds Thermal Science

[10] Markov V , Sokolova V , Rzhavtsev A and Eliseev I 2019
Research of factors affecting thickness of the covering obtained by electro contact sintering of composite materials Journal of Physics: Conference Series

[11] Kalinin G Y, Mushnikova S Y, Fomina O V, Teplenicheva A S, Zotov O G and Sokolov S F 2011 Physical simulation of thermomechanical treatment of a high-strength nitrogen-bearing austenitic steel and its effect on the microstructure and properties of the steel Russian Metallurgy (Metally) Mar 1;2011(3):202-8. DOI: 10.1134/S0036029511030086

[12] Puzryakov A F 2003 Theoretical foundations of plasma spraying. (Moscow: MSTU) 358 p.

[13] Makeenko A, Larionova T, Klimova O, Galkin V, Starykh R and Tolochko O 2017 Synthesis of Gadolinium-Yttrium-Aluminum Garnet Activated with Cerium by Spraying High Aqueous Salt Solutions. Key Engineering Materials. Jan 28;721. DOI: 10.4028/www.scientific.net/KEM.721.267

[14] Koltsova T S, Nasibulin A G, Shamshurin A, Shakhov F, Mikhailov V 2015 Hybrid Aluminum Composite Materials Based on Carbon Nanostructures. Materials Science Jan 1;21(3):372-5 DOI: 10 5755/j01.ms.21.3.7355

IOP Conf. Series: Materials Science and Engineering **786** (2020) 012025 doi:10.1088/1757-899X/786/1/012025

[15] Chubinskii A N , Tambi A A , Teppoev A V , Semishkur S O and Bakhshieva M A 2014 Physical nondestructive methods for the testing and evaluation of the structure of wood-based materials. Russian Journal of Nondestructive Testing,
[16] Ged'O V M , Markov V A , Sokolova V A , Parfenopolo G K and Chernyh L G 2018 Research of composite coverings wear resistance for forestry machines by electro contact sintering IOP Conference Series: Materials Science and Engineering,

Influence of the tilt of the sprung elements of the construction of car cranes on cargo stability

O N Mekhonin[1,*], R V Shchetkin [1] and K G Pugin [1,2]

[1] Federal State Budgetary Educational Institution of Higher Education «Perm National Research Polytechnic University», Perm, Russia

[2] Federal State Budgetary Educational Institution of Higher Education «Perm State Agro-Technological University named after Academician D.N. Pryanishnikov», Perm, Russia

[*]E-mail: onm13@yandex.ru

Abstract. This article describes the key features of determining the coefficient of cargo stability of automotive cranes on the base chassis, taking into account the roll of sprung structural elements. The possible factors affecting the error of the cargo stability calculation are identified, and the ways of solving this problem are proposed.

1. Introduction

The main requirement for lifting machines is their safety. That is why in the design and manufacture of lifting machines special attention is given to this issue. Automobile crane is a hoisting machine consisting of a base chassis or vehicle (car) and a crane-manipulator unit (CMU). The specificity of automotive cranes is that in most cases their design and production is carried out under the individual requirements of a particular customer, and the number of combinations of different models of CMU, basic vehicles or chassis reaches several hundred thousand. The possibility of manufacturing automobile cranemanipulator with one or another of the CMU, on the other, or other automobile chassis or vehicle is determined by calculation on the basis of allowable loads on the axle of the vehicle and conditions of the cargo sustainability. [3]. And each case requires a theoretical study of the product, sometimes taking considerable time from the overall manufacturing process. That is why it is especially important to use a calculation method that allows you to reflect accurately the technical features of the designed unit of equipment. After all, in modern production conditions, much attention is paid to the consumption of materials used, as well as the amount of total labor and energy costs.

2. Determination of the coefficient of cargo stability of automotive cranes.

Here and further, the term cargo stability is understood as the ability of an automobile crane-manipulator to counteract the tipping moments that arise when lifting cargo. One of the main indicators both on safety, and on use of technical potential of load-lifting equipment is the coefficient of cargo stability determined by the equation 1 [2].

$$K_s = \frac{M_{hf}}{M_{op}} \geq 1{,}15 \quad (1)$$

where M_{hf} – moment of holding forces, M_{of} – moment of overturning forces.

However, it should be noted that the current method of calculating the cargo stability does not take into account the design features of the applied suspension of the basic chassis of automobile cranes. In automotive cranes, unlike cranes, there is a distinctive feature – they are not fully raised on the retractable hydraulic supports, and retractable supports are raised only until the unloading of the chassis suspension, while the wheels remain in contact with the support surface of the ground. Therefore, when lifting the load, there is often a roll of the truck crane in the direction of lifting the load as a result of compression of the elastic suspension elements of the base chassis (Figure1).

Figure 1. Roll of the automobile crane-manipulator with separation of the front left retractable hydraulic support when carrying out tests on cargo stability.

In the current method of calculation in the construction of the support contour of the crane is taken over the passage of the tipping ribs through the elastic suspension elements (springs), it does not take into account the possible displacement of the tipping ribs as a result of compression of the elastic elements of the suspension and the roll of the sprung structural elements of the base chassis.

3. Research objectives and the scheme of the experiment.

The authors of this study set the task to analyze the possible effect of the displacement of the edges of the truck crane rollover, which occurs due to the compression of the elastic suspension elements of the base chassis when lifting cargo. For this purpose, on the example of a biaxial base chassis with a dependent spring suspension, we consider the possible effect of the displacement of the tilting edge on the final value of the load stability coefficient, since it is the dependent suspension that is the most common for currently produced automotive crane manipulators. For the theoretical calculation of the coefficient of load stability, we construct a scheme of the support circuit of the lifting machine (Figure2).

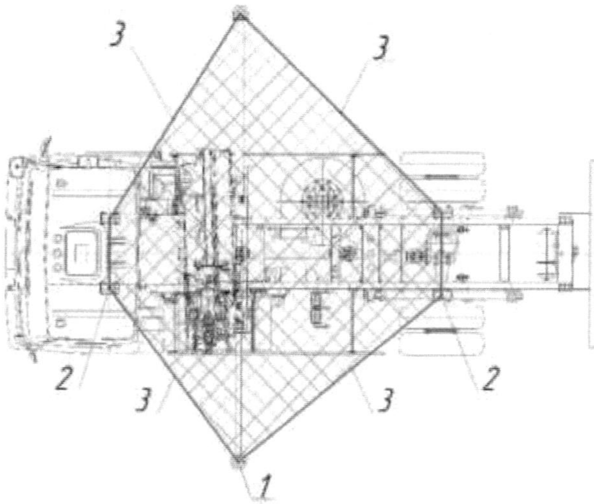

Figure 2. The reference circuit of the automobile crane-manipulator, where:
1 – reference platform outriggers, 2 – bearing point of the spring, 3 – fin rollover, 4 – foot contour of the car.

To conduct a theoretical experiment it can be suggested that when lifting a load the boom position perpendicular with respect to the edges of a rollover occurs the actual offset of this edge to the axis of the CMU (Figure3).

IOP Conf. Series: Materials Science and Engineering **786** (2020) 012026 doi:10.1088/1757-899X/786/1/012026

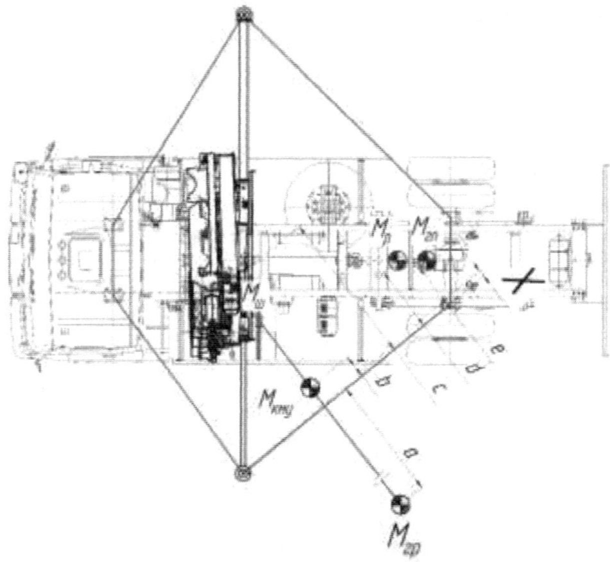

Figure 3. The proposed scheme of the experiment, where X is the displacement of the point of passage of the rib tipping at the point of attachment of spring suspension, [mm]; M_{gr} – coordinate of the center of gravity of the cargo; M_{kmy} – coordinate of the center of gravity of the CMU; M_r – coordinate of the center of gravity of the subframe; M_{gp} – coordinate of the center of gravity of the loading platform; a, b, c, d, e – the distance from the center of gravity of these elements to the edges of a rollover [mm].

As a consequence of the reduced shoulder of the support frame of the crane, and thus decreases and the final value of the factor of cargo sustainability. The displacement value X [mm] is a value that depends on many factors: the rigidity of the base chassis suspension, the dimension and type of tires used, the bending of the subframe of the working equipment and the chassis frame. For determining the ratio of the load resistance for each of the estimated provisions were identified overturning and restraining moments in the following formulas:

$$M_{of} = M_{gr} \cdot a \text{ [t·m]} \quad (2)$$

$$\mathrm{M}_{hf} = \mathrm{M}_{\mathrm{KMY}} \cdot b + \mathrm{M}_{\mathrm{sh}} \cdot \mathrm{c} + \mathrm{M}_p \cdot d + \mathrm{M}_{\mathrm{gr}} \cdot e \text{ [t·m]} \quad (3)$$

The following calculation results were obtained for the automotive crane-manipulator based on KAMAZ 43253 chassis with VM10L74 CMU (see Table 1).

Table 1. Tilting edge offset (X) from 0 to 300 mm.

Offset X, [mm]	Tipping moment, [t·m]	Holding moment, [t·m]	Stability factor	Δ [%]
0	8,68	13,74	1,58	0
50	8,7	13,55	1,56	1,7
100	8,72	13,35	1,53	3,3
150	8,74	13,16	1,5	4,9
200	8,75	12,97	1,48	6,5
250	8,77	12,78	1,45	8
300	8,78	12,59	1,43	9,5

According to the results obtained, the displacement of X in the range of 0-50 mm can increase an error of the final value of the coefficient of cargo stability up to 1.7%, and within 100-150 mm from 3.3 to 4.9%. The maximum displacement step of X was assumed to be 300 mm, since in this case a larger displacement is unlikely and may result in a change of the cargo stability coefficient for more than 10%. Whith such deviations automobile crane-manipulator would be not sustainable for conducting full-scale dynamic and static testing, and would require additional work that was not identified. The linear graph of the dependence of the change of the value of the cargo stability coefficient depending on the displacement of X is shown in figure 4.

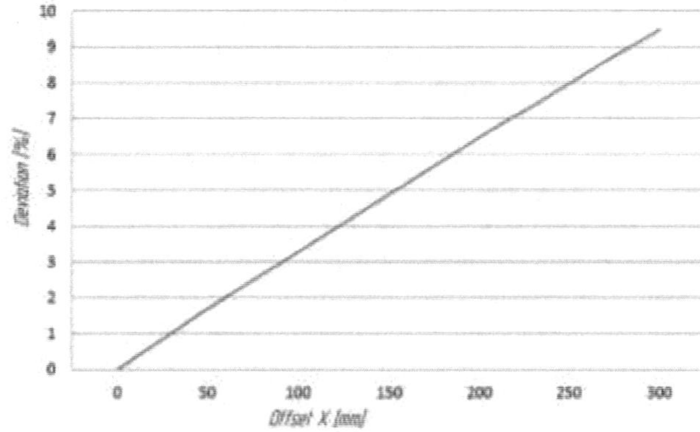

Figure 4. Graph of dependence of the error of the cargo stability coefficient value of the displacement of X.

There is also a linear dependence of the decrease in the value of the stability coefficient the increase in the displacement of X (see Figure 5).

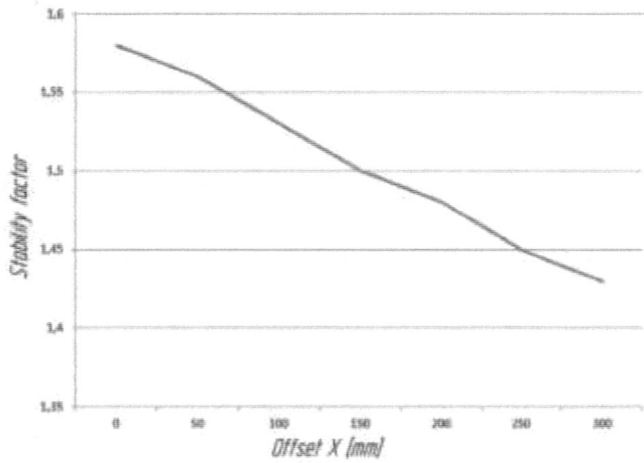

Figure 5. Graph of the stability factor value versus the displacement X value.

4. Conclusion

After analyzing the results of this experiment, it can be concluded that the displacement of the reference contour point can significantly affect the final value of the cargo stability coefficient for automotive cranes. This points out to the need of further study of the effect of the suspension stiffness and the roll of the spring-loaded structural elements of the base chassis on the possible displacement of the rollover ribs. Taking into account the conditions of modern production, it is necessary to develop and further include in the current methodology for calculating the stability of additional correction factors that allow the most accurate simulation of the behavior of the crane when lifting cargo. Thus, the study of this direction of calculation method development aimed to improve the accuracy of determining the cargo stability of automotive crane manipulators.

References

[1] Shchetkin R V 2015 features of construction of reference circuits of automobile cranes-manipulators with spring suspension in the calculation for stability Industry and security, 2 20-3

[2] Mekhonin O N 2018 Features of determination of cargo stability of automobile cranes-manipulators Materials of the all-Russian scientific and practical conference of young scientists, postgraduates, students and schoolchildren (with international participation) "Chemistry Ecology. Urbanism", Perm, Russia, April 19-20 2018 - Perm: publishing house of PERM national research Polytechnic University. 454-8

[3] Burckhardt, M 1993 Fahrwerktechnik: Radschlupf Regelsysteme Wurzburg: Vogel, — 432

[4] Zorin V A , Baurova N I 2009 Improving the safety of road construction machinery and equipment // Science and technology in the road industry, 1 39-40

[5] Tuzelbaev A G 2017 Calculation and tests for cargo resistance of automobile crane-manipulator crane RK 30002K on KAMAZ chassis 43118-0003078-46 Technical regulation in transport construction, 5

[6] Pugin K G 2012 Development and current state of the road construction industry: studies. manual Perm: publishing House of Perm. NAT. research. Polytechnic. UN-TA – 193

[7] Lagerev I A 2012 Dynamics of three-link hydraulic cranes Bryansk: BSTU. – 196

[8] Zhadanovsky B V 2016 Organization of stability of hoisting vehicles in the construction industry. Vestnik MGSU **5** 52-8

[9] Federal rules and regulations "Rules of industrial safety of hazardous production facilities, which are used lifting facilities " Approved by the order of Rostekhnadzor 533 from 12 11.2013

Structural schemes analysis of road and construction machines

A G Saveliev[1,*] and V A Mikhailovskaya[2]

[1]Moscow Automobile and Road Construction State Technical University (MADI), 64, Leningradsky prospect, Moscow, Russia

[2]Moscow State Technical University named after Bauman (MSTU), 5, 2-Baumanskaya Street, Moscow, Russia

[*]E-mail: prof.saveliev@yandex.ru

Abstract. The paper develops a methodology for computation external movability of mechanisms with two-sided and one-sided ties. These mechanisms are often used in the design of road-building machines. Also this article provides with the some of examples of structural schemes of bulldozers, excavators, motor graders and loaders. Describing the main motions of the operating equipment, here is given some example of using different types of hinges.

1. Introduction

Track type tractors, excavators, loaders and motor graders are widespread machines due to their majority of application.

All these machines can be united in groups by structural features.

The first group – machines that have as operating equipment blade. Machines in this group have operating equipment, that are design for cutting and moving engaged material, for example track type tractors with tilted and not tilted blades, motor graders. Track type tractors equipment consist of rods couplings, like straight frames, that attached blade with undercarriage. Also there are one or several rods for misalignments compensation. And this rods take horizontal loads.

The second group – machines that have buckets type of operating equipment. The thing that characterizes them is all motions of the operating equipment are for the filling out or unloading. Excavators and loaders have such equipment. Loaders bucket is attached to the carriage of machine due to hinged boom. Excavators buckets can have more operating motions in comparison with the loaders, bulldozers and motor graders (Figure 2).

2. Goal

The purpose of this article is to develop methods, which will help engineers to create new structural schemes of mechanisms according to working motions that these mechanisms perform and according to tie that these mechanisms have in their design.

IOP Conf. Series: Materials Science and Engineering **786** (2020) 012027 doi:10.1088/1757-899X/786/1/012027

3. Analytical model

Track type tractors with tilted equipment have one distinction – the possibility of rotating in the horizontal plane. This motion is possible due to frame, that pivotally connected to frame and to undercarriage. As an example, motor graders have a draw bar, more often joined by spherical pivot with a main frame.

A circle member is rotatably attached to a draft frame. A blade is mounted to blade support, that has tilt frame. The tilt frame is attached to the legs for pivotally movement of the blade support and the blades movement relative to the circle frame. Also this construction allows a sideshifting of the blade. Figure 2. shows possible movement of blades operating equipment. Possible movements of blades and buckets are:

- In vertical plane – to lift and to down blade or bucket, this movement can do track type tractors without tilted blade and with rigid attachment of frame with blade, for example DZ-42G (motion V_1, along axis Z).
- Additionally lateral motion is possible in the vertical plane of X axis. It is the usual movement for motor graders, the track type tractors, loaders, excavators-manipulators (motion V_2).
- Also the equipment rotation around Z axis is possible in the horizontal plane. It is typical motion for the motor graders, track type tractors, loaders and excavators (motion V_3).
- Equipments rotation in vertical longitudinal plane (relatively Y axis) is necessary to change the angle of cutting edge. It is also typical movement for graders, track type tractors, loaders and excavators (motion V_4).
- Additionally equipments sideshifting in the horizontal plane in lateral direction (along Y axis) is possible. It is typical motion for motor graders and for some track type tractors (motion V_7).
- Movement forward or backward is possible by the movement of the machine or by movement of the operating equipment. (motion V_5, along axis X).

Excavator equipment (Second group)	Loader equipment (Second group)	Motor grader equipment (First group)
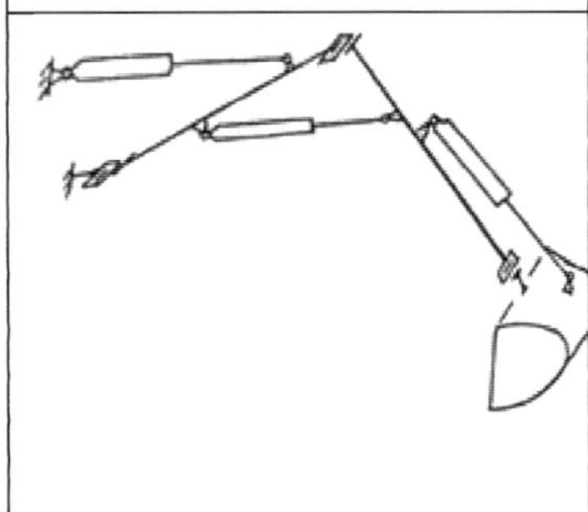	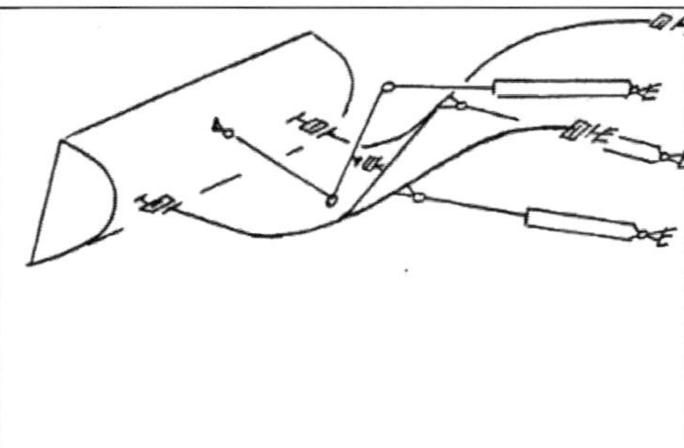	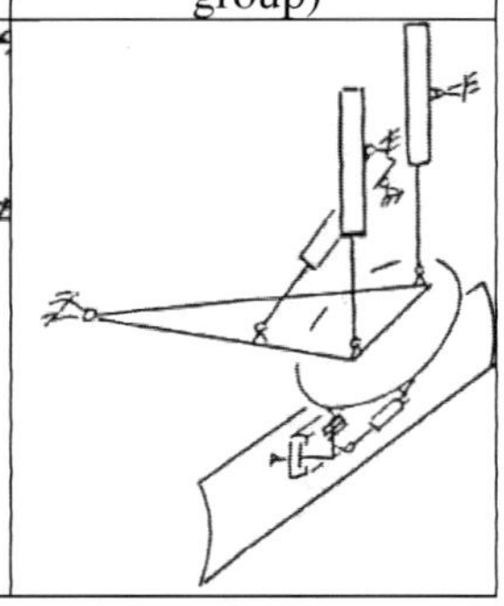

Figure 1. Operating equipments, schematic drawings.

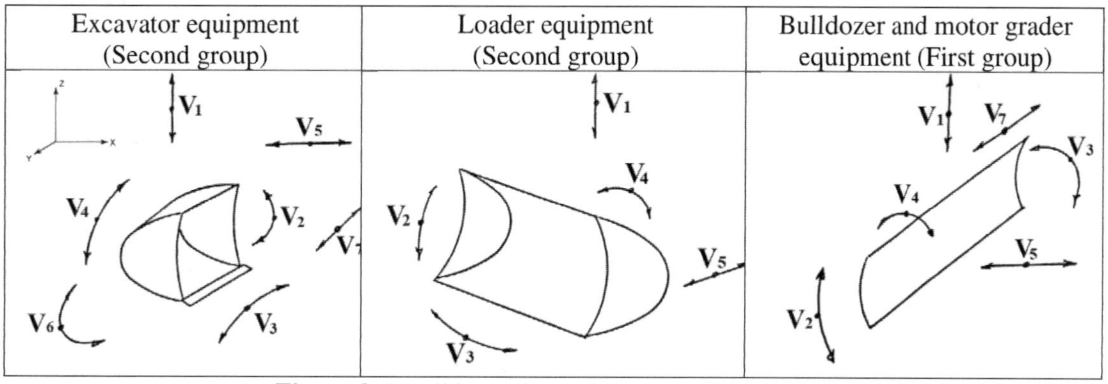

Figure 2. Possible motions of the operating equipment.

Thus, all six motions of blade that is possible have been described. All possible motions of the equipment are revealed due to analysis of technical materials and constructions of well-known machines. Connections of machine operating equipment elements are implemented through kinematic pairs. There are revolute and prismatic joints, cylindrical joints and in rare cases Hooke's joint, spherical joint with translation motion. Kinematic pairs, that have been used in design of road and construction machines, are presented in Figure 7.

As was proved in the Reshetov's paper [1], machines will be able to move their operating equipment, if the design of machines has no redundant constraints. Otherwise in the design of operating equipment will appear self-breaking internal forces, which will have value, that will exceed value of forces, that appear due to loads on the blade or other operating equipment. Moreover, if operating equipment has redundant constraints, the movements of this equipment will be possible only due to clearances in the joints. And the value of these joints will set the range of movements of operating equipment.

You can measure the number of redundant constraints by using the equation of Chebyshev-Malishev [1]:

$$W = 6 * n - 5 * P_y - 4 * P_y - 3 * P_y - 2 * P_y - W_{\mathrm{M}}, \quad (1)$$

W – a number of external movability;

P_i – a number of kinematic pairs of certain class;

W_M – a number of local movability.

Below there are some variants of operating equipment schemes for bulldozer, excavator and motor grader.

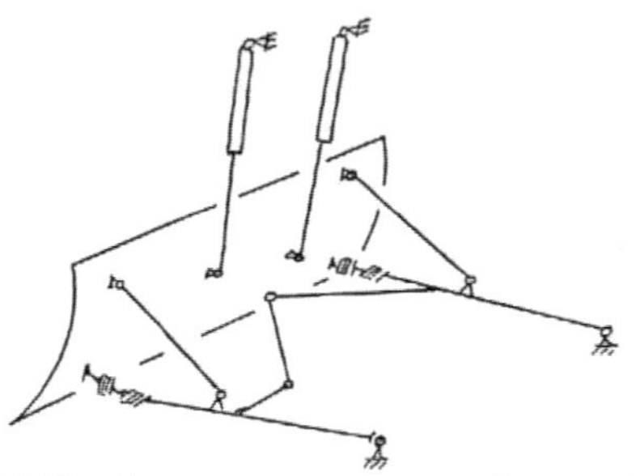

Figure 3. Track type tractors operating equipment without tilted blade, schematic drawing.

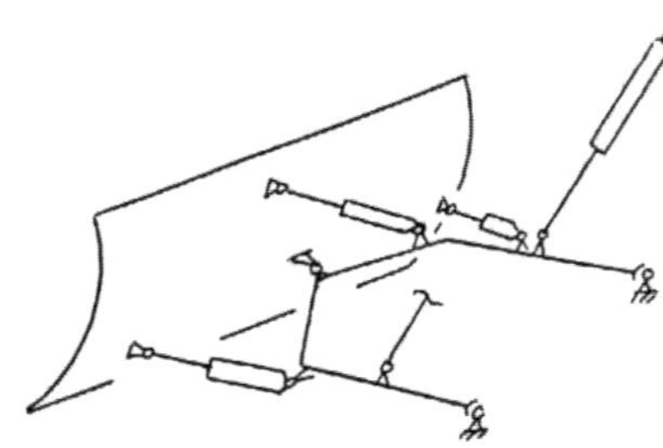

Figure 4. Track type tractors operating equipment with tilted blade, schematic drawing.

IOP Conf. Series: Materials Science and Engineering **786** (2020) 012027 doi:10.1088/1757-899X/786/1/012027

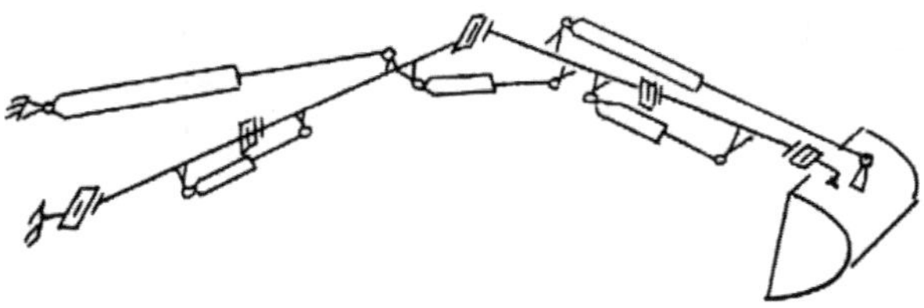

Figure 5. Excavators operating equipment with jointed elements of boom and stick, schematic drawing.

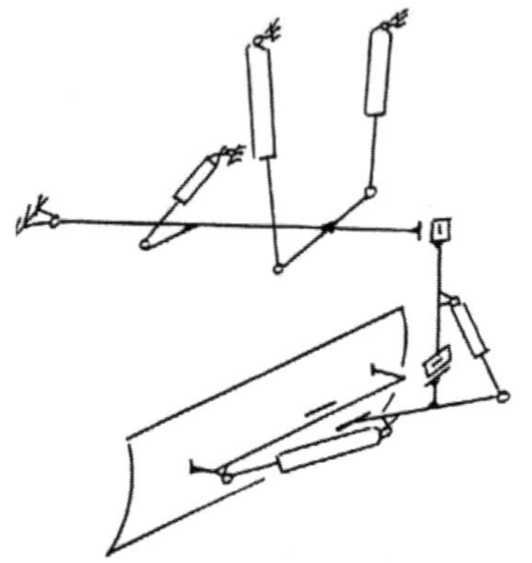

Figure 6. Motor graders operating equipment, schematic drawing.

Classification	Scheme
Revolute joint	
Prismatic joint	
Cylindrical joint	
Hooke's joint	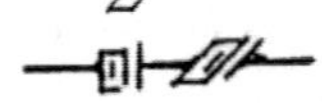
Spherical joint	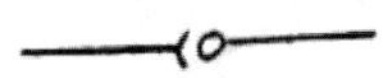
Spherical with longitudinal motion joint («ball in pipe»)	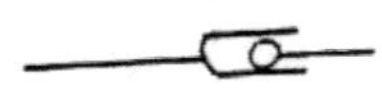

Figure 7. Kinematic pairs, that are used in the design of construction and road machines.

Local movability (relative motion) will appear, when the rods are attached by spherical joints, therefore the relative rotation of the rods around their own axis appears. And this axis passes through this spherical joints.

Most part of machines, that have been using in a construction area, has redundant constraints. In the L. N. Reshetov's book was proved, that existence of redundant constraints causes great harm for machines design. If redundant constraints are in design of machine, then it will better to make parts of the design with more accuracy, because this redundant constraints can make assemble of part impossible, reduce a value of possible loads on the blade, reduce performance of a machine, due to high level of friction in joints, and also increase the weight and overall sizes.

In addition to this, redundant constraints can reduce a functionality of operating equipment and reliability.

Modern operating equipment of road machinery should have high performance, low metal intensity, high reliability and low cost of production.

Nowadays road and construction machines use hydraulic systems for realizing their operating functions, for example to change the position of the motor graders blade, vertical hydrocylinders are engaged or if it is necessary to sideshift the circle and draft frame of a motor grader then inclined cylinder should be engaged.

If design of a machine has additional degrees of freedom, that are technologically proved, it will expand possibilities of operating equipment and will increase a performance.

Date about limits in the functionality of motor graders operating equipment are in the Table 1.

IOP Conf. Series: Materials Science and Engineering **786** (2020) 012027 doi:10.1088/1757-899X/786/1/012027

Table 1.

Country	Company, Power, kW	Blade sideshift right and left, mm	Maximum lift and depth of blade, mm	Blade tip range, °
USA	CAT, motor grader 160 K, 139 kW	943 851	452 790	40; -5
USA	John Deere, motor grader 770 G, 138 kW	683 683	490 No data	42; -5
China	Shantui, motor grader SG18-3, 132 kW	No data No data	440 500	44; -1
Turkey	Hidromek, motor grader HMK 600MG, 140 kW	No data no data	525 660	46
Russia	Road Machinery Factory, motor grader DM-14.0, 132 kW	800 800	No data 500	No data

Similar data can be gotten about other operating equipment of road machines.

The equation Chebysheva-Malisheva has gotten from the condition, that the body has six degree of freedom in space: three – revolute and three – prismatic. But, in certain cases, it is better to take into consideration the direction of possible movement of operating equipment and direction of holding bonds, for example, in case of using one side types of bonds like ropes in rope excavators.

$$W = 12\,n - 11\,P_{11} - 10\,P_{10} - 9\,P_9 - 8\,P_8 - 7\,P_7 - 6\,P_7 - 5\,P_5 - 4\,P_4 - 3\,P_3 - 2\,P_2 - P_1 - Wm, \quad (2)$$

where P_{10} – revolute pair;

P_8 – cylindrical pair or Hooke's hinge;

P_6 – spherical pair;

P_4 – type of pair like «ball in pipe»;

P_2 – kinematic connection like rod with several spherical hinge;

P_{11} – ratchet gear;

P_9 – cylindrical pair, that cannot rotate or translationally move any side (a combination of ratchet gear and prismatic pair);

P_7 – ratchet gear with universal joint (spherical pair with pin, if the stroke of pin has settled);

P_5 – «ball in the tube», if the stroke of ball has settled (ball on the bottom of the pipe);

P_3 – ratchet translational pair on the flat;

P_1 – cam; one sided rod: from one side is spherical joint, from another – pair P_5; or connection through a cable.

IOP Conf. Series: Materials Science and Engineering **786** (2020) 012027 doi:10.1088/1757-899X/786/1/012027

Coefficient 12 at the beginning of the equation considers the possibility of translation and rotation motion in all sides. Kinematic pairs with even number of coefficients – possibility to move in all sides, with odd number of coefficients – have no possibility to move in any direction.

Equations (1) and (2) let quickly define existence or absence of redundant constraints or degrees of freedom, but which kinematic pairs or bodies have redundant constraints in the design of machines by these equations is impossible to define. Also you cannot define whether a system will be instantly changeable. Therefore, if you want to avoid mistakes in creating new schemes of road machines, you have to consider certain movability and bonds. Balance equation is:

$$n\,(w_{z+} + w_{z-} + w_{y+} + w_{y-} + w_{x+} + w_{x-} + f_{z+} + f_{z-} + f_{y+} + f_{y-} + f_{x+} + f_{x-}) - (K_1 S_{z+} + K_2 S_{z-} + K_3 S_{y+} + K_4 S_{y-} + K_5 S_{x+} + K_6 S_{x-} + K_7 S_{wz+} + K_8 S_{wz-} + K_9 S_{wy+} + K_{10} S_{wy-} + K_{11} S_{wx+} + K_{12} S_{wx-}) = n(w_{i+-} + f_{i+-}) - K_j S_j,$$

Where w_{i+-} – degree of freedom of rotation every part in all sides around every axis;

f_{i+-} – degree of freedom of translational motion every part in all sides along every axis;

n – number of bodies (elements of construction);

K_j – number of bonds in every translational or rotational bonds, that named S_j.

You can represent all kinematic pairs or joints as a set of ties, for example, ratchet gear, that have rotating degree of freedom w_{z+}, you can present like:

$$S_{z+} + S_{z-} + S_{y+} + S_{y-} + S_{x+} + S_{x-} + S_{wz-} + S_{wy+} + S_{wy-} + S_{wx+} + S_{wx-}\,,$$

Rotational pairs, that have rotational degree of freedom in both sides around axis OZ you can present w_{z+} and w_{z-} as:

$$S_{z+} + S_{z-} + S_{y+} + S_{y-} + S_{x+} + S_{x-} + S_{wy+} + S_{wy-} + S_{wx+} + S_{wx-}.$$

Similarly, any kinematic pair can be represented.

There are particular cases, when instantly changeable scheme appears and you can consider this cases, if each link has its own coordinate system. For example, if longitudinal axis of rods, that are connected by spherical hinges, match with each other, than will appear instantly changeable scheme. Thus, you can consider particular cases, when you define mutual arrangement of ties coordinate systems. It means, that ties, for example, $S_{y'+}$ and $S_{y'-}$; $S_{y''+}$ and $S_{y''-}$ (Figure 8) eliminate degrees of freedom along the OY axis in both sides, because axis O'Y' and O''Y'' of ties system coordinate, when it intersects with the axis OX of the main coordinate system, it forms the arm, that named 1.

Also you should take into consideration the direction of the action of external forces. For example, there is a beam, hinged fixed by one end and freely leaned by another end (Figure 9). Ties of system: $S_{z'+} + S_{z'-} + S_{y'+} + S_{y'-} + S_{x'+} + S_{x'-} + S_{w'y+} + S_{w'y-} + S_{w'x+} + S_{w'x-}$ – of revolute joint A; $S_{y''+}$ – of carriage B. Rotational moment of the force of gravity G is balanced, because the pair of ties $S_{y'+}$ and $S_{y''+}$ forms the arm, if this pair of ties in intersection with flats, that formed by axis of main coordinate system. If the force G acts in the opposite direction, then there will not be a holding pair of ties and redundant mobility will appear.

This approach is also important, for example, in the design of bulldozer units, when the ground ties are set vertically down and horizontally in all directions from the adhesion forces of a machine with the ground. This allows you to exclude from the design of a machine redundant breakout connection, especially when the bulldozer unit is operating on the rough surface or when dozer is operating at different levels and at different angles to the horizontal plane.

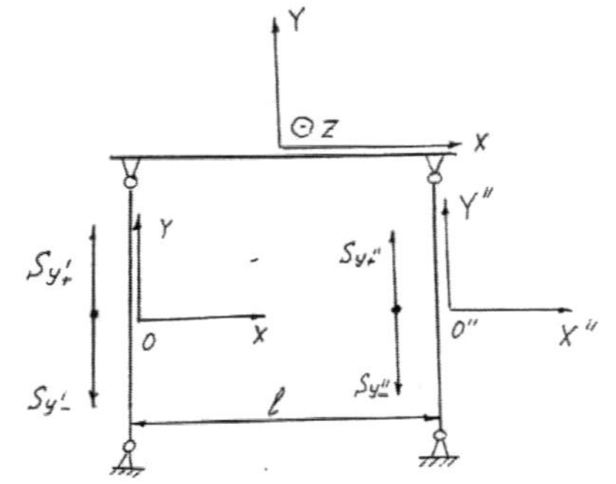

Figure 8. Scheme of hinged four links mechanism.

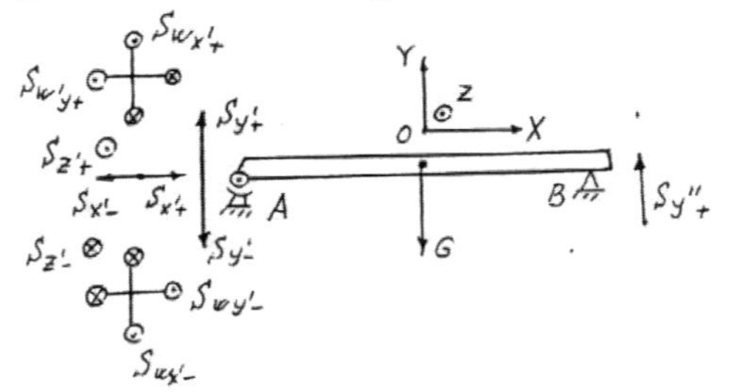

Figure 9. Scheme of bar attachment.

Therefore to ensure a fixed position of the construction and road machines relative to the base surface and fixed position of the working equipment relative to the machine, it is not necessary to have twelve one-sided links. It is necessary to identify the possible directions of external forces of construction and road machine and to ensure links that would counteract possible external forces. For example, for a road roller, a vertical link down Sy- is not necessary. For a blade of a bulldozer with a non-tilted blade in the traditional version, it is necessary to have all twelve one-sided links, because external force on the blade can be applied at any point and directed in any direction. Existence of more than twelve one-way links leads to the appearance of additional stresses in the working equipment of machines, and the incorrect arrangement of the links leads to excessive mobility of the construction and road machine.

4. Example

Let's consider as an example working equipment of track type tractor DZ-110A (Figure 10). The blade has following movabilities: $w_{z+} + w_{z-} + w_{y+} + w_{y-} + w_{x+} + w_{x-} + f_{z+} + f_{z-} + f_{y+} + f_{y-} + f_{x+} + f_{x-}$. Hydrocylinder AB sets the links $S_{y+} + S_{y-}$. Rods CD and EF set the links $S_{wx+} + S_{wx-}$ and $S_{wz+} + S_{wz-}$, because there is a shoulder for the moment C'E'. Rods KL and MN set the links $S_{wy+} + S_{wy-} + S_{z+} + S_{z-}$, rod QP – the links $S_{x+} + S_{x-}$. There are no redundant links in the picture below and tilt of blade can be set without any extra stresses in the design.

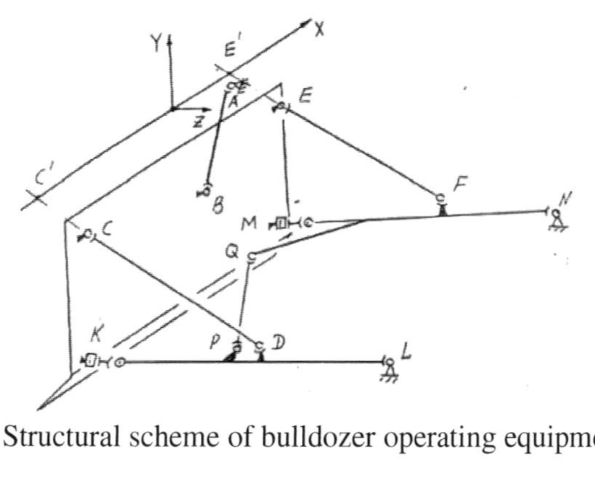

Figure 10. Structural scheme of bulldozer operating equipment DZ-110A.

5. Conclusions

Thus, in order to increase the efficiency of construction and road machines with a significant simplification of the structure and reduction of production costs, it is necessary to analyze their structural schemes. And then decide which links apply in the design one-sided or both-sided. To extend the functionality of construction and road machines by eliminating redundant links in the design, it is better to use in the design the number of links less than the number of links that external forces set on machine.

It is necessary to create the way to formalize analyzing and synthesizing new structural schemes of the construction and road machines in order to speed up this work. At the same time, it is possible to accelerate the memorization and analysis of structural schemes by creating database of already existing structural schemes. It could speed up searching for new ones and comparing them with each other.

In the design of machines it is necessary to use schemes without redundant links. Otherwise, self-extinguishing forces will occur even before the application of the workloads.

References

[1] Reshetov L N 2000 Self-stabilizing mechanisms – 2nd ed – M : Mechanical Engineering, 364

[2] Chew M , Shen S N T and Issa G F 1995 /Kinematic Structural Syntesis of Mechanisms Using Knowledge-Based Systems *Journal Mechanical Design* ASME **117/98**

[3] Adamszyk E and Kozen Z Struktursynthese und Stabilitatsanalyse der Oberbauten von Schaufelfadgeraten Konstruktion/Wissenschaft/Forschung, 5/86-27-40

[4] Mruthyunjaya T S , 2003 Kinematic structure of mechanisms revisited, Mechanism and Machine Theory, 279-302

[5] R L Norton 2004 Design of Machinery, McGraw and Hill, New York

[6] Mavroidis C , Roth B 1995 New and Revised Overconstrained Mechanisms *Journal of Mechanical Design* ASME **117/75**

[7] Liu T and Yu C -H 1995 Identification and Classification of Multi-Degree-of Freedom and Multi-Loop Mechanisms *Journal of Mechanical Design* ASME **117/104**

[8] Huafeng Ding 2015 Automatic structural synthesis of planar mechanisms amd its application to creative design. Doctor thesis, university of Duisburg-Essen, Germany

Improving the efficiency and mobility of urban housing maintenance services

D O Burkeev[1], A I Romanova[1,*], A A Murafa[1], O V Maksimchuk[2] and A V Voronin[3]

[1]Kazan State University of Architecture and Engineering 1, Zelenaya St., Kazan, Republic of Tatarstan, 420043, Russian Federation

[2]Volgograd State Technical University 1 Academicheckaya St., Volgograd, 400074, Russian Federation

[3]Tyumen Industrial University 38 Volodarskogo St., Tyumen, 625003, Russian Federation

* E-mail: aisofi@kgasu.ru

Abstract. Important areas of development in the field of housing construction are to energy efficiency, improvement of the system of operation and maintenance of buildings, transparency in the operational performance of buildings for owners and investors. To obtain such benefits, it is necessary to develop a strategy for implementation of intelligent technologies and smart engineering systems to improve energy efficiency, automate operational processes and create an 'intelligent' profile of a building.The article presents a study of energy efficiency of buildings and justifies smart management of real estate objects, in order to save resources and timely respond to the decline in their operational qualities.In the modern conditions of market economy, when most of the housing stock is privately owned, it is necessary to create a modern economic and organizational mechanism for overhauls, reconstruction and modernization of building with the use of BIM technologies.The result of the overhaul and reconstruction works is the restoration and improvement of the operational characteristics of buildings. An overhaul or a reconstruction is advisable if the technical condition of the building has been compromised and effective operation and maintenance are impossible.

1. Introduction

Today operation and maintenance of residential buildings are impossible without innovative intelligent technology. Gault (2018) defined and measured innovation in all sectors of economy. Appio, Lima and Parutis (2019) wrote about technological forecasting and social changes in reference to Smart Cities. Arundel, Bloch and Ferguson (2019) and Marconi and others (2016) highlighted such priorities as promoting innovation in the public sector and aligning the measurement of innovation with political objectives.

Rauter and others (2018) and Pavlatos and Kostakis (2018) reviewed open innovation and its impact on economic efficiency and innovation activity. According to Caragliu and Del Bo (2019) innovative Smart Cities directly influence the Smart City policies and urban innovations.

In Russia, a number of large cities positioning themselves as Smart, such as Moscow, St. Petersburg and Kazan, have developed a property management system at the local level, however, there is no uniform national standard for this process.

In order to improve the efficiency of residential real estate management in modern economy, it is necessary to categorize buildings according to legal criteria and functional purpose at the overhaul stage and also to be able to plan the operational performance of a building at the construction stage by creating a Building Information Model and using intelligent technologies in design, construction and operation. Recent work in this field focused on dynamic characteristics of smart grid technology acceptance (Park, Kim and Yong 2017) and introduction of the Internet of Things for development of smart buildings (Jia and others 2019).

The life cycle of a building has, tentatively speaking, three stages, namely, design, construction, and operation. There is also a fourth stage, that is reconstruction or demolition (including the change of functional purpose), closely interconnected with the operation.

At the first and second stages, the terms of reference are developed; the design and estimate documentation is drawn; BIM modeling is performed; investment analysis of the construction project is carried out if necessary; the investor, the general contractor and the other participants in the investment and construction processes are determined; and a physical object, having a certain functional purpose and a set of consumer qualities and properties, is created.

At the third stage, the good condition of all structural elements and systems of the building are maintained and the adjacent territory is improved or developed. Any building, regardless of the form of ownership and location, must be under constant maintenance to ensure normal functioning during the entire life cycle, which can reach one hundred and more years (Jiang 2012; Xie 2011; Romanova and others 2015).

The operation stage includes 'consumption' of the building; that is, in other words, the use of its engineering systems, equipment, adjacent territory, etc., for specific purposes. The main task in this case is to ensure trouble-free operation of all structural elements of the building with strict observance of sanitary and hygienic standards, proper use of engineering equipment, maintenance of optimum temperature and humidity conditions, compliance with maintenance and repair plans, servicing and improvement of adjacent territories, etc.

Li and others (2018) showed the implementation of architecture, technologies and systems into a Smart Home. Gao and others (2017) addressed the theory and practice of intelligent collection of data on the operation of residential buildings by the example of modern technologies in smart district heating system.

Wetzel and Thabet (2015) addressed the use of a BIM-based framework to support safe facility management processes. Sepasgozar and others (2019) confirmed the relevance of implementing citizen-centric technology in developing smart cities and created a model for predicting the acceptance of urban technologies. As early as five years ago Fazli and others (2014) stated the effectiveness of building information management (BIM) in project management.

The duration of trouble-free operation of the building and its systems can differ. The standard service life of the foundation, the walls and the main bearing elements is regarded as the standard service life of the building. The service life of some individual elements can be 2–3 times less than the useful life of the building. It is necessary to replace its elements or their parts to continue using the building comfortably and safely for its intended purpose in the course of the entire service life. Also, the engineering systems and the structural elements require setup, restoration and failure prevention over the entire service life of the building. Operation of buildings to full wear is often impossible, therefore, works are needed to compensate for normal wear. If planned maintenance is not performed on time, premature failure of the structure can occur.

It should be noted that buildings and constructions are the main consumers of heat. Government regulation in the field of thermal performance of building is prioritized in most countries of the world. Such requirements root in rational use of non-renewable natural resources, environmental protection, reduction of greenhouse effect and minimization of emissions of carbon dioxide and other harmful substances into the atmosphere. For example, thermal imaging allows localizing the places of heat loss and assessing the condition of exterior wall envelope in order to timely protect it from the adverse effects of the external environment.

In the Russian Federation, energy audit of buildings is voluntary. The proportion of buildings surveyed is about 5%, most of which have been surveyed in preparation for the overhaul of the property.

Today, available are various structural elements, engineering systems and building materials, of foreign and domestic manufacture, with different performance indicators, service intervals and cost of maintenance. When estimating the construction costs and making the investment project analysis, the cost of operation is often calculated incorrectly, and the owner of the building often lacks additional funds for operating expenses, which leads to a considerable loss in performance of the building in the first five years. The regulatory technical documentation that sets forth the rules of design, construction and operation of buildings is continuously amended. However, such amendments are mainly aimed at reducing energy losses and energy consumption of buildings by changing the layout of the building, adding facade insulation, introducing energy-saving technologies (most often by reducing the overall energy consumption of electrical installations, lighting devices, etc.) and they pertain to construction and operation of smart buildings.

Despite the fact that information and digital technologies, including voice recognition systems, artificial intelligence and robotics, are on the rise in Russia, the overwhelming majority of the country's population still take utility meter readings manually and report them monthly to the utility companies. This creates inconveniences for residents who have to keep track of utility meter readings (utility meters are often installed in hard-to-reach places), and, obviously, taking meter readings by hand does not help store and systemize such important data. Since in Russia there is still no generally accepted system of intelligent metering that would collect, store and use meter readings, valuable data gets lost, while they can be used not only for basic statistical purposes but also for energy saving studies, decision-making, design of real estate, improvement of existing housing, development of BIM technologies and in other areas.

This paper discusses how to use BIM-modeling not only at the stage of design but also the stage of operation, taking into account all engineering systems and structural elements with given service intervals, strength properties, stress-strain behavior and other characteristics and providing for automation of operational processes by installing various fire-detection systems, security systems, gas and vapor sensors, anti-icing systems for roofing, etc., to minimize human factor and ensure maintainability. During BIM modeling, it is recommended to use energy-efficient design and construction technologies, to minimize operating expenses by automating the maintenance of systems, installed in the building, to reduce the heat and power consumption whenever possible, and to evaluate the possibility of introducing models of using renewable energy sources.

The purpose of the study is to scientifically substantiate the improvement of building operation quality, to conduct a practical survey of buildings in order to determine their operational characteristics, and to define the conditions for the development of smart innovations in the operation of real estate.

2. Materials and Methods

Regardless of the functional purpose and the form of ownership, buildings and structures are the main energy consumers due to the availability of systems and equipment that converts this energy into comfortable and safe living conditions for owners or temporary tenants.

At the same time, the requirements to the utility services are often unobserved, for example, the air temperature in living rooms, the temperature and pressure of the hot water, the illumination and the heat supply in common areas and utility services rooms, etc. fail to meet the requirements of regulatory documents. In addition, if the scheduled maintenance and repairs are not performed in due time, utility networks, technological equipment and structural elements of the building wear out faster. Construction errors and violations, made during the erection phase, lead to deterioration in the performance of the building. An example of such violation is incompliance with the requirements to external insulation of the facade. This violation is often found both in apartment buildings and single-family houses. Blow-through slits and cracks appear, forming cold bridges. Sometimes, poor-quality materials are used or insulating techniques are ignored. When using modern eco-friendly materials for individual housing construction, for example, laminated veneer lumber (LVL), the construction technology is often violated. The consequences are significant heat losses during the operation, mold and fungi on the walls and comfortable microclimate in the house. In some cases, in the absence of proper quality control, such violations result in the loss of bearing capacity of the structural elements of the house.

At the design stage, simultaneously with designing in effective thermal protection, the use of energy resources should be rationalized by increasing the efficiency of engineering systems and equipment and reducing the consumption of heat and power with the help of automated equipment management systems.

Unfortunately, in individual housing construction, future home owners apply simplified procedures for development and approval of the house design, technical supervision of the quality of construction and installation works, acceptance and commissioning of the house because the current procedures, established by law of the Russian Federation in the field of construction, allow it.

Some people build their houses relying on their own resources, without any design or quality control at all stages of the construction works, or contract incompetent construction organizations. This can result in decreased strength characteristics of the building, poor environmental performance and energy efficiency, deteriorated microclimate in the house and even danger to life and health of the residents if the requirements of regulatory and technical documentation in the field of construction are seriously violated.

The main task of the building envelope is to provide comfortable living conditions and protection from adverse climatic conditions; what is also important is to ensure the natural illumination indoors.

Generally, Russian experience in construction can be characterized by low interest in the construction of individual houses. In the Soviet and post-Soviet period, standard projects for individual housing construction had been issued in mass circulation, and there had been also recommendations for developers. The government had set the task of using secondary resources for production of building materials, such as production wastes. It has been planned that the construction of individual houses would reach 30% of housing development in the country. However, priorities changed and it became necessary to build apartment buildings on a massive scale. Today, in the Russian Federation, apartment buildings make up over 75% of the new housing development (the figure may vary depending on the source).

Every year, the number of housing grows due to the support from the federal target programs. Increasing the availability of land plots and providing them with the necessary infrastructure made it possible to actively develop individual housing construction using modern durable, environmentally friendly and energy efficient materials.

For example, the current housing construction program in the Republic of Tatarstan envisaged commissioning of 2.403 million square meters of housing in 2018. Also in 2018, 87 houses with 9,711 thousand apartments (527 thousand square meters) were to be built under the program of government housing stock under the President of the Republic of Tatarstan.

The ratio of the number of apartment buildings to individual real estate in the Russian Federation is 86% to 14%, respectively.

Today, the share of cast-in-situ, brick-built and stone houses is growing, while the number prefabricated concrete panel houses and wooden houses, built during the Soviet times, is decreasing, thanks to housing renovation programs. Such old buildings are demolished due to their low energy efficiency and high cost of reconstruction of the main structural elements in order to maintain them in good condition. Most residential buildings erected in the Soviet and post-Soviet times, including individual houses, were built with poor thermal insulation of the envelope.

Since thermal imaging inspection is an integral part of smart operation of a building, the thermal imaging survey was performed and included the following activities:

- visual inspection;
- instrumental examination of external envelope of the building and its engineering systems (thermal imaging survey);
- collection, systematization and analysis of the obtained data;
- development of measures to improve energy efficiency and energy saving, assessment of resource savings;
- assessment of the economic efficiency of implemented and proposed activities.

Thermal imaging of the studies buildings was made with the FLIR E4 thermal imaging camera (refer to manufacturer's web-site for specifications and certificates: https://www.flir.eu/support/products/e4#Overview).

The following devices can also be used for such thermal imaging surveys: heat flux density meter, hygrometer, anemometer, data logger with temperature sensors, or thermometers, pyrometer, manometers, power quality analyzer, light meter, etc., depending on the purpose and objectives of the survey.

Obviously, the intellectualization of a building becomes possible only with regular monitoring of the utility systems.

The transition to smart buildings by creating a BIM model of the existing buildings and its engineering systems can be carried out not only in new construction projects, but also in periodic maintenance, overhauls, reconstruction of buildings and structures, taking into account the structural features of buildings, their historical and cultural value, etc.

The introduction of intelligent technologies and systems should be carried out in several stages, depending on the depreciation of the building and its engineering systems, the form of ownership, the location, the level of accessibility for all categories of citizens, and other indicators. However, a complete BIM model must be created during the next repair works, with brake down to sections requiring repair. Thus, the building owner can have a real understanding of the condition of the property, the amount of funds needed to maintain and repair the utility systems in the coming years, taking into account the existing performance indicators and the target indicators to be obtained with the intellectualization.

In Russia, low-rise construction is currently attracting increasing attention from both consumers and government agencies, developers, investors, architects and design studios. A privately owned single-family house is a dynamically developing type of private property and a commodity in market operations. A private house is not only a residential building, it also incorporates ancillary buildings, a land plot, etc. The quality of a low-rise building includes many properties (individual indicators) characterized by the particular complexity of their interrelations and a multi-level structure. There are the following system problems and shortcomings in the field of quality management of low-rise construction:

- construction activities are not adequately controlled, since they do not ensure full compliance with the established requirements of regulatory and technical documentation in force in the territory of the Russian Federation;

- use of materials that do not meet the requirements of regulatory and technical documentation in force in the territory of the Russian Federation, violation of the requirements of transportation and storage of building materials;
- violation of construction techniques and errors in installation of structures of buildings;
- violation of the requirements for the maintenance of buildings and structures.

In the context of tough market competition in the construction industry, many construction companies choose low-quality building materials, with reduced durability and environmental performance, in order to reduce the cost of construction.

To ensure a comfortable microclimate in the residential buildings, it is necessary that the temperature of the external envelope would be about the same as the air temperature in the room, otherwise discomfort is possible due to heat dissipation by the external envelope and formation of condensate on the internal wall, which can lead to mold and fungi growth. In this case, it is necessary to take into account that the temperature of the internal walls depends not only on the selected material and thermal insulation, but also on the quality of construction and installation works. In the presence of construction joints, improperly mounted structural elements and improperly installed facade insulation, windows and doors, etc., there appear cold bridges that let the heat escape.

Often, violations of the requirements for installation of building structures are hidden, and visual inspection cannot identify them. In this case, special equipment, such as a thermal imager, is recommended.

The Federal Law No. 261-FZ On Energy Saving and Improving Energy Efficiency sets forth the main objectives of an energy survey of a building:

- obtaining objective information on the volume of used energy resources;
- determination of the figures for energy efficiency;
- determination of the potential of energy saving and improvement of energy efficiency;
- development of measures of energy saving and improvement of energy efficiency and their cost assessment.
-

3. Results

The obtained thermal images show the defects of external envelopes of the studied buildings of various types, forms of ownership and functional purpose (Figs. 1, 2).

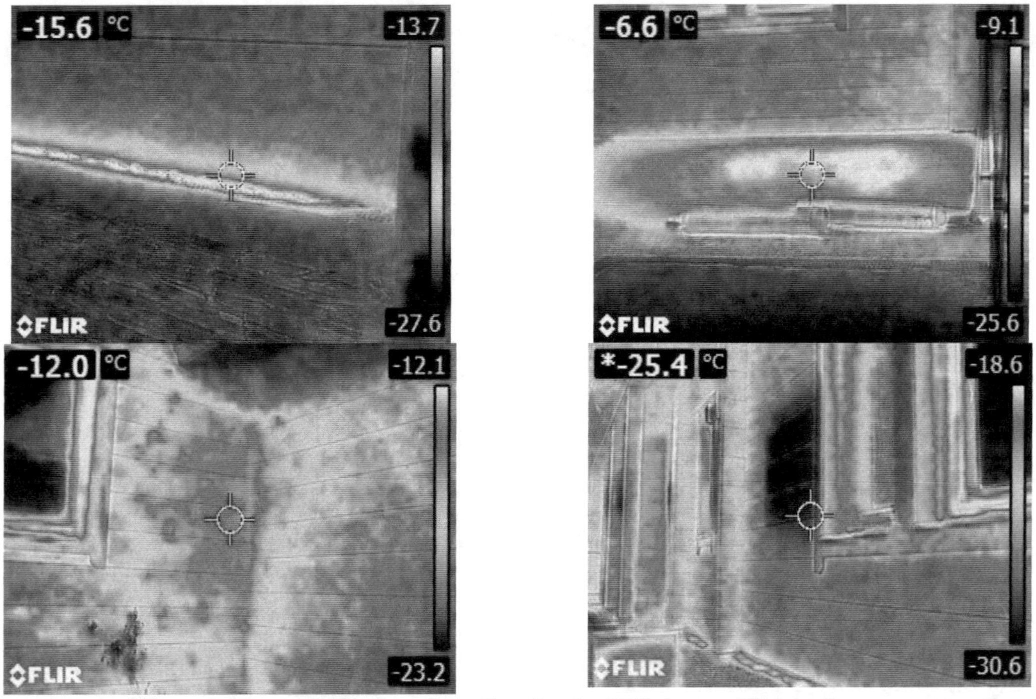

Figure1. Thermal images of the facade of a LVL house (Object 1).

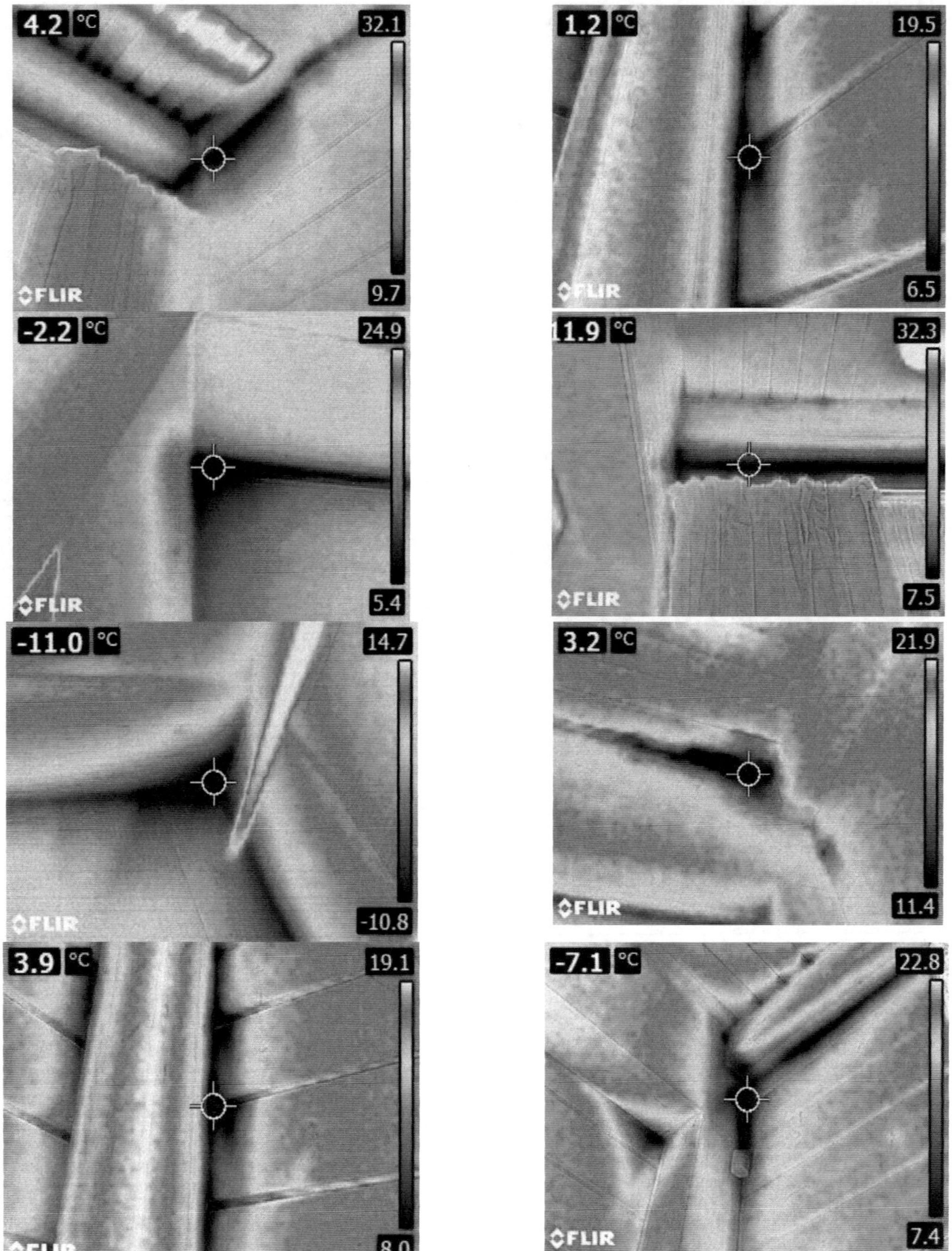

Figure2. Thermal images of the envelope made inside a LVL house (Object 2).

It was found that during the construction and installation works at Object 1, the installation technology of windows and doors structures was compromised; the construction joint between the foundation of the house and the first row of logs was not insulated; and the construction technology of the LVL house was compromised, namely, laying of sealant in-between the logs.

At Object 2, gross violations of the construction technology of assembling a LVL house were found, namely:

- the sealant in-between the logs had low quality and has become useless over time;
- due to repeated violations of the construction technology, the house has subsided unevenly; some beam structures got 'suspended' on the 'portals', which were decorated with substandard stone, without any subsidence elements. Now the beam structures rest on the columns also without any possibility to compensate the subsidence of the house. Due to this, the beams are suspended and, as a result, some rows of logs lack proper support. The joints between the logs along the entire perimeter of the house have opened up;
- there are no jacks to compensate the subsidence of the house;
- in the frame of the house on the second floor and in the double-height part of the ground floor, the insulation is compromised; poor-quality materials are used, namely, the insulant and the tape; the insulation was mounted with gross violations of the construction technology.

Due to the numerous violations of the construction technology, mold and fungi has appeared in the freeze-and-thaw action areas; the windows also thaw and freeze inside. A 1.2-2 m thick ice built-up was found on the roof.

Another object surveyed was an apartment building with a bearing frame made of cast in-situ reinforced concrete, having 182 apartments. The energy efficiency class of the building has not been defined yet. The material of the exterior wall envelope was lime brick masonry with insulation.

The conducted survey revealed gross violations of the construction technology in brickwork:

- brickwork joints were not sealed during masonry works;
- low quality insulation was used;
- brickwork of the internal walls was done with violations and the insulation does not fit tightly to the walls;
- windows and doors were installed with crude errors;
- heating lines were not installed in the staircases;
- doors on the staircase landing are wooden, without insulation; gaps and joints were not sealed.

The thermal imaging survey of five objects (four wooden low-rise LVL houses from 400 to 1800 m^2 assembled by different construction teams and one high-rise apartment building) showed that the quality of the work performed on the construction site was not controlled properly. Gross violations were made in the construction and installation works, the materials used did not meet the requirements of the regulatory technical documentation, and the operation of these building has been conducted with gross violations as well, all summing up to the following consequences:

- mold and fungi appeared on the walls and the comfortable microclimate in the building was compromised;
- increased heat loss through the exterior wall envelope resulted in excessive energy consumption;
- in the absence of due attention to the violations, the carrying capacity of some structural elements, or of the building as a whole, may have been compromised.

As a rule, in case of poor quality construction and installation works, or in case of violations of the requirements of the regulatory and technical documentation, the areas of excessive moistening and freeze-and-thaw action should be detected by means of visual inspection during the operation phase. Also, in case of violations described above, the dew point may be shifted inside the building, leading to the condensation on the walls, destruction of insulation or walls, and growth of mold and fungi, which has a negative effect on the health of the residents. If the carrying capacity of the outer walling during the operation is compromised due to the above factors, overhaul, reconstruction, or demolition of the building becomes necessary.

To ensure the standard quality and energy efficiency at all stages of the life cycle of real estate objects, a construction organization must increase the labor productivity of workers, improve the quality of construction and installation works performed by them, and ensure saving of construction materials based on the improved labor productivity of workers, efficient use of working time, mechanical aids and material resources, and availability of modern equipment and means of non-destructive control. All of these are examples of effective property management.

At the same time, the demand for electricity, water and natural gas is rapidly growing in the 21st century. Since their cost is high, closer attention must be paid to how we use natural resources and whether we are doing anything to use them more effectively to reduce consumer costs and, of course, to protect the environment.

Unlike the United States of America and other Western countries, Russia began thinking about energy saving measures only recently. The law on energy saving projects was passed only at the end of 2009, and it served as a starting point for the development of improved accounting infrastructure (AMI, also known as the 'smart metering system').

However, this good first step aggravated the situation at first: homeowners associations (HOA), property management companies (PMC), and real estate owners found themselves absolutely unprepared for the amount of data they suddenly had to deal with. Utility consumers were fined if they were late in providing utility meter reading to their HOA or PMC.

Such amount of work both for utility consumers (who had to take meter readings manually), and HOAs or PMCs (who has to process and store this data somehow) is extremely difficult, if not impossible, without suitable technology. Unfortunately, today there is still little progress in this matter.

We interviewed employees of HOAs and PMCs in the city of Kazan. The majority of respondents feel that they really struggle with processing of the collected data but they do not fully support the idea of switching to smart metering systems. However, it should be noted that:

- utility consumers would like to simplify the collection of utility meter readings, find a way to control electricity, water and heat consumption and, therefore, save money.
- engineering companies that are seeking new markets would like to introduce new technologies as early as the construction stage, because it can help not only save resources and reduce construction costs, but also attract more technically savvy customers, i.e. homeowners and tenants who want to save on their utilities by using smart technologies.

Smart metering technology automates the process of taking meter readings and makes it easy to regularly provide data to utility companies and to accurately invoice consumers. Such systems also benefit consumers since the readings are taken every 15 minutes to 1 hour (depending on how the setup) and it allows consumers to monitor the consumption of electricity, water and heat during the day and night. It can change the rules of the game when it come to maintenance problems. For example, if a consumer monitors water consumption and sees that water is used from 2 to 4 am, while the household is usually asleep, this indicates a leak. In general, smart metering systems allow households to use energy and water more responsibly and help consumers make a positive contribution to the environmental protection. Smart meters can be used with home power management systems, such as web-based tools, that can be installed in the building. Smart meters can display power consumption, help find ways to save energy and money, and allow users to remotely adjust thermostats or turn off appliances.

To achieve the ambitious goals set in the innovative concept of energy saving in Russia, it is necessary:

1) to conduct expert review of building design projects in terms of energy efficiency;
2) control the quality of construction works.

The second factor is especially important in construction of low-rise individual houses that do not fall under the control of state supervisory authorities.

This approach seems especially useful in planning and performing maintenance and repair of buildings. It provides for using proven, modern, high-quality materials, which guarantee of long service life of a building and significantly increases the transparency of operation of a building. The proposed automation of operation processes can decrease the number of failures of elements and systems.

4. Discussion

The intellectualization of operation processes based on the utility automation with smart metering technology can improve the efficiency of the above mechanism.

At the moment there are three main reasons that prevent the implementation of AMI:

1) only the owner (owners) of the building can decide which meter to install on their property. If this basic yet crucial step is not taken, all benefits associated with the use of the smart metering system will be delayed;

2) the second reason is of psychological nature; although it takes a lot of time to take meter readings manually and then report this data to HOA or PMC, many consumers are used to doing it, and they may be reluctant to learn how to use new meters and understand the advantages of transition to AMI;

3) the third reason explains why HOAs do not the initiate the transition to smart metering systems. Housing managers understand that, while AMI brings tremendous benefits to consumers, it will replace employees who don't want to lose their jobs with devices. However, rightsizing (using fewer people but more devices) can obviously help HOAs and PMCs to reduce their operating costs.

5. Conclusion

The introduction of smart metering technology forms a much-needed basis for digitizing housing infrastructure and metering the natural resources that people consume daily. Thus, AMI helps reduce operating expenses, and the financial factor is crucial in any decision making process. However, there are several other benefits that are convenient for consumers who monitor the consumption of utilities:

1) the automated process prevents human errors in talking the readings;

2) synchronized databases immediately transmit data to utility providers and utility bills are generated electronically, instead of using paper, which is another environmental benefit;

3) smart metering systems are extremely helpful in obtaining more accurate readings for statistical purposes and help prevent power outages;

4) since in Russia starting from 1 January 2017 PMCs are responsible for paying to utility providers for the provided electricity, heat and water, smart accounting technologies can help employees of PMCs and HOAs to monitor readings, analyze utility consumption, compare billing periods for a particular consumer, track payments, prevent non-payments, etc.;

5) as any modern technology, AMI makes data processing less time consuming. Building operation and maintenance costs can be significantly reduced.

Finally, considering all the years of development and implementation of AMI in the USA and following their experience, consumers in Russia can be confident that the transfer to digital means of collecting and storing data on the use of utilities will bring obvious benefits, not to mention the fact that it completely correlates with Federal Law No. 261 "On energy saving and improvement of energy efficiency" and with the national concept of digitalization of Russia.

References

[1] Bakhareva, O V , Romanova, A I , and others 2016 On the Building Information Modeling of Capital Construction Projects Market Development *Journal of Internet Banking and Commerce* **21** S3

[2] Jiang, F (2012) The Study of the Relationship between House Price and Price Tolerance in China from the Perspective of Systems Engineering Systems Engineering Procedia, **5** 74–80

[3] Marconi, N , & others 2016 Manufacturing and economic development: The actuality of Kaldor's first and second laws Original Research Article Structural Change and Economic Dynamics **37** 75–89

[4] Murafa, A A , Karimova M 2018 Utility Services Automatization through the Use of Smart-metering Technology International Scientific Conference 'S M A R T Polis: Sustainability, Management, Architecture, Renovation, Technologies Kazan, November 08-10/ 2018', 128–31

[5] Romanova, A I 2015 Increased construction and service quality products in terms of self-regulation Proceedings of Kazan State Architectural University **2**, 330–337

[6] Romanova, A I , Zagidullina, G M , Afanasyeva, A N and Hkairetdinov, R S 2015 Experience in the Region to Increase the Availability of Housing Services *Mediterranean Journal of Social Sciences*, **6(4)** 549–54 Doi:10 5901/mjss 2015 v6n4s2p549

[7] Xie, Y 2011 Research on the Land Scale Control Model of Public Housing Construction in China: An Example of Harbin Procedia Engineering, e **15** 5121-5

[8] Zagidullina, G M , Romanova, A I & others (2013) Indicative Model of Socio-Economic Development of Small Towns. *World Applied Sciences Journal: IDOSI Publications*, **24** 350–7 DOI: 10 5829/idosi wasj.2013.24.03.13203

[9] Zagidullina, G M , Romanova, A I and others 2013 Peculiarities of Housing Construction Development in the Region. Middle-East *Journal of Scientific Research*, **16** 490–495

[10] Appio, F P , Lima, M , Paroutis S (2019) Understanding Smart Cities: Innovation ecosystems, technological advancements, and societal challenges96. *Technological Forecasting and Social Change*, **142** 1-14 doi.org/10.1016/j.techfore.2018.12.018

[11] Rauter, R , Globocnik, D , Perl-Vorbach, E and Baumgartner, R J 2018 Open innovation and its effects on economic and sustainability innovation performance. *Journal of Innovation & Knowledge*, open access, In Press, Corrected Proof. doi.org/10.1016/j.jik.2018.03.004Get

[12] Pavlatos, O and Kostakis, H 2018 Management accounting innovations in a time of economic crisis. *The Journal of Economic Asymmetries*, **18** e00106. doi.org/10.1016/j.jeca.2018.e00106

[13] Arundel, A , Bloch, C , Ferguson, B 2019 Advancing innovation in the public sector: Aligning innovation measurement with policy goals. Research Policy, **48(3)**, 789–8 doi org/10.1016/j.respol.2018.12.001Get

[14] Gault, F 2018 Defining and measuring innovation in all sectors of the economy Research Policy, 47(3) 617-22 doi org/10.1016/j.respol.2018.01.007Get

[15] Hsi-PengLu, Chiao-Shan Chen, Hueiju Yu Technology roadmap for building a smart city: An exploring study on methodology. Future Generation Computer Systems, Available online 13 March 2019, In Press, Accepted Manuscript. doi.org/10.1016/j.future.2019.03.014

[16] Sepasgozar, Samad M E , Hawken, S , Sargolzaei, S and Foroozanfa, M 2019 Implementing citizen centric technology in developing smart cities: A model for predicting the acceptance of urban technologies. *Technological Forecasting and Social Change*, **142** 105–16 doi.org/10.1016/j.techfore.2018.09.012

[17] Min Li, Wenbin Gu, Wei Chen, Yeshen He, Yannian Wu, Yiying Zhang (2018). Smart Home: Architecture, *Technologies and Systems. Procedia Computer Science*, **131** 393–400 doi.org/10.1016/j.procs.2018.04.219Get

[18] Caragliu, A and Chiara F Del Bo 2019 Smart innovative cities: The impact of Smart City policies on urban innovation. Technological Forecasting and Social Change, **142**, 373–83.doi.org/10.1016/j.techfore.2018.07.022Get

[19] Park, C , Kim, H , Yong, T 2017 Dynamic characteristics of smart grid technology acceptance. Energy Procedia, **128** 187–93 doi.org/10.1016/j.egypro.2017.09.040Get

[20] LinGao, Xuyang Cui, Jiaxin Ni, Wanning Lei, Tao Huang, Chao Bai, Junhong Yang 2017 Technologies in Smart District Heating System. Energy Procedia, **142** 1829–34 doi.org/10.1016/j.egypro.2017.12.571Get

[21] Jia, M , Komeily, A , Wang, Y , Srinivasan, R S 2019 Adopting Internet of Things for the development of smart buildings: A review of enabling technologies and applications. *Automation in Construction,* **101** 111–126. doi.org/10.1016/j.autcon.2019.01.023

[22] Wetzel, E M , Thabet, W Y 2015 The use of a BIM-based framework to support safe facility management processes. *Automation in Construction*, **60** 12–24 doi org/10.1016/j.autcon.2015.09.004

[23] Fazli, A , Fathi, S , Enferadi, M H , Fazli, M , Fathi, B Appraising Effectiveness of Building Information Management (BIM) in Project Management *Procedia Technology,* 16 1116–25

IOP Conf. Series: Materials Science and Engineering **786** (2020) 012029 doi:10.1088/1757-899X/786/1/012029

Evaluation of savings on operational services from innovative energy-saving materials

A I Romanova*, A V Berval, D O Burkeev, R S Safina, E V Ilyina and D A Egorov

Kazan State University of Architecture and Engineering 1, Zelenaya St., Kazan, Republic of Tatarstan, 420043, Russian Federation

* E-mail: aisofi@kgasu.ru

Abstract. Innovative infrastructure projects, as a rule, are aimed at creating or some modification of infrastructure objects, which are characterized by large-scale, high degree of capital intensity, their long-term implementation, freezing investments for a long period due to a long payback period. The most fertile post for the implementation of infrastructure projects is construction.A high share of the economic effect is explained by the fact that the modern world is focused on increasing the efficiency of each innovation, thereby saving on the costs of natural resources, which affects environmental stabilization and the rational use of the resources received.In this aspect, investments in green building can be considered particularly advantageous, allowing us to talk about the double effect of innovations in construction: here we can also reduce the environmental burden on nature with the help of new technological solutions at the stage of development and application of new environmentally friendly materials (effect for the manufacturer), and comfort for a consumer of highly environmentally friendly construction products in an environmentally friendly area (effect for the buyer) in a large metropolis.

1. Introduction

Modern problems of the implementation of innovative materials in green building P.G. Grabovy (2012), O.V. Maksimchuk, O.A.Baulina (2017), etc. Energy technologies and materials have been of interest to modern researchers for more than ten years. So, Pronin S.A. (2001) introduces the concept of innovation in the technical features of green building.

Rauter and others (2018) and Pavlatos and Kostakis (2018) reviewed open innovation and its impact on economic efficiency and innovation activity. A.N. Larionov (2008, 2009, 2011, 2012) argues that environmental friendliness and economic efficiency are factors in the investment attractiveness of housing construction.

At the third stage, the good condition of all structural elements and systems of the building are maintained and the adjacent territory is improved or developed. Any building, regardless of the form of ownership and location, must be under constant maintenance to ensure normal functioning during the entire life cycle, which can reach one hundred and more years (Jiang 2012; Xie 2011; Romanova and others 2015).

IOP Conf. Series: Materials Science and Engineering **786** (2020) 012029 doi:10.1088/1757-899X/786/1/012029

Li and others (2018) showed the implementation of architecture, technologies and systems into a Smart Home. Gao and others (2017) addressed the theory and practice of intelligent collection of data on the operation of residential buildings by the example of modern technologies in smart district heating system.

Thus, an important and urgent task in the field of environmental construction is the implementation of a package of measures for the rational use of energy resources.

2. Materials and Methods

This article discusses the main parameters of the effectiveness of the application of environmental construction technologies (mainly resource saving) and the formation of the economic effect of the use of these technologies.

3. Results

Conditional calculations were carried out on the individual residential building basis (author's vision and approach). An average area of house is 100 sq. m, a family of four lives in the house. The average electricity consumption per month is presented in Table 1.

Table 1. Average electricity consumption of an individual residential building of 100 sq. m area for a family of four (authors vision and approach).

#	Household appliances list	Consumer Power, W	Use mode	Consumption, kW
1	Fridge, economy class	100-300	constantly	60
2	Automatic washing machine, 6kg	2000-2500	4-6 times per week (mainly mode - standard wash at 40 degrees, sometimes up to 80 deg.).	30
3	TV - 2pcs	80-140	TV is turned on for 5-6 hours a day	35
4	Electric stove	2000	4 burners, 2 of which are used constantly, oven	62
5	Exhaust hood	250	unstable	2
6	Microwave	600-1200	often used for heating, sometimes for rapid cooking (breakfast)	20
7	Notebook - pc, Mobile Phone 3 pcs, Wi-fi router DVD player	300-500	computer and router are always on, the remaining equipment is used from time to time	15
8	Kettle Thermo + to 5L	3000	It stands on a heated 85 degrees	12
9	Iron / vacuum cleaner	400-2000	3-4 hours a week	12
10	Landline phone	1	(On charging tube)	0.7
11	Watches - Service	1	(With vibration and city lights)	0.7
12	Camcorder, camera	12	recharging the batteries 2-4 times a month	3
13	Lighting	700-1500	4 bedrooms, bathroom, toilet, hall, kitchen	60
	Total monthly electricity consumption			312.4

A family of four consumes 312.4 kW of electricity for household needs. The use mode of household appliances is various. For example, a fridge is used constantly, an iron and a vacuum cleaner are used 3-4 times a week on average.

Table 2 shows the electricity tariffs for the period from 2009 to 2018. There is a cost increase dynamic of 1 kW by 0.16 rubles in a year on average.

Table 2. The electricity tariffs for 2012-2018 in the Republic of Tatarstan.

Year	2012	2013	2014	2015	2016	2017	2018
The cost of 1 kW / rubles	2.06	2.23	2.33	2.43	2.88	2.99	3.20
Cost increase, rubles	-	0.17	0.1	0.1	0.45	0.11	0.21

Consequently, a family of four pays 999.68 rubles per month on average for consumed electricity per month in 2018.

To switch to solar panels, it is necessary to carry out some preparatory works: average monthly consumption in the summer and winter, determine capital investments, and determine how much our facility will depend on an alternative energy source.

We select a solar battery for an individual residential building, with an area of 100 sq. m, with an average monthly energy consumption of 312.4 kW.

Daily consumption of 312.4 / 30 = 10.41 kWh.

Define the maximum and minimum insolation values for the Kazan city are given in Table 3.

Table 3.The average monthly level of insolation (solar radiation) in Kazan.

#	Month	The solar radiation level indicator, kWh / sq. m
1	January	0.68
2	February	1.44
3	March	2.82
4	April	4.29
5	May	5.52
6	June	5.93
7	July	5.72
8	August	4.49
9	September	2.86
10	October	1.51
11	November	0.83
12	December	0.54

The table shows that the highest value of the solar radiation was in June, and the smallest one was in December.

Let us determine by the formula (1) the amount of energy that a solar battery can generate each month of the year.

$$W = k * P * E \quad (1)$$

P – the power of the panel, kW;

E – insolation value, kWh /sq. m

To determine an energy amount that solar panel can generate, you need to know the power of the panel. When analyzing the solar panel market, we found that 500W panels have the greatest demand. We calculate the energy amount that this panel can allocate.

The calculation of the energy amount that the solar panel can generate is presented in Table 4.

Table 4. An energy amount calculation that a solar panel can generate

#	Month	The solar radiation level indicator, kWh / sq. m	P, kW	W, kW
1	January	0.68	0.5	0.34
2	February	1.44	0.5	0.72
3	March	2.82	0.5	1.41
4	April	4.29	0.5	2,145
5	May	5.52	0.5	2.76
6	June	5.93	0.5	2,965
7	July	5.72	0.5	2.86
8	August	4.49	0.5	2,245
9	September	2.86	0.5	1.43
10	October	1.51	0.5	0.755
11	November	0.83	0.5	0,415
12	December	0.54	0.5	0.27

Table 4 shows that the solar panel works most efficiently from March to September since in these months there is a high insolation rate.

Based on this calculation, presented in table 4, you can determine the need for solar panels for each month of the year. To do this, daily consumption is divided by the obtained values. The requirement calculation for solar panels is presented in Table 5.

Table 5. Calculation for solar panels requirements.

#	Month	Daily intake, kW	W, kW	N, pcs
1	January	10.41	0.34	30.62
2	February	10.41	0.72	14.46
3	March	10.41	1.41	7.38
4	April	10.41	2.145	4.85
5	May	10.41	2.76	3.77
6	June	10.41	2.965	3.51
7	July	10.41	2.86	3.64
8	August	10.41	2.245	4.63
9	September	10.41	1.43	7.28
10	October	10.41	0.755	13.79
11	November	10.41	0.415	25.09
12	December	10.41	0.27	38.56

At the initial stage, it is proposed to purchase 5 panels, since such an amount will be enough to completely replace energy needs in the summer, and excess electricity will be accumulated and consumed if necessary.

For installation, we select a set of solar modules CNH 500-W-72P with power parameters - 500 W and voltage - 48 V. The set description of solar panels is given in Table 6.

Table 6. Kit description of solar battery CNH 500-W-72P.

#	Accessories	Price per unit, rubles	Quantity, pcs	Cost, rubles
1	Solar modules 500W / 48V	16 500	5	82 500
2	Batteries DELTA GX 12-200	29 300	1	29 300
3	Uninterruptible power supply 24 / 48V	18 500	1	18 500
4	The inverter IC-48-1500	13 100	1	13 100
	Total			143 400

We calculate the energy amount that the selected solar panel set can generate. The calculation of the energy amount that the selected CNH 500-W-72P kit can generate is shown in Table 7.

Table 7. An energy amount calculation that a kit of solar panel CNH 500-W-72P can generate.

#	Month	N, pcs	W_1, kW	W_5, kW
1	January	5	0.34	1.7
2	February	5	0.72	3.6
3	March	5	1.41	7.05
4	April	5	2.145	10.725
5	May	5	2.76	13.8
6	June	5	2,965	14.825
7	July	5	2.86	14.3
8	August	5	2.245	11.225
9	September	5	1.43	7.15
10	October	5	0.755	3.775
11	November	5	0.415	2.075
12	December	5	0.27	1.35

Based on this calculation, we can calculate the savings from the use of solar panels. The calculation of savings is presented in Table 8.

Table 8. Calculation of savings when using work out a kit CNH 500-W-72P.

#	Month	W_5, kW	Daily energy savings, kW
1	January	1.7	8.71
2	February	3.6	6.81
3	March	7.05	3.36
4	April	10.725	-0.315
5	May	13.8	-3.39
6	June	14.825	-4.415
7	July	14.3	-3.89
8	August	11.225	-0.815
9	September	7.15	3.26
10	October	3.775	6.635
11	November	2.075	8.335
12	December	1.35	9.06

Based on this calculation, it is clear that the most efficient use of solar panels occurs in the summer period (April, May, June, July, August). In these months, it is proposed to completely switch to the use of solar panels and then we will save 1562 kW per year. (312.4 kW * 5 month).

In dynamics, the cost of electricity will grow by 16 kopecks per year, this can be seen in Figure 1.

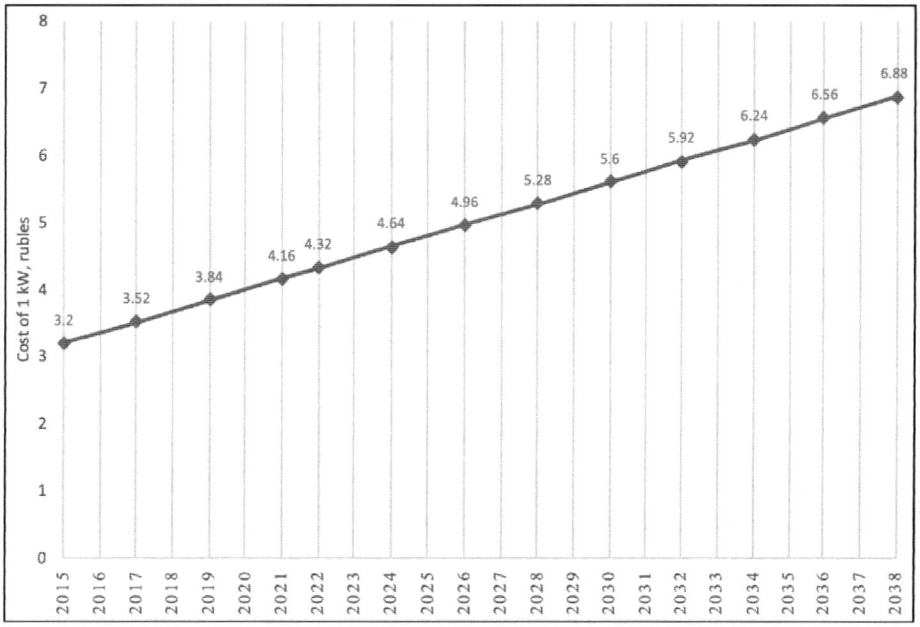

Figure 1. Rising cost of electricity in dynamics until 2038.

After analyzing the development dynamics of the cost of 1 kW of electricity, we can calculate the economic effect of using solar panels in the summer (April, May, June, July, August). Electricity consumption for a year without using a set of solar panels is 3 748.8 (312.4 kW * 12 months). When using the selected set, the energy consumption per year is reduced by 1 562 kW per year and will be 2186.8 kW, based on this, it is possible to calculate the energy savings. The calculation of savings when using the CNH 500-W-72P kit is presented in Table 9.

Table 9. Calculation of savings by the solar panel kit usage.

Year	Electricity consumption per year, kW	Consumption cost, rubles	Electricity consumption per year when installing solar panels, kW	Consumption cost, rubles	Savings by the solar panel kit usage, rubles
2015	3748.8	11996	2186.8	6997.8	4998.4
2016	3748.8	12596	2186.8	7347.6	5248.3
2017	3748.8	13196	2186.8	7697.5	5498.2
2018	3748.8	13796	2186.8	8047.4	5748.2
2019	3748.8	14395	2186.8	8397.3	5998.1
2020	3748.8	14995	2186.8	8747.2	6248
2021	3748.8	15595	2186.8	9097.1	6497.9
2022	3748.8	16195	2186.8	9447	6747.8
2023	3748.8	16795	2186.8	9796.9	6997.8
2024	3748.8	17394	2186.8	10147	7247.7
2025	3748.8	17994	2186.8	10497	7497.6
2026	3748.8	18594	2186.8	10847	7747.5
2027	3748.8	19194	2186.8	11196	7997.4
2028	3748.8	19794	2186.8	11546	8247.4
2029	3748.8	20393	2186.8	11896	8497.3

Year	Electricity consumption per year, kW	Consumption cost, rubles	Electricity consumption per year when installing solar panels, kW	Consumption cost, rubles	Savings by the solar panel kit usage, rubles
2030	3748.8	20993	2186.8	12246	8747.2
2031	3748.8	21593	2186.8	12596	8997.1
2032	3748.8	22193	2186.8	12946	9247
2033	3748.8	22793	2186.8	13296	9497
2034	3748.8	23393	2186.8	13646	9746.9
2035	3748.8	23992	2186.8	13996	9996.8
2036	3748.8	24592	2186.8	14345	10247
2037	3748.8	25192	2186.8	14695	10497
2038	3748.8	25792	2186.8	15045	10747
2039	3748.8	26392	2186.8	15395	10996

From the calculation, we can see that the amount of savings increases each year, due to the increase in the cost of 1 kW. By 2025, the selected kit will be used to cover the initial cost of acquiring it, but do not forget that the first 25 years the solar panel is 100% in use, and 80-90% in the next 20 years, therefore, the cost in subsequent years it will only increase, which can be seen in Figure 2.

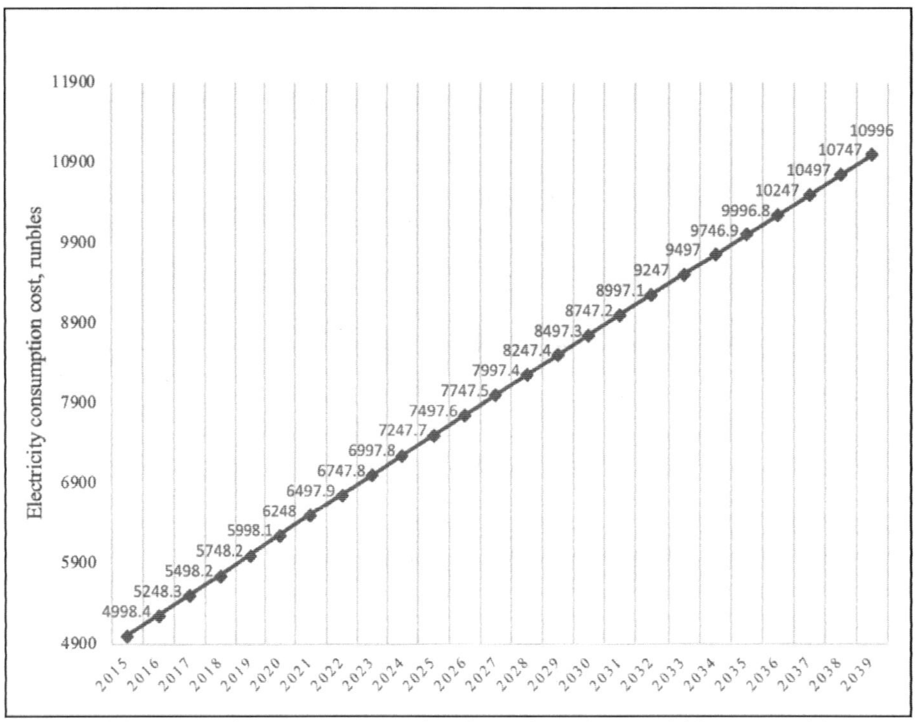

Figure 2. The rising cost of energy consumption from the solar panel kit usage with a cumulative total.

We consider the calculation of energy savings and cash costs during the usage of the energy-saving lamp for100 sq. m. area house. The calculation bases on the fact that the lamp is turned on 6 hours a day (180 hours per month). It is also believed that 1 energy-saving lamp of 20 W in terms of light output is equal to an incandescent lamp of 100 W. A comparative analysis of an incandescent lamp and a compact fluorescent lamp is presented in Table 10.

Table 10. Comparative analysis of incandescent and compact lamps fluorescent tubes.

Name	Incandescent lamp	Compact fluorescent lamp
The cost of the lamp, rubles	10	150-200
The light output of the lamp, kW	0.1	0.02
Operating time, hour	1000	8000-15000
Operating period lamp when used 180 hours per month, month	1000 / 180 = 5.6	8000 / 180 = 44.4
Operating period lamps when used 180 hours per month, year	0.5	3.7
The cost of 1 kW / h, rubles	3.2	3.2
Cost of electricity per month, rubles	3.2 * 180 * 0.1 = 57.6 57.6 + 10 = 67.6	3.2 * 180 * 0.02 = 11.52 11.52 + 180 = 191.52
Energy costs for the first year, rubles	3.2 * 180 * 12 * 0.1 + 10 * 2 = 711.2	3.2 * 180 * 12 * 0.02 + 180 = 318.24
Savings by the use of energy-saving lamps for the first year, rubles	711.2 – 318.24 = 392.96	
Energy costs for the second year, rubles	3.2 * 1.05 * 180 * 12 * 0.1 + 10 * 2 = 745.76	3.2 * 1.05 * 180 * 12 * 0.02 = 145.152
Savings by the use of energy-saving lamps for the second year, rubles	745.76 – 145.152 = 600.608	
Energy costs for the third year, rubles	3.36 * 1.05 * 180 * 12 * 0.1 + 10 * 2 = 762.048	3.36 * 1.05 * 180 * 12 * 0.02 = 152.409
Savings by the use of energy-saving lamps for the third year, rubles	766.048 – 152.409 = 609.639	
Energy costs for three years, rubles	711.2 + 745.763 + 762.048 = 2219.011	318.24 + 145.152 + 152.409 = 615.801
Savings by the use of energy-saving lamps for three years, rubles	2219.011 – 615.801 = 1603.21	

The cost of a standard incandescent lamp varies from 5 to 10 rubles. For the calculation, we took the average price of 10 rubles. The cost of an energy-saving lamp varies from 150 to 250 rubles. It is 25 times more expensive than an incandescent lamp.

The operating time of a standard incandescent lamp is 1000 hours on average. The operating time of a compact fluorescent lamp is about 8000-15000 hours. For calculation simplicity, we took the minimum time set by the manufacturer.

Based on these data and the fact that we use a lamp for 180 hours a month, we can calculate the approximate operating time of each lamp. This is six months for an incandescent lamp and almost 4 years for an energy-saving lamp.

When calculating energy consumption per month, we see that the cost of energy-saving lamps is almost 2.9 times higher than that of an incandescent lamp. At first glance it is ineffective, but when calculating for a year, the savings from using energy-saving lamps are 393 rubles.

Since the energy-saving lamp has a service life of 3.7 years, the savings from its use will increase, with each year and the third year of operation it will be 1603 rubles per lamp.

The calculation presented in Table 11 is carried out during the operation of the lamps during the year.

Table 11. Cost calculation during the operation of lamps for a year.

#	The name of a room	The number of lamps in the room, pcs	Costs of using an incandescent lamp per year, rubles	Costs of using an energy-saving lamps per year, rubles
1	Living room №1	5	3556	1591.2
2	Living room №2	3	2133.6	954.72
3	Living room №3	2	1422.4	636.48
4	Living room №4	3	2133.6	954.72
5	Bathroom	3	2133.6	954.72
6	Restroom	3	2133.6	954.72
7	Hallway	3	2133.6	954.72
8	Kitchen	5	3556	1591.2
	Total		19202.4	8592.48

To carry out this calculation, the average number of lamps in an individual residential building was determined. An economic benefits analysis was made for using energy-saving lamps after a year of service. For a year of using energy-saving lamps, it is possible to save 10610 rubles with an average price of one lamp of 180 rubles. In previous calculations, we determined that the average service of an energy-saving lamp is 3-3.5 years. The cost calculation of working for both types of lamps is in Table 12.

Table 12. Cost calculation for the lamp operation for three years.

#	The name of a room	The number of lamps in the room, pcs	Costs of using an incandescent lamp per year, rubles	Costs of using an energy-saving lamps per year, rubles
1	Living room №1	5	11095	3079
2	Living room №2	3	6657	1847.4
3	Living room №3	2	4438	1231.6
4	Living room №4	3	6657	1847.4
5	Bathroom	3	6657	1847.4
6	Restroom	3	6657	1847.4
7	Hallway	3	6657	1847.4
8	Kitchen	5	11095	3079
		Total	59913	16627
		Overall savings: 59913 – 16627 = 43286		

For 3 years of operation of energy-saving lamps, it is possible to save 43288 rubles, with an initial investment of 4860 rubles (27 lamps * 180 rubles).

Another innovative development of our time is energy-saving windows. By purchasing energy-saving windows, the buyer receives a large number of advantages: you can more effectively keep heat in the house, in the winter they return up to 90% of the heat waves from heating devices; have improved thermal insulation characteristics, thereby helping to keep warm in the house and save on heating; reduce exposure to ultraviolet rays: in the summer the room does not overheat, paintings, wallpapers and other surfaces do not fade; have high light transmission: the room is also light, as if ordinary windows were installed in it; on such windows, the formation of condensate is practically reduced to zero (due to the optimal surface temperature of the glass unit)

Energy-saving when switching to energy-saving windows occurs in three directions:

1) in winter, heating costs are reduced;
2) in the summer, air conditioning costs are reduced;
3) reduced lighting costs in the daytime. [3]

In an individual house with an area of 100 sq. m, there are six window openings, it is proposed to replace wooden windows with double glazing with six-chamber ones with a two-chamber double-glazed window 52 mm thick with argon filling (4M1-20Ar-4V1-20Ar-I4).

The area of one window is 1.4 * 1.7 = 2.38 sq. m, the total glazing area is 14.28 sq. m (2.38 * 6).

To calculate the effectiveness of replacing wooden windows, it is necessary to calculate the heat loss through the windows using the formula (2):

$$Q = {S*dT}/{R} \quad (2)$$

Q – heat loss through the windows, W * h;

S – total glazing area, sq. m;

dT – temperature difference between indoor and outdoor, °C;

R – the value of thermal resistance between the interior and the street, sq. m * °C / W.

The temperature regime inside the house for calculation we take +20 °C. This mode is the most comfortable for humans. The average monthly air temperature is presented in Table 13.

Table 13. Monthly average air temperature.

#	Month	Temperature outside, °C
1	January	-13,5
2	February	-13,1
3	March	6,5
4	April	3,7
5	May	12,4
6	June	17
7	July	19,1
8	August	17,5
9	September	11,2
10	October	3,4
11	November	-3,8
12	December	-10,4

We need the monthly average air temperature to calculate the temperature difference between the indoor and the street. It was said earlier that it is necessary to replace plastic windows with energy-saving ones to see how much heat loss will decrease. Table 14 shows the thermal resistance values for wooden and energy-saving windows.

Table 14. Thermal resistance values.

#	Name	R_o, sq. m*°C/W
1	Double glazed plastic window	0.4
2	Six-chamber window with a double-chamber double-glazed window 52 mm thick with argon filling (4M1-20Ar-4V1-20Ar-I4)	0.945

Table 14 shows that in energy-saving windows, the value of thermal resistance is 2 times higher. Table 15 shows the calculation of heat loss through plastic and energy-saving windows.

Table 15. Heat loss calculation for standard plastic and energy-saving windows.

#	Month	Monthly average air temperature, °C	dT, °C	Q_{window}, Wh	
				Standard plastic window	4M1-20Ar-4V1-20Ar-I4
1	January	-13.5	33.4	1192.38	504.71
2	February	-13.1	33.1	1181.67	500.18
3	March	6.5	26.5	946.05	400.44
4	April	3.7	23.7	846.09	358.13
5	October	3.4	23.4	835.38	353.6
5	November	-3.8	23.8	849.66	359.64
7	December	-10.4	30.4	1085.28	459.38

Table 15 shows that heat loss through plastic windows is higher than through energy-saving. Table 16 shows heat losses per day and per month through plastic and energy-saving windows.

Table 16. Heat loss calculation for plastic and energy-saving windows.

#	Month	Q_{window}, watt * day		Q_{window}, watt * month	
		Standard plastic window	4M1-20Ar-4V1-20Ar-I4	Standard plastic window	4M1-20Ar-4V1-20Ar-I4
1	January	28617.12	12113.07	858513.6	363392
2	February	28360.08	12004.27	850802.4	360128
3	March	22705.2	9610.667	681156	288320
4	April	20306.16	8595.2	609184.8	257856

Table 16 shows the reduction in heat loss when replacing plastic windows with energy-saving ones and reveals a general tendency to save in value terms. The calculation of heat loss reduction is presented in Table 17.

Table 17. The calculation of heat loss reduction

№ p/p	Mounth	Q_{window}, kW * month		Heat loss reduction, kW	Savings from reducing heat loss, rubles
		Standard plastic window	4M1-20Ar-4V1-20Ar-I4		
1	January	858,5136	363,392	495.1216	1584.39
2	February	850,8024	360,128	490.6744	1570.16
3	March	681,156	288,32	392.836	1257.08
4	April	609,1848	257,856	351.3288	1124.25
5	October	601,4736	254,592	346.8816	1110.02
5	November	611,7552	258,944	352.8112	1128.99
7	December	781,4016	330,752	450.6496	1442.08
			Total	2880.3	9216.97

Heat loss reduction in January of 495 kW, which in cost terms is equal to 1584.389 rubles, in March, heat losses are reduced from 681.156 kW to 288.32 kW, which in cost terms is 1257.08 rubles. Replacing plastic windows with energy-saving ones per year results in a decrease of 2880.3 kW, which in value terms is equal to 9216.97 rubles. Figure 3 shows what savings appear monthly.

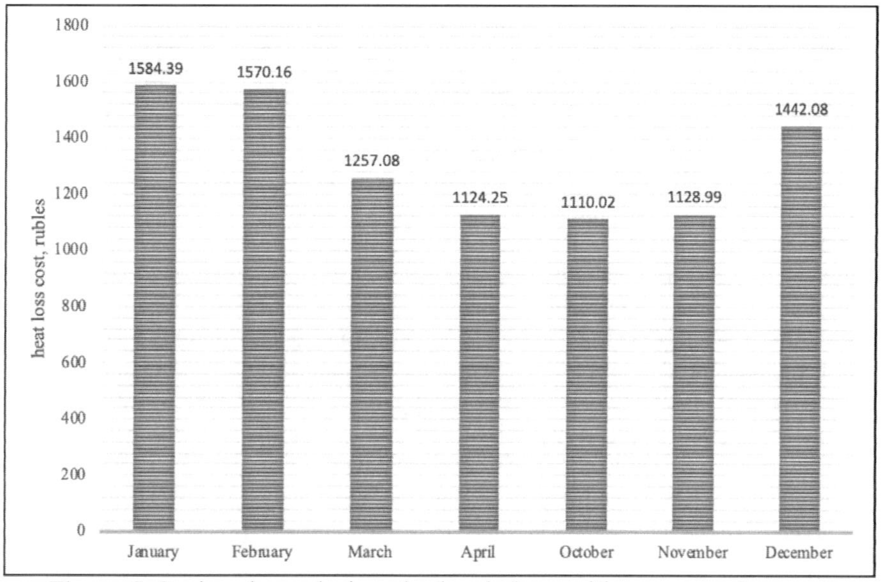

Figure 3. Savings by replacing plastic windows with energy-saving ones.

In this research, it was proposed to replace conventional plastic windows with more energy-efficient ones, whose air resistance coefficient is higher, replace fluorescent lamps with energy-saving lamps and calculate the acquisition and savings from the use of solar panels.

It shows the obvious savings from work and services on energy efficiency in housing construction, which allows it to be transferred to the rank of "Green Building".

4. Conclusion

Today, all regions are interested in the development of environmental construction and the rational use of natural resources. A high share of innovations in construction is associated with the permanent task of the state and the ongoing interest of private investors in solving housing problems of the population. Demographic growth in the world entails a solution to the problem of providing housing for all residents. The situation in the construction market has been the high cost of the facilities being built, and they are trying to solve this problem with the help of innovative projects. Cost reduction, quality growth, improved the ergonomic effect, and increased safety is the main directions of the construction industry. The solution of local problems, such as natural disasters, special climatic conditions are also fundamental in several countries. However, the creation of any mass-scale product in any industry requires the construction of a facility where products will be manufactured and stored, which is why the construction innovation market dominates others.

So, Green Building is necessary for the continued existence of mankind, as the population increases every year and with it, the needs increase, which, in turn, will lead to an increase in the production and consumption of electricity.

References

[1] Baskakova A P , Danilova N I and Scheklein S E 2002 Energy saving and increasing the efficiency of energy use in buildings and structures Ural State Technical University, 328

[2] Grabovy P G and Starovoitov A S 2012 Innovative construction - energy efficiency and environmental friendliness Real Estate: Economics, Management **2**, 68-71

[3] Larionov A N 2009 Methodological approaches to the development of environmental housing programs Business Education Right Bulletin of the Volgograd Institute of Business **9**, 68-80

[4] Larionov A N 2008 Modern Trends in the Development of Ecological Housing *The Economic Revival of Russia* **2**, 10-17

[5] Larionov A N 2011 Environmental friendliness and economic efficiency as factors of the investment attractiveness of housing construction Real Estate: Economics, Management **1**, 14-7

[6] Larionova Yu V 2012 To the question of the formation of a system of technical regulation in ecological housing construction Economics of Education **4**, 129-43

[7] Maksimchuk O V , Baulina O A and Klyushin V V 2017 "Smart" living as one of the aspects of the formation of a "smart" city Sociology of the City, Volume **1** 61-77

[8] Pronin S A 2001 Energy-saving technologies and materials Building Materials, Equipment, Technologies of the 21st century, 17

[9] Rauter R , Globocnik D , Perl-Vorbach E and Baumgartne, R J 2018 Open innovation and its effects on economic and sustainability innovation performance *Journal of Innovation & Knowledge*, Available online, open access, In Press, Corrected Proof doi org/10 1016/j jik 2018 03 004Get

[10] Pavlatos O and Kostakis H 2018 Management accounting innovations in a time of economic crisis *The Journal of Economic Asymmetries*, **18** November 2018, e00106 doi org/10 1016/j jeca.2018.e00106

[11] Jiang F 2012 The Study of the Relationship between House Price and Price Tolerance in China from the Perspective of Systems Engineering Systems Engineering Procedia, **5** 74–80

[12] Romanova A I 2015 Increased construction and service quality products in terms of self-regulation. Proceedings of Kazan State Architectural University **2** 330–3

[13] Romanova A I , Zagidullina G M , Afanasyeva A N & Hkairetdinov R S 2015 Experience in the Region to Increase the Availability of Housing Services. *Mediterranean Journal of Social Sciences* **6(4)** 549–54 Doi:10.5901/mjss.2015.v6n4s2p549

[14] Xie Y 2011 Research on the Land Scale Control Model of Public Housing Construction in China: An Example of Harbin Procedia Engineering, 15 5121-5

[15] Zagidullina G M , Romanova A I and others 2013 Indicative Model of Socio-Economic Development of Small Towns World Applied Sciences *Journal: IDOSI Publications*, **24** 350–7 DOI: 10 5829/idosi wasj 2013 24 03 13203

[16] Zagidullina G M , Romanova A I and others 2013 Peculiarities of Housing Construction Development in the Region Middle-East *Journal of Scientific Research*, **16** 490–5

The method of synthesis of new structural schemes of additional working equipment of the motor grader with the use of graphs and matrices

V A Mikhailovskaya

Moscow Automobile and Road Construction State Technical University (MADI), 64, Leningradsky prospect, Moscow, Russia

E-mail: valeriyamikhailovskaya7@gmail.com

Abstract. The article explores existing methods for the synthesis of structural schemes, but special attention is paid to a method based on the theory of graphs and matrices. The paper presents structural schemes of the additional working equipment of the motor grader and matrices of kinematic pairs, and graphs of fastening of elements of the structural scheme obtained by this method. The authors compare possible variants of these schemes and describe their advantages and disadvantages. The researchers suggest some test results conducted on a grader. The article calculates the stress-strain state, taking into account experimental data, and two design options for additional equipment of the grader.

1. Introduction

It is difficult to imagine modern life without the use of road construction equipment, which makes it possible to comprehensively mechanize the most difficult construction works, such as building, repairing and cleaning roads, preparing foundation ditches for building houses, clearing trees from territories, and developing high-strength soils etc.

Graders are mainly designed for planning and profiling, so the work performed by the grader is an important stage in the construction of roads. In addition to their primary purpose, they can also perform other operations when using additional working equipment. For example, clearing territories can be carried out much more efficiently if we use a bulldozer blade and an additional side blade [1].

Design of the working equipment of the grader, taking into account the structural schemes with the best characteristics, is a difficult task for modern industry engineers. Often, the development of new equipment is based on examples of existing structures and calculation methods, but such way hardly helps to find the optimal design. A design with the most favorable characteristics for these conditions and tasks can be selected when all possible structural schemes are obtained. Therefore, the search for structural schemes of the working equipment of the grader is an important and urgent task today.

The aim of the paper is to demonstrate the method which helps to obtain new structural schemes of the additional working equipment of the motor grader, which will expand the range of options for the

IOP Conf. Series: Materials Science and Engineering **786** (2020) 012030 doi:10.1088/1757-899X/786/1/012030

equipment of the motor grader in real constructions. Moreover, the article describes some designs in order to identify their advantages and disadvantages and prospects for further application in practice.

2. Materials and methods

A significant part of the machines has redundant connections. The work of L. Reshetov proves that the presence of redundant connections does great harm, since they require precision in manufacturing, and eventually make assembly of machines impossible, reduce the load capacity of the machine, reduce efficiency due to high friction forces in the joints, increase mass and dimensions, which leads to reduced productivity and reliability of the machine [2].

There are a large number of methods for structural synthesis of mechanisms: the Frank method, a method based on graph theory, a method for obtaining new structural schemes by adding Assur groups, a method based on grouping theory, etc. [3]. To obtain new structural schemes of the additional working equipment of the grader, we will use the method based on graph theory.

To determine the presence or absence of redundant connections, we use the Somov-Malyshev formula:

$$W = 6 * n - 5 * P_5 - 4 * P_4 - 3 * P_3 - 2 * P_2 - 1 * P_1 - W_m, \quad (1)$$

where

n – number of elements in the structural scheme;

P_i – the number of kinematic pairs of the i-th class;

W_m – number of mobility points.

Along with the known types of constructions for attaching the dozer blade to the grader, there are some others, the disadvantages and advantages of which will be discussed later in the article.

The same scheme can be performed with different kinematic pairs, which will affect both the performance of this design and reliability.

One of the most popular designs among manufacturers is a dozer blade mounted on four rods, where the two lower blades are fixed to each other by means of a reinforcing plate that prevents lateral loads from affecting the hinges. The blade is raised and lowered by a hydraulic cylinder, which is attached either to the blade or to the reinforcement plate with the head of the rod, both options are used in practice.

Figure 1. An example of a bulldozer equipment when the blade is attached to the blade.

Figure 2. An example of a bulldozer equipment when the blade is attached to a reinforcement plate.

IOP Conf. Series: Materials Science and Engineering **786** (2020) 012030 doi:10.1088/1757-899X/786/1/012030

3. Model

Possible structural schemes for this type of additional construction equipment:

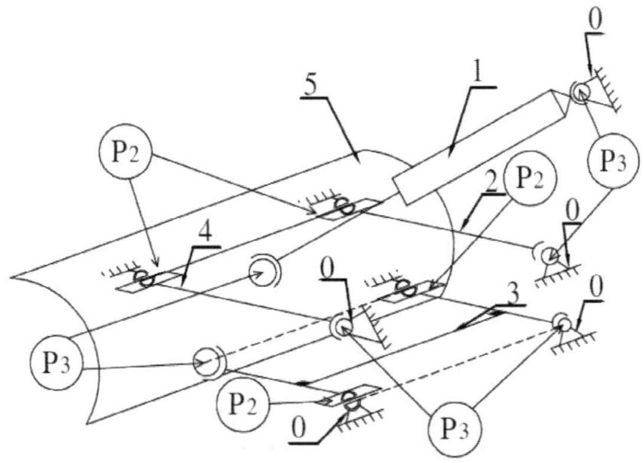

Figure 3. Structural scheme for bulldozer equipment with hydraulic cylinder mount on the blade, option 1.

Let's check availability of redundant connections by the formula:

$$W = 6 * 5 - 3 * 6 - 2 * 4 - 4 = 0$$

Let's make a matrix and a graph for this structural scheme.

	0	1	2	3	4	5
0		P_3	P_3	P_5	P_3	
1	P_3					P_3
2	P_3					P_2
3	P_5					P_5
4	P_3					P_2
5		P_3	P_2	P_5	P_2	

Figure 4. The matrix of kinematic pairs of the structural scheme in Figure 3.

The compilation of such matrices makes it possible to more clearly and quickly assess whether the created structural scheme differs from existing ones. Moreover, it is easier to store it in a computer in this form, creating entire databases of such matrices. In addition, looking at the obtained matrix of kinematic pairs, we can immediately assess whether the structural scheme contains redundant connections or not, because the matrix of kinematic pairs of the structural scheme that does not contain excess mobilities is symmetric with respect to the main diagonal. The graph for the structural scheme is as follows:

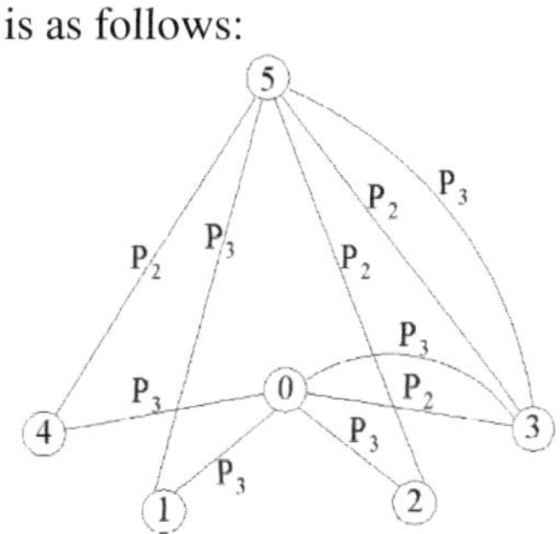

Figure 5. The graph of connections of the elements of the structural scheme in Figure 3.

Looking at this graph, we can conclude that link 3 has several links with both link 5 and link 0. Due to manufacturing errors that are inevitable in production, the axis of attachment of link 3 to links 5 and

IOP Conf. Series: Materials Science and Engineering **786** (2020) 012030 doi:10.1088/1757-899X/786/1/012030

0 may not coincide, therefore it was necessary to carry out one of the fastenings with axial movement, thus allowing the structure to self-mount.

It is possible to reduce this graph, the third-class kinematic pair in conjunction with the second-class kinematic pair gives a fifth-class pair.

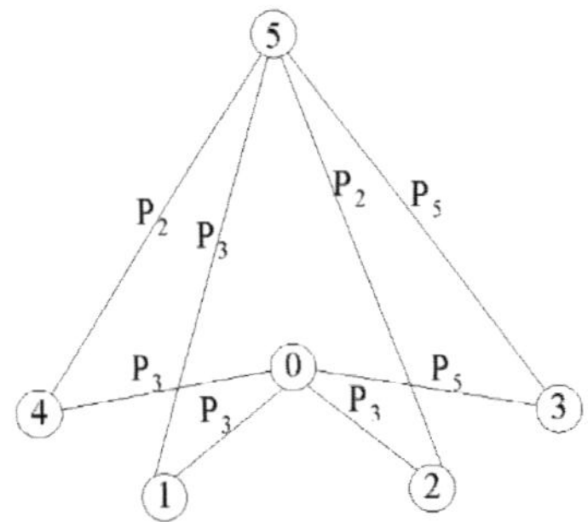

Figure 6. Reduced graph for the structural scheme in Figure 3.

Using graph theory, we can also simplify the process of tracking new structural schemes and even automate it using software [4].

An example is the bulldozer equipment mounted to the motor grader through three rods and a hydraulic cylinder.

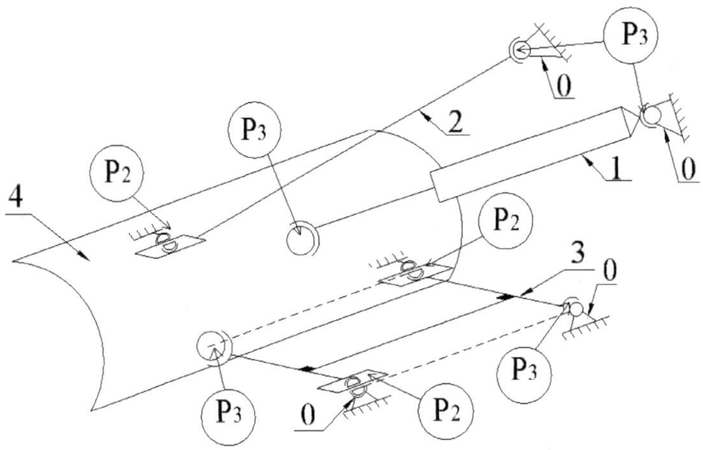

Figure 7. Structural scheme for bulldozer equipment with hydraulic cylinder mount on the blade, option 2.

Let's check the presence of redundant connections by the formula:

$$W = 6 * 4 - 3 * 5 - 2 * 3 - 3 = 0$$

	0	1	2	3	4
0		P_3	P_3	P_5	
1	P_3				P_3
2	P_3				P_2
3	P_5				P_5
4		P_3	P_2	P_5	

Figure 8. The matrix of kinematic pairs for the structural scheme in Figure 7.

IOP Conf. Series: Materials Science and Engineering **786** (2020) 012030 doi:10.1088/1757-899X/786/1/012030

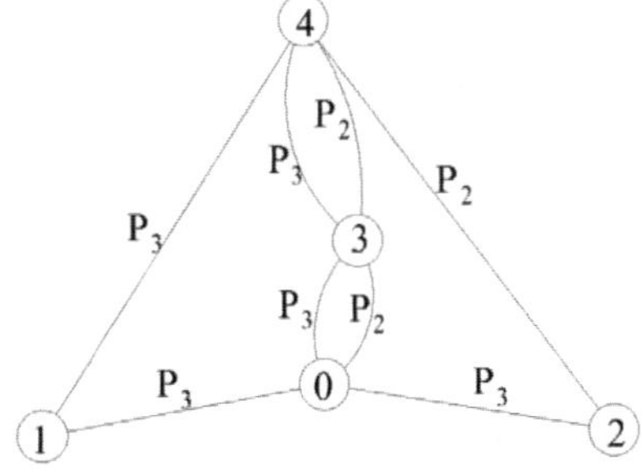

Figure 9. The graph corresponding to the structural scheme in Figure 7.

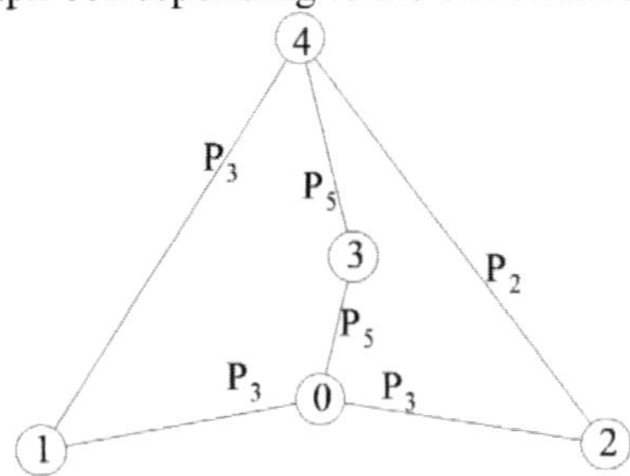

Figure 10. The reduced graph for the structural scheme in Figure 7.

Today, there is a large number of designs that are not inferior to the existing ones in terms of performance and manufacturing costs, but they are not so popular among the already familiar parallelogram suspension. Let's consider the design and structural schemes of some of them.

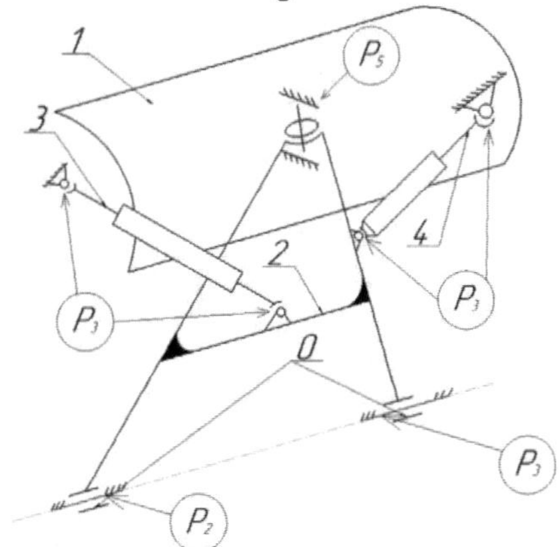

Figure 11. Structural scheme of bulldozer equipment suspended on a triangular frame, option 1.

	0	1	2	3	4
0			P5	P3	P3
1			P5		
2	P5	P5		P3	P3
3	P3		P3		
4	P3		P3		

Figure 12. The matrix of kinematic pairs for the structural scheme in Figure 11.

Let's check the presence of redundant connections by the formula:

$$W = 6 * 4 - 3 * 5 - 2 * 1 - 5 * 1 - 2 = 0$$

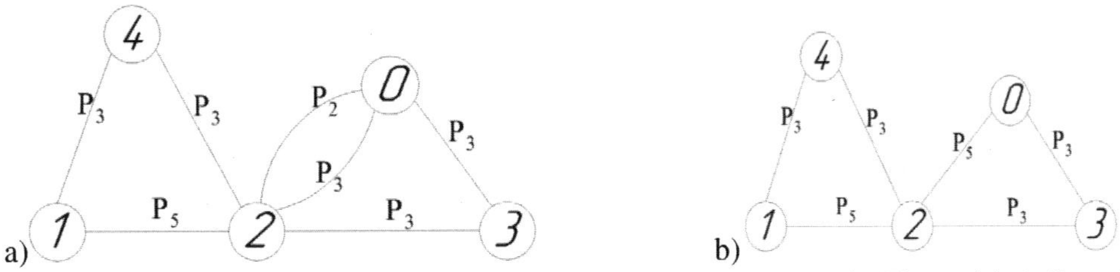

Figure 13. Connection graphs of the elements of the structural scheme in Figure 11 a) Complete graph b) Compressed graph.

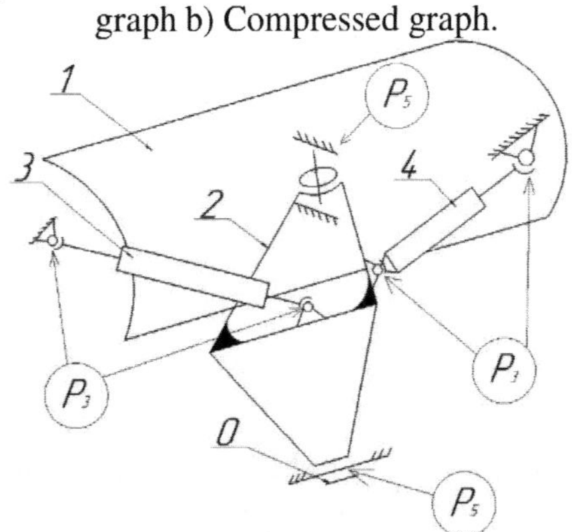

Figure 14. The structural diagram of the bulldozer equipment suspended on a frame in the form of a rhombus.

a)

	0	1	2	3	4
0			P5	P3	P3
1			P5		
2	P5	P5		P3	P3
3	P3		P3		
4	P3		P3		

b)

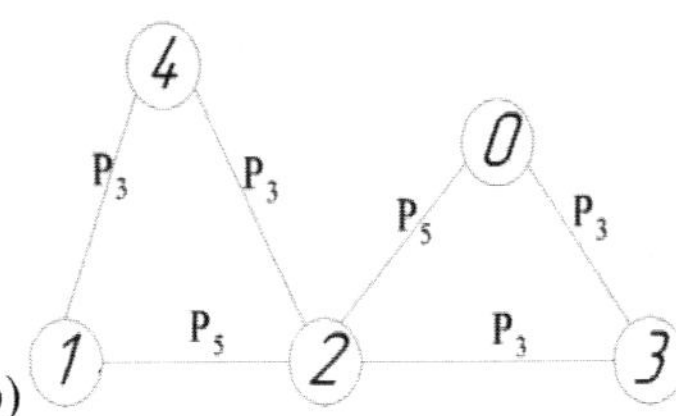

Figure 15. a) A matrix of kinematic pairs; and b) a graph of the connection of elements for the structural scheme in Figure 14.

Considering the scheme in Figure 11, we can see that when attaching the triangular frame with the back side, we obtain a new scheme. Let’s look at it, calculate the stresses obtained by loading both design options in the SolidWorks software and compare it.

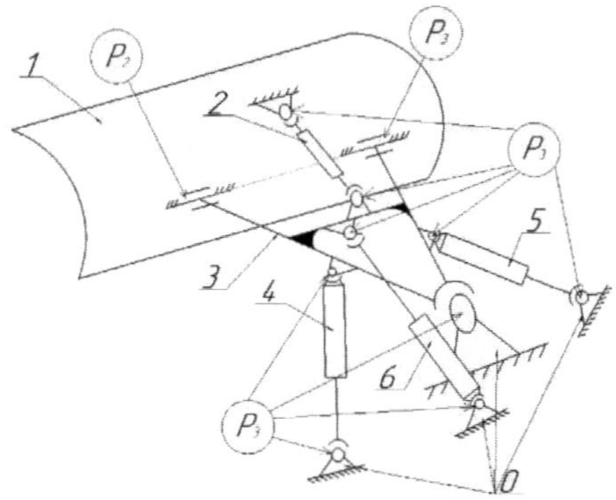

Figure 16. Structural scheme of bulldozer equipment suspended on a triangular frame, option 2.

	0	1	2	3	4	5	6
0				P3	P3	P3	P3
1			P3	P5			
2		P3		P3			
3	P3	P5	P3		P3	P3	P3
4	P3			P3			
5	P3			P3			
6	P3			P3			

Figure 17. The matrix of kinematic pairs for the structural scheme in Figure 16.

Let's check the presence of redundant connections by the formula:

$$W = 6 * 6 - 3 * 10 - 2 * 1 - 4 = 0$$

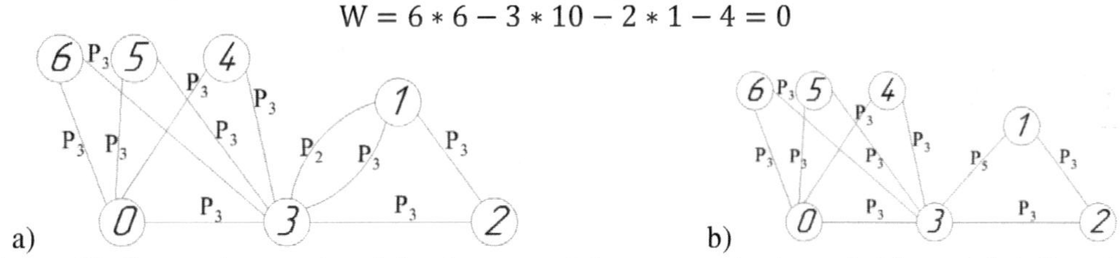

Figure 18. Connection graphs of the elements of the structural scheme in Figure 16 a) Complete graph b) Compressed graph.

There was an experiment at the MADI University training ground to determine the stresses in the metalwork of the motor grader at additional calculated positions, in which the motor grader abutted the edge of the main working equipment of the dump into an insurmountable obstacle. At the same time, strain gauges were glued at control points of the metal structure, according to which stresses were recorded in the metal structure. The experiment was carried out taking into account various calculated provisions of the grader. As a result of the experiment, it was found that the maximum traction force on the dump of the grader at one of the calculated positions is 49850 N. Therefore, under a real existing load of this value, we calculated the models of additional equipment of the grader.

IOP Conf. Series: Materials Science and Engineering **786** (2020) 012030 doi:10.1088/1757-899X/786/1/012030

Figure 19. One of the points (blue mark) on the metal structure of the grader, where the strain gauge was glued. In the illustration, a spinal beam is at the attachment point of the hydraulic cylinders.

Calculation of the design model of the bulldozer equipment corresponding to the structural scheme in Figure 16 showed maximum stresses equal to 154 MPa, which is less than the yield strength, so we can conclude that the design has retained its performance under this load.

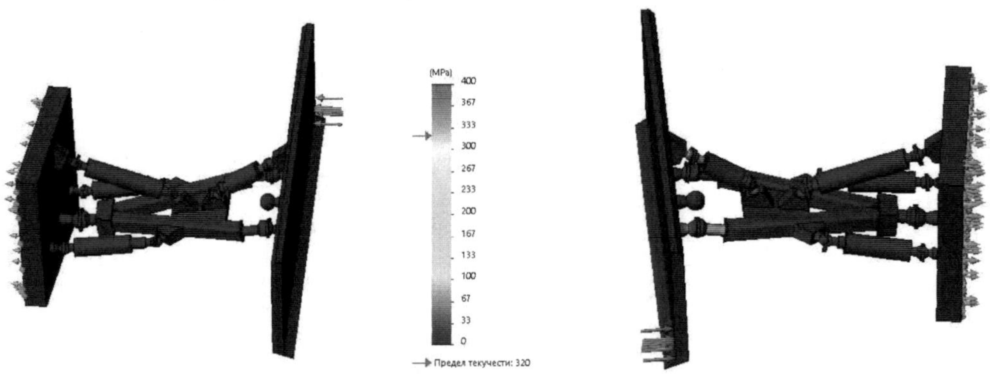

Figure 20. The stress state of the metal structure of the bulldozer equipment when a load of 49850 N is applied to the edge of the blade.

Calculation of the structure corresponding to the structural scheme in Figure 11, showed maximum stresses three times higher than the values obtained in the calculation of the previous design. The calculation showed that the thin spots are the fastening of the hydraulic cylinders for raising / lowering and turning the blade.

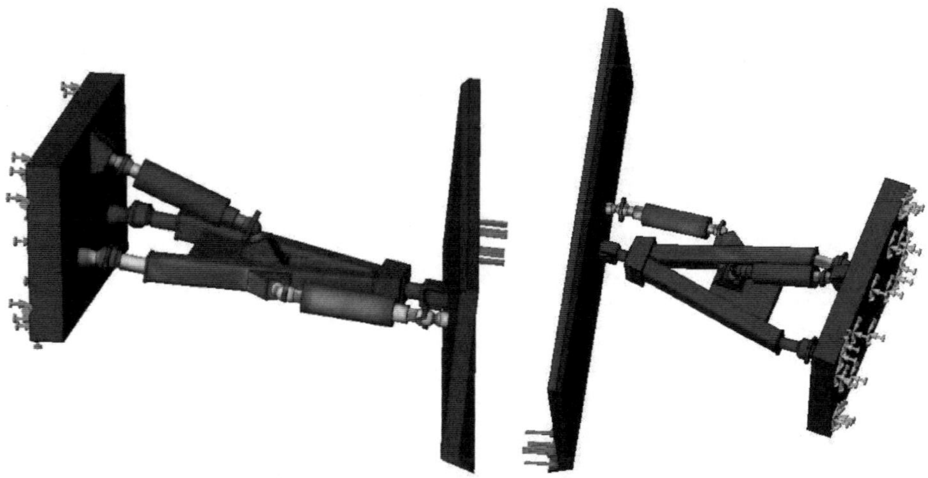

.**Figure 21.** Stresses arising in the metal structure corresponding to the scheme in Figure 11.

The calculation of both models was carried out at the same load, which was applied in both cases to the edge of the blade. Thus, the second design is recommended for use on light graders.

4. Conclusion

1. The use of structural schemes without redundant connections allows to reduce the stress-strain state of the metal structures of the grader.
2. The use of graph and matrix methods for synthesizing new structural schemes allows accelerating the identification of structural schemes without redundant connections.
3. Using experimentally identified real powers on the working body of the grader, we can determine the actual design position of the grader, thereby reducing its stress-strain state and increasing the strength in some nodes of the metal structure, which will increase its productivity by increasing the working effort on the working bodies.
4. The presented method for the synthesis of new structural schemes in conjunction with traditional methods of calculation forms a powerful and comprehensive tool for designing new designs of working equipment of the grader.

References

[1] Saveliev A G 2000 Obosnovanie parametrov strukturnykh skhem i sterzhnevykh sistem rabochego oborudovaniya dorozhno-stroitelnykh mashin: dis ... d-r tekhn. nauk: 05.05.04 364
[2] Reshetov L N 1985 Samoustanavlivayushchiesya mekhanizmy – 2-e izd. – M : Mashinostroenie, 272
[3] Huafeng Ding 2015 Automatic structural synthesis of planar mechanisms amd its application to creative design Doctor thesis, university of Duisburg-Essen, Germany
[4] Pucheta M A 2008 Computational methods for design and synthesis of planar mechanisms Doctor thesis, Argentina (in Eng)
[5] Pennestri E and Belfiore N P 2015 On Crossley's contribution to the development of graph based algorithms for the analysis of mechanisms and gear trains, Mechanism and Machine Theory, 92-106
[6] Glazunov V A 1991 Sintez prostranstvennykh rychazhnykh mekhanizmov parallelnoy struktury i razrabotka metodov ikh analiza: dis . d-r tekhn. nauk: 05.02 18. – Moskva, 311
[7] Mruthyunjaya T S 2003 Kinematic structure of mechanisms revisited, Mechanism and Machine Theory, 279-302

[8] Voloboev V G 2003 Metodologiya optimalnogo vybora parametrov elementov rabochego oborudovaniya zemleroynykh i zemleroyno-transportnykh mashin: dis .. d-r. tekhn. nauk: 05.05 04. – Omsk, – 351

[9] Shevchenko V , Chaplygina O and Beztsennaya Zh 2015 Methods to determine measures providing a motor-grader road-holding ability//*International Scientific Journal* «Machines Technologies Materials», 78-83 (ISSN: 1313-0226)

The deformation-topographic method for the study of the tribotechnical indicators of the working bodies of rotary excavators

Y I Gustov[1,*] , I V Gadolina[2], A A Yushkov[3]

[1]Moscow State University of Civil Engineering, Russia, 129337 Moscow, 26 Yaroslavl highway

[2]Blagonravov Mechanical Engineering Research Institute of the Russian Academy of Sciences, Russia, 101000 Moscow, 4 M. Kharitonyevskiy Pereulok

[3]Baumetall Grupp, Russia,105120 Moscow, 10 Nizhnyaya Syromyatnicheskya str

*E-mail: gustov.u.i@mgsu.ru

Abstract. The article describes the deformation-topographic method, developed on the basis of hypsographic and sclerobatic modeling of the microrelief of friction surfaces. Hypsographic modeling allows us to represent the microrelief of roughness of friction surfaces in the form of a normalized coordinate system "relative reference line - relative approximation" t_p–$\mathcal{E}$. Sclerobatic modeling allows to reproduce the diagrams of the distribution of microhardness along the normal to the friction surface of details and in the potential area of their worn layer.

1. Introduction

We consider the partial values of the micro-wear D_a (the share of depressions) and the micrometall D_m (the share of protusions) in the normalized coordinate system t_p–$\mathcal{E}$ (see Figure 1). The area of partial micro-wear is located above the curve of the reference line L_γ (hypsogram), the share of the partial micrometall is located below the line L_γ. For each partial value, we established the centers of gravity C_a and C_m, respectively, through the pairs of medians. The line connecting the centers of gravity — the bicentroid L_β — intersects the curve of the reference line L_γ at the pole P.

IOP Conf. Series: Materials Science and Engineering **786** (2020) 012031 doi:10.1088/1757-899X/786/1/012031

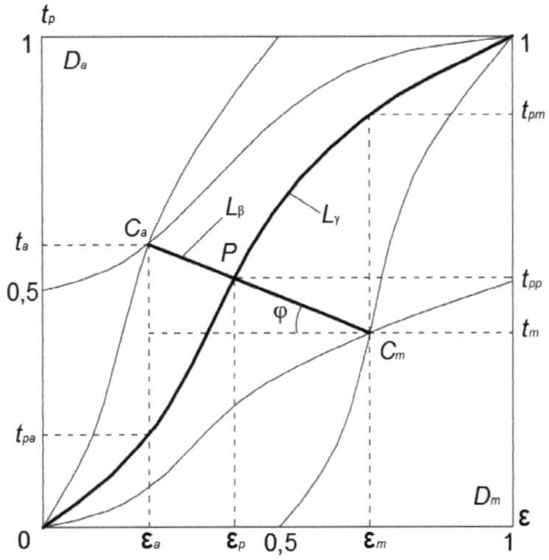

Figure 1. normalized coordinate system "relative reference line - relative approximation" t_p–Ɛ.

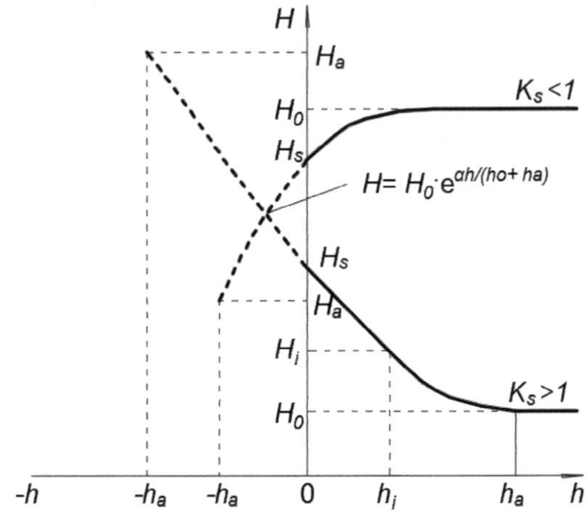

Figure 2. Sclerobatic model of wear.

Rotary excavators are important part of the mining fleet. Their working bodies (buckets, teeth, etc.) could wear out intensively during the operation in different soil conditions [10, 13]. This leads to a significant decrease in the performance of excavators and the associated complex of machines and equipment [2, 3].

Thus, the increase in abrasive wear resistance and durability of the working bodies of rotary excavators seems to be an important issue [9, 11, 12]. The solution of related issues requires the study of other important tribological indicators associated with the wear resistance [4, 5, 6].

2. Methods

The partial values of microwear and micrometals are determined by the relationship of bicentroid segments to the length of bicentroids:

$$D_a = C_aP/C_aC_m,\ D_m = PC_m/C_aC_m = 1 - D_a. \quad (1)$$

Using the partial values, we can determine the coefficient of motion friction (sliding) by the dependence:

$$f = (1 - D_a)^{1/Da} = D_m^{1/Da}, \quad (2)$$

From which, with unknown values of D_a and D_m, but with a known coefficient of friction, we can find the partial microwear D_a by the method of selection and find the partial micrometall D_m by (1). The coefficient of incomplete friction force of the rest (coefficient of adhesion) is calculated from the tilt angle of the bicentroid:

$$F_p = \mathrm{tg}\ \varphi = \mathrm{tg}[(t_a/t_m)/(Ɛ_m - Ɛ_a)], \quad (3)$$

where $Ɛ_a$, t_a and $Ɛ_m$, t_m – coordinates of the centers of gravity of D_a and D_m (see Figure 1).

By expressions (2) and (3), we can estimate the indicator of the adhesion of the detail surface to the ground $P_c = f_p/f$.

The partial values, determined by (2) with the method of selection with known values of the friction coefficient, make it possible to determine the degrees of tribo-deformation hardening (or softening) of the friction surfaces K_s and wear products K_a (see Figure 2).

The degree of hardening K_s is calculated by the formula:

$$K_s = H_s/H_0 = (D_{m\gamma}/D_m)^{Dm/Da} = (0{,}618/\ D_m)^{Dm/Da}, \quad (4)$$

where H_0, H_s – respectively, the initial (technological) and operational microhardness of the detail surface; $D_{m\gamma}$ – harmonic value of the partial micrometall in the golden proportion system [7].

The degree of tribodeformational hardening of wear products is estimated based on the following dependance:

at $K_s>1$ (hardening of the friction surface)

$$K_a = H_a/H_o= K_s[K_s^{0.5}+ (K_s - 1)^{0.5}], \quad (5)$$

at $K_s<1$ (softening of the friction surface)

$$K_a = K_s[K_s^2- (K_s -1)^2], \quad (6)$$

where H_a – limit microhardness of friction surface.

Friction temperatures of friction surfaces T_S and wear products T_a are calculated by the formulas:

$$T_s= T_0 + \Delta T_s= T_0 + \ln K_s/\lambda, \quad (7)$$

$$T_a= T_0 + \Delta T_a= T_0 + \ln K_a/\lambda, \quad (8)$$

where T_0 – ambient temperature; ΔT_s and ΔT_a – the temperature increment of the friction surface and wear products due to friction and wear of the details respectively; $\lambda=2.15\cdot10^{-3}$, 1/°C – average temperature coefficient for steels and alloys.

Efficiency coefficients of the friction bodies (detail - abrasive) can be determined by the expression:

$$\eta = 1/(1+2fL_\gamma/t_{pp}), \quad (9)$$

where t_{pp} – relative reference line at the pole approximation ε_p (see. Figure 1), which is calculated depending on

$$t_{pp} = 1/(1+2D_a). \quad (10)$$

With established practical equality $2f= t_{pp}$ expression (9) takes the form

$$\eta=1/(1+L_\gamma). \quad (11)$$

Hypsogram L_γ length can be measured instrumentally (for example, with an odometer) or calculated by the formula

$$L_\gamma=1.4142 (D_{max}/D_{min})^{DaDm}, \quad (12)$$

where $D_{max}=D_a$ at $D_a> D_m$, $D_{min}=D_a$, at $D_a< D_m$.

To assess the wear resistance of materials of the working bodies of excavators, we use a complex criterion in the form of:

$$\varepsilon =1/ D_a f \eta. \quad (13)$$

3. Results

Basing on the known coefficients of sliding friction of the teeth of rotary excavators [3, 4, 5], we obtained the following values of tribotechnical indicators (table 1).

Table 1. Tribotechnical indicators of rotary excavator teeth.

Soil type	f	D_a	K_s	K_a	ΔT_s, °C	ΔT_a, °C	L_γ	η, %	ε
Peat	0,25	0,5	1,236	1,975	98,55	316,5	1,4142	41,4	19,32
Loam	0,3	0,32	0,816	0,516	94,58	307,7	1,66	37,5	27,78
Sludge	0,25	0,5	1,23	1,975	98,55	316,5	1,4142	41,4	19,32
Clay	0,35	0,09	0,020	0,019	1819,5	1838,0	1,709	36,9	86,03
Clay sandstone	0,3	0,32	0,816	0,516	94,58	307,7	1,666	37,5	27,78

According to the table values of tribotechnical indicators, we can make the following conclusions:

1. During the operation in peat soil and in sludge, the teeth experience tribodeformational hardening of the friction surface (Ks = 1.236) and wear products (Ka = 1.975). The corresponding friction temperatures are ΔTs = 98.55 and ΔTa = 316.5 oC. The efficiency (η = 41.4%) corresponds to the values typical for dry friction conditions, around 40% [7, 8].
2. During the operation in loam and in clay sandstone, the teeth experience triboedeformational softening (Ks = 0.816 and Ka = 0.516). The corresponding friction temperatures are of the surface and wear products are 94.58 and 307.7 oC, respectively. In this case, the efficiency of the friction pair "tooth-soil" is 37.5%, which corresponds to the conditions of dry friction.

3. Operation in clay conditions leads to significant tribodeformational softening of surface layers and teeth wear products (Ks = 0.02 and Ka = 0.019), accompanied by frictional heating to temperatures of 1522 and 1542 ° C, respectively. The efficiency is equal to 36.9%.

4. Comparison of the wear resistance criterion values Ɛ shows the greatest abrasive wear resistance of teeth when working in clay (Ɛ = 86.03), the smallest - in peat and sludge (Ɛ = 19.32); intermediate value (Ɛ = 27.78) - in loam and in clay sandstone. The abnormal value (Ɛ = 496.28) confirms the conclusion about the absence of friction between the teeth in the siltstone.

Basing on the table values of the partial values D_a and D_m=1-D_a, we can construct a normalized system of relative reference lines and approximations t_p – Ɛ based on the approximating function.

$$t_p = Ɛ^{Da/Dm}. \quad (14)$$

Figure 3 shows the t_p – Ɛ system for an excavator when operating in peat and sludge with a friction coefficient of 0.25 and a partial microwear and micrometall values of 0.5 each.

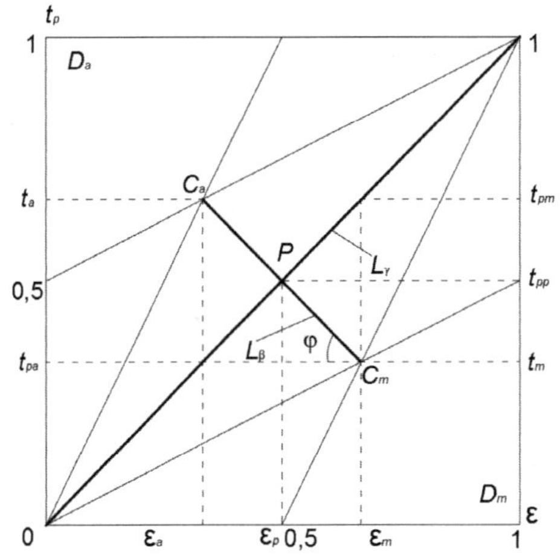

Figure 3. t_p – Ɛ system for an excavator when operating in peat and sludge.

Relying on this system, we can note the following:

1. At equal partial values, the dependence of relative reference lines on relative approximations (hypsogram) is linear.
2. The coefficients of the centers of gravity of the partial microwear, the pole and the partial micrometall are fixed by the values $Ɛ_a$ = 0.333 t_a = 0.666; $Ɛ_p$ = 0.5 t_{pp} = 0.5; $Ɛ_m$ = 0.666 t_m = 0.333.
3. The length of the bicentroids is $L_β$ = 0.4709, the length of the hypsogram is $L_γ$ = 1.4142. The relation $γ=L_γ/L_β$=3,0 is a constant of the system.
4. The tangent of the bicentroid slope (φ = 45 °) equals to 1.0, characterizes the friction coefficient - the coefficient of incomplete rest friction force (f_p). The coefficient of sliding friction is f = 0.25. The ratio f_p/f = 4.0.
5. The degrees of tribodeformational hardening of friction surfaces and wear products are K_s = 1.236 and K_a = 1.975, the corresponding frictional temperatures are $ΔT_s$ = 98.5 and $ΔT_a$ = 316.5 °C. The efficiency at the tooth contacting with the soil is η = 41.4%. A complex criterion for wear resistance is characterized by the value Ɛ = 19.32.
6. The values of the tribotechnical indicators obtained from the t_p – Ɛ system (see Figure 2) coincide with the values of Table 1.

Figure 4 shows the t_p – Ɛ system for teeth during operation in loam and clay sandstone. The analysis of this system allowed us to get the following tribotechnical indicators (Table 2).

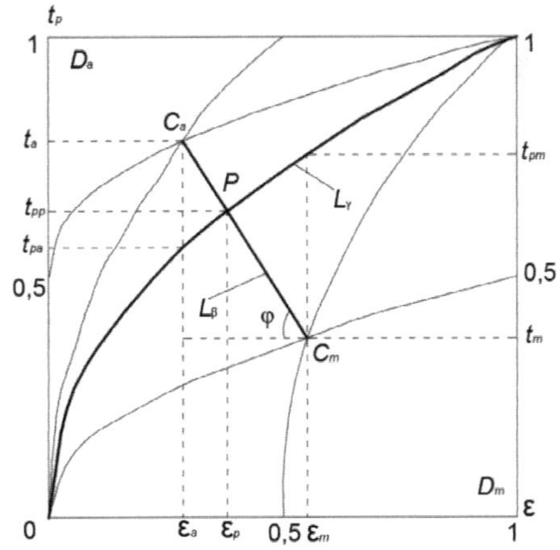

Figure 4. $t_p - \varepsilon$ system for an excavator bucket teeth when operating in loam and clay sandstone.

Table 2. Tribomechanical characteristics of the teeth during the operation in loam and clay sandstone.

K_s	K_a	D_a	ε_a	t_{pa}	t_a	ε_p	t_{pp}	ε_m
0,835	0,56	0,35	0,295	0,275	0,785	0,385	0,64	0,565

t_{pm}	t_m	L_p	L_γ	tgφ	f	η, %	ε
0,765	0,375	0,486	1,50	1,52	0,294	40,0	23,65

Comparison of the values of tables 1 and 2 shows satisfactory coincidence of values D_a (0,32 and 0,35) with discrepancy Δ=8,6 %; friction coefficient f (0,30 and 0,294) at Δ=2,0 %; hypsogram length L_γ (1,666 and 1,50) at Δ=10,7%; degree of tribodeformational softening K_s (0,816 and 0,835) at Δ=2,3 % K_a (0,516 and 0,566) at Δ=8,5 %; complex wear resistance criterion ε (27,78 and 23,65) at Δ=17,5 %.

The results from the table 2 allow us to determine the friction fatigue t by the formula:

$$t = \ln(1/\varepsilon_a) / \ln(\varepsilon_p/\varepsilon_a), \tag{15}$$

The value of friction fatigue is 4.58.

This value indicates the wear of the teeth on the multi-cycle fatigue mechanism. [14, 15].

Test of the wear mechanism by the established dependences shows that the low-cycle fatigue mechanism is possible with the condition:

$$L_\gamma = 1{,}4142+(0{,}05 - 0{,}1D_a)^{0,5} \tag{16}$$

It is performed (see Table 2) with a slight divergence of the values L_γ (Δ=2,4%). The divergence of L_γ from Table 1 is Δ=11,8 %.

Basing on the conditions (15) and (16), we can assume that the wear of teeth in these soils occurs by the mixed mechanism of low-and multi-cycle fatigue.

Figure 5 illustrates $t_p - \varepsilon$ system for teeth during operation in clay. In this case we obtained the following values of tribotechnical indicators (Table 3).

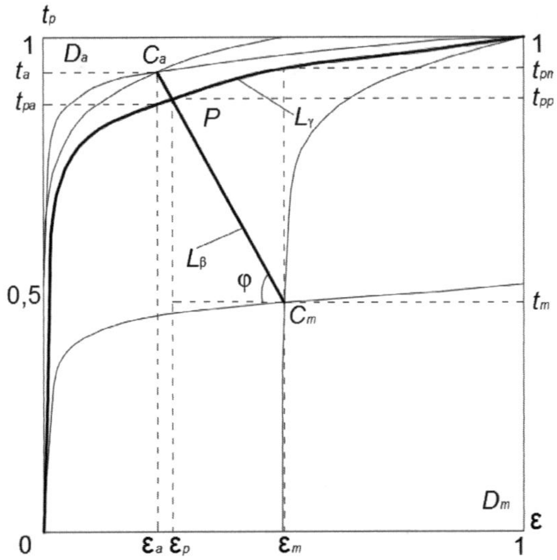

Figure 5. $t_p - \varepsilon$ system for an excavator bucket teeth when operating in clay.

Table 3. Tribomechanical characteristics of the teeth during the operation in clay.

ΔT_s	ΔT_a	D_a	ε_a	t_{pa}	t_a	ε_p	t_{pp}	ε_m	t_{pm}
1522	1544	0,104	0,23	0,86	0,92	0,26	0,87	0,505	0,93

t_m	K_s	K_a	L_β	L_γ	$tg\varphi$	f	η, %	ε
0,57	0,04	0,036	0,53	1,71	1,272	0,348	36,9	74,88

A comparison of the calculated (see table 1) and graphic (see figure 5) values shows the following:

1. The values of the partial micro-wear Da = 0.09% and Da = 0.104 almost coincide (discrepancy Δ = 13.5%), the same with the length of the hypsogram Lγ = 1.709 and Lγ = 1.71 (Δ = 0.06%); friction coefficient f = 0.35 and f = 0.348 (Δ = 0.57 °); Efficiency (Δ = 0%); complex wear resistance criterion ε = 86.03 and ε = 74.88 (Δ = 14.9%).
2. Indicators of tribodeformational softening Ks = 0.02 and Ka = 0.019 are of the same order as the graphic values Ks = 0.04 and Ka = 0.036. The corresponding friction temperatures have the values ΔTs = 1819.5 ΔTa = 1838 ° C and ΔTs = 1522 ΔTa = 1542 ° C with discrepancies Δs = 19.5% and Δa = 19.2%.
3. The frictional fatigue is t = 12, that indicates the wear of the teeth by the mechanism of multi-cycle fatigue. Verification by the Lγ hypsogram and partial microwear condition confirms it.

$$1{,}71=1{,}618+(0{,}025-0{,}1\cdot 0{,}104)^{0{,}5}=1{,}74\ \Delta=1{,}7\%.$$

The value of L_γ = 1.71 and the calculated L_γ = 1.74 with a discrepancy Δ = 1.75% confirms the conclusion about the wear mechanism.

4. Conclusion

1. The deformation-topographic method of studying the microrelief of friction surfaces makes it possible to determine the principal tribomechanical parameters for a given coefficient of sliding friction of the working bodies of ground-moving equipment in various soil conditions.
2. Relying on the initial coefficient of sliding friction, we can determine the partial values of micro-wear and micrometals. They are basic indicators of friction surfaces microtopography for calculating the estimated tribomechanical characteristics, establishing wear mechanisms and criteria for wear resistance of teeth.

3. By means of the proposed approximating power function, we reproduce the curve of the reference line in the normalized coordinate system "relative reference line - relative approximation". Its characteristic indicators help to find the target tribotechnical values.
4. Comparison of tribotechnical indicators calculated by a given friction coefficient with the same indicators of the normalized coordinate system showed their satisfactory numerical coincidence. This confirms the validity of the application of the deformation-topographic method.
5. The paper shows the mechanisms of wear and complex criteria for wear resistance of the teeth of rotary excavators during operation in different soil conditions.

References

[1] Y I Gustov 2011 Triboengineering of construction equipment and machinery: Monograph/ Mosk state un-t of civ ing - M : MGSU, 192 (in Russian)
[2] I V Gadolina P A Pobegaylo, D Yu Kritsky and L Papic 2019 Refinement of the engineering practice of evaluation of the wear rete of excavator working tool // *Reliability*, **1** 18-23 (in Russian)
[3] L Papich 2016 Interval Estimation of the Availability Factor of the Bucket – wheel Excavator Based on Bootstrap Modeling // *Journal of Machinery Manufacture and Reliability* **45(6)** 531-7
[4] D Dowson 1997 Progress in tribology: an historical perspective – In: New Directions in Tribology/ Ed by I M Hutchings Buru St Edmunds and London , MEP, 4-20
[5] N Barimeyer 2013 Positive displacement: Transport of metal particles by means of negative pressure // *Bulk Solids Handling* **31-1** 28-31
[6] Hu J 2006 Shape Characterisation of Concrete Aggregate [Text] // *Image Anal Stereol* **28** 43-53
[7] H Czichos A systems approach to the science and technology of friction, lubrication and wear Tribology series, 1 Amsterdam – Oxford – New York – 1978
[8] G Biresaw and K Mittal 2013 Surfactants in Tribology Volume 3 PDF Taylor Francis Group, LLC , CRC Press, **XVII**
[9] K Grigoroudis and D J Stephenson Modelling low stress abrasive wear *Wear* 1997 T **213** **1-2** 103-11
[10] Q K Luu and D Söffker Dynamics and control of the bucket-wheel excavator part i: dynamic modeling of bucket-wheel boom In collection: Proceedings of the ASME Design Engineering Technical Conference 2014
[11] P Hrabel, R Choteborsky and R Medlin 2006 Influence of Overlay Layers on the Abrasive Wear *Research in Agricultural Engineering* **52(3)** 115–22
[12] G W Stachoviak and A W Batchelor 2005 Laboratory abrasive wear tests: investigation of test methods and alloycorrelation Engineering Tribology 3rd ed Oxford: Elsevier, 831
[13] F Borik and W G Scholz 1971 Gouging abrasion test for materials used in ore and rock crushing: part II Effect of metallurgical variables on gouging wear, J *Mater* **6** 590–605
[14] H Sin, N Saka and N P Suh 1980 Abrasive wear mechanists and the grit size effect Massachusetts Institute of Technology, Cambridge, Mass 02139 (U S A) Fundamentals of Tribology 493-518
[15] P J Mutton 1980 High stress abrasion testing of wear resistant materials, *Tech Bull* , **24(1)** Broken Hill Proprietary, Melbourne Res Lab , Clayton, Victoria, Australia, 38–44

IOP Conf. Series: Materials Science and Engineering **786** (2020) 012032 doi:10.1088/1757-899X/786/1/012032

Durability of bridge asphaltic concrete pavements under temperature loads

I V Grishyn*, G P Ivanov and R A Kayumov

Kazan State University of Architecture and Engineering, 420043, Kazan, Zelenaya str., 1, Russia

* E-mail: iliyahu6357@gmail.com

Abstract. This article presents the major results obtained during the research of the effect of temperature factors on the formation of cracks in asphalt concrete pavements of bridges. The main problem was divided into a number of parts: determination of temperature field in pavement; creation of an engineering design model for determining the stress-strain state caused by temperature deformation; determination of asphaltic concrete rheological characteristics; introduction of criteria for assessing the durability of asphaltic concrete pavement.

1. Introduction

The main objective of the studies is to determine the stress-strain state of the asphalt concrete pavement of bridges caused by the difference in the thermal expansion-compression coefficients of the hard base (reinforced concrete slab or orthotropic metal slab) and asphalt concrete pavements interconnected by a system of protective-adhesive and waterproofing layers, and their durability. As a basis for research, we took asphalt concrete pavement of an orthotropic plate of a metal bridge, with a single protective-adhesive layer made of a material similar in its properties to bitumen. In general, the design was considered as a multilayer plate.

2. Temperature field of a multilayer plate

We assumed that the temperature varies only in thickness. Initially, we attempted to obtain an analytical approximate solution using Fourier series. However, in the case of multilayer plates, this approach led to the fact that the solution did not have physical sense. Moreover, for single-layer plates, this method is perfectly suitable and was used to verify the results of determining the temperature field in ANSYS, which implements the finite element method (FEM) [1]. The verification showed a difference in the results of the analytical approach and the FEM for a single-layer plate of less than 4% and we decided to use ANSYS for a multi-layer plate.

We took the data on temperature changes for December, January and February as temperature boundary conditions (this period has the highest rates of temperature change) in 2006. The initial temperature conditions assumed a uniform temperature distribution. We used a flat 4-node PLANE55 element (Quad 4node 55), as in the test task. An example of a temperature field is shown in Figure 1.

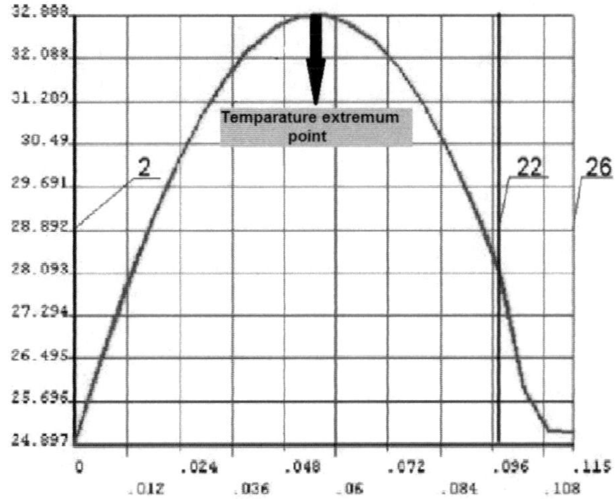

Figure 1. The temperature field of the bridge canvas:
point 2 - free boundary of asphalt concrete; point 22 - the contact plane of asphalt concrete and the protective-adhesive layer; point 26 - the free surface of the metal flooring of the orthotropic plate.

Analyzing Figure 1, we can conclude that the stress-strain state is influenced by the temperature drop across the thickness of the asphalt concrete. We found that the stiffeners of the orthotropic plate do not significantly affect the temperature field of the multilayer plate. According to the calculation results in ANSYS, depending on the thickness of asphalt concrete, the maximum values of the temperature difference in thickness were: for asphalt concrete of 0.1 m thick - 5.5 °C; for asphalt concrete of 0.15 m thick - 7.5 °C; for asphalt concrete of 0.2 m thick - 8.9 °C.

3. Stress-strain state model of multilayer plate

We considered 3 calculation models describing the stress-strain state of a multilayer plate [2]:

- A model using the hypothesis of complete deformations uniform in height, with stresses determined by the following equation (uniform deformation hypothesis):

$$\sigma_i(y) = E_i \cdot c - E_i \cdot \alpha_i \cdot \Delta T_i(y), \tag{1}$$

- A model using the hypothesis of linearly varying heights of the entire multilayer deformation plate, with stresses determined by the following equation (inclined hypothesis):

$$\sigma_i(y) = E_i \cdot (a \cdot x + b) - E_i \cdot \alpha_i \cdot \Delta T_i(y), \tag{2}$$

- A model using the hypothesis of linearly varying deformations for each layer (the hypothesis of a broken line):

$$\sigma_{zz}^{(i)}(z,y) = E_i \cdot \varepsilon_{zz}^{(i)}(z,y) - E_i \cdot \alpha_i \cdot \Delta T_i(y), \tag{3}$$

$$\varepsilon_{zz}^{(i)}(z,y) = \frac{\partial u^{(i)}(y,z)}{\partial z} = \grave{U}_i(z) \cdot \frac{-y + y_{i+1}}{h_i} + \grave{U}_{i+1}(z) \cdot \frac{y - y_i}{h_i}, \tag{4}$$

where:
i – number of layer;
$\sigma_i(y)$ – stress in the i-th layer, MPa;
E_i – the elastic modulus of the material of the i-th layer, MPa;
α_i – coefficient of linear thermal expansion-compression, $\frac{1}{°C}$;
$\Delta T_i(y)$ – function representing the temperature difference in the i-th layer, °C;
c – value of deformations uniform in height;
a, b – sought coefficients for the hypothesis of the inclined;

$\grave{U_i}(z)$ – displacements of the boundary of the layers i and i-1, specified in the form of approximating functions satisfying the boundary conditions;

$h_i = y_{i+1} - y_i$ – thickness of the i-th layer, m.

The hypotheses of inclined and broken lines are shown more clearly in Figure 2.

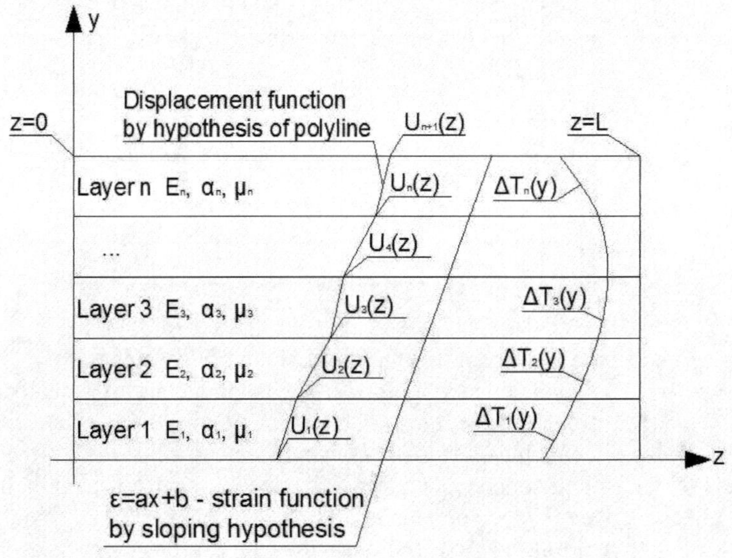

Figure 2. Hypotheses of inclined and broken lines.

The results obtained by the above hypotheses were compared with the results obtained in ANSYS. In this case, the selection criteria were the difference between the results obtained using the hypothesis and the FEM results and the simplicity of the equations obtained using the hypothesis. In addition, using FEM, we preliminary analyzed the factors affecting the choice of a hypothesis for the calculation model and made the following conclusions:

1. The temperature difference across the thickness of the asphalt concrete does not significantly affect its thermally stressed state. As a consequence, we further assumed that the temperature is distributed uniformly throughout the structure and is equal to the ambient temperature.
2. The normal stresses of asphalt in the transverse and longitudinal directions differ slightly.
3. Normal stresses of asphalt concrete increase from the edge of the plate to its center.
4. An increase in the stiffness of the base by a factor of 1000 does not lead to a significant increase in stresses. For this reason, we concluded that the models under development are applicable not only for metal bridges with an orthotropic plate, but also for reinforced concrete bridges, where the stiffness of the base, taking into account the walls of the main beams, can be higher.

As a result of the comparison, we concluded that the hypothesis of the inclined and the broken line hypothesis correspond to a numerical solution with a sufficient degree of accuracy. Basing on the simplicity criterion, we should choose the hypothesis of the inclined. A comparison of the hypothesis of the inclined with the uniform deformation hypothesis, shown in Table 1, allows us to conclude that the simplest uniform deformation hypothesis is best suited to determine the thermally stressed state.

Table 1. A comparison of the hypothesis of the inclined with the uniform deformation hypothesis.

	The difference between the hypothesis of uniform deformations from the FEM	The difference between the hypothesis of the inclined and the FEM
Difference in maximum stresses at various geometrical parameters of an orthotropic plate	8.4 %	12 %
The difference in maximum stresses with an increase in the elastic modulus of a metal of an orthotropic plate by 1000 times	10 %	10 %
The difference in maximum stresses with an increase in the elastic modulus of asphalt concrete by 10 times	1.5 %	2.5 %
The difference in maximum stresses with a decrease in the elastic modulus of asphalt concrete by 10 times	4 %	4 %
The difference in maximum stresses with an increase in the thickness of asphalt concrete by 0.05 m	2 %	7 %
The difference in maximum stresses with a decrease in the thickness of asphalt concrete by 0.03 m	6 %	6 %

4. Building a creep model for asphalt concrete

To describe the creep of asphalt concrete, the equation of the stiffness theory was used in the following form [3, 4]:

$$\dot{\varepsilon}_{cr}(t) = \frac{\sigma^{k(T,\sigma)}}{{}^{1}/_{(c(T)\cdot s^4(\sigma,T))} \cdot \varepsilon_{cr}^{\alpha}(t)} \tag{5}$$

where: $\dot{\varepsilon}_{cr}(t)$ – creep relative strain rate;

σ – stress, MPa;

t – time, sec;

T – temperature, °C.

$k(T,\sigma)$, $c(T)$, $s^4(\sigma,T)$, α – coefficients, which in the general case depend on temperature and stresses, which are selected from the condition of best approximation to the experimental data.

Since during the experiment the stresses remained constant, integrating expression (5) we obtain the expression of creep deformations in explicit form:

$$\varepsilon_{cr}(t) = \sqrt[\alpha+1]{(\alpha+1)\cdot c(T)\cdot \sigma^{k(T,\sigma)}\cdot t}\cdot s(\sigma,T), \tag{6}$$

To obtain creep curves from fine-grained dense asphalt concrete grade B-I, we took standard samples with a diameter and height of 71.4 mm. Samples were inserted into a lever specially made for this, which was located under a canopy on the street (Figure 3), where under conditions of variable temperature equal to the ambient temperature, a compressive load of various sizes was applied to the samples. As a result, the following dependences were obtained [5], graphically shown in Figures 4, 5 (experimental curves are shown at a variable temperature, on average equal to the specified):

Stresses 0-1.75 MPa:

$$\varepsilon_{cr}(t) = \sqrt[4]{4\cdot(350+310e^{-0.16T})\cdot 1.75^{(-15.5-15.5e^{-0.04}\ )}\cdot t}\,\frac{\sigma}{1.75\cdot 71.2} \tag{7}$$

1) Stresses 1.75-3.5 MPa:

$$\varepsilon_{cr}(t) = \sqrt[4]{4 \cdot (350 + 310e^{-0.16T}) \cdot \sigma^{(-15.5-15.5e^{-0.04}\ )(1-0.171(\sigma-1.75))} \cdot t^{\frac{(1+33.71(\sigma-1.75))}{71.2}}} \tag{8}$$

2) Stresses 3.5-5 MPa:

$$\varepsilon_{cr}(t) = \sqrt[4]{4 \cdot (350 + 310e^{-0.16T}) \cdot \sigma^{(-15.5-15.5e^{-0.04T})(0.7-0.133(\sigma-3.5))} \cdot t^{\frac{60}{71.2}}} \tag{9}$$

3) Stresses >5 MPa:

$$\varepsilon_{cr}(t) = \sqrt[4]{4 \cdot (350 + 310e^{-0.16T}) \cdot \sigma^{(-1\ .5-15.5e^{-0.04}\ )(0.7-0.133\cdot1.5)} \cdot t^{\frac{60}{71.2}}} + \\ + \frac{(0.034+3.14\cdot10^{-3}T+8.43\cdot10^{-5}T^2-1.17\cdot10^{-7}T^3-2.31\cdot10^{-8}T^4)}{3}(\sigma-5) \tag{10}$$

Figure 3. Asphalt concrete test lever.

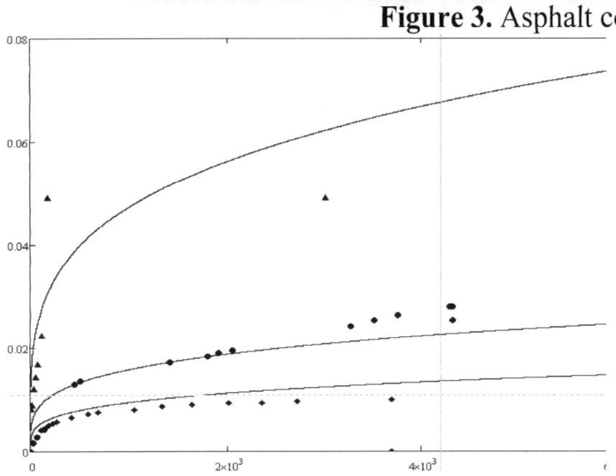

▲ – for -3 °C, ● – for -11 °C, ◆ – for -13 °C

Figure 4. Creep curves at 5 MPa.

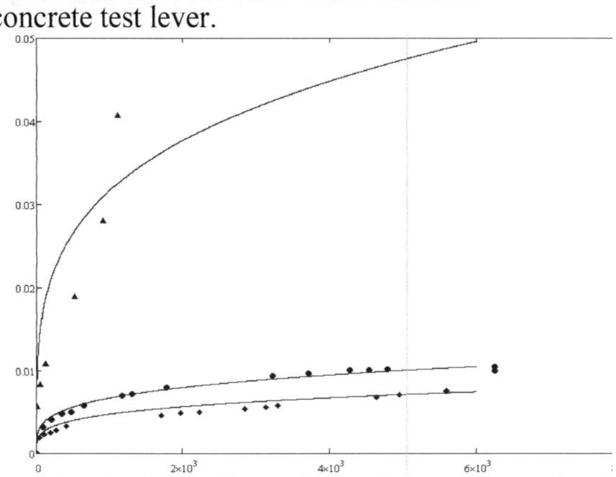

▲ – for 21.8 °C, ● – for -1.5 °C, ◆ – for -6 °C

Figure 5. Creep curves at 1.75 MPa

In order to determine asphalt concrete creep strains under tension, an asphalt concrete bend experiment was performed on beams (Figure 6) and a conversion coefficient was found that allows us to determine tensile creep strains, knowing the creep strains under compression [6]:

$$k(\varepsilon_{cr}^{\mathrm{p}}) = \frac{1+1.5\cdot 10^{17}\cdot 2.3\cdot(\varepsilon_{cr}^{\mathrm{p}})^6}{1+1.5\cdot 10^{17}\cdot(\varepsilon_{cr}^{\mathrm{p}})^6}, \quad (11)$$

Multiplying (7-10) by (11), we obtain creep strains under tension at any stress level and temperature.

Figure 6. Bending experiment on asphalt concrete beams.

5. The criterion for assessing the durability of asphalt concrete bridge pavements under thermal effects

Given the above, the full deformations will be expressed as follows:

$$\varepsilon = \varepsilon_{\mathrm{el}} + \varepsilon_T + \varepsilon_{cr}, \quad (12)$$

Where :ε_{el} - elastic deformations;
ε_T - thermal strains;
ε_{cr} - creep strains.

From equation (12), we can determine the thermally stressed state, taking into account creep. But since temperature stresses act for a long time, the durability of the pavement is defined as its long-term strength using the theory of linear damage accumulation, the possibility of using which is mentioned in [7]:

$$P = \sum_{i=1}^{n} \frac{dt_i}{\tau_i} \quad (13)$$

where: i – time interval number;
dt_i – the duration of the i-th time interval, sec;
τ_i –durability in the i-th time interval, under the assumption that during the time dt_i the stresses act constantly, sec;
n – total number of time intervals.

When P value becomes equal to one, destruction occurs. The durability τ can be determined, for example, by Zhurkov or Bartenev equations. In the absence of a sufficient amount of experimental data, we used the following data:

- [8] presents data for fine-grained asphalt concrete B-I in the temperature range from -40 °C to 0 °C:

$$\tau = \tau_0 \exp\left(\frac{u_0-\gamma}{RT}\right), \quad (14)$$

where: τ – durability of the material, under the action of constant stress σ, sec;
T – temperature, K;
$R = 8{,}31 \cdot 10^{-3}$ – universal gas constant, $\frac{kJ}{mol\cdot K}$;

$\tau_0 = 10^{-13}$ – constant corresponding to the period of oscillation of kinetic units ($\tau_0 = 10^{-12} \div 10^{-13} c$);

$u_0 = 89{,}85$ – effective activation energy of the destruction process, $\frac{kJ}{mol}$;

$\gamma = 60{,}58 \cdot 10^{-1}$ – structural coefficient characterizing the sharpness of the decrease in the activation energy of fracture with increasing stress, $\frac{kJ}{mol \cdot MPa}$.

- in [9], data is presented for sandy asphalt concrete type G in the temperature range from -15 ° C to +50 ° C

$$\tau = 5{,}9 \cdot 10^{-2} \cdot \sigma^{-(1{,}8+0.2\sigma)} \exp\left(\frac{43{,}2}{RT}\right) \quad (15)$$

In equation (15), stresses σ are taken in $\frac{kg}{cm^2}$, and durability is obtained in seconds. The change in ambient temperature can be set in the following form:

$$T = \sum_{i=1}^{2} a_i \sin\left(\frac{\pi \cdot t \cdot 2}{p_i} + \frac{\pi}{2}\right) \quad (16)$$

where: $a_1 = 30°C$, $a_1 = 5\,°C$ - the amplitude of the annual and daily average temperature fluctuations.

$p_1 = 1\,year$, $p_2 = 1\,day$ - periods of annual and daily temperature fluctuations.

By setting the temperature load in accordance with (16), taking into account (12), (1), (7) - (11) and substituting the obtained values in (13), it is possible to obtain the time after which the process of destruction of asphalt concrete pavement begins. The result of applying this procedure showed that the data obtained in this way do not correspond to the service life observed in the practice of building asphalt concrete pavements. Figures 7 and 8 show the stress variation curve during the year and the damage accumulation curve for the year for laying asphalt concrete at +25 °C. We can see that the onset of destruction is projected in less than a year.

Most likely this is a consequence of the lack of experimental data on the durability of asphalt concrete and the possible phenomenon of "healing" of damage in asphalt concrete at positive temperatures. Therefore, we decided to introduce a hypothesis: damage accumulation occurs only below a certain limit {X}, called the damage accumulation limit.

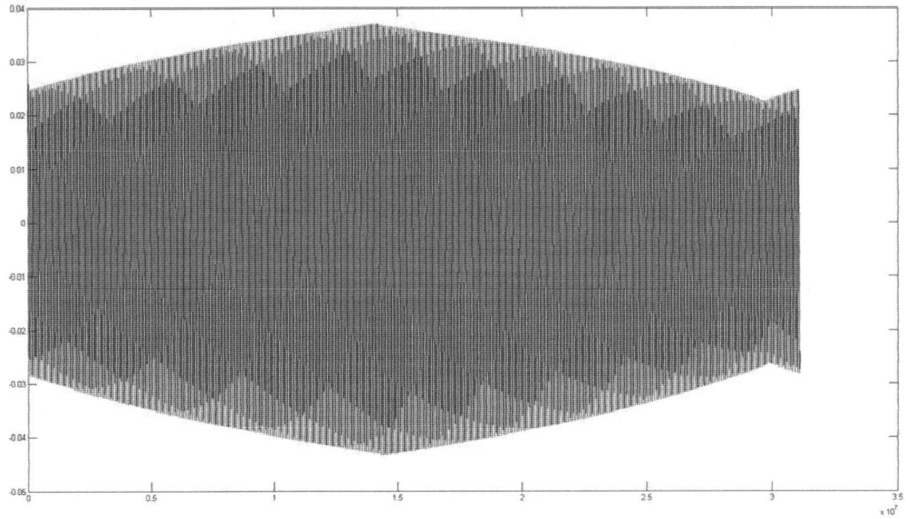

Figure 7. Curve of stress variation (MPa) over time, taking into account creep.

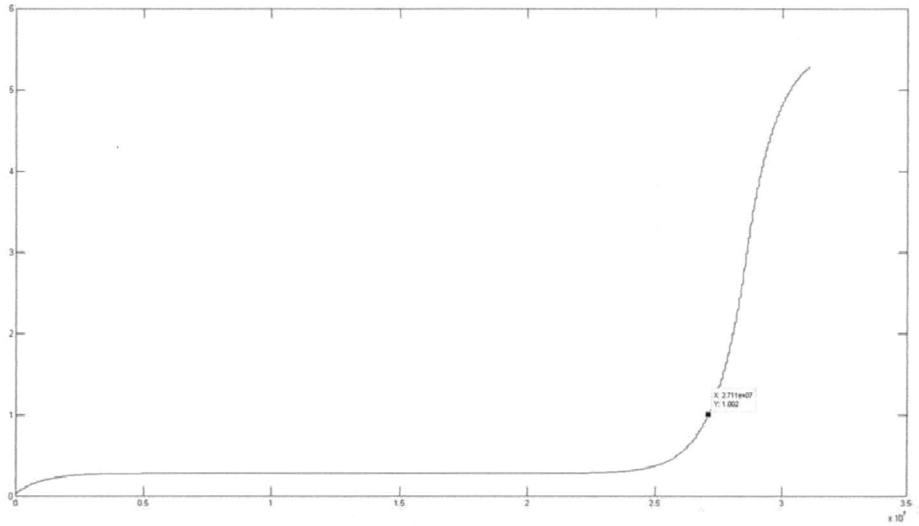

Figure 8. Curve of accumulation of damage for the year.

Based on observational data on asphalt concrete pavements of bridges and roads, we can conclude that cracks caused by temperature changes do not appear earlier than the autumn-winter period, i.e. at temperatures below + 10 °C. At {X} = + 10 °C and when laying asphalt concrete at + 30 °C, it turns out that *P* reaches only the value of 0.33 in a year, i.e. destruction will begin in 3 years (Figure 9, 10). Thus, at the moment, it is possible to predict the start time of the destruction of asphalt concrete bridge pavements only with certain assumptions that require verification.

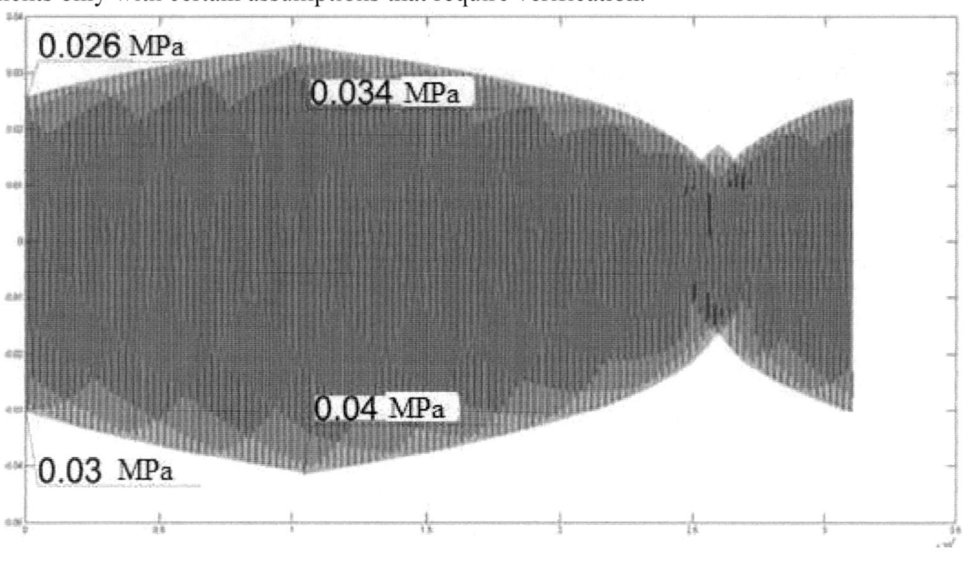

Figure 9. The curve of stress variation (MPa) over time, taking into account creep at {X} = + 10 °C.

IOP Conf. Series: Materials Science and Engineering **786** (2020) 012032 doi:10.1088/1757-899X/786/1/012032

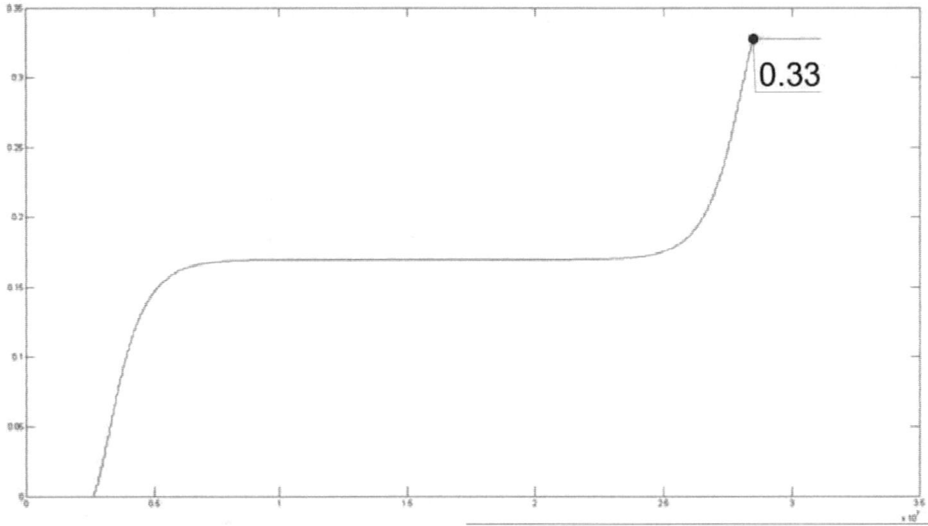

Figure 10. The curve of damage accumulation for the year at {X} = + 10 °C.

6. Conclusion

1) Numerical experiments have established that the real rate of change in ambient temperature does not lead to large temperature differences in the layer of asphalt concrete (not more than 5 ° C). This, in turn, does not lead to a significant difference in the thermally stressed state of asphalt concrete in comparison with the case of a uniform temperature distribution (not more than 6% of the stresses in the elastic setting).

2) Numerical studies with FEM have established that the thermal stress state of asphalt concrete of the bridge pavement corresponds to the stress state of a multilayer beam with a uniform distribution of total deformations along its thickness, which is typical for the operation of centrally stretched elements.

3) Numerical studies with the FEM found that even a multiple increase in the stiffness of the base of a multilayer plate leads to an increase in stresses in asphalt concrete of not more than 10% in an elastic setting.

4) New experimental data on the creep of asphalt concrete have been obtained. Based on the analysis of experimental data, we found dependencies that describe the creep of asphalt concrete at various stresses in the temperature range from -30 ° C to +30 ° C. These equations can be used to calculate taking into account the creep of roads, bridges and other structures with asphalt concrete pavement.

5) At present, there is insufficient experimental data to predict the beginning of the destruction of asphalt concrete bridge pavements. However, we can introduce simplifying hypotheses based on observational data on the destruction of asphalt concrete pavements.

6) To accurately predict the onset of failure of asphalt concrete pavements, it is necessary to obtain more accurate data on the creep of asphalt concrete under cyclic loads, its durability and the presence or absence of the phenomenon of “healing” of damage at positive temperatures.

References

[1] Grishin I V , Kayumov R A and Ivanov G P 2012 Determination of temperature field in orthotropic slab pavement // *Proceedings of KGASU* **3(21)** "KazGASU" 156-66

[2] Grishin I V , Kayumov R A and Ivanov G P 2011 About pavements of metal bridges with orthotropic slab under temoerature loading // *Proceedings of KGASU* **3(17)** "KazGASU" 171-8

[3] Malinin N N 1975 Applied theory of plasticity / N N Malinin - M : Mashinostroeni 400

[4] Rabotnov Yu N 1966 Creep of structural elements / Yu N Rabotnov M Nauka 752
[5] Grishin I V , Kayumov R A and Ivanov G P 2013 Experimental research of asphaltic concrete rheology under various temperatures // *Proceedings of KGASU* **2(24)** "KazGASU" 99-107
[6] Grishin I V , Kayumov R A and Ivanov G P 2013 Experimental research of asphaltic concrete tensile creep // Collection of reports of the international scientific-practical conference "Innovative materials, technologies and equipment for the construction of modern transport facilities", BSTU, Belgorod 133-43
[7] G Kiryukhin 2005 Designing the Composition of Asphalt Concrete and Methods for Testing It / G N Kiryukhin // Auto Roads and Bridges: Overview inform / "*INFORMAVTODOR*" **6**
[8] Nikolskiy Yu E 1971 Investigation of low-temperature properties of asphalt concrete with bitumen of different viscosity *Works of Soyuzdornii* **44** 77-87
[9] Kiryukhin G N 1978 On the issue of long-term durability of asphalt concrete *Works of Syzdornia* **99** 31-8

IOP Conf. Series: Materials Science and Engineering **786** (2020) 012033 doi:10.1088/1757-899X/786/1/012033

Evaluation of the efficiency of roundabouts in modern cities

A S Gavaev

Tyumen Industrial University, 625000, Tyumen, Volodarskogo Street, 38

E-mail: gavaev@list.ru

Abstract. The article presents the results of studies that allow to reduce time delays by determining the optimal parameters of roundabouts based on the established dependencies of road characteristics on the geometric parameters of road intersections and traffic flows. In analytical studies, we identify the most effective types of configurations of the road intersection at the same level and establish the dependence of the average vehicle speed on the total traffic intensity for these configurations. Moreover, the paper establishes the dependence of the average speed on the intensity of traffic flow at different radii of the central guide island of the roundabout and the dependence of the average speed on the radius of the central guide island with different bandwidth of the carriageway on the approach to the roundabout.

1. Introduction

The high rate of motorization in major cities around the world is steadily leading to an increase in the number of road congestion. The problem is especially acute in the intersection zone, in the areas of merging and interweaving of traffic flows, where the direction of movement changes. At these points, the parameters of traffic flows deteriorate - the connection speed decreases and travel time increases due to traffic delays and traffic imperfections [15]. This leads not only to a decrease in the comfort of travel and quality of transport services [1], but also to an increase in ambient air pollution by products of incomplete combustion of fuel due to an unstable driving regime, frequent forced stops and congestion of cars at intersections [10]. The population of the city is constantly exposed to noise and exhaust gases [2]. An increase in the intensity of traffic and pedestrian flows directly affects road safety [6]. Ensuring fast and safe movement of vehicles in modern cities requires the use of a range of architectural, planning and organizational measures [3].

Existing guidelines are aimed at increasing the efficiency of using roundabouts, namely, reducing the number of conflict situations in traffic [4], preventing accidents and reducing the severity of their consequences at roundabouts [8].

Roundabouts belong to intersections at which the movement of vehicles takes place in a circle, which unloads the transport hub. However, the number of accidents at such intersections is higher than at a standard four-way intersection. Traffic accidents at such intersections rarely lead to serious consequences, injuries and deaths [7, 14, 16]. But at the same time, the number of small incidents here is quite large, which obviously reduces the throughput of city streets.

2. The main part

To evaluate the efficiency of roundabouts we used micromodelling as the main research method [9]. To forecast traffic flows, we should assess the main parameters of the traffic conditions - speed of movement, level of loading, time spent [5].

Modeling of intersections [11] was carried out with different variants of traffic organization: a four-way intersection (X-intersection) and a roundabout.

When changing the configuration of the intersection from a four-way two-lane intersection to a roundabout, the traffic parameters for traffic flows change in a positive direction [13]. So the average delay time is reduced by 43%, and the average speed increases by 28%.

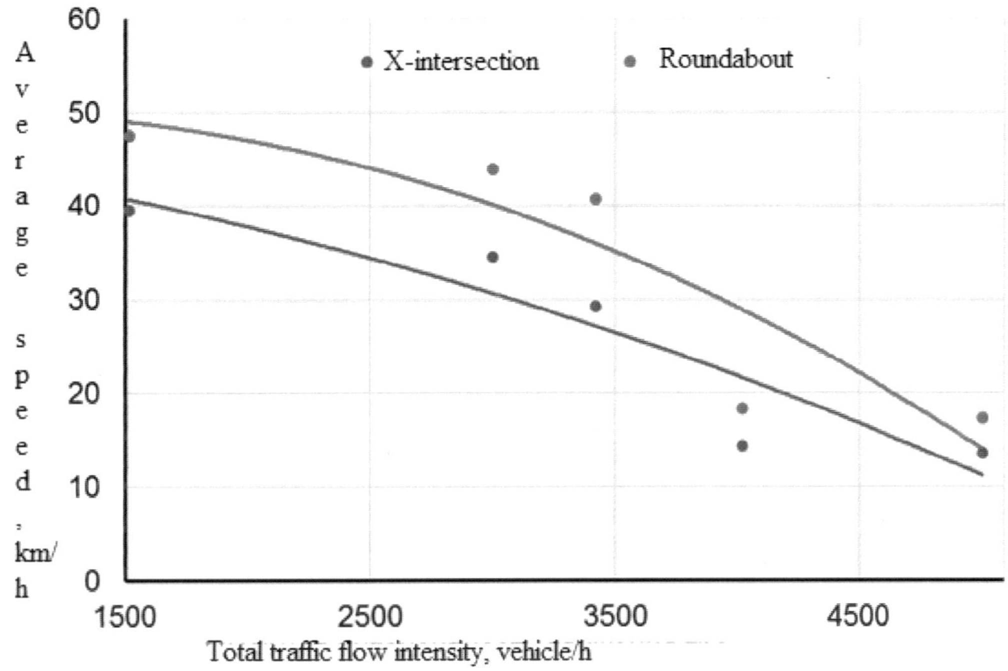

Figure 1. Dependence of the change in average vehicle speed on the total traffic intensity for various intersection configurations.

With an increase in the total intensity of traffic, the average speed of traffic decreases both at the four-way intersection and at roundabouts. However, the average speed at the roundabouts is higher than at the four-way intersection. Therefore, roundabouts are more efficient.

Research aimed at improving the organization of traffic was implemented using the computer simulation program PTV Vissim 11, based on the laboratory of transport modeling of the Tyumen Industrial University [8]. The software allows to build an intersection, simulate the movement of traffic and pedestrian flows along them and evaluate measures to improve traffic and the appropriateness of their implementation.

The intensity of traffic flows in both the main and secondary directions was used as input parameters [12].

The dependences of changes in traffic parameters, namely, the average speed of movement, on the radius of the central guiding island of a roundabout for different traffic intensities and lanes at the approach to the roundabouts are shown in Figures 1-4.

The radii of the central guiding island - 40, 60, 80, 100 meters, were considered based on the classification of the guiding central island and its types.

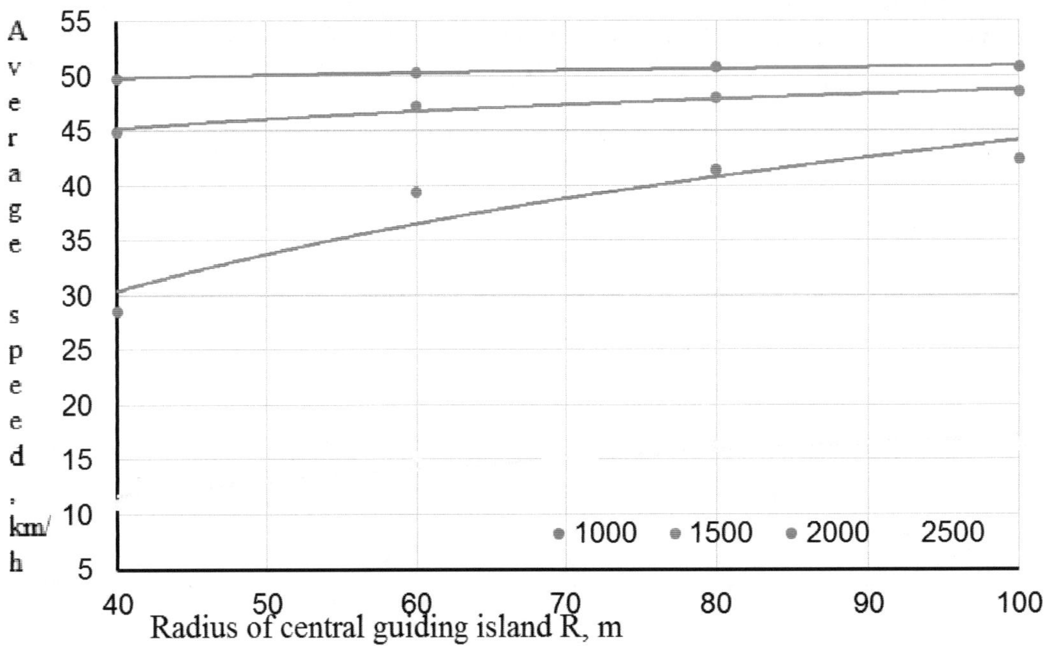

Figure 2. The dependence of the change in average speed on the radius of the central guide island with a bandwidth of 2/2 and traffic intensity of 1000, 1500, 2000, 2500 vehicles per hour.

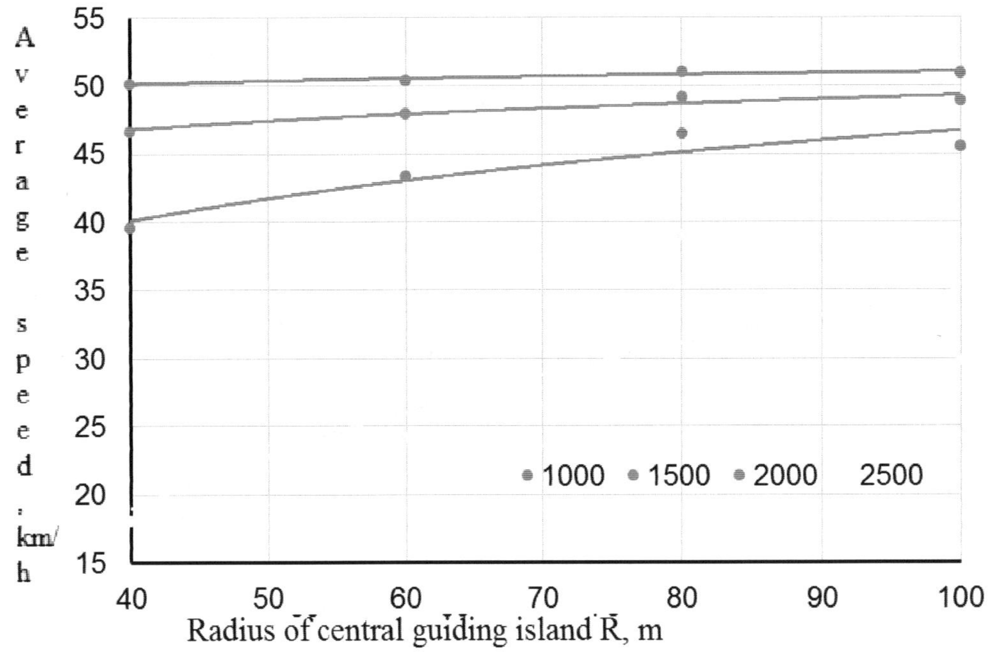

Figure 3. The dependence of the change in average speed on the radius of the central guiding island with a bandwidth of 4/2 and traffic intensity of 1000, 1500, 2000, 2500 vehicles per hour.

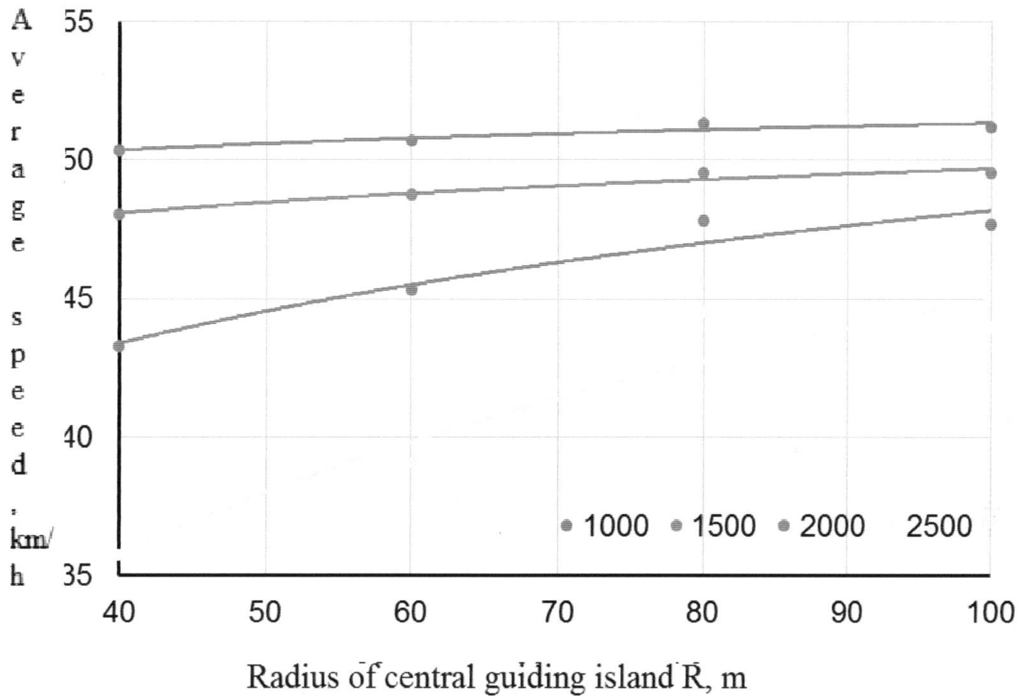

Figure 4. The dependence of the change in average speed on the radius of the central guide island with a bandwidth of 4/4 and traffic intensity of 1000, 1500, 2000, 2500 vehicles per hour.

3. Conclusion

At an intensity of less than 2000 vehicles per hour and an increase in the radius of the central guide island of the roundabout from 40 m to 100 m, the average speed does not change significantly.

With increased intensity and an increase in the radius of the central guide island of the roundabout, the average speed increases significantly, logarithmically.

The revealed patterns allow us to conclude that the potential for using roundabouts in cities with high traffic intensity exceeds the capabilities of the most common four-way intersections.

References

[1] Anisimov I A, Gorbunova A D, Burakova L N Email Author and Burakova, O D 2018 Evaluation of the effectiveness of the adaptive traffic management system for its development and interaction with electric transport IOP Conference Series: Earth and Environmental Science **177(1)** International Conference on Sustainable, ICSC 2018

[2] Basbas S, Papanikolaou A 2009 Evaluation of a sustainable urban transport system through the use of the transecon methodology. *International Journal of Sustainable Development and Planning* **4(1)** 18-34

[3] Chebotareva A A, Chebotarev V E and Danilina E I Transport and logistics in a modern metropolis: New opportunities. International *Journal of Recent Technology and Engineering* **8(1)** 2247-51

[4] Chuprov S, Viksnin I, Kim I and Nedosekin G 2019 Optimization of autonomous vehicles movement in urban intersection managementsystem. Conference of Open Innovation Association, FRUCT 60-6

[5] Efimova T B, Haitbaev V A and Pogorelova E V Intellectual Algorithms for the Digital Platform of "Smart" Transport. In: Ashmarina S, Mesquita A, Vochozka M (eds) Digital Transformation of the Economy: Challenges, Trends and New Opportunities Advances in Intelligent Systems and Computing, **908**

[6] Ertman S, Ertman J and Zakharov D 2016 Adaptation of urban roads to changing of transport demand E3S Web of Conferences **6(21)** International Conference on Sustainable Cities, ICSC

[7] Gubaidullina T, Ivanova N, Absalyamova S and Yerina T Analysis of national strategies for sustainable development with regard to fundamental conceptual premise *In Journal of Physics: Conference Series* **1141(1)** 012018

[8] Karmanov D, Zakharov D and Fadyushin A 2018 Evaluation of changes in traffic parameters for various types of traffic signal regulation Transportation Research Procedia 13th International Conference on Organization and Traffic Safety Management in Large Cities, SPbOTSIC **36** 274-280

[9] Kolesov V I, Danilov O F and Petrov A I 2018 Specific features of goal setting in road traffic safety IOP Conference Series: Earth and Environmental Science **177(1)** 3rd International Conference on Sustainable

[10] Larin O, Mavrin V and Almetova Z Simulation modeling for the evaluation of conflicts at stops of the urban route network Transportation Research Procedia 13th International Conference on Organization and Traffic Safety Management in Large Cities **36** 411-7

[11] Makarova I, Shubenkova K, Mavrin V, Gabsalikhova L, Sadygova G and Bakibayev T Problems, risks and prospects of ecological safety's increase while transition to green transport Advances in Intelligent Systems and Computing **879** 172-1804

[12] Panteleeva M and Borozdina S 2019 Mathematical model of evaluating the quality of "smart city" transport interchanges functioning E3S Web of Conferences **97**, International Scientific Conference on Construction the Formation of Living Environment

[13] Safiullin R, Kerimov M, Afanasyev A and Marusin A A 2018 model for justification of the number of traffic enforcement facilities in the region Transportation Research Procedia 13th International Conference on Organization and Traffic Safety Management in Large Cities, SPbOTSIC **36** 493-99

[14] Vasilyeva E and Sazonova T 2018 Evaluation of the effectiveness of the adaptive traffic management system for its development and interaction with electric transport IOP Conference Series: Earth and Environmental Science 3rd International Conference on Sustainable, **177(1)**

[15] Zhang J and Shi T Spatial 2019 analysis of traffic accidents based on WaveCluster and vehicle communication system data Eurasip Journal on Wireless Communications and Networking **1**

[16] Zhankaziev S 2017 Current Trends of Road-traffic Infrastructure Development Transportation Research Procedia **20** 731-9

Development of a finite-element model for bench testing of road fence rack

I V Demiyanushko[1], I A Karpov[1], P S Mikheev ,V S Nadezhdin1[1,*], O V Titov[1], B T Tavshadze[1] and J Awrejcewicz[2]

[1]Moscow Automobile and Road Construction State Technical University (MADI), 64, Leningradsky prospect, Moscow, Russia

[2]Lodz University of Technology, Poland

*E-mail: nadezhdin@niimech.ru

Abstract. The article compares the results of mathematical modeling by the finite element method (FEM) with the results of bench tests of steel racks of a road fence. The object of the study was a road fence rack made of steel 3, with a cross section in the form of a channel (No. 12). As a result of mathematical modeling, we obtained the dependences of the force in the rack on the movement, as well as the energy absorption of the racks. The obtained parameters were compared with the empirical results of field tests.

1. Introduction

In an accident, as a result of the interaction of the vehicle with the road fence, the vehicle elements and the fence deform and the trajectory of the vehicle changes. We should note that road barriers are a rigid structure, designed for the set energy absorption value provided by GOST [1]. Therefore, when installing them, we must take into account that collisions with barriers can lead to more serious consequences than leaving a road in the absence of barriers.

The aim of the proposed work was the development of a finite element model of bench tests of a road fence rack. We chose a road rack with a height of 750 mm with a cross-section in the form of a channel (No. 12) as an object of study, because it is one of the most used in the elements of road construction [2].

Full-scale bench tests were carried out for the selected rack in accordance with GOST [3]. Bench tests of the road fence racks were carried out on a specialized bench, which allows applying static or shock load, created by a horizontally moving mass (loading element) at a given value of the loading speed, to the object of study. The purpose of full-scale tests was to determine the nature of the deformation and strength characteristics when applying a static load.

2. The main part

When conducting field tests for a specialized laboratory bench, we imposed a number of requirements:

1) the height of the application of the load from the level of the lower plane of the rack or soil plate should be in the range from 300 to 600 mm;
2) the load on the fence rack should be at least 50 kN;
3) the rack loading speed (loading element speed) should be in the range from 1 to 80 mm / min;
4) the stroke of the loading element must be at least 500 mm.

The loading element must be absolutely rigid. During the tests, we measured the force and deformation at the contact point of the loading element and the fence rack. According to its results we built the load diagram "Force - movement of the rack at the contact point " (Figure 1.). The diagram allows to calculate the amount of energy absorption, numerically equal to the area of the figure under the curve of the diagram. According to the calculation results, the energy absorption was 6.3 kJ.

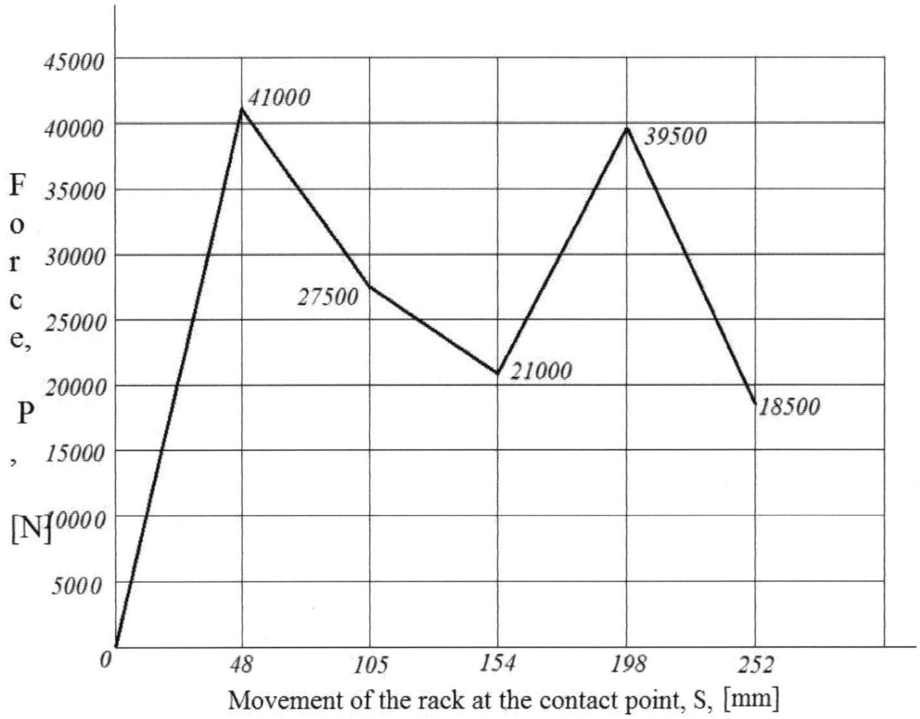

Figure 1. "Force - movement of the rack at the contact point".

In order to reduce material costs for multiple expensive full-scale tests, we implemented mathematical modeling of the loading of the road fence racks using the FEM. Solid-state geometric and finite element (FE) models of the object of study were created without simplifications (Figure 3.).

The thickness of the rack of the road fence is much less than its geometric dimensions, so the rack was modeled in the form of a shell [4-16].

According to the results of the FE analysis performed in the LS-Dyna software package, according to the existing methods [4-13], the diagram "Force - movement of the rack at the contact point" was obtained again (Figure 4.). For the new diagram, according to the calculation results, the energy absorption value was 6.15 kJ.

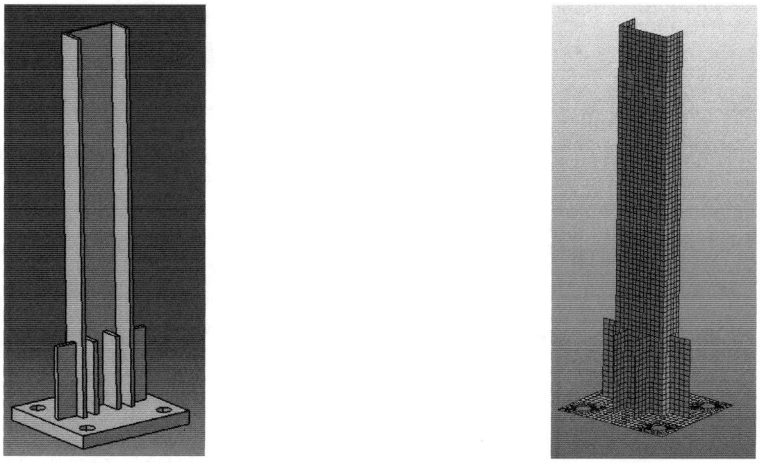

(a) (b)

Figure 2. Rack models: (a) –geometric model; (b) – FE model.

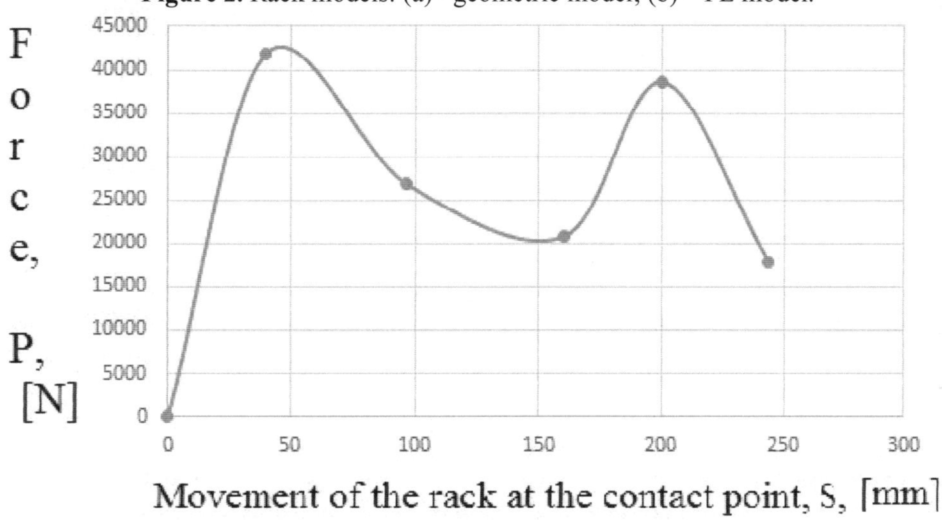

Figure 3. "Force - movement of the rack at the contact point".

As can be seen from the diagram (see Figure 3), with a force of 41900 N, plastic deformation of the rack occurs.

Comparing the results obtained in mathematical modeling of the tests of the road fence racks with the results of full-scale bench tests, we can conclude that the energy absorption value is almost the same. Consequently, mathematical modeling is an equivalent substitute for a full-scale (expensive) bench test, with a calculation error Δ :

$$\Delta = \frac{E_э - E_м}{E_э} \cdot 100 = \frac{6{,}3 - 6{,}15}{6{,}3} \cdot 100 = 2{,}38 \quad (1)$$

where E – energy absorption value obtained during a full-scale experiment, E - energy absorption value obtained by the results of mathematical modeling.

Figure 4 shows a comparative analysis of the operation of the rack in full-scale and virtual experiments.

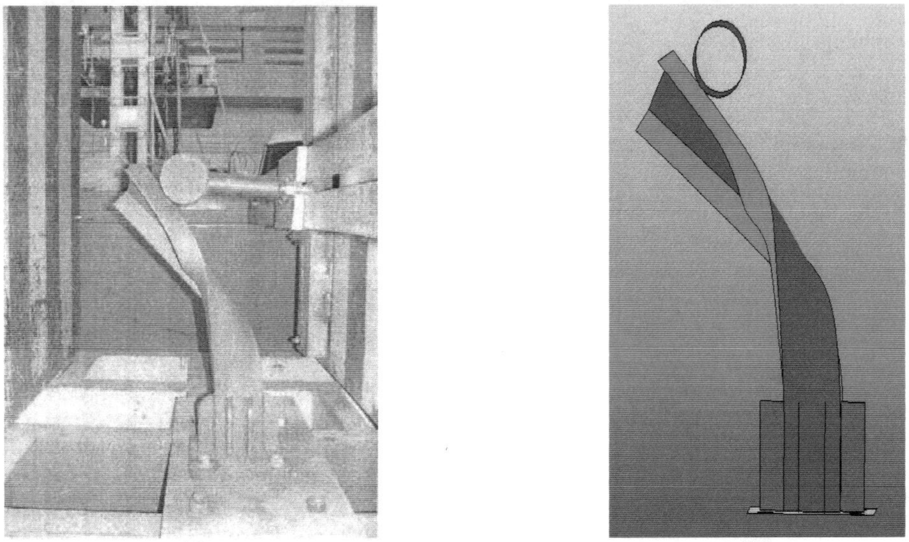

(a) (b)

Figure 4. Comparison of the results of the racks of the road fence: (a) - bench tests of the racks; (b) - virtual rack tests.

3. Conclusion

Based on the results of the work, we can make the following conclusions:

1) The value of energy absorption obtained as a result of modeling converges with the value of energy absorption obtained during field tests, with an error of 2.38%;

2) The time of implementation (preparation + conduct) of the experiment during modeling is much lower than the time spent on the full-scale experiment (preparation + conduct);

3) Material costs for design development (taking into account further modifications and their tests) during a virtual experiment will be significantly lower than material costs for conducting full-scale bench tests. The difference in material costs will increase with an increase in the number of research objects.

References

[1] GOST R 52607-2006 Tekhnicheskie sredstva organizatsii dorozhnogo dvizheniya. Ograzhdeniya dorozhnye uderzhivayushchie bokovye dlya avtomobiley. Obshchie tekhnicheskie trebovaniya.

[2] I V Demiyanushko and Obshchev A G 2011 Razrabotka tekhnologii ustroystva sovremennykh energopogloshchayushchikh dorozhnykh ograzhdeniy bezopasnosti // *Vestnik Permskogo gosudarstvennogo tekhnicheskogo universiteta* Okhrana okruzhayushchey sredy, transport, bezopasnost zhiznedeyatelnosti **1** 150-9

[3] GOST 33129-2014 Dorogi avtomobilnye obshchego polzovaniya Ograzhdeniya dorozhnye Metody kontrolya

[4] I V Demiyanushko 2017 Primenenie chislennykh metodov nelineynoy dinamiki k resheniyu problem stolknoveniy pri naezde transportnykh sredstv na dorozhnye barery bezopasnosti // V sbornike: *Nelineynaya dinamika mashin* - sbornik **IV** Mezhdunarodnoy SHkoly-konferentsii molodykh uchenykh 56-64

[5] I V Demiyanushko and Karpov I A 2013 Modelirovanie naezda avtomobilya na stoyku dorozhnogo ograzhdeniya // *Transportnoe stroitelstvo* **10** 16-9

[6] Karpov I A 2014 Razrabotka konechno-elementnykh modeley trosovykh dorozhnykh ograzhdeniy s ispolzovaniem programmnogo kompleksa LS-DYNA // *Avtomobil Doroga Infrastruktura* **2** 1

[7] I V Demiyanushko, I A Karpov and V S Nadezhdin 2019 Computational simulation and experimental study of cable for cable barriers // Part of the Mechanisms and Machine Science book series Mechan Machine Science, **73** IFToMM WC: Advances in Mechanism and Machine Science 2883-91

[8] I V Demiyanushko, I A Karpov and B Tavshavadze 2019 Applications FEM Analysis in Researches Reliability and Dynamics of Road Safety Barriers at Collisions With Cars // Part of the Mechanisms and Machine Science book series *Mechan Machine Science*, **73** IFToMM WC: Advances in Mechanism and Machine Science 1651-8

[9] Demiyanushko I 2017 Developments of non-linear dynamics FEM simulation of the impact performance of road safety barriers with use of experimental validation of models / Demiyanushko i , karpov I , Tavshavadze B // Mathematical and Numerical Aspects of Dynamical System Analysis 14th International Conference «Dynamical systems theory and application» **11-14** 165-74

[10] Reid and John D 2010 Development of advanced FE material model for cable barrier wire rope , Lincoln, Nebraska, U S Department of transport

[11] Nelson and Ryan, et al 2005 Evaluation of alternative cable anchor design and three-cable guardrail adjacent to steep slope Transportation research report **TRP-03-155-05** Lincoln, Nebraska: Midwest Roadside Safety Facility, University of Nebraska-Lincoln

[12] Hitz and Rebecca, et al 2008 Design and evaluation of a low-tension cable guardrail end termainal system Transportration Research report **TRP-03-131-08** Lincoln, Nebraska: Midwest Roadside safety Facility, University of Nebraska-Lincoln

[13] Ferdous MR, Abu-Odeh A, Bligh RP, Jones HL and Sheikh NM 2011 Performance limit analysis for common roadside and median barriers using LS-DYNA *Int J Crashworth* **16(6)** 691–706

[14] Gutowski M, Palta E, Fang H 2017 Crash analysis and evaluation of vehicular impacts on W-beam guardrails placed on sloped medians using finite element simulations Adv Eng Softw **117** 88–100

[15] Neves R R , Fransplass H and Langseth M 2018 Performance of some basic types of road barriers subjected to the collision of a light vehicle Braz Soc Mech Sci **40** 274

[16] Ana P C Freitas, Rui F M Silva and João M P Dias Multibody and Finite Element Models for the Design of Motorcyclist's Roadside Protections // III European Conference on Computational Mechanics 768

Optimizing the location for the enterprise recycling the forestry production waste

N A Tyurin, L Ya Gromskaya*, T S Antonova, V A Kalyashov and V N Artemiev

St. Petersburg State Forest Technical University, 5, Institutsky per., 194021, St. Petersburg, Russian Federation

*E-mail: gromskaya.stl@gmail.com

Abstract. The article discusses the optimization of the location of the enterprise for recycling the wood waste from logging and wood processing industries. A promising area of waste recycling is the production of wood-fuel pellets as an alternative renewable energy source. The raw material supply of pellet production means a lot of collection activities, in which transport costs play a decisive role. Sources of raw materials and consumers of finished products are distributed over vast territories. The paper proposes an integrated approach to solving the problems of the territorial location of the recycling plant based on logistic methods and the use of geographic information systems. The paper presents a mathematical model and a technique for optimizing the territorial distribution of pellet production by the criterion of the minimum transportation costs for raw materials. The developed methodology takes into account the existing infrastructure. It involves the full use of sources of raw materials, and also allows us to choose the best option for the location of the enterprise, taking into account the social and economic possibilities of construction. The technique has great versatility, which allows it to be used to place enterprises of various fields of activity. It clearly reflects the results of the problem solution, which simplifies the decision-making process on the location of enterprises.

1. Introduction

The current level of development of logging and the woodworking industry is accompanied by the formation of a large amount of wood waste both at the stage of logging and at the stage of its processing. A promising direction for the disposal of such waste is the involvement of wood biomass in the form of wood-fuel pellets in the fuel and energy complex as an alternative environmentally friendly renewable energy source.

Location is one of the key factors affecting the competitiveness and profitability of any processing enterprise, which necessitates the development of methodological approaches to determining and justifying the place of their rational territorial distribution.

The main specific feature of pellet production enterprises is the crucial role of raw materials, their quality indicators and the territorial distribution of sources. Raw materials for such enterprises require a lot of collection activities, i.e. it is of a geographic nature, especially when using logging waste. Most of wood waste is dispersed over a large territory, which has a low transport density and a very low

IOP Conf. Series: Materials Science and Engineering **786** (2020) 012035 doi:10.1088/1757-899X/786/1/012035

development of forest infrastructure (primarily forest roads), the industrial utilization of wood waste requires a serious feasibility study. However, effective scientific methods for the territorial distribution of such enterprises with respect to sources of raw materials so far do not exist. To solve such problems of enterprise location, we advise to use universal geographic information systems and mathematical-cartographic modeling.

The purpose of the study is to develop a mathematical model and methodology for optimizing the territorial distribution of pellet production according to the criterion of minimum transportation costs for raw materials.

2. Statement of the problem of choosing the location of production

On one fragment of the map it is necessary to place one enterprise for the recycling of waste from timber industry with an installed capacity of *M* tons per year of finished products. There is data on *n* free sources of raw materials and their reserves Q_i in m^3 i = 1 ... n in the studied region. We know the output coefficients of finished products v_i for raw materials of the i-th source, m^3/t. It is necessary to determine the optimal location of the enterprise by the criterion of minimum freight transport work for the delivery of raw materials, taking into account the existing transport infrastructure of the region. At the same time, the condition must be met - the location of the enterprise must be in a village or town.

3. Mathematical model and algorithm for locating a waste recycling enterprise

The methodology for the application of geographic information technologies in the solution of the problem of locating enterprises for processing waste from forestry production consists of the following sequence of steps:

1) Creating a vector map of the study area (forestry, administrative region, region). When creating a map, we must use a raster substrate to create the correct map, and also have reliable information about the geographical coordinates of the created map and its projections.

2) Collection and analysis of data on existing and potential sources of wood raw materials, their location and possible supply volumes.

3) Creation of an electronic database using the most preferred software. It is recommended to use tables created in the MS Excel table processor environment and saved in dbf format.

4) Application of information from the database to the map using the geocoding operation.

Let's find the coordinates of the center of gravity of unused waste from the timber industry of the region x^c, y^c. The criterion for the search for the center of gravity of the region's reserves is the minimum transport work for transporting waste from sources to the center of gravity. The objective function of finding the coordinates of the center of gravity is as follows:

$$R = \sum_{i=1}^{n} Q_i \sqrt[2]{(x^c - x_i)^2 + (y^c - y_i)^2} = \min \tag{1}$$

where Q_i – possible (unused) volumes of wood waste in the i-th source, m^3; n - is the number of raw materials in the region; x_i, y_i - coordinates of the centroid of stocks of the i-th source.

The obtained coordinates of the center of gravity of regional stocks of timber industrial waste allow us to obtain a basic, initial solution to the problem of locating a recycling enterprise. The main drawback of the obtained solution is not taking into account the transport infrastructure existing in the region. All transportation is provided through a direct air line (hypotenuse) and it is assumed that there is a transport connection of each source with the center, which, as a rule, is not possible even in regions with developed transport infrastructure. To take into account the transport infrastructure when searching for the location of the enterprise, we use the obtained initial reference solution of the coordinates of the center of gravity of the regional waste stocks and the geoanalysis tools in the following sequence.

Let's map the obtained coordinates x^c, y^c of the gravity center of the stocks of regional timber waste from the region and calculate the elongation coefficients k_i of the transportation of raw materials from the centroids of their stocks x_i, y_i to the reference solution x^c, y^c of the enterprise location taking into account the existing transport infrastructure. The elongation coefficient is determined by the formula

$$k_i = \frac{l_i^t}{l_i^v} \tag{2}$$

where l_i^t – distance of waste transportation from the i-th source to the center of gravity of the reserves x^c, y^c over the existing transport network, km; l_i^v - the air distance from the centroid of the i-source reserves to the center of gravity of the reserves x^c, y^c.

We determine the coordinates of the optimal lcoation of the recycling enterprise x^o, y^o according to the criterion of minimum transport work according to the following objective function

$$R = \sum_{i=1}^{n} k_i \cdot Q_i \sqrt[2]{(x^o - x_i)^2 + (y^o - y_i)^2} = \min \tag{3}$$

The new obtained solution on the location coordinates of the enterprise will already take into account the existing transport infrastructure, but it assumes the full use of raw materials in the region and does not take into account some social and economic possibilities of building the enterprise at the coordinates x^o, y^o (lack of settlements, heat and energy, labor). To account for these factors, we proceed to the third stage of the algorithm for finding the optimal location of the enterprise.

Using geoanalysis tools (buffering), we find the nearest (closest *m*) to the point with coordinates x^o, y^o settlements that satisfy the requirements for the possibility of building a processing plant and measure their coordinates x_j, y_j, $j=1...m$ in a GIS project. The coordinates will be promising options for locating a processing plant there. To determine the optimal one, we proceed to the fourth stage of the algorithm.

For each j-th variant of enterprise location, the problem of optimizing the supply of raw materials from the i-th source q_i is solved by the criterion of minimum transport work. Objective function

$$R_j = \sum_{i=1}^{n} k_{ij} \cdot q_{ij} \sqrt[2]{(x_j - x_i)^2 + (y_j - y_i)^2} = \min \tag{4}$$

Limited by

$$\sum_{i=1}^{n} \frac{q_{ij}}{v_i} = M \quad , i=1,2...n, j=1,2...m \tag{5}$$

$$q_{ij} \leq Q_i \quad , i=1,2...n, j=1,2...m \tag{6}$$

$$q_{ij} \geq 0 \quad , i=1,2...n, j=1,2...m \tag{7}$$

where R_j – optimal transport work on the delivery of raw materials of the j-th variant of the enterprise location, m^3 * km; q_{ij} - the volume of supply of raw materials from the i-th source to the *j* location of the enterprise, m^3; k_{ij} is the elongation coefficient for the transportation of raw materials from the i-th source to the j-th variant of the enterprise location taking into account the transport infrastructure (2); *M* - the capacity of the processing enterprise for the production of finished products, tons; v_i - the coefficient of output of finished products for raw materials of the i-th source, m^3 / t .; Q_i - the possible volume of supply of raw materials of the i-th source, m^3

The objective function (4) determines the total freight work for the transportation of raw materials. Expressions (5) and (6) determine the balance between the required volume of raw materials and the presence of its reserves at collection points. Restriction (7) reflects the condition of non-negative supply volumes of raw materials.

The final version of the territorial location of the enterprise is selected according to the criterion of minimum transport work from j options $j=1,2...m$. From m places, we choose a place for which the value of R_j is minimal. As a result, we get the address of the location of the enterprise, after which we should evaluate the feasibility of locating the enterprise in the selected village. If it is impossible to place the enterprise in this place, then it is excluded from consideration and the place with the smallest sum from the remaining options is determined. This procedure is performed until a solution is found.

The problem of the optimal location of the pellet production plant can be solved using the non-linear programming method in the MS Excel add-on "Solution Search" and geographic information technologies in the following sequence:

1. By solving the optimization model (1), we find the coordinates of the optimal position of pellet production without taking into account the existing transport network.
2. The obtained coordinates of the station x_0, y_0 are mapped and the elongation coefficients are calculated taking into account the use of existing roads of the region for the delivery of logging waste according to formula (2).
3. Once again, we calculate the optimal station coordinates using the optimization model (3) and map the point of the position of the pellet plant now taking into account the existing transport network of the region.
4. Near the point of the found optimal position of pellet production, all settlements are considered for the possibility of placing a pellet production plant. For each location of the enterprise, the problem of optimizing the supply of raw materials by the criterion of the minimum transport work (4-6) is solved and the final decision is made.

The performed machine experiments confirmed the reliability of the proposed methodology, which can be recommended for optimizing the territorial location of wood waste recycling enterprises according to the criterion of the minimum transport work for the delivery of raw materials.

4. Conclusion

The developed technique has great versatility, which allows it to be used to locate the enterprises in various fields of activity. The methodology also cartographically clearly reflects the results of problem solution, which simplifies the decision-making process on the location of enterprises.

References

[1] Vasilev N I 2011 Pellety i toplivnye brikety - progressivnye vidy tverdogo biotopliva *Energosberezhenie* **4** 1420

[2] Sevastyanova S N 2009 Bioenergetika Drevesnye (toplivnye) granuly/S N Sevastyanova//*Vestnik OGU* Orenburg, **10(104)** 133- 8

[3] Salminen E O, Borozna A A, Tyurin N A 2010 Lesopromyshlennaya logistika Uchebnik dlya studentov vyssh uchebn zaved, obuchayushchikhsya po napr podgot «Tekhnologiya lesozagotovitelnykh i derevopererab proizv » Sankt-Peterburg, Moskva, Krasnodar: «Lan», – 352

[4] Tyurin N A 2016 Geoinformatika na promyshlennom transporte Uchebnoe posobie SPb : SPbGLTU, – 106

[5] Mikhaylov R A 2012 Modeli optimalnogo razmeshcheniya skladskikh kompleksov s uchetom razlichnogo potrebitelskogo sprosa naseleniya v zadache logisticheskogo upravleniya tovarodvizheniem: dis kand tekhn nauk: 05 13 10, Voronezhskaya gosudarstvennaya lesotekhnicheskaya akademiya Voronezh, – 150

[6] Arranz J I, Miranda M T, Montero I, Sepulveda F J Rojas C V 2014 Characterization and combustion behaviour of commercial and experimental wood pellets in South West Europe : Fue, 199-207

[7] Di FulvioaF Comparison of energy-wood and pulpwood thinning systems in young birch stands [Text] / Fulvio Di Fulvioa, Anders Kroonb, Dan Bergstroma, Tomas Nordfjella // Scandinavian

[8] Grigorev I, Khitrov E, Kalistratov A, Bozhbov V and Ivanov V 2014 New approach for forest production stocktaking based on energy cost // International Multidisciplinary Scientific GeoConference Surveying Geology and Mining Ecology Management, SGEM 14, 407-14

[9] Gubaidullina T, Ivanova N, Absalyamova S and Yerina T 2018 Analysis of national strategies for sustainable development with regard to fundamental conceptual premise In Journal of Physics: Conference Series **1141(1)** 012018 IOP Publishing

[10] Lehtikangas P 2001 Quality properties of pelletised sawdust, logging residues and bark Biomass & Bioenergy **20** 351-60

[11] McKendry P 2002 *Energy production from biomass* (part1): overview of biomass Biores Technol **83** 37-46

[12] Persson T 2012 Tree stumps for bioenergy - harvesting techniques and environmental consequences [Text] // *Scandinavian Journal of Forest Research*, **27(1)** 705-8

Effective system of learning for engineers

M F Guskova[1,*], P F Sterlikov[2] and F F Sterlikov[3]

[1]Russian University of Transport (RUT), 9 Obraztsova str, 127994 Moscow, Russia

[2]A.I. Yevdokimov Moscow State University of Medicine and Dentistry, 20/1 Delegatskaya str, 127473 Moscow, Russia

[3]K.G. Razumovsky Moscow State University of technologies and management (the First Cossack University), 73 Zemlyanoy Val str, 109004 Moscow, Russia

*E-mail: oet2004@yandex.ru

Abstract. The article analyzes the problem of the formation of a system of intensive computerized learning (SICL), based on the consistent use of didactic foundations for creating effective learning systems. The paper generalizes experience of using SICL to create training courses for students of technical areas and specialties. The paper uses the author's experience of implementing an additional professional educational program for advanced training and professional retraining of higher education teachers "Development of the content of professional education, improving its quality in inextricable connection with science". The creation of SICL for academic discipline begins with the statement of the didactic task and continues with the formation of the didactic system (a system of methods, means and forms of training). The didactic system (DS) takes into account the conditions of the didactic task and the laws of learning. The theory and methodology of SICL are described for three groups of stages of learning: lecture (presentation), work (bringing to the required level of mastery of educational issues) and final (bringing to the required level of mastering of the content of topics, sections and parts of the academic discipline). The stated positions can be used to create a computerized textbook that forms the basis of SICL.

1. Introduction

In recent decades, a large-scale modernization of education has been taking place in the Russian Federation. It aims to create new principles for the formation of a competent and highly qualified human potential, as an important factor in innovative economic development and the country's geopolitical competition. Activities carried out by the state in this direction are reflected in the National Education Project (2019-2024). The project is subordinate to the development of the educational space in our country. It is aimed at ensuring the global competitiveness of Russian education and becoming one of the 10 leading countries in the world in the quality of general education. [5,7]

The implementation of the National Project involves the implementation of several main areas of development of the education system, including: updating its content, creating the necessary modern infrastructure.

Achieving high learning outcomes has always been the main requirement for our universities. And the results are formed, first of all, by the competence and motivation of teachers, their knowledge of the methods of effective training systems and ensuring the availability of these systems for students.

2. Formation of the competencies of an engineer in the field of construction, road, lifting and communal machines

In our opinion, the list of the main competencies of a modern teacher should include a set of professional and pedagogical knowledge, skills and possessions, which allows him to be in demand and participate in practical research in the field of construction, road, lifting and communal machines, as well as in the development of intensive systems of training, and in the implementation of such systems in the educational process for students of construction specialties and areas. [3]

3. Creation of a system of intensive computerized learning (SICL)

The problem of creating effective learning systems has interested the pedagogical community all the times. This issue is considered as one of the most significant for modern education in Russia. The creation of effective learning systems is also of interest to higher school teachers. There are many studies on this subject. [2]

The most progressive solution in creating effective learning systems is contained in the work of the research team under the guidance of A.A. Zolotarev. The authors developed a concept that highlights the theory and methodology of systems of intensive computerized learning (SICL). The purpose of the concept is to offer the developers of the educational process, and, first of all, teachers and practitioners, a theoretical basis applicable in any subject area, which combines the best of the well-known author's pedagogical schools and their own experience into an effective holistic system that allows students to form competencies. [6]

The creation of a system of intensive computerized learning (SICL) for each academic discipline is formed by certain sequential actions:

1. First, the didactic task is carefully developed and set.
2. After that, the didactic system is formed.

The teacher (as a researcher and as a developer) should be able to set a didactic task:

- set learning goals,
- form a structured content of the discipline,
- determine the degree of readiness of students to study the discipline,
- take into account the risks of organizational and material-technical groups,

Then the teacher designs, develops and implements the didactic system:

- a system of methods,
- system of tools,
- a system of forms of learning,

adequate to the laws of teaching and the conditions for the implementation of the pedagogical process.

This basic idea can be represented in the form of a written summary (Figure 1).

IOP Conf. Series: Materials Science and Engineering **786** (2020) 012036 doi:10.1088/1757-899X/786/1/012036

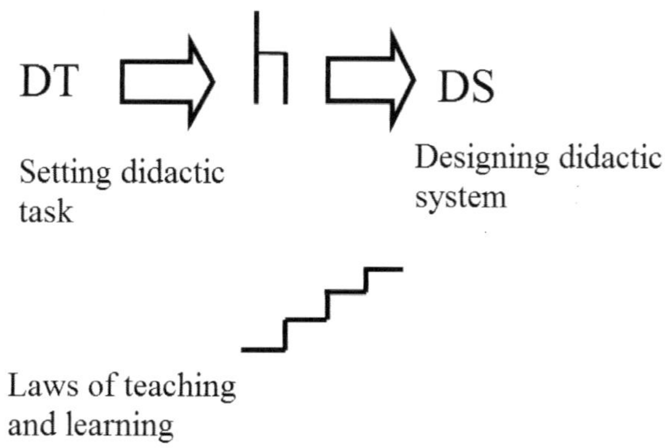

Figure 1. The main idea of intensive information learning.

The symbol "h" means the student's chair, in which the teacher should be mentally, designing the training system (technology) in accordance with the laws (stages shown in steps) of learning. Each of the nine stages has its own didactic system (Table 1.).

Table 1. The relationship between the stages of learning (teaching) and the elements of the didactic system.

Elements of the didactic system	Lecture stages		Work stages				Final stages		
	Initial	Key	Actions				Final generalization	Control	Corrections and adjustments
			With full support system	With reduced support system	Without external support, slow	Without supports, free			
	1	2	3	4	5	6	7	8	9
Methods of learning	The system of teaching methods for lecture stages		The system of teaching methods for work stages				The system of teaching methods for final stages		
Tools of learning	A set of training tools for lecture stages		A set of training tools for work stages				A set of training tools for final stages		
Forms of learning	System of forms of teaching at lecture stages		System of forms of teaching at work stages				System of forms of teaching at final stages		

The group of lecture stages (presentation) includes the first (initial) and second (key) stages. At these stages, the student studies the content in the volume of the educational topic, including several educational questions. The level of assimilation of the content at these stages demonstrates only the ability of the student to begin reproduction of information based on all the necessary teaching tools.

The group of work stages (internalization) includes stages from the third to the sixth inclusive. Here, on the background of the learned topic, each individual educational question is studied. The level of mastering the question increases with the transition from stage to stage and at the last stage - the sixth - reaches the ability to perform an action in the mental form at the skill level, i.e. automated, without supports.

The group of final stages includes stages from the seventh to the ninth. The seventh stage is devoted to the final generalization. It is similar to the work stages, but it does not apply to individual issues of the topic, but to the topic as a whole. This stage may contain several sub-steps necessary to ensure mastery of the learner to the level of ability to perform a complex action at the level of proficiency. [4]

In the SICL concept, emphasis is placed on the phased assimilation of the learning content as the main pattern of learning.

The completion of the formation of SICL provides the passage of two steps in the formation of the didactic system, which corresponds to the condition of the stated didactic task and the laws of learning. At the first step, an a priori didactic system (DSa) is developed, and at the next, a real (DSr) system is developed.

In the process of formation, the a priori didactic system (DSa) does not take into account the restrictions imposed by the condition of the didactic task on the material base and the organization of training. Learning tools are only named at this stage.

The real didactic system (DSr) provides the limitations of the material base and the organization of training. This section is especially relevant for the training of engineering personnel in the field of construction, road, lifting and communal machines. Learning tools are defined here. It can be landfills, laboratories, stands, which will allow in practice to master competencies. Such objects can either belong to an educational organization or be used on a contractual basis. [1]

When designing a DSa, the learning tool are just named; when developing DSr, they are formed. With such a step-by-step procedure for the formation of the didactic system, it is possible to fix the difference between the "ideal" structure, which corresponds to modern knowledge about the pedagogical system, and the actual one, which is possible under the specific restrictions of pedagogical practice. This allows us to outline a plan for moving from a real system to a priori.

4. Structure of a system of intensive computerized learning (SICL)

This concept is included in the SICL as the first section. The second section of the SICL defines the content of the regulations for the process of setting the didactic task. Here, on the example of a certain topic, the sequence of steps of the SICL developer is presented, which includes: the formulation of the learning objectives; collection, selection and structuring of the content of the educational material; setting the levels of assimilation of educational material and requirements for the initial level of competence of students; the procedure for accounting for the inclusion of restrictions that are imposed on the input design data of the didactic system. [1]

The third and fourth sections demonstrate the theory of designing didactic systems, both at the lecture (presentation) and at the work stages of training, in addition, they prescribe the methodology for their development and use in the educational process. The third and fourth sections begin by examining the features of the formulation of didactic tasks.

The paragraphs of the sections that relate to the design of DSa should be completed with tables reflecting the relationship of the stages of training and elements of these systems.

The points of development of the DSr contain information on the nature, functions and technologies of developing training tools, and in addition, the types of classes within which these didactic systems can be implemented are defined.

A feature of the fifth section of SICL is its construction. It differs from the third and fourth sections because it is the peculiarity of the formulation of the didactic task at the final stage that determines the corresponding specific types of classes for each of the stages. These include a final lecture and a business game - both of these types of classes are presented as integral didactic systems.

The sixth section of SICL provides a summary table of the relationship between the stages of training and the elements of the didactic system. Based on it, an algorithm for designing a holistic didactic system adequate to the conditions of the didactic task is constructed. The section presents options for a training system that allows you to implement the developed SICL and thereby ensure the required quality of training.

The seventh section gives an idea of a computerized textbook, where using a specific example SICL is implemented in its unique didactic capabilities. Such a textbook is the core of SICL, which reflects the current state of the most important elements of the content of training - theoretical knowledge and knowledge derived from them about the algorithms for performing actions. In addition, the computerized textbook is considered the core of SICL because the system of training testing is included in it, which guarantees the unconditional formation of competencies among students. The creation of computerized textbooks will allow educational institutions to efficiently solve one of the key social and economic problems facing modern Russia: accelerated modernization of education in all its forms (full-time, part-time, part-time), and especially when distance learning technologies are used, including open learning education. [7]

5. Conclusion

The paper generalizes the experience of using SICL to create training courses for students of technical areas and specialties. The author's experience of implementing an additional professional educational program for advanced training and professional retraining of higher education teachers "Development of the content of professional education, improving its quality in inextricable connection with science" was used. The creation of SICL for academic discipline begins with the statement of the didactic task and continues with the formation of the didactic system (a system of methods, means and forms of training). The didactic system (DS) takes into account the conditions of the didactic task and the laws of learning.

The article presents theory and methodology of SICL for three groups of learning stages: lecture (presentation), work (bringing to the required level of mastery of educational issues) and final (bringing to the required level of mastering of the content of topics, sections and parts of the academic discipline). The stated positions can be used to create a computerized textbook that forms the basis of SICL.

References

[1] Abrameshin A E and Sterlikov F F 2010 O sisteme intensivnogo informatizirovannogo obucheniya *Ekonomicheskie nauki* **72** 321-5

[2] Gulyaev V N , Ivanov E S , Loginov I P and Khasanov R Sh 2017 K voprosu o podgotovke nauchno-pedagogicheskikh kadrov na sovremennom etape razvitiya vysshey voennoy shkoly *Voennyy akademicheskiy zhurnal* **2 (14)** 126-30

[3] Guskova M F and Sterlikov F F 2017 Development of the principles of the quality management system which based on the economic theory of value: Proceedings of the 2017 International Conference "Quality Management, Transport and Information Security, Information Technologies", IT and QM and IS 2017 422-4

[4] Guskova M F , Sterlikov P F and Sterlikov F F 2015 Sovremennye informatsionnye obrazovatelnye tekhnologii: teoriya i praktika *Voprosy ekonomiki i prava* **87** 31-5

[5] Sakhapov, R L and Absalyamova, S G 2015 The use of telecommunication technologies in education network In Proceedings of 2015 12th International Conference on *Remote Engineering and Virtual Instrumentation* (REV) 14-7 IEEE

[6] Zolotarev A A 2008 Teoriya i metodika sistem intensivnogo informatizirovannogo obucheniya (didakticheskie osnovy sozdaniya effektivnykh sistem obucheniya) M : RITS MGIU

[7] Zolotarev A A 2001 Sovremennye obrazovatelnye tekhnologii v kontekste intensivnogo informatizirovannogo obucheniya M : MOSU 17-8

[8] National project «Obespechenie globalnoy konkurentosposobnosti rossiyskogo obrazovaniya, vkhozhdenie Rossiyskoy Federatsii v chislo 10 vedushchikh stran mira po kachestvu obshchego obrazovaniya»

IOP Conf. Series: Materials Science and Engineering **786** (2020) 012037 doi:10.1088/1757-899X/786/1/012037

Methods of recruiting of mobile repair services and maintenance of machines performing reclamation works

A S Apatenco[1,*] and N S Sevryugina[2]

[1]Moscow Russian State Agrarian University named K.A. Timiryazev, Moscow, str. Timiryazevskaya, 49, Russia

[2]Moscow state university of civil engineering, Moscow, Yaroslavskoye shosse, 26, Russia

*E-mail: a.apatenko@rgau-msha.ru

Abstract. The specifics of reclamation works are associated with difficult conditions for their mechanization. Seasonality of the work requires an increase in the efficiency of using machines, preferably with the absence of their failures. Usually, equipment failures in melioration organizations are eliminated at the place of machine's work (units) by the on-site repair teams equipped with mobile repair workrooms (MRWs). Cost analysis of the units operation in the technological complex shows that there is an optimal amount of MRWs for performing repair and technical impacts. The authors develop a technique for selecting mobile repair workrooms and their completeness with repair equipment. It justifies the effectiveness of the inclusion of the service department to eliminate machine failures when performing technical works of mobile repair and maintenance services, basing on a car chassis, with special equipment and accessories infrastructure.

1. Introduction

In Russian Federation, works in the field of urban planning and environmental engineering are carried out in difficult climatic conditions. Territorial diversity is characterized by land use zones both with insufficient or uneven precipitation regimes, frequent droughts and hot dry winds, and areas of excessive moisture. The expansion of the urban sphere leads to a change in the ecosystem, and nowadays the issue of reducing the nature's burden through the development of land reclamation becomes urgent. Raising the standard of living of the population by reducing the territories of the natural complex is already unacceptable; it is required to actively reorient the industrial land plots (often abandoned or naturally depleted) into the category acceptable for organized or private agriculture, through land's melioration. Reclaimed land, occupying 6% of the arable land, produces up to 65% of vegetable production and potatoes, all rice, about 20% of feed for livestock and other products. The structure of the state management of the land-reclamation complex of the Russian Federation is represented by 80 sub-departments of the Ministry of Agriculture of the Russian Federation in the field of land reclamation with a total staff of more than 17 thousand people [1].

Construction of modern land reclamation is based on the use of a large number of various means of mechanization. The most important task of the technical policy in this area is the complex mechanization of work. The solution of this task requires a continuous increase in the fleet of reclamation, construction and road machines, increase in their productivity and reliability, and improvement of utilization and support of working population. Statistical studies show that a third of the fleet of specialized land-reclamation equipment and vehicles standing on the balance sheet in institutions have a wear rate of more than 75%, and operation of more than 50% of the fleet exceeds the standard operating time.

Under these conditions, an important role is played by technical support measures aimed at maintaining equipment in constant readiness due to operational and service optimization [2–9]. The analysis showed that equipment failures in meliorative institutions are eliminated at the place of work of the machines by the forces of field repair teams equipped with mobile repair workrooms (MRW) [10, 11]. Currently, there are many models of MRWs with different configuration and cost [12]. Various techniques have been developed for their selection, which basically boil down to determining the required number of MRWs. So, recommendations to determine the number of MRWs for servicing logging machines were developed based on the comparison of losses due to machine downtime, the untimely repairments and the cost of maintaining mobile repair workroom [13]. In studies [14, 15], it is noted that when designing a fleet maintenance system with MRWs, in addition to determining their number it is necessary to select the optimal set of process equipment, its place the workroom and select the vehicle chassis. The proposed methods don't pay attention to the issues of functional affiliation, operating conditions and the nature of technological operations, their cyclical nature and complexity, although these factors are key in assessing the causes of the failures of machines, their systems, components and assemblies.

The purpose of the research is to develop a technique for selecting mobile repair workrooms and their completeness while eliminating machine failures during reclamation work.

2. Materials and methods

We investigated the flow of failures and the costs of operating units of technological complexes when performing reclamation works of a specialized organization. The technical equipment included standard complexes with the following units: KF-2.8 brush cutter, MP-15 loader collector, BDM-2.5 land-reclamation harrow, BDT-3.0 heavy disk harrow, PV-1.5 wood residues, planner VP-8, skating rink water filling ZKVB-1,5A, seeder APP-2,8. These machines are widely used in the performance of reclamation works.

3. Results and discussion

The experience of repair and technical impacts (RTI) in the elimination of machine failures at land reclamation facilities has shown that the use of several mobile repair workrooms is the most effective. In this case, the MRW is considered as a service department of a land reclamation institution, including personnel performing repair and maintenance impacts, repair and process equipment (RPE), and repair materials [16].

To determine the amount of RTI, we developed a mathematical model that describes the minimum cost of operating units of technological complexes, based on the amount of work done, taking into account the losses from machine downtime due to sudden failures.

$$Y_{(a)} = \frac{C_1 M_{0(a)} + C_2\left(M - M_{0(a)}\right) + C_3 S}{M W_0\left(1 - K_{n(a)}\right)} \to \min \quad , \qquad (1)$$

where: Y(a) – minimum cost of operating units of technological complexes, rubles / ha.; C1 – unit downtime losses, rubles / hour; C2 – direct costs of operating one unit, rubles / hour; C3 – the cost of maintaining service posts using MRW, rubles/hour; Mo(a) – average number of inoperative units, pcs.; (M-Mo(a)) – average number of operating units, pcs.; S - the number of service posts using MRW, pcs.;

Kn(a) – aggregate downtime factor; Wo – hourly average operational productivity of one technological complex, hectare / ha, M – total number of units, pcs.

For machine complexes that perform reclamation works, we established that, depending on the availability of RTI using MRW, the specific operating costs vary along an extreme curve and have their own optimum (Figure 1).

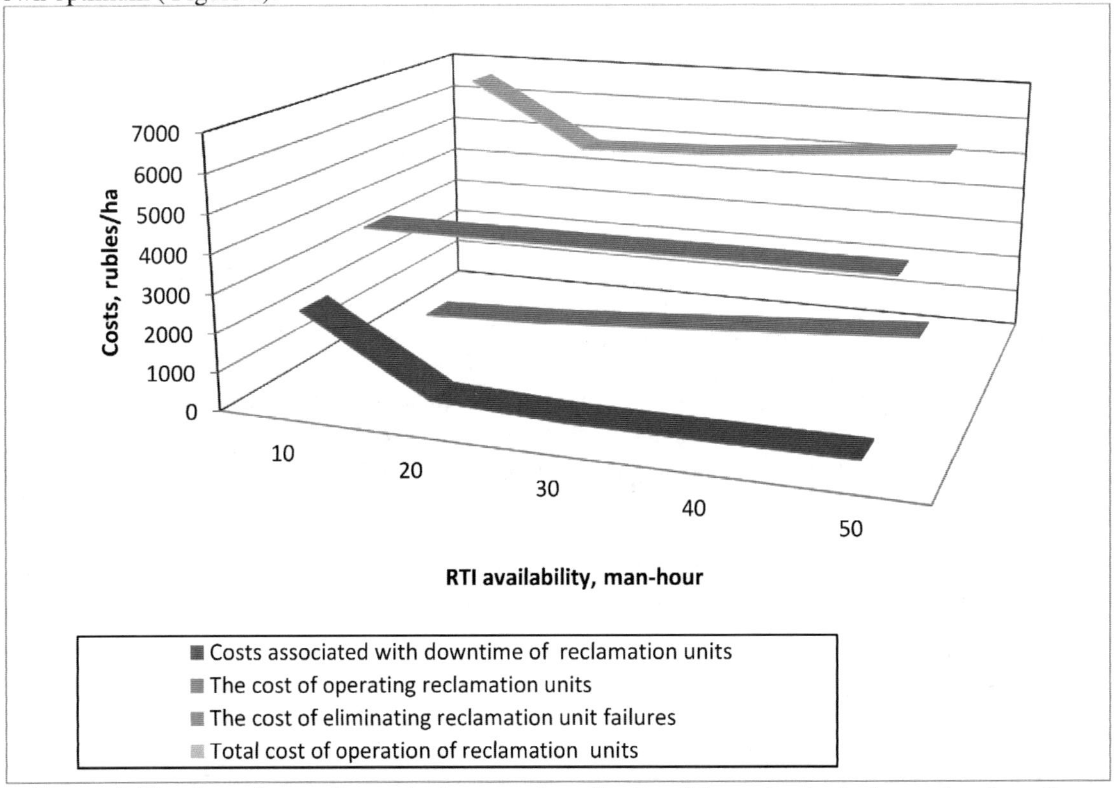

Figure 1. The dependence of costs in the operation of units of the technological complex from the availability of repair and technical impact using MRW.

Analysis of the dynamics of changes in cost components shows that the increase in costs for eliminating failures is associated with the attraction of additional service points (MRWs). Increasing the availability of RTI using MRWs leads to a reduction in losses from the downtime of reclamation units for technical reasons. With a minimum availability of RTI, the unit cost of operation is of the greatest importance, which is associated with an increase in losses from the downtime of reclamation units for technical reasons [16]. Analysis of the dependence of costs in the operation of units of technological complex from the availability of RPE shows that there is an optimal amount of RPE when performing repair and technical impacts. Thus, using the proposed model, it is possible to determine the required number (S) of MRWs, taking into account the optimal number of RTI.

However, the effectiveness of repair and technical impacts largely depends on the configuration of the MRW repair and process equipment. The nomenclature of the RTE depends on the list of repairs being carried out and is determined by the structure of the machine failures. Based on the analysis of the flow of failures, it has been established that the elimination of malfunctions of the units of technological complexes for the performance of reclamation works is mainly associated with disassembly-assembly, mounting and adjusting, diagnostic, welding, washing and lubricating works. Therefore, MRWs must be equipped with a disassembly-assembly tool, diagnostic welding and fitting equipment, washing and filling installations.

IOP Conf. Series: Materials Science and Engineering **786** (2020) 012037 doi:10.1088/1757-899X/786/1/012037

In general, the equipment of the MRW set with the necessary equipment of i-name (equipment identification) $N_{eq.id.}$ is determined as follows:

$$N_{eq.id.} = \frac{Qi}{F_{Wt} m_{sp} K_f}, \tag{2}$$

where: Q_i - the complexity of the i-type of work, man-hour (determined on the basis of the analysis of the flow of failures, the standards of troubleshooting labor intensity); F_{Wt} – daily working time fund of MRW, hours; m_{sp} - the number of service personnel (repair technicians) in the repair workroom (excluding mechanics-drivers of repaired cars); K_f - equipment utilization coefficient (takes into account the time spent on the preparation and adjustment of equipment, tools and accessories). Daily fund of working time depends on the duration of the works by repair workroom per day when the reclamation units perform their tasks. To correct the results of calculating the required amount of MRW equipment, it is necessary to take into account the demand for certain types of work provided by certain equipment [17-18]. The specified number of sets of technological equipment is determined by the expression:

$$N^F_{eq.id.} = \frac{N_{eq.id}}{K_{dem}}, \tag{3}$$

where: $N^F_{eq.id.}$ – number of sets of equipment; K_{dem} - demand ratio on the types of work provided by the set. K_{dem} for individual sets of equipment, based on the existing set of mobile repair workrooms, is determined

$$K_{dem} = \frac{F_{Wti}}{\Sigma F_{Wti}}, \tag{4}$$

where: F_{Wti} - daily worktime fund of mobile repair workroom that performs i-th type of work, man-hour, for land reclamation units is accepted, within 6-8 man-hours;

$\sum F_{Wti}$ - total daily fund of sets, for performance of i types of work, man-hour.

Basing on the proposed methodology for troubleshooting at technological complexes performing reclamation work, we recommend a mobile repair workroom based on a car chassis in the following configuration: lathe, gas welding equipment, automatic welding, sandpaper, drilling machine, crane, oxygen and propane cylinders, synchronous generator, two two-level beds, two household cabinets, a table, two autonomous heaters, an intercom, a mini-kitchen.

Calculations show that the use of MRW on the basis of the Ural 4320-58 chassis as part of the service of the operating organization, while eliminating the failures of machines belonging to the reclamation complexes, will reduce the unit operating costs by 20-30%.

4. Conclusion

1. The failures of equipment in melioration organizations, as a rule, are being repaired on-site by forces of on-site repair teams equipped with mobile repair workrooms (MRW).
2. Cost analysis in the operation of units of the technological complex shows that there is an optimal amount of MRWs when performing repair and technical impacts.
3. A technique has been developed to select mobile repair workrooms and their completeness with repair and technological equipment while eliminating machine failures during reclamation work.
4. To eliminate machine failures when performing reclamation works as part of the service department of the Federal State Budgetary Enterprise “Spetsmeliovodkhoz” management”, it is recommended to use a mobile repair workroom based on the chassis Ural 4320-58.

IOP Conf. Series: Materials Science and Engineering **786** (2020) 012037 doi:10.1088/1757-899X/786/1/012037

References

[1] The results of the 2014-2017 of the federal target program "development of land reclamation of agricultural land in Russia to 2014-2020 years: inform Ed FGBNU «Rosinformagroteh», Moscow, 2018
[2] V V Nedoluzhko and A N Novikov, J 2011 *World of Transport and Technological Machines* **1** 33-38
[3] R L Sakhapov, R V Nikolaeva, M H Gatiyatullin and M M Makhmutov 2018 Modeling of regional transport and logistics systems, *J of Physics*: Conference Series **1141(1)** 012133
[4] N I Baurova, V A Zorin and V M Prikhodko 2017 Information model of the state of the technical system, J *All materials Encyclopedic reference* **6** 11-16
[5] O Poddaeva, A Kubenin and D Gribach 2018 Measures of Improving the Accuracy of the Calculation of Energy Efficiency and Energy Saving of Construction Transport Infrastructure, J Advances in *Intelligent Systems and Computing* **692** 490-497
[6] V A Zorin, N I Baurova and A A Pegachkov 2019 Assessment of products risks of mechanical engineering by results of diagnosing, J *Periodicals of Engineering and Natural Sciences* **7(1)** 287-293
[7] N S Sevryugina 2012 Theory of technical safety formation of the complete transport and technological machines life cycle: monograph Belgorod: Publishing house of BGTU 179
[8] N Sevryugina 2017 J MATEC Web of Conf RSP 117 00151
[9] P D Kapyrin and N S Sevryugina 2018 J IOP Publishing Ltd 365 042018
[10] A S Apatenko 2014 Analysis of the causes of downtime and failures of aggregates to perform kulturtehnicheskih of works, J *Machinery and equipment for the village* **2** 14–17
[11] Ch A Abdulmazhidov 2013 Integrated application of the kanaloochistitelnyh machine Bulletin of the Moscow State Agroengineering University n V P Goryachkina **3(59)** 28-32
[12] Yu V Bashkirtsev, I G Golubev, V V Bykov and M I Golubev 2017 Type and operation of technological equipment service companies FGBNU «RIAMA», Moscow
[13] V A Makuev, V I Panferov, Y A Shamarin and V M Korneev 2014 The basics of creating branded service system forest and agricultural machinery *Vestnik of Moscow State University forest-Forest bulletin* **18(2)** 10-11
[14] P N Tarasenko 2014 Designing mobile repair units Belarusian National Technical University, Minsk
[15] A V Naumov, A S Tetenkin and A A Perevertov 2012 Improvement of technological process of recovery machines by applying new models of mobile maintenance and repair tools J *New materials and technologies in mechanical engineering* **16** 133-6
[16] A S Apatenko 2015 Improving the system of technical operation during import substitution of machines for reclamation work, J Environmental Engineering 2 74-7
[17] A S Apatenko and V A Evgrafov 2014 Optimization of the security units of reclamation technological complexes in repair and technical effects, J *Technique and equipment for the village* **8** 41–44
[18] A S Apatenko 2013 Acquisition of a fleet of machines for watering peatlands, taking into account unplanned failures, J *Technique and equipment for the village* **12** 36-38

Optimizing the number of anti-skid device for vehicle

R L Sakhapov* and M M Makhmutov

Kazan State University of Architecture and Engineering, 420043, Kazan, Zelenaya str., 1, Russia

* E-mail: rusakhapov@gmail.com

Abstract. We can image the shift work of the vehicle in the form of: a) the movement of the unit from the park to the object of work; b) work on the object; c) the movement from the object to the park. Depending on the condition of the roads, as well as the weather conditions of spring and autumn, the driving wheels of the vehicle can be equipped with the hooks just before leaving, on the road or directly at the work site. In any case, it is necessary to spend a certain amount of the shift time for installation and disassembly of the anti-skid device.

1. Introduction

The characteristics of the undercarriage systems of transport vehicles significantly affect the technical level. The advantages of wheeled propulsion units are obvious compared to crawlers. They are distinguished by their low specific metal consumption and the possibility of using them on highways. Wheeled vehicles provide the best working conditions for the driver and service personnel.

Fundamentals of the wheels rolling theory were laid in the works of such scientists as M. G. Bekker [3], I. P. Ksenevich [4], M. M. Makhmutov [5, 6], N.K. Mazitov [9], R. L. Sakhapov [7, 8] and others.

Their works note that the towing properties of vehicles are enhanced by:

1) the use of the entire mass of the tractor;
2) an increase in tire contact with the soil;
3) the development of new design solutions;
4) the development of new technological solutions.

The solution to this issue will be reduced to the determination of such number of anti-skid devices, which will ensure the maximum performance of the vehicle in these operating conditions [2, 3].

Let's present the shift performance of a vehicle operating without devices, as

$$W_{Ш} = 0{,}36\,T_O B V_T (1 - \delta_{Ш}) \quad (1)$$

Then the shift performance of the vehicle working with devices will be determined as:

$$W_C = 0{,}36 (T_O - z_C T_C) B V_T (1 - \delta_K), \quad (2)$$

where V_T – theoretical speed, m/s; T_O – basic working time, hrs; B – vehicle width, m; T_C – time required for installation and disassembly of the anti-skid device, hrs.

Consequently, the change in shift performance will be:

$$\Delta W_{CM} = 0,36\ B\ V_T\left[T_O(\delta_Ш - \delta_K) - z_C T_C(1-\delta_K)\right]. \tag{3}$$

With the increase in the number of anti-skid devices on the driving wheel, the expression in square brackets of the optimization model changes, having an extremum point [4].

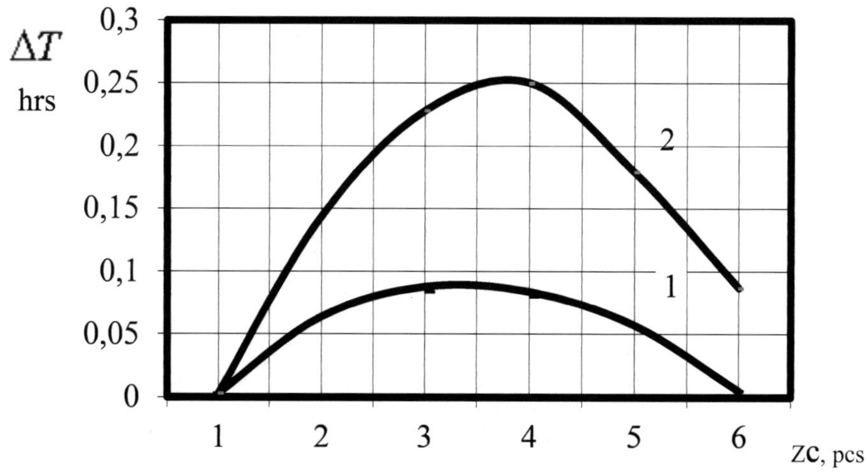

Figure 1. Curves of the optimization of anti-skid devices number on a wheeled vehicle, class 0.6 - 0.9 at: $\delta_Ш$ = 0,25; P_K = 11,8 kN; $[\sigma]$ = 200 MPa; T_O = 5,0 hrs; T_C = 0,2 hrs; τ_{cp} = 1,6 kN/m; $K_{T1} = 1,5 \cdot 10^6$ N/m^3; 1 – h_C = 0,20 m, l_C = 0,05 m; 2 – h_C = 0,10 m, l_C = 0,10 m.

For the full slip mode of the propulsor, equipped with removable hooks, the model of their quantity optimization, according to the increment factor of the shift time, will receive the following form:

$$\Delta T = (1-\delta_K)\cdot(T_O - z_C T_C) \tag{4}$$

For general case we have:

$$\Delta T = T_O(\delta_Ш - \delta_K) - z_C T_C \cdot (1-\delta_K) \tag{5}$$

The resulting model of optimization of the number of antiskid devices graphically represents a curve with an extreme value (Figure 1).

2. Materials and methods

The results of the study showed that diamond-shaped intersections at two levels are the safest intersection types. First, the diamond-shaped intersection is simple in layout and provides good visibility for drivers, which greatly reduces the likelihood that the driver can choose the wrong direction to enter or exit. Second, the ramps of the diamond-shaped intersection at different levels have a direct form. Curvy ramps have a higher degree of risk than direct ones. Third, in most cases at the diamond-shaped intersections a minor road passes over the main road.

The research plan includes:

- development of experimental research methods in accordance with GOST 7057-2001;
- the choice of objects and places for experimental studies in accordance with GOST 20915-75;
- determination of research plans and the optimal limits of change of the considered factors based on GOST 24026-80;
- preparation of tools for operation, and determination of their measurement errors;

- conduction of experimental studies and registration of measured factors (parameters) in accordance with GOST 30745-2001, GOST 3481-79;
- preparation and processing of the obtained experimental material in accordance with GOST 8.207-76.

Figure 2 shows the purpose, objectives of research and ways to achieve them.

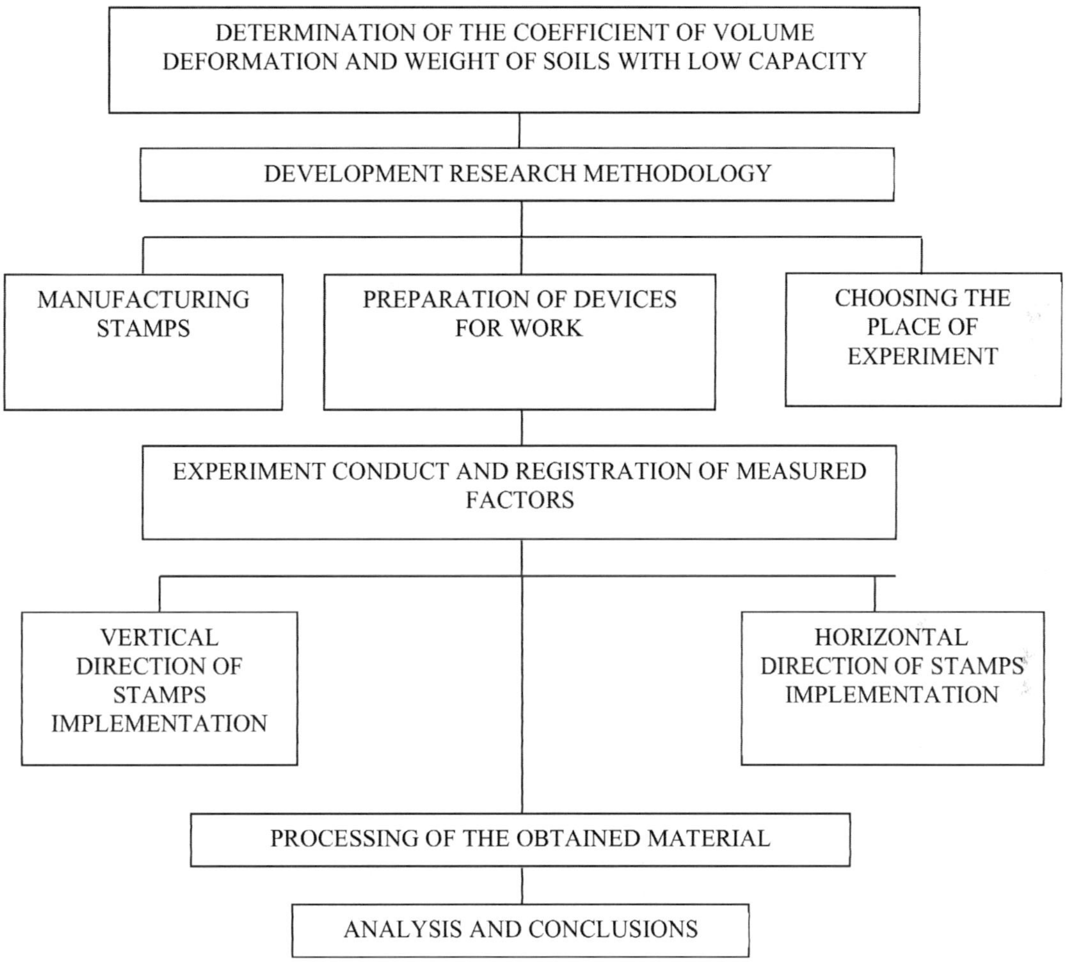

Figure 2. Program of experimental studies.

3. Discussion

With the rising number of hooks, we observe an increase in the shift time to a certain value, after which it starts to decrease. The maximum value of the increment in these operating conditions of the unit will determine the optimal number of devices on the wheel [7].

In order to get the mathematical model of the optimal number of anti-skid devices on the propulsor from the expression (2), we take the derivative by equating it to zero and substituting the model (3), we get:

$$\Delta T^{/} = \begin{pmatrix} T_O\delta_{Ш} - T_O\delta_{Ш} + T_O z_C(\delta_{Ш} - \delta_{CШ}) \times \dfrac{\arccos\left(\dfrac{R_K}{R_K - h_C}\right)}{180} - \\ - z_C T_C + z_C T_C \delta_{Ш} - z_C^2 T_C(\delta_{Ш} - \delta_{CШ}) \times \dfrac{\arccos\left(\dfrac{R_K}{R_K - h_C}\right)}{180} \end{pmatrix}^{/}.$$

Consequently,

$$T_O \cdot (\delta_{Ш} - \delta_{CШ}) \cdot \frac{\arccos\left(\dfrac{R_K}{R_K - h_C}\right)}{180} - T_C + T_C\delta_{Ш} -$$

$$- 2 z_{opm} T_C \cdot (\delta_{Ш} - \delta_{CШ}) \cdot \frac{\arccos\left(\dfrac{R_K}{R_K - h_C}\right)}{180} = 0$$

$$z_{opm} T_C \cdot (\delta_{Ш} - \delta_{CШ}) \cdot \frac{\arccos\left(\dfrac{R_K}{R_K - h_C}\right)}{90} =$$

$$= T_O \cdot (\delta_{Ш} - \delta_{CШ}) \cdot \frac{\arccos\left(\dfrac{R_K}{R_K - h_C}\right)}{180} - T_C + T_C\delta_{Ш}.$$

From here we find Z_{opt}:

$$z_{opm} = \frac{90 \cdot \left[T_O \cdot (\delta_{Ш} - \delta_{CШ}) \cdot \dfrac{\arccos\left(\dfrac{R_K}{R_K - h_C}\right)}{180} - T_C \cdot (\delta_{Ш} - 1) \right]}{T_C \times (\delta_{Ш} - \delta_{CШ}) \times \dfrac{\arccos\left(\dfrac{R_K}{R_K - h_C}\right)}{90}}.$$

Therefore, after the transformations we get:

$$z_{opm} = \frac{0{,}5 \cdot T_O \cdot (\delta_{Ш} - \delta_{CШ}) \cdot \dfrac{\arccos\left(\dfrac{R_K}{R_K - h_C}\right)}{180} - 90 \cdot T_C \cdot (1 - \delta_{Ш})}{T_C \cdot (\delta_{Ш} - \delta_{CШ}) \cdot \dfrac{\arccos\left(\dfrac{R_K}{R_K - h_C}\right)}{90}}. \quad (6)$$

Analysis of this model shows that the optimal number of anti-skid devices is influenced by the vehicle's basic operation time, device installation and disassembly time, slip coefficients of the wheel with and without hooks [8].

Let's determine the optimal number of hooks on the propulsor for the next levels and intervals of variation of significant factors (Table 1).

IOP Conf. Series: Materials Science and Engineering **786** (2020) 012038 doi:10.1088/1757-899X/786/1/012038

Table 1. Levels and factors of interval variation.

Levels	Sign	Factors			
		T_o, h	T_c, h	$\delta_{\text{Ш}}$	δ_K
Lower	- 1	5	0,1	0,15	0,05
Central	0	16	0,2	0,30	0,10
Upper	+ 1	7	0,3	0,45	0,15
Variation Interval	ΔX	1	0,1	0,15	0,05

4. Results

The calculated values of the optimal number of hooks on the propulsor according to model (5) showed the following values (Table 2).

Table 2. Calculated values of the optimal number anti-skid devices according to the criterion of traction properties.

T_o, h	T_c, h			$\delta_{\text{Ш}}$			δ_K		
	0,1	0,2	0,3	0,15	0,30	0,45	0,05	0,10	0,15
5	17,9	5,4	1,2	-21,9	5,4	9,3	6,8	5,4	3,0
6	22,9	7,9	2,9	-19,4	7,9	11,8	9,3	7,9	5,5
7	27,9	10,4	4,6	-16,9	10,4	14,3	11,8	10,4	8,0
T_c, h	T_o, h			$\delta_{\text{Ш}}$			δ_K		
	5	6	7	0,15	0,30	0,45	0,05	0,10	0,15
0,1	17,9	22,9	27,9	-4,4	22,9	26,8	24,4	22,9	20,5
0,2	5,4	7,9	10,4	-19,4	7,9	11,8	9,3	7,9	5,5
0,3	1,2	2,9	4,6	-24,4	2,9	6,8	4,3	2,9	0,5
$\delta_{\text{Ш}}$	T_o, h			T_c, h			δ_K		
	5	6	7	0,1	0,2	0,3	0,05	0,10	0,15
0,15	-21,9	-19,4	-16,9	-4,4	-19,4	24,4	-2,2	-19,4	-
0,30	-19,4	7,9	10,4	22,9	7,9	2,9	9,3	7,9	5,5
0,45	-16,9	11,8	17,3	26,8	11,8	6,8	12,2	11,8	11,3
δ_K	T_o, h			T_c, h			$\delta_{\text{Ш}}$		
	5	6	7	0,1	0,2	0,3	0,15	0,30	0,45
0,05	6,8	9,3	11,8	24,4	9,3	4,3	-2,2	9,3	12,2
0,10	5,4	7,9	10,4	22,9	7,9	2,9	19,4	7,9	11,8
0,15	3,0	5,5	8,0	20,5	5,5	0,5	19,4	5,5	11,3

With an increase in the basic operating time of the vehicle, regardless of the time of installation and disassembly, the optimal number of hooks increases at a central level of variation of the factors in a linear relationship. At $T_C = 0.1$ hrs, the increase in maintenance by 1 hour increases Z_{opt} by 5 pcs., At $T_C = 0.2$ hrs - by 2.5 pcs., at $T_C = 0.3$ hrs - by 1.7 pcs.

With an increase in the basic operating time of the vehicle, regardless of the size of wheel slip, the optimal number of hooks increases at a central level of variation of the factors in a linear relationship. However, when $\delta_{\text{Ш}} = 0.15$, Z_{opt} has negative values, which means that it is inexpedient to use hooks in these operating conditions. At $\delta_{\text{Ш}} = 0.30 \dots 0.45$ increase in T_O for 1 hour, increases Z_{opt} by 2.5 pcs.

With an increase in the basic operating time of the vehicle at the central level of factors variation, regardless of the wheel slip value δ_K, the optimal number of hooks increases in a linear relationship and the increase in maintenance T_O for 1 hour increases Z_{opt} by 2.5 pcs.

After studying the effect of installation and disassembly time of the anti-skid device on the optimal number of hooks, we should note that regardless of the basic operating time of the vehicle, with an increase in T_C, Z_{opt} decreases along the curve with negative acceleration (Figure 3). At the central level of factors variation, with an increase in T_C in the range of 0.1 ... 0.2 hrs, Z_{opt} decreases by 34.5%, for T_O = 5 hours - by 30.2%, for T_O = 7 hours - by 37, 2%. For the range of T_C 0.2 ... 0.3 hrs, Z_{opt} is reduced by 22.2; 36.7 and 44.2% respectively.

With an increase in the time of installation and disassembly of the device at the central level of factors variation, regardless of the value of the wheel slip $\delta_Ш$, the optimal number of hooks decreases along the curve with a negative acceleration. However, when $\delta_Ш$ = 0.15, $Z_{opt\,t}$ has negative values, which means that it is inexpedient to use hooks in these operating conditions. At = 0.30 ... 0.45, the increase in the vehicle in the range from 0.1 ... 0.2 hours reduces Z_{opt} by 15 pcs., and in the T_C range from 0.2 ... 0.3 hrs to 5 pcs.

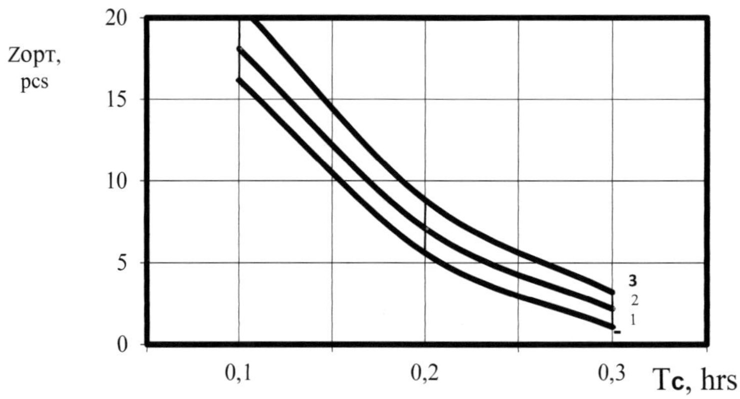

Figure 3. Change of the optimal number of anti-skid devices from the time of installation and disassembly of the device and the basic operating time of the vehicle:
1 – T_O = 5 hrs; 2 – T_O = 6 hrs; 3 – T_O = 7 hrs.

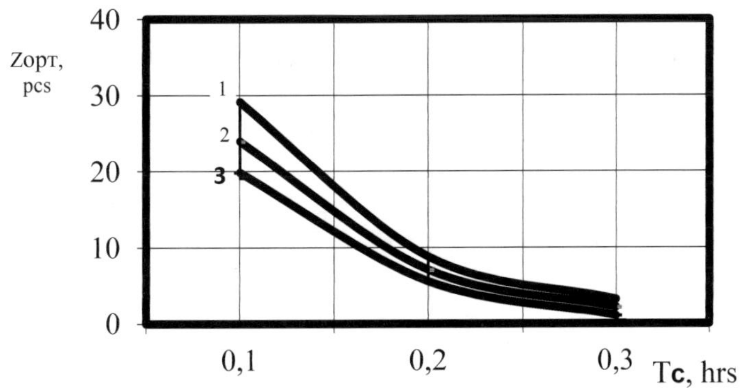

Figure 4. The influence of the installation and disassembly time of the device and wheel slip factor δ_K on the optimal number of the hooks:
1 – δ_K = 0,05; 2 – δ_K = 0,10; 3 – δ_K = 0,15.

With an increase in the disassembly time of the device, at the central level of factors variation, regardless of the the wheel slip value δ_K, the optimal number of hooks decreases along the curve with negative acceleration (Figure 4). At δ_K = 0.05 ... 0.15 with an increase in the T_C in the range from 0.1 ... 0.2 hours, Z_{opt} is reduced by 15 pcs., And in the T_C range from 0.2 ... 0.3 hours - by 5 pcs.

After studying the influence of wheel slip on the optimal number of hooks, we should note that at the central level of factors variation at T_O = 5 hours, it is inexpedient to equip the vehicle with hooks. For T_O at 6 ... 7 hours, the expediency of equipment arises at high values of propulsor slipping and with an increase of $\delta_Ш$ in the range of 0.30 ... 0.45 Z_{opt} increases by 4 pcs.

With an increase in wheel slippage at a central level of factors variation, regardless of the time of installation and disassembly of the device, the optimal number of hooks increases along a curve with a negative acceleration (Figure 5). However, at low values of $\delta_Ш$ = 0.15, it is inexpedient to equip the vehicle with hooks. With an increase of $\delta_Ш$ in the range 0.30 ... 0.45 at T_C = 0.1 h, Z_{opt} increases by 1.2 times, at T_C = 0.2 hours - 1.5 times, and at T_C = 0.3 hours - in 2.3 times.

With an increase in wheel slip, at a central level of factors variation, regardless of the value of wheel slip, equipped with hooks δ_K, the optimum number of hooks decreases along a curve with negative acceleration (Figure 5). At δ_K = 0.05 with an increase of $\delta_Ш$ in the range 0.30 ... 0.45, Z_{opt} is increases by 3 pcs., for δ_K = 0.10 - by 4 pcs., for δ_K = 0.15 - by 5 pcs.

After studying the influence of the slip of the equipped wheel δ_K on the optimal number of hooks, we should note that at the central level of factors variation at T_O = 5 hours, with an increase of δ_K in the range from 0.05 to 0.10, Z_{opt} decreases by 20.5%; at T_O = 6 h - by 15.1%; at T_O = 7 h - by 11.9%. With an increase of δ_K in the range from 0.10 to 0.15, Z_{opt} decreases by 44.4; 30.7% and 23.1% respectively. With an increase of δ_K in the range from 0.05 to 0.10, depending on the time of

IOP Conf. Series: Materials Science and Engineering **786** (2020) 012038 doi:10.1088/1757-899X/786/1/012038

installation and disassembly of the device, Z_{opt} decreases in a linear relationship and equals to 2 pcs on average.

With wheel slippage value $\delta_{Ш}$ = 0.15, at the central levels of factors variation, Z_{opt} has negative values, therefore, it is not advisable to equip the propulsor with removable hooks in these operating conditions. With an increase of δ_{K} in the range from 0.05 to 0.10, with $\delta_{Ш}$ = 0.30, Z_{opt} decreases by 15.1%; for $\delta_{Ш}$ = 0.45 - by 3.2%. With an increase of δ_{K} in the range from 0.15 to 0.10, Z_{opt} decreases by 30.3 and 4.2%, respectively.

One of the factors that increase the productivity of the vehicles (as shown by the models [8]) and, consequently, the range of expediency of removable hooks usage, is the time required for installation and disassembly of the anti-skid device [9]. The value of this factor depends on the mass of the removable device, convenience and the connection of its attachment to the wheel rim.

We have developed and manufactured prototypes of removable anti-skid devices with minimal attachment on the wheel rim.

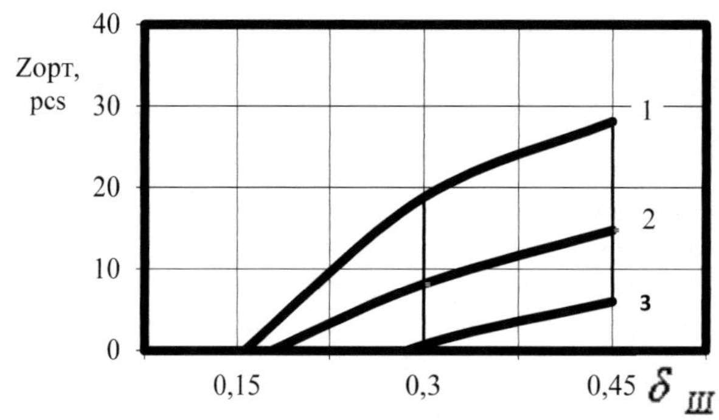

Figure 5. Influence of slipping coefficient $\delta_{Ш}$ on the optimal number of anti-skid devices depending on the time of installation and disassembly of the device: 1 – T_C = 0,1 hrs; 2 – T_C = 0,2 hrs; 3 – T_C = 0,3 hrs.

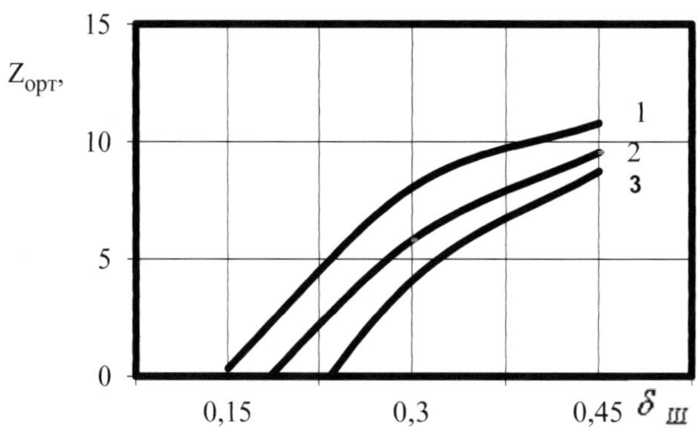

Figure 6. Change the optimal number of anti-skid devices from $\delta_{Ш}$ and δ_{K} : 1 – δ_{K} = 0,05; 2 – δ_{K} = 0,10; 3 – δ_{K} = 0,15.

IOP Conf. Series: Materials Science and Engineering **786** (2020) 012038 doi:10.1088/1757-899X/786/1/012038

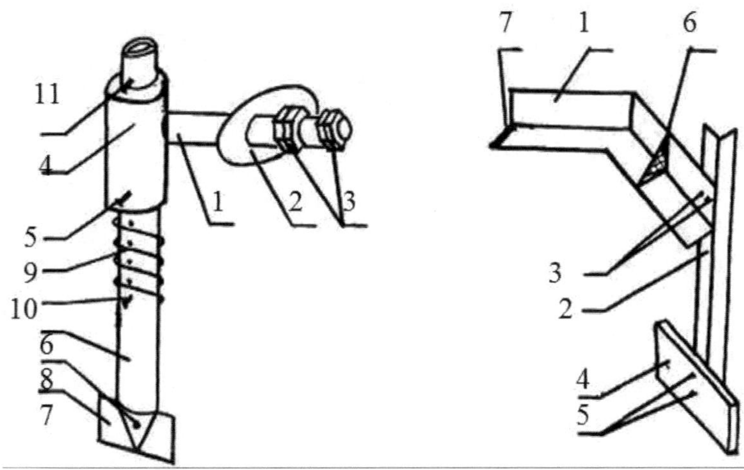

a) b)

Figure 7. Anti-skid devices for traction class:
a – 0,9…2,0; b – 0,6.

The anti-skid device for tractiob class 0.9 ... 2.0 consists (Figure 7) from the axis 1, fastened to the wheel rim disk by nuts 3 and fixing the position of the cheek 2, the guide sleeve 4, with the bolt 5, the adjustable stand 6 with the holes 10; hook 7, fixed in the lower groove of the rack using a bolt connection 8, and spring 9.

A device for wheel propulsors consists of a bracket 1 (Figure 7), a pillar 2 fixed to the bracket with the help of bolts 3; a hook 4 connected to the bottom of the rack by a bolt connection 5. To increase the resistance of the bracket, stiffener 6 is welded to bending and torsion moments.

5. Conclusion

Mitigation of dynamic loads on the wheel disk is achieved by a rubber gasket 7, worn on one of the edges of the bracket.

To attach the device to the wheel rim, it is enough to unscrew one of the disc nuts and, aligning it with the hole in the bracket, fix the device. On average, the installation and disassembly of the device takes 8 ... 12 minutes.

Thus, this analysis of the model for optimizing the number of removable hooks, depending on the operating time of the vehicle, installation and disassembly of the anti-skid device, soil properties by the criterion of traction characteristics of the propulsor, shows its suitability in engineering calculations.

References

[1] Makhmutov M M 2003 Optimizatsiya chisla ustroystv protivoskolzheniya na kolesnom dvizhitele Traktory i selkhozmashiny **10** 11

[2] Sakhapov R L , Makhmutov M M and Zemdikhanov M M 2017 Eksperimentalnaya model vliyaniya vlazhnosti i tipa pochv na udelnoe soprotivlenie sdvigu // *Izvestiya KGASU* – M **1(39)** 286-93

[3] Makhmutov M M 2011 Povyshenie tyagovo-stsepnykh svoystv kolesnykh dvizhiteley mashinno-traktornykh agregatov // RGAZU Moskva

[4] Makhmutov M M 2007 Vliyanie ustroystv protivoskolzheniyana pochvosberegayushchie svoystva dvizhiteley// *Traktory i selkhozmashiny*, **1** 9 10

[5] Makhmutov M M 2012 Rezultaty eksperimentalnykh issledovaniy buksovaniya kartofeleuborochnogo agregata // *Vestnik RGAZU* – M **13** 73-8

[6] Makhmutov M M , Makhmutov M M , Gadzhiev P I , Slavkin V I , Ramazanova G G , Navruzova L G , Bykovskiy V S , Tepikin A I and Alimov A S Patent 2536610 RF Sposob povysheniya

tyagovo-stsepnykh svoystv dvizhitelya v usloviyakh mezhduryadya i ustroystvo dlya ego osushchestvleniya (RU 2536610) 15 08.2012; publ. 27 12.2014 bul 36

[7] Sakhapov R L , Makhmutov M M , Zemdikhanov M M and Ismagilov A G 2016 Opredelenie raskhoda topliva stroitelnoy mashiny i ee ekologicheskaya otsenka // v sbornike trudov konferentsii Aktualnye napravleniya nauchnykh issledovaniy XXI veka: teoriya i praktika 4 **5-4(25-4)** 136-40

[8] Sakhapov R L and Makhmutov M M 2016 Teoriya funktsionirovaniya mekhanicheskoy energii v stroitelnykh mashinakh // *Mekhanizatsiya stroitelstva,* t **77(8)** 17-21

[9] Sakhapov R L , Nikolaeva R V , Gatiyatullin M H and Makhmutov M M 2016 Modeling the dynamics of the chassis of construction machines *Journal of Physics: Conference Series* **738(1)** 012119

The relationship of microrelief indicators of the friction surfaces of mobile interfaces of construction equipment

Y I Gustov*, S Y Sletov and P E Shishkov

Moscow State University of Civil Engineering, Russia, 129337 Moscow, 26 Yaroslavl highway

*E-mail: moidm@mgsu.ru

Abstract. The paper presents the relationships of microrelief indicators of the friction surfaces of mobile interfaces of a hydraulic excavator equipment and passenger elevator. The authors use partial micro-wear and micrometall as initial indicators of microrelief. These indicators are used to determine important operational tribomechanical characteristics. The article proposes an approximating function of the curve of the reference lines for the entire domain of relative approaches.

1. Introduction

The microrelief of friction surfaces can be represented by the normalized coordinate system "relative reference line - relative approximation" (Figure 1).

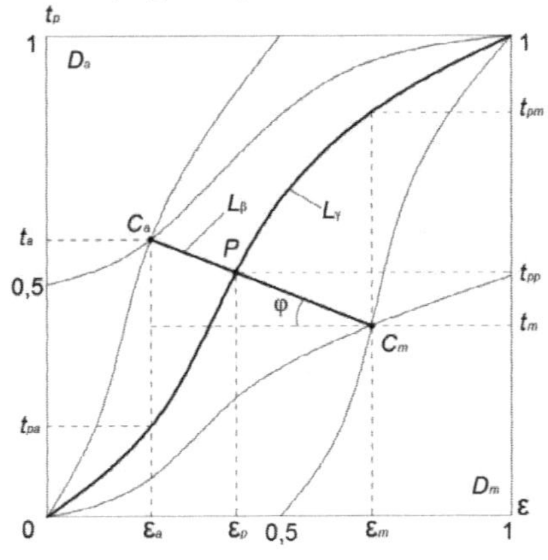

Figure 1. Calculated curve of the reference line of the friction surface of the passenger elevator.

Studies of this system allowed us to establish a number of dependencies between tribomechanical indicators of worn surfaces. At the same time, as the initial indicators, we consider the partial values of micro-wear and micrometall, corresponding to the fractions of micro-hollows and micro-protrusions of the profile on a certain profilographic base.

The theoretical and applied interest is caused by the relation of the initial (basic) indicators and their derivatives, which shows the self-organization (synergism) of the microrelief during friction and wear of surfaces.

The aim of the research is to establish the relationship between the indicators of the tribomechanical system "relative reference line - relative approximation" and to determine the operational characteristics of the used friction units.

To achieve this goal, we planned the following main tasks:

1) The determination of the pole coordinates (ε_p, t_{pp}), the centers of gravity of the partial values of micro-wear (ε_a, t_a) and micro-metal (ε_m, t_m), as well as the relative values of the reference lines (t_{pa}, t_{pm}).
2) The determination of the length of the reference lines curve (L_γ).
3) The determination of the degree of tribodeformational hardening (softening) of friction surfaces (K_s) and wear products (K_a).
4) The estimation of frictional temperatures of friction surfaces (ΔT_s) and wear particles (ΔT_a).
5) The determination of friction coefficients of motion and incomplete friction force of rest.
6) The evaluation of the efficiency of friction units (η).
7) Approximation of the curve of the reference lines of friction surfaces.

2. Methods

We investigated the friction surfaces of hinge pairs of hydraulic excavator working equipment and traction sheaves of passenger elevator. We took the profilograms of worn surfaces by means of a profilograph - a profilometer PERTH-O-METER DWR-LP (Germany). Basing on the profilograms, we constructed a normalized coordinate system "relative reference lines - relative approximations" $t_p - \varepsilon$ (see Figure 1).

The curve of the reference lines L_γ – a pegellinie - divides the coordinate system $t_p - \varepsilon$ into two parts. The upper part characterizes the partial micro-wear D_a (the share of hollows), the lower part - the partial micrometall D_m (the share of protusions) of the friction surface relief. The centers of gravity of the partial values C_a and C_m are determined by the intersection of the corresponding pairs of medians. The line connecting the centers of gravity - a bicentroid - crosses the line L_γ at the pole P.

The fundamental partial values of D_a and D_m can be determined by the methods of base reference lines, geometric areas, bicentroid segments, and pole approximations [1-3].

The method of the base reference lines can be described as relation of the sum of the reference lines of the profilogram protusions on all the approximations to the sum of the base lengths. In this case, the partial micrometal D_m and the microwear D_a are equal to:

$$D_m = \sum_1^n t_{pi}/n\,,\; D_a = 1 - D_m, \tag{1}$$

where $t_{pi} = \Delta x_i/L_\delta$ – relative reference line, n – number of approximations, L_δ – profilographic base length.

The method of the geometric areas can be described in the following way: the partial values D_a and D_m are represented by a set of geometric figures (for example, squares, triangles, etc.), which sum of the areas in relation to the area of the $t_p - \varepsilon$ system is determined by the values of D_a and D_m.

The method of bicentroid segments can be described as the division of segments before (C_aP) and after (C_mP) the pole by the bicentroid length $L_\beta = C_aC_m$:

$$D_a = C_aP/C_aC_m\,,\; D_m = C_mP/C_aC_m. \tag{2}$$

The method of pole approximation consists of equating the partial micro-wear D_a to the numerical value of the pole approximation ε_p, i.e.:

$$D_a = \varepsilon_p,\; D_m = 1 - \varepsilon_p, \tag{3}$$

The established correspondence (3) indicates the possible relations of tribomechanical indicators during the self-organization of the microrelief.

In accordance with the goal of the study, we tested and analyzed the following relationships and dependencies.

We found the pole coordinates by the relations:

$$\varepsilon_p = D_a,\ t_{pp} = D_a^{D_a/D_m}. \tag{4}$$

We found the coordinates of the center of gravity of a partial micrometall by expressions:

$$\varepsilon_m = D_a^{D_a},\ t_a = D_a D_m^{D_a}. \tag{5}$$

We found the coordinates of the center of gravity of the partial microwear by dependencies:

$$\varepsilon_a = 0{,}565\varepsilon_p,\ \varepsilon_a = \varepsilon_p \varepsilon_m,\ t_a = 0{,}5\left(1 + \varepsilon_a^{\frac{Da}{Dm}}\right). \tag{6}$$

The relative values of the reference lines were calculated by the formulas:

$$t_{pa} = \varepsilon_a{}^{D_a/D_m}, t_{pp} = \varepsilon_p{}^{D_a/D_m}, t_{pm} = \varepsilon_m{}^{D_a/D_m}. \tag{7}$$

The length of the curve of the reference lines was determined by the dependence:

$$L_\gamma = 1{,}4142(D_{max}/D_{min})^{D_a \cdot D_{max}}, \tag{8}$$

where $D_{max} = D_a$ at $D_a > D_m$, $D_{min} = D_a$ at $D_a < D_m$.

The above linear indicators for (2) - (8) were determined by micrometry with a reference value of 0.5 mm with a scale bar and a KU-A brand odometer.

We determined the tribodeformational hardening of the friction surface K_s and wear products K_a by the formulas:

$$K_s = H_s/H_0 = [0{,}618/D_m]^{Dm/Da}$$

$$\text{at } K_s > 1 \text{ (hardening) } K_a = H_a/H_0 = K_s[K_s^{0,5} + (K_s-1)^{0,5}] \tag{9}$$

$$\text{at } K_s < 1 \text{(softening) } K_a = H_a/H_0 = K_s[K_s^{2} + (K_s-1)^{2}] \tag{10}$$

where H_0 – initial metal microhardness , H_s – friction surface microhardness, H_a - microhardness of wear products.

The frictional temperatures of the friction surface T_s and wear products T_a are estimated by dependencies:

$$T_s = T_0 + \Delta T_s = T_0 + \ln K_s / \lambda \tag{11}$$

$$T_a = T_0 + \Delta T_s = T_0 + \ln K_s / \lambda \tag{12}$$

where T_0 - ambient temperature, ΔT_s and ΔT_a - temperature increment, $\lambda = 2{,}15 \cdot 10^{-3}$ 1/ ^{0}C – average temperature coefficient for steel and fusions [4].

The coefficient of friction of movement (slip) is determined by the expression:

$$f = D_m^{1/Da} = (1-D_a)^{1/Da}. \tag{13}$$

The coefficient of static friction is estimated by the slope of the bicentroid (see figure 1):

$$fp = tg\varphi = (ta-tm)/(\varepsilon m - \varepsilon a). \tag{14}$$

The efficiency was estimated by the formula:

$$\eta = 1/(1 + L_\gamma). \tag{15}$$

The curve of the reference lines $L\gamma$, depending on its type, can be approximated by the following partial functions [5]:

$$tp = \varepsilon,\ tp = b\,\varepsilon^{v} \tag{16}$$

and general function:

$$t_p = \varepsilon^{\varepsilon} \varepsilon^{\frac{p}{1-\varepsilon_p}} / (1 + \varepsilon_p - \varepsilon)\mathcal{E}^{1-\varepsilon} \tag{17}$$

3. Results

Figure 2 shows the normalized coordinate systems “relative reference line - relative approximation” t_p - $\mathcal{E}$ of worn hardened pins (a) and bushings (b) of the hinge of the working equipment of a hydraulic excavator.

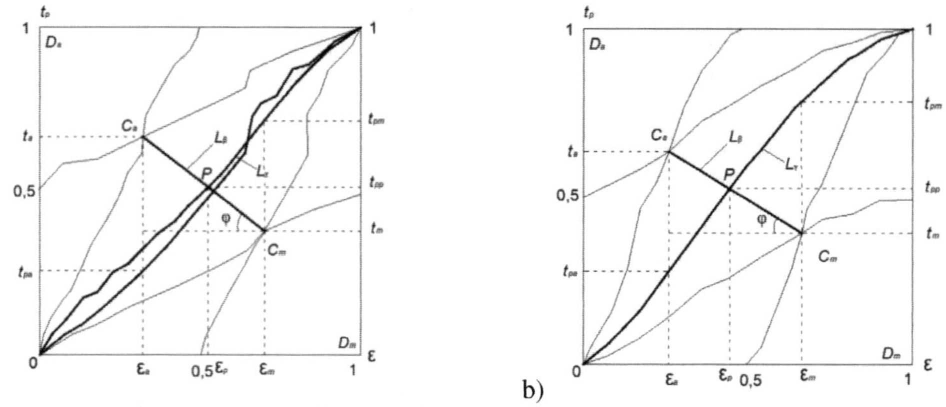

Figure 2. Coordinate system t_p - $\mathcal{E}$ of worn hardened pins (a) and bushings surface microrelief.

The initial partial sizes of microwear D_a and micrometall D_m and their derivative tribomechanical indicators are given in Table 1.

Table 1. Partial and derivative tribomechanical indicators of worn hardened parts of a hydraulic excavator hinge.

Detail	Da	*f*	Ks	Ka	ΔTs, °C	ΔTa, °C	*f*p	ta	tm	εm	ε a	Lγ	Lβ	tpa	tpp	tpm	η, %
Pin	0,50	0,25	1,236	1,974	98,5	316,3	0,714	0,625	0,35	0,695	0,31	1,45	0,47	0,255	0,485	0,72	40,8
Bushing	0,565	0,229	1,31	2,229	125,6	372,8	0,524	0,592	0,353	0,724	0,272	1,508	0,48	0,184	0,469	0,657	39,9

Figure 3 shows the coordinate systems t_p - ε of a worn-out borated pin (a) and a hardened bushing (b) of a hydraulic excavator hinge.

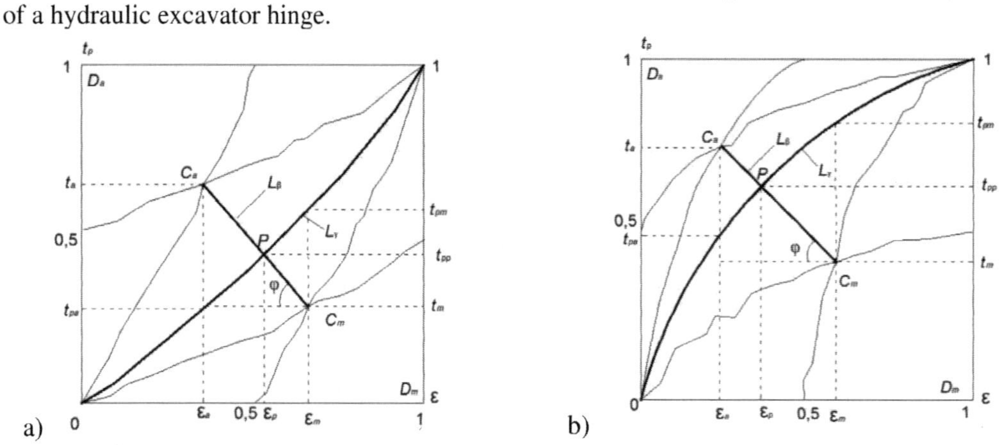

Figure 3. The coordinate system t_p - ε of the microrelief of a worn-out borated pin (a) and a hardened bushing (b) of a hydraulic excavator hinge.

Table 2 presents the initial partial values of D_a and D_m and their derived tribomechanical indicators.

Table 2. Tribomechanical indicators of worn-out borated pin and hardened bushing.

Detail *	Da	tpp	Lβ	Ks	Ka	f	t	Lγ	tgφ	ΔTs, ^0C	ΔTa, ^0C	η, %	ΔTsa, ^0C
P-1	0,460	0,52	0,476	1,167	1,738	0,262	2,16	1,472	0,609	60,0	215,0	40,5	137,5
P-2	0,545	0,53	0,470	1,291	2,163	0,236	2,14	1,48	0,625	99,4	300,0	40,3	199,7
P-4	0,535	0,445	0,470	1,484	2,847	0,239	2,47	1,45	0,699	153,6	406,0	40,8	279,8
P-5	0,455	0,535	0,470	1,163	1,772	0,263	2,51	1,49	0,608	58,8	211,0	40,0	134,9
B-4	0,365	0,640	0,485	0,849	0,630	0,297	3,66	1,50	0,957	63,7	180,0	40,0	121,85
B-5	0,525	0,520	0,475	1,269	2,09	0,242	2,30	1,47	0,773	92,7	286,8	40,4	189,75
B-7	0,320	0,620	0,500	0,816	0,571	0,300	3,00	1,50	0,647	79,1	217,9	40,0	148,5

*P – pin, B -bushing

The graphs of the approximating functions (17) agree with the experimental curves of the reference lines (see figures 2 and 3). Knowing this, it is possible to construct a normalized coordinate system t_p - ε without profilographing the friction surfaces, but using, for example, the known friction coefficient.

For the pair "traction rope - traction elevator pulley" with the coefficient of sliding friction $f = 0.276$ [3], we obtained the following values of the reference lines *tpi* with given relative approaches ε *i* (table 3). To determine the partial micro-wear $D_a = \varepsilon p$, we used the expression (13). Using the selection method with the known coefficient of friction we got $D_a = 0.41$ and $D_a/D_m = 0.695$.

Table 3. Values of the relative reference lines of the approximating function.

ε i	0,1	0,2	0,3	0,4	0,5	0,6	0,7	0,8	0,9	1,0
tpi	0,158	0,281	0,403	0,526	0,648	0,763	0,864	0,945	0,994	1,0

The graph of the calculated curve of the reference lines of the friction surface of the traction pulley, built according to the data of Table 3, is presented in Figure 3. The resulting coordinate system t_p - ε is characterized by the indicators given in Table 4.

Table 4. Indicators of the coordinate system t_p - **ε** of the traction pulley friction surface.

indicators	ε a	ε p	ε m	tpa	tpp	tpm	ta	tm	Lβ	Lγ	Ks	Ka	η, %
calculated	0,260	0,415	0,645	0,36	0,545	0,795	0,665	0,385	0,480	1,50	1,08	1,43	40,0
experimental	0,275	0,410	0,600	0,40	0,536	0,730	0,694	0,400	0,450	1,54	1,07	1,39	39,4
Δ,%	5,8	1,2	7,5	11,1	1,7	8,8	4,4	3,9	6,7	2,7	0,9	2,9	1,5

Comparison of calculation system (tp - ε) indicators with the experimental values for (3) - (7) showed satisfactory coincidence (see the differences of Δ in Table 4).

Thus, the partial values of microwear and micrometal and their derivative ratios make it possible to determine the main operational tribological characteristics of friction surfaces. According to the proposed approximating function, it becomes possible to build the calculated curve of the reference lines from the known coefficient of friction of motion (slip).

4. Conclusion

1) The basic indicators of the "relative reference line - relative approximation" system are the partial values of micro-wear and micrometall. On their basis, we can determine the operational tribomechanical characteristics of the mobile interfaces of machines and equipment.

2) The partial values of micro-wear and micrometall and the derived tribomechanical characteristics expressed through them testify to the self-organization of the process of friction and wear.

3) The proposed approximating general function of the curve of the reference lines is in qualitative and quantitative agreement with the experimental curves in the entire domain for determining the relative approximations.

4) The calculated tribomechanical indicators are in satisfactory agreement with the experimentally obtained characteristics. The developed method makes it possible to predict service tribomechanical

IOP Conf. Series: Materials Science and Engineering **786** (2020) 012039 doi:10.1088/1757-899X/786/1/012039

indicators by the profile of the friction surface, as well as by the known or given coefficient of motion friction

References

[1] Gustov Y I 2011 Tribotekhnika stroitelnykh mashin i oborudovaniya : Monografiya /Mosk gos stroy un-t – M : MGSU, 192
[2] Gustov Y I Voronina I V Orekhov A A 2011 Metodologiya issledovaniya tribomekhanicheskikh pokazateley stroitelnoy tekhniki / *Mekhanizatsiya stroitelstva* **8** 10-2
[3] Gustov Y I , Gustov D Y , Voronina I V 2015 Tribotekhnicheskie pokazateli rabochikh organov i sopryazheniy stroitelnoy tekhniki // Sbornik dokladov Mezhdunarodnoy nauchno-tekhnicheskoy konferentsii «Interstroymekh –Kazan» Izdatelstvo KGASU, 68-73
[4] A P Babichev , N A Babushkina, A M Bratkovskiy 1991Fizicheskie velichiny : Spravochnik / et al -M : Energoizdat 1232
[5] A V,CHichinadze , E D Braun, N A Bushe i dr 2001 Osnovy tribologii (trenie , iznos , smazka) Uchebnik dlya tekhn Vuzov / M : Mashinostroenie 664

International Scientific Conference Interstroymeh - 2019 (ISM - 2019) IOP Publishing
IOP Conf. Series: Materials Science and Engineering **786** (2020) 012040 doi:10.1088/1757-899X/786/1/012040

Open gear train as a source of oscillations of the construction machines mechanisms

V I Skel

Moscow State University of Civil Engineering, Russia, 129337 Moscow, 26 Yaroslavl highway

E-mail: skelvi@mgsu.ru

Abstract. The article considers kinematic, force and parametric characteristics of an open cylindrical gear train of the rotation mechanism of an excavator as sources of oscillations of mechanisms with a tooth frequency. The author builds a two-mass dynamic model with an internal perturbation in the form of a variable gear ratio and taking damping into account. The paper shows that the main perturbations can be the difference between the basic pitches of the gear and the wheel and the error of the involute profile of the side surface of the teeth. These errors occur during the manufacture of a gear pair and due to wear of the teeth. The errors in the relative position of the gear and wheel teeth, the variable tooth hardness with an overlap ratio greater than one, friction in the engagement for open gear train of a large module and manufactured by 8 ... 9 degrees of accuracy, play a significantly smaller role in the excitation of oscillations. The effect of the considered sources of oscillations should be taken into account in the resonant modes of the mechanism operation.

1. Introduction

Open gear trains are widely used in the mechanisms of rotation of construction excavators and cranes, as well in some models of concrete mixers, mills and other similar machines. Questions of the dynamics of construction machines and their mechanisms remain relevant [1]. The peculiarity of open gear trains is that usually they do not have a casing that isolates the transmission from the external environment; the gear ratio is usually more than recommended for one pair of wheels of closed gear circuits; the engagement module is relatively large: it can be much larger than 4 ... 5 mm (for powerful machines up to 60 ... 80 mm); wheel diameter over 1000 mm; degree of accuracy (kinematic, smoothness of work and contact patch) is usually no better than 8 ... 9th; It is difficult to ensure normal transmission lubrication conditions. Most often, open gears are made with an involute tooth profile, which is important due to the constructive absence of a single frame for gear and wheel bearings (low sensitivity of the involute gear to small errors in the center distance). All these features are largely interrelated and quite often one follows from the other. It is especially relevant for the accuracy and wear of open gear trains. In a closed drive of construction machines and equipment, gear trains may also have a gearing module of at least 5 mm [2]. The issues of this article may be interesting for such mechanisms as well.

There are many analytical and experimental works [3, 4, 5, 6, 7, 8, 9, 10] on the dynamics of gear trains which describe and identify main causes of dynamic loads. These include: the presence of the difference between the basic pitches of the gear teeth and the wheel, the error of the involute tooth profile due to manufacturing errors and wear, variable stiffness of the gearing with one-pair and two-pair gearing, friction in the gearing, the variability of the instantaneous gear ratio. It is also possible to note the studies indicating the nonlinearity of oscillations of gears and their role as an internal vibration exciter of oscillations of the entire mechanism.

This paper aims to study an open gear train as an internal source of oscillations in the drive mechanisms of machines of the construction industry.

2. The main part

We conducted this study on the example of the mechanism of rotation of the excavator on a two-mass dynamic model (Figure 1) with elasticity, damping and internal disturbance *i*: kinematic, force or parametric, in different modes of motion. This model is a non-free non-conservative holonomic system with a stationary kinematic connection. Let us explain the possible internal perturbations for the system considered in the study.

We took into account the kinematic perturbation from the action of the difference between the basic pitches of the gear and wheel, the error of the involute tooth profile, non-parallelism and skewing of the relative position of the axes of the gear and the open gear train wheel. Force perturbation was investigated by the influence of friction force in the gearing. Parametric perturbation was represented by a variable stiffness of the link when the overlap coefficient is greater than one.

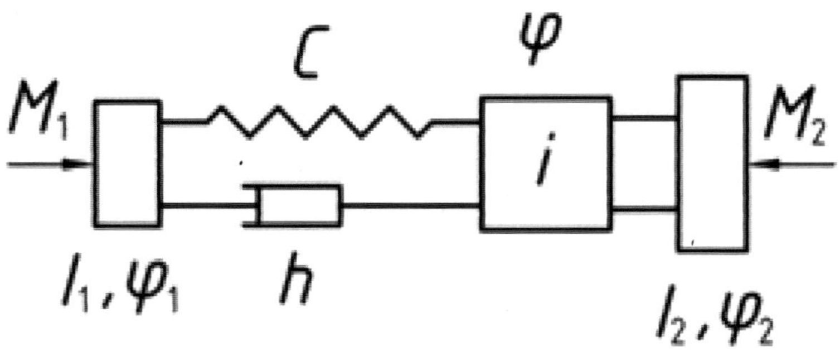

Figure 1. Generalized dynamic model of the mechanism of rotation of the excavator.

Figure 1 shows: J_1 - moment of inertia of the engine; J_2 - moment of inertia of the turntable; C - stiffness of the single-bucket excavator turning mechanism; h – generalized coefficient of viscosity in the mechanism (mainly structural damping); M_1, M_2 - driving moment and moment of resistance; φ_1, φ_2 - coordinates of movement of the masses J_1, J_2; i - gear ratio of the crown pair; φ - additional coordinate – movement of the massless gear.

Let's consider the above disturbances in more detail.

The difference in the basic pitch of the gear teeth and the wheel f_{pb} is the main reason why the teeth of the gearing wheels engage or go out of it outside the engagement line. When this occurs, the so-called edge or middle blows happen [1]. The contact of involute teeth outside the engagement line leads to a variable instantaneous gear ratio (Figure 2) with a constant value of the average gear ratio, and is an internal source of perturbation of oscillations.

Figure 2 shows: φ_{1k} - edge gearing angle, counted on the angle of rotation of the gear φ_1; $\varphi_{1\tau}$ – the end of the rotation on the angular pitch of the gear teeth; f_{u1} - the greatest error of the gear ratio at edge gearing; f_{u2} - gear ratio error when contacting involute profile sections; τ_1 – angular tooth pitch; i – gear ratio; U_{cp} – average per wheel rotation gear ratio; the dotted line of the graph shows the change in gear ratio at $f_{pb}< 0$, the solid line – at $-f_{pb}> 0$. Figure 2 graphs are built without regard to compliance of the elements involved in the transfer of load.

The measurements of the basic pitch of the gear wheels of a large module of 10 and 36 mm showed that for this wheel the error of the main pitch almost always retained its positivity or negativity.

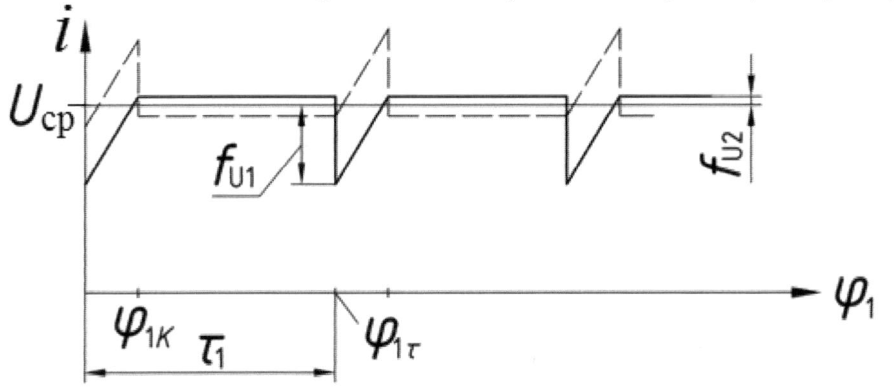

Figure 2. The change in gear ratio due to the difference of the basic pitch of gear teeth and wheels f_{pb}..

The error of the involute teeth profile also leads to a variable instantaneous gear ratio and a corresponding perturbation of the system. For excavators and other machines operating in a dusty atmosphere, the distortion of the side surface profile of the teeth of open gears due to wear may be significantly greater than manufacturing tolerances. We can note that the edges of the teeth wear out first. This leads to an increase in the basic pitch of the wheels and a corresponding disturbance. Experiments show that fluctuations can increase by 1.5 ... 2 times with worn teeth of gearing wheels compared to the new teeth. At the same time, there is evidence that the wear of the teeth has almost no effect on the magnitude of oscillations [11]. The effect of the gear manufacturing errors on the smoothness of the transmission is considered in [12].

We assessed the non-parallelism and skewness of the relative position of the axles of the gear and the wheel of the open gear train using the error of the gear ratio relative to the gear number of the transmission. Numerical estimates of this error for 12 degrees of accuracy give values less than 0.0002. This value is significantly less than the relative error in the gear ratio resulting from the difference in the basic pitch of the gear teeth and the wheel. We made the similar conclusion when evaluating other violations of the relative position of the wheels of an open gear pair: displacement of the wheels along and across the center line, non-perpendicularity of the plane of the wheels to their axes and non-parallelism of the teeth of the spur gears to their axes.

The force perturbation of the friction force in the gear happens due to two factors: 1) in an involute spur transmission, the friction force in the engagement changes its direction when the point of contact of the teeth along the engagement line passes through the engagement pole; 2) during the movement of the point of contact of the teeth along the engagement line, the shoulder of the friction force action changes relative to the supports.

For excavator rotation mechanisms, the reduced flexural stiffness of the output gear shaft, taking into account the compliance of the supports, is usually an order of magnitude higher than the torsional stiffness of the mechanism. Therefore, we can consider the influence of friction forces only on torsional oscillations. We also assume that the coefficient of friction in the process of teeth engagement on the excitation of oscillations with a tooth frequency does not change significantly, and therefore we take the coefficient of friction $f = const$ for this analysis. Another assumption is that the moment of friction in the gear supports is almost constant and quite small and can be neglected.

The effect of friction was taken into account by the relation of the gear ratio of an open gear pair by the moment i_M to the gear number of this pair U. The gear ratio of the moment $i_M = M_2/ M_1$, where M_1

and M_2 – moments occurring on the gear and the wheel, with the friction forces. We found these moments from the equilibrium condition of the gear and the wheel. We should also note that we performed an experiment on an excavator with two drives of the turning mechanism, when one gear and wheel were well lubricated and the second pair was without lubrication. As a result, there was practically no difference in the oscillations of the lubricated and non-lubricated teeth on the recording of moments on the shafts. A similar result is noted in [13].

Variable stiffness at one-pair and two-pair engagements can lead to parametric oscillations. In this case, the oscillations of open gears were described by the Hill equation. For the considered mechanisms, there is a sufficient level of the natural decrement of oscillations in the system and it does not lose stability. Numerous studies, for example [4], conclude that the average specific stiffness of a single-pair link is estimated by the value of C_0 = 1390 MPa and two-pair – C_d = 2700 MPa. However, these values and the corresponding calculations show that the stiffness of the teeth is more than 10 times greater than the stiffness of the shafts of the turning mechanism. Therefore, in the reduced stiffness C of the calculated dynamic system (Figure 1), the stiffness of the engagement plays a small role and is not a source of oscillation.

Preliminary studies of the dynamics of the rotation mechanism of the excavator show that it is characterized by a repetitive mode of acceleration-braking-stopping. In this mode, when a rotary platform accelerates, a transition resonance may occur in the rotation mechanism. A resonance, if it occurs, does not have time to develop to significant values. And noticeable fluctuations in the turning mechanism occur only at resonance. Therefore, the main studies were performed for the resonant mode, for which the single-frequency model of small oscillations is justified with obligatory consideration of damping. The perturbation from the gear transmission was represented by the variable gear ratio according to figure 2. We assumed that the disturbance amplitude from the difference of the basic pitch is equal to the disturbance amplitude from the teeth profile errors and these disturbances were represented by the two-term decomposition of the curve in figure 2 in a Fourier series. The nature of the equations in dimensionless coordinates was:

$$\begin{cases} \ddot{x}_1 + b\dot{x}_1 + x_1 = -\mu \cdot \phi(x_1, x_2, \dot{x}_1, \dot{x}_2, \gamma, b) + a_1 \\ \ddot{x}_2 = \mu a_2 \cdot \phi(x_1, x_2, \dot{x}_1, \dot{x}_2, \gamma, b) + a_3 \end{cases},$$

where $\phi(x_1, x_2, \dot{x}_1, \dot{x}_2, \gamma, b)$ - a function depending on the coordinates x1 and x2, their velocities, the damping properties of the system and the phase of the perturbation γ, μ – relative amplitude of kinematic perturbation, a1, a2, a3 – constant components depending on the inertial, stiffness and force (driving moments and resistance) system parameters.

We studied the nonlinear equations of motion by graphic methods in the stroboscopic phase plane [14].

3. Results

The results of studies of the adopted dynamic model are quite logical and consistent with the results of field tests. For example: 1) in relation to the mechanism of rotation of the excavator with internal kinematic perturbation, the oscillations decrease after reaching a maximum; 2) the maximum amplitude of the arising oscillations depends on the amplitude of the internal kinematic perturbation, the ratio of the oscillating masses and the magnitude of the attenuation coefficient 3) depending on the magnitude of the attenuation coefficient, the oscillations can take the form of beats or be almost periodic (as one of the masses tends to infinity, the oscillations become periodic); 4) resonant modes of operation should be avoided: the main forced oscillations occur with the tooth frequency of an open gear; 5) with existing norms on the kinematic accuracy of large-module tooth wheels, the main kinematic perturbation is the difference between the basic pitch of the engaging wheels and the error in the profile of the side surface of involute teeth; 6) if we consider the difference between the basic pitch of the engaging wheels as the main cause of the kinematic perturbation, then the practical overlap ratio of the open gear pair must exceed one by at least on the relative angle of rotation of the wheels with the edge gearing; 7) studies have shown that for single-bucket excavators, when the reduced value of the moment of inertia of the turntable is more than twice the inertia moment of the rotor of the engine, the excitation of resonant oscillations only by friction forces is unrealistic (the estimate was carried out with a friction coefficient of not more than 0.23, a gear ratio of 10 and logarithmic decrement of oscillations on average 0.23). Design-wise, the reduction of possible resonant oscillations can be obtained by using special dampers, vibration absorbers or due to the characteristics of the engine.

4. Conclusion

Usually, the issues of the dynamics of gear wheels are solved by modeling methods, for example, in the works [6, 7, 15, 16]. In this article, we chose an original model, which along with simplicity has certain universal properties for studying oscillations of mechanisms with gear wheels. Recommendations to reduce resonant oscillations relate to improving the accuracy of manufacturing gears, damping oscillations in one way or another, and changing the shape of teeth, for example, using flanking, non-involute tooth profile and non-circular gear wheels [17].

References

[1] Sharapov R R , Kaytukov B A and Stepanov M A 2017 Nekotorye problemy dinamiki i nadezhnosti stroitelnoy tekhniki *Mekhanizatsiya stroitelstva* T **78(7)** 5-8

[2] Kaytukov B A 2019 Razrabotka unifitsirovannykh mekhanizmov privoda betonosmesiteley prinuditelnogo deystviya / B A Kaytukov , V I Skel, P O Goryaynova // Vestnik BGTU im V *G SHukhova* **4** 151-3

[3] Petrusevich A I , Genkin M D and Grinkevich V K 1956 Dinamicheskie nagruzki v zubchatykh peredachakh s pryamozubymi kolesami Izd ANSSSr

[4] Kovalev V A 1966 Dinamikazubchatykhperedach International congress on the theory of machines and mechanisms, 2-nd, Zakopane

[5] Mathematical Models Used in Gear Dynamics-a Review H NevzatOzguvent and D R Houser *Jollmal of Sollnd and Vibration* 1988 **121(3)** 383-411

[6] GRZEGORZ LITAK, MICHAEL I FRISWELL Dynamics of a Gear System with Faults in Meshing Stiffness NonlinearDynamics 2005 **41** 415–21

[7] Kalinin D V and Temis YU M 2015 Modelirovanie nelineynykh kolebaniy tsilindricheskikh zubchatykh peredach aviatsionnykh privodov *Vestnik Samarskogo gosudarstvennogo aerokosmicheskogo universiteta* Tom **14(3)** 193-202

[8] Rashid Sharapov and Vitaly Vasiliev 2017 Analysis of the spectrum distribution of oscillation amplitudes of the concrete mix at shock vibration molding MATEC Web of Conferences Volume 117 RSP 2017 – XXVI R-S-P Seminar 2017 Theoretical Foundation of Civil Engineering, Warsaw, Poland, August 21-5

[9] Plavelsky E P and Sharapov R R 2017 The Problems of Dynamics of Wheeled Vehicles with Flow Able Building Cargo International Conference on Industrial Engineering, ICIE **206** 86-92

[10] Plavelsky E and Sharapov R 2018 Modeling of Movement the Flowable Building Load in the Operating Vessel of Vehicle International Scientific Conference Energy Management of Municipal Transportation Facilities and Transport EMMFT 2017 AdvancesinIntelligent Systemsand Computing, 692 Springer, Cham

[11] Deryuga I F 1970 Issledovanie izgibnoy prochnosti iznoshennykh zubev evolventnykh zubchatykh koles Avtoreferat dissertatsii na soiskanie uchenoy stepeni kandidata tekhnicheskikh nauk (05 161) / Tomskiy politekhn in-t im S M Kirova - Tomsk, 24

[12] Issledovanie vliyaniya pogreshnostey zubchatykh koles na plavnost raboty peredachi Zabelin D A Vestnik Polotskogo gosudarstvennogo universiteta Seriya V: *Promyshlennost Prikladnye nauki* 2009 **2** 39-43

[13] Kalinin D V and Temis YU M 2018 Analiz vliyaniya sil treniya v zatseplenii na dinamicheskie nagruzki v zubchatykh peredachakh *Izvestiya vuzov Mashinostroenie*, **6(699)** 32-43
[14] T K Hayasi 1968 Nelineynye kolebaniya v fizicheskikh sistemakh - M : Mir, 432
[15] Dynamic Model of Spur Gear Pair with Modulation Internal Excitation Zhong Wang, Lei Zhang, Yuan-Qing Luo, and Chang-Zheng Chen International Journal of Rotating Machinery Volume 2017, Article ID 1264904, 8 pages https://doi org/10 1155/2017/1264904
[16] Gearbox Simulation Models with Gear and Bearing Faults Endo Hiroaki and Sawalhi Nader, Mechanical Engineering Department
[17] Karpov, Oleksii, Nosko, Pavlo, Fil, Pavlo, Nosko, Oleksii and Olofsson, 2017 Prevention of resonance oscillations in gear mechanisms using non-circular gears Mechanism and Machine Theory 114 1-10 10 1016/j mechmachtheory 2017 03 010

IOP Conf. Series: Materials Science and Engineering **786** (2020) 012041 doi:10.1088/1757-899X/786/1/012041

The method of synthesis of new structural schemes of additional working equipment of the motor grader with the use of graphs and matrices

M L Rakhmanov[1,2,*], A G Saveliev[1], V A Mikhailovskaya[1] and M A Kosov[1]

[1]Moscow Automobile and Road Construction State Technical University (MADI), 64, Leningradsky prospect, Moscow, Russia

[2]Moscow Aviation Institute (National Research University), Moscow, Russia

[*]E-mail: domrml@yandex.ru

Abstract. The paper considers a new approach to determining the external noise of road construction and earthmoving machines. The article shows that the existing method gives overestimated measurement results for earth moving vehicles. It was also revealed that due to the fact that the known methods for determining the external noise of road construction machines do not give a real picture of the external noise emitted by the machines, there are no devices for measuring the real noise of the vehicles. The authors provide the results of experimental studies performed at the plant of JSC "Bryansk Arsenal".

1. Introduction

Constantly increasing requirements for road construction and earthmoving machines (RCM and EM), especially in terms of their safety, ergonomics and environmental friendliness, also require the development of new, more advanced methods for determining the parameters of machines that affect these indicators.

One of the main parameters affecting labor safety is the level of external noise emitted by machines. The limit value of external noise affecting workers located in the working area is determined by GOST 12.1.003-83 and is equal to 80 dBA based on an 8-hour working day.

The World Health Organization estimates that the noise level at which the risk of hearing damage is minimal is 75 dBA at an 8-hour working day.

Since July 1, 1995, the Russian Federation introduced mandatory certification of road construction and earthmoving machinery, which should check the level of external noise.

The methods for determining external noise according to GOST 27717-88 and GOST 28975-91 suggest measuring external noise when simulating working movements or operating the machine at maximum speed and only on a spherical measuring surface where noise sensors are located, which does not provide reliable data on external noise of the machine.

The existing methods for determining the external noise of road construction and earthmoving machinery do not imply the possibility of changing the external noise emission by the machine during operation and do not take into account the accumulation of the effect of noise on workers in the working area with unstable operating mode and location of the machine. Moreover, they can't identify the areas around the entire trajectory of the machine, where external noise is maximum.

Inaccuracies in the methods for determining the external noise emitted by road construction and earthmoving machinery can lead either to the danger of harmful effects on the human body or to unreasonably high material costs for noise reduction.

Therefore, there is a need to develop a methodology that allows to increase the accuracy of determining the external noise emitted by road construction and earthmoving machinery, which provides an objective assessment of machines taking into account the technology of work.

In relation to road construction and earthmoving machinery the Russian Federation applies GOST 27717-88, GOST 28975-91 and RD 22-219-88. There are also standards ISO 4872, 1585, 7216, 6393, 4871 as well as national standards SAE J1805, J1174, J1372, J88, J1262, J1008 VDI 3749 and others, which practically repeat the well-known ISO standards. All known techniques involve the use of measuring surfaces in the form of a hemisphere or parallelepiped.

As a result of the analysis of existing methods for determining the external noise of road construction and earthmoving machinery, we drew some conclusions.

1. The use of hemispherical measuring surfaces and measuring surfaces in the form of a parallelepiped does not allow us to determine the external noise of the road construction and earthmoving machinery, since there is no way to surround the entire trajectory of the movement of cyclic machines during operation, and obviously the entire trajectory of the movement of a machine with continuous working movement.
2. Measurement of external noise only at maximum engine speeds without taking into account the external real load on the working body leads to distorted results for measuring external noise of the road construction and earthmoving machinery.
3. Among the existing methods for determining noise, only the SAE J1166 standard proposes the complete elements of the duty cycle for cyclic machines and the conditions for performing work for continuous machines.
4. Known methods for determining the external noise of road construction and earthmoving machinery do not give a real picture of the external noise emitted by the machines.
5. The development of new, more accurate methods is possible based on the analysis of real work processes performed by machines.

The aim of the work is to develop a methodology that would improve the accuracy of determining the external noise emitted by road construction and earthmoving machinery and which could provide an objective assessment of machines taking into account the technology of the work performed.

2. Analytical model

All road construction and earthmoving machines with continuous working movements can be divided into two groups:

- machines having a low working speed, which the working area allows the personnel to be there all the time, for example, pavers, road construction materials spreaders;
- machines having a working speed that does not allow personnel to be in the working area, for example, motor graders, wheeled bulldozers (when paving paths and roads).

The working cycle time of a machine performing cyclic working movements can be determined by the formula:

$$\mathrm{T_c = T_{px} + T_{xx} + T_{oct}} \tag{1}$$

where: T_{px} – time of operating run; T_{xx} – time of blank run; T_{oct} – time of stops at the beginning of blank and operating run. The working cycle time can also be determined by another formula:

$$T_c = {}^{S_p}/_{V_p} + {}^{S_{xx}}/_{V_{xx}} + T_{stp}, \tag{2}$$

where: S_p, S_{xx} – lengths of blank and operating run; V_p, V_{xx} – velocities of blank and operating run. Formula (1) can be presented in expanded form for various machines. For example, for loaders:

$$T_c = 4 * T_{pp} + T_n + T_p + T_m + T_g + T_x \tag{3}$$

where: T_{pp} – gear shift duration; T_n – loading duration; T_p – unloading duration; T_m – loader maneuvering duration; T_g, T_x – duration of cargo transportation and blank run. Similar formulas can be obtained for other machines with cyclic working movements. During working and idling, stops at the beginning of blank and operating runs, the operating modes of the machine can be different and therefore the levels of noise emitted by the machine will also be different.

As a result of a theoretical analysis of noise emission of road construction and earthmoving machinery with cyclic working movement, it was obtained that the emitted noise consists of:

- noise emitted by the machine before performing the duty cycle when shifting gears - L_1;
- noise emitted by the machine when making the operating run - L_2;
- noise emitted by the machine before performing blank run when shifting gears and turning on the reverse - L_3;
- noise emitted by the machine when making the blank run - L_4.

When performing the duty cycle with a bulldozer (Figure 1), we determined the equivalent noise level at the control point. It was assumed that the control point at which it is necessary to determine the equivalent noise level is located on the longitudinal axis of the bulldozer and the axis of the direction of movement. The distance from the control point to the center of the bulldozer at the beginning of the cycle is R_0.

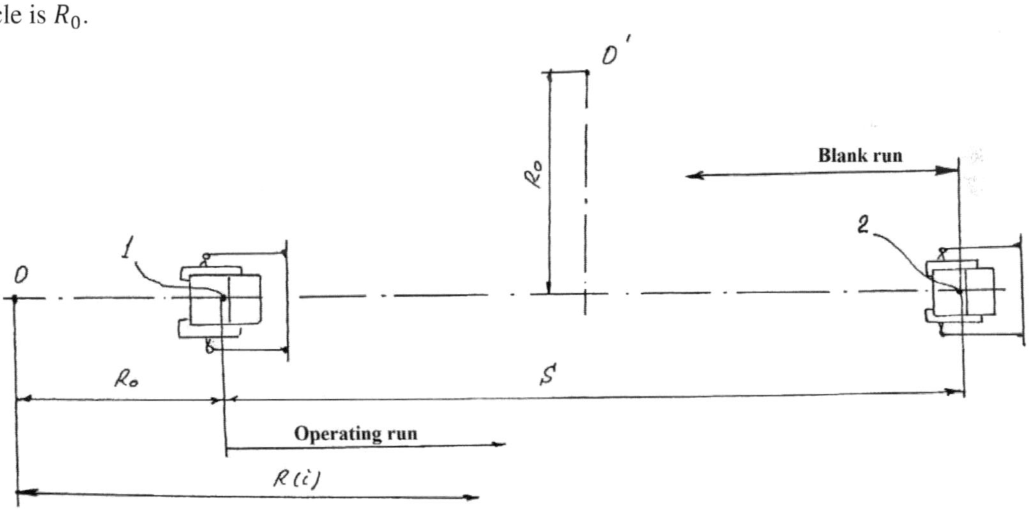

Figure 1. Scheme of the working cycle of the bulldozer. O, O′ - control points; 1 - the start point of the operating run; 2 - the starting point of the blank run.

Noise L_2; when performing the operating run decreases, since the distance R_i from the center of the bulldozer to the control point increases. The current value of the noise level can be determined by the formula:

$$L_i = L_0 - 20 * lg\frac{R_i}{R_0}, \tag{4}$$

where L_0 – noise level at the control point at the beginning of the operating run. The current distance value is determined by the formula:

$$R_i = R_0 + V_{px} * T_{px} \tag{5}$$

where: V_{px} – the current speed of the bulldozer when performing the operating run; T_{px} - the current time of the operating run.

The equivalent noise level for the entire operating run is determined as the average noise level according to the formula:

$$L_{\text{ekb}} = 10 * lg\left[\frac{1}{T_2}\int_0^{T_1} 10^{0,1*L_1(t)}dt\right], \tag{6}$$

The noise level at time t is determined by the formula:

$$L_2 = L_{0px} - 20 * lg\frac{R(i)}{R_0}, \tag{7}$$

where: L_{0px} – noise level when the bulldozer is at point 1 (Figure 1); R_0 – distance from measuring point to point 1; R(i) – current distance from the measuring point to the moving bulldozer.

Final formula (6) has the form:

$$L_{\text{ekb2}} = L_{0px} - 10 * lg\frac{S+R_0}{R_0}, \tag{8}$$

where $S = V_{px} * T_2$.

When performing the blank (reverse) run of the bulldozer, the distance R will be determined by the formula:

$$R = S - V_{xx} * t_{xx} + R_{0,} \tag{9}$$

where: t_{xx} – current time of the blank run; S – length of blank run equal to the operating time.

The equivalent noise level of a bulldozer when idling for time T is determined by the formula:

$$L_{\text{ekb4}} = L_{0xx} + 10 * lg\left[\frac{R_0^2}{V_{xx}*T_4} * \left(\frac{1}{S-V_{xx}*T_4+R_0} - \frac{1}{S+R_0}\right)\right], \tag{10}$$

Accordingly, for the current moment

$$L_{\text{ekb4}} = L_{0xx} - 10 * lg\frac{(S-V_{xx}*t_{xx}+R_0)*(S+R_0)}{R_0^2}, \tag{11}$$

where L_{0xx} – the noise level at the control point when the bulldozer is at the near point at the blank run. The equivalent noise level for a full cycle covers four operating modes and is determined by the formula

$$L_{\text{ekb}} = 10 * lg\frac{\sum_{i=1}^{4} T_i*10^{0,1*L_{0xx}}}{\sum_{i=1}^{4} T_i}, \tag{12}$$

Considering the noise emission of a bulldozer only when it is working and idling, the formula for determining the equivalent noise level for the operating run at the current time t_{px} has the form:

$$L_{\text{ekb2}}(t_{px}) = L_{0px} + 10 * lg\left[\frac{R_0^2}{V_{xx}*t_{px}} * \left(\frac{1}{R_0} - \frac{1}{V_{px}*t_{px}+R_0}\right)\right] \tag{13}$$

At the current moment of time at the blank run:

$$L_{\text{ekb}}(t_{xx}) = 10 * lg\frac{T_2*10^{0,1*L_{\text{ekb2}}(T_1)}+t_{xx}*10^{0,1*L_{\text{ekb4}}(t_{xx})}}{T_2+t_{xx}}, \tag{14}$$

where: L_{ekb2} – equivalent noise level over the entire operating run, i.e. during T_2; L_{ekb4} – equivalent noise level at the current time t_{xx} of the blank run, defined as

$$L_{\text{ekb4}}(t_{xx}) = L_{\text{ekb}}(t_{xx}) = 10 * lg\frac{T_2*10^{0,1*L_{\text{ekb2}}(T_1)}+t_{xx}*10^{0,1*L_{\text{ekb4}}(t_{xx})}}{T_2+t_{xx}}, \tag{15}$$

when the loader performed the operation (Figure 2), it was obtained that when the loader moves from point 1 to point 2, the equivalent level of external noise during this movement is determined by the formula:

$$L_{\text{ekb1-2}}(t) = L_0 + 10 * lg\left[\frac{R_0^2}{V_{xx}*t*\sin\alpha} * (arctg\frac{V*t+R_0*\cos\alpha}{R_0*\sin\alpha} + L_{\text{ekb1-2}}(t)\right] = I, \tag{16}$$

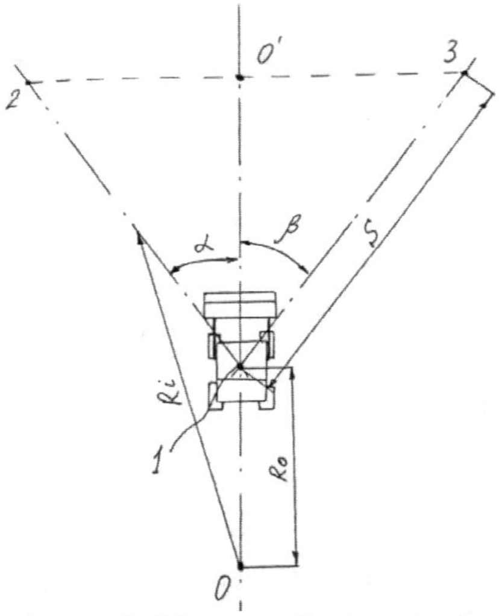

Figure 2. Loader duty cycle scheme. O, O′ - control points; 1 - the start point of the operating run; 2 - loading point; t. 3 - point of discharge.

When moving from point 2 to point 1 (when performing a reverse run), the equivalent level of external noise of the loader during this movement is determined by the formula:

$$L_{\text{ekb}1-2}(t) = L_0 + 10 * lg\left[\frac{R_0}{V*t*\sin\alpha} * \left(arctg\frac{V*t-S-R_0*\cos\alpha}{R_0*\sin\alpha} + arctg\frac{S-R_0*\cos\alpha}{R_0*\sin\alpha}\right)\right], \quad (17)$$

As shown by preliminary tests of road construction and earthmoving machinery with continuous working movement, machines having a low working speed and which working area allows personnel to be there constantly, emit different noise in different directions within the working area at the same distance from the machine. Therefore, the equivalent noise level affecting an employee can be determined by the formula:

$$L_{\text{ekb}} = 10 * lg\frac{1}{\sum_i T_i} * \left[\sum_i T_i * 10^{0,1*L_{0i}}\right], \quad (18)$$

where: L_{0i} – values of the equivalent noise level acting on an employee located at the i-th point, during his stay at the i-th point; T_i – time spent by the employee at the i-th point of his working area.

As a result of the analysis of theoretical dependencies, we found that existing methods overestimate the level of external noise affecting a motionlessly located object, which depends on the distance of movement of road construction and earthmoving machinery with cyclic working movement, less than 6 dBA or more on the level of external noise, determined by existing methods. The expected value of the level of external noise of road construction and earthmoving machinery, having continuous working movement, also differs from the level of external noise determined by existing known methods.

3. Experiment, results and conclusions

For experimental determination of external noise, we used both standard methods, described in current GOST 28975-91, and the newly developed method, protected by a patent for the invention according to application No. 95121519/03. According to the new method of measuring external noise emitted by road construction machines, a measuring surface with noise sensors located on it surrounds the entire working trajectory of the cyclic movement of the machine. One of the tested machines was the DZ-198 motor grader manufactured by "Bryansk Arsenal" JSC.

Figure 3. Motor grader DZ-198 produced by JSC "Bryansk Arsenal" at operation during testing.

Some calculation results and experimental data of external noise levels for the 1840 mini loader manufactured by Case are shown in Table 1 and shown in Figures 4, 5, 6. As a result of preliminary measurements, we found that the background noise was 51 dBa, and the noise level of the mini-loader, measured at a distance of 4 m from the machine at the beginning of the working cycle, was L_o = 79 dBA, and the average speed of the mini-loader during its duty cycle was 5 km / h.

Table 1. Theoretical and experimental data on the level of external noise for a mini-loader 1840 manufactured by Case Initial data: $L_o = 79$ dBa, $S = 10$ m, $R_0 = 4$ m, $V = 5$ km / h, $\alpha = 30°$.

Time, [sec]	Theoretical data, [dBa]	Experimental data, [dBa]	Data discrepancy, [dBa]	From point to point
0,0	79,00	79,0	0,00	1 to 2
0,6	78,13			1 to 2
1,2	77,39			1 to 2
1,8	76,75			1 to 2
2,4	76,19			1 to 2
3,0	75,69	75,6	0,09	1 to 2
3,6	75,24			1 to 2
4,2	74,83			1 to 2
4,8	74,45			1 to 2
5,4	74,11			1 to 2
6,0	73,78	73,8	0,02	1 to 2
6,6	73,50			2 to 1
7,2	73,27			2 to 1
7,8	73,10			2 to 1
8,4	72,97			2 to 1
9,0	72,90	72,7	0,20	2 to 1
9,6	72,90			2 to 1

Table 1.

Time, [sec]	Theoretical data, [dBa]	Experimental data, [dBa]	Data discrepancy, [dBa]	From point to point
10,2	72,96			2 to 1
10,8	73,11			2 to 1
11,4	73,37			2 to 1
12,0	73,78	74,2	0,42	2 to 1
12,6	74,13			1 to 3
13,2	74,27			1 to 3
13,8	74,31			1 to 3
14,4	74,29			1 to 3
15,0	74,24	74,8	0,56	1 to 3
15,6	74,17			1 to 3
16,2	74,08			1 to 3
16,8	73,99			1 to 3
17,4	73,89			1 to 3
18,0	73,78	74,2	0,42	1 to 3
18,6	73,69			3 to 1
19,2	73,60			3 to 1
19,8	73,53			3 to 1
20,4	73,47			3 to 1
21,0	73,43	72,8	0,63	3 to 1
21,6	73,41			3 to 1
22,2	73,42			3 to 1
22,8	73,48			3 to 1
23,4	73,59			3 to 1
24,0	73,78	74,2	0,42	3 to 1
		Max. discrepancy	0,63	

Let's look at the graphs of the theoretical and experimental equivalent noise levels according to the data in Table 1.

IOP Conf. Series: Materials Science and Engineering **786** (2020) 012041 doi:10.1088/1757-899X/786/1/012041

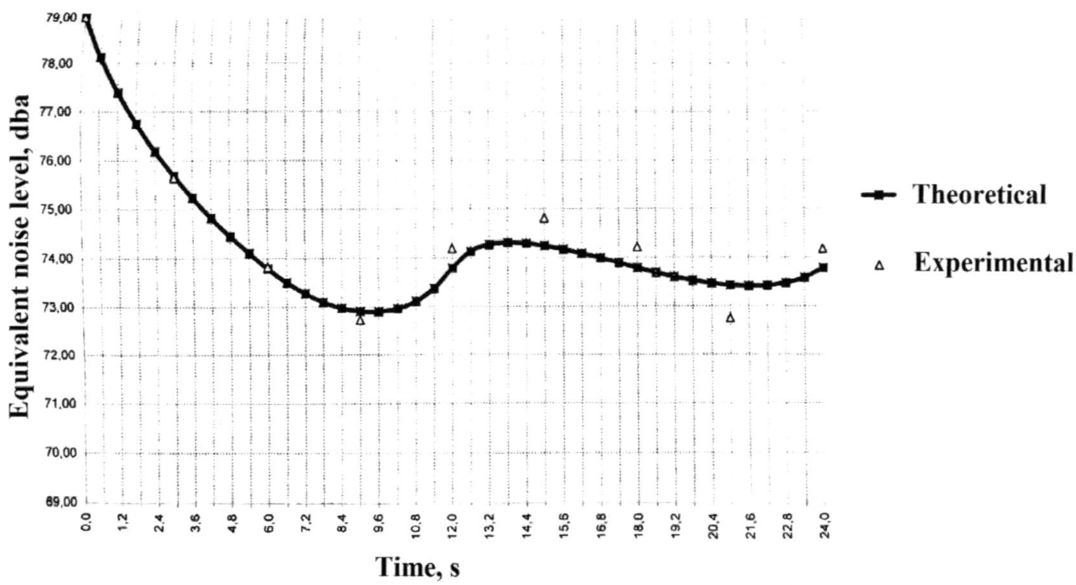

Figure 4. Graphs of theoretical and experimental noise levels for Case 1840 mini-loader.

Table 2. Initial data: $L_o = 79$ dBa, $L'_0 = 79,29$ dBa, $S = 10$ m, $R_0 = 4$ m, $R'_0 = 8,66$ m, $V = 5$ km/h, $\alpha = 150°$. The control point is located between the loading and unloading points.

Time, [sec]	Theoretical data, [dBa]	Experimental data, [dBa]	Data discrepancy, [dBa]	From point to point
0,0	72,29	72,3	0,01	1 to 2
0,6	72,74			1 to 2
1,2	73,23			1 to 2
1,8	73,75			1 to 2
2,4	74,31			1 to 2
3,0	74,88	74,4	0,48	1 to 2
3,6	75,43			1 to 2
4,2	75,92			1 to 2
4,8	76,30			1 to 2
5,4	76,53			1 to 2
6,0	76,64	76,0	0,64	1 to 2
6,6	76,72			2 to 1
7,2	76,85			2 to 1
7,8	76,98			2 to 1
8,4	77,07			2 to 1
9,0	77,10	77,5	0,40	2 to 1
9,6	77,07			2 to 1
10,2	76,99			2 to 1
10,8	76,89			2 to 1
11,4	76,77			2 to 1
12,0	76,64	76,0	0,64	2 to 1
		Max. discrepancy Blank run	0,64	

IOP Conf. Series: Materials Science and Engineering **786** (2020) 012041 doi:10.1088/1757-899X/786/1/012041

Time, [sec]	Theoretical data, [dBa]	Experimental data, [dBa]	Data discrepancy, [dBa]	From point to point
0,0	77,06	77,1	0,04	2 to 1
0,6	77,47			2 to 1
1,2	77,78			2 to 1
1,8	77,96			2 to 1
2,4	77,99			2 to 1
3,0	77,89	77,9	0,01	2 to 1
3,6	77,70			2 to 1
4,2	77,46			2 to 1
4,8	77,19			2 to 1
5,4	76,91			2 to 1
6,0	76,64	77,5	0,86	2 to 1
		Max. discrepancy	0,86	

The graphs reflecting the data in Table 2 are presented below.

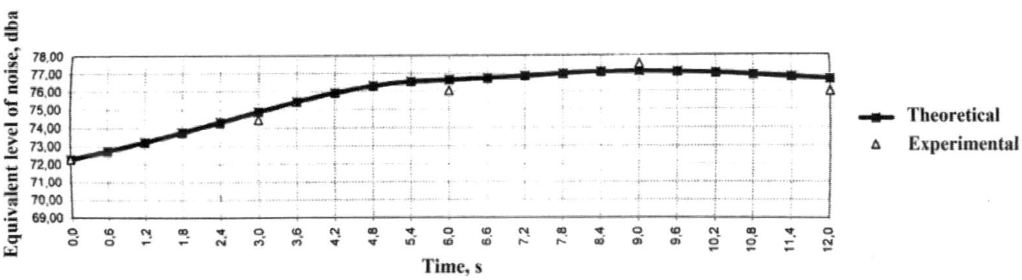

Figure 5. Graph of theoretical and experimental noise levels of the 1840 mini-loader for a full cycle (control point between the loading and unloading points).

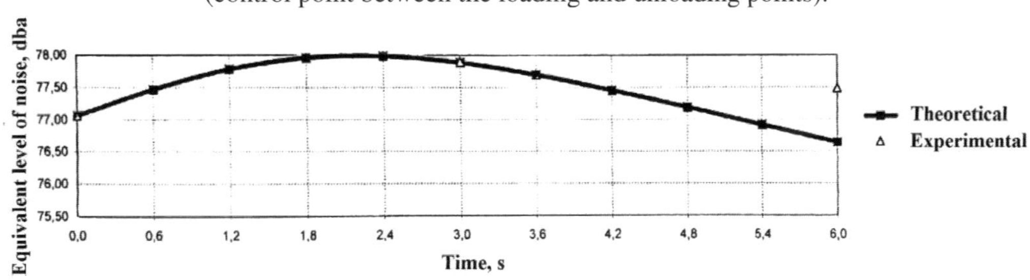

Figure 6. Schedule of theoretical and experimental noise levels of the 1840 mini-loader per blank run (reference point between the loading and unloading points).

References

[1] Rakhmanov M L and Saveliev A G 1995 Sposob izmereniya vneshnego shuma, izluchaemogo dorozhno-stroitelnymi i zemleroynymi mashinami. Zayavka 95121519/03(037794) na vydachu patenta Rossiyskoy Federatsii na izobretenie. Prioritet ot 20.12.1995

[2] Rakhmanov M L and Saveliev A G 1995 Sposob izmereniya vneshnego shuma, izluchaemogo dorozhno-stroitelnymi i zemleroynymi mashinami i ustroystvo dlya ego osushchestvleniya. Zayavka 95121898/03(037811) na vydachu patenta Rossiyskoy Federatsii na izobretenie. Prioritet ot 20.12.1995. Reshenie Goskomizobreteniy RF o priznanii izobreteniem ot 04.02.1997

[3] Rakhmanov M L and Saveliev A G 1996 Novye sposoby izmereniya vneshnego shuma // Stroitelnye i dorozhnye mashiny 6
[4] Rakhmanov M L 1997 Novye metodiki izmereniya vneshnego shuma, izluchaemogo dorozhno-stroitelnymi i zemleroynymi mashinami // *Mekhanizatsiya stroitelstva* **5**
[5] Rakhmanov M L 1997 Teoreticheskoe opredelenie ekvivalentnogo urovnya vneshnego shuma, izluchaemogo zemleroyno-transportnymi mashinami // *Mekhanizatsiya stroitelstva* **11**
[6] ISO 6393-85 Acoustic - Measurement of airborne noise emitted by earth-moving machinery – Method for determining compliance with limits for exterior noise – Stationary test condition
[7] Elliot S J and Nelson P A 1994 Active noise control//Noise/News International **2(2)** 75-98
[8] Goyder H G D 1980 Methods and Applications of Structural Modeling from Measured Structural Frequency Response Date/*Journal of Sound and Vibration* **68(2)** 209-230

Implementation of laboratory works on the modern laboratory equipment

A N Drozdov* and E R Belkina

Moscow State University of Civil Engineering, Russia, 129337 Moscow, 26 Yaroslavl highway

*E-mail: drozdovan@mgsu.ru

Abstract. The Department "Mechanization of Construction" of Moscow State University of Civil Engineering continues to work in the implementation of the concept of development of educational disciplines in terms of the widespread use of computers and software in the educational process. This paper determines the main characteristics of two types of impact machines: electric hammers with vacuum magnetic angle valve and impact wrenches using NATIONAL INSTRUMENTS equipment and, in particular, the NATIONAL INSTRUMENTS ELVIS workstation.

1. Introduction

It is much more rational to use the existing equipment of National Instruments, and NI ELVIS workstation, which allows you to significantly expand the scope and volume of laboratory works. Along with the typical radio measuring devices (millimeter, a generator of standard signals and an oscillograph) it contains useful, but rare in hardware form devices as a meter of frequency characteristics, a generator of arbitrary waveforms, a character graph, etc. We use the following equipment:

National Instruments Workstation, Lab VIEW graphical programming environment. Blocks-modules are made in metal cases with a schematic circuit diagram of the studied device attached to the front panel with internal connections made according to the scheme. Radio components, circuit elements mounted on specialized large connectors in the measures of power plugs, are installed in sockets on the front panel of the module. This design clearly represents the principles of the functioning of the studied devices and provides an easy and reliable change of parts even during operation. Power and radio meters are also connected to special sockets on the front panel of the unit.

2. The main part

The main studied parameter is the impact impulse. Its form and characteristics determines the efficiency of the implementation of working processes: breaking down and drilling rocks for hammers and perforators and tightening threaded connections for wrenches. To determine the impact impulse, we used methods of pressure determination in the pneumatic working chamber of the hammer and perforator and the method for determining the deformation on the spindle of the wrench [1-10].

1. In cooperation with JSC INTERSKOL, we prepared a prototype model of electric hammer M-10 / 1100E, providing the possibility of installing a piezometric sensor to measure the pressure in the vacuum magnetic angle valve working chambers, allowing to estimate the energy of a single machine impact and the trajectory of the impactor.

Figure 1. Sensor location on the hammer M-10/1100E.

We used a PS2001-250 sensor with a sensitivity of 250 mV / bar, with positive polarity, with a measured range from 0.01 to 25 MPa.

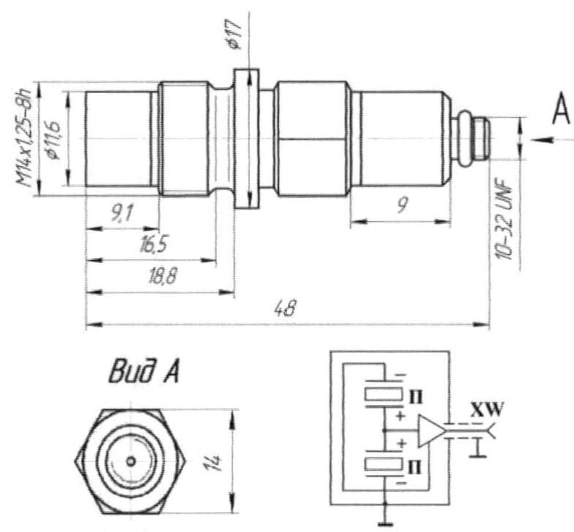

Figure 2. Appearance and wiring diagram PS2001-250.

This experimental stand was used to study the method of determining the energy of a single impact, which is the main to this type of machines. Let us describe a method for estimating a single impact energy.

To determine the energy and speed characteristics, as well as to identify the movement of an impactor as a function of time, we use B.V. Sudnishnikov theorem. This theorem is a geometric interpretation of the solution of the differential equation of motion of a body of mass m along the x axis under the action of a force R (t). As a result of integration from 0 to T, we obtained the solution in the form:

$$x = \frac{1}{m}\int_0^T (T-t)R(t)dt + \dot{x}_0 T$$

It can be interpreted as: the displacement of mass m in time T in the direction of any axis is equal to the static moment of the impulse area formed at the same time by the projection of force on this axis relative to the straight line t = T divided by the mass m plus the projection of the initial velocity ($\dot{x}_0$) multiplied by the time T. The figure shows an experimentally obtained graph for determining the operating parameters of pneumatic impact machines.

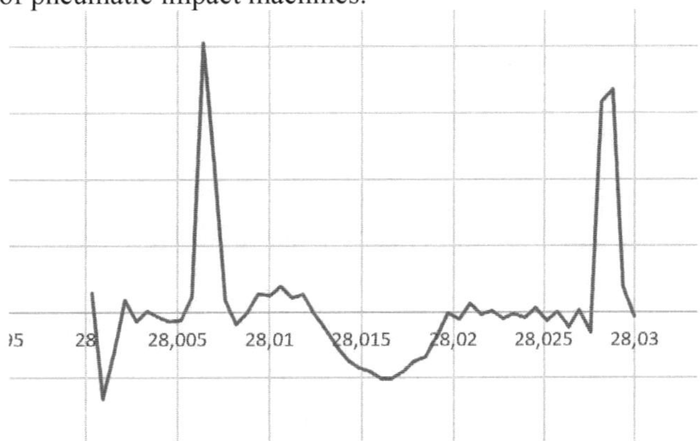

Figure 3. Impulse oscillogram.

2. To test wrenches to assess the impact of the impulse, we used the method of strain gauging. The sensors are located on the torsion bar 2 (Figure 3).

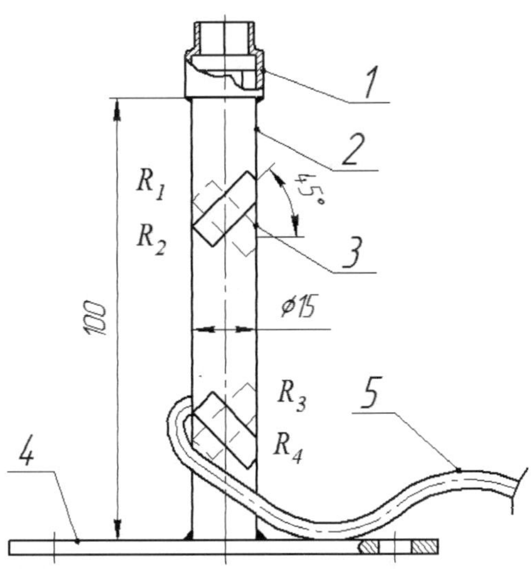

Figure 4. Sensors location.

As a result of experimental studies we found:

- the applicability of this method to determine the main parameters of impact impulses (shape, maximum amplitude, duration) in the study of impact wrenches;
- parameters of shock impulses for nut drivers of IE 3101;

- a relationship between the maximum impulse loads of the wrenches in the 2 studied modes, which makes it possible, in theoretical determination of the value, to select the calculated load for the colliding parts;

In addition to the above, the use of the described method is convenient for determining the energy parameter of wrenches - the energy of an impact impulse, which is proportional to the expression

$$\int_0^{T_{имп}} \varepsilon^2(t)dt,$$

derived from the impulse waveform.

3. Conclusion

This work allows us to modernize the formulation of laboratory works and makes it possible to provide identification of the results of the experiment and theoretical models developed at the department and listed in the references [11-22].

References

[1] Drozdov A N and Stepanov V V 2016 Matematicheskie modeli ruchnyh mashin dlya stroitelno-montazhnyh rabot s primerami realizatsii M : NIU MGSU, 152
[2] Drozdov A N and Stepanov V V 2013 Elektricheskie udarnye gaykoverty Dinamika M : MGSU
[3] Drozdov A N and Stepanov V V 2017 Gidroimpulsnye gaykoverty Monografiya M : MGSU, 84
[4] Drozdov A N and Nemkov S A 2014 Stend dlya izmereniya energii udara metodom konechnyh skorostey dlya ruchnyh mashin udarno-vrashchatelnogo deystviya // *Mekhanizatsiya stroitelstva* **12** 4
[5] Sudnishnikov B V and Esen N N 1965 Elementy dinamiki mashin udarnogo deystviya Novosibirsk SO AN SSSR, 84
[6] Tihonov A N , Batuev V N and Drozdov A N 2017 Elektroprivod stroitelnogo mekhanizirovannogo instrumenta M : NIU MGSU, 244
[7] Drozdov A N and Stepanov V V 2014 Analiz mekhanizmov gidroimpulsnyh gaykovertov // *Mekhanizatsiya stroitelstva* **11** 3
[8] Drozdov A N and Stepanov V V 2016 Matematicheskie modeli ruchnyh mashin dlya stroitelno-montazhnyh rabot s primerami realizatsii M : NIU MGSU, 152
[9] Drozdov A N and Nemkov S A 2014 Stend dlya izmereniya energii udara metodom konechnyh skorostey dlya ruchnyh mashin udarno-vrashchatelnogo deystviya // *Mekhanizatsiya stroitelstva* **12** 4
[10] Tihonov A N , Batuev V N and Drozdov A N 2017 Elektroprivod stroitelnogo mekhanizirovannogo instrumenta M : NIU MGSU, 244
[11] Manzhosov V K 2006 Modeli prodolnogo udara Ulyanovsk : UlGTU, 160
[12] Kudryavtsev E M 2013 Sistemy avtomatizirovannogo proektirovaniya mashin i oborudovaniya: Ucheb Dlya vuzov - M :ASV 379
[13] Sakhapov, R L and Absalyamova, S G 2015 The use of telecommunication technologies in education network In Proceedings of 2015 12th International Conference on Remote Engineering and Virtual Instrumentation (REV) 14-17 IEEE
[14] Kudryavtsev E M 2018 Kompyuternoe modelirovanie, proektirovanie i raschet elementov mashin i mekhanizmov Ucheb Posobie dlya vuzov M : ASV, 328
[15] Drozdov A N , Nemkov S A and Stepanov V V 2016 Opredelenie ratsionalnyh parametrov kompressionno-vakuumnogo udarnogo mekhanizma ruchnogo perforatora dlya snizheniya vibratsionnoy nagruzki rabotnika «*Bezopasnost truda v promyshlennosti*» **10** 56-9
[16] Belkina E R and Navasardyan A G Eksperimentalnoe issledovanie rabochih protsessov v elektricheskih molotkah

[17] Belkina E R Eksperimentalnoe issledovanie rabochih protsessov v udarnyh gaykovertah
[18] Drozdov A N 2019 Theoretical and empirical methods of estimating energy and force characteristics of impact wrenches Journal of Civil Engineering and Architecture, USA, (ISSN 1934-7359)
[19] A N Drozdov, V V Stepanov, S A Nemkov and D V Fedorov 2014 Utochnennaya model kompressionno-vakuumnogo udarnogo mekhanizma ruchnyh elektricheskih perforatorov Samara, Mezhdunarodnaya nauchno-tekhnicheskaya konferentsiya
[20] Drozdov A N 2012 Stroitelnye mashiny i oborudovanie Moskva, Academa
[21] Absalyamova S G and Sakhapov R L 2014 The usage of telecommunicational technologies in the integration of universities and business Proceedings of 2014 11th International Conference on Remote Engineering and Virtual Instrumentation, REV 21-3
[22] Sakhapov R L and Absalyamova S G 2015 The use of telecommunication technologies in education network Proceedings of 2015 12th *International Conference on Remote Engineering and Virtual Instrumentation*, REV 2015 **12** 14-7

The research of resistance to snow cutting and moving with an auger of a small-sized rotary-auger snowplow

M F Zakirov

Department of Building Materials, Mechanization and Geotechnics, Kalashnikov Izhevsk State Technical University, Studencheskaya Street 7, Izhevsk, 426069, Russia

E-mail: maratzf@yandex.ru

Abstract. Small-sized rotary-auger snowplows are being used to remove snow from pavements, areas near houses and in cramped conditions. To design and operate such snowplows, it is necessary to know the resistances that arise in the process of interaction of the working bodies with snow. The article describes the design of the laboratory stand "Ground channel" and the order of the experimental study to determine the resistance to snow cutting and moving with a screw of a small-sized snowplow. The paper presents the values of the total resistance on the screw feeder, depending on the frequency of rotation of the working body and the thickness of the cut snow layer.

1. Introduction

For most regions of the Russian Federation, the issues of winter maintenance of roads and territories remain relevant for almost half of the calendar year. Timely and regular works on snow cleaning and combating winter slippery conditions ensures uninterrupted and safe movement of vehicles and pedestrians. Particular attention should be paid to the areas adjacent to apartment buildings, public buildings and institutions, public transport stops, etc., where traffic is quite busy, especially in the morning and evening hours. Snow removal from local areas is mainly done by wipers manually, who do not complete the entire scope of work in a timely manner every time. Snow is a continuously changing environment, the mechanical properties of which change thousands of times or more [1,2]. The influence of atmospheric effects and the occurrence of snow cover, the intensive movement of pedestrians and vehicles during snowfall or after it, lead to a rapid compaction of snow [3,4]. Cleaning the surface from the compacted and icy snow mass leads to a significant increase in the labor and energy intensity of the process [5], as well as the complication of the snow removal technology.

2. The main part

The winter maintenance of sidewalks, near-house territories and intra-block drives using small-sized snowplows differs from the winter maintenance of highways by a number of technological and organizational features. Small-sized rotary-auger snowplow machines are being used to clean freshly fallen snow with a density of 100 ÷ 400 kg/m3. The working body of such snowplow consists of an auger feeder and a blade rotor mounted on a uniaxial self-propelled chassis (Figure 1). The auger allows to create a more compact and maneuverable design of a rotary-auger snowplow compared to a milling rotary snowplow, which is equipped with a multiple run milling cutter. Such mill has a large cutting surface, which leads to an increase in the energy consumption of the snow-cutting process for the feeder [6], and this is undesirable for small-sized snowplows having engines with a capacity of up to 10 kW. With the help of two independent power take-off shafts, torque is transmitted from the engine to the working body and the driving wheels. During the stepwise movement of the snowplow, the auger feeder, having the right and left direction of the turns, cuts and moves the snow from the edges to the central part, where the vane rotor is mounted in a common space behind the feeder. High-speed rotation of the rotor impeller causes snow to fall in a given direction by means of an exhaust nozzle. Such designs of small-sized snowplows are quite productive and are capable of throwing snow in the shortest possible time, and also have small dimensions and weight, which simplifies the control and maneuvering in cramped conditions.

Figure 1. Small-sized rotary-auger snowplow.

Many methods for calculating the basic parameters of snowplows are based on the most significant resistances that occur during the operation of the machine [7], and have a number of specific assumptions, due to the complexity of the processes occurring during the interaction of working bodies with snow [8,9]. Resistances that occur on the working bodies of small-sized low-speed snowplows with a small width are not researched enough yet, and differ sharply from high-speed snowplows based on trucks and tractors in many specific indicators [10].

Solution of the problem of calculation and design, determination of the optimal parameters and forecasting the development of the working bodies of construction, material handling, road, utility machines and equipment remains one of the topical directions of scientific research. The choice of design parameters and the assignment of operating modes of the working bodies of machines under specific operating conditions cannot be made without a comprehensive study of the process of interaction between the working body and the developed medium [11]. Scientific studies of the working bodies of machinery and equipment are carried out mainly in three areas:

1. mathematical modeling using software;
2. physical modeling on special laboratory stands;
3. reproduction of working processes in natural conditions on the structures of the working bodies of existing machines.

Due to the lack of data on resistance to snow cutting and movement with an auger of a small-sized snowplow [12,13], we measured horizontal, vertical and lateral components of the force of resistance to snow cutting and movement by an auger feeder mounted on a soil trolley at the laboratory stand “Ground channel” (figure 2) [14]. Single-pass auger with a diameter of 0.2 m is made in the form of an axial belt blade of metal sheet, welded to a shaft with a diameter of 0.06 m, with a width of 0.6 m and a screw pitch of 0.2 m, while the angle of elevation of the helix of the outer cutting edge auger is 17.7 °. The working body is driven into rotation by an electric motor using a toothed-belt transmission located on the left. A welded metal subframe, which is bolted to the intermediate frame of the laboratory stand, is provided for the installation of the drive, attachments and protective case. On both sides of the subframe, there are quick-release bearing supports for fastening the shaft of the studied working body.

Figure 2. Laboratory stand "Ground channel" with an auger.

The intermediate frame with the installed working equipment is attached to the hinged frame of the undercarriage of the laboratory stand with the help of seven strain gages: four horizontal, two vertical and one longitudinal. The readings from strain gauges were taken using an analog-to-digital converter LTR-212 (L-Card) connected to a laptop and recorded with the PowerGraph software. We used Delta frequency converter to carry out an experimental study at various speeds of rotation of the auger, and the ATV 2500LB automobile traction winch to move the trolley at a speed of 0.23 km/h along the rails of the laboratory bench.

The experimental study took place in March 2019 in natural conditions on wet snow with an average density of 400 kg/m3 at an ambient temperature of -5 ... 0 ° C. The task of the study was to determine the resistance to snow cutting and movement by the auger, depending on the frequency of rotation of the working body and the thickness of the cut snow layer. Before each experiment, the channel of the laboratory stand was filled with snow, and its surface was leveled with a grading rail. The parallelogram attachment of the hinged frame to the undercarriage of the laboratory stand allowed the working body to be buried to the required depth using screw mechanisms. The thickness of the cut snow layer was set using a metal ruler and was 25, 50, 75 and 100 mm. After checking the receiving signals from each strain gauge link, we began their calibration, followed by a switch to the data recording mode. The required frequency of the electric current was set on the frequency converter, and the drive of the working device was turned on with a rotation frequency in the range of 300-700 rpm. Then the traction winch was connected to drive the undercarriage of the laboratory bench. In the course of the experiment, horizontal, vertical and longitudinal components of resistance to snow cutting and movement by an auger were recorded in the form of oscillograms (Figure 3) taking into account the calibration coefficients of each tensometric link.

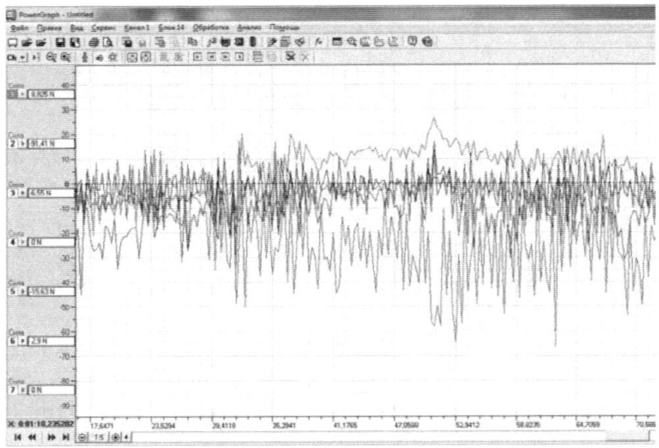

Figure 3. The characteristic form of the original oscillogram.

After the auger feeder cleaned the targeted section, the drive of the undercarriage and the auger was turned off, and the recorded data was saved on the laptop. The undercarriage returned to its original position at the beginning of the channel, and the cycle was repeated anew, starting with the loading of snow into the channel of the laboratory stand, its compaction and leveling. The experiments were repeated as many times as necessary, according to the plan of the experiment.

The results of the experimental study were processed by the average peak values of the resistance to snow cutting and movement by the auger during the steady movement of the undercarriage, which corresponded approximately to the middle part of the movement path. Based on the horizontal, vertical and longitudinal components of the resistance forces, we determined the total resistance that occurs when the auger is in operation at the current time. After processing all the oscillograms obtained in the course of laboratory experiments, we received the graphs of the values of the total resistance on the auger feeder of a small-sized snowplow, depending on the rotation frequency of the working body and the thickness of the cut snow layer (Figure 4).

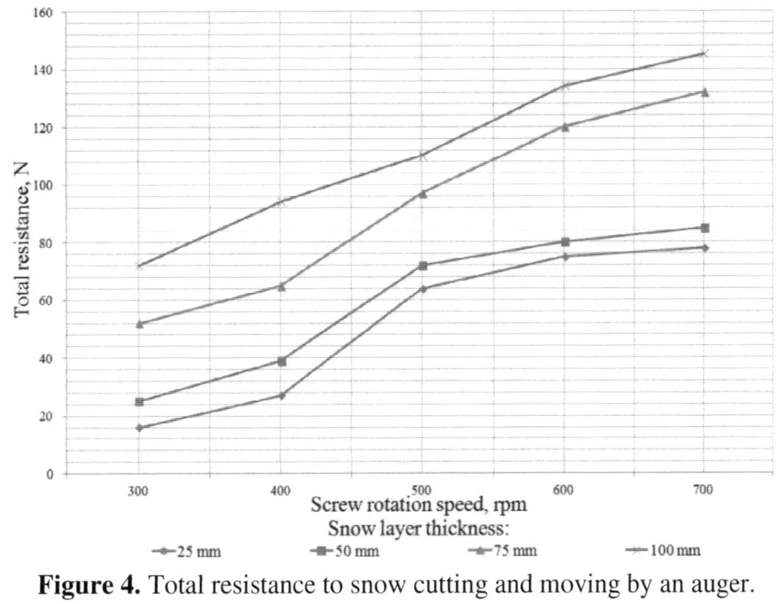

Figure 4. Total resistance to snow cutting and moving by an auger.

3. Conclusion

Based on the obtained results, we can conclude that the depth of the working body leads to an increase in the total contact length of the cutting edge of the auger with the developed medium. Therefore, with an increase in the thickness of the cut snow layer, the total resistance on the auger feeder grows mainly due to the horizontal and vertical components responsible for cutting snow from the array. Moreover, the amount of snow mass, simultaneously moving along the shaft of the auger, increases as well, which leads to a slight increase in the longitudinal component of the resistance force. It should be noted that in real constructions of small-sized snowplows the longitudinal component is almost zero during snow removal due to the right and left execution of the auger turns, which balance each other. We noted that for large thicknesses of the cut snow layer (75 and 100 mm), additional resistance occurs on the shaft of the auger feeder, which partially compresses the snow during the stepwise movement of the snowplower.

In the course of the experimental study, we found that at the speed of movement of the undercarriage of 0.23 km/h, the rotational speed of the working body in the range of 300-400 rpm is hardly enough to completely remove a 100 mm thick layer of snow. This is due to insufficient performance of the auger feeder. To ensure effective cleaning of the surface from a high layer of snow with small total resistances and a low auger rotation speed, it is necessary to reduce the translational speed of the snowplow. We should take into account that reducing the working speed of movement leads to a decrease in the performance of the machine as a whole [15]. On the other hand, work at high rotational speeds requires a lot of power to drive an auger feeder, especially when the snow layer is of small thickness, when the total resistance on the working member increases 2-4.9 times for the snow layer 100 and 25 mm respectively. The ratio of the optimal speed of movement of the base chassis and the frequency of rotation of the auger at different thicknesses of the snow layer is a task for further research in this direction.

4. References

[1] Mellor M 1964 Properties of snow US Cold Regions Research and Engineering Laboratory Cold regions science and engineering Hanover NH Pt III Sect At

[2] Kuroiwa D, Mizuno Y and Takeuchi M 1967 Micromeritical properties of snow *Physics of Snow and Ice: proceedings* **1(2)** 751–72

[3] Voitkovskiy K F Golubev V N Lapteva N I Troshkina Ye S Ushakova L A and Pavlov A V 1975 Mass transfer and metamorphism in snow cover International Association of Hydrological Sciences Publication 114 (Symposium at Grindelwald 1971 – snow Mechanics) 16–24

[4] Yosida Z 1957 Physical studies on deposited snow III Mechanical properties Low Temp Sci **11** 1–41

[5] Uesugi S 1998 Laborsaving technology of a rotary snowplow *Yuki* **30** 33–38

[6] Aleshkov D and Sukovin M 2017 Aerodynamic characteristics of the milling and rotary snowblower feeder in the loading gate area *International review of mechanical engineering* **11(9)** 701–708

[7] Zakirov M F 2015 Study of the influence of the screw step on the feeder drive power of the small rotary snow plow *Intellektualnye sistemy v proizvodstve* **2(26)** 56–57

[8] Yosida Z 1980 Theoretical studies on motion of snow kicked up by a snowplow *Low Temp Sci Ser* A **35** 47–65

[9] Korchagin P A, Teterina I A and Rahuba L F 2018 Improvement of human operator vibroprotection system in the utility machine *Journal of Physics: Conference Series* **944(1)** 12–59

[10] Yosida Z 1974 Theoretical studies on snow removal by a plough. Flow-type kick up of snow caused by a plough moving at high speeds *Low Temp Sci Ser* A **32** 39–70

[11] Nixon W and Frisbie T 1993 Field measurements of plow loads during ice removal operations Iowa department of transportatiton

[12] Zakirov M F and Baranchik V P 2014 Determination of the resistance to cutting of snow by screw of small rotary snow plow Interstroimeh (Samara: SGASU) 25–28

[13] Zakirov M F 2016 Determination of power demand on the drive of feeder of compact rotary snowplow Arhitekturno-stroitelnyj i dorozhno-transportnyj kompleksy: problemy, perspektivy, novacii (Omsk: SibADI) 348–352
[14] Zakirov M F, Menshikov R A and Shuklin I V 2016 The equipping laboratory stand with the executive tools and the research of small-sized snow blowers feeders *Intellektualnye sistemy v proizvodstve* **1(28)** 52–54
[15] Zakirov M F 2015 Determination of the optimal moving speed of rotary snow plow by method of analysis of the fourth coordinate of working process Interstroimeh – 2015 (Kazan: KGASU) 98–102

Improving the level of traffic service on the road network of cities

A I Solodkiy* and N V Chernikh

Saint Petersburg State University of Architecture and Civil Engineering (SPbGASU), Russia, 190005 Saint Petersburg, 4 Vtoraya Krasnoarmeiskaya

* E-mail: asolodkiy@mail.ru

Abstract. The problems associated with the formation of congestion and the low quality of transport service continue to be relevant for modern megacities. One of the parameters that evaluates the effectiveness of the organization of traffic, is the level of service traffic. The article provides foreign experience of using this parameter. The need to improve the level of traffic service on the road network by using a set of measures is noted as the main problem. The authors give a classification of event types. The paper notes that only the integrated application of the measures specified in the classification, taking into account the restrictions imposed and the existing traffic conditions, has the greatest effect.

1. Introduction

Problems of organization and road safety in large cities are becoming increasingly relevant due to the continuous growth of motorization and the lack of development of urban transport infrastructure. Congestions that happen everywhere on the road network cause the need to improve traffic and road management, and the use of technologies that provide a designated level of service in key urban areas with an appropriate level of road safety for road transport.

Systemic congestion leads to a highly time-consuming travel of citizens both on individual and on public transport. The level of comfort and safety of these trips is low [1]. One of the parameters that evaluates the effectiveness of the organization of road traffic is the level of service of road traffic, which is the ratio of the average vehicle speed to the speed of vehicles in free traffic conditions [2]. This provision is enshrined by the Decree of the Government of the Russian Federation dated November 16, 2018 N 1379 "On approval of the rules for determining the main parameters of road traffic and keeping their records", which was developed pursuant to Federal Law No. 443-FZ "On the Organization of Road Traffic in the Russian Federation and on amendments to certain legislative acts of the Russian Federation".

2. The main part

In foreign sources, in particular [3], the level of service (LOS) is an indicator describing traffic conditions in a traffic flow, as a rule from the point of speed of movement and travel time, freedom to maneuver, comfort, and convenience.

Level of Service is a concept introduced to relate the quality of transport service to a given flow speed, used to determine how well a vehicle works from the perspective of the traveler [4], [5], [6], and also from the position of drivers, transit passengers, cyclists and pedestrians - the so-called multi-modal level of service [7]. Usually there are six service levels, and each is assigned a letter from A to F, where service level A represents the best working conditions, and service level F is the worst.

Level of service in urban areas on the streets and roads with regulated traffic is significantly different from the level of service on countryside roads or during unregulated traffic [8, 9]. There is a following procedure for measuring the level of service on a section of city street:

- determination of the study area;
- determination of peak hourly volume and peak hourly ratio;
- determination of the speed of free flow;
- classification of the type and class of a city street;
- determination of operating time, delay time and, therefore, determination of the average speed of movement during rush hours and hours of the lowest network load;
- data analysis to determine the level of service

The criteria for determining the level of service of a city street are based on the class of a city street and the average speed of movement. If the demand at any point exceeds the capacity of the segment, then the calculated average speed of movement may not be a good indicator of the level of service. The above stages describe the determination of the city street type and the average speed of movement, and using the Table 1, we can determine the level of service of an urban street [10].

Table 1. Level of Service for different categories of US urban streets.

Urban street class	I	II	III	IV
Level of service		Average Travel Speed (km/h)		
A	>72	>59	>50	>41
B	>56-72	>46-59	>39-50	>32-41
C	>40-56	>33-46	>28-39	>23-32
D	>32-40	>26-33	>22-28	>18-23
E	>26-32	>21-26	>17-22	>14-18
F	≤26	≤21	≤17	≤14

The definition of the level of service is completely based on the average speed of movement, delay time, travel time, type and class of an urban street. For urban streets, a 'C' level of service is required as a minimal, according to Highway Capacity Manual 2010. This means that for different urban streets the average speed varies from 40-56 km / h for streets of class I to 23-32 km / h for streets of type IV.

Regarding the level of service for the urban streets in the Russian Federation, the Ministry of Transport of the Russian Federation signed the Order No. 114 "On Approval of the Road Traffic Monitoring Procedure" on April 18, 2019. This document provides the gradation of average vehicle speeds in accordance with the level of service for different categories of streets (see Table 2).

Table 2. Level of service for different categories of RF urban streets.

Urban street class	Main streets of citywide value of regulated traffic (2nd and 3rd class)	Main streets of district value	Streets and local roads	Streets and local roads in production areas
Level of service	Average Travel Speed (km/h)			
A	60-50	>85	40-34	50-43
B	50-40	85-67	34-27	43-34
C	40-30	67-50	27-20	34-25
D	30-24	50-40	20-16	25-20
E	24-18	30-40	16-12	20-15
F	<18	<30	<12	<15

Nowadays the Russian Federation urgently needs to develop methodological approaches to determining the main parameters of road traffic, namely the level of service, to comply with the provisions and standards enshrined in the above regulatory documents.

Important task is to increase the level of traffic service on the road network. For this porpose the authors have developed a method for increasing the level of service for road traffic in urban conditions, taking into account the probabilistic characteristics of the traffic flow and a combination of various factors influencing traffic conditions. The use of the proposed method provides an increase in the efficiency of functioning of the urban road network (URN) by implementing a set of measures for improvement of the transport infrastructure elements, organization and traffic management in large cities.

The first step is to determine the core network of city streets, which is the main area for the task of raising the level of road traffic service. To determine the possibility of including a street in the core network, we propose an algorithm for the formation of the street road network (see Figure 1). The selection criteria are the correspondence and traffic intensity of URN [9-12].

For each city street (or street section) claiming to be included in the core road network, we should determine the proportion of correspondences to the main streets of city-wide significance (by SP 42.13330.2016) and when this share is more than 30% - the street is included for further consideration.

The next stage calculates the coefficient of intensity of movement of the core network K_{Noc} (1). The value of the specified coefficient for each city street is compared with the weighted average traffic intensity of a large city $N_{cp.vz}$.

$$K_{N_{oc}} = \frac{k\,N_i}{N_{cp.vz.}} \tag{1}$$

Where N_i – traffic intensity on the sections of the considered route, vehicles/day; k – correction factor (1,2-1,5); $N_{cp.vz.}$ – weighted average traffic intensity, vehicles/day.

IOP Conf. Series: Materials Science and Engineering **786** (2020) 012044 doi:10.1088/1757-899X/786/1/012044

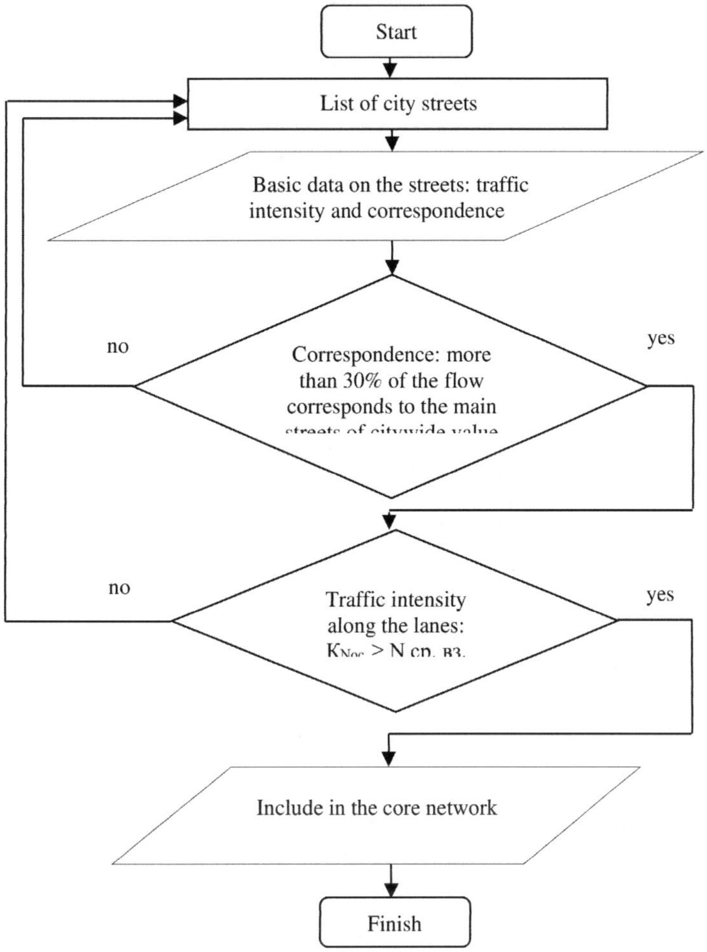

Figure 1. Algorithm for the formation of the core road network.

Thus, the core network of streets and roads is formed. First of all, the list of city streets included in the core network is checked for compliance with the assigned level of service.

If the level of service complies with the assigned for each category of city street - the city network is functioning properly and does not require additional adjustments or measures to reorganize the traffic. The values of the service level that do not comply with the designated ones indicate the need to identify areas that limit the speed of movement on the considered network. The algorithm for increasing the level of traffic service for each element of the core network is following: (see Figure 2).

IOP Conf. Series: Materials Science and Engineering **786** (2020) 012044 doi:10.1088/1757-899X/786/1/012044

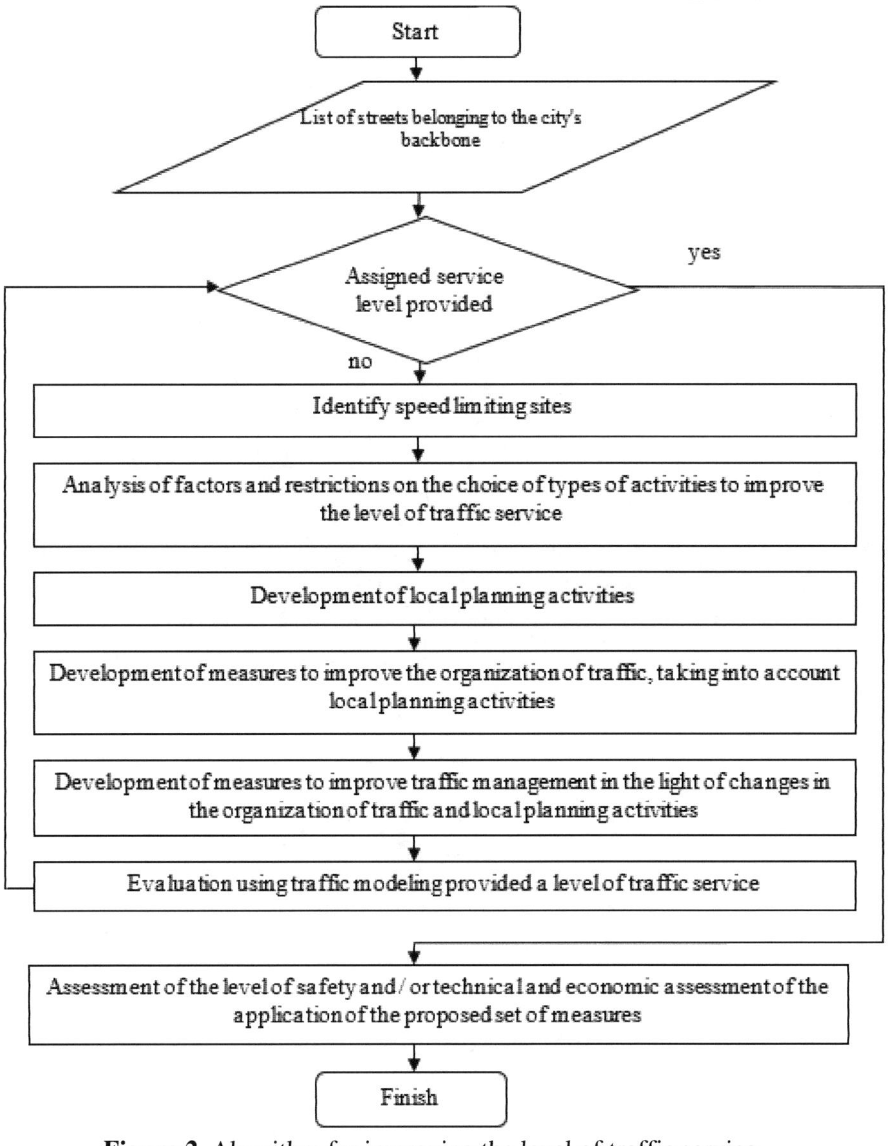

Figure 2. Algorithm for improving the level of traffic service.

Identification of problem areas is based on the analysis of the graph of the speed of movement on the specific route. First of all, it is necessary to include areas with a significant reduction in the speed of movement.

For problem areas, we should carry out a detailed analysis of the existing traffic conditions and its organization and factors that limit the possibility of using these or other measures. The following conditions and limitations are considered:

- spatial (territorial) restrictions;
- the intensity of movement of the various participants;
- change in traffic intensity;
- the demand for turning maneuvers in various directions;
- the existing traffic organization in the considered section of the street;

- existing methods of traffic control at the intersections of the considered section of the street;
- the organization of public transit stops;
- organization of parking space;
- economic restrictions;
- the time of implementation of the proposed activities.

Taking into account the above factors, we can develop measures to improve the level of traffic service, which can be divided into three groups (see. Figure 3):

1. Development of transport infrastructure;
2. Organization of traffic;
3. Movement control.

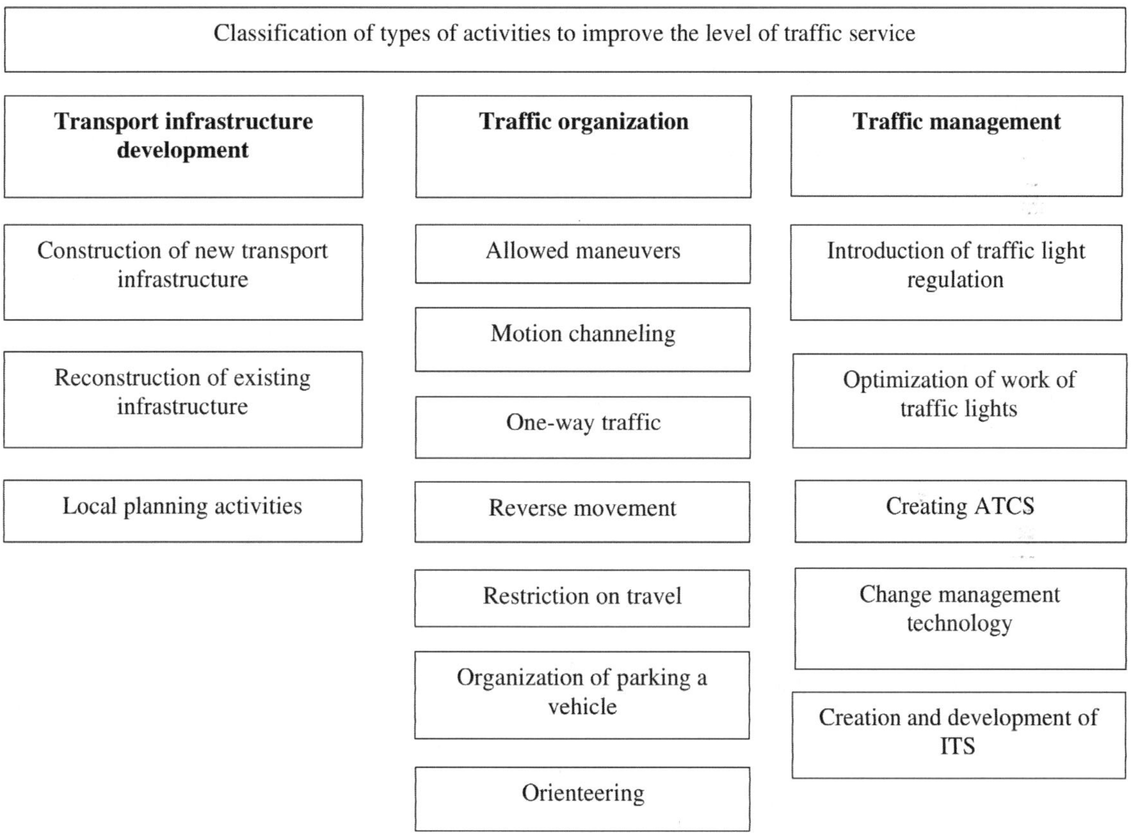

Figure 3. Classification of types of activities to improve the level of traffic service.

The biggest limitations and influence on subsequent decisions come from the measures for the development of transport infrastructure. Since the proposed methodology is focused on increasing the level of service for the movement on the existing URN, local planning measures were taken for consideration, as well as the measures to improve the organization of traffic and its management. If it is impossible to ensure a given level of service for road traffic by the specified set of measures, a transition to reconstructive measures of the URN elements is necessary.

The greatest effect is provided by the complex application of these measures. The choice of a complex of measures depends on the existing conditions, the factors listed above and restrictions.

The development of measures and the assessment of their effectiveness should be carried out with the use of transport modeling. The development of measures is carried out for each haul and the intersection of the considered route (a section of a street, a complex of streets).

After identifying areas that limit traffic speed, we advise to prepare recommendations for improving the level of service.

The sequence of the development of activities:

- analysis of factors and restrictions on the choice of types of activities to improve the level of traffic service;
- development of local planning activities;
- development of measures to improve the organization of traffic, taking into account local planning activities;
- development of measures to improve traffic management, taking into account the proposed changes in traffic management and local planning activities.

The proposed measures should serve as a basis for improving the level of service of the urban network; for this purpose, it is vital to make an assessment using transport modeling.

If the assigned level of road traffic service is reached, the proposed set of measures is accepted and a feasibility study of its use is made.

If the assigned level of traffic service is not reached, the cycle of development of a set of measures is repeated. If it is impossible to ensure the assigned level of road traffic service due to a set of measures, including local planning activities, measures to improve traffic organization and management, it is necessary to reconstruct the URN elements.

3. Conclusion

The complex of the developed techniques and algorithms allows to carry out a research of the conditions of road transport movement and will contribute to the solution of an important task - to increase the level of service and the quality of functioning of the traffic movement on the road network of large cities.

References

[1] A Solodkiy 2016 Problems of solving transport infrastructure of large cities of Russia and their solutions (on the example of St Petersburg), Collection of works of the international scientific-practical conference "Transport planning and modeling", 136-45
[2] Information on http://base garant ru/72108806
[3] Highway Capacity Manual Washington, D C 2010
[4] Dr Tom V 2014 Mathew Transportation Systems Engineering, *IIT Bombay*, **21.1-21.5**
[5] Regional Transportation Plan 2040 «What Moves You», 391-395
[6] 2014 Guide to Level of Service (LOS) Target Setting for Highway Assets College of Engineering University of Wisconsin-Madison Madison, Wisconsin, 173
[7] 2008 National cooperative highway research program «Multimodal Level of Service Analysis for Urban Streets» Washington, D C 122 p
[8] Absalyamov T 2015 The influence of cultural and sport mega-events on sustainable development of the city *Procedia-Social and Behavioral Sciences*, **188** 197-201
[9] Gubaidullina, T , Ivanova, N , Absalyamova, S and Yerina, T 2018 Analysis of national strategies for sustainable development with regard to fundamental conceptual *premise In Journal of Physics: Conference Series* **1141(1)** 012018 IOP Publishing
[10] R Babit, V Sharma and A K Duggal 2016 Level of service concept in urban roads International *Journal of Engineering Science Invention Research & Development* **III(I)** 44-8
[11] N Chernikh 2019 Problems of forming the core network of major cities, // Collection of reports of the international conference «Education The science Production»
[12] Sakhapov R L and Absalyamova S G 2013 Integration of universities and business as the condition of formation of the innovative economy 2013 International Conference on Interactive Collaborative Learning, ICL 2013 189-91

Determination of rational cutting parameters while drilling intermediate by ring

V P Baranchik[1,*] and M F Zakirov[2]

[1]Udmurt State University, 1 Universitetskaya St., 426034 Izhevsk, Russia

[2]Department of Building Materials, Mechanization and Geotechnics, Kalashnikov Izhevsk State Technical University, Studencheskaya Street 7, Izhevsk, 426069, Russia

*E-mail: barvp@mail.ru

Abstract. The article analyzes the problem of reducing the cost of changing bits during deep drilling of directional wells due to the rational choice of operational parameters in the process of sinking rocks that alternate in strength. The authors propose a technology to control drilling parameters on the basis of preliminary information on the limits of rock strength over sections of the well, which allows to increase the bit flight and reduce the energy intensity of the drilling process.

1. Introduction

Analysis of research in the field of improvement of cutting drilling tools indicates the impossibility of creating structures that do not change during deep drilling and are optimal for the entire range of drilled rocks [1-8]. The bit is repeatedly replaced during the drilling process, which increases the time for tripping, decreases the average rate of penetration and economic indicators. A rational solution to the problem is an adaptive chisel, automatically adapting to changes in the physical and mechanical properties of the rock. Self-regulation of the tool depending on the physicomechanical properties of the face should be based on the implementation of the optimal parameters of the rock cutting process.

Modern ideas on the mechanics of destruction of soils and rocks are based on the existence of defects in the structure of the rock and the anisotropy of the physicomechanical properties. A cutting angle that intensifies the tension and shear in the compressed volume of the rock stands out among the numerous factors that influence the resistance of soils and rocks to cutting [6]. It is known that soils and rocks practically do not resist stretching and collapse along sliding lines from shear forces [6].

The fundamentals of the theory of cutting and digging the soil, laid by academician V.P. Goryachkin, received rapid development in the writings of many of his followers, which resulted in the specific resistance to cutting and digging of soils of varying degrees of strength, determined experimentally in relation to dump and bucket working bodies. The work of V.V. Sokolovsky, which developed the basis of the theory of limiting equilibrium of a granular medium with adhesion, takes a special place in the studies of the interaction of working bodies with the environment. The disadvantage of his theory is the idea taken from soil mechanics: soil destruction under passive pressure on a fixed retaining wall imitating a moving working body. However, the calculation of the soil destruction zones according to V.V. Sokolovsky gave impetus to the development of graph-analytical methods of A.N. Zelenin [6] and E.I. Berestov [3].

The physical and mathematical model of cutting soil and rock while drilling a well with varying physico-mechanical characteristics was developed with the aim of using the same bit on soft and hard rocks in order to increase the bit's flight and reduce the total length of tripping, i.e. increase bit versatility. The model consists of three parts:

- determination of the axial and circumferential forces on the bit;
- definition of a family of optimal combinations of limit normal and tangential stresses;
- algorithm of control of regime parameters of drilling.

2. Determination of axial and circumferential force on the bit

In relation to the cutting bit, the cutting angle is the angle of inclination of the front face of the tool to the borehole wall. It is convenient to present the diagram of the forces acting on the cutter by turning the well along the diameter d, and replacing the cutter with a slider with an inclined blade (Figure 1). Such a scheme is similar to the interaction scheme of a screw-nut threaded pair. The elevation angle α of the screw surface of the cutter blade corresponds to the cutting angle, and the screw pitch h corresponds to the tool feed per revolution. The cutter removes chips of length πd and thickness h under the action of axial force P in one revolution. The force R of interaction between the cutter and the rock is deflected from the normal to the helical surface by an angle φ of external friction of the cutter on the rock. The circumferential force T is related to the axial force P by the ratio:

$$\frac{T}{P} = tg(\alpha + \varphi) \tag{1}$$

Let's express the forces:

$$T = \tau_{\text{в}} \cdot \pi \cdot d \cdot h,$$

$$P = \sigma_{c} \cdot \frac{\pi \cdot d^2}{4}, \tag{2}$$

where: τ_v, σ_c – core material strength limits for shear and compression.

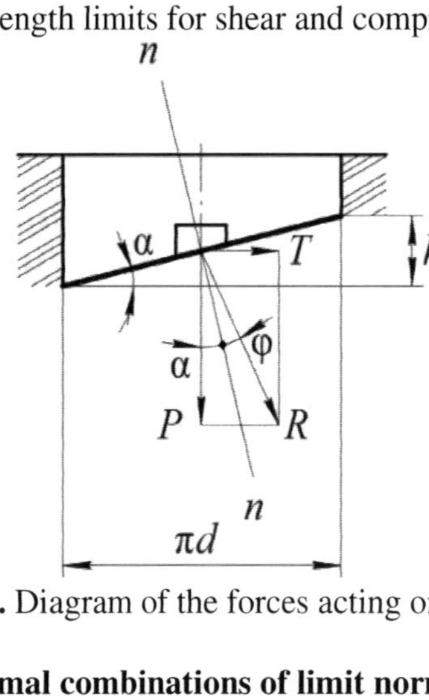

Figure 1. Diagram of the forces acting on the cutter.

3. Definition of a family of optimal combinations of limit normal and tangential stresses

The family of optimal combinations of limit normal and tangential stresses is determined using the theory of limit stress states of connected environments with the joint action of normal and tangential stresses [6].

The ratio of normal and tangential stresses σ/τ for connected environments is determined by the Coulomb's law:

$$\tau = c_0 + \sigma \cdot tg\,\rho,$$

where: c_0 – specific grip;
ρ – internal friction angle.
In the case of limiting stresses (destruction by drilling), taking into account expressions (2), we have:

$$\frac{\tau_в}{\sigma_c} = tg\,\rho + \frac{c_0}{\sigma_c} \tag{3}$$

In relation (3), the angle of internal friction ρ is unknown. We define it by applying the theory of the ultimate stress state of the soil and rocks in the form of a diagram (Figure 2). The diagram is based on two points σ_c and $\tau_v = c_0$, obtained during laboratory tests of cores for uniaxial compression and net shear, respectively. The tangent to the Mohr limit circles of compression and shear limits the region of rock destruction under the combined action of normal and tangential stresses. The points of the circle 3 determine the family of optimal ratios τ_v/σ_c, at which only tensile and shear stresses are present in the drilled rock.

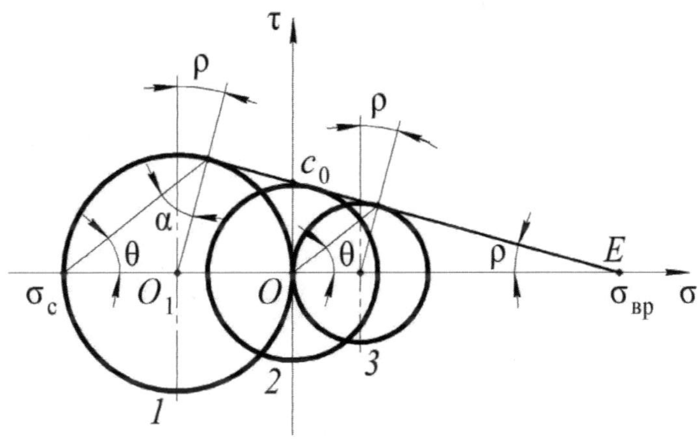

Figure 2. Diagram of limit stress state of rock:
1, 2, 3 – limit circles of stresses of compression, shear, and tensile, respectively.

The optimum cutting angle is determined by the ratio:

$$\alpha = 90° - (\rho + \theta)$$

where: ρ – internal friction angle;
θ – shear angle.

$$tg\,\rho = 1 - \left(\frac{2 \cdot c_0}{\sigma_c}\right)$$

$$\theta = 45° - \frac{\rho}{2}$$

4. Algorithm of control of regime parameters of drilling

Optimal control of the drilling regime is possible only by changing the cutting angle. With an unregulated bit, only approximate adherence to the coordinates belonging to circle 3 is possible in order to prevent compression forces in the drillable volume. If possible, it is necessary to maintain the balance using the following relations arising from the diagram (Fig 2):

$$\frac{T}{P} = tg(\alpha + \varphi) = \frac{4 \cdot h \cdot \left(1 - \frac{2 \cdot c_0}{\sigma_c}\right)}{d} \qquad (4)$$

With the change of c_0/σ, the axial force P, the supply of flushing fluid affecting φ, chisel h feed can act as regulating parameters. Such an almost optimal (rational) regulation of the process parameters will allow to pass the problem area by the same bit, i.e. increase the bit flight and reduce the energy intensity of the well drilling process.

Economic analysis and decision-making on the effectiveness of savings on tripping operations with the possible risk of reducing the mechanical drilling speed should be made taking into account the specific mining and geological conditions.

5. Conclusion

We proposed a technology to control drilling parameters on the basis of preliminary information on the limits of rock strength over sections of the well, which allows to increase the bit flight and reduce the energy intensity of the well drilling process.

References

[1] Artemev K A 1975 O vozmozhnosti ispolzovaniya teorii predelnogo ravnovesiya sypuchey sredy dlya opredeleniya soprotivleniya gruntov rezaniyu i kopaniyu // *Issledovaniya i ispytaniya dorozhnyh i stroitelnyh mashin:* Sb trudov SibADI **53(6)** 3-8
[2] Balovnev V I 1981 Modelirovanie protsessov vzaimodeystviya so sredoy rabochih organov dorozhno-stroitelnyh mashin M : VSH, 336
[3] Berestov E I 2008 Soprotivlenie gruntov rezaniyu: Monografiya / E I Berestov, A P Smolyar Mogilev: Belorus -Ros un-t, 179
[4] Baranchik V P , Zakirov M F and Kibardina N A 2014 Opredelenie ekvivalentnyh protsessov rezaniya anizotropnyh sred pri ispolzovanii smennyh rabochih organov mashiny *// Intellektualnye sistemy v proizvodstve* **1(23)** 22-23
[5] Vetrov YU A 1983 Zakonomernosti energoemkosti rezaniya gruntov // *Gornye, stroitelnye i dorozhnye mashiny*: resp mezhvedomstv nauch -tekhn sb **36** 7-14
[6] Zelenin A N 1969 Laboratornyy praktikum po rezaniyu gruntov M : VSH , 310
[7] Sokolovskiy V V 1960 Statika sypuchey sredy 3-e izd M : Fizmatgiz, 243
[8] Shchemelev A M and Berestov E I 1980 Analiticheskoe opredelenie skola struzhki grunta // Issledovanie i ispytanie stroitelnyh i dorozhnyh mashin Omsk, 68-78

Methods of the building of the dynamic models of the drive machines loads with the use of system monitoring data (on example of the acceleration of the drive of the drill tool)

A V Ashcheulov

Institute of machine building, metallurgy and transport, Saint-Petersburg Polytechnic University, Russia, Saint-Petersburg, Polytehnicheskaya street 29,

E-mail: studgidro@gmail.com

Abstract. The article considers the sequence of modern machines' monitoring systems data processing. The aim is to build the mathematical models of the complex dynamic processes of the interaction between working bodies of the machines with external ambience. On the example of drilling machines on the starting and accelerating modes of the drilling tool, the authors obtained the dependencies of the torque on the output shaft of the rotation drive of the drill string depending on the frequency and rotation acceleration, and the depth of drilling related to the rock strength.

1. Introduction

Mathematical modeling of machines is a common method of studying their dynamic processes of motion: acceleration/deceleration, automatic control, and special high-speed operations. Usually these studies are carried out to synthesize the optimal parameters of control devices, to determine the properties of a system under extreme conditions, and so on. Scientists usually use a systematic approach [1] for compiling mathematical models of machines. This approach presents a machine as a system consisting of separate subsystems: the main engine, transmission, and a working body interacting with the environment. To control the machine they use a control subsystem that imitates the actions of an operator or an ACS subsystem — an automatic control system. In general, the scheme of such a system can be depicted as in Figure 1. For a particular machine and depending on the research problem to be solved, the blocks of subsystems can be represented by subsystems of a smaller order, up to individual elements - devices.

IOP Conf. Series: Materials Science and Engineering **786** (2020) 012046 doi:10.1088/1757-899X/786/1/012046

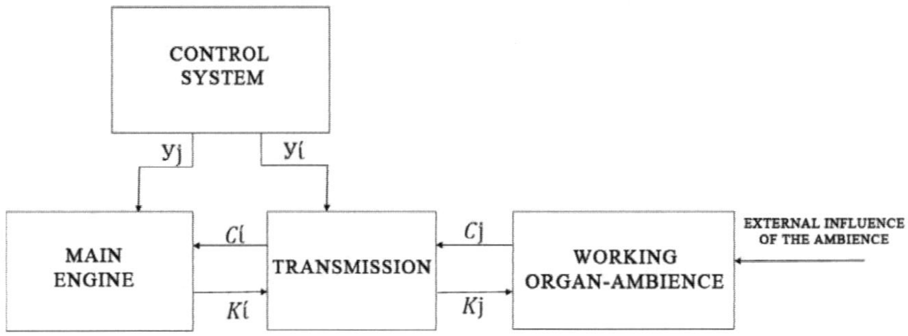

y_j y_i – CONTROLLING INFLUENCES;

C_j C_i – A POWER PARAMETERS;

K_j K_i – AN KINEMATIC PARAMETERS.

Figure 1. A Block scheme of machine systems.

The method of element-node structures proposed in [2, 3] and developed in the work of the author [4] is becoming more widespread in mathematical modeling of machines. In this method, the calculated model of the machine is represented by a dynamic system with inputs and outputs, internal and feedback parameters. The equations are written explicitly, as a function of real time. The final system of equations is a mixed system of nonlinear differential and algebraic equations of a large order, which is solved by numerical methods [4]. If the mathematical models of the main engine (internal combustion engine or electric motor) and transmission elements are represented in the technical literature quite well and their properties are well-known, then modeling the subsystem "working body - environment" has problems. This is due to the fact that most of the equations of load on the working bodies of machines known in the literature [5, 6, 7, etc.] refer to the statics of machines. In these equations, the loads on the working bodies do not depend on the time of the process, on the speed of movement of the mechanism, on the time-varying parameters of the environment. It is not possible to use such equations in mathematical modeling of the dynamics of machines.

There is a need for mathematical models, where a change in the kinematic parameters of the machine system would lead to a change in the load parameters in real time. A graphic interpretation of the required "working body – environment" models for some machines is shown in Figure 2. There are no such models for specific types of machines in the technical literature. In this case, most researchers of the dynamics of machines are limited to the consideration of simplified models where the loads on the working bodies are modeled as typical signals used in the theory of automatic control: a single jump; linear loading, harmonic function, i.e. without feedback from changes in the kinematics of the working body. The use of typical loading signals of mechanisms does not always give the optimal solution for the required parameters of the drives and the ACS of the machine system.

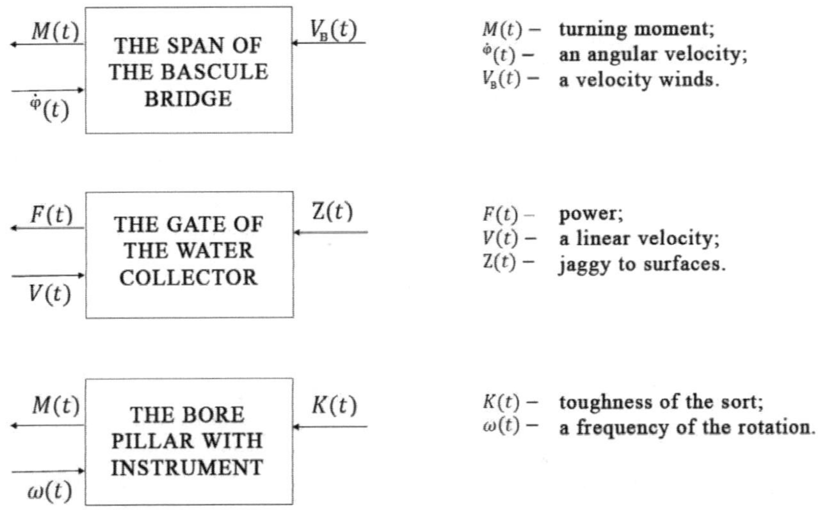

Figure 2. Some variants of the accounting subsystem schemes «working body - environment».

Due to the complexity of the physical processes occurring in the subsystem "working body-environment", direct theoretical mathematical modeling of this link is an extremely time-consuming task. Empirical methods based on experiments in bench conditions are also difficult because of the complex similarity of the inertial masses of the working body, complex imitation of environmental parameters, high cost and labor intensity of bench studies [8].

2. Concept headings

The author's methodology [4] proposes to use data from monitoring systems of machine operation at the design stage of machines. These monitoring systems are installed on modern machines for diagnostics of parameters, for example, the so-called "black boxes" on airplanes. Previously, such systems, due to their high cost, were installed only on especially important machines, but now, with the reduction in the cost of electronic components, many machines are equipped with monitoring systems: mechanisms of movable bridges, sluices, gates of hydraulic dams, quarry excavators, gantry cranes, combined road machines, various kinds of technological plants, drilling rigs, etc. Data from these systems is usually used only after machine crashes and serves as a source of information on determining the causes of these events. In some cases, these systems are used for administrative control over the operation of machines and mechanisms and are aimed at ensuring the norms of proper and safe operation. But this data can also be used in the development of mathematical dynamic models of the working body interacting with the environment.

The data of the monitoring systems of machines consists of digital records of the values of various physical parameters in real time: displacements, forces, pressures, temperatures, stresses, etc. The developers of the monitoring systems usually display this data on the operator's screen in the form of graphs that look like a cardiogram. But working with graphics is inconvenient and there is a need for numerical data. Numerical data is not always possible to obtain because the developers of monitoring systems do not always make them available. If such difficulties arise, it is recommended to contact the developers of monitoring systems. The obtained data requires system archiving and conversion to typical data formats. In addition, it is necessary to carry out data rationing on various grounds, for example, technological operations, climatic conditions, and environmental properties. Then we can get the required mathematical dependencies by the methods of mathematical processing, including statistical methods.

3. Results

As an example, let us consider the solution of a problem for drilling rigs, which bores oil and gas wells to a depth of 3000–3500 m. This corresponds to a class of installations with a lifting capacity of 250–320 tons. According to the classification of drilling rigs, such technical characteristics have fixed and cluster installations (mobile, on a rail track, for relatively short distances up to 1.5 km). Drilling rigs are machines with continuous action that combine the processes of rotation and supply of the working body (the drill string with a drilling tool). When a drilling tool interacts with the rock, a torque is generated at the output shaft of the drive mechanism - the moment of resistance, which value depends on the strength of the soil, the speed of rotation and the feed rate of the drilling tool. Similar processes take place on milling machines, multi-bucket excavators and many other machines.

A feature of modern designs of drilling rigs is a movable drive rotation of the drill string. Previously, the drive was placed on the drilling table permanently, but in modern designs it goes up and down (like a shuttle) with a winch along the guide beam. Such systems are called top drive systems (TDS). The developers of this equipment are mainly foreign firms. Russian drilling companies also began to use this tool, since it can significantly increase drilling productivity. According to import substitution programs in the Russian Federation, similar domestic TDSs are being created [9, 10, 11]. This work is a continuation of previously performed research [12] within the framework of the federal target program "Research and development in priority areas of the development of the scientific and technological complex of Russia for 2014–2020". St. Petersburg Polytechnic University of Peter the Great, together with the industrial partner JSC SKB PN, commissioned and sponsored by the Ministry of Science and Higher Education of the Russian Federation, carries out the project "Developing and researching the principles of controlling the curvature vector of a small diameter well with the help of rotary controlled systems" (Agreement No. 075-15-2019-1403 of 06/20/2019. T & V RFMEFI57517X0138).

TDSs are equipped with monitoring systems, which records has the parameters of the drive operation processes are recorded for the entire period of drilling of each particular well. Figure 3 shows a fragment of one of these records, which shows some time-varying parameters of the TDS: the rotational speed V of the drill string, the torque M on the output shaft, the vertical displacement GS of the drill string. In this case, the recording of V and GS parameters is a result of direct measurement, and the recording of M values is the result of measurements of indirect parameters, such as current strength (electromechanical drive) or pressure (hydraulic drive), and calculations that are performed by the monitoring system controller. Such records are the initial data for subsequent analysis. However, not all monitoring systems of modern machines issue records in this form. There is no unification yet. Each monitoring system is an individual system of its developer and to obtain records in the form shown in Figure 3 the system may require some additional computational transformations.

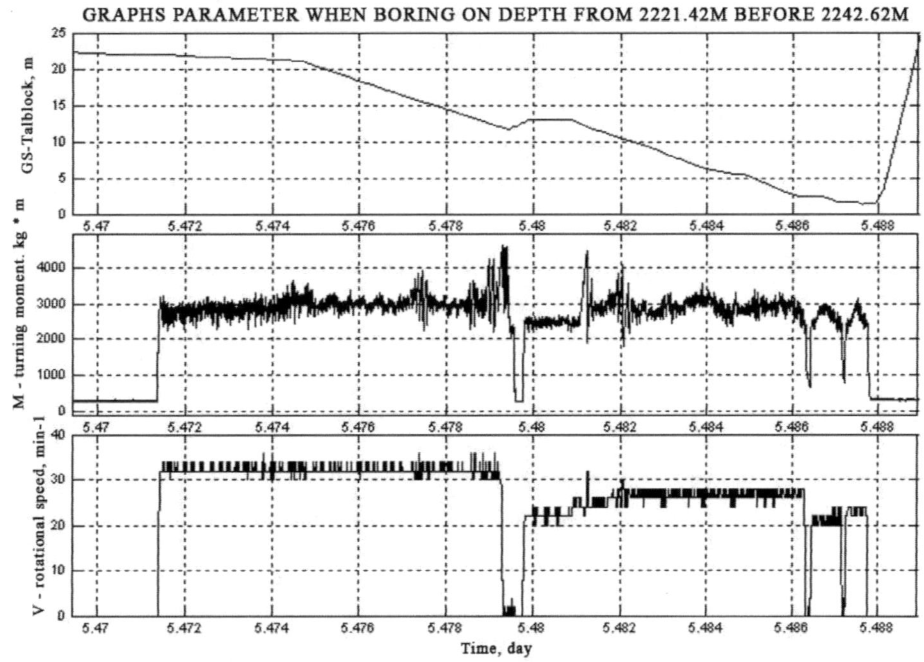

Figure 3. Fragments of the TDS monitoring systems recording.

To solve a specific task, we selected only parts of transient processes of acceleration of the drill string from the records of general processes (for other tasks we can use the other areas). In the selected areas, we eliminated random parameter outliers and smoothed the parameters associated with the "noise" of the measuring instruments. As a result of this preparatory work, we formed a database of transients of rotation frequency and torque for different depths of drilling. We can look at the example of selected transient functions in Figure 4. One well can have from 150 to 300 processes of acceleration of the drill string depending on the length of the used drilling pipes. Due to the use of different type of drilling tool, the data bank for one well was divided into two parts: data for depths up to 700 m and data from 800 to 3500 m.

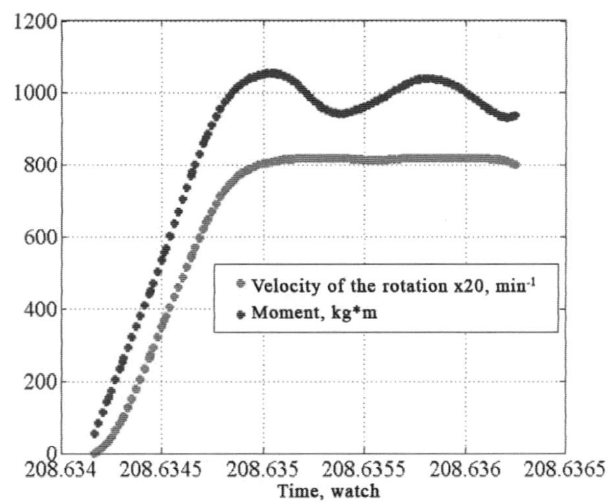

Figure 4. An example of transient process record of the acceleration of the drill string.

Studies have shown that it is not enough to display the torque function during acceleration of a drill string as a simple polynomial function of the speed of rotation of the drill string, as suggested by the authors of [13]. The method of controlling the speed of most machines, including on drilling rigs, is remote manual. The rate of increase of rotational speed, i.e. acceleration, may be different. Therefore, based on [14], the polynomial function should be represented as a sum of terms, which will include the acceleration of the rotation of the drill string.

In the general case, we used the following dependence for approximation

$$\widetilde{M}(\omega,\dot{\omega}) = a\omega^h + b\omega + c\omega^2 + d\dot{\omega}^k + e\dot{\omega} + f\dot{\omega}^2 + g, \tag{1}$$

where, $\widetilde{M}$ – approximation moment, $\omega, \dot{\omega}$ – the speed and acceleration of rotation of the drill string, respectively, $a, b,\ c,\ d,\ e,\ f,\ g,\ h,\ k$ – dependency parameters.

We determined the coefficients of the approximating dependence (1) sequentially for each individual section of the acceleration of the drill string. These coefficients were calculated by solving the problem of multidimensional minimization of the standard deviation of the values of the approximating dependence $\widetilde{M}$ on the processed source data $\overline{M}$:

$$\min \frac{1}{N}\sum_{i=1}^{N}\left(\widetilde{M}_i - \overline{M}_i\right)^2 =$$
$$\min_{a,b,c,d,e,f,g,k,h} \frac{1}{N}\sum_{i=1}^{N}\left(a\omega^h + b\omega + c\omega^2 + d\dot{\omega}^k + e\dot{\omega} + f\dot{\omega}^2 + g - \overline{M}_i\right)^2, \tag{2}$$

where N – the number of values of the moment in each individual section of acceleration.

The calculations were performed in Matlab R2012a.

To solve this problem, we used the method of Nelder-Mead unconditional minimization, since it has a high rate of convergence and does not require the calculation of the gradient. The method consists of sequential movement and deformation of an irregular multidimensional simplex in the direction of the optimum point. The coefficients were found with a relative error of the values of the function and arguments of 10^{-5}. The sought approximating dependence (or coefficients optimization) for the moment was obtained in several successive stages. After each minimization for a set of optimized parameters, one or several coefficients weakly dependent on depth were taken as constant values and excluded from the optimized parameters at further optimization stages.

As a result of processing the initial data at depths of drilling from 800 to 3200 m, it was found that the coefficients of degrees h, k take the same little varying values $h = k = 0.55$. As a result, they were assumed constant and equal $h = k = 0.5$. Parameters $c,\ d,\ e,\ f$ do not depend on the depth of drilling and can be assumed as constant: $c = -2.1 \cdot 10^{-3}$, $d = -0.454$, $e = 1.14 \cdot 10^{-4}$. Parameters a, b, g depend on the depth of drilling and should be considered as functions. Therefore, the dependence (1) can be written in the form:

$$\widehat{M}(\omega,\dot{\omega},h) = \tilde{a}(h)\sqrt{\omega} + \tilde{b}(h)\omega - 2.1 \cdot 10^{-3}\omega^2 -$$
$$-0.454\sqrt{|\dot{\omega}|} + 1.14 \cdot 10^{-4}\dot{\omega} + 7.28 \cdot 10^{-11}\dot{\omega}^2 + \tilde{g}(h) \tag{3}$$

where the dimension of the moment $\widehat{M}$ - «kg*m», and the rotational speed ω - «min^{-1}».

Despite its little value, we can't ignore the parameter $f = 7.28 \cdot 10^{-11}$ because it brings a significant correction to the overall result at the beginning of the acceleration section, where the acceleration is large and the moment is small. Parameters a, b, g could be found from the equations

$$\tilde{a}(h) = 1.21 \cdot 10^{-5}h^2 + 2.69 \cdot 10^{-3}h + 1.08, \tag{4}$$
$$\tilde{b}(h) = 1.15 \cdot 10^{-4}h^2 - 1.89 \cdot 10^{-1}h + 1.82, \tag{5}$$
$$\tilde{g}(h) = 2.11 \cdot 10^{-4}h^2 - 2.22 \cdot 10^{-1}h + 1.46, \tag{6}$$

where the dimension of drilling depth h – «m».

4. Discussion

The results were analyzed by comparing the initial experimental dependencies of the torque and the calculated dependences of the approximating moment (3), received from the coefficients obtained through the approximating dependences (4), (5), (6).

IOP Conf. Series: Materials Science and Engineering **786** (2020) 012046 doi:10.1088/1757-899X/786/1/012046

To quantify the quality of the approximation, we used the relative error introduced as the standard deviation of the approximating moment from the initial moment:

$$\hat{E}_{rel} = \sqrt{\frac{1}{N}\sum_{i=1}^{N}\left(\frac{\hat{M}_i-\bar{M}_i}{\bar{M}_i}\right)^2}, \quad (7)$$

The analysis showed that at those specific drilling depths for which the values of the parameters *a*, *b*, *g* were calculated before obtaining the approximation equations (4), (5), (6), the coincidence of the calculated dependences with the experimental ones is very good. The relative error is literally a few percent. This is illustrated on Figure 5.

If we use the dependences (4), (5), (6) and carry out the analysis at arbitrary depths of drilling, then the accuracy of the obtained results decreases. For example, figures 6, 7, 8 for three sections of arbitrarily chosen depths h = 790 m, h = 1440 m, h = 2120 m show the corresponding illustrations of the analysis results.

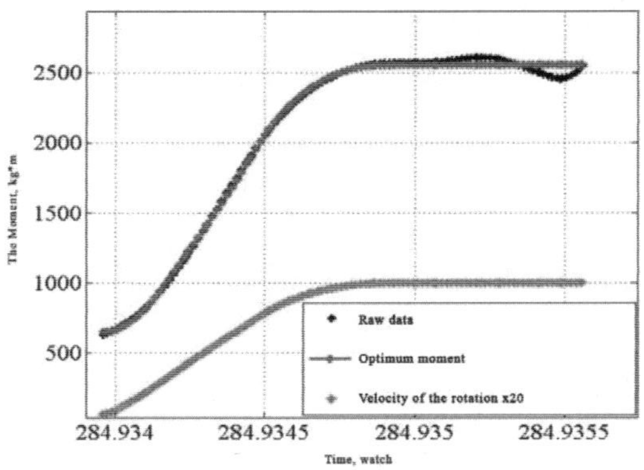

Figure 5. A comparison of approximating moment with initial moment at the depth h=3200 m.

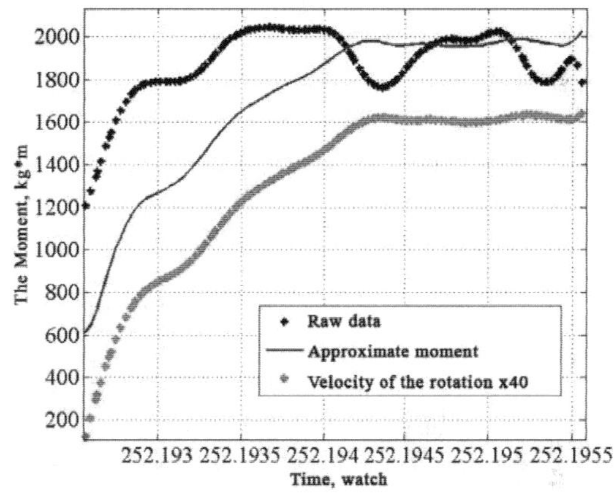

Figure 6. A comparison of approximating moment with initial moment at the depth h=790 m.

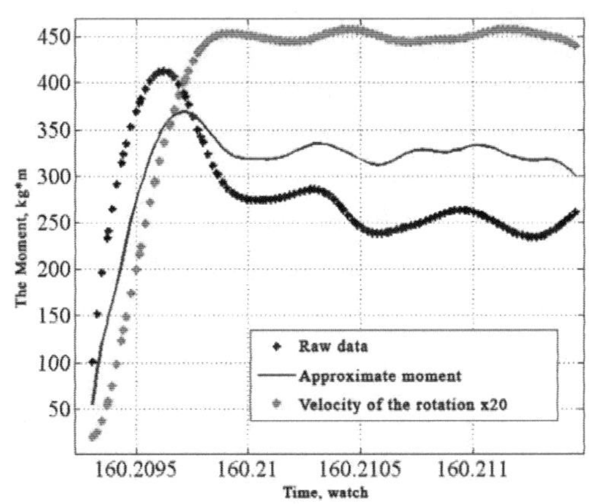

Figure 7. A comparison of approximating moment with initial moment at the depth h=1440 m.

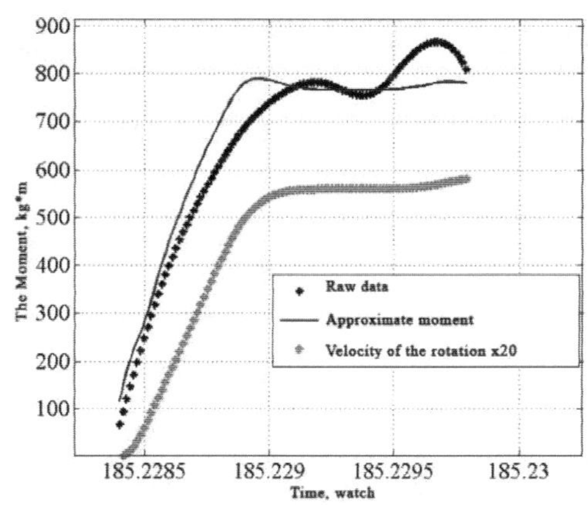

Figure 8. A comparison of approximating moment with initial moment at the depth h=2120 m.

The values of the specified error for the three considered sections at depths $h = 790$ m, $h = 1440$ m, $h = 2120$ m equal respectively $\widehat{E}_{rel} = 19\%$, $\widehat{E}_{rel} = 25\%$, $\widehat{E}_{rel} = 14\%$.

More details of the research results are presented in [15].

5. Conclusion

We advise researchers to use digital data of completed production processes on modern machines equipped with electronic parameter monitoring systems. This data can be used to solve many engineering problems, including building of new dynamic models of systems-machines, opening the way to the construction of digital twins of complex technology.

With the help of such monitoring systems, we developed and tested a method for constructing empirical differential dependencies for the "Working Body – Environment" machine link on the example of the acceleration mode of drilling tools for drilling rigs with top drive systems. The torque on the drill string during start-up and acceleration of the drive depends not only on the rotational speed and depth of drilling associated with the rock strength, but also on the acceleration of rotation. According to the proposed method, we can construct mathematical models of dynamic processes of other modern machines.

The resulting mathematical dependencies can be used in mathematical models of drilling rigs to simulate the dynamic processes of the drives and their control systems. Created data archives allow us to build mathematical models of other dynamic processes of drilling rigs.

References

[1] Balovnev V I 1981 Modelirovanie protsessov vzaimodeystviya so sredoy rabochih organov dorozhno-stroitelnyh mashin - M : Vysshaya shkola, – 335

[2] Y G Berengard and M M Gaytsgori, E Y 1980 Malinovskiy Raschet i proektirovanie stroitelnyh i dorozhnyh mashin na EVM/ Pod red E Y Malinovskogo – M : Mashinostroenie, - 216

[3] I I Bazhin, YU G Berengard and M M Gaytsgori i dr 1988 Avtomatizirovannoe proektirovanie mashinostroitelnogo gidroprivoda– M : Mashinostroenie, – 312

[4] Ashcheulov A V 2007 Metodologiya proektirovaniya gidravlicheskih podemnyh mekhanizmov razvodnyh mostov// dissertatsiya na soiskanie uchenoy stepeni doktora tekhnicheskih nauk/ Sankt-Peterburgskiy politekhnicheskiy universitet Sankt-Peterburg

[5] Vetrov Y A 1971 Rezanie gruntov zemleroynymi mashinami - M : Mashinostroenie– 360

[6] Zelenin A N 1968 Osnovy razrusheniya gruntov mekhanicheskimi sposobami - M : Mashinostroenie, – 376

[7] Bulatov A I , Proselkov Y M and SHamanov S A 2003 Tekhnika i tekhnologiya bureniya neftyanyh i gazovyh skvazhin: Ucheb dlya vuzov – M : OOO «Nedra-Biznestsentr» – 1007

[8] Berestov E I 2008 Soprotivlenie gruntov rezaniyu: monografiya Mogilev: Belorus - Ros Un-t, – 179

[9] Ashcheulov A V , SHestopalov A A , Horoshanskiy A E and Kochetkov A V 2015 Osobennosti raboty burovyh ustanovok s sistemoy verhnego privoda // *Mezhdunarodnyy nauchno-tekhnicheskiy i proizvodstvennyy zhurnal* «Himicheskoe i neftegazovoe mashinostroenie» - M: **2**, 14-7

[10] A V Ashcheulov, A A SHestopalov and A A Lobachev 2016 Analiz dinamicheskoy nagruzhennosti silovogo verhnego privoda burovyh ustanovok *Himicheskoe i neftegazovoe mashinostroenie*: Moskva, **3** 12-5

[11] Ashcheulov A V 2016 Optimization of Regulator Parameters of the Automatic Regulation System of the Rotation Frequency of the Valve Engine of the Top Drive System of Bore Installation *International Journal of Applied Engineering Research* **11(22)** 11055-9

[12] Sozdanie ekonomichnogo verhnego elektroprivoda dlya mobilnyh burovyh ustanovok: otchet o PNI v 5 tomah / FGAOU VO SPbPU; lead: Ashcheulov A V SPb , 2014-2017 – **GR114110570105**

[13] TSuprikov A A , CHerednichenko V G , Kritskaya V G and Kritskaya L M 2011 Analiz protsessa razrusheniya porody dolotom pri burenii skvazhin Nauchnyy zhurnal KubGAU, **74(10)**
[14] J A Nelder and R Mead 1965, *Computer Journal*, **7**
[15] Lobachev A A 2017 Issledovanie nagruzhennosti elementov reduktora sistemy verhnego privoda// dissertatsiya na soiskanie uchenoy stepeni kandidata tekhnicheskih nauk po spetsialnosti 05 02 02/ Sankt-Peterburgskiy politekhnicheskiy universitet Petra Velikogo Sankt-Peterburg

IOP Conf. Series: Materials Science and Engineering **786** (2020) 012047 doi:10.1088/1757-899X/786/1/012047

The influence of the ratio of the vibration device's two stages' imbalances' static moments with asymmetric oscillations on the system amplification factor

M D Gerasimov[1,*], N D Vorobev[1], V G Ryazantsev[1] and Otakulov O Kh[2]

Belgorod State Technological University named after V.G.Shukhov, 46 Kostyukova street, Belgorod, 308012, Russian Federation

Ferghana Polytechnic Institute, Uzbekistan

*E-mail: mail_mihail@mail.ru

Abstract. Vibration devices with asymmetric oscillations can consist of two or more stages of vibrators with directed vibrations. Component stages of a vibration device with asymmetric oscillations have multiple frequencies of rotation of the imbalanced shafts. The multiplicity of rotation frequencies of the imbalanced shafts of the vibrator's stages with directed vibrators in a vibration device with asymmetric oscillations is imperative. In this case, the parameters of the total asymmetric oscillation depend entirely on the ratio of the static moments of the imbalances of the stages included in the vibratory device. The simplest vibration device with asymmetric oscillations is a device consisting of two stages of vibrators with directed vibrations. The article is devoted to the question of the influence of the ratio of the vibration device's two stages' imbalances' static moments with asymmetric oscillations on the system amplification factor.

1. Introduction

Vibration device with asymmetric oscillations, formed on the basis of two or more successively installed imbalance vibrators with directed vibrations [1,2,3,4], as a rule, has a multiple value of the rotation frequency of the imbalance shafts of individual stages' vibrators:

$$\omega_1 : \omega_2 : \omega_3 : ... : \omega_n = 1 : 2 : 3 : ... : n. \quad (1)$$

where ω_1, ω_2, ω_3, ..., ω_n – the angular velocity of the successively installed imbalance shafts of a vibration device with asymmetric oscillations, consisting of n stages [c^{-1}].

A vibration device with asymmetric oscillations generates driving force F, where the components magnitudes F_p and F_x act in opposite directions and obey the relation [5,6,7,8]:

$$Fp = k_d \cdot F_x \quad (2)$$

where F_p - component of the driving force acting in the direction of useful work;

F_x - component of the driving force acting in the direction of the idle;

k_d - the amplification factor of the vibration system, which characterizes the ratio of the components of the driving force acting in opposite directions: in the direction of useful work and in the direction of idle, respectively.

Let's set the values of the static moments of the first and second stages imbalances as ($M_1=m_1r_1$) and ($M_2=m_2r_2$), respectively. The ratio F_p/F_x, and the value of the amplification factor k_d will depend on the ratio $(m_1r_1)/(m_2r_2)$. Let's set the value of each stage imbalance's the center of gravity offset as equal to one, $r_1=r_2=1{,}0$. Then we have an opportunity to obtain the dependence of the change in the amplification factor of a vibration system with asymmetric oscillations on the ratio of the static moments of the first and second stage imbalances, by changing only the mass ratio (m_1/m_2) of the first and second stage imbalances:

$$k_d = f\left(\frac{M_1}{M_2}\right) \quad (3)$$

where $M_1=m_1r_1$ - static moments of imbalance of the first stage of the vibration device with asymmetric oscillations;

$M_2=m_2r_2$ - static moments of imbalance of the second stage of the vibration device with asymmetric oscillations.

2. Methods

We developed a program for addition of the harmonic oscillations, which is being used to study the influence of the static moments ratio of the vibration device's two stages' imbalances with asymmetric oscillations on the system amplification factor [9,10,11].

3. The main part

The output parameters of the total asymmetric oscillation are controlled by the frequency of the forced oscillations, the mass and the radius of the center of gravity offset of the imbalances. An example of sampling the initial parameters are given in Table 1.

Table 1. An example of the formation of the initial parameters for the calculation with a change in the ratio $(m_1\!:m_2)$.

Adopted Initial Parameters	Examples of initial parameters sample $(m_1\!:m_2)$					
	6:1		7,175:1		10:1	
№ of vibrator	1	2	1	2	1	2
Mass (kg)	6	1	7.175	1	10	1
Radius(cm)	1	1	1	1	1	1
Initial phase (degrees)	0	0	0	0	0	0
Speed (turn/min)	500	1000	500	1000	500	1000
	The value of the parameter used in the calculations					
R (m)	0.01	0.01	0.01	0.01	0.01	0.01
φ_0 (rad)	0.00	0.00	0.00	0.00	0.00	0.00
ω (1/sec)	52.36	104.72	52.36	104.72	52.36	104.72

As a result of the calculation, for each adopted mass ratio of imbalances $(m_1\!:m_2)$, we plot the change in the driving force of each stage and the total driving force of asymmetric oscillations. Figure 1 shows the result for the ratio of the masses of the imbalances of the first and second stages $(m_1\!:m_2)$ equal to (6: 1).

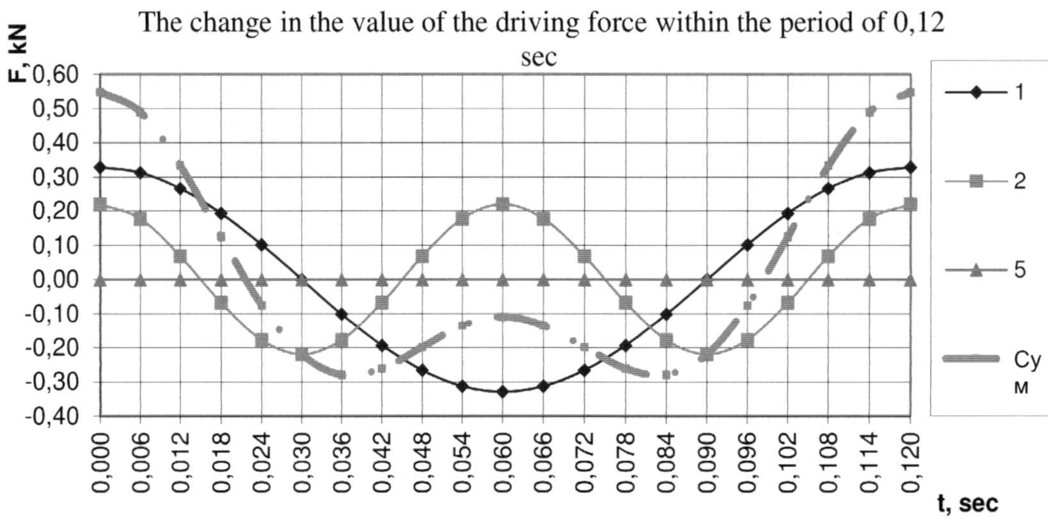

Figure 1. The change in the value of the driving force of a vibrator of first stage (1), second stage (2) and total driving force (Cym) with the ratio of first and second stage imbalance masses ($m_1 : m_2$) equals to (6:1).

Figure 1 shows that the value of the total driving force on the time interval of 0.03 ... 0.09 sec takes different values, and therefore, has a different value of the amplification factor k_d: $k_{d0,030}$, $k_{d\,036}$, $k_{d\,0,042}$ ... $k_{d0,09}$.

We have two options for using the value of amplification factor in the design calculations of vibration devices with asymmetric oscillations with two stages of directed vibrators:

- by the smallest value of the amplification factor within one period. In this case, at $t = 0.036$ seconds or at $t = 0.084$ seconds, which is the same;
- according to the average value of the amplification factor within one period on a segment of 0.03 ... 0.06 seconds.

For each adopted ratio $m_1 : m_2$, we determined the value of the amplification factor and summarized the results of the calculations in Table. 2

Table 2. Summary table of system's amplification factor.

	Ratio of imbalances of the first and second stages $m_1 : m_2$						
	5:1	6:1	7,175:1	10:1	13:1	15:1	17:1
kd0.03	2.24999	2.499986	2.793732	3.499968	4.249949	4.749935	5.249918
kd0.036	1.882391	1.964569	2.049228	2.212998	2.343753	2.413813	2.473671
kd0.042	2.155701	2.099621	2.049174	1.967972	1.914999	1.890008	1.870249
kd0.048	3.203999	2.76395	2.44604	2.042576	1.831669	1.743248	1.677687
kd0.054	5.924195	4.04817	3.114769	2.231258	1.86247	1.722609	1.623892
kd0.06	9	5	3.519685	2.333333	1.888889	1.727273	1.615385
kd. average	4.069379	3.062716	2.662104	2.381351	2.348622	1.89939	1.852177

The calculation clearly shows that the minimum value of the amplification factor k_d is present at any ratio of the masses of the imbalances of the first and second stages. However, for each ratio of the masses of the first and second stage vibrators, the minimum value of the amplification factor is reached at a certain moment of the oscillation period. This dependence can be represented by the equation or graphically (Figure 2).

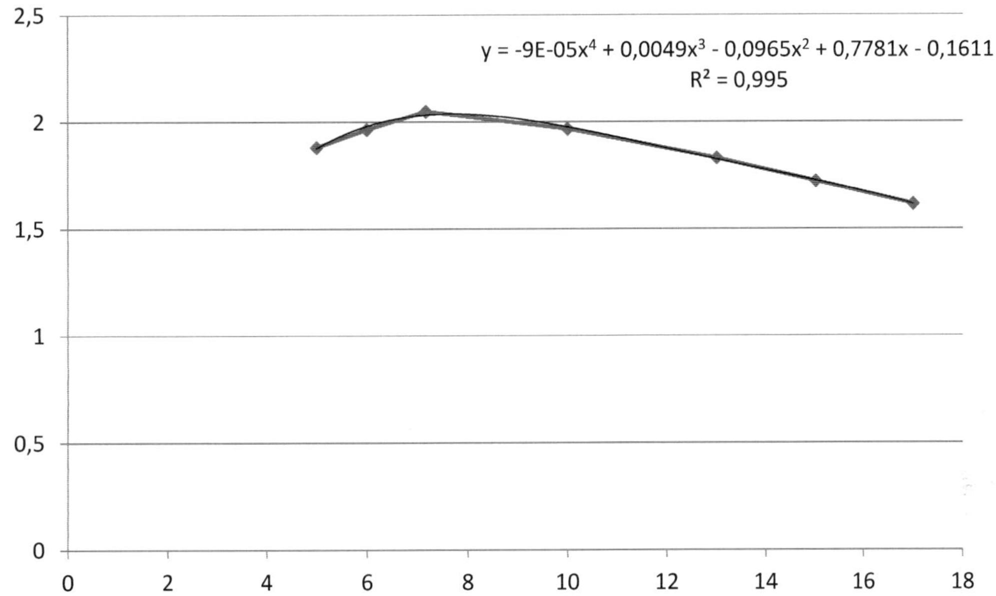

Figure 2. The change in the minimum value of the system amplification factor (y) with asymmetric oscillations from the mass ratio of the imbalances $(m_1\colon m_2)$ of first and second stages (x).

The equation will be:

$$k_d = -9E - 05x^4 + 0.004 \cdot x^3 - 0.096 \cdot x^2 + 0.778 \cdot x - 0.161 \tag{4}$$

where x – mass ratio of first and second stage imbalances, $m_1\colon m_2$.

Similarly, we can get the equation for the dependence of the average value of the amplification factor on the value of the mass ratio of the imbalances $(m_1\colon m_2)$ of first and second stage.

$$k_{d.cp} = 0{,}001 \cdot x^4 - 0.057 \cdot x^3 + 1.004 \cdot x^2 - 7.525 \cdot x + 23.05 \tag{5}$$

Thus, we got the equations describing the influence of the ratio of the vibration device's two stages' imbalances' static moments with asymmetric oscillations on the system amplification factor.

4. Conclusion

The obtained equations (4) and (5) can be used in the design of a vibration device with asymmetric oscillations, consisting of two stages of imbalanced vibrators with directed vibrations. With a given mass ratio of imbalance vibrators of the first and second stages, we can determine the project value of the amplification factor. Conversely, given the value of the amplification factor of a vibration system with asymmetric oscillations, it is possible to determine the rational mass ratio of the first and second stage imbalances.

References

[1] A Kleibl and Ch Heichel 2010 Patent US 7,804,211. B06B 1/16. Vibration Generator. Prior Publication Data US2009/0243410A1

[2] Krupenin V L and Veprik A M 1996 Vibroconductors equipped with impact elements and distributed vibroimpact systems // Proceedings of the 2-nd European Nonlinear Oscilations Conference V 1 Czech Prague: CTU, 229-34

[3] V I Babitsky and V L Krupenin 2001 Vibration of Strongly Nonlinear Discontinuous Systems - Berlin Heidelberg, New York: Springer-Verlag, 404

[4] Vibration machines in the construction and production of construction materials: Reference book 1970/ Under edited by V A Bauman and others – M : Mashinostroenie 548-50

IOP Conf. Series: Materials Science and Engineering **786** (2020) 012047 doi:10.1088/1757-899X/786/1/012047

[5] Blekhman I I 2013 Theory of vibration processes and devices Vibration mechanics and vibration technology - SPb , Ore and Metals Publishing House, - 640

[6] An Introduction to Mechanical Engineering Michael Clifford (editor) Richard Brooks, Kwing-So Choi, Donald Giddings, Alan Howe, Thomas Hyde, Arthur Jones and Edward Williams University of Nottingham Part 2 480 Hodder Education, An Hachette UK Company, 338 Euston Road, London NW1 3BH.

[7] Gerasimov M D , Glagolev S N , Gerasimov D M and Mkrtychev O V 2015 Determination Of The Driving Force's Asymmetry Factor And The Vibrostand's Work's Analysis *// International Journal of Applied Engineering Research,* **10(24)** 45392-8 http://www ripublication com/Volume/ijaerv10n24 htm

[8] Gerasimov M D , Mkrtychev O V , Glagolev S N , Gerasimov D M and Latyishev S S 2016 Method of Determination of Vibrating Screens Oscilation's Amplitude in a Characteristic Point for Plane Motion ARPN *Journal of Engineering and Applied Sciences* **11(20)** 12295-301 www arpnjournals com

[9] Gerasimov MD and Stepanischev V A 2014 Theoretical studies of asymmetric driving force of vibratory pile drivers / *Fundamental research* **5(3)** 473-476

[10] Gerasimov MD 2016 Addition of oscillations in vibration exciters // *Bulletin of Belgorod State Technological University* **3** 116-122

[11] Gerasimov MD, Vorobev ND, Gerasimov DM and Stepanishchev V A 2017 Study of a multi-stage vibration mechanism with directional oscillations of each step // *Bulletin of the Belgorod State Technological University* **3** 88-932

IOP Conf. Series: Materials Science and Engineering **786** (2020) 012048 doi:10.1088/1757-899X/786/1/012048

Ensuring the safety of lifts in operation

M A Stepanov* and and B A Kaytukov

Moscow State University of Civil Engineering, Russia, 129337 Moscow, 26 Yaroslavl highway

*E-mail: masmias@mail.ru

Abstract. The article discusses the issues of improving the safety of automobile lifts during operation. A safety assessment is proposed to be carried out using complex reliability indicators that include the loss coefficient. After conducting a cluster analysis of a car lift, the authors determined the failure rates and recovery time for each element. The paper determines the dependence of the recovery intensity on the waiting time as well. In accordance with the contribution of each element, there article calculates the safety level of the lift. Taking into account the statistical data, we determine the contribution of each element of the elevator to risk-safety, which corresponds to the percentage of output applications of the lift's individual elements. According to the obtained dependences, the safety level of the elevator is determined from the lift's service time. The proposed technique will allow determining the safety level of the lift during certain periods of service to reduce downtime and increase the efficiency of use.

1. Introduction

Nowadays in construction and other works at height, lifts are used when it is necessary to transport goods and deliver people. Their functional purpose is not limited to the construction site, they are used for storage, repair and redevelopment.

Safety of the lifts is estimated by the criticality of events that are set for each specific case on the basis of expert analysis. For systems with a high degree of risk and complexity it is recommended to use a probabilistic risk analysis [1-12]. Safety assessment can be done by using complex reliability properties (reliability, durability and maintainability), quantitatively represented by formulas, including the loss factor as an indicator of a separate property [13].

$$K_j = \frac{1}{\left(1 + \frac{T_{pi}}{T_{\text{нар}}}\right)}, \tag{1}$$

where T_{pi} -recovery time of i-th element;

$T_{\text{нар}}$ - time between failures.

Lifts are complex technical systems, so the danger during operation can arise in the form of failure of their structural elements. The main causes of failures of lifts are: design flaws, the quality of the elements manufacture, and the low level of their maintenance. Ensuring the safety of the lifts operation with the adjustment of the resource components for a certain period of service is an important task.

In order to solve this problem, it is necessary to analyze the faults and identify the elements with "weak" reliability. Then it is possible to work out the proposals for improving the safety level of the lifts on this basis.

2. The main part

Improving the level of safety requires addressing the issue of raising the technical level of the lifts manufacturing, the quality of their maintenance and repair. In addition, it is necessary to use the lift to its intended purpose, and to carry out maintenance and diagnostics in a timely manner. Analysis of various structures of automobile lifts showed that they are equipped with the following safety elements: critical load limiter; gravitational locking system of rotation and boom lift with non-exposed supports; a device for blocking the lifting of supports with the working equipment raised; a device protecting the outriggers from spontaneous extension when working with the cradle; system of an emergency stop of movement when controlling from a working platform. During operation, this equipment requires careful maintenance, monitoring and high-quality diagnosis. These measures make it possible to promptly determine the causes of faults in the lifts, which helps to reduce downtime and operating costs. Creating a diagnostics system for lifts can significantly reduce downtime for technical reasons and speed up troubleshooting. [14,15]

Safety is the absence of unacceptable risk, where the required level is achieved by specialists with a certain recovery rate. Solving such problems rationally, we propose to use the method of hierarchical clustering, which can clearly provide a solution with the construction of a tree diagram-dendogram.

In addition, the method of clustering elements of lifts allows us to gather the nodes into clusters for functional purposes. On the basis of statistical data on the operation of automobile hydraulic lifts that worked at various sites in Moscow, the following elements were established from the position of reliability: internal combustion engine (ICE), hydraulic system, running gear.

Automobile hydraulic lift is a special type of equipment that is widely used in construction, therefore, the safety of operation and service life are important factors. The elements of general functionality in automotive hydraulic lifts are: internal combustion engines, clutch, starter, cooling system, valve, hydraulic motor, hydraulic cylinder, highway, pump. Let's consider the dependence of the intensity of recovery (repair) of lifts μ on the waiting time before troubleshooting $T_{ож}$ .The results of the analysis of some elements of the lifts are presented in Table 1.

Table 1. Summary statistical data on the recovery of the elements of automotive hydraulic lifts for one year.

Recovering element		Number of applications, (count)	Recovery time, (hrs)
ICE	Clutch	4	28,3
	Starter	2	3,1
	Internal combustion engine	7	22,8
	Carburetor	1	4
	Cooling system	1	25
Hydrosystem	Hydraulic motor	4	148
	Track	6	230
	Control valve	2	158
	Valve	2	108
	Pump	1	99
	Hydraulic motor	1	168
Chassis	Muffler	1	8,4
	Springs	1	67
	Tyre covers	1	5,5
	Rear bumper	1	9
	Tank leakage	1	3,87
	Wing welding	1	4,1

According to the data of the table, the dependence of the recovery intensity of the lifts on the waiting time for recovery (repair) is constructed (Figure 1) with least squares approximation:

$$\mu = 2{,}601 T_{\text{ож}}^{-0{,}322}, \tag{2}$$

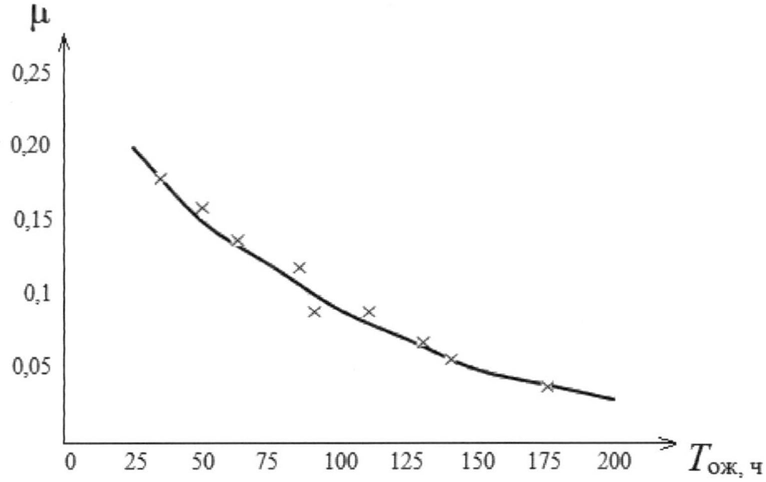

Figure 1. Dependence of the recovery intensity of the lifts μ on the waiting time for recovery (repair) $T_{\text{ож}}$.

According to the operational data, the automobile hydraulic lift works in 1.5 shift mode at a construction site for a year. According to the passport of this lift, the service life is 20 years, then the number of working hours for the entire service life will be 46 thousand hours.

Taking into account these sources, the failure rate of a hydraulic lift for one year is determined by the formula [16]:

$$\lambda = \frac{m(\Delta t)}{N \cdot \Delta t}, \tag{3}$$

where λ -intensity of the lift failures;

$m(\Delta t)$ - number of failures;

N - number of elements failed during the period Δt.

Taking into account the statistical data, we determine the contribution b_i of each element of the elevator to risk-safety, which corresponds to the percentage of output applications of the lift's individual elements. The rank of repair costs is determined by the dependence [17]:

$$r_i = \frac{T_{ожi} + T_{pi}}{\sum_i^n (T_{ож} + T_p)}, \tag{4}$$

where $T_{ожi}$ - waiting time before eliminating the continuity of the i-th element;

T_{pi} - recovery time of the i-th element;

$T_{ож}$ - total waiting time for all items.

We can estimate the danger risk with the formula:

$$P_i = b_i \cdot r_i, \tag{5}$$

The safety level of one lift operation, taking into account the contribution to it of elements, is estimated by the formula:

$$E = \sum (1 - P_i) = 0,821, \tag{6}$$

For the elements of the lift (table), the failure rate (λ) is determined by the formula (2): for the internal combustion engine - $3,76 \cdot 10^{-4}$; hydraulic system - $3,52 \cdot 10^{-4}$; the chassis - $3,25 \cdot 10^{-4}$, and the recovery rate (μ) is determined by the formula [17]:

$$\mu = \frac{1}{\frac{T_{ож} + T_{рем}}{N}}, \tag{7}$$

Where $T_{рем}$ - time of the element's recovery.

According to the formula (6), for ICE - μ=0,052, hydraulic system – 0,017; chassis – 0,036.

Fig. 2 shows the dependences of the ratio of the failure rate (λ) to the recovery rate (μ) on the number of applications ($K_з$) for repairs for ICE - 1, the hydraulic system - 2 and the chassis - 3, where $\Psi = \frac{\lambda}{\mu}$ is the ratio of the failure rate to the recovery rate.

Based on the obtained dependencies, we determine the safety level of the lift from the service life. Data on the operation of automotive hydraulic lifts at different sites allows us to establish the dependence of the number of failures of individual elements of the lifts ($K_{от}$) on the service life (T), the equations of the approximating curves, which look like:

For ICE - $K_{от} = 0,71T^2 - 5,1T + 16$; hydraulic system - $K_{от} = 0,63T^2 - 4,2T + 8,5$ and chassis - $K_{от} = 0,58T^2 - 5,71T + 12,1$.

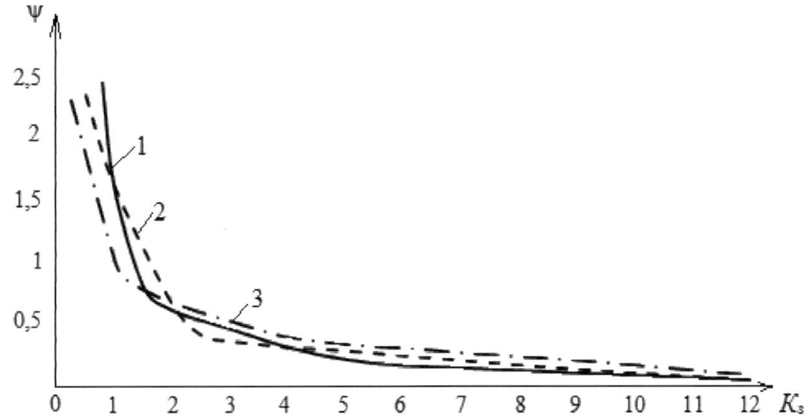

Figure 2. Dependences of the ratio of the failure rate (λ) to the recovery rate (μ) on the number of applications ($K_з$) for repairs for ICE - 1, the hydraulic system - 2 and the chassis – 3.

Considering the average service life of internal combustion engines, hydraulic systems and chassis, we can determine the number of failures for different service lives, which makes it possible to rationally carry out restoration work and systematize the maintenance and diagnostics. This approach allows us to assess the level of safety of lifts at various points in operation and to identify the qualitative and quantitative indicators of the level of safety.

3. Conclusion

The proposed method for assessing the level of safety of automotive lifts can reduce downtime and increase their efficiency.

Moreover, it can help to plan the diagnosis of the main elements of the machine and prepare a stock of sets of equipment for the repair and replacement of failed elements.

References

[1] Ancev V Yu , Tolokonnikov A S , Gorynin A D 2013 Automation of the failure risk estimation of cargo cranes // *Izvestiya TulGU Technical science* **7(1)** 214-20

[2] Zorin V A 2005 Performance bases of the engineering systems: the textbook for high schools M : OOO Magister-Press, 536

[3] Zorin V A, Dauhello V A , Sevryugina N S 2009 Safety requirements for the land transportation systems: a textbook Belgorod: BelGTU, 186

[4] Polyakov Yu I 2008 Theory of danger manifestations: the main strategy tasks of the engineering safety regulation by technical standards // *Bulletin of the Scientific Center* **2** 154 – 64

[5] Sevryugina N S 2012 Theory of technical safety formation of the complete transport and technological machines life cycle: monograph Belgorod: Publishing house of BGTU 179

[6] Sevryugina N S , Stepanov M A , Mechiev A V 2017 General methodological approach to the risk assessment of the structural safety of elevators // *Mechanization of construction* **78(4)** 24-9

[7] Stepanov M A , Mechiev A V 2016 Modeling of the elevator park maintenance system // *Scientific Review* **3** 27-32

[8] Telichenko V I , Zavalishin S I , Khlystunov M S 2005 Global risks and new safety hazards of the important construction sites of the megalopolis // Book of reports of the thematic scientific and practical conference "City construction complex and the safety of citizens" - Moscow: MGSU, 211-8

[9] Baurova N I , Zorin V A , Prikhodko V M 2015 Description of scenarios of transition of material from an operable to inoperable state using an equation of fold catastrophe theory Polymer Science Series D **8(1)** 1-5

[10] Baurova N I , Zorin V A , Prikhod'ko V M 2015 Influence of structural defects in carbon fibers on their sensor properties evaluated using scanning electron microscopy Fibre Chemistry **46(5)** 283-7

[11] Baurova N I , Zorin V A , Prikhodko V M 2015 Description of scenarios of transition of material from an operable to an inoperable state Polymer Science Series D **8(1)** 1-5

[12] kov A A , Roytman V M , Shilova L A 2016 Model of stability of life support systems in emergency situations International Journal of Applied Engineering Research **11(3)** 1666-9

[13] Sevryugina N S , Stepanov M A Vertical transport: resource by the criterion of safety Vertikalnyy transport: korrektirovka resursa po kriteriyu bezopasnosti» Inzhenerno-stroitelnyy zhurnal **72017** 23-35 doi: 10 18720/MCE 75 3

[14] Rashid Sharapov, Vitaly Vasiliev 2017 Analysis of the spectrum distribution of oscillation amplitudes of the concrete mix at shock vibration molding MATEC Web of Conferences 117 (2017) RSP 2017 – XXVI R-S-P Seminar 2017 Theoretical Foundation of Civil Engineering, Warsaw, Poland, August 21-5

[15] Plavelsky E P , Sharapov R R 2017 The Problems of Dynamics of Wheeled Vehicles with Flow Able Building Cargo International Conference on Industrial Engineering, ICIE **206** 86-92

[16] Stepanov M A , Kaytukov B A, Releabity and Safaty of Transpportation of Use IPICSE-2018 https//org 10 1051201875 103 013

[17] Stepanov M A , Kaytukov B A , Mechiev A V Opredelenie urovnya bezopasnosti liftov pri ekspluatatsii Podemno-transportnoe delo **6(1018)** 7-10

Experimental studies of the dependence of the heterogeneity coefficient of the mixture on the technological parameters in the mixer with a variable working chamber

N M Lozovoi*, L V Radinskaya and S Y Lozovaya

Belgorod State Technological University named after V.G.Shukhov, 46 Kostyukova street, Belgorod, 308012, Russian Federation

*E-mail: lozwa@mail.ru

Abstract. Mixing bulk materials is widespread in various industries: construction, agro-industrial complex, chemical technology, energy industry and many others. At the same time, the task of preparing homogeneous mixtures is associated with a number of difficulties, such as a wide range of changes in the physicomechanical properties of the processed materials, requirements for the quality and composition of the product, productivity, energy and metal consumption, etc.

1. Introduction

In road construction, when they use quicklime or burnt lime waste as additives, cohesive soil, before mixing with lime, is moistened to the condition of 0.7–0.8 of the moistness at the soil yield line. If the process requires using mixed bulk additives, then they should be added before the soil is moistened. Before processing mixtures with an anionic emulsion, we should add a mixture of 1–2% lime-powder and 2–4% cement [1–3], which should be pre-mixed with each other.

Tools and methods of intensive mixing allow us to seriously revise the basic recipes of modified mixtures and their production technology, while increasing the impact of modifying additives on the physicomechanical and technological parameters of the prepared mixtures leads to significant savings in expensive components [4-5].

The task of increasing the efficiency of devices used in the preparation of bulk mixtures inclined to adhesion and agglomeration, segregation by the physicomechanical properties of particles (size and density), with a large volume ratio of components (1:10 or more), make it necessary to constantly improve the known types of mixing equipment and create new ones.

Today, the production of certain types of materials for construction purposes is impossible without the use of mixing equipment capable of providing the necessary level of homogeneity of the mixture. The system of additives is used in the compositions of dry construction mixtures to regulate their technological properties. A number of components are introduced in small quantities (0.05-0.5%), but their influence on the formation of the properties of mortar mixes and solutions is extremely large. They form their own structural network in the aqueous phase or interact with the dispersed phase and maintain

IOP Conf. Series: Materials Science and Engineering **786** (2020) 012049 doi:10.1088/1757-899X/786/1/012049

the stability of the system, enhance the anti-sedimentation effect, increase the plasticity of the system, provide the necessary level of tixotropic properties.

Nowadays, modifying additives having a complex composition are supplied for sale in a ready-for-use form. Therefore, a mixer capable of high-quality mixing and homogenizing powders from the initial components, differing from each other in particle size (from micron fractions to 5 mm) and density (from 0.1 g / cm3 to 4.0 g / cm3) is the main device of technological cycle of production of dry mixes and their components.

2. Experimental installation of the mixer with a variable working chamber

One of the possible ways to improve the mixing equipment for bulk materials is the use of mixer designs with the ability to control and regulate the movement of particles inside the mixing chamber, which will provide the necessary movement of particles, and the stable achievement of the required quality of the mixture [5-9]. Therefore, the ability to control the process of mixing, simplicity of design, quick and easy adjustment of the mixer to work with different bulk materials is an urgent issue. This can be implemented using the mechanism of deformation of thin-walled elements in mixers to increase degrees of freedom and to impact the material controlling the movement of the mixture and the corresponding mixing processes to obtain high-quality multicomponent mixtures with desired characteristics [6, 9].

We conducted the studies using an experimental test bench (Figure 1, a) and a laboratory setup (Figure 1, b). The experimental stand is based on the modular principle, i.e. the stand base remains unchanged, and individual modules are being replaced. This made it possible to simplify the formulation of experiments and to conduct them on a unified basis according to an invariable method.

The experimental stand allows us to change the diameter, length of the variable chamber over a wide range, adjusting the drive rotation frequency within the required limits, and thus allows investigating the process of mixing materials in the mixer with the variable working chamber using multifactorial experiment methods. The stand consists of a bed where all the main nodes are installed. The working chamber is fixed with one side to the rear support, which, depending on the length of the working chamber, can move along the frame. The movement to the deforming elements is transmitted from the actuator, providing stepless adjustment of movement.

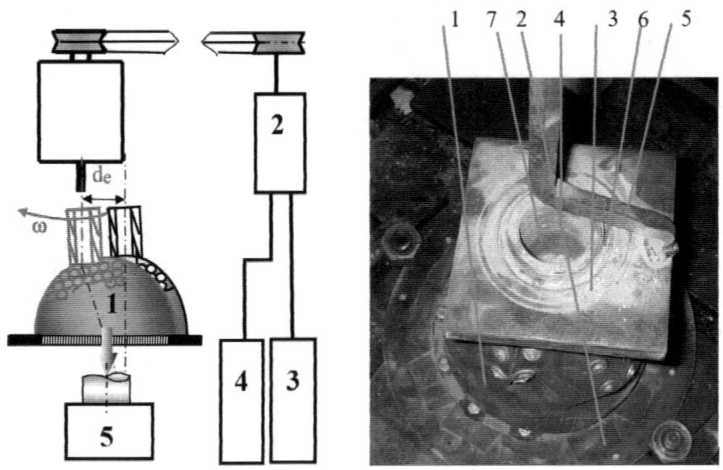

Figure 1. Equipment for experimental studies:
a - assembled experimental stand: 1 – device under research; 2 - drive unit; 3 - tachometer; 4 - control and measurements unit; 5 –bunker for final product;
b - laboratory setup: 1 – camera; 2 – base; 3 – ring; 4 – pin; 5 – drive ; 6 – bearing unit; 7 – loading hole; 8 – gate; 9 – shoulder blades; 10 - holes in the base for unloading material; 11 – gate for unloading.

The laboratory setup consists of a chamber 1 with a volume of 3 liters, fixed on the base 2 by a ring 3. The chamber 1 is fastened to the pin 4 with collars 4, through the drive 5 and the bearing unit 6, the rotation of the pin is transmitted from the spindle of the drilling machine. The loading is carried out through the hole 7, unloading through the gate 8. Blades 9 are attached to the pin 4, holes 10 are made in the base 2 for unloading the material. Their width is 6 mm and their size is justified by the fact that, if necessary, the chamber is filled with grinding bodies of size 6,5 mm and more and the device can work as a mill of superfine grinding.

The quality of mixing materials in devices with variable working chambers (the coefficient of heterogeneity of the mixture K_n) is influenced by the following factors:

$$K_n = f\,(\omega_c, L, D, d_{ch}, d_e, d_c, K_3, \Delta, t, K, R_l, j); \quad (1)$$

where ω_c – pin rotation speed around eccentric vertical axis, c^{-1}; D - diameter of the base of the working chamber, m; d_{ch} - particle size of the base material, m; d_e – eccentricity of the pin rotation around the vertical axis, m; d_C – pin diameter, m; K_3 - the fill coefficient of the chamber with mixed material; L – height of the working chamber, m; Δ - deformation coefficient of the working chamber in height; t - mixing time of mixture components, sec; K – blend ratio, %; R_l – radius of blades arrangement on the pin; j – number of blades.

We experimentally established that the mixing process takes place the more intensively, the greater the deformation coefficient of the variable chamber. It depends on the radius of the blades, since the blades should not be in contact with the walls of the chamber during the moving. The chamber has the shape of a half-bar, in which the height must be equal to the diameter of the base and the diameter of the pin is equal to 1/5 of the diameter of the base (by construction). The filling coefficient of the variable chamber with material, by analogy with existing mixers, was assumed to be 0.75% (2.25 l) of the chamber volume (3 l). Due to the problem of mixing the basic components of dry construction mixtures with different physical and mechanical characteristics, and due to the novelty of the mixing device, we chose sand as the main basic material with particle sizes of 1.25-2.5 mm, brand cement 350, to determine the device's performance under the proposed conditions. The blend ratio of the mixture ranged from 90 / 10-60 / 40% [3-5].

Based on the estimating tests, the mixing time ranged from 1 minute to 7 minutes, the number of revolutions of the pin ranged from 100 to 300 rev/min (pin rotation frequency 10.5 - 31.5 c^{-1}). With a number of revolutions per minute less than 100 with a minimum mixing time, the heterogeneity coefficient was 60%, and with a speed of more than 300 rpm and a maximum mixing time of 7 min, the heterogeneity coefficient was 75.7%, which can be explained by the fact that the cement particle has higher dispersion than sand particle and under the action of centrifugal forces sand particles fly to the walls of the chamber. It causes a separation into fractions, which increases the coefficient of mixture's heterogeneity. In this case, the eccentricity value was taken as maximum - 0.06 m.

At the experiments, we chose the following levels of variation to obtain a regression equation for the dependence of the mixture heterogeneity coefficient on the technological parameters (mixing time t, the blend ratio K, expressed in density ρ and drive speed n) (Table 1).

For K_n conversion formulas will be

$$x_1 = \frac{t-240}{110}; \quad x_2 = \frac{K-25}{9}; \quad x_3 = \frac{n-200}{60}. \quad (2)$$

In coded form, the regression equation for mixing sand and cement is

$$Y = 12{,}01 - 1{,}3x_1 - 12{,}52x_2 + 2{,}3x_3 + 1{,}13x_1x_2 + 5{,}2x_1x_3 - 1{,}1x_2x_3 - \\ - 2{,}43x_1^2 + 7{,}24x_2^2 + 6{,}1x_3^2 \quad (3)$$

IOP Conf. Series: Materials Science and Engineering **786** (2020) 012049 doi:10.1088/1757-899X/786/1/012049

Table 1. Intervals and levels of variation of independent variables.

Levels	*t*, s		*K*, %		*n*, rpm	
	Natural	Coded	Natural	Coded	Natural	Coded
Main	240	0	75/25 (1225 kg/m^3)	0	200	0
Variation intervals	110	-	11/9 (45 kg/m^3)	-	60	-
Upper	350	+1	66/34 (1270 kg/m^3)	+1	260	+1
Lower	130	-1	84/16 (1180 kg/m^3)	-1	140	-1
Top point	420	1,682	60/40 (1300 kg/m^3)	1,682	300	1,682
Bottom point	60	-1,682	90/10 (1150 kg/m^3)	-1,682	100	-1,682

A negative sign at x_1 and x_2 shows that with an increase in mixing time and the proportion of filler in the mixture (sand), the heterogeneity coefficient decreases, with an increase in the number of revolutions of the pin, the heterogeneity coefficient increases.

Substituting the conversion formulas (2) into equation (3), we have:

$$K_n = 306,2 - 0,2t - 5,5K - 1,2n + 0,001tK + 0,0008tn - 0,002Kn - \\ - 0,0002t^2 + 0,09K^2 + 0,002n^2 . \quad (4)$$

Using the formulas, we determined the significance of factors $t = 8\%;\ \mathrm{K} = 78\%;\ \mathrm{n} = 14\%$. The blend ratio in the mixture has the greatest influence on the value of the coefficient of heterogeneity of sand with a particle size of 1.2 - 2.5 mm with brand cement 350 in a mixer with a variable working chamber.

We built three-dimensional figures, showing the dependence of the heterogeneity coefficient on the main factors. Figure 3 shows the most typical shapes reflecting the values of the heterogeneity coefficient of the mixture from the main factors. Any point lying on a particular surface shows what values of time, blend ratios and the number of revolutions of the pin are required for the desired coefficient of heterogeneity of the mixture.

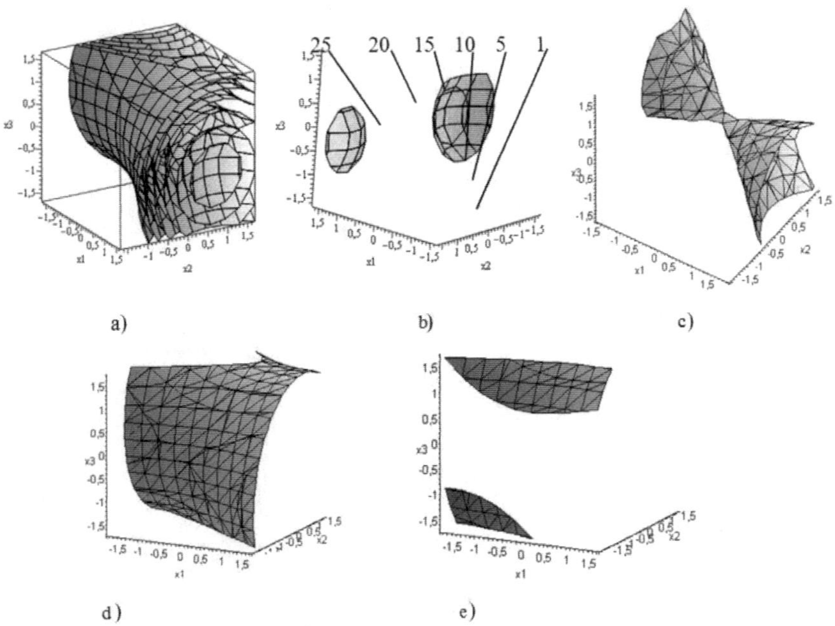

Figure 2. Graphs of the dependence of the heterogeneity coefficient on the main factors: a - the surfaces displaying change of coefficient of heterogeneity; b - K_n=1%; c - K_n=6,8%; d - K_n=30%; e - K_n=60,9%.

Analysis of three-dimensional figures (Figure 2, a, b) and graphs (Figure 3) shows that to obtain the heterogeneity coefficient of the mixture K_n = 1%, we can use two modes of mixing:

• .

The presence of a double mode of mixing is typical in obtaining mixtures with an heterogeneity coefficient of up to 6% (Figure 3, c, 5).

Figure 3, d, shows that the heterogeneity coefficient of the mixture is K_n = 61% with mixing time up to 240 seconds,pin speed from 100 to 140, blend ratio about 80/10. Moreover, this also applies to mixing modes with mixing time to 350 s, the number of revolutions of the pin from 250, the blend ratio of about 84/16. This can be explained by the fact that, firstly, sand and cement have a big difference in particle size and weight. In this case, the rotation frequency of the pin and the blend ratio are too small, so the cement falls in the voids between the sand particles. Secondly, in the case of a larger number of revolutions and mixing time, the cement particles are distributed closer to the chamber walls, which also reduces the homogeneity of the mixture.

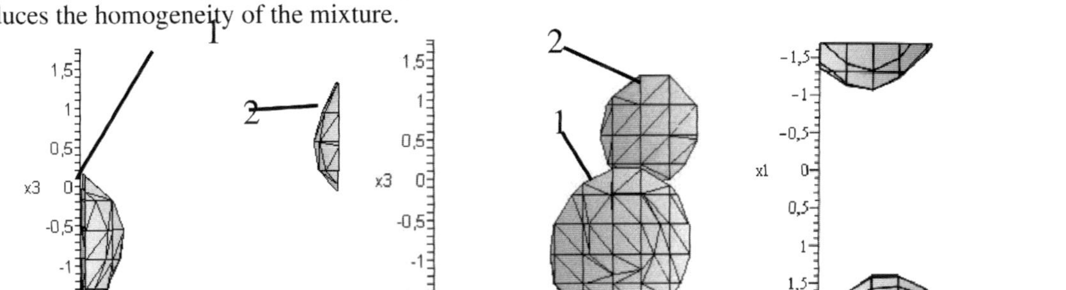

Figure 3. Surfaces reflecting the dependence of *Kn*=1% on technological parameters.

Here we can assume that in this device, with a smaller difference in the particle size of sand and cement, the smallest homogeneity coefficient can be obtained with a pin speed of up to 300 rpm, which is 5-10 times less than in known high-speed devices and with a shorter mixing time. The reason is that the particles of the material are mixed in the cross section by rotating the pin with the blades, and in the longitudinal section due to the deformation of the chamber, i.e. the number of degrees of freedom of mixing the material is maximum.

Figure 4 shows nomograms of the dependence of the mixture heterogeneity coefficient on two parameters with a fixed third. The areas highlighted in color show at which technological parameters the heterogeneity coefficient corresponds to: 1-5% (red); 5-10% (green); 10-15% (blue).

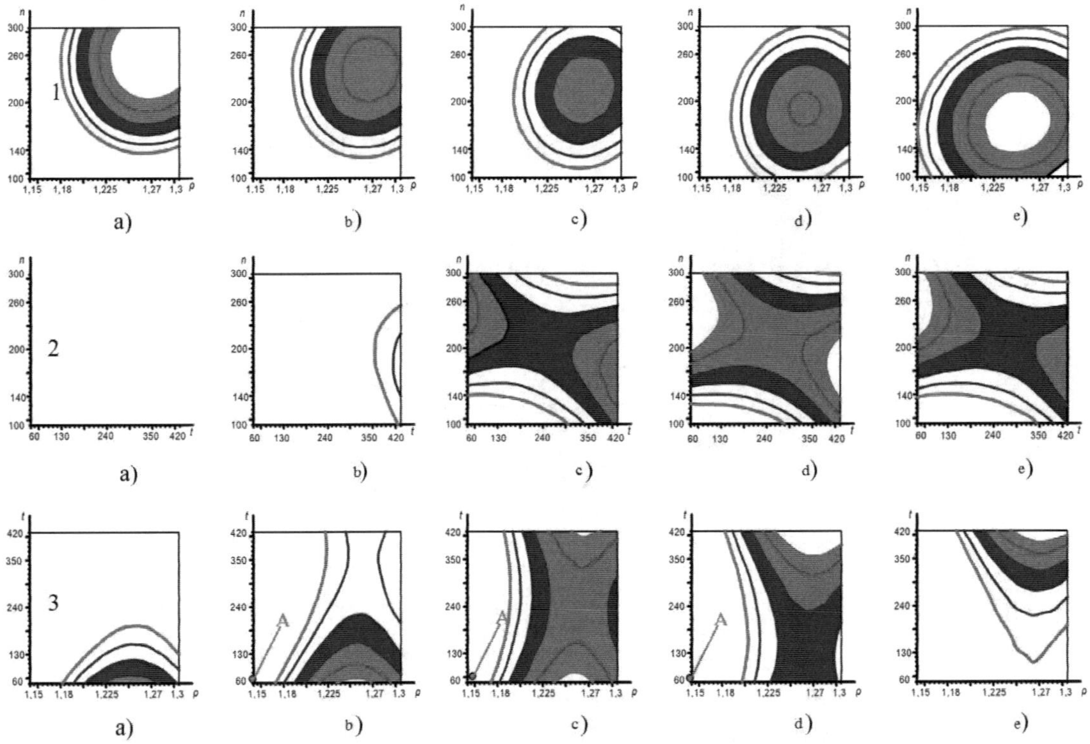

Figure 4. The change in the coefficient of heterogeneity of the mixture in the range from 1 to 25% depending:

1) on pin speed and density: a) *t*=60s; b) *t*=130s; c) *t*=240s; d) *t*=350c; e) *t*=420s;

2) on pin speed and time: a) ρ=1150 kg/m^3; b) ρ=1180 kg/m^3; c) ρ=1225 kg/m^3; d) ρ=1270 kg/m^3; e) ρ=1300 kg/m^3;

3) on time and density: a) *n*=100 rpm; b) *n*=140 rpm; c) *n*=200 rpm; d) *n*=260 rpm; e) *n*=300 rpm – Kn=1-5%; – Kn=5-10%; – Kn=10-15%; – Kn=15%; – Kn=20%; – Kn=25%.

3. Conclusion

The presence of a double blending mode is typical when obtaining blends with a heterogeneity factor of up to 6%: the first mode - when the number of revolutions of the pin is from 210 to 300 rpm, the mixing time is 60-100s and the percentage blend ratio is from 66/34 to 60/40 (ρ = 1225 ... 1300 kg / m^3); the second mode - when the number of revolutions of the pin is from 130 to 210 rpm, the mixing time is 380 - 420 s and the percentage blend ratio is from 66/34 to 60/40 (ρ = 1225 ... 1300 kg / m^3).

References

[1] Metodicheskie rekomendatsii po stroitelstvu osnovaniy dorozhnyh odezhd s ispolzovaniem svyaznyh gruntov, ukreplennyh mineralnymi ili organicheskimi vyazhushchimi s dobavkami PAV i promyshlennyh othodov/Soyuzdornii - 1985 39

[2] GOST R 52129-2003 Poroshok mineralnyy dlya asfaltobetonnyh i organomineralnyh smesey Tekhn usloviya - Izd ofits ; Vved 01 10 2003; Vved vpervye - M : Gosstroy Rossii, FGUP TSPP, 2004 - 33

[3] Laskorin B N 1984 Bezothodnaya tekhnologiya pererabotki mineralnogo syrya - M : Nedra, - 334

[4] SHirina N V 2009 Stroitelnye rastvory Proshloe i nastoyashchee [Monografiya - Belgorod: Izd-vo BGTU, – 219

[5] S Yu Lozovaya, N M Lozovoj and N V Shirina 2013 Influence of Processing Characteristics of Mixer with the Changing Camera on Quality of Heat-Insulating Plasters on the Basis of Modifying Agent and Anthropogenic Raw Materials/ *World Applied Sciences Journal* **24(10)** 1404-10, ISSN 1818-4952 © IDOSI Publications, 2013 DOI: 10 5829/idosi wasj 2013 24 10 7021, Published: Sep 15 1404-10

[6] Lozovaya S YU , Lozovoy N M , Uvarov V A , Ryadinskaya L V , Sharapov R R , Ryadinskaya L V and Sharapov R R 2016 Studying changes in design and technological parameters of the grinding and mixing devices with cylindrical working chambers deformable in cross section International *Journal of Pharmacy & Technology* **8(4)** 24733-24747 URL: http://www ijptonline com/wp-content/uploads/2017/02/24733-24747 pdf

[7] Tarshis, M YU 2008 K raschetu smesitelya sypuchih materialov so slozhnym dvizheniem elastichnoy rabochey kamery / M YU Tarshis // Izv VUZov, «*Himiya i himicheskaya tekhnologiya*» **51(8)** 75 -7

[8] Tarshis, M YU 2003 Novye apparaty s elastichnymi rabochimi elementami dlya smeshivaniya sypuchih sred Teoriya i raschet: YAGTU, – 84

[9] S Yu Lozovaya, N M Lozovoy, L V Ryadinskaya, I B Kostina, O N Satler, L V Logvinova and E I Ziborova 2018 Experimental studies of energy characteristics of mixer with deformable camera/ International Conference “Complex equipment of quality control laboratories” IOP Conf Series: *Journal of Physics*: Conf Series **1118(012024)** IOP Publishing doi:10 1088/1742-6596/1118/1/012024

IOP Conf. Series: Materials Science and Engineering **786** (2020) 012050 doi:10.1088/1757-899X/786/1/012050

Road safety strategies in the Russian Federation for 2018–2024 in Omsk Region

M G Simul* and S M Porkhacheva

Siberian State Automobile and Highway University, Russia, 644080 Omsk, 5 Mira avenue

*E-mail: simul79@yandex.ru

Abstract. The paper explores the causes of accidents and ways to reduce their number through the use of targeted programs to improve traffic safety. The authors justify the choice of target indicators used in the development of such programs. The article gives an example of the development of a set of measures for one of the sections of Omsk city.

1. Introduction

The safety indicators of the transport process, primarily road traffic, do not comply with the world standards. The number of deaths in traffic accidents per 1,000 vehicles is 4 times higher than in developed countries [1]. Modern federal programs targeted to reduce road accidents, as a rule, are focused on a long period of implementation. Each program establishes annual target indicators, the fulfillment of which should ultimately ensure the achievement of the final target national indicator [1, 2].

2. The main part

Road traffic is a multifactorial structure, so it is necessary that the complex of control measures covers all the constituent elements of the problem in a specific territorial unit (region, city, district). Therefore, one of the effective mechanisms for reducing road hazard is the use of an integrated complex approach in ensuring road safety. One of the examples of an integrated approach is the development of targeted programs. This method of reducing road accidents is enshrined at the legislative level and is presented as a priority principle for ensuring road safety. The entire list of activities provided in the framework of integrated programs is aimed at achieving a single goal - improving the road safety [2]. A feature of target programs is a clear target orientation, characterized by quantitative targets. In accordance with the criteria for grouping targeted programs by time of goal achievement, they identify short-term, medium-term, and long-term target programs. The problem of reducing road accidents requires a lot of attention, therefore, to achieve positive results, we should apply long-term targeted programs in the first instance.

At the implementation stage, the target integrated program has the form of a targeted system that evolves over time due to the introduction of the executors required to carry out the relevant activities.

After reaching the goal, the program, as a system, can either end its existence or move to the class of functioning systems that support the achieved goal. A distinctive feature of this method of improving road safety is the development of quantitative target indicators. Thus, the positive effect is achieved due to the structured presentation of the program goal in the form of a system of interrelated subgoals (Figure 1).

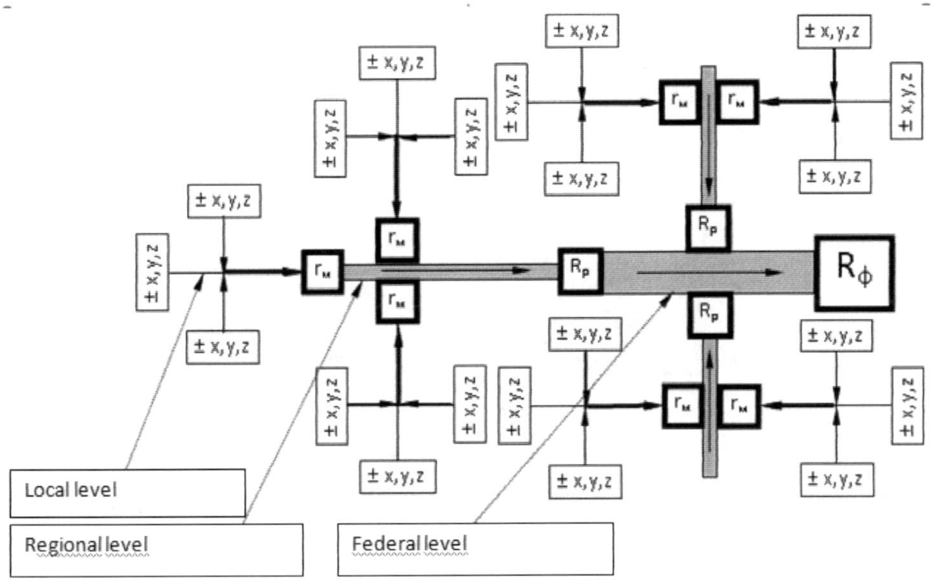

Figure 1. Scheme of the formation of the federal final target indicator of road accidents reduction where r_M, R_p, R_f - respectively, the results of negative and positive nature at the local, regional and federal levels; x, y, z - factors that have either a positive impact on the accident rate or a negative.

The issues of developing of targeted programs for specific territorial units are described in [3,4]. Indicators of the road safety conditions in Omsk for 2016-2018 are given in Table 1.

Table 1. Absolute accident indicators in Omsk for 2016 - 2018.

Accident indicator	2016	2017	2018
Number of accidents, count	2318	2257	2300
Deaths, humans	54	71	77
Wounded, humans	2925	2789	2870

For the improvement road safety in different regions of the Russian Federation, they analyze the causes and conditions associated with accidents, and then propose comprehensive measures to reduce accident rates. Examples of such proposals are covered in the proceedings of scientific conferences, where specialists from different federal subjects exchange their experience and analyze the use of modern smart systems, both in Russia and abroad. Such examples of proposals could be seen in [5, 6, 7, 8, 9]. The work [10-12] presents a model of a Smart City. The model is based on traffic police data on the vehicle fleet, the number of accidents, and the number of dead and injured. The key indicators are calculated on the basis of this model.

To reduce the number of accidents we propose the following steps:

1. To perform an analysis of accidents for several years across the entire road network of the study area.
2. To select and rank traffic accidents according to stability and type.
3. To select a group of the most stable places for the concentration of traffic accidents.
4. To study the designated areas in order to identify hazards: limited visibility, dangerous speeds, poor state of technical means of traffic control.

5. To develop a set of measures aimed at eliminating the identified hazards and to achieving the specified goals

To show an example of the measures development, we consider one of the sections of Omsk, which is shown in Figure 2. The proposed accident reduction technology has been developed on its basis.

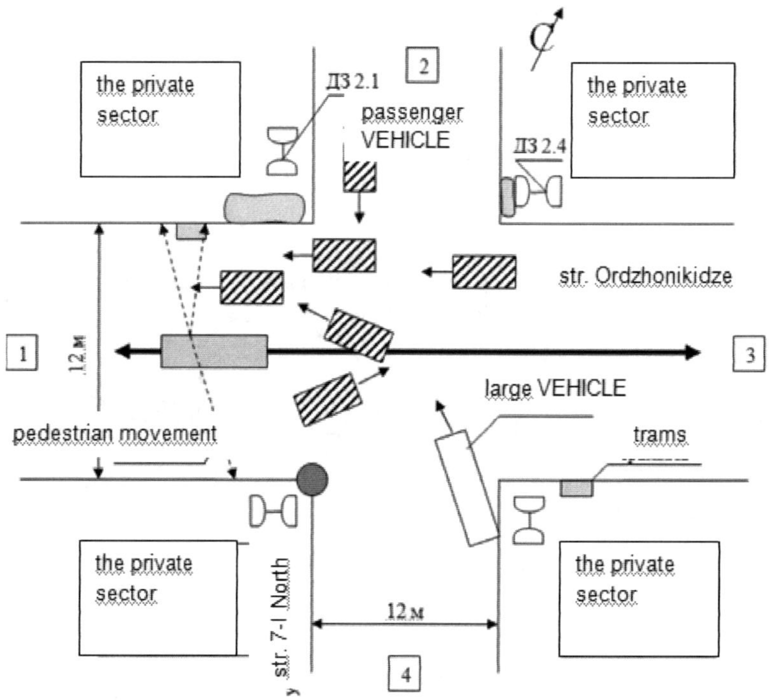

Figure 2. Scheme of the intersection of Ordzhonikidze and 7th North street in Omsk.

Let's look at the data on the dynamics of accidents at this site (Table 2).

Table 2. Dynamics of accidents in the areas of Herzen and Ordzhonikidze streets, Omsk.

№	Accident concentration place	Dangerous areas	Accident rate dynamics 2014-2017
1	Herzen str – Oktyabrskaya str	Oktyabrskaya str., № 48, №65	9-20-18-28
2	Herzen str – 7th North str	-	18-22-26-30
3	Herzen str – 24th North str	24th North str, № 216	12-12-14/3-28/5
4	Ordzhonikidze str – Oktyabrskaya str	№ 47, № 56	0-18-18-29
5	Ordzhonikidze str – 7th North str	-	13-23-22-16

We suggest the following activities:

- To move a tram stop from the intersection down to the block for at least 30 meters on the sides 1 and 3;
- To arrange the order of priority movement for drivers by installation of road signs: 2.1 "Main road" on the side 4; 2.4. "Yield" on the side 1; 5.15.1 - "Directions of movement along the lanes" on all the sides;
- To increase the duration of "green" cycles in the traffic light cycle;
- To eliminate defects of the road surface in the conflict zone of the intersection;
- To move the pedestrian crossing down to the block along with the public transport stop;
- To install pedestrian barriers to prevent pedestrians from entering the road.

In order to select an indicator suitable for assessing the actual impact on the traffic conditions of a specific road network during the implementation of municipal programs, and associated with the generally accepted target index *Nn*, we carried out a study of quantitative relations of generally accepted indicators of road accidents among themselves:

Nn,- number of deaths,

Nid – number of accidents with deaths or injuries;

NMVC – number of accidents with only material damage.

We used the following sample for statistical analysis: about 500 thousand traffic accidents (469700) for a long period (15 years) on the road network of the city of Omsk. Based on the analysis, we made the following conclusions.

1. Between the average annual indicators: there is a relation between the number of accidents with dead *Nn* and the number of accidents with injured *Nid* within certain limits. The same conclusion was made in relation to the number of accidents with injured *Nid* and the number of accidents with only material damage *NMVC*.

2. The entire street-road network of a big city can be decomposed into several groups of streets and roads with a stable range of the annual accident rate. This conclusion can be used when planning the sequence of activities for the systematic reduction of accidents across the city. For example, it was revealed that for the city of Omsk, traffic incidents occur on 850 streets out of a roughly 1,300 streets in total. We selected the most dangerous streets from the general list of emergency streets, which is about 12% of the total number of streets with the accidents. In 15 years, there were about 350 thousand traffic accidents on these streets, which is 77% of the total number of traffic accidents in the city. The most emergency streets are also divided into the groups according to the quantitative indicator of the accident rate.

3. Some of the emergency streets have places of the accident concentration with a small dispersion of accidents by year. Usually, these places have permanent (stationary) or temporary factors and conditions that contribute to the dangerous behavior of drivers and pedestrians and create conflict situations (N_{ks}). These factors and conditions can be visually identified and eliminated by specially selected events. This conclusion indicates the possibility of systematic reduction of the number of accidents in specific places of accident concentration. We can assume that, by eliminating the identified hazards in a particular area, the average annual number of accidents will be reduced by at least 50%.

3. Conclusion

The proposed approach makes it possible to systematically reduce the annual number of accidents on a specific street-road network due to the targeted selection of measures to eliminate specific hazards and conditions that cause road accidents at specific locations of their concentration. This approach makes it possible to indirectly influence the program indicator “the number of people died in road accidents for the reporting year” and ensure the fulfillment of the indicators established by federal regulatory documents. Moreover, a prerequisite is to conduct a systematic preventive work with various social groups.

References

[1] On approval of the Transport Strategy of the Russian Federation for the period up to 2020: Order of the Ministry of Transport of the Russian Federation of 12 05 2005, N 45 - [Electron resource] // IS “Consultant Plus” - Access mode: www consultant ru

[2] The program for improving road safety in 2013–2020: the federal targeted program, approved by an order of the Government of the Russian Federation 1995 - [Electron resource] // IS “Consultant Plus” - Access mode: www consultant ru

[3] Salnikov V A and Ryabokon YU A 2008 Formirovanie tselevykh pokazateley munitsipalnykh programm s idetom Federalnoy programmy po snizheniyu dorozhnoy avariynost //Organizatsiya i

bezopasnost dvizheniya v krupnykh gorodakh: Sbornik dokladov vosmoy mezhdunarodnoy konferentsii «Organizatsiya i bezopasnost dvizheniya v krupnykh gorodakh / SPb gos arkhit -stroit un-t SPb 460

[4] Porkhacheva S M and Simul M G 2016 Problemy obespecheniya bezopasnosti dvizheniya na UDS goroda // Problemy kachestva i ekspluatatsii avtotransportnykh sredstv: Organizatsiya avtomobilnykh perevozok i bezopasnost dorozhnogo dvizheniya [Tekst]: materialy XI mezhdunar zaochn naid -tekhn konf 15 march 2016 , Penza – Penza: PGUAS, **526** 335-8

[5] Road safety rules URL: https://ru wikipedia org

[6] Dergunov, S A 2014 Dorogi budushchego – dorogi peremen Innovatsii v nauke: sbornik statey po materialam KHKHKH mezhdunarodnoy naidno-prakticheskoy konferentsii – Novosibirsk 95-103

[7] Absalyamov T 2015 The influence of cultural and sport mega-events on sustainable development of the city *Procedia-Social and Behavioral Sciences*, **188** 197-201

[8] Novikov A 2017 Adaptation capacity of the traffic lights control system (TSCS) as to changing parameters of traffic flows within intellectual transport systems (ITS) Transportation Research Procedia 455-62

[9] Gubaidullina, T , Ivanova, N , Absalyamova, S , & Yerina, T (2018, December) Analysis of national strategies for sustainable development with regard to fundamental conceptual premise *In Journal of Physics: Conference Series* **1141(1)** 012018 IOP Publishing

[10] Kolesov V I 2019 Parametricheskaya identifikatsiya modeli obobshchennogo zolotogo secheniya v sfere dorozhnoy bezopasnosti // Organizatsiya i bezopasnost dorozhnogo dvizheniya: materialy KHII Natsionalnoy naidno-prakticheskoy konferentsii s mezhdunarodnym idastiem 2 t / otv red D A Zakharov – Tyumen: TIU, 317 -22

[11] Sakhapov R and Nikolaeva R 2017 Economic aspects of traffic safety administration Transportation Research Procedia 578-83

[12] Sakhapov R L , Nikolaeva R V , Gabdullin T R and Makhmutov M M 2015 Motor roads as a factor of the economic potential of the republic of Tatarstan *Procedia Economics and Finance* **24** 606

Ensuring the safety of the transshipment process in river ports

O V Leonova*, A S Veretennikova and E G Veretennikov

Moscow State Academy of Water Transport, Moscow, Novodanilovskaya quay, 2, building 1, Russia

* E-mail: pptmir@mail.ru

Abstract. The paper considers the issues of ensuring the safety of the transshipment process in river ports. The authors propose an approach to integrated safety assessment of the transshipment process in river ports taking into account various types of risk. The article analyzes the accidents and injuries during transshipment works in river ports. The authors establish that injuries in river ports are mainly due to operational reasons i.e. due to the presence of a significant number of cranes that have completed their standard service life. The paper studies the influence of the operating time on the static and fatigue strength characteristics of the metal structure of gantry cranes after 30-40 years of operation. The research results show that long-term operation don't cause significant changes in mechanical characteristics, with static and fatigue strength.

1. Introduction

The transshipment process in river ports is a set of technological operations for the movement of goods, using special equipment and ensuring the required safety.

Nowadays, during the transshipment process in the port, the most attention is paid to the control of the transshipment equipment, with special attention required for the steel structures - since they cannot be completely replaced under operating conditions in case of resource depletion. Bearing elements of metal structures are subject to static, dynamic and other types of loads and require evaluation of their performance according to the criteria of static and cyclic strength.

Evaluation of the metal constructions performance of the transshipment equipment allows to conduct a quantitative assessment of the technical risk of operation based on physical-mechanical analysis methods using the" load-strength" model. This is an integral part of a comprehensive assessment of the safety with various types of risk of the transshipment process in river ports.

2. The various impacts on the safety of the transshipment process

There are many hazardous things that could happen during the transshipment process: technical system or device failure, pollution or destruction of the ecological system, harm to health, death of a person or a group of people, material damage from the hazards that have occurred or an increase in safety costs.

We can identify the following sources of danger during the transshipment process: transshipment equipment, cargo, the technological process, weather conditions and the human factor (Figure 1).

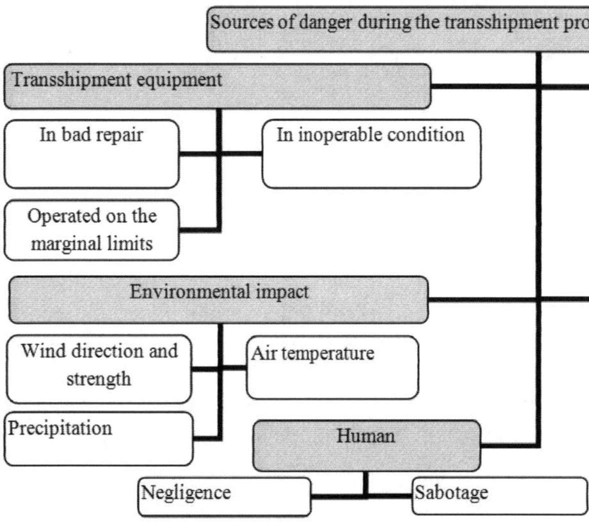

Figure 1. Block diagram of the sources and factors affecting the safety of the transshipment process.

These factors are characterized by the undesirable events, each of which may occur in relation to a particular risk object. The relation of objects of risk and undesirable events allows us to distinguish individual, technical, environmental, social and economic risks. Each type of risk determines the typical sources and risk factors, classification and characterization.

In case of safety violation of the transshipment process in the river port, the objects of negative factors influence may be: a person or a group of people, the environment, technical facilities and equipment, and cargo. We made recommendations on the need to assess various types of risk for the listed objects of impact (Table 1).

Table 1. Recommendations on the need to assess various types of risk depending on the object of impact.

Object of impact	Types of risk				
	Individual	Technical	Ecological	Social	Economic
Person or group of people	++	—	—	+	+
Environment	—	—	++	—	++
Technical buildings and equipment	—	++	—	—	++
Cargo	—	—	—	—	+
Note:					
++ - evaluation is required					
+ - evaluation recommended					
— - evaluation is not required					

3. The analysis of accident rate during the transshipment works in ports

To assess the nature of the accidents in ports, we analyzed statistical data and generalized the causes of accidents and injuries that have occurred over a number of years during the operation of port transshipment equipment. The results of the analysis are shown in Figure 2.

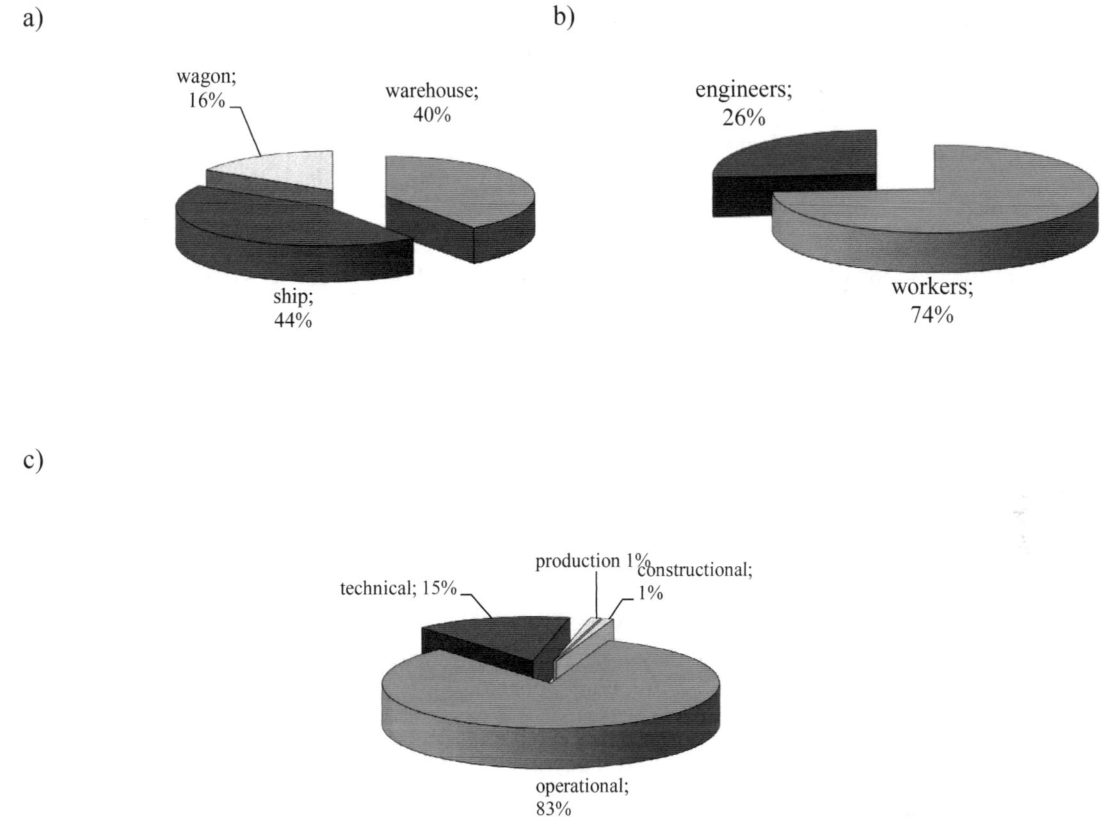

Figure 2. The results of the analysis of accidents during transshipment works in ports:
a) Classification of operations of the transshipment process by injury rate
b) Classification of accidents by employee categories
c) Classification of equipment failure by its nature.

The diagrams show:

a) Types of transshipment operations do not have a major impact on injuries during the transshipment process (Figure 2a).

b) Unqualified actions of workers cause accidents and injuries 3 times more than actions of engineers and technicians (Figure 2b). At the same time, a big role is played by the reasons related to the unsatisfactory supervision of the technical condition of the equipment, often due to the presence of a significant amount of transshipment equipment that work despite the end of its service life in the river ports. This necessitates a study of the impact of the technical state of the equipment. In particular, on the carrying capacity of the steel structure.

c) Injuries are caused mainly by operational reasons, while the level of accidents for operational reasons significantly exceeds the level of accidents for technical reasons (Figure 2c). This also confirms the need to conduct research on the operated equipment, and also makes it possible not to take into account production and constructional reasons at this stage of risk assessment.

Studies show that the impact of operating time contributes to the mechanical and fatigue characteristics of the material in a non-obvious way [1]. In this regard, we carried out the studies and clarified its impact on:

- mechanical characteristics of the metal structures material samples of gantry cranes, which main indicators are are the values of tensile strength (σ_B), yield strength (σ_T), relative elongation (δ), relative narrowing (ψ), impact toughness (KCV);
- fatigue resistance characteristics, which main indicators are the endurance limit values (σ-1), the fatigue curve slope (m);
- crack resistance characteristics, which main indicators are the values of crack growth rate (v), stress intensity factor (K), parameters of the Paris equation (C) and (n).

4. The impact of operating time on mechanical characteristics with static strength and toughness

The impact of operating time on the above characteristics was investigated for the steel St3 and St38b2 used for gantry cranes.

We took the samples from metal structures of gantry cranes "Albatros 10 / 20-32 / 16" and "Ganz 6 / 5-32 type N", which have been operated in the ports of Russia for 30-40 years.

We carried out the tests to determine the mechanical characteristics in two stages: tensile tests and tests for impact bending.

Tests for tensile with operating time were carried out on a tensile testing machine IM-4R. The results of tensile test samples are presented in Table 2 [2].

Table 2. Mechanical characteristics of samples made of steel St38b2 and St3 with and without operating time.

Crane, material	Operating time, years	Metal construction element	Yield strength бT, MPa	Temporary resistance бB, MPa	Relative elongation δ%
Albatros 10/20-32/16 (St38b2)	40	Counterbalance	297	562	23,5
			300	535	25,0
	40	Jib	280	441	21,5
			286	450	22,0
	40	Jib	261	418	22,5
			232	412	27,0
	40	Portal	253	412	34,0
			255	416	34,0
	40	Trunk	255	369	32,0
			259	371	33,0
Medium	40		267,8	438,6	27,45
St38b2 [3]	without operating time		220-240	380-470	25
Ganz 6/5-32 type N (St3)	30	Column	244	427	33,0
			247	429	32,0
	30	Jib	287	462	31,
			322	477	28,0
Medium	30		275	448,75	31
St3 [3]	without operating time		250	380-490	26

We compared the mechanical characteristics (samples with operating time and without operating time) using the methods of mathematical statistics. A check by the Fisher and Student criteria showed that the samples belong to the same general population [4]. The results of tensile tests show that the operating time of 30–40 years do not cause changes in the values of the tensile strength σB, the yield strength σT, the relative elongation δ of the steel St3 and St38b2.

Tests for bending impact were carried out on a PSW-300 pendulum scraper in accordance with [4] on samples with a V-type concentrator. Figures and Tables 3 and 4 give the results of testing with and without operating time for impact toughness [2].

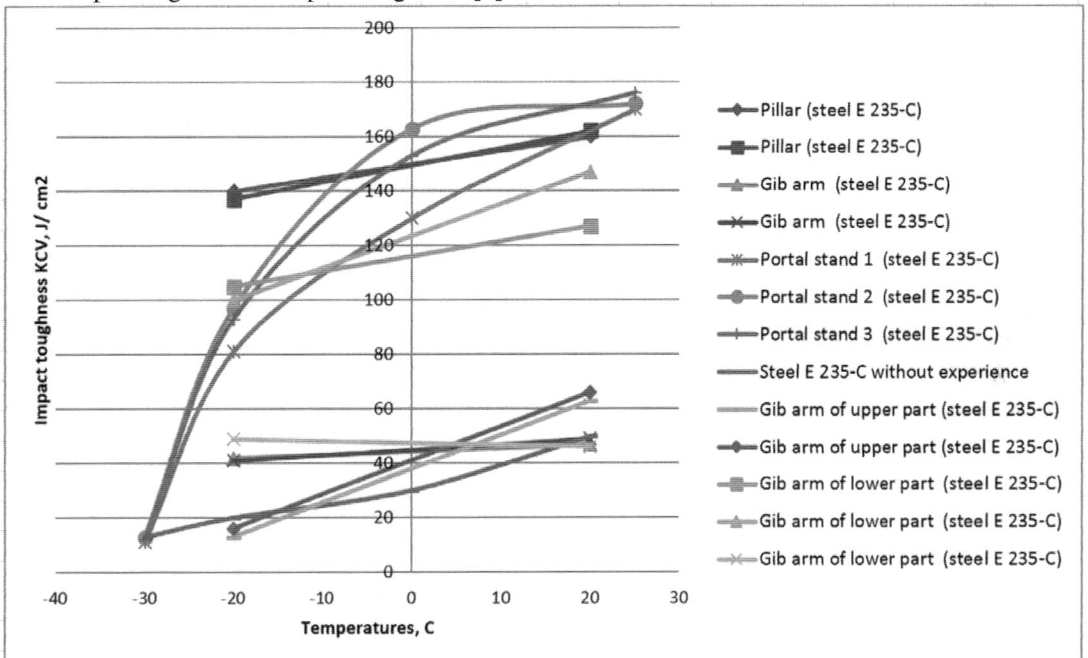

Figure 3. Change of impact toughness depending on the temperature for steel St3.

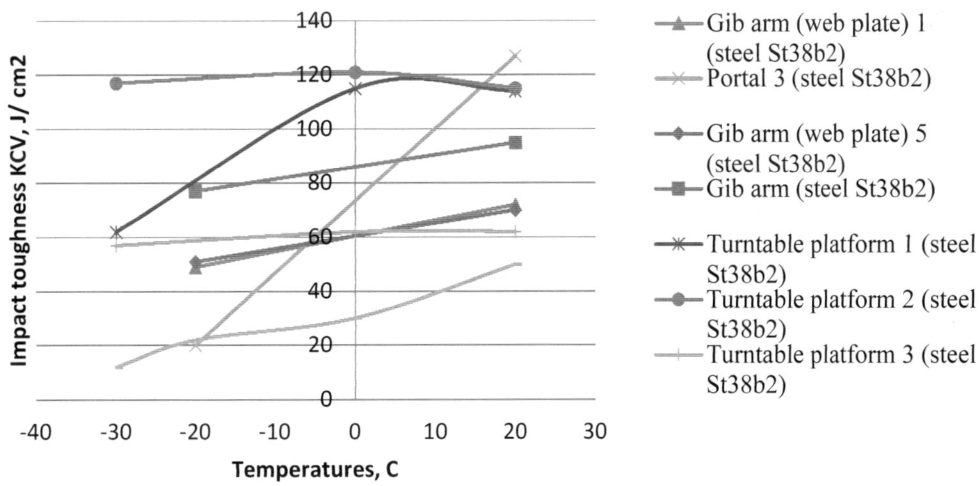

Figure 4. Change of impact toughness depending on the temperature for steel St38b2.

Table 3. Comparison of test results for the determination of impact toughness for St3.

t, °C	Impact toughness KCV, J/cm²							
	Operating time							
	30 years				29 years			0 years
	Column (section 1)	Column (section 1)	Column (section 4)	Column (section 4)	Portal foot (section 1)	Portal foot (section 2)	Portal foot (section 3)	Samples of steel without operating time
25					170	172	176	
20	160	162	47	49				50
0					130	163	153	30
-20	140	137	42	41	81	97	93	20
-30					11	13	12	15

Table 4. Comparison of test results for the determination of impact toughness for St38b2.

t, °C	Impact toughness KCV, J/cm²							
	Operating time							
	40 years				30 years			0 years
	Portal (section 3)	Jib (section 1)	Jib (section 5)	Counterbalance	Turntable (section 1)	Turntable (section 2)	Turntable (section 3)	Samples of steel without operating time
20	127	72	70	65	114	115	62	50
0					115	121	62	30
-20	20	49	51	47				22
-30					62	117	57	12

Tables 3 and 4 show that the impact toughness values of samples with an operating time of 30–40 years have a variation caused by the fact that samples were taken from metal structures with different levels of tension in operation. In this regard, it is possible to make a preliminary conclusion that the effect of the level of loading of the metal structure influences the measurements of impact toughness. To assess the impact of operation time on the values of impact toughness, we need an additional research.

5. Impact of operation time on fatigue strength and durability

Fatigue tests were carried out on samples that were cut from the trunk of the gantry crane "Albatros 10 / 20-32 / 16" and the jib of the gantry crane "Ganz 6 / 5-32 type N". The test results are summarized in table 6.

Table 5 . The results of fatigue tests of samples of steel St38b2 and St3 with the operating time.

Sample number	Tension Σ_a, MPa	Number of cycles, *N* Steel St38b2, Operation time 40 years	Steel Ct3, Operation time 29 years
1	316	$1{,}62 \cdot 10^5$	$1{,}7 \cdot 10^5$
2	277	$3{,}12 \cdot 10^5$	$2{,}1 \cdot 10^5$
3	238	$9{,}43 \cdot 10^5$	$5{,}1 \cdot 10^5$
4	200	$2{,}63 \cdot 10^6$ passed the test base	$2 \cdot 10^6$ passed the test base

View of the fatigue curve for samples taken from elements of a metal structure with operating time of 30 and 40 years, presented in figure 5 [4].

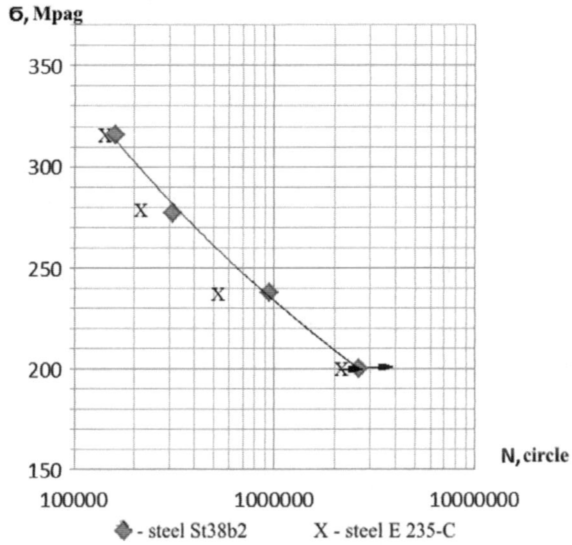

Figure 5. Fatigue curve for steel St38b2 with 40 years in operation.

As we can see from Table 6, the fatigue limit of steel St38b2 and St3 based on $2 \cdot 10^6$ is 200 MPa, i.e. the fatigue resistance characteristics of St38b2 steel with a life of 40 years and St3 steel with a life of 30 years in operation remained at the same level.

6. Impact of operation time on the characteristics of crack resistance

Tests on cyclic crack resistance of the samples were carried out on the “Amsler HFP” pulsator and on the “SCHENCK” PC-400S universal electrohydraulic machine. Figure 6 and 7 show the speed of fatigue fractures growth in the metals in operation [2].

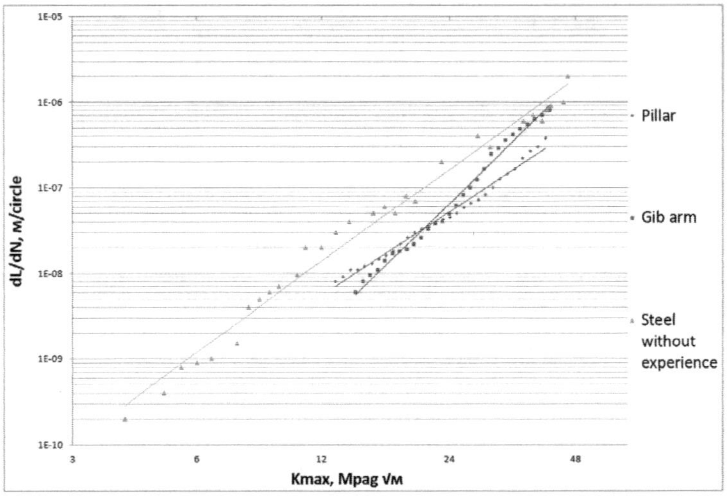

Figure 6. Speed of the fatigue fractures growth in steel E 235-C with 30 years in operation.

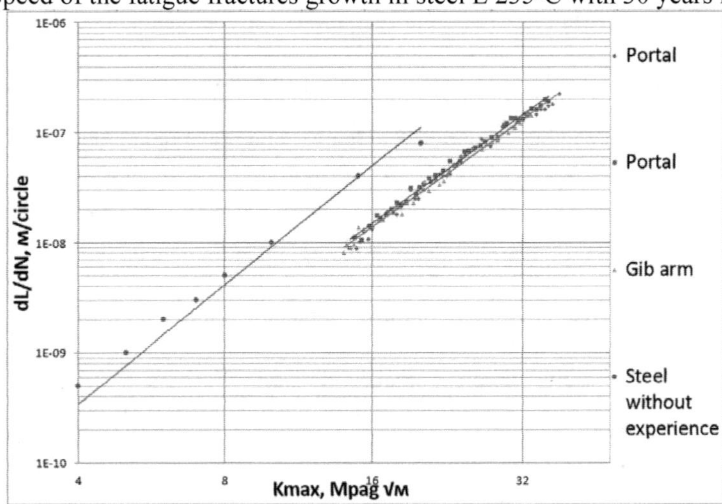

Figure 7. Speed of the fatigue fractures growth in steel St38b2 with 40 years in operation.

According to the obtained experimental data of crack growth rates, corresponding to different values of ΔK, we calculated the coefficients "C" and "n", which are characteristics of the cyclic crack resistance of the material under given test conditions:

$$da/dN = C\ \Delta K^n \quad (1)$$

Table 6. Characteristics of the cyclic crack resistance of samples of steel St38b2 and St3.

Crane	Steel	Object	Characteristics of cyclic crack resistance		Coefficient of determination
			C	n	
Albatros 10/20-32/16	St38b2	Portal	$1{,}66 \cdot 10^{-12}$	3,26	0,989
Albatros 10/20-32/16	St38b2	Portal	$1{,}73 \cdot 10^{-12}$	3,27	0,992
Albatros 10/20-32/16	St38b2	Jib	$1{,}87 \cdot 10^{-12}$	3,22	0,989
Ganz 6/5-32 type N	St3	Jib	10^{-14}	4,83	0,989
Ganz 6/5-32 type N	St3	Column	$2 \cdot 10^{-12}$	3,25	0,988

Comparison of the kinetic diagrams of fatigue destruction and characteristics of cyclic crack resistance of metal portal and jib of a gantry crane after 40 years of operation with the metal of a gantry crane after 30 years of operation showed that metal after 40 years of operation is characterized by higher values of fatigue cracks growth rate in the region of low values of the linear section of a diagram - from 14 to 25 MPa√m, - up to 1.5-2.5 times. With increasing ΔK, this difference weakens and the rates of crack growth become hardly distinguishable.

The results of tests on cyclic crack resistance showed that the kinetic diagrams of fatigue failure of the metal of all series on St38b2 steel are located close to each other. For samples of steel St3, we need to conduct additional research.

7. Conclusion

The paper provides a block diagram of the sources and factors affecting the safety of the transshipment process and recommendations on the assessment of various types of risk depending on the object of impact.

Moreover, this article analyzes the causes of accidents in the river ports during the transshipment process.

The study presents the results of research on the evaluation of the influence of operation time on the mechanical characteristics, characteristics of fatigue strength and durability of St3 and St38b2 steels with an operating time of 30 ... 40 years.

The materials obtained as a result of this research allow to conduct a quantitative assessment of the technical risk of the metal constructions of the transshipment equipment on the basis of physical and mechanical methods of analysis using the load-strength model [5-8].

8. Discussion

Research results show that operation for 30 ... 40 years don't cause significant changes in the mechanical characteristics, characteristics of crack resistance and fatigue. This is mainly due to the fact that the metal structures of gantry cranes are designed with large safety margins.

However, it is necessary to pay special attention to the study of the influence of operation time on the mechanical characteristics, characteristics of crack resistance and fatigue of details and assemblies, depending on their loading during operation.

To improve the reliability of port transshipment equipment, it is necessary to ensure a high level of maintenance, repairs and compliance with the rules of safe operation.

For a comprehensive assessment of the safety of the transshipment process in river ports, in accordance with the recommendations of Table 1, it is also necessary to take into account the results of a quantitative assessment of the environmental risks, as well as the results of an individual risk assessment. The obtained results must be compared with an acceptable risk that combines technical and economic aspects.

References

[1] Pustovoy V N 2012 Problemy bezopasnoy ekspluatatsii portovykh kranov v usloviyakh sverkhnormativnoy ekspluatatsii // Podemnye sooruzheniya Spetsialnaya tekhnika **7-8(126)** 11– 3

[2] Veretennikova A S and Leonova O V 2015 Otsenka vliyaniya narabotki na prochnostnye kharakteristiki metallokonstruktsiy portalnykh kranov // Materialy Mezhdunarodnoy nauchno-tekhnicheskoy konferentsii «Interstroymekh-2015»/ Otv redaktor R L Sakhapov – Kazan, Izdatelstvo kazansk gos arkhitekt -stroit un-ta, 27-34

[3] RD 212-0093-88 Pravila Remont metallicheskikh konstruktsiy gruzopodemnykh kranov //Transport, 1989 72

[4] Leonova O V , Petrova I M , Veretennikov E G , Veretennikova A S 2015 Otsenka vliyaniya narabotki portalnykh kranov na ustalostnye kharakteristiki stali St38b2 // *Rechnoy transport (XXI vek)* **3(74)** 36-7

[5] Leonova O V , Veretennikov E G , Smirnova A S 2011 Obosnovanie bezopasnoy ekspluatatsii portalnykh kranov // *Rechnoy transport* **4(52)** 78 - 81

[6] Veretennikov E G 2011 Obosnovanie bezopasnoy ekspluatatsii portalnykh kranov na osnove teorii riska: Avtoreferat dis kand tekhn Nauk – M : – 20
[7] Sakhapov R , Nikolaeva R 2017 Economic aspects of traffic safety administration Transportation Research Procedia 578-83
[8] Sakhapov R L , Nikolaeva R V , Gabdullin T R , Makhmutov M M 2015 Motor roads as a factor of the economic potential of the republic of Tatarstan *Procedia Economics and Finance* **24** 606

IOP Conf. Series: Materials Science and Engineering **786** (2020) 012052 doi:10.1088/1757-899X/786/1/012052

A method for constructing standard-size series of quick couplers

E V Duganova[1,*], N S Sevryugina[2] and P D Kapyrin[2]

[1]Belgorod State Technological University, Belgorod, str. Kostukova, 46, Russia

[2]Moscow state university of civil engineering, Moscow, Yaroslavskoye shosse, 26, Russia

*E-mail: crumb1202@mail.ru

Abstract. The article analyzes the effectiveness of the use of multifunctional machines when performing work in the field of urban planning and environmental design. It has been established that unification ensures a reduction in the cost of equipment operation and allows to increase the rationality of using the working bodies according to their functional characteristics. The authors evaluated the typical nomenclature rows of quick couplers. It has been revealed that the variety of quick couplers offered by manufacturers does not allow to determine a common design feature, and to use it for unification for various classes of machines. In this regard, the authors take its weight to the mass of the base machine during the unification of standard-size series, as a main parameter. The paper develops a technique for unification of quick couplers. Then, the authors modify the mathematical model for estimating the cost of standard sizes of quick couplers according to the exponents and proportionality coefficients. They propose a parametric unification for 14 rows of the main parameter of the quick coupler. Finally, the paper develops a model for estimating the range of parametric applicability of various quick coupler constructions.

1. Introduction

The constructive diversity of special machines is functionally justified by technological processes that should be performed with the least expenditure of both material and time resources. [1-4] At the same time, operations for performing works in the field of urban planning for landscaping are typical processes, which can be improved by creation of a new technique, new working bodies or modernization of existing ones [5-7]. The unification process allows to solve the problem of minimizing the costs of technological equipment by establishing a rational range of components of the machine, criterionally and functionally identical in the given parameters.

The typical hierarchical structure of the basic machine with different nomenclature of working bodies allows to set the level of unification of structural elements:

$$К_{у.э} = \frac{Q_{у.э}}{Q_э}, \tag{1}$$

where $Q_{у.э}$ – number of unified construction elements;

$Q_э$ – number of standard sizes of construction elements.

The tendency of design development for machines ensuring technological processes in the sphere of urban planning and environmental engineering follows the path of model assembly, which also requires the expansion of the nomenclature of unified elements. To reduce the cost of machines production we should assess the prefabrication, which is quantified as:

$$К_{сб} = \frac{U}{U + D}, \tag{2}$$

where U – the number of assembly units in the machine design;

D – the number of details that make up the nodes and assembly units.

The task of unification, which is being solved at the design stage, emphasizes the shift of attention to increasing the functionality of machines by applying unified structural elements in the design of working equipment, i.e. the use of quick couplers as an additional element in the structure of the working equipment. [9-10].

Quick couplers are universal devices that allow to modernize machines on a functional basis, transferring them to the multifunctional category according to the working bodies used. Development of standard-size series for quick couplers will ensure the creation of techno-economically optimal nomenclature, built on the basis of parametric series of the main parameter.

2. Analysis of quick couplers design

Quick couplers have found wide application in various fields and in particular in the field of urban planning and environmental engineering.

According to their purpose, quick couplers provide a quick change of working equipment without the involvement of additional working personnel.

The key element of the design is the mechanism for gripping the working body and the method of its fixation for reliable operation of the joint.

For example, Wedgelock Equipment Ltd offers a reduced weight design that is reliable and controlled both hydraulically and manually. To increase the nomenclature, quick couplers are available with various types of fastening finger centers, with elements of safe lifting of loads. Bob-tach offers a frame design that provides automatic gripping of attachments.

The design of quick couplers can be divided according to the method of attachment: using holes, hooks, or a combination of them; and type of connection: with manual, semi-automatic and automatic.

The analysis showed that the variety of quick couplers proposed by the manufacturers does not allow to determine the general design feature, and to use it as unified for various classes of machines [8-11].

3. Development of methods for unification of quick couplers

The stages of work on the unification of quick couplers for machines operated in the field of urban planning and environmental engineering include an analysis of standard-sized series of designs from world manufacturers.

Using the method of vertical unification, we set the main and basic parameters of the unification of the quick coupler. For quick couplers, the main parameter is the indicator of the ratio of quick couplers weight relative to the mass of the base machine. The rationale for this decision is the significance of this parameter, since it determines the overall size of the quick couplers, affects the stability of a multi-functional machine, has the stability of a physical value and is technologically linked to other key parameters and economic indicators of the quick couplers.

This main parameter does not depend on the pre-operational stages of the life cycle of the quick coupler.

The main parameters also include the geometric characteristics: the axial distance of the connecting dimensions and the specific permissible load perceived from the elements of the working equipment and the working body.

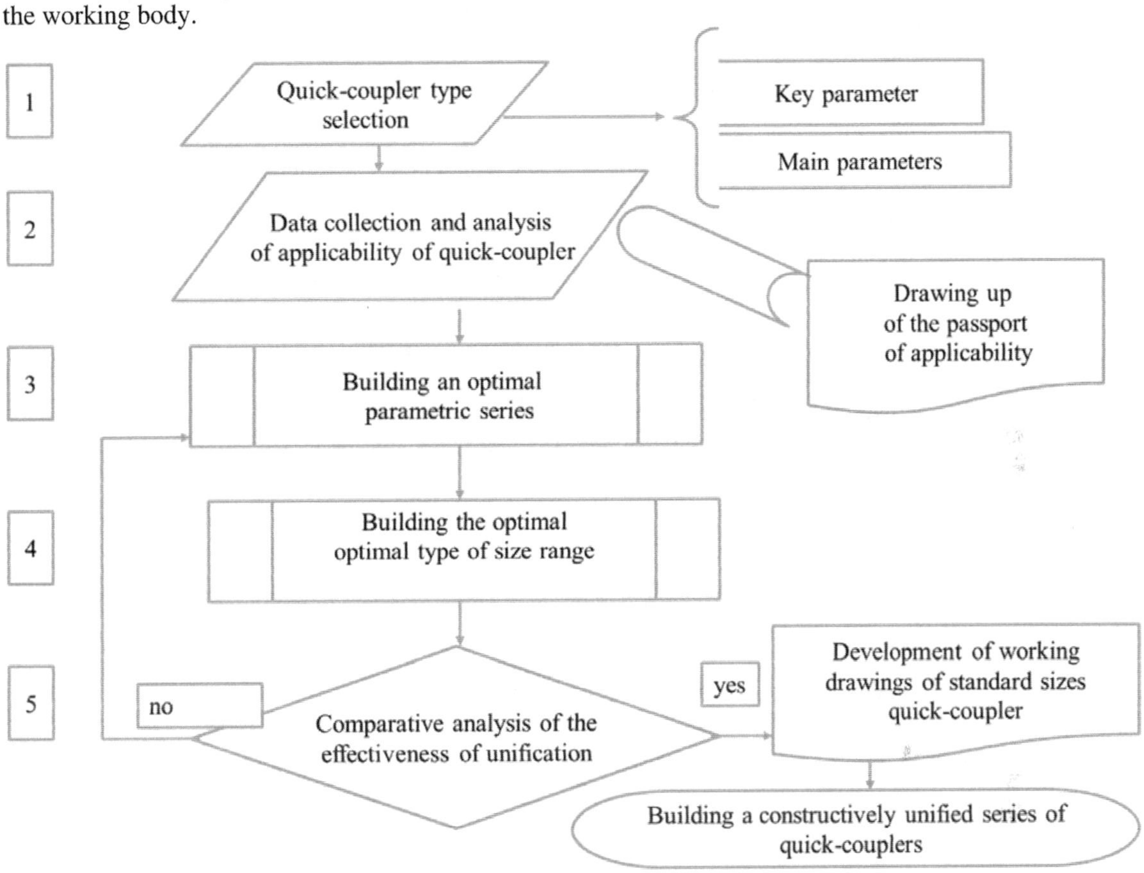

Figure 1. Algorithm of quick coupler unification model.

4. Modeling

The basic qualifying attribute of the methods for constructing standard series is the nature of the comparison of series and the type of modal analysis. There are elementary, classical and non-classical methods for constructing standard-size series.

After the performed analysis, we choose the method of complete sequential search through the series as a whole as the most applicable for constructing a parametric series of quick couplers. The search for the optimal series is carried out by comparing the total costs of implementing the life cycle stages on an annualized basis according to the output programs of all the members of three adjacent rows constructed by the series of preferred numbers R10 - R20 - R40.

Evaluation of the constructive diversity of quick couplers types produced in different countries confirmed the need to develop methods of unification, because the manufacturers indicate that the shape of the part does not correspond to the form adopted for unification [9-10].

The evaluation of the functional efficiency of quick-coupler resulted in the widespread use of hydraulic quick-coupler with two hooks, providing automatic connection with the working body. The classic design is with the mounting and non-mounting plates, which share the range of application of the working body of the base machine.

For the compilation of the unification model, we adopted design parameters for quick couplers of both versions.

We propose to use the method of limited unification and optimization of the standard nomenclature of quick couplers.

The calculation of the cost of standard sizes of quick couplers ($C_i^{'}$) is carried out on the basis of regression processing of the database of standard-sized quick couplers series with the definition of the exponent and the proportionality coefficient:

$$C_i^{'} = K_i A_i^u V^v \left(\frac{1}{R}\right)^w P^x N_i^n, \tag{3}$$

where K_i – coefficient of proportionality, takes into account the type of a detail and design implementation;

A_i – the main parameter;

R – geometric characteristic of the mount, as the main parameter;

V – axial parameters of quick coupler according to the characteristic of stability of base model;

P – metal intensity of quick coupler, in monetary equivalent to metal intensity of base model;

N_i – the annual program for the release / modernization of quick couplers;

u, v, w, x, n - the exponents calculated by the method of regression analysis of the corresponding initial statistical aggregates.

The modal analysis with a number of transformations made it possible to obtain a logarithmically linear model of relation of influence parameters:

$$\lg C_i^{'} = \lg K_i + u \lg A_i + v \lg V + w\left(\frac{1}{R}\right) + x \lg P + nLgN_i \,, \tag{4}$$

As mass and geometric characteristics of quick couplers parameters are set as key, then to determine the cost of metal quick coupler it is necessary to take into account, along with the value of the main parameter, the value of the axial connecting size.

Analyzing the method of the quick coupler installation, we found a multiplicative effect of the dependence of the main parameter — mass on the parameters of height, width and length of the quick coupler as a whole, geometry of connecting elements and stability of the base machine, etc. Accordingly, in order for the ratio $\frac{m_{q-k}^u}{m_m}$ to be within the recommended limits, it is required to set the value of the axial mounting dimensions b.

In this case, the mathematical model (3) will have a modified form:

$$C'_i = K \cdot \left(\frac{m_{q-k}^u}{m_m}\right) \cdot b^v \cdot \left(\frac{1}{R}\right)^w \cdot P^x \cdot N_i^n, \tag{5}$$

where $\frac{m_{q-k}^u}{m_m}$ – the transformed expression of the main parameter, with known values of the masses;

b – cross-axle mounting dimensions of quick coupler providing stability of the base model.

Analysis of the applicability data provides a reduction of the range of sizes for each of the selected types of quick couplers by a combination of the main parameters of the quick coupler and the base machine, i.e. their masses.

Solving the production problems of manufacturing quick couplers, one should take into account the costs of service life according to the annual production programs of each of the selected basic i-th sizes, counting changes in the cost of individual structural elements, costs of technical re-equipment and technological reconstruction [12-13].

IOP Conf. Series: Materials Science and Engineering **786** (2020) 012052 doi:10.1088/1757-899X/786/1/012052

As shown above, the modification of mathematical models makes it possible to transfer the unification problem to the level of a multi-criteria solution of the problem, combining and simultaneously optimizing the initial series of sets of specific numerical values of the main parameters of the details [14-15].

5. Parametric unification

Analysis of the design characteristics of quick couplers from different manufacturers allowed us to establish the quantitative ratio of the values of the main parameter. We distinguish 14 rows of the rational range of the unification of the main parameter from 160 to 1000 kg.

We built a histogram of the field of application of quick couplers on the base model based on the main parameter.

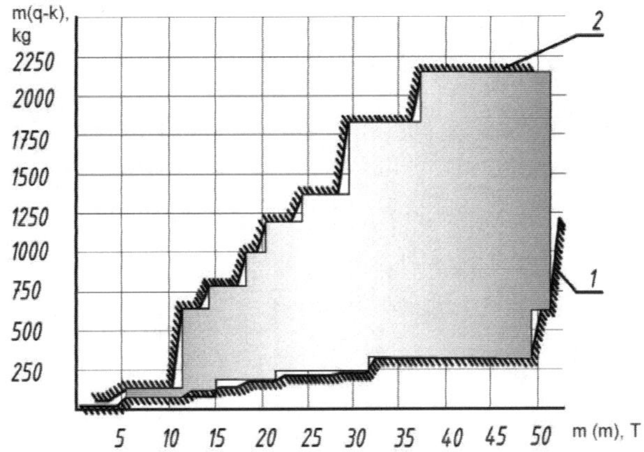

Figure 2. The area of combination of quick couplers' main parameters used with machines of urban planning and environmental management with the rationing of the main parameter of quick coupler according to the maximum (2) and minimum (1) values within the limits of the tolerance region.

The infogram shows that 90% of the rows have the structure of regular geometric progressions (stepped geometric rows with thickening to a larger parameter value), and 10% - the structure of "irregular" stepped geometric progressions.

Taking into account the step view of the row - the first three mass numbers in the R10 row, and the rest in the R20 row, the optimal standard-sized quick coupler row is set according to the main parameter, with the following mass numbers: 160/180/200/250/280/320 / 360/400/450/500/560/630/710/800.

IOP Conf. Series: Materials Science and Engineering **786** (2020) 012052 doi:10.1088/1757-899X/786/1/012052

6. The practice of assessing parametric applicability of quick couplers

The compilation of a table of optimal standard-sized rows of quick couplers is carried out taking into account the adopted limitations in the standard-sized series of all values of the optimized row of the main parameters and the unification range.

As a result, a model was obtained for estimating the range of parametric applicability of various designs of quick couplers, taking into account the consistency of the values of the main and basic parameters (Table 1).

Table 1. The range of applicability of quick couplers on the level of consistency of the values of the main and basic parameters.

Main parameter	Basic parameteres			
Mass of quick coupler, kg	Mass of base machine, tons	Geometry of connection, mm	Length, mm	Height, mm
160	3-5	230 - 320	960	480
200	5-7	270 - 350	1000	490
250	7-9	290 - 380	1030	500
280	9-12	320 - 430	1050	510
320	12-16	340 - 450	1070	530
360	16-19	370 - 480	1100	550
400	19-25	400 - 500	1130	570
450	25-30	430 - 540	1150	590
500	30-35	460 - 570	1170	620
560	35-40	500 - 620	1200	640
630	40-45	550 - 670	1230	650
710	45-50	590 - 750	1250	660
800	50-55	600 - 780	1270	665

7. Conclusion

We analyzed the quick couplers types for constructive and functional diversity;

We complied the methodology for developing the standard series of quick couplers;

The main parameter is justified as the ratio of the weight of the quick coupler to the weight of the base machine, and the basic parameters are the geometric characteristics of the connecting elements of the quick couplers as with the working equipment and the working body;

We introduced a modified model for calculating the cost of a quick coupler, taking into account the main and main parameters;

We got an optimal number of quick coupler with mass values: 160/180/200 / for the R10 series and 250/280/320/360/400/450/500/560/630/710/800 for the R20 series.

We developed a model for estimating the range of parametric applicability of various quick coupler designs, taking into account the consistency of the values of the main and basic parameters.

8. Discussion

The practice of applying basic methods of maintaining the functionality of machines throughout the entire operational period shows their temporal limitations due to rapid obsolescence. Scientific and technical support allows us to quickly make adjustments to the basic techniques, taking into account innovative trends in the development of both the mechanical component of the structure and the drive elements: electrical, hydraulic, and digital control systems, in particular, the possibility of flexible adjustment of the standard-size rows of machine components unification, reducing operating costs and increasing functionality [14-18].

References

[1] A Volkov, V M Roytman and L A Shilova 2016 Model of stability of life support systems in emergency situations Intern *J of Applied Engineering Research* **11(3)** 1666-9

[2] I V Dunichkin, A S Kovaleva and I V Tashlykova 2018 Approaches to the assessment of energy potential of renewable energy sources in the territory of Russia *J Power equipment Autonomous system* **1(1)** 15-27

[3] N Bilgin, T Dincer, H Copur and M Erdogan 2004, *J Tunnelling and Underground Space Technology* **19(6)** 629–36

[4] V V Nedoluzhko, A N Novikov 2011 *J World of Transport and Technological Machines* **1** 33-8

[5] A S Apatenko 2015 Improving the system of technical operation during import substitution of machines for reclamation work, J Environmental Engineering **2** 74-7

[6] A S Apatenko 2013 Influence of the service life of machines on their operational reliability when performing reclamation work, J *Technique and equipment for the village* **10** 4-6

[7] A S Apatenko 2013 Acquisition of a fleet of machines for watering peatlands, taking into account unplanned failures, J *Technique and equipment for the village* **12** 36-8

[8] N S Sevryugina, E V Prokhorova and E A Volkov 2014 Fast coupling device (utility model) Pat Of the Russian Federation **146194**, IPC E02F3 / 28 Applicant and Patentee: Belgor state tehnol un-t - 2012153300/03; decl 12/10/2012

[9] E V Duganova, N A Zagorodniy, D N and A S Korneyev 2018 Studying quick coupler efficiency in working attachment system of single-bucket power shovel *IOP Conference Series: Materials Science and Engineering* **327(4)** 042026

[10] N Sevryugina 2017 J *MATEC Web of Conf RSP* **117** 00151

[11] P D Kapyrin and N S Sevryugina 2018 J *IOP Publishing Ltd* **365** 042018

[12] F K Abdrazakov, D G Goryunov 2004 The optimal distribution of technology - the basis for the stable development of production, *J Mechanization of construction* **1** 8-10

[12] N I Baurova, V A Zorin and V M Prikhodko 2017 Information model of the state of the technical system, J *All materials Encyclopedic reference* **6** 11-16

[13] O Poddaeva, A Kubenin and D Gribach 2018 Measures of Improving the Accuracy of the Calculation of Energy Efficiency and Energy Saving of Construction Transport Infrastructure, J *Advances in Intelligent Systems and Computing* **692** 490-497

[14] A V Shchelgunov 2018 The main problems of import substitution in the domestic engine-TV construction in relation to low-power generating sets, *J Power and energy equipment Autonomous systems* **1(2)** 46-51

[15] V A Zorin, N I Baurova and A A Pegachkov 2019 Assessment of products risks of mechanical engineering by results of diagnosing, J *Periodicals of Engineering and Natural Sciences* **7(1)** 287-93

[16] R L Sakhapov, R V Nikolaeva, M H and M M Makhmutov 2017 Modeling of traction-coupling properties of wheel propulsor, J *of Physics: Conference Series* **936(1)** 012033

[17] Sakhapov R , Nikolaeva R 2017 Economic aspects of traffic safety administration Transportation Research Procedia 578-83

IOP Conf. Series: Materials Science and Engineering **786** (2020) 012053 doi:10.1088/1757-899X/786/1/012053

The network patterns characteristics research as one of the factors of traffic congestion

O V Saprykina*, E A Akhmedova and K I Vinogradov

Samara State Technical University 443100, Molodogvardeyskaya street, 244, Samara, Russia

*E-mail: olga_grineva_@mail.ru

Abstract. The article presents the planning structures analysis method of urbanized areas base on the geometric indicators of graph theory. For research, we developed a software package written in the Python programming language. The methodology of the analysis consists of identifying areas with a planning structure of a certain type in urban area and evaluating the selected area according to several calculated indicators. According to the results of the study, hypotheses were made about the impact of the type of planning structure on the level of traffic load of the city's road network.

1. Introduction

Modern urban science identifies several basic types of planning structures - regular, scenic, radial-circular, combined. The indicated types of planning structure schemes appeared in ancient times and continue to be used in modern city-planning projects in explicit or modified form. At present, in the theory of urban planning, the question of the optimal choice of the type of planning structure when designing territories, based on their prospects for the development of the transport network, remains relevant.

This paper presents some criteria for comparing the types of planning structures, calculates the values of the criteria for each type and compares the characteristics of the planning structures in the context of the modern city fabric. The study examines the basic types of planning structures implemented in twelve cities of the world. The twelve examples considered are obviously insufficient number to form valid conclusions, therefore the objective of this study is limited to hypothesizing. However, the above method of analyzing the planning structures may be applicable for a full-scale study in order to prove the hypotheses put forward.

To conduct the research we developed a software package written in the Python programming language. Additionally, we used third-party libraries OSMnx, GeoPandas, NetworkX. We used maps of cities downloaded from OpenStreetMap. The following cities were studied: Adelaide (Australia), Bologna (Italy), Vancouver (Canada), Vienna (Austria), Yekaterinburg (Russia), Yokohama (Japan), Lyon (France), Manchester (United Kingdom), Samara (Russia) , Barcelona (Spain), St. Louis (USA), Frankfurt (Germany). The choice of cities is due to the following criteria: the city is a large industrial center, it is not a capital city (except for the city of Vienna), it has an extensive transport network and continues active urban development.

2. Selection of study areas in cities

The methodology of the analysis consists of identifying areas with a planning structure of a certain type in urban tissue and evaluating the selected area according to several calculated indicators. In order to solve the problem in each of the twelve cities, two sites were allocated with a regular and scenic planning structure. Since the radial-circular structure is extremely rare as a layout of individual sections of the city, examples with this structure were found only for two cities. To illustrate the characteristics of the planning scheme of the combined type, large-scale areas are considered, covering almost the entire territory of the city.

The size of the selected areas was limited to an area of no more than 10,000 square meters. Plots are selected in the vicinity of the central district of the city. For each site, the software selected a planning scheme, for which the basic and design characteristics of road networks are analyzed. The calculation of the characteristics was carried out in terms of the unit area of the urbanized territory.

An example of the selected area and the corresponding selected planning scheme for the city of Vancouver is presented in Figure 1.

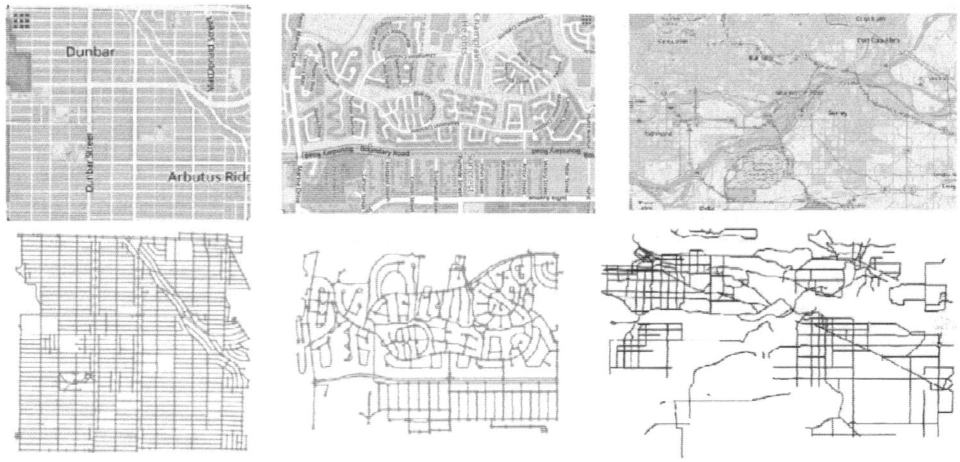

a b c

Figure 1. Vancouver city planning schemes (a) regular structure; (b) scenic structure; (c) combined structure.

3. Congestion level of the cities

The congestion level is the main factor determining the delayed movement of vehicles. Increasing the congestion level leads to a decrease in the speed of the traffic flow. The load limit is usually reached on city highways during peak hours. Table 5 shows the average values of the congestion level for the studied cities, according to the Tomtom Traffic Index [1].

Table 1. Congestion level of the road networks

City	Congestion Level, %	City	Congestion Level, %
St. Louis	13	Frankfurt	28
Yokohama	21	Samara	28
Vienna	23	Manchester	30
Bologna	24	Lyon	33
Barcelona	25	Yekaterinburg	34
Adelaide	27	Vancouver	39

According to the given data, the most congested transport networks are owned by the cities of Vancouver, Yekaterinburg, Lyon, Manchester (the congestion level is more than 30%). The less busy cities are St. Louis, Yokohama, Vienna.

We should also note cities with high levels of transport service and comfort of living. Among the cities under consideration, according to the rating of the international consulting company Mercer, for 2017 there are Vienna (2nd place), Vancouver (3rd place), Adelaide (6th place), and Frankfurt (7th place in Europe) [2]. The Mercer index is based on the results of an annual comparative study of 450 cities in the world. Data evaluation is carried out according to 39 factors grouped into 10 categories, including the level of transport service quality (in the category "Public services and transportation / electricity, water, public transportation, traffic congestion, etc.).

Let's consider the characteristics of the planning structures and transport networks of these cities and try to identify the differences inherent in cities with high quality transport service.

4. Main geometric characteristics of the transport network

As the main characteristics of the transport network, we considered the following indicators: the length of the transport network (km), the average length of the streets (km), the number of sections of the road network, the number of intersections. The obtained indicators show a general picture of the development of the transport network for each section, and also are the basis for calculating the more complex integral characteristics of transport networks.

The length of the transport network is the total length of all sections of the road network. It is measured along the street axis in one direction, regardless of the number of lanes [3]. Graphs of the obtained indicators of the transport network for various types of planning structures are presented in Figure 2.

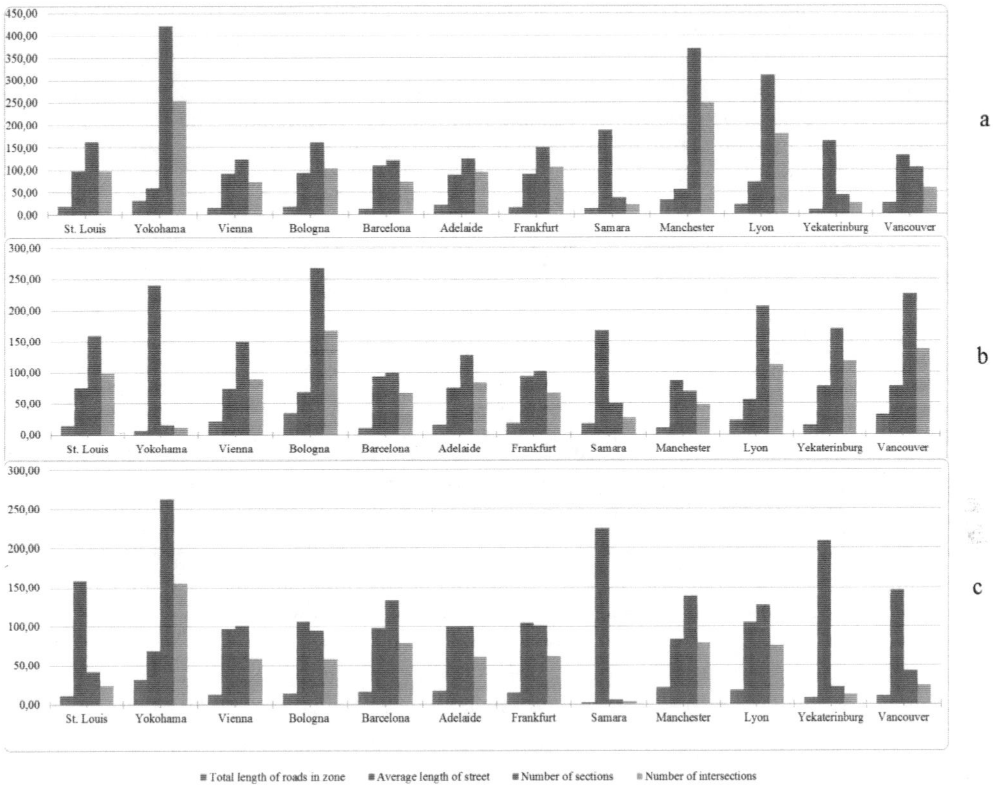

Figure 2. Main indicators of the transport network for various types of planning structures (a) regular structure; (b) scenic structure; (c) combined structure.

Analyzing the graphs of indicators of transport networks on a regular layout, we can see that for the cities of Yokohama, Manchester and Lyon there is a large number of sections of the road network and intersections with low values of the average street length and the total length of roads, which indicates an increased density of the transport network on considered sites. Regular and combined planning of the cities of Yekaterinburg, Vancouver and Samara is characterized by a high value of the average length of the street, which can adversely affect the traffic load of these cities. Considering the regular planning structures of the cities of St. Louis, Barcelona and Lyon, we can see that the geometry of the quarter is close to the square, but this fact does not affect the values of the main indicators of transport networks.

Areas of the territory with a scenic layout of the cities of Yokohama and Samara are distinguished by an increased value of the indicator of the total length of roads. It is worth noting that in both cities the scenic planning structure on the considered sites was formed in relatively recent times with unregulated development of the territory; in the other cities the scenic parts were formed in the Middle Ages. The scenic layouts of Vancouver and Adelaide are characterized by a high degree of curvature of the roads, but this fact does not affect the picture of the main characteristics.

Studying the combined planning structures of the cities of Vienna, Bologna, Frankfurt and Adelaide we obtained a similar picture of the main indicators, determined by the correlated values of the average length of the street and the number of road network sites, we note that in these cities the combined planning is closest in its configuration to the radial-circular one.

5. Calculated transport network characteristics

To calculate the characteristics of the transport network, we represent the transport network as a connected planar graph G, having a certain number of vertices n and edges m. As the estimated characteristics of the transport network we consider the density of the road network, network connectivity, and Engel coefficient.

The density of road network is one of the main indicators of the development of the city's road network and is determined by the ratio of the total length of streets and roads (L) to the built-up area of the territory (S) according to the following formula [4]:

$$D = \frac{L}{S} \tag{1}$$

The density of the network reflects the transport security of the territory and is measured by the number of kilometers of the network per 1000 km^2 of urbanized territory.

Increasing the density of the transport network increases the value of the indicator of service of the population of the city, but worsens the comfort of living. According to the current town planning standards (TSN 300-301-2002), the recommended density of transport networks in cities is accepted in the range of 4–5.5 km / km^2 [5].

The rank of network connectivity is the number of network links that can be removed (or blocked) without breaking the network into isolated fragments. Thus, the rank of network connectivity determines the number of possible alternative routes on the transport network [6].

A network is called connected if for any two of its vertices *n1* and *n2* there is a path from the vertex *n1* to the vertex *n2*.

A connected network component is a certain subset of points such that for any two points of this set there is a path from one to another, and there is no way from the point of this set to a point that does not belong to this set.

Let a graph of a network *G* with *n* vertices and *m* edges contain *k* connected components. Then the rank of the connectivity of the network *R* is determined by the Euler formula:

$$R(S) = m - n + k \tag{2}$$

When presenting a road network as a graph, similar terms from graph theory can be used. In this case, the network connectivity rank is the cyclomatic number of the graph.

The zero value of the rank of connectivity between two points means the absence of a path between them. If the rank of the connection is equal to one, then the route has no alternative, i.e. there is only one path between two points.

Low network connectivity increases the likelihood of congestion and eliminates the use of smart transport systems, so one of the significant tasks of network design is to increase the connectivity rank [7]. The rank of connectivity should also be taken into account in the management of transport infrastructure [8]. Modern studies of the dependence of the connectivity on the accident rate on the road network determine that higher connectivity is associated with lower accident rates with the participation of pedestrians and cyclists [9, 10].

The meshedness coefficient (Gamma index) is an important property that evaluates the complexity and structure of the transport network with cycles of arbitrary lengths [11-13].

The entire network is represented through a set of elementary loops. The following elementary cycles are distinguished: triangular (length is equal to three), regular (length 4), hexagonal (length 6). The meshedness coefficient is determined by the formula:

$$M = F/F_{max} \tag{3}$$

where *F* – number of faces (without outer face) of a planar graph,

N – number of vertices,

K – number of edges by Euler formula ($F = K — N + 1$).

Fmax – maximum possible number of faces in a connected planar graph having the same number of vertices *N* and edges K_{max}= 3*N* — 6. Thus, F_{max} = 2*N* — 5.

The value of the meshedness coefficient can take a value in the range [0,1]. In this case, the value zero corresponds to tree structure, and the value one is a connected graph.

IOP Conf. Series: Materials Science and Engineering **786** (2020) 012053 doi:10.1088/1757-899X/786/1/012053

Thus, the meshedness coefficient and the complexity of the structure are directly proportional. The Engel coefficient is interpreted as an indicator of public service and is used to assess the provision of a territory with a transport network. Calculated by the following formula:

$$k_E = \frac{L}{\sqrt{S \cdot H}} \tag{4}$$

where L – length of transport network in the region, km; S – territory space, km^2; H – population, thousands people.

The calculation of the Engel coefficient was made only for the urban area as a whole, which is due to the lack of accurate population data for sites with regular and scenic planning structures. The values of the estimated indicators of transport networks obtained in the framework of the study are presented in Figure 3.

We can see on the presented graphs that there is no correlation between the density of road network, the meshedness coefficient and the type of planning structure. However, for the planning structure of the combined type, the rank of connectivity is an order of magnitude higher than for the planning structures of other types, which indicates an increased reliability of the transport network on the combined structure.

The cities of St. Louis, Manchester, Adelaide, Vancouver are leading in terms of the high Engel coefficient, which means a decent level of accessibility to the transport infrastructure in these cities. In the planning structures of these cities, an increased number of diagonal links are formed, forming the shortest paths of the peripheral points with the central regions of the city. It can also be concluded that the Engel coefficient or transport accessibility does not affect the level of traffic in the transport networks.

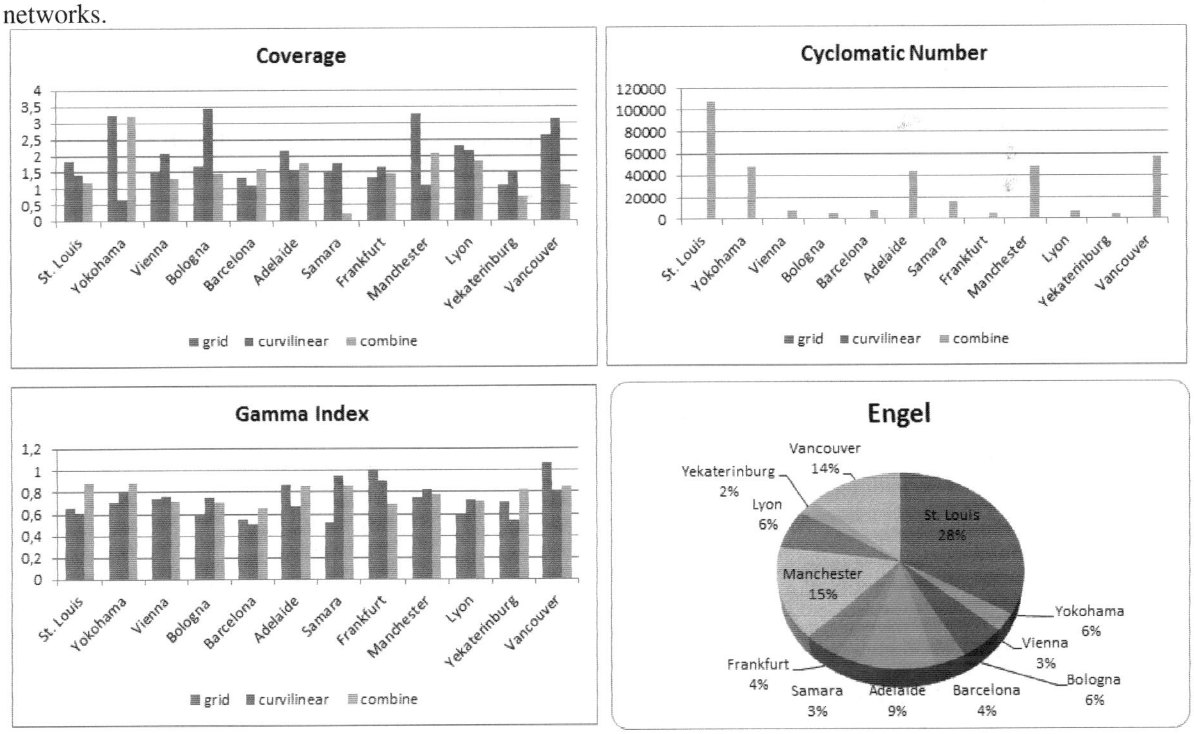

Figure 3. Calculated indicators of transport networks.

6. Conclusion

This paper presents a method for analyzing the planning structures of urbanized areas based on geometric indicators of graph theory. The proposed method is based on grapho-analytical technique, which is one of the key tools for assessing and analyzing the urban development situation of an urbanized space.

The proposed method allows to:

- explore the existing or emerging planning structure in a specific area of urbanized territory;
- calculate objective indicators of transport networks on the presented planning structures;
- based on the calculated indicators, form an understanding of the quality of transport infrastructure and the level of transport accessibility at the studied site;
- make correct and reasoned design decisions when choosing a planning structure based on knowledge of the patterns of their operation.

The limited number of considered cities makes it impossible to draw an objective conclusion about the dependence of geometric indicators on the type of planning structure. Nevertheless, the proposed hypotheses indicate the necessary direction of research.

According to the results of the study, the following hypotheses were proposed:

- the main geometric indicators of the planning structure do not depend on its type;
- there is no correlation between the road network density, the meshedness coefficient and the type of planning structure;
- for a planning structure of the combined type, the rank of connectivity is an order of magnitude higher than for planning structures of other types;
- an increase in the number of diagonal links leads to an increase in transport accessibility;
- transport accessibility does not affect the level of congestion of transport networks;
- trunk networks of cities that have a radial-circular configuration, have similar values of average street length and number of sections of road network.

The direction of further research should be related to the justification of the proposed hypotheses. For these purposes, it is necessary to study planning structures for a larger number of urbanized sites in various cities of the world with the involvement of Big Data technologies.

References

[1] Measuring Congestion Worldwide, Information on: http:// www tomtom com/ (viewed on 12 01 2018)
[2] Quality Of Living City Ranking, Information on: https://mobilityexchange mercer com/
[3] Bulavina L V 2013 Design and evaluation of the transport network and route system in cities [Proectirovanie i ocenka transportnoy seti i marshrutnoy system v gorodah], Yekaterinburg: Ural Federal University, 48 (in Russian)
[4] Mihailov A U 2004 Modern trends in design and reconstruction of street-road networks of cities [Sovremenye tendencii proectirovaniay i rekonstrukcii ulichno-dorojnih setey gorodov] Novosibirsk: Science, 267 p (in Russian)
[5] Database of estimated norms TSN / Normative TSN 300-301-2002 Information on: www complexdoc ru (viewed on 12 01 2018) (in Russian)
[6] Deviatov, M M and Volkov I M 2006 The methodology of the highways functional classification for the modernization purpose [O metodologii funkcionalnoy klassifikacii avtomobilnykh dorog dlia celey ikh modernizacii], *Roads and bridges*, **16(2)** 30-41 (In Russian)
[7] Vukan R Vuchic 1999 Transportation for livable cities, New Brunswick, NJ: Center for Urban Policy Research, Rutgers University
[8] Iliasov B G 2000 Automation of complex objects management in critical situations with the help of an intelligent information system [avtomatizaciay upravlenia slojnimi objectami v criticheskih

situaciah s pomoschiu intelectualnoy informacionoy sistemi], Samara: ICCS RAS, 105-10 (in Russian)

[9] Zhang Y, Bigham J, Li Z, Ragland D and Chen X 2012 Associations between Road Network Connectivity and Pedestrian Bicyclist Accidents, Paper presented at TRB

[10] Gubaidullina, T , Ivanova, N , Absalyamova, S and Yerina, T 2018 Analysis of national strategies for sustainable development with regard to fundamental conceptual premise *In Journal of Physics: Conference Series* **1141(1)** 012018 IOP Publishing

[11] Sakai, H and Naya, K 1992 Relationships between α and β-indices for measuring network connectivity and relationships between the development of forest-road networks and these indices, **70** 245–50

[12] Sakhapov R and Nikolaeva R 2017 Economic aspects of traffic safety administration Transportation Research Procedia 578-83

[13] Sakhapov R L , Nikolaeva R V , Gabdullin T R and Makhmutov M M 2015 Motor roads as a factor of the economic potential of the republic of Tatarstan Procedia Economics and Finance **24** 606

Analysis of contact parameters of large tires with soil for earthmoving machinery

V A Zhulay*, V L Tyunin, N M Volkov and D N Degtev

Voronezh State Technical University, Voronezh, Russia, 84 October's 20 anniversary street

* E-mail: zhulai@vgasu.vrn.ru

Abstract. The article demonstrates the equations for finding the contact width parameters of a large tire with a deformable support surface, such as the radius and angle of the sidewall arc, the sidewall width in the contact area, the track width under the wheel center and at any point of contact. For some large-size tires, the paper discovers graphical dependences of the track width at any point of contact on the internal tire pressure, the vertical wheel load and the tire structure. The authors analyzed the results of these dependencies.

1. Introduction

Earth-moving machines (EMM) separate the soil from the massif and move it. They are equipped not only with a powerful working body, but also with transport running equipment (bulldozers, scrapers, motor graders) [1].

The EMM uses two types of undercarriage equipment: wheeled and crawlers. Wheeled vehicles have several advantages relative to crawlers, namely: high resource; the possibility of developing high transport speeds; significant reduction in metal construction, operating costs, etc. In addition, when relocating, wheeled equipment allows the car to move via public roads.

The parameters of the contact of the wheel propulsion (WP) with the supporting surface determine the main characteristics of the EMM operation such as traction force, rolling resistance, performance and fuel efficiency [2].

Works [3, 4] confirm that the questions of studying the width of a tire profile on the ground are relevant, since EMM during the operation mainly move on soils of different strength. In this regard, the following method was developed to determine the width of the EMM tire profile when driving on the ground [5].

2. The main part

We took the model described in [3, 6, 7] as a basis and made the following assumptions: the wheel moves with constant linear V_{WP} and angular ωK speed along a straight-line trajectory; the pneumatic tire is deformed in the radial direction (only those of its elements that are within the contact area *nm* (Figure 1)); a wheel with a pneumatic tire is represented: outside the contact area, in the form of a cylinder with a width of B_p with radius r_0, in the contact area with associated cylinders with radiuses R_1 and R_2 ($R_2 > R_1$), whose centers are located on a vertical axis passing through the center of rotation of the wheel.

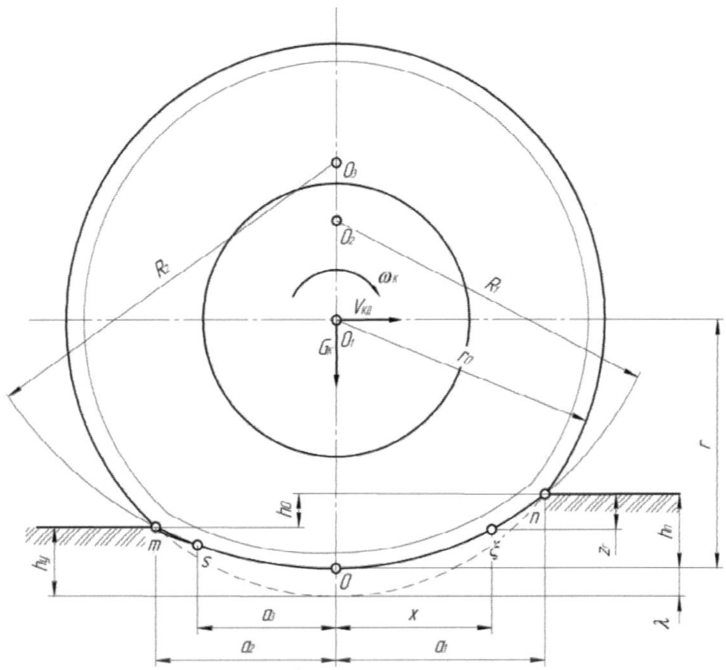

Figure 1. Diagram of the interaction of the wheel propulsion with a deforming support surface.

Total soil deformation h can be determined by the formula:

$$h_{\Pi} = \frac{a_1^2}{2r_0}\left(\frac{K_1}{C_1 + K_1}\right), \quad (1)$$

where a_1 – the length of the loading area of the contact area; r_0 – the radius of the undeformed tire; K_1 – the deformation coefficient of the pneumatic tire during compression; C_1 – the coefficient of complete deformation of the soil..

The amount of soil deformation at any point of contact $z_g(x)$ is determined from the expression:

$$z_{\Gamma}(x) = h_{\Pi} - \frac{x^2}{2R_i}, \quad (2)$$

where x – the current value of the coordinate of the point ξ; R_i – is the current value of the radius in the contact area (in the loading zone $R_i = R_1$, in the unload zone $R_i = R_2$).

Expressions for determining these values depending on the value of the variable x in the contact area are presented in the source [5].

Side wall arc angle is:

$$\alpha_{Б\xi}(x) = 2\sqrt{6\left[1 - \frac{2H_{\xi}(x)}{U - B_{ПР}}\right]}, \quad (3)$$

where $H_\xi(x)$ – tire profile height at the point ξ (Figure 1).

The radius of the side wall is:

$$r_{Б\xi}(x) = \frac{U - B_{ПР}}{2\alpha_{Б\xi}(x)}. \quad (4)$$

IOP Conf. Series: Materials Science and Engineering **786** (2020) 012054 doi:10.1088/1757-899X/786/1/012054

Sidewall width in the contact zone:

$$b_{Б\xi}(x) = r_{Б\xi}(x)\left[1 - \cos\left(\frac{\alpha_{Б\xi}(x)}{2}\right) - \frac{1}{2}\left(\sin\left(\frac{\alpha_{Б\xi}(x)}{2}\right) - \frac{z_{Г}(x)}{r_{Б\xi}(x)}\right)^2\right]. \quad (5)$$

Track width at any point of contact:

$$B_{К\xi}(x) = B_{ПР} + 2b_{Б\xi}(x) \quad (6)$$

Using this technique, we got graphical dependences of the track width $B_{К\xi}(x)$ on the value of variable x for tire of 16.00-24 model YA-140 (static modulus of total deformation of soil E_1 = 5 MPa, internal tire pressure P_ω = 0.3 MPa, vertical load on the wheel $G_К$ = 45 kN) (Figure 2).

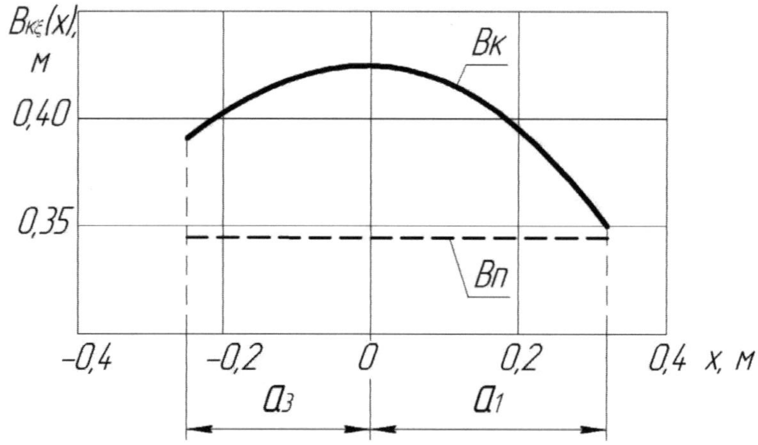

Figure 2. The dependence of the width of the gauge on the value of the variable x for tires of 16.00-24 YA-140 model.

When the tire comes into contact with the supporting surface (Figure 2), the track width increases from the value of the tread width to the maximum value under the center of the wheel. Having passed the center, the width of the track begins to decrease to the value at point a_3, at which the tire comes out of contact with the supporting surface. The difference between the width of the contact $B_{К\xi}(x)$ and the width of the tread B_p for tires 16.00-24 on loose soil at x = 0 is 24.8%, and at x = a_3 is13.7%.

According to this method, we received the graphical dependences of the effect of the internal pressure in the tire (Figure 3), the vertical load on the wheel (Figure 4) and the design of the tire (Figure 5) on the gauge width.

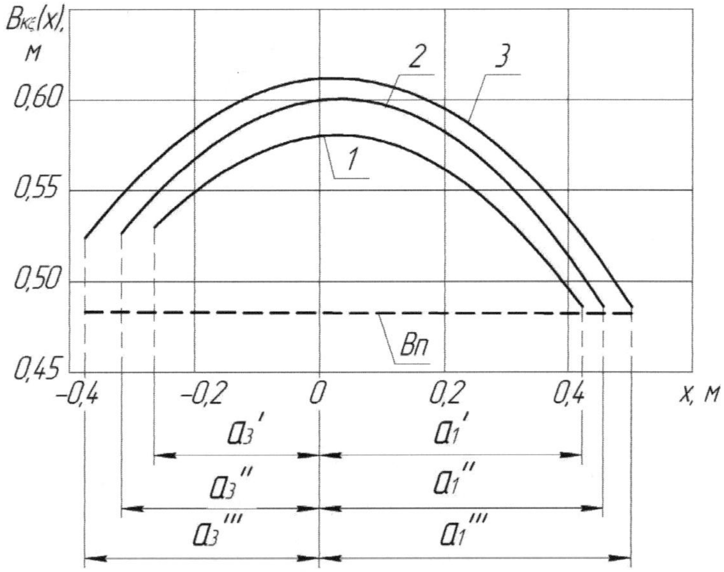

Figure 3. The dependence of the internal pressure in the tire 21.00-28 mod. DF-27 on loose soil: 1 – P_ω = 0,4 MPa, 2 – P_ω = 0,3 MPa, 3 – P_ω = 0,2 MPa.

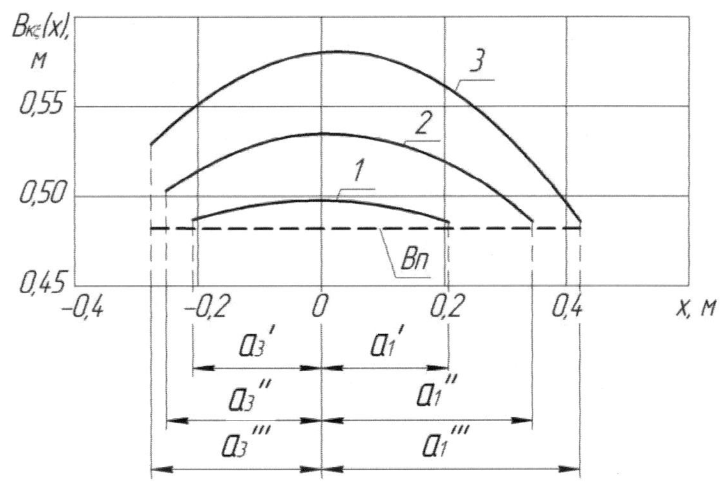

Figure 4. The dependence of the vertical load on the wheel 21.00-28 mod. DF-27: 1 - G_K = 15 kN, 2 - G_K = 50 kN, 3 - G_K = 85 kN.

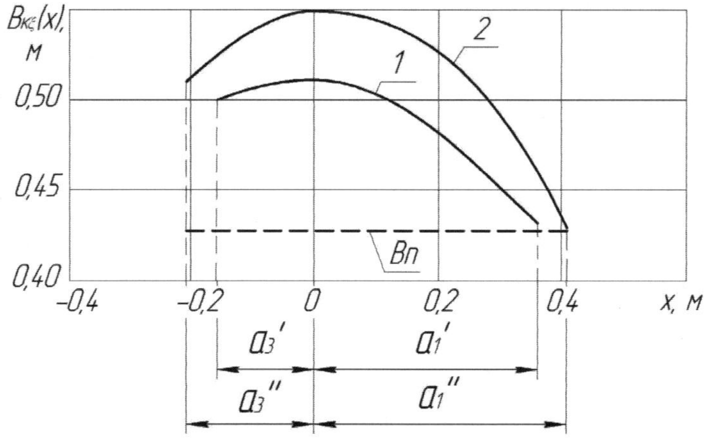

Figure 5. Dependence of the tire design 18.00-25 mod. F-27 on loose soil:
1 - diagonal tire, 2 - radial tire.

After analyzing the presented graphical dependencies, we can draw the following conclusions. A decrease in the internal pressure in the tire P_ω from 0.4 to 0.2 MPa (Figure 3) leads to an increase in $B_{K\xi}(x)$ at $x = 0$ by 7.8%. When the vertical load on the wheel decreases from 85 to 15 kN (Figure 4), the $B_{K\xi}(x)$ decreases by 14.4% at $x = 0$ and decreases by 12% at $x = a_3$. The value of $B_{K\xi}(x)$ in the diagonal construction is smaller than the radial one (Figure 5) at $x = 0$ by 7.2%, and at $x = a_3$ by 2.8%.

3. Conclusion

The parameters of the contact area of a large-sized tire with the soil affect the basic characteristics of the WP EMM, therefore it is necessary to determine and take into account the width of the contact, and it is necessary to conduct experimental studies [8-10] to confirm the dependencies that were obtained analytically.

References

[1] V I Balovnev, S N Glagolev and R G Danilov i dr 2011Mashiny dlya zemlyanyh rabot: konstruktsiya, raschet, potrebitelskie svoystva: v 2 kn Kn 1 Ekskavatory i zemleroyno-transportnye mashiny: uchebnoe posobie dlya vuzov / pod obshch red V I Balovneva – Belgorod: Izd-vo BGTU, – 401

[2] Zhulay V A 2016 Otsenka toplivnoy ekonomichnosti samohodnyh zemleroyno-transportnyh mashin / V A ZHulay, V L Tyunin, A V Krestnikov // *Mekhanizatsiya stroitelstva* **77(8)** 27-31

[3] Ulyanov N A 1962 Osnovy teorii i rascheta kolesnogo dvizhitelya zemleroynyh mashin / N A Ulyanov – M : Mashgiz, 208

[4] Ageykin YA S 1972 Vezdekhodnye kolesnye i kombinirovannye dvizhiteli / M : Mashinostroenie, – 184

[5] Zhulay V A 2015 Opredelenie parametrov oblasti kontakta pnevmoshiny zemleroyno-transportnoy tekhniki s deformiruemoy opornoy poverhnostyu / V A ZHulay, V L Tyunin // *Stroitelnye i dorozhnye mashiny* **8** 51-4

[6] Ulyanov N A 1982 Kolesnye dvizhiteli stroitelnyh i dorozhnyh mashin: Teoriya i raschet /N A Ulyanov – M : Mashinostroenie, 279

[7] Vasilenko A V 2000 Metodika rascheta tyagovyh kachestv kolesnogo dvizhitelya s krupnogabaritnymi pnevmaticheskimi shinami zemleroyno-transportnyh mashin : dis kand tekhn nauk / Vasilenko Andrey Vladimirovich – Voronezh, 220

[8] Zhulay V A 2015 Komplekt mobilnoy tsifrovoy apparatury dlya obespecheniya provedeniya eksperimentalnyh issledovaniy zemleroyno-transportnyh mashin *Mekhanizatsiya stroitelstva* **8(854)** 14-6
[9] Sharipov L H 1986 Issledovanie tyagovyh i ekspluatatsionnyh pokazateley avtogreydera DZ-146 / L H SHaripov, YU M Buzin, V A ZHulay // *Stroitelnye i dorozhnye mashiny* **10** 4-5
[10] Sakhapov R L , Nikolaeva R V , Gatiyatullin M H and Makhmutov M M 2017 Modeling of traction-coupling properties of wheel propulsor *Journal of Physics: Conference Series* **6** "6th International Conference on Mathematical Modelling in Physical Sciences, IC-MSQUARE 2017" 012033

Synthesis of the six-part spatial mechanism

A G Mudrov* and R L Sakhapov

Kazan State University of Architecture and Engineering, 420043, Kazan, Zelenaya str., 1, Russia

*E-mail: Alexmudrov42@rambler.ru

Abstract. The synthesis of flat and spatial mechanisms is of a great importance in the development of various devices, including robotics, automated lines, etc. These questions are set out in the works of Russian scientists (on the structure of the mechanisms, the synthesis) as well as in the foreign science. The purpose of this article is to synthesize a six-part spatial mechanism to reproduce the linear function of the slave link, which takes place at certain angles of rotation of the drive and slave links in devices with linear functions. The authors used the method of uniform best approximation of functions, graphical method, analytical theory of infinitesimal flat displacements of a rigid body, quadratic approximations, and analytical theory of a finite number of positions of a solid body.

1. Introduction

The synthesis of mechanisms is a section of the theory of mechanisms and machines, which examines the methods for designing a mechanism scheme for given kinematic and dynamic properties.

The synthesis of flat and spatial mechanisms is of a great importance in the development of various devices, including robotics, automated lines, etc. These questions are set out in the works of Russian scientists (on the structure of the mechanisms [1,2], the synthesis [3,4,5]) as well as in the foreign science [6,7,8]. Considering the synthesis, we used the analysis of the mechanisms according to P. L. Chebyshev (method of uniform best approximation of functions), graphical method, analytical theory of infinitesimal flat displacements of a rigid body, quadratic approximations, and analytical theory of a finite number of positions of a solid body. Along with this, the general questions of the theory of mechanisms are explained in the following sources [9,10,11,12].

The synthesis of spatial mechanisms with only rotational hinges has its own specifics. First, it is necessary to solve the problem of the mechanism existence, i.e. it is necessary to determine the linear and angular parameters of the links and the relationship between them, the turnability of the links. Then it's time to solve one or another synthesis problem. Professor P. G. Mudrov made a great contribution to the synthesis of spatial mechanisms with rotational hinges [13].

It should be noted that the kinematics of the output link of the spatial mechanisms with rotational hinges depend primarily on the angular parameters of the links (the angles of crossing of the axes of the hinges of the links). From the linear parameters, the length of one of the links is set according to needs, based on the design of the hinge nodes, while the other links are determined by the structural formulas.

IOP Conf. Series: Materials Science and Engineering **786** (2020) 012055 doi:10.1088/1757-899X/786/1/012055

Additional information on spatial mechanisms can be obtained in [14-18].

The purpose of this article is to synthesize a six-part spatial mechanism to reproduce the linear function of the slave link, which takes place at certain angles of rotation of the drive and slave links in devices with linear functions.

2. The main part

The analysis of spatial mechanisms with rotational hinges showed that a mechanism with the following structural parameters [13] (Figure 1) can be used to reproduce a linear function on some part of the rotation of the slave link with its full turnability:

$$\left.\begin{aligned} &\alpha_2 = 180^0 - 2\alpha_6,\ \ell_2 = 0,\ \alpha_3 = \alpha_5 - \alpha_4,\ \ell_3 = \ell_5 - \ell_4,\ \alpha_4 = 180^0 - \alpha_6, \\ &\ell_4 = \ell_6,\ \frac{\ell_1}{\sin\alpha_1} = \frac{\ell_5}{\sin_5} = \frac{\ell_6}{\sin\alpha_6} \end{aligned}\right\} \quad (1)$$

The angle of rotation of the slave link mechanism is

$$\left.\begin{aligned} \psi_5 = {} & \arcsin\frac{(\cos\alpha_6 - \cos\alpha_1)\sin\varphi_1}{1-\cos\alpha_1\cos\alpha_6 - \sin\alpha_1\sin\alpha_6\cos\varphi_1} + \arcsin\frac{(\cos\alpha_6 + \cos\alpha_1)\sin\varphi_1}{1-\cos\alpha_1\cos\alpha_6 - \sin\alpha_1\sin\alpha_6\cos\varphi_1} + \\ & + \arcsin\frac{-(\cos\alpha_6 + \cos\alpha_5)\sin\varphi_1}{1-\cos\alpha_5\cos\alpha_6 - \sin\alpha_5\sin\alpha_6\cos\varphi_1}. \end{aligned}\right\} \quad (2)$$

Where α_6- angle of crossing of the axles of the frame pivots (lies between 65^0 and 85^0);

α_5- the angle of crossing of the axles of the driven crank joints (lies from (213^0-α_6) to (183^0-α_6)).

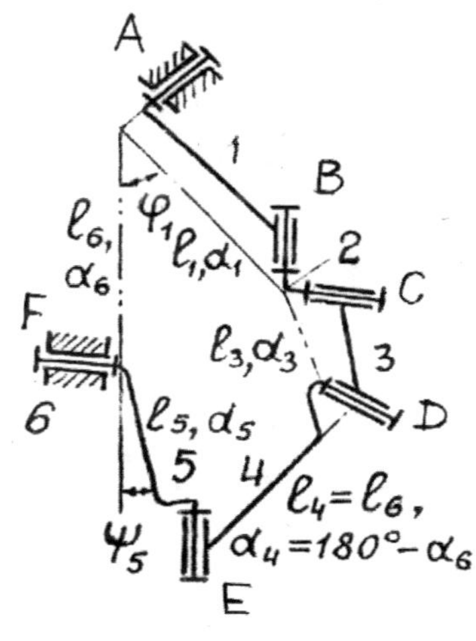

Figure 1. Scheme of the mechanism.

The graph of dependence of the rotation angle of the driven crank on the angle of rotation of the leading one is shown in Figure2. As can be seen from the graph, the curve is symmetric with respect to point a, therefore, the segment of the line is cb=2ab, and angle is $\varphi_p=2\varphi'_P(\psi_P=2\psi'_P)$.

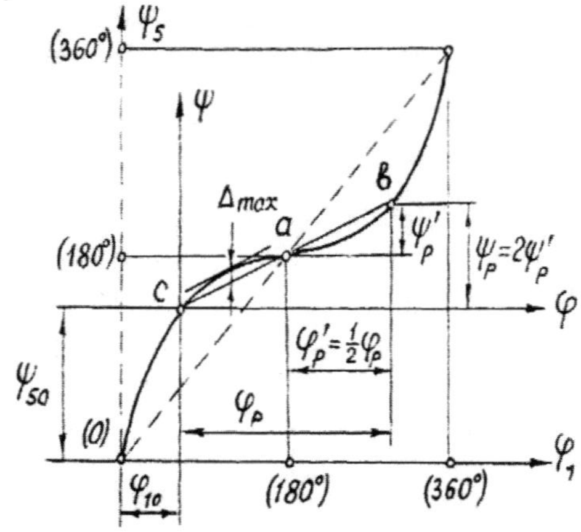

Figure 2. Graph of the slave (driven) link movement.

We know the angle φ_P (φ_P=180^0…300^0), and the relation of φ_P and ψ_P, i.e.

$$K = \frac{\psi_P}{\varphi_P} = \frac{\psi_P^{'}}{\varphi_P^{'}} = 0,1...0,5.$$

It is necessary to synthesize the mechanism so that the maximum deviation Δ_{max} of the curve $\psi_5=f(\varphi_1)$ in the segment **cb** does not exceed the specified value determined by the relative error.

$$\Delta\psi^0 = \frac{\Delta_{max}}{\psi_P} 100\% \leq \Delta\psi.$$

Considering the accuracy let's assume $\Delta\psi$ = (0,2…0,5)%.

Additionally, it is necessary to create an equation relating the input and output parameters of the synthesis mechanism. The equation is derived by solving a vector equation (Figure 3),

$$\overline{AB} = \overline{AF} + \overline{FE} + \overline{ED} + \overline{DC}. \quad (3)$$

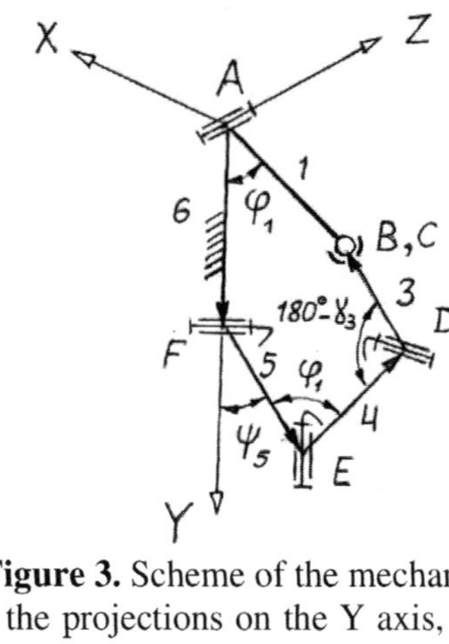

Figure 3. Scheme of the mechanism.

Solving the vector equation in the projections on the Y axis, substituting the value of unit vectors and taking into account that

$$\left.\begin{aligned} &\varphi_1 = 180^0 + \varphi_P^{'} = 180^0 + \varphi_P/2,\ \psi_5 = 180^0 + \psi_P^{'} = 180^0 + K\varphi_P/2,\ \frac{\ell_1}{\sin\alpha_1} = \frac{\ell_5}{\sin\alpha_5} = \frac{\ell_6}{\sin\alpha_6}, \\ &\sin\gamma_3 = \frac{-(\cos\alpha_6 - cos\alpha_6)\sin\varphi_1}{1+\cos\alpha_5\cos\alpha_6 - \sin\alpha_5\sin\alpha_6\cos\varphi_1}, \cos\frac{(1+\cos\alpha_5 cos\alpha_6)\cos\varphi_1 - \sin\alpha_5\sin\alpha_6}{1+\cos\alpha_5\cos\alpha_6 - \sin\alpha_5\sin\alpha_6\cos\varphi_1}, \end{aligned}\right\}$$

and

$$\left.\begin{aligned}&\ell_1\cos\varphi_1=\ell_6+\ell_5\cos\psi_5-\ell_6(\cos\varphi_1\cos\psi_5-\cos\alpha_5\sin\varphi_1\sin\psi_5+\\&+(\ell_5-\ell_1)\left\{\begin{aligned}&\left[\cos\alpha_5\sin\varphi_1\cos\gamma_3-(\sin\alpha_5\sin\alpha_6-\cos\alpha_5\cos\alpha_6\cos\varphi_1)\sin\gamma_3\right]\sin\psi_5-\\&-(\cos\varphi_1\cos\gamma_3-\cos\alpha_6\sin\varphi_1\sin\gamma_3)\cos\psi_5\end{aligned}\right\}.\end{aligned}\right\}\quad(4)$$

the angle α_1 is determined

$$\alpha_1=\arcsin\frac{\sin\alpha_6\left[(1-\cos\frac{\varphi_P}{2}\cos\frac{K\varphi_P}{2})(1+\cos\alpha_6\cos\alpha_5)+\sin\frac{\varphi_P}{2}\sin\frac{K\varphi_P}{2}(\cos\alpha_6+\cos\alpha_5)+\sin\alpha_6\sin\alpha_5(\cos\frac{\varphi_P}{2}-\cos\frac{K\varphi_P}{2})\right]}{(\cos\frac{K\varphi_P}{2}-\cos\frac{\varphi_P}{2})(1+\cos\alpha_6\cos\alpha_5+\sin\alpha_6\sin\alpha_5\cos\frac{\varphi_P}{2})-\sin\alpha_6\sin\frac{\varphi_P}{2}\left[\begin{aligned}&\sin\frac{K\varphi_P}{2}(\sin\alpha_6\cos\alpha_5\cos\frac{\varphi_P}{2}-\\&-\cos\alpha_6\sin\alpha_5)+\sin\alpha_6\sin\frac{\varphi_P}{2}\cos\frac{K\varphi_P}{2}\end{aligned}\right]}\quad(5)$$

Maximum deviation Δ_{max} of the curve will be written as

$$\left.\begin{aligned}\Delta_{max}&=\arcsin\frac{(\cos\alpha_6-\cos\alpha_1)\sin\varphi_k}{1-\cos\alpha_1\cos\alpha_6+\sin\alpha_1\sin\alpha_6\cos\varphi_k}+\arcsin\frac{(\cos\alpha_6+\cos\alpha_1)\sin\varphi_k}{1+\cos\alpha_1\cos\alpha_6+\sin\alpha_1\sin\alpha_6\cos\varphi_k}+\\&+\arcsin\frac{(\cos\alpha_6+\cos\alpha_5)\sin\varphi_k}{1+\cos\alpha_5\cos\alpha_6+\sin\alpha_5\sin\alpha_6\cos\varphi_k}-K\varphi_k\end{aligned}\right\}\quad(6)$$

The angle φ_k is found from the equality of the partial derivative $\frac{d\psi_5}{d\varphi_1}$ to the slope K, i.e.

$$\left.\begin{aligned}\frac{d\psi_5}{d\varphi_1}&=\frac{\cos\alpha_6-\cos\alpha_1}{1-\cos\alpha_1\cos\alpha_6-\sin\alpha_1\sin\alpha_6\cos\varphi'_k}+\frac{\cos\alpha_6+\cos\alpha_1}{1+\cos\alpha_1\cos\alpha_6-\sin\alpha_1\sin\alpha_6\cos\varphi'_k}-\\&-\frac{\cos\alpha_6-\cos\alpha_5}{1+\cos\alpha_5\cos\alpha_6-\sin\alpha_5\sin\alpha_6\cos\varphi'_k}=K,\end{aligned}\right\}$$

where $\varphi_k'=180^0+\varphi_k$.

The sequence of the synthesis is as follows:

We set the working section of the angle φ_p of the driving crank, as well as the extreme upper values of the coefficient K and the angles α_5 and α_6. The angle α_1 is determined by the formula (5), with the check of the restriction $\alpha_1 \geq 3^0$. Then we calculate the objective function $\Delta\psi^0$, which is compared with the lower value of $\Delta\psi$, and the condition

$$\Delta\psi^0 \leq \Delta\psi \text{ or } \Delta\psi_{i+1}{}^0 - \Delta\psi_i{}^0 \geq 0. \quad (7)$$

If this condition is not met, then the values of the angles α5 and α6 are sequentially reduced by a certain amount and the process is repeated until one of the conditions (7) is obtained.

If all values of K, α5 and α6 are enumerated, and condition (7) is not met, then the value of $\Delta\psi$ increases by a certain amount and the process is repeated again until the required condition (7) is fulfilled.

After determining K, α1, α5 and α6, we find the other parameters of the mechanism (α2, α3, α4, linear parameters) using the above formulas (1).

The size of the working, straight-line section of the driven crank turn is determined by ψp = Kφp.

The beginning of new system's ψ0φ coordinates (and point C – beginning of the straight-line section) ψ50 = 1800- 0,5ψp , φ10 = 1800- 0,5φp.

Example. We need to synthesize a mechanism, which reproduces linear function on the segment determined by the angle φp=1800, with the accuracy Δψ=0,5% and K=0,5…0,1.

Let's accept the values α6=750, α5=1100 and K=0,25. Then we determine the angle α1 using (5)

$$\alpha_1 = \arcsin \frac{\sin 75^0 \begin{bmatrix} (1-\cos\frac{180^0}{2}\cos\frac{0,25\cdot180^0}{2})(1+\cos 75^0\cos 110^0)+ \\ +\sin\frac{180^0}{2}\sin\frac{0,25\cdot180^0}{2}(\cos 75^0+\cos 110^0)+ \\ +\sin 75^0\sin 110^0(\cos\frac{180^0}{2}-\cos\frac{0,25\cdot180^0}{2}) \end{bmatrix}}{\begin{matrix}(\cos\frac{0,25\cdot180^0}{2}-\cos\frac{180^0}{2})(1+\cos 75^0\cos 110^0+ \\ +\cos\frac{180^0}{2}\sin 75^0\sin 110^0)-\sin 75^0\sin\frac{180^0}{2}\begin{bmatrix}\sin\frac{0,25\cdot180^0}{2}(\sin 75^0\cos 110^0\cos\frac{180^0}{2}- \\ -\cos 75^0\sin 110^0)+\sin 75^0\sin\frac{180^0}{2}\cos\frac{0,25\cdot180^0}{2}\end{bmatrix}\end{matrix}} == \arcsin\frac{0,039658}{0,700063} = 34,506^0.$$

Determining angle φ_k, after intermediate value conversions we can see φ_k equals:

$$\varphi_k = \arccos x_1 = \arccos 0,632664 = 50,75^0,$$

$$\varphi_k = \arccos x_2 = \arccos - 0,695432 = 134,06^0,$$

$$\varphi_k = \arccos x_3 = \arccos - 0,445834 = 116,47^0.$$

Since the angle φ_k must lie within $0\ldots90^0$, then the angle $\varphi_k=50,75^0$ will be valid. Now, by expression (6), Δ_{max} is determined as

$$\Delta_{max} = \arcsin\frac{(\cos 75^0-\cos 34,5^0)\sin 50,75^0}{1-\cos 75^0\cos 34,5^0+\sin 75^0\sin 34,5^0\cos 50,75^0} + \arcsin\frac{(\cos 75^0+\cos 34,5^0)\sin 50,75^0}{1+\cos 75^0\cos 34,5^0+\sin 75^0\sin 34,5^0\cos 50,75^0} -$$

$$-\arcsin\frac{(\cos 75^0+\cos 110^0)\sin 50,75^0}{1+\cos 75^0\cos 110^0+\sin 75^0\sin 110^0\cos 50,75^0} - 0,25\cdot 50,75^0 = 0,410507^0.$$

The relative error will be

$$\Delta\psi^0 = \frac{\Delta_{max}}{K\varphi_P}100\% = \frac{0,410507}{0,25\cdot180^0}100\% = 0,91\%.$$

The obtained value of the relative error is greater than the required $\Delta\psi = 0.5\%$, i.e. $\Delta\psi 0 > \Delta\psi$, therefore it is necessary to enter the correction and accepting the other values of the parameters, repeat the calculation.

Having accepted $\alpha_5=108^0$ and $K=0,225$, we leave the angle $\alpha_6=75^0$ the same.

Counting the values gave the following final results.

$$\alpha_1 = \arcsin\frac{0,039388}{0,070148} = 34,18^0,\ \varphi_k = 42,63^0,\ \Delta_{max} = 0,202909$$

The relative error will be

$$\Delta\psi^0 = \frac{\Delta_{max}}{\varphi_P}100\% = \frac{0,202909}{0,225\cdot180^0}100\% = 0,5\%.$$

The error corresponds to the specified.

Thus, the parameters of the synthesized six-part spatial mechanism have the following meanings:

$\alpha_1 = 34,18^0$, $\alpha_2 = 180^0 - 2\cdot 75^0 = 30^0$, $\alpha_3 = 108^0 - 34,18^0 = 73,81^0$, $\alpha_4 = 180^0 - 75^0 = 105^0, \alpha_5 = 108^0, \alpha_6 = 75^0$,

$\ell_2 = 0, \ell_4 = \ell_6 = 100\text{мм}, \ell_1 = \ell_6 \sin 34,18^0/\sin 75^0 = 58,16\text{мм}, \ell_5 = \ell_6\sin 108^0/\sin 75^0 = 98,46\text{мм}, \ell_3 = \ell_5 - \ell_4 = 40,3\text{мм}.$

3. Conclusion

The six-part spatial mechanism with rotational hinges with the specified parameters ensures the reproduction of a linear function with sufficiently high accuracy and can be recommended for practical use in engineering, for example, in robots or manipulators.

The synthesis of spatial mechanisms with rotational joints is carried out in a certain sequence. First of all, the question of the conditions for turning the links of the mechanisms is solved by analyzing their average gear ratio. If it is zero, then the slave link will be the balance bar, if one - it will be a crank, if two, it will be the crank that makes two turns per master drive, etc.

The multi-functionality of the mechanisms also generates various problems of their synthesis that can be solved in one way or another.

In addition to reproducing a linear function, we can solve some other problems with the synthesis of mechanisms, for example: to ensure the movement of a slave link with a given variable law, characterized by a given degree of non-uniform rotation; synthesis of mechanisms for the given positions of the master and slave units; synthesis according to the elliptic law of motion of the slave link; ensuring the movement of an excavator shovel from the condition of undermining of soil; synthesis by axial displacement of the working shaft of the screw mixer from the middle position, etc.

References

[1] Pozhbelko V I 2014 Metod sistemnogo strukturnogo analiza podvizhnosti ploskih i prostranstven¬nyh mekhanizmov v opredelennom prostranstve ih dvizheniy / V I Pozhbelko // *Teoriya mekhanizmov i mashin* **1** 70-80
[2] Hrostitskiy A A 2011 Osobennosti struktury i geometrii prostranstvennogo shestizvennogo mekha¬nizma s izbytochnymi svyazyami Sovremennoe mashinostroenie – Sankt-Peterburg: SPbGPU, 399 - 409
[3] Artobolevskiy I I 1933 Sintez i kinematicheskoe issledovanie sfericheskih mekhanizmov -*Trudy instituta selskohozyaystvennogo mashinostroeniya*, **1**
[4] Artobolevskiy I I , Levitskiy N I 1959 Sintez ploskih mekhanizmov M : Fizmatgiz, -1084
[5] Yarullin M G 2013 Sintez strukturnyh modifikatsiy mekhanizma Bennetta Trudy KNITU, 271 - 80
[6] Brunnthaler, K , Schrocker, H-P , Husty, M 2005 A New Method for the Synthesis of Bennett Mechanisms / University Innsbruck, Austria // Proceedings of CK2005, International Workshop on Computational Kinematics – Cassino, 53–61
[7] Perez, A and McCarthy J M 2003 Dimensional Synthesis of Bennett Linkages, *Transactions of the ASME*, **125** 98-104
[8] Oliveira, Jr A A , Carvalho and J C M 2007 Modeling of the Bennett's linkage as leg of a mobile robot, 12 th IFToMM World Congress, Besancon 1- 6
[9] Dimentberg F M 1982 Teoriya prostranstvennyh sharnirnyh mekhanizmov – M : Nauka Glavnaya redaktsiya fiziko-matematicheskoy literatury, 336
[10] 100 let mekhanizmu Bennetta 2004/ Materialy mezhdunarodnoy konferentsii po teorii mekhanizmov i mashin – Kazan: RITS «Shkola», 292
[11] Nastase, A and Bocioaca, R 2004 Utilisarea Programului AutoCAD Pentru Generarea Configuratiilor mechanismului Bennett, Simposionul national cu participare international Proieectarea Asistata de Calculator, Brasov 237-40
[12] Jingfang, Liu, Yueqing, Yu, Zhen, Huang and Xiao'ou Huang 2013 General Order Principle for Multi-Bennett Linkages, *Chinese jornal of mechanical engineering*, **26(1)** 1-7
[13] Mudrov P G 1976 Prostranstvennye mekhanizmy s vrashchatelnymi parami – Kazan: Izd- vo Kazanskogo gosudarstvennogo universiteta, –264 s
[14] Mudrov A G and Jarullin M G 2003 From the Bennett mechanism to the differential gear Educational book Kazan, Kazan University press, – 92
[15] Chen, Y and Baker, E J 2004 *Using a Bennett linkage as a connector between other Bennett loops,* Proc IMechE, **219** 177-85
[16] Mudrov A G 2003 Prostranstvennye mekhanizmy s osoboy strukturoy /-RITS «Shkola», Kazan, 300

[17] Sakhapov R L , Nikolaeva R V , Gatiyatullin M H and Makhmutov M M 2017 Modeling of traction-coupling properties of wheel propulsor Journal of Physics: Conference Series 6 "6th International Conference on Mathematical Modelling in Physical Sciences, IC-MSQUARE 012033
[18] Sakhapov R L and Absalyamova S G 2018 Fourth industrial revolution and the paradigm change in engineering education International Scientific Conference on Energy, Environmental and Construction Engineering (EECE-2018) electronic edition "MATEC Web of Conferences" 12003

IOP Conf. Series: Materials Science and Engineering **786** (2020) 012056 doi:10.1088/1757-899X/786/1/012056

On complex formation in chemical copper plating processes from solutions with organophosphorus ligands

R M Mukhametshina*

Kazan State University of Architecture and Engineering, 420043, Kazan, Zelenayastr., 1, Russia

*E-mail: rumya211@yandex.ru

Abstract. Nowadays the processes of chemical vapor deposition of metals are widely used in mechanical engineering, both for the production of materials with desired functional properties and for decorative purposes. One of the most widely used chemical vapor deposition processes is chemical copper plating. The paper aims to study the complexation of copper (II) with nitrilotrimethylene phosphonic acid and to identify the possibility of using nitrilotrimethylene phosphonic acid as ligands of copper (II) ions for obtaining stable solutions of chemical copper plating. The formation of copper (II) complexes with nitrilotrimethylene phosphonic acid has been studied by the electron paramagnetic resonance (EPR) method. The paper proposes a method for processing EPR data – measurements to calculate the distribution of complex forms in the form of dependence = dP/dpH – f(pH), where P – the EPR spectrum parameter, which seems promising for describing paramagnetic systems characterized by complex equilibria with small changes in the magnetic parameters relative to the line width. The significance of the results for the engineering industry is that the authors studied the new materials based on copper complexes with nitrilotrimethylene phosphonic acid. The article establishes compositions and calculates the stability constants of the complexes at various pH of the medium. The research also shows the possibility of obtaining high-quality metal coatings based on copper (II) complexes with nitrilotrimethylene phosphonic acid.

1. Introduction

It is hard to imagine modern development of the engineering industry without the use of the achievements of chemistry in the field of the synthesis of new chemicals and materials. In recent decades, metal-polymer materials formed by a chemical method have been widely used in mechanical engineering to produce a variety of protective and decorative coatings [1-4]. The advantages of chemical deposition, unlike other methods of coating (electroplating, cladding, thermal spraying, metallization in vacuum) are the possibility of obtaining a uniform coating on products of complex configuration, availability of materials, simplicity of equipment, low energy costs.

In solutions of chemical vapor deposition of metals, the nature of the complex-forming ligand plays an important role for ensuring the required stability of solutions. The stability of solutions is related to the strength of the formed complex ions and their spatial structure. In recent years, organophosphorous

IOP Conf. Series: Materials Science and Engineering **786** (2020) 012056 doi:10.1088/1757-899X/786/1/012056

complexones have attracted a lot of attention from researchers. They are widely used in various areas of the economy, including mechanical engineering [5,6].

The literature has information about the prospects for the use of organophosphorus ligands in the processes of chemical deposition of metals [7-10]. One of the most interesting representatives of complexing compounds with phosphonic groups is nitrilotrimethylene phosphonic acid (NTP). The combination of a high basicity nitrogen atom with phosphonic groups in a ligand molecule causes the formation of both stable mononuclear and strong polynuclear complexes with most metal ions. The availability of acid synthesis opens up wide possibilities for its practical use [11,12].

The existing solutions of chemical deposition of metals have insufficient stability and therefore the issues of increasing their stability and reusability remain highly relevant [13–15]. In this regard, the study of the complex formation processes in solutions of chemical copper plating based on nitrilotrimethylene phosphonic acid is becoming important in order to optimize the technological processes of metal plating and determine the methods of controlling these processes.

2. Methods

To study the complex fromation of copper with nitrilotrimethylenephosphonic acid, we used the electron paramagnetic resonance (EPR) method [16–18]. The electron paramagnetic resonance method is one of the most sensitive to changes in the structure and composition of the first coordination sphere of the complexing metal. The main parameters of the EPR spectra are the intensity and width of the lines, the g-factor and the constant of the hyperfine interaction, which depend on the symmetry of the nearest surrounding of the central ion and on the nature of the ligand.

This paper, basing on the analysis and processing of the parameters of the EPR spectra of liquid and frozen solutions, discovers the probable structure of the resulting complex ions, and calculates the equilibrium constants of complex formation. The EPR spectra were recorded on an RE-1306 spectrometer equipped with a magnetic field calibrator assembled on the basis of a magnetic induction meter III-I-I. The samples were placed in calibrated capillaries of glass, which did not give an EPR signal in the working region of magnetic fields. The EPR spectra were recorded at 298 and 77 K in the three-centimeter wavelength range. In some cases, we took measurements at elevated temperatures (up to 390 K).

3. Results and discussion

Six-basic nitrilotrimethylenephosphonic acid has the following structure:

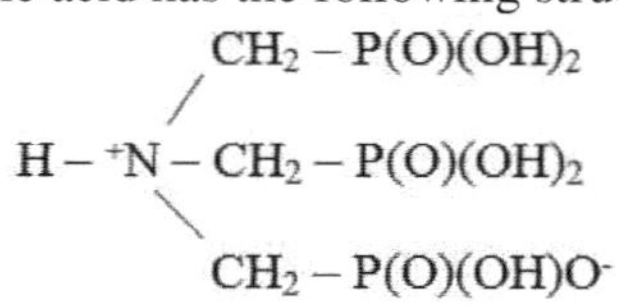

The study of complexation processes in the system Cu(II) – NTP – H_2O was carried out by the EPR method. Figure 1 shows the typical EPR spectra in different pH ranges, taken at room temperature in solutions with a metal ligand ratio of 1:2. In the pH range 0.5 - 13, copper ions form a series of coordination compounds with NTP. In the complexation processes, NTP exhibits the function of a polydentate ligand at the expense of the nitrogen atom and the oxygen atom of the phosphonic groups, which are able to coordinate with the metal ion. In a strongly acidic enivronment (pH <0.5), the EPR spectrum has one wide line with g-factor equal to 2.19, which is typical for the copper aquacomplex. With increasing pH, the shape and parameters of the spectra change, which is associated with changes in the composition and symmetry of the nearest coordination sphere of the central ion.

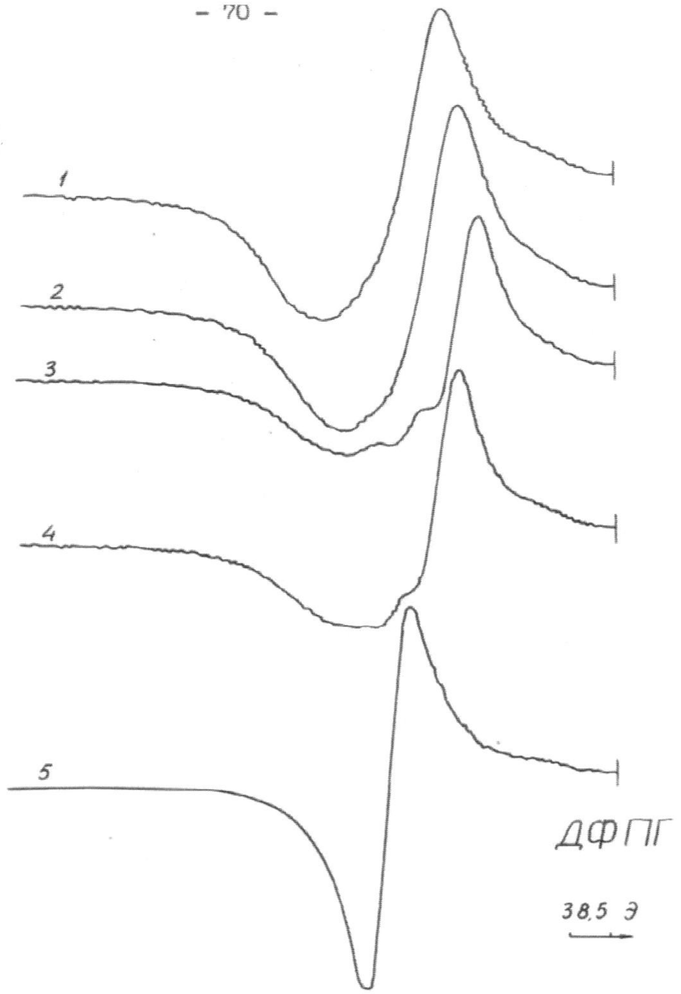

Figure 1. EPR spectra of copper (II) complexes with NTP in liquid solutions depending on pH: 1 – 2,09; 2 – 4,02; 3 – 5,47; 4 – 10,3; 5 – 12,48; 6 – 13,25;
$C_{Cu(II)}$ = 0,02 mol/l; C_{NTP} = 0,04 mol/l.

An increase in the pH of the environment of more than 0.5 causes a decrease in the g-factor value. This indicates the enhancement of the planar bond of the copper (II) ion with the ligand and the entry of the NTP molecule into the equatorial plane of copper. Depending on the pH of the solution, NTP, like other polybasic acids, could be in various dissociating forms. According to the literature data [1], the EPR spectrum observed up to a pH value of 3.5 should be attributed to the formation of a protonated complex $[CuHL]^{-}$ (Figure 1,1), where NTP acts as a dual-ligand. The entry of the ligand into the equatorial plane of the central ion leads to a weakening of bonds with axial water molecules. In this regard, with a further increase in pH, the coordination of NTP occurs in the axial position of the complex and a spectrum is observed in the pH range 3.5–5.0 (Figure 1, cr. 2) can be attributed to the following protonated complex $[CuH_2L]^{2-}$, where two coordination sites in the inner sphere of the central ion are occupied by the oxygen donor atoms of the double-deprotonated phosphonic NTP groups. The intensity of absorption of this type of complex is maximum at pH 4. At pH 4.5, the formation of a mono-impregnated complex $[CuHL]^{3-}$ begins, which intensity increases as the pH rises and reaches its maximum value at pH 5.5 (Figure 1, 3). The character of the spectrum changes, a hyperfine structure appears from the nuclei $Cu^{63/65}$. During the transition to the fully deprotonated complex $[CuL]^{4-}$

(Figure 1, 4), significant changes occur in the EPR spectrum, which may be caused by a strong distortion of the symmetry of the nearest coordination sphere of the central ion. The complex $[CuL]^{4-}$ prevails in a wide range of pH 7.5 - 11.5. With further increase in pH, we detected a new type of spectrum (Figure 1, 5), which is basic in the pH range of 11.5 - 13. The appearance of this spectrum can be explained by the occurrence of a hydroxyl group in the inner sphere of the central ion. Thus, the EPR spectrum observed in the pH range > 11.5 should be attributed to the formation of a new complex particle - the mixed $[Cu(OH)L]^{5-}$ hydro complex; the axial position is occupied by the nitrogen atom and the hydroxyl ion. An increase in pH more than 13 leads to a spectrum with a clearly defined ultrathin structure consisting of four components from copper nuclei. This is caused by the fact that at pH> 13 there is a transition of Cu (II) hydroxocomplexes in the $[Cu(OH)_4]^{2-}$complex. A similar type of spectrum with similar values of parameters was obtained in [19] in the study of copper (II) - tartrate complexes.

The magnetic parameters obtained from the spectra of liquid solutions (Table 1) show an initial increase in the hyperfine interaction with the Cu (II) nucleus and a decrease in g_{iso} in the series $[CuH_3L]^-$, $[CuH_2L]^{2-}$, $[CuHL]^{3-}$ followed by a sharp drop in the hyperfine interaction (HFI) with the nucleus Cu(II) for $[CuL]^{4-}$and complete disappearance for $[Cu(OH)L]^{5-}$. This can be explained by the initial control of the complex with the formation of a structure that is maximally close to a tetragonal elongated octahedral for $[CuHL]^{3-}$ with the subsequent transition to the structure of a trigonal bipyramid, which is accompanied by a sharp decrease in HFI with the Cu (II) nucleus. The formation of $[CuL]^{4-}$ causes a change in the geometry of the complex and a strong mixing of $d_{x^2-y^2}$ and d_{z^2}) conditions. The formation of the $[Cu(OH)L]^{5-}$ complex with an even stronger ligand OH^-compared to H_2O leads to a trigonal compression along the $N \rightarrow Cu \rightarrow OH^-$ axis and an additional decrease in HFI with the nucleus Cu (II). Removing the degeneracy of the electronic levels by the mechanism of the dynamic Jahn-Teller effect leads to a complete averaging of the hyperfine structure from Cu (II) and the spectrum represents a single structureless line with a width of ~ 45 Oe.

Table 1. The isotropic (g_{iso}, A_{iso}) and anisotropic ($g_{\parallel}$, $g_{\perp}$), ($A_{\parallel}$, $A_{\perp}$) parameters of the EPR spectra in the system Cu(II) – NTP – H_2O, $C_{Cu(II)}$ = 0,02 mol / l; C_{NTP} = 0,04 mol / l.

Complex ion	g_{iso}	A_{iso}, [Oe]	$g_{\parallel}$	$g_{\perp}$	$< g >$	$A_{\parallel}$, [Oe]	$A_{\perp}$, [Oe]	α^2
$[Cu(H_2O)_6]^{2+}$	2,19	-	2,404	2,08	2,19	133	-	0,84
$[Cu(H_3L)(H_2O)_4]^-$	2,18	45	-	-	-	-	-	-
$[CuH_2L(H_2O)_3]^{2-}$	2,17	-	-	-	-	-	-	-
$[CuHL(H_2O)_3]^{3-}$	2,17	55	-	-	-	-	-	-
$[CuL(H_2O)_2]^{4-}$	2,18	50	$g_x = 2,329$ $g_y = 2,14$ $g_z =$ 2,05		2,175	130	-	0,79
$[CuL(OH)]^{5-}$	2,165	45	-	-	-	-	-	-
$[Cu(OH)_4]^{2-}$	2,134	78,3	2,272	2,061	2,131	190	32	0,86

In the research of complexation reactions, the spectra of frozen solutions (when there is an anisotropy of the g-factor and the HFI constants) are of particular interest, since the anisotropic components are more sensitive to the environment of the paramagnetic ion than their average values. An analysis of the EPR spectra taken in the Cu(II) – NTP – H_2O system at 77 K allowed us to determine their parameters, which are listed in Table 1. Lowering the temperature of the solution and freezing at 77 K leads to stability in the entire pH range of four kinds of complexes. Magnetic parameters obtained from the spectra of frozen solutions confirm the conclusions of the analysis of the isotropic parameters of the EPR. In the acidic pH range, the EPR spectra of copper (II) complexes in frozen solutions are anisotropic lines with resolved components of the hyperfine structure in a parallel orientation. At low pH values, the aquacomplex $[Cu(H_2O)_6]^{2+}$ is thermodynamically stable. The range of its existence in vitrified solutions expands to pH 3.5. At pH 4, we observe the formation of the $[CuL]^{4-}$ complex; it is stable in the pH 4–11 range and is characterized by a strong rhombic nature of the magnetic parameters (Table 1), which is also typical for trigonal bipyramidal complexes. The maximum dentate equals to four is easily realized at low temperatures, as evidenced by the presence of the complex form $[CuL]^{4-}$at sufficiently low pH values of the solution. The$[CuH_3L]^-$, $[CuH_2L]^{2-}$, $[CuHL]^{3-}$ complexes are probably characterized by a rapid proton exchange, which is indirectly confirmed by the absence of these forms in the EPR spectra of frozen solutions. The broadening of the lines leads to a weak distinguishability of these forms in the EPR spectra. In the pH range above 11, the EPR spectra of copper (II) complexes have a complex shape with poorly resolved components of the hyperfine structure. In the pH range of 11.5 - 12.75, the $[CuL(OH)]^{5-}$ complex accumulates, and at pH 12.75, along with this complex, the $[Cu(OH)_4]^{2-}$ axial symmetry complex begins to accumulate as well, which spectrum has a well-resolved hyperfine structure in the region of parallel and perpendicular orientation.

We estimated the covalence parameter α^2 (the density of the unpaired electron on the central atom), which makes it possible to qualitatively estimate the degree of covalence of the metal – ligand bond (Table 1). The value α^2 is calculated by the formula [20]:

$$\alpha^2 = \frac{A_{\parallel}}{P} + (g_{\parallel} - 2{,}0023) + \frac{3}{7}(g_{\perp} - 2{,}0023) + 0{,}04, \quad (1)$$

where P - constant characterizing the hyperfine interaction in a free uncomplexed central atom, $P = 0{,}036$ cm^{-1} for copper (II) complexes. The values $g_{\parallel}$, $g_{\perp}$, $A_{\parallel}$ and $A_{\perp}$ determined from spectra taken at liquid nitrogen temperature.

It turned out that in the deprotonated complex the bond of copper (II) with the ligand is the most covalent. NTP, acting as a tetradentate ligand in a deprotonated complex, leads to a stronger bond in the equatorial plane, which is reflected in a decrease in the parameter α^2. Values of g – factors ($< g >$) calculated by the equation:

$$< g > = \frac{g_{\parallel} + 2g_{\perp}}{3} \quad (2)$$

slightly differ from the g - factors determined from the EPR spectra of liquid solutions (Table 1). This suggests that the symmetry of the nearest environment of copper (II) is preserved in the solution.

Thus, based on the data obtained, the following equilibria can be used as the basis for the mathematical model of equilibrium processes in the system Cu (II) - nitrilotrimethylenephosphonic acid:

$$Cu^{2+} + H_6L \leftrightarrows [CuH_3L]^- + 3H^+ \quad (3)$$

$$Cu^{2+} + H_6L \leftrightarrows [CuH_2L]^{2-} + 4H^+ \quad (4)$$

$$Cu^{2+} + H_6L \leftrightarrows [CuHL]^{3-} + 5H^+ \quad (5)$$

$$Cu^{2+} + H_6L \leftrightarrows [CuL]^{4-} + 6H^+ \quad (6)$$

$$Cu^{2+} + H_6L + H_2O \leftrightarrows [Cu(OH)L]^{5-} + 7H^+ \quad (7)$$

$$Cu^{2+} + 4H_2O \leftrightarrows [Cu(OH)_4]^{2-} + 4H^+ \quad (8)$$

Basing on the EPR data and using *CPESSP* (Chemical Parametres of Equilibria in Solutions with Solid Phases) [21] software, we determined the equilibria constants complex stability constants (Table 2).

Table 2. Equilibrium and stability constants of copper (II) complexes with nitrilotrimethylenephosphonic acid.

Complex	*lgKp*	*lgβ*
$[CuH_3L]^-$	-1,61	5,25
$[CuH_2L]^{2-}$	-4,92	8,08
$[CuHL]^{3-}$	-9,59	10,86
$[CuL]^{4-}$	-16,27	16,07
$[Cu(OH)L]^{5-}$	-28,08	18,26
$[Cu(OH)_4]^{2-}$	-37,1	18,9

In the pH range of 0 - 10, we used the experimental dependence $< H >= f(pH)$ (figure 2) for calculations, where H is the magnetic field strength corresponding to the position of the inflection point for the high-field component of the EPR spectrum. In the case when changes in the magnetic parameters of the spectra occur within the width of the line, it can be assumed that $< H >$ of the selected point of the spectrum is determined by a linear combination of characteristic values for the particles present in equilibrium with a weight proportional to their concentration. An increase in pH more than 11 leads to a dramatic change in the shape of the spectrum associated with the formation of the complex form $[Cu(OH)L]^{5-}$ and the value of the $< H >$ parameter at the same time ceases to be a linear combination of the value of H_i of individual complexes. However, the difference in the widths of the lines leads to the fact that the overlay of the lines does not significantly affect the amplitude of the line related to $[CuL(OH)]^{5-}$, which made it possible to use the dependence $I = f(pH)$ in the pH 10 region - 13.5 (Figure 3) , where I is the intensity of the EPR signal, for the processing.

The search for satisfied convergence of experimental and calculated data was carried out during the search for the minimum of the functional:

$$F_{max} = \sum_{n=1}^{N}\left(x_{exp} - x_{calc}\right)^2, \quad (9)$$

where x_{exp} – experimental value of properties – H or I; x_{calc} – its calculated version.

For the system Cu(II) – NTP in the range of pH 0 - 13.5, in the presence of a twofold excess of the ligand with respect to the metal ion in the region of pH> 11, deprotonation of the complex form $[CuL]^{4-}$takes place with the formation of the hydroxocomplex $[Cu(OH)L]^{5-}$, which with a further increase in the alkalinity of the environment (pH> 13.1) decomposes with the formation of a copper complex $[Cu(OH)_4]^{2-}$. During the calculation procedures, we obtained the following values of the stability constants: for $[Cu(OH)L]^{5-} - 18{,}25 \pm 0{,}04$ and $[Cu(OH)_4]^{2-} - 18{,}9 \pm 0{,}08$.. The latter constant has good convergence with the literature data [1].

IOP Conf. Series: Materials Science and Engineering **786** (2020) 012056 doi:10.1088/1757-899X/786/1/012056

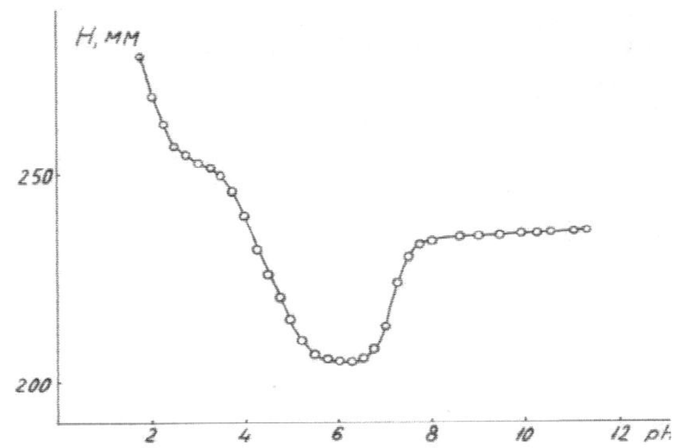

Figure 2. The change in the magnitude of the magnetic field (H) of the high-field component of the EPR spectrum depending on pH: $C_{Cu(II)}$ = 0.02 mol / l; C_{NTP} = 0.04 mol / l.

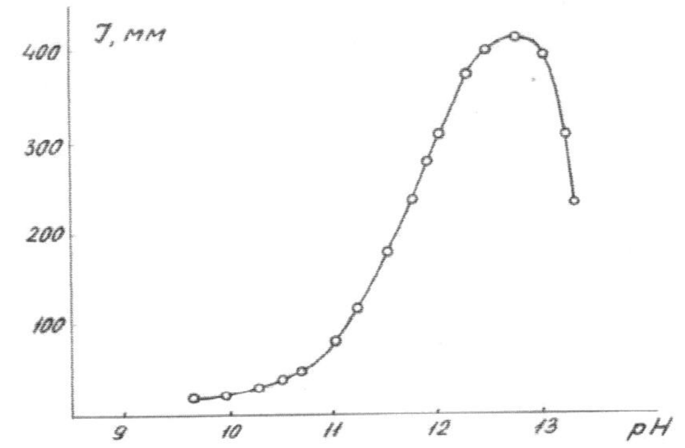

Figure 3. The change in the intensity of the EPR signal depending on the pH in solutions containing 0.02 mol/l Cu(II) and 0.04 mol/l NTP.

Figure 4 shows the magnetic resonance data as a differential dependence representing the ratio of the change in the position of the characteristic point of the EPR spectrum to the change in the pH of the solution.

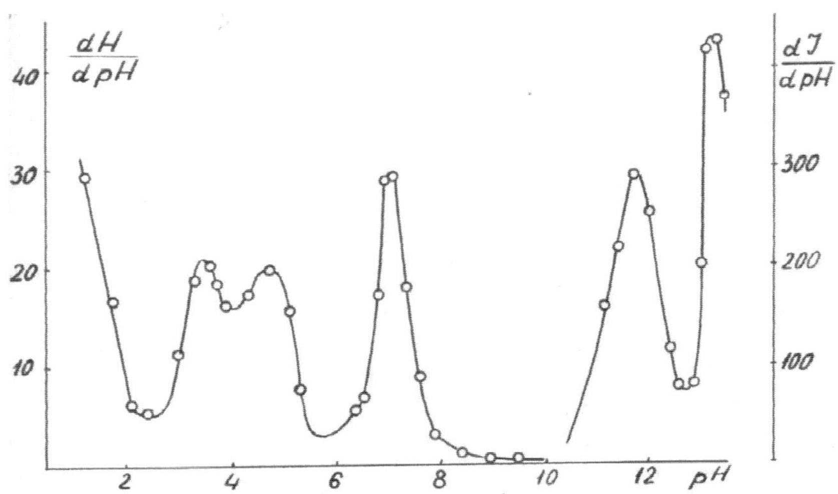

Figure 4. Change in $\frac{dP}{dpH}$ and $\frac{dI}{dpH}$ depending on pH; $C_{Cu(II)}$ = 0,02 mol/l; C_{NTP} = 0,04 mol/l.

Comparison of the obtained dependence with the distribution of complex forms showed that the areas of maximum accumulation of individual complexes correspond to minimal changes in the position of the high-field component of the spectrum, and sharp maxima correspond to the boundaries of the existence of the complexes. The region of existence of $[CuH_2L]^{2-}$ and the boundary of the forms $[CuH_2L]^{2-}$ and $[CuHL]^{3-}$ are poorly expressed due to the proximity of the formation constants of these complexes. The regions of existence of$[Cu(H_2O)_6]^{2+}$, $[CuH_3L]^{-}$, $[CuHL]^{3-}$, $[CuL]^{4-}$ and $[Cu(OH)L]^{5-}$ on the differential diagram are clearly shown. It seems to us that this type of dependence, namely $\frac{dP}{dpH}$ – *f*(pH), where P is a parameter of the EPR spectrum, is promising for describing paramagnetic systems characterized by complex equilibria with relatively small changes in the magnetic parameters relative to the width of the line.

IOP Conf. Series: Materials Science and Engineering **786** (2020) 012056 doi:10.1088/1757-899X/786/1/012056

4. Conclusion

The electron paramagnetic resonance technique was studied using copper (II) complexes with nitrilotrimethylene phosphonic acid. The study of complexing properties suggests that stable complex compounds in a wide range of pH values. Based on the cumulative study of liquid and frozen solutions with a consistent change in pH (pH change step of 0.25), we obtained data on the composition and stability of copper (II) complexes with NTP formed in the pH range 0 - 13 established in the region of their existence and structure. The obtained information is of significant interest in regard with the practical effects of solutions, copper (II) complexes and NTP for chemical deposition without the use of toxic electrolytes.

References

[1] Ignatenko N L Danyushina G A , Serbinovskiy M YU , Otych N A , Loginov V T and Levintsev V A 2008 Poluchenie metallopolimernyh kompozitsionnyh pokrytiy himicheskim osazhdeniem // Izv Vuzov Sev -Kavk Region Tekhn Nauki Spetsvypusk Kompozitsionnye materialy 48-50

[2] Vasileva N A 2015 Kompozitsionnye metallopolimernye pokrytiya, formiruemye himicheskim metodom // *Izv Vuzov Sev -Kavk Region Tekhn* Nauki **4** 133-7

[3] Burkat G K , Fujimura T , Dilator V Y , Rollover E A and Veretennikova M V 2005 Preparation of composite electrochemical nickel-diamond and iron- diamond coatings in the presence of detonation synthesis nanodiamonds // Diamond and Related Materials 14(11-12) 1761-4

[4] Meeraker J A 2002 On the mechanism of electroless plating: 1 Oxidation of formaldehyde at different electrode surfaces// *J of Appl Electrochemistry* **11** 387-93

[5] Ivanov V V and Shcherbakov I N 2008 Povyshenie ekspluatatsionnyh svoystv detaley avtomobiley putem naneseniya kompozitsionnyh nikel-fosfornyh pokrytiy // *Izv Vuzov Sev -Kavk Region Tekhn Nauki* **3** 113-5

[6] Saxena V , Uma Rani R and Sharma A K 2010 Studies on ultra high solar absorber black electroless nickel coatings on aluminum alloys for space application // *J Appl Electrochem* 40(2) 333-9

[7] Gamer P U , Mukhametshina R M and Guseva L G 1987 Himicheskoe mednenie iz rastvorov s fosfororganicheskimi kompleksonami // V sb : Tez dok IX Vsesoyuznoy nauchno-tekhnicheskoy konferentsii po elektrohimicheskoy tekhnologii “Galvanotekhnika - 87” 260

[8] Collins L I and Perbins P G 1977 Constants and Salt Formation by 1-Hydroxyethyl-1,1-diphosphonic Acid (HEDP) // *J Appl Chem Biotechnol* **27(12)** 651-61

[9] Mukhametshina R M , Korshin G V , Guseva L G and Gamer P U 1987 Osobennosti razryada kompleksov medi(II) s fosfororganicheskimi kompleksonami // V sb : Himicheskaya fizika i fizicheskaya himiya CHast II, Tallin 130

[10] Rizkalla E N , Laki M T and Ismail M I 1980 Metal chelates of phosphonate-containing ligands V Stability of some 1-hydroxyethane-1,1- diphosphonic acid metal chelates // Talanta 27(9) 715-9

[11] Kabachnik M I , Medved T YA and Dyatlova N M 1974 i dr Fosfororganicheskie kompleksony // Uspekhi himii 43(9) 1554-74

[12] Dyatlova N M , Temkina V YA and Popov K I 1988 Kompleksony i kompleksonaty metallov M : Himiya, 544

[13] Uhanova S I 1983 Stabilizatory rastvora himicheskogo medneniya // *Obmen opytom v radiopromyshlennosti* **2** 12-5

[14] Sakhapov, R L , Nikolaeva, R V , Gatiyatullin, M H , Makhmutov, M M 2016 Risk management model in road transport systems // *Journal of Physics: Conference Series* **738(1)** 012008

[15] Sakhapov, R L , Nikolaeva, R V , Gatiyatullin, M H and Makhmutov, M M 2017 Modeling of traction-coupling properties of wheel propulsor // *Journal of Physics: Conference Series* **738(2)** 012115

[16] Pentin YU A and Volkov L V 2006 Fizicheskie metody issledovaniya v himii M : Mir 683

[17] Brustolon M and Gifmello E 2009 Eltctron Paramagnetic Resonance A Practitioner's Toolkit – Hoboken: John Wiley s Sons I nc , 553
[18] Vishnevskaya G P , Safin R SH , Muhametshina R M i dr 1986 Izuchenie kompleksov medi(II) s oksietilidendifosfonovoy kislotoy metodom EPR// *Koord himiya* **12(11)** 1473-7
[19] Schoenberg L N 1971 The Structure of the Complexed Copper Species in Electroless Copper Plating Solutions // *J Electrochem Soc* **118(10)** 1571-6
[20] Vishnevskaya G P , SHapnik M S and Safin R SH 1981 Issledovanie kompleksov medi(II) s etilendiamintetrauksusnoy kislotoy i etilendiaminom metodom EPR // *Neorganicheskaya himiya* **26(10)** 2781-5
[21] Salnikov YU I , Devyatov F V , ZHuravleva N E , Golodnitskaya D V 1984 Kompleksoobrazovanie kobalta(II) i nikelya(II) s limonnoy kislotoy // *Neorganicheskaya himiya* **29(9)** 2273-6

IOP Conf. Series: Materials Science and Engineering **786** (2020) 012057 doi:10.1088/1757-899X/786/1/012057

Metal plating based on copper (II) complex compounds with hydroxyethylidene diphosphonic acid

R M Mukhametshina

Kazan State University of Architecture and Engineering, 420043, Kazan, Zelenaya str., 1, Russia

E-mail: rumya211@yandex.ru

Abstract. The materials used in mechanical engineering must ensure proper work of the details and meet the requirements of the minimum complexity of its manufacture and efficiency. In this regard, the issues of improving the technological processes of obtaining metal coatings are of particular importance. This paper studies the possibility of using copper (II) complexes with hydroxyethylidene diphosphonic acid to obtain high-quality copper coatings by the method of electron paramagnetic resonance. The author are determines stability constants of the complexes and their regions of existence. Moreover, the article investigates the effect of stabilizing additives on the technological characteristics of solutions of chemical copper plating. Due to the obtained results, the author developed a highly stable solution of chemical copper plating based on copper (II) complex compounds with hydroxyethylidene diphosphonic acid.

1. Introduction

A promising direction for the development of mechanical engineering, including the transport system, is the improvement of technological processes of applying metal layers on the surface of products. In recent decades, the chemical method of coating has become popular, in particular, for the metallization of dielectrics. Metallized plastics have wear resistance, chemical resistance, heat resistance and mechanical strength, have a decorative appearance and are resistant to light. Due to these properties, metallized plastic is widely used in mechanical engineering.

One of the important issues of obtaining metal coatings is the choice of complexing agents for the preparation of technological solutions. At present, a large number of complex compounds for the deposition of metals have been developed and introduced into production [1-3]. However, the search for new, more efficient ligands, which help to obtain coatings with given functional properties, is an urgent task and constantly attracts the attention of researchers.

To optimize and develop promising technological processes for obtaining metal coatings, experimental studies of the complex formation processes are necessary. From this point of view, it is of considerable theoretical and practical interest to study the patterns of complexation of complex copper (II) ions with hydroxyethylidene diphosphonic (HEDP) acid.

The choice of copper (II) hydroxyethylidene diphosphonic acid as a ligand can be explained, on the one hand, by its high complexing properties. HEDP is a polybasic acid that can serve as a complexone in a wide pH range. The compound is non-toxic and is available on an industrial scale. One of the objectives of the study was to discover the possibility of using HEDP as ligands of copper (II) ions in chemical copper plating solutions, since the solutions of chemical copper plating currently used in practice have insufficient stability, and the complexing agent plays an important role in providing the required stability of the solutions. In connection with the problem of low-waste production, increasing the stability of chemical copper plating solutions and developing new efficient technological processes is an urgent task.

2. Methods

To research the complexation processes in the studied systems, we used the electron paramagnetic resonance (EPR) method, which is one of the most sensitive to changes in the structure and composition of the first coordination sphere of the complexing metal. The fundamentals of the EPR method and its application for the study of complex formation reactions and the state of paramagnetic ions in solutions are described in a number of papers [4–7].

The main parameters of the EPR spectra are the intensity and width of the lines, the g-factor and the constant of the hyperfine interaction, which depend on the symmetry of the nearest environment of the central ion and on the nature of the ligand.

This paper, based on the analysis and processing of the parameters of the EPR spectra of liquid and frozen solutions, explores the probable structure of the resulting complex ions, and calculates the equilibrium constants of complex formation. The EPR spectra were recorded on an RE-1306 spectrometer equipped with a magnetic field calibrator collected on the basis of a magnetic induction meter III-I-I. The samples were placed in calibrated capillaries of glass, which do not give an EPR signal in the working region of magnetic fields. The EPR spectra were recorded at 298 and 77 K in the three-centimeter wavelength range. Measurement of the g-factors of the studied complexes was carried out by comparison with the g-factor of DPPH.

The determination of the density of the unpaired electron on the central atom was carried out according to the formula:

$$\alpha^2 = \frac{A_{\parallel}}{P} + (g_{\parallel} - 2{,}0023) + \frac{3}{7}(g_{\perp} - 2{,}0023) + 0{,}04, \qquad (1)$$

where P - constant characterizing the hyperfine interaction in a free uncomplexed central atom. The values $g_{\parallel}$, $g_{\perp}$, $A_{\parallel}$ and $A_{\perp}$ determined from spectra taken at liquid nitrogen temperature. The calculation of the equilibrium constants of complexation was carried out using the software *CPESSP* (Chemical Parametres of Equilibria in Solutions with Solid Phases) [8].

We determined the quality of the coatings using surface micrographs and X-ray structural analysis. X-ray diffraction studies were performed on a Dron-3 diffractometer in CuK_{α} radiation. The study of the surface morphology of the coatings was performed on an electron microscope. The deposition rate of copper was determined by the gravimetric method, and the stability of solutions of chemical copper plating was evaluated by the “Trigger” method [9].

3. Research of the complex formation processes in the system Cu(II) - HEDP - H2O

In the study of metal deposition processes, it is necessary to know the composition and stability of the resulting complexes, their structure, and the symmetry of the nearest environment. The literature data on the complexation of copper (II) with hydroxyethylidene diphosphonic acid is inconsistent and does not provide a complete picture of the composition and stability of the resulting complexes. In addition, the study of complex formation equilibria was carried out in the pH range up to 10. In this connection, it seems necessary to study the complex formation processes of Cu (II) with HEDP in a wider pH range in order to develop promising technological processes for the deposition of copper coatings.

Five-basic hydroxyethylidenediphosphonic acid (HEDP, H_5L) has the structure:

IOP Conf. Series: Materials Science and Engineering **786** (2020) 012057 doi:10.1088/1757-899X/786/1/012057

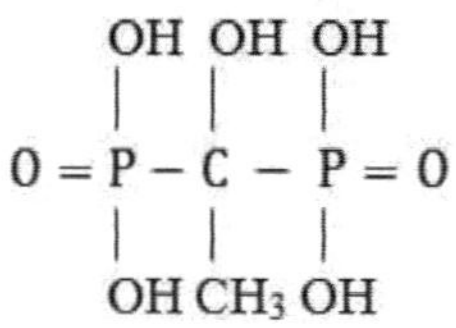

To study the composition and structure of Cu (II) complexes with HEDP in aqueous solutions with an acidity varying from 0.5 to 13, we used the EPR method [10]. The parameters of the EPR spectra finely react to changes in the composition of the coordination sphere of the metal-complexing agent. A consistent increase in pH during the transition from acidic to alkaline (the step of pH change was 0.2 ... 0.5) in the system Cu(II) - HEDP - H_2O allows us to clearly record systematic and reproducible changes in the parameters and form of the EPR spectra during the formation of various forms complexes (table 1).

Table 1. Parameters of the EPR spectra of copper (II) complexes with HEDP in vitrified solutions (77K) at various pH values.

pH	Parameters of EPR spectra			
	$g_{\|}$ ($\mp$0,003)	$g_{\perp}$ ($\mp$0,003)	$A_{\|}$ ($\mp$1), Oe	g_o ($\mp$0,003)
0,5	2,420	2,057	127	-
1-5,5	2,430	2,065	119	-
1-6	2,430	2,066	124	-
1-2	-	-	-	2,19
6-9	2,400	2,053	128,5	-
6-7,5	-	-	-	2,17
10-12	2,390	2,045	133,5	-

In the system Cu (II) - HEDP - H_2O at pH 0.5 for excess of ligand and in the range of pH 0.5 ... 1.3 for lack of ligand, the shape and parameters of the EPR spectra correspond to the copper aquacomplex. Above of these pH values we can note the participation of various forms of HEDP in complexation with copper (II). In the acidic pH range, we can observe an anisotropic spectrum with components of a hyperfine structure resolved in parallel orientation, characterized by a decrease in the constant $A_{\|}$ and an increase in $g_{\|}$ and $g_{\perp}$ ($A_{\|}$ = 119 Oe, $g_{\|}$ = 2.430, $g_{\perp}$ = 2.065) compared to similar values of parameters for aqua copper. Changes in the parameters of anisotropic EPR spectra for complexes formed in the acidic region indicate the hardening of axial bonds of copper with ligands. Based on our data, we can assume that in the acidic region under conditions of lack of ligand, a complex of the composition $[CuH_3L]^0$ is formed, and with an excess of ligand - a complex of the composition$[Cu(H_3L)_2]^{2-}$. The intensity of the spectra corresponding to this complex is maximum at pH 3. In the pH range 3.5 ... 5.5, the formation of insoluble complexes occurs. With an excess of ligand, starting from pH 6, the intensity of the EPR signal in solution increases dramatically, which indicates the formation of new types of soluble complexes. For a pH value of 9, $[Cu(H_2L)_2]^{4-}$ with an anazotropic spectrum parameters prevail in the solution: $g_{\|}$ = 2,400; $g_{\perp}$ = 2.053; $A_{\|}$ = 128.5, and the range of pH 10 ... 12 forms complexes $[Cu(HL)_2]^{6-}$, which give anazotropic spectra with the following parameters: $g_{\|}$ = 2.390; $g_{\perp}$ = 2.045; and $A_{\|}$ = 133.5. A decrease in $g_{\|}$, $g_{\perp}$ and an increase in $A_{\|}$ in the spectra of these complexes in comparison with similar parameters of the spectra of the aquacomplex and the $[CuH_3L]^0, [Cu(H_3L)_2]^{2-}$complexes indicate an increase in planar bonds of copper with ligands. The intensity of the EPR spectrum due to the accumulation of the complex particle$[Cu(HL)_2]^{6-}$ is maximum at pH 11, and then the signal intensity decreases from pH 11.5, which may indicate the formation of binuclear complexes. Since the literature data on the composition and stability of the resulting complex ions in the alkaline region (pH 9… 13) is not available, we performed the simulation of equilibrium processes in the Cu (II) - HEDP - H_2O system based on the EPR data. Using the CPESSP software we processed the experimental dependence $I = f(pH)$, where I is the intensity of the EPR signal, and obtained the formation equilibrium constants and the stability constants of the identified complex forms. The calculation results are shown in Table 2.

Table 2. The equilibrium constants of formation (K_p and stability (β) of copper (II) complexes with HEDP (L).

Complex	lgK_p	$lg\beta$
$[Cu(HL)_2]^{6-}$	-29,5	16,44
$[Cu_2(HL)_2]^{4-}$	-20,04	25,9
$[Cu_2(HL)L]^{5-}$	-28,71	29,23
$[Cu_2L_2]^{6-}$	-39,52	30,42
$[CuL]^{3-}$	-21,65	13,32
$[Cu_2L(OH)_2]^{3-}$	-33,25	29,72

4. The study of chemical deposition of copper from solutions based on HEDP

The literature has just a limited information on the use of organophosphorus ligands, including HEDP, in metal plating processes. Practically, there is no information about the properties and quality of the coatings obtained, it says only about the high stability of the solutions. Since we studied the complexation of copper (II) with HEDP, it is of practical interest to research the possibility of obtaining copper coatings from solutions based on HEDP. It should be noted that this ligand is non-toxic and available on an industrial scale.

We chose the process of chemical copper plating for the study. It is currently widely used in practice to obtain an electrically conductive copper sublayer and an independent copper coating on dielectrics. However, popular in the production of chemical copper plating solutions have insufficient stability and not always the coating turns out to be of good quality. In addition, the ligands used (tartrate, trilon B) are scarce. Depending on the composition and structure of the complex being formed, it is possible to obtain different rates of copper recovery on the catalytic surface of the dielectric and metal, different solution stability and different physicochemical properties of the depositions.

In the study of the complex formation of copper (II) with HEDP, we found that in strongly alkaline solutions copper (II) ions are present mainly in the form of $[Cu_2L_2]^{6-}$ in the Cu (II) - HEDP -H_2O system .

We determined the stability constants of the resulting complexes: $[Cu_2L_2]^{6-} - lg\beta$ = 30.42. If we compare copper complexes with HEDP with trilonate complexes ($lg\beta$ = 21.1), traditional in solutions of chemical copper plating, it turns out that copper complexes with HEDP are the most stable, which probably causes the high stability of copper plating solutions based on HEDP.

The experiments were carried out with copper-plating solutions containing (mol/l): $CuSO_4 \cdot 5H_2O$ – 0,1; ligand (HEDP) - 0.2; formaldehyde - 0.3; $NaOH$ – up to pH 12.8. The coatings were deposited at a temperature of 25^0C for 0.5 hours on samples of ABS-plastic, previously degreased, etched and activated by well-known methods [11]. The loading density was 2 dm^2/l. The deposition rate was determined by the weight gain, and the stability of the solutions was determined by the "Trigger" method [9]. The following data were obtained: deposition rate - 1.5 μm / h; stability - 47.9 min; the appearance of the coating is light pink.

Experiments have shown that uniform, fine-crystalline and dense depositions of copper, with good adhesion to the base, can be obtained from solutions with HEDP, although the speed of deposition is not very high. The peculiarities of the copper plating process in a solution with HEDP are its high stability, good coating quality of light pink color resembling electroplating copper. It is known that the properties of copper deposits, as well as the properties of metallic coatings of a different nature, are determined by their structure. Therefore, after the deposition from solutions of chemical copper plating, the depositions were subjected to phase-structural studies.

Figure 1 shows the line X-ray diffraction patterns of annealed copper samples (as a reference) and coating from HEDP solutions. They show that practically pure copper with a lattice period of 0.3607

IOP Conf. Series: Materials Science and Engineering **786** (2020) 012057 doi:10.1088/1757-899X/786/1/012057

nm is deposited from a solution based on HEDP. However, both solutions have growing crystals oriented along the 111 face.

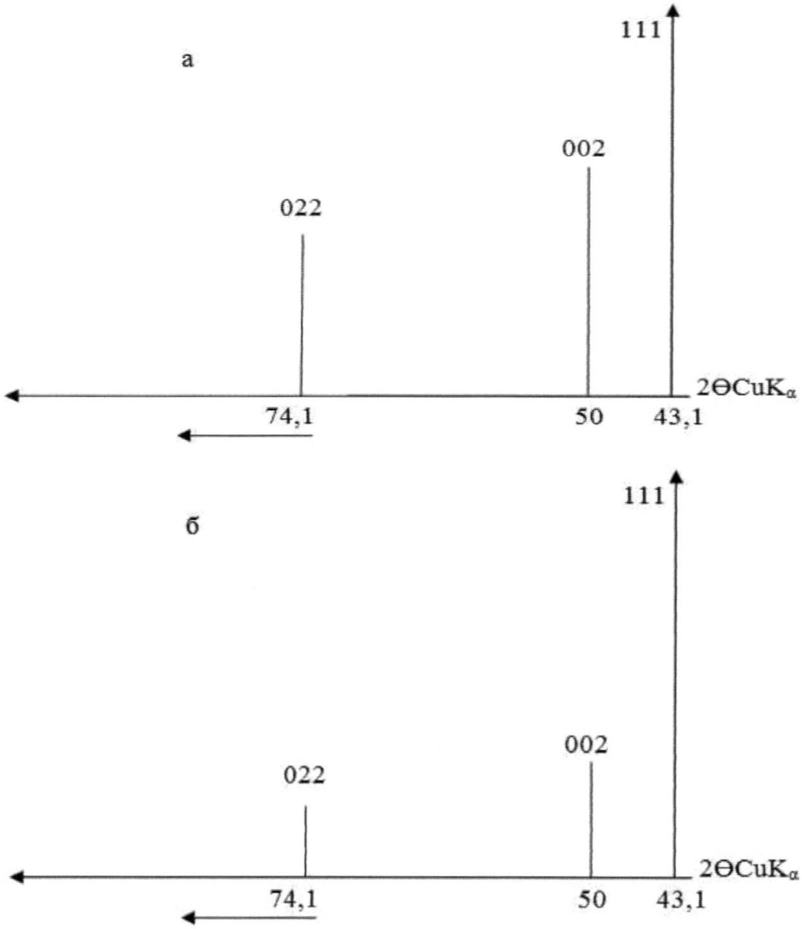

Figure 1. X-ray diffraction patterns: a – annealed copper; b – deposition of copper from solutions of chemical copper plating based on HEDP.

Radiographs also show some small peaks, corresponding to impurities of copper (I) oxide and copper (II) oxide. The magnitude of these peaks was approximately at the noise level and it was not possible to quantify the content of copper oxides on the surface or in the bulk of the coating.

We also carried out an electron microscopic study of the surface structure of copper deposits. The data obtained indicate that the surface of copper is an even crystalline deposition.

The problem of stability and repetitive use is relevant for a solution of chemical copper plating. Therefore, we also tested the effect of some additives. When choosing a stabilizing additive, we relied on the experience of stabilization of chemical metallization solutions. The literature tells that organic compounds containing a trivalent nitrogen atom and bivalent sulfur, capable of adsorption bonding to the metal surface (as a result of formation of a chemical bond between lone electron pairs and free d - orbitals of the metal), are the most promising for stabilizing solutions of chemical copper plating. In addition to electronic factors, the effectiveness of the additive is determined by its stereochemistry — the ability of adsorbed additive molecules to block a large number of active sites of the catalyst [12–15].

We evaluated the effect of additives of various types on the process of copper (II) reduction from complexes with HEDP: I - 1-ethylthiobenzylphosphonic acid; II, III - ammonium salts of 1-ethylthiobenzylphosphonic acid; IV - morpholine; V - N-phenylamide-1-phenyl-1-N-

IOP Conf. Series: Materials Science and Engineering **786** (2020) 012057 doi:10.1088/1757-899X/786/1/012057

morpholinomethanesulfonic acid (Table 3). We should note that additive IV is recommended in the literature [16-18] as a stabilizer for copper plating solutions, and the effect of other additives was tested for the first time. We investigated the effect of additives on the stability of the solutions, the deposition speed and the appearance of the coatings (Table 3).

Table 3. Effect of additives on the process of copper plating.

Additive, [10⁻⁵mol/l]	Deposition speed, [μm/h]	Stability, [min]	The appearance of the coating
I	1,5	49,1	light pink
II	1,5	41,4	light pink
III	1,5	42,7	light pink
IV	1,2	48,5	light pink
V	1,4	49,8	light pink

Studies have shown that additives have different effects on the process of copper plating. Additives I and V are of the greatest interest. Additives II – IV have little effect on the stability of the solution and have virtually no effect on the deposition rate. If additives I and V are used as stabilizing substances, solution stability can be achieved.

On the basis of the stabilizing additive I (I - ethylthiobenzylphosphonic acid) in the Cu(II) – L – CH_2O system, where L is HEDP, we carried out experiments to identify the optimal conditions for the chemical deposition of copper. As a result of studying the influence of the concentration of the main components, pH and temperature on the quality of the copper deposition, as well as the deposition speed and the stability of the solution, we selected the optimal solution composition [mol / l]:

Copper (II) sulfate	0,1
Hydroxyethylidenediphosphonic Acid	0,2
Formalin	0,3
I – ethylthiobenzylphosphonic acid	$1 \cdot 10^{-5}$
Sodium hydroxide	up to pH 12,8

The study of the influence of the concentration of the additive on the deposition speed and the stability of the solution showed that the optimal concentration lies in the range $1 \cdot 10^{-5} - 2{,}5 \cdot 10^{-5}$ mol / l. Raising the concentration of the additive above these values increases stability, but at the same time the speed of copper deposition decreases sharply. A solution of this composition at a temperature of 18-25^{0}C can give a pink, dense, fine-crystalline depositions of copper on dielectric products.

In the study of the complex formation of copper (II) with HEDP, we found that in strongly alkaline solutions copper (II) ions are present mainly in the form of $[Cu_2L_2]^{6-}$ in the $Cu(II) - HEDP - H_2O$ system. We determined the stability constants of the resulting complexes: $[Cu_2L_2]^{6-} - lg\beta$ =30,42. If we compare copper complexes with HEDP with trilonate complexes ($lg\beta = 21{,}1$), which are traditional in solutions of chemical copper plating, it turns out that copper complexes with HEDP are the most stable, which probably leads to high stability of copper plating solutions based on hydroxyethylidenediphosphonic acid.

5. Conclusion

We used the electron paramagnetic resonance method to study the formation of copper (II) complexes with hydroxyethylidene diphosphonic acid. Then we determined the stability constants of the complexes and their regions of existence. Moreover, the study shows the possibility of obtaining high-quality copper coatings from solutions of chemical copper plating based on HEDP. X-ray structural analysis of the samples showed that there is a deposition typical for pure copper with a slight difference in the broadening of the peaks, which is associated with the grain sizes of copper crystals. The research established that hydroxyethylidene diphosphonic acid forms stable complex compounds with copper (II) ions and is a promising ligand for obtaining high-quality metallic coatings.

References

[1] Dyatlova N M , Temkina V YA and Popov K I 1988 Kompleksony i kompleksonaty metallov M : KHimiya, 544

[2] Kabachnik M I , Medved T YA , Dyatlova N M 1974 i dr Fosfororganicheskie kompleksony // *Uspekhi khimii* **43(9)** 1554-74

[3] Schoenberg L N 1971 The Structure of the Complexed Copper Species in Electroless Copper Plating Solutions // *J Electrochem Soc* **118(10)** 1571-6

[4] Vertts D and Bolton D 1975 Teoriya i prakticheskie prilozheniya metoda EPR : Per s angl – M : Mir – 548 s

[5] Pentin YU A and Volkov L V Fizicheskie metody issledovaniya v khimii M : Mir - 683s

[6] Rakitin YU V , Larin G M and Minin V V 1993 Interpretatsiya spektrov EPR koordinatsionnykh soedineniy M : Nauka 400s

[7] Brustolon M and Gifmello E 2009 Eltctron Paramagnetic Resonance A Practitioner's Toolkit – Hoboken: John Wiley s Sons I nc , 553

[8] Salnikov YU I , Devyatov F V , ZHuravleva N E and Golodnitskaya D V 1984 Kompleksoobrazovanie kobalta(II) i nikelya(II) s limonnoy kislotoy // *Neorganicheskaya khimiya* **29(9)** 2273-6

[9] Ukhanova S I 1983 Stabilizatory rastvora khimicheskogo medneniya // Obmen opytom v radiopromyshlennosti **2** 12-5

[10] Vishnevskaya G P , Safin R SH , Mukhametshina R M i dr 1986` Izuchenie kompleksov medi(II) s oksietilidendifosfonovoy kislotoy metodom EPR / // *Koord khimiya* **12(11)** 1473-7

[11] SHalkauskas M and Vashkyalis A 1985 KHimicheskaya metallizatsiya plastmass – L : KHimiya 143

[12] Kryukova N P , Bolotin S N and Panyushkin V T 2003 Issledovaniya kompleksoobrazovaniya v sisteme med(II) – L – gistidin – D – ornitin metodom spektroskopii EPR // Akademiya nauk Seriya khimiya **5** 1060-3

[13] Collins L I and Perbins P G 1977 Constants and Salt Formation by 1-Hydroxyethyl-1,1-diphosphonic Acid (HEDP) // J Appl Chem Biotechnol **27(12)** 651-61

[14] Rizkalla E N , Laki M T and Ismail M I 1980 Metal chelates of phosphonate-containing ligands V Stability of some 1-hydroxyethane-1,1- diphosphonic acid metal chelates // *Talanta* **27(9)** 715-9

[15] Mukhametshina R M , Korshin G V , Guseva L G and Gamer P U 1987 Osobennosti razryada kompleksov medi(II) s fosfororganicheskimi kompleksonami // V sb : KHimicheskaya fizika i fizicheskaya khimiya CHast II, Tallin 130

[16] Gamer P U , Mukhametshina R M and Guseva L G 1987 KHimicheskoe mednenie iz rastvorov s fosfororganicheskimi kompleksonami // V sb : Tez dok IX Vsesoyuznoy nauchno-tekhnicheskoy konferentsii po elektrokhimicheskoy tekhnologii "Galvanotekhnika - 87", Kazan, 260

[17] Sakhapov, R L , Nikolaeva, R V , Gatiyatullin, M H , Makhmutov 2016 M M Risk management model in road transport systems // *Journal of Physics: Conference Series* 738(1) 012008

[18] Sakhapov, R L , Nikolaeva, R V , Gatiyatullin, M H and Makhmutov, M M 2017 Modeling of traction-coupling properties of wheel propulsor // *Journal of Physics: Conference Series* **738(2)** 012115

IOP Conf. Series: Materials Science and Engineering **786** (2020) 012058 doi:10.1088/1757-899X/786/1/012058

Exploring the effect of thermoplastic composition on light emission

R L Sakhapov*, R F Kashipov and M H Gatiyatullin

Kazan State University of Architecture and Engineering, 420043, Kazan, Zelenaya str , 1, Russia

*E-mail: rusakhapov@gmail.com

Abstract. Currently, reflective coloring agents are used for road markings. To obtain the best result from the use of the light-reflecting effect of the marking, we propose to use luminophore (light-emitting powder), by adding it to the thermoplastic composition heated at about 200 ° C. It can ensure the best visibility in the dark, in the conditions of rain and snowfall. The effect of the use of this technology will be visible both in urban and rural areas, where it can surely be an excellent alternative in case of the absence of stationary lighting. This will give the driver a good view of the route, even with the lights off. By analyzing the available and relevant information and conducting the necessary research and experimentation, the main goal will be to obtain the best results and a rational proportion of components in the manufacture of experimental samples that provide the maximum light-emitting effect.

1. Introduction

The glow effect of the luminophore in the dark helps a driver to concentrate solely on the road when the opposite vehicle blinds with high lights. In addition, driving in the dark, especially over long distances, is associated with psycho-emotional stress, fatigue and drowsiness for the driver. At night, the human body should rest and recover, and the attention and reaction of the driver at night are noticeably different from the driving conditions in the daytime. Road markings, glowing in the dark, not only enhance the attention of drivers, but even make the trip exciting. The use of this technology on the structural elements of bridges or other road structures, especially in dangerous places on the road, will reduce the number of accidents. The lack of visibility of road markings is the most frequent disadvantage of the road leading to road accidents [1,12].

The principle of operation of the technology is quite simple: the surface, which absorbs and accumulates the energy of light from natural or artificial sources, begins to glow in the darkness. The maximum intensity of the glow could be observed for a few hours, followed by its gradual decrease. At night, luminous lines, writings and other symbols serve as an additional measure of the security system on difficult sections of motorways. Moreover, on sports fields and cycle tracks this light provides an opportunity for exciting competitions even at night. The new luminous direction in the field of road arrangement opens:

IOP Conf. Series: Materials Science and Engineering **786** (2020) 012058 doi:10.1088/1757-899X/786/1/012058

- huge earning opportunities for road companies;
- unique opportunities for real energy savings;
- real opportunities to ensure safe traffic on the roads;
- excellent opportunities to enjoy the beauty of the night trails.

Figure 1. Luminous markings on a highway.

The most "problematic" parameter is the specific coefficient of retroreflection. Insufficient values of this important characteristic of the road marking, which determines its visibility in the dark time with reflected light from the headlights of vehicles, are often recorded both on the new marking (during the acceptance control) and during operation. It is also necessary to note another disadvantage of the markings, directly related to this parameter - an uneven value of the elements spread on the area, first of all applied by hand markings [2].

2. History and origins

The word Luminophore comes from the Latin "lumen" - "light" and the Greek "phoros" - "bearing." Since ancient times, people know the glow of natural luminophores - some insects, minerals, rotting wood. Scientists were able to understand the nature of such a glow - luminescence - only relatively recently, with the development of atomic physics. After all, luminescence is explained by the phenomena occurring in the atom of matter: excitation of the atom and then its return to a stable energy state; during this transition light is emitted [3].

Excitation of an atom can cause ultraviolet and x-ray radiation, electric field, chemical reactions. Luminophores transform all these different types of energy into light. In some luminophores, the color of the glow is ultraviolet, in others - blue, green, yellow, red. By combining various luminophores, we can get a spectrum of daylight. These mixtures are usually used in fluorescent lamps.

Organic luminophores are used to create luminous colors, road signs. Luminophores are also used for the analysis of various substances, the manufacture of luminous devices, in the extraction of minerals, in biology.

Photoluminophores are a type of luminophores that have the properties of preserving the stored energy upon excitation, and its return, with the possession of its own afterglow of a certain duration after the cessation of excitation in the form of light radiation in the visible, ultraviolet or infrared region. A very wide list of compounds belongs to this class of luminophores. Scientists distinguish both natural photoluminophores and artificially synthesized [7].

3. Luminophore characteristic

Luminophore is a substance that is capable of converting the energy absorbed by it into light radiation [4]. All luminophores by their chemical nature are divided into inorganic (most of which belong to crystallophosphorus) and organic. Self-luminescence of inorganic phosphors is caused in most cases by the presence of extraneous cations, which are contained in small amounts (up to approximately 0.001%) [5].

The most common types of luminophores on the market are photoluminophores and electroluminophores. They are a mixture of a very complex composition: strontium aluminate activated by europium, dysprosium, yttrium. Their chemical formula is: ($SrAl_2O_4$):Eu, Dy, Y.

Aluminate-based photoluminophores are inert to aqueous and solvent media, resistant to various irradiations and have an afterglow of up to 20 hours.

Luminophores are inherently non-toxic, fire and explosion safe, and have no dangerous radioactive radiation.

Photoluminophore is a non-hazardous substance, hazard class of the components is 4. Luminophore does not pollute the environment and has high stability of chemical reactions [5].

The comparative characteristics of zinc-based luminophores and aluminates of strontium luminophores of long afterglow are shown below. The DLO-7D brand and zinc phosphor were taken as the basis.

Glowing in the dark luminescent pigment (Luminophore), is characterized by its peculiarity, ability to absorb the energy of natural and artificial light sources and emit it in the form of a visible glow in the dark. DLO-7D brand pigment consists of very thin crystals, which basically have the molecule of strontium aluminate $SrAl_2O_4$:Eu, Dy, radically different from ordinary phosphorescent pigment, which is based on zinc sulfide or radioisotopes with their self-luminosity properties.

Some advantages and differences of luminophore pigments:

- The luminescence period in the dark is 50 times longer than that of a conventional pigment based on ZnS.
- Activation by waves of different lengths (200-450 nm), but the best result is obtained with activation energies above 350 nm [6].
- It does not pollute the environment and has high chemical stability.
- Absence of hazardous and radioactive substances.
- The initial brightness of the afterglow is at least 10 times longer than that of radioluminescent and photoluminescent pigments.
- Increased luminescence and afterglow with increased activation time.
- Excellent weather and light resistance of the pigment.

The DLO-7D luminophore (Figure 2) continues to accumulate and preserve light energy, not reaching the saturation point, much longer than the classic phosphorescent pigment.

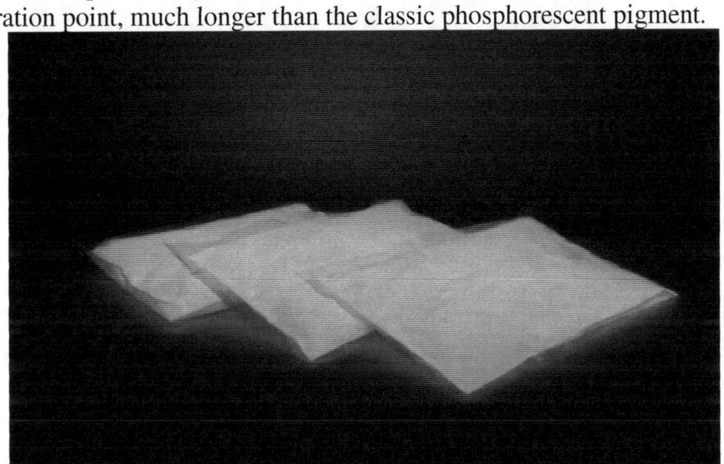

Figure 2. Luminophore DLO-7D.

4. Afterglow characteristics

Brightness, afterglow and color diffusion.

When activated, the most effective energy saturation can be obtained when the unprotected pigment is exposed to the directional ultraviolet ray (UV) of the sun, a halogen lamp, a gas discharge lamp, and other light sources rich in ultraviolet light.

On the other hand, ordinary fluorescent lamps are rich in ultraviolet light, and then quick activation is possible if we add pigment next to them. The brightness of the afterglow is also proportional to the intensity of the ultraviolet contained in the activating light and the activation time.

The upper curve shows the characteristics of the afterglow of the strontium aluminate luminophore $SrAl_2O_4$:Eu, Dy in comparison with the pigment based on zinc sulfide.

The measurements were taken from shielding silky surfaces that were activated by a 200-lux light source for 4 minutes.

As can be seen from the graph (Figure 3), $SrAl_2O_4$:Eu, Dy is 10 times brighter and has approximately 10 times longer afterglow than the pigment based on Zn: Cu sulfide.

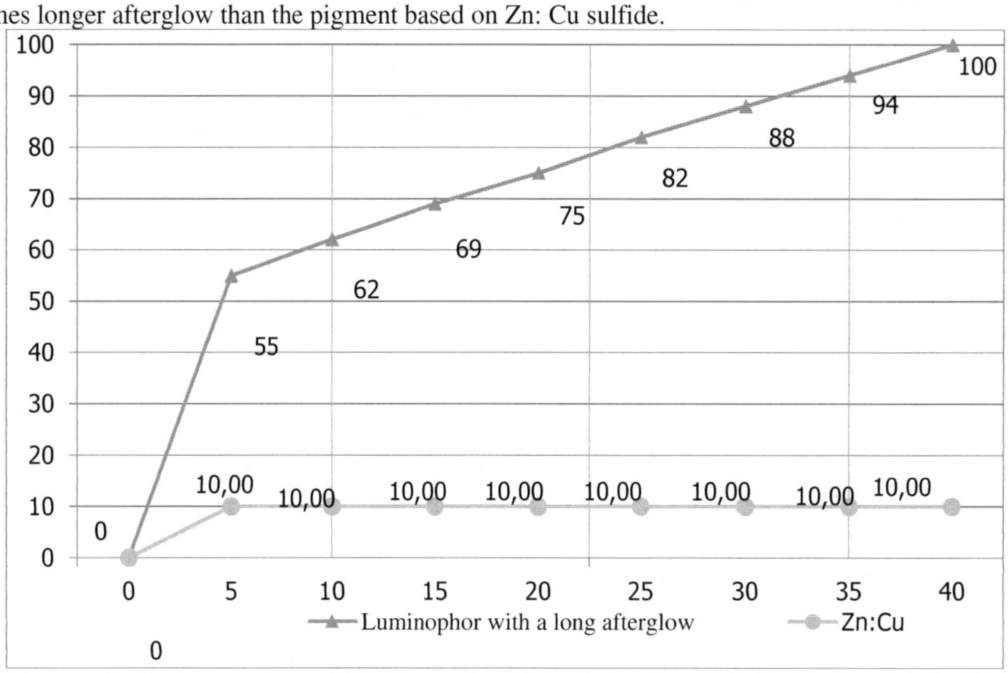

Figure 3. Afterglow Efficiency.

The activation time and brightness of the afterglow of $SrAl_2O_4$:Eu, Dy continues to accumulate and preserve light energy (without reaching the saturation point) much longer than the classical phosphorescent pigment. The graph (Figure 3) shows the results when both types of pigments were activated using D65, a standard source of daylight intensity of 200 Lux.

5. Retroreflective materials

The most important property of the marking is its good visibility in bright light, and in dark nights, and in heavy rain, and in cloudy weather, and in the light of the sun, and in the light of headlights. Due to this property, it is possible to reduce the number of road traffic accidents at night by 30%.

Thermoplastics designed for road marking on asphalt pavements are made by dry mixing of the following components: binders, fillers, pigments, processing aids [13].

Good visibility is achieved by using retroreflective materials in road markings, for example, reflective glass beads [14]. They are not used independently as marking material, but only in combination with it, to increase the visibility of marking, especially at night, rainy and cloudy weather.

Their reflective property is based on the ability of glass beads, which are located on the surface of the marking material, to refract the light coming from the headlights of the car and reflect it at a different angle so that it gets into the eyes of the driver.

There are three methods of application:

1. The introduction of the marking material in the amount of 10 - 20%;
2. Sprinkling onto freshly applied markings in an amount of 200 - 300 g / m^2;
3. A combination of these two methods at the same time.

To guide the light of the headlights, refracted by the bead and reflected from its inner surface, into the driver's eyes to the greatest extent, the bead must be half the height above the marking surface, and its surface should be free from marking material. In this case, it is also fairly well retained by the marking material. But if the bead protrudes from the marking layer by more than a half, it could easily be hit by the wheel of a car. If the bead is immersed in a layer of marking material more than a half, the amount of reflected light will decrease [13,14].

6. Research results

The purpose of the study was to study the effect of luminophore on thermoplastics, i.e. in obtaining a rational ratio of substances that provide the maximum retroreflective effect in conditions of complete absence of light.

To determine the rational ratio of luminophore to thermoplastic, we drew up an experimental plan.

During the study, we calculated various ratios of substances:

Table 1. Calculated ratios of luminophore and thermoplastic.

Luminophore, g	Thermoplastic, g
20	200
40	200
60	200
80	200
100	200
120	200
140	200

The results of the dependence of the illumination time on the radiation level of the samples in the ratios 20/200, 40/200, 60/200, 80/200, 100/200, 120/200, 140/200 are shown in Figure 4.

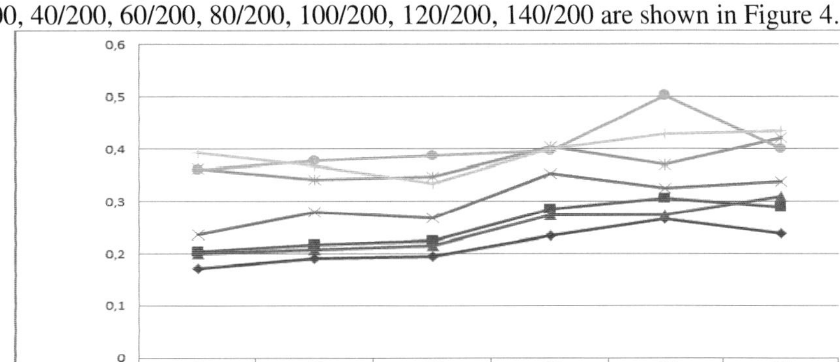

	5	10	15	20	25	30
20/200	0,172	0,19	0,194	0,234	0,266	0,238
40/200	0,204	0,216	0,224	0,284	0,306	0,288
60/200	0,2	0,208	0,214	0,274	0,274	0,31
80/200	0,236	0,28	0,268	0,352	0,324	0,338
100/200	0,362	0,342	0,348	0,404	0,37	0,42
120/200	0,36	0,378	0,386	0,396	0,502	0,4
140/200	0,392	0,368	0,334	0,4	0,428	0,434

Figure 4. The dependence of the illumination time (Y) on the radiation level (X).

The results of the dependence of the decay time on the radiation level of the samples in the ratios 20/200, 40/200, 60/200, 80/200, 100/200, 120/200, 140/200 are shown in Figure 5:

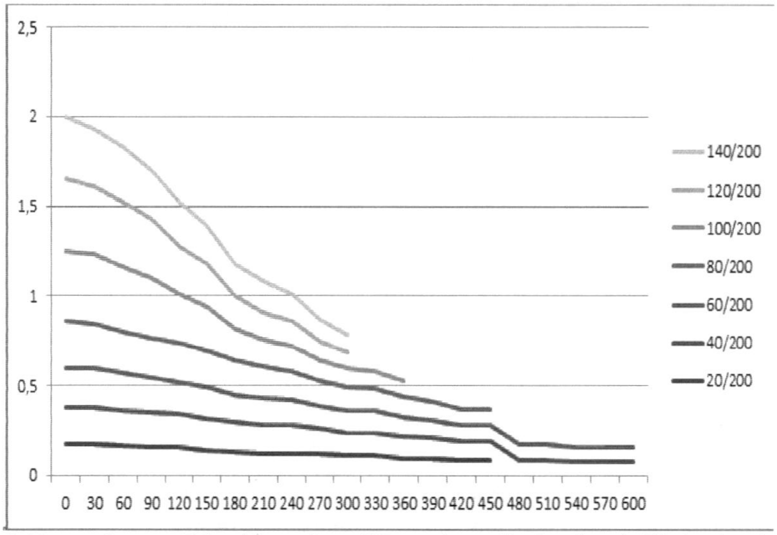

Figure 5. Dependence of the decay time (Y) on the radiation level (X) (activation 30 s).

As a result of the study and processing of the obtained indicators, we determined that the characteristics of the experimental samples with a high luminophore content practically did not differ.

They also have increased brightness compared to other experimental samples, but the duration of the glow decreases faster.

Thus, samples with 60 grams and 80 grams of luminophore will be rational in terms of component ratio and efficiency.

7. Conclusion

As a result of research on the properties of thermoplastics, we determined the most effective indicators of the luminophore glow in thermoplastics.

We determined and processed the luminosity indices for a certain period of time, luminosity indices during the complete decay of prepared thermoplastic samples with different ratios of thermoplastic and luminescent powder.

The application of the above technology has many advantages compared with the technology of conventional thermoplastics:

1. Creating a favorable aesthetic appearance in the conditions of the city, rural areas, etc.
2. Ensuring safe traffic on roads in the absence of light or in the presence of sections of roads with low light.
3. The absence of negative effects on the environment and human health.
4. Low material costs compared to conventional thermoplastics.
5. The application of this technology is possible with a road marking machine, which will improve the performance of the equipment and increase its efficiency and road safety.

References

[1] GOST R 51256-2011 "Technical means of traffic management. Road markings. Classification. Technical requirements"
[2] GOST 32953-2014 “Public roads Road markings Technical requirements"
[3] Kazankin O N and others 1975 "Inorganic phosphors " Ed “Chemistry”, LO
[4] *Chemical Encyclopedia*, **2** 1990211-222
[5] Zvereva EM, Manashirov O Ya and Golota AF 2006 New class of multifunctional phosphors based on strontium aluminates // Herald of the Southern Scientific Center of the Russian Academy of Sciences **2(4)** 31-37

[6] Zvereva EM, Manashirov O Ya and Golota AF 2006 Development of a new type of phosphors with enhanced functionality based on strontium aluminates // Proceedings of the VIII international conference "Opto-, nanoelectronics, nanotechnologies and microsystems" Ulyanovsk 193

[7] Zvereva EM, Golota AF and Manashirov O Ya 2008 Study of phase and structural transitions in the SrAl2O4 – CaAl2O4 binary system // *Proc report All-Russian Conference "Solid State Chemistry and Functional Materials - 2008"* Yekaterinburg, October **21-24** 145

[8] Sakhapov R L and Kashipov R F 2018 Phosphor applications in road marking // *Technique and technology of transport* **3(8)** 6 URL: http://transport-kgasu ru/files/N8-06BDD318 pdf

[9] Sakhapov R L and Absalyamova S G 2015 New priorities of industrial policy and a paradigm shift in engineering education News of the Kazan State University of Architecture and Engineering **3** 221-9

[10] Absalyamova, S G and Absalyamov, T B 2015 Remote employment as a form of labor mobility of today's youth *Mediterranean Journal of Social Sciences*, **6(1 S3)**, 227

[11] Sakhapov, R L and Absalyamova S G 2015 The use of telecommunication technologies in education network In Proceedings of 2015 12th International Conference on Remote Engineering and Virtual Instrumentation (REV) 14-7 IEEE

[12] Nikolaev R V 2016 Proposals for the design of roads, taking into account the perception of the driver of road conditions News of the Kazan State University of Architecture and Engineering **2(36)** 252-8

[13] https://www.studioroosegaarde.net/info.

[14] https://www.heijmans.nl/nl/nieuws/eerste-pilot-met-glowing-lines-oss.

[15] https://revolution.allbest.ru/manufacture/00492273_0.html.

[16] http://snab.ru/lkm/02/13.html.

IOP Conf. Series: Materials Science and Engineering **786** (2020) 012059 doi:10.1088/1757-899X/786/1/012059

Formation of a system for assessing the renovation of monolithic buildings

D Topchiy[1] and A Bolotova[2]

[1] Moscow State University of Civil Engineering, Yaroslavskoe shosse, 26, Moscow, 129337, Russia
[2] Moscow State University of Civil Engineering, Yaroslavskoe shosse, 26, Moscow, 129337, Russia

E-mail: 89161122142@mail.ru, bolotova63@mail.ru

Abstract. Reliability and quality of monolithic construction works are determined by analyzing the system of preparation for implementation of certain technological operations, establishing and ensuring the order, sequence and timetable of works, and supply of all types of resources. Periodic inspections of compliance with construction norms and regulations are held in the course of construction and installation works. As a follow-up on such inspections, prescriptive orders on rectification of detected nonconformities and violations within a specified time limit are issued. There is no provision for the time consumed by corrective action in the calendar plan, and consequently the actual duration of construction as often as not exceeds the scheduled term to a great extent. In this connection, there is a task of assessing the level of impact of violations on the efficiency of functioning and the organizational quality of production processes. Analytic research into this problem allows for a conclusion on the need for development of a mathematical model to be used in the assessment of duration of construction taking into account the random pattern of occurrence of violations.

1. Introduction

Organizational and technological reliability is understood as the ability of technological, organizational, and managerial solutions to ensure attainment of a target result of construction operations under the conditions of random disturbances that are intrinsic to construction as a complex stochastic system. Assessment of the level of organizational and technological reliability of monolithic construction projects requires quantitative characteristics. The main criteria of organizational and technological reliability are duration, cost and quality of construction [1-3]. Collection of information and registration of random events that have an impact on the functioning of a production process in monolithic construction operations include recording the time of occurrence of defects and failures [7-8], their duration, causes and circumstances of loss of productivity, deviations from the schedule of implementation of the planned scope of works, completion of technological stages and stages of works.

2. Materials and methods

The type of distribution of the quantity of violations is determined as follows. There is a list of types of violations numbered by indexes i=1, 2,..., m. For the purpose of modelling the time of elimination of violations we assume that the quantity of violations ξ_i, of each type i=1, 2,..., m is a random value. In order to derive the type of distribution of random values ξ_i let us fix the time interval ΔT – (mean) inspection frequency. Let us divide the ΔT interval into s equal parts $\Delta T_s=\Delta T/s$ long and designate by p_{is} the probability of occurrence of at least one i violation during the fixed time phase. Then the probability of an opposite event, namely, not a single i violation occurring in the fixed time phase, is $q_{is}=1-p_{is}$, with the p_{is} probability being irrespective of the number of the time phase. It means the presence of binominal distribution at the ΔT segment, and the probability of occurrence of precisely k violations is equal to $b\ \ (k)=C^k p^k q^{s-k}$. Proceeding from a natural assumption that sp_{is} is limited for all s, it appears that $sp_{is}\to\lambda_i$, with $s\to+\infty$, where λ_i is a constant value that differs for each index i=1, 2, …, m. In this case, with $s\to+\infty$, $b\quad _{is}(k)\to\pi_i(k)$ [4-6], where

$$\pi_i(k) = \frac{\lambda_i^k}{k!}e^{-\lambda_i} \tag{1}$$

Thus, the random value ξ_i at the ΔT segment is distributed by the Poisson law with λ_i parameter.

For the purpose of assessment of λ_i, 1, 2,…, m parameters, let us use statistical data on the indicators under review based on analysis of construction control operations. Information has been mainly obtained from results of systemic analysis and statistical processing of actual data on conformities and violations detected by applied research of Scientific Research and Testing Center "MGSU STORY-TEST" in monitoring the quality of construction projects from 2012 to 2016 [2-5, 12]. Prescriptive orders issued by construction supervision authorities specify the names of violations, type of the works performed, deadline for elimination, and the issue date of the prescriptive order. All violations of construction regulations are divided into three groups: violations of requirements of design documentation, violations of requirements of normative documents, and violations of requirements of technological regulations. Nonconformities in production processes have various degrees of impact on the overall construction of the project. There is no provision for the time consumed by corrective action in the calendar plan, and consequently the actual duration of construction as often as not exceeds considerably the scheduled term.

Let us fix index i=1, 2,…, m and find mathematical expectation $M\xi_i$ of random value ξ_i. We obtain directly from the mathematical expectation:

$$M\varsigma_i = \sum_{k=0}^{+\infty} k\pi_i(k) = \sum_{k=0}^{+\infty} k\frac{\lambda_i^k}{k!}e^{-\lambda_i} = \lambda_i e^{-\lambda_i}\sum_{k=1}^{+\infty}\frac{\lambda_i^{k-1}}{(k-1)!} = \lambda_i e^{-\lambda_i}e^{\lambda_i} = \lambda_i \tag{2}$$

By virtue of the law of large numbers, mathematical expectation $M\xi_i$ of random value ξ_i is equal to the frequency limit of occurrence of i violation in the sample, i.e. the ratio limit:

$$\lambda_i = \frac{N_i}{N}, \text{with } N \to +\infty, \tag{3}$$

where N_i is the quantity of i violations in the sample; N is the overall number of inspections held.

Thus we can assess λ_i parameter using equation (3).

Probability assessment for extending the duration of construction due to any violations requires an assessment of $D\xi_i$ dispersion of random value ξ_i for each index i=1, 2,…, m. For this purpose let us first find the mathematical expectation of random value ξ_i^2:

$$\begin{aligned} M\varsigma_i^2 &= \sum_{k=0}^{+\infty} k^2\pi_i(k) = \sum_{k=0}^{+\infty} k^2\frac{\lambda_i^k}{k!}e^{-\lambda_i} = \lambda_i e^{-\lambda_i}\sum_{k=1}^{+\infty}\frac{k\lambda_i^{k-1}}{(k-1)!} = \\ &= \lambda_i e^{-\lambda_i}\Big(1+\sum_{k=2}^{+\infty}\Big(1+\frac{1}{k-1}\Big)\frac{\lambda_i^{k-1}}{(k-2)!}\Big) = \lambda_i^2 e^{-\lambda_i}\sum_{k=2}^{+\infty}\frac{\lambda_i^{k-2}}{(k-2)!} + \lambda_i e^{-\lambda_i}\sum_{k=1}^{+\infty}\frac{\lambda_i^{k-1}}{(k-1)!}) \\ &= \lambda_i^2 e^{-\lambda_i}e^{\lambda_i} + \lambda_i e^{-\lambda_i}e^{\lambda_i} = \lambda_i^2 + \lambda_i \end{aligned} \tag{3.1}$$

We obtain from the last equation and equation (2) by virtue of determination of dispersion:

$$D\varsigma_i = M\varsigma_i^2 - \left(M\varsigma_i\right)^2 = \lambda_i + \lambda_i^2 - \lambda_i^2 = \lambda_i \tag{4}$$

Then we should assess the time consumed by elimination of violations at a certain job. Let us use a calendar plan of a construction project that consists of n jobs numbered by indexes j=1, 2,…, n. Each j job is characterized by the time of its commencement t^0, completion t^1, and duration T_j, where

$$t_j^1 > t_j^0 \geq 0, T_j = t_j^1 - t_j^0 \tag{5}$$

for all j=1, 2,…, n, with $t^0 = 0$

For each index j=1, 2,…, n, we draw up a list of B_j indexes (types) of violations that may occur in the execution of j job. Each index $i \in B_j$ is assigned compatibility indicator a_i equal to 0 in the event that one or more violations with index i can be eliminated irrespective of another or other violations (in parallel), and equal to 1, if these violations must be eliminated one-at-a-time (in sequence). Let us designate the set of indexes with a_i=0 as C_j, and the set of indexes $i \in B_j$ with a_i=1 as E_j,

We know τ_i, the time of elimination of a single i violation, for each index i=1, 2,… . With that in mind and proceeding from the above-mentioned values for $M\xi_i$ and $D\xi_i$, we can assess the time consumed by the elimination of violations in the course of execution of each job as follows.

Let us fix j job and divide it into time intervals $[t_{j0}, t_{j1}]$, $[t_{j1}, t_{j2}]$, …, $[t_{j,lj-1}, t_{jlj}]$ each being ΔT long,

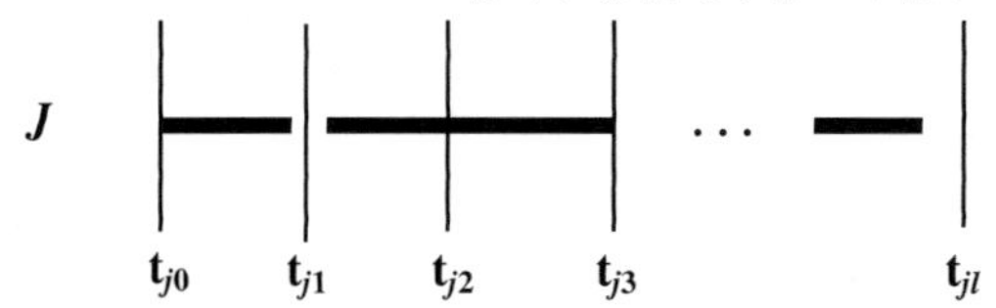

Fig. 1. An extract from a job schedule with j job broken down by ΔT intervals.

The number of i violations at each of the intervals $[t_{jr}, t_{jr+1}]$, r=0, 2, …, l_j-1, where $i \in B_j$, is a random value ξ_i, which has a Poisson distribution with mathematical expectation and dispersion equal to λ_i, (see (2) and (4)). We calculate the time consumed by elimination of these violations by proceeding from the following natural assumptions. With regard to violations capable of being eliminated in parallel, i. e. with $i \in C_j$, the time of their joint elimination is equal to the maximum duration of elimination of each of them. Regarding violations that can be eliminated in a consecutive manner only, i.e. with $i \in E_j$, the time of their joint elimination is equal to the total time consumed by elimination of all possible violations.

To assess the time spent on parallel elimination of violations in the course of performing a separate work, we fix the index j=1, 2, ..., n and first, we estimate the total time spent on elimination of all violations with indices $i \in C_j$ (which allow parallel elimination) on the interval ΔT_j work.

For each index $i \in C_j$ from equality (1) we have: $\pi_i(0) = \frac{\lambda_i^0}{0!} e^{-\lambda_i} = e^{-\lambda_i}$. Consequently, the probability of occurrence of at least one i violation on the interval ΔT is equal to $\pi_i(k > 0) = 1 - e^{-\lambda_i}$. Since these violations can be eliminated sequentially, the time spent on their cumulative elimination is a random variable τ_{0i}, with the distribution function $F_i(t)$, taking the value $\pi_i(0) = e^{-\lambda_i}$ when $t < \tau_i$, and - the value 1 for $t \geq \tau_i$, that is

$$F_i(t) = \begin{cases} e^{-\lambda_i} \text{ where } t < \tau_i \\ 1 \text{ where } t \geq \tau_i \end{cases} \tag{6}$$

With the distribution function (6) for all indices $i \in C_j$, is easy to obtain the distribution function $F_{j0}(t)$ of a random variable τ_{j0} - the time spent on the elimination of all such violations in the j work as the product of $F_i(t)$ distribution functions for all $i \in C_j$:

$$F_{j0}(t) = \prod\{F_i(t), i \in C_j\} \tag{7}$$

Indeed, since $\tau_{j0}=\max\{\tau_{0i}: i \in C_j\}$, then by virtue of the independence of the random variables ξ_i for all $i \in B_j$, for arbitrary $x > 0$, we have:

$$F_{j0}(t) = P(\omega: \tau_{j0}(\omega) \le t) = P(\omega: max\{\tau_i(\omega), i \in C_j\} \le t) = P(\cap \{\{\omega: \tau_i(\omega) \le t\}, i \in C_j\})$$
$$= \prod \{P(\omega: \tau_i(\omega) \le t), i \in C_j\} == \prod \{F_i(t), i \in C_j\}.$$

Equalities (6) and (7) allow us to estimate the expectation and variance of a random variable τ_{j0} as follows

We enumerate the duration τ_i eliminate violations $i \in C_j$ in ascending order: $\tau_{i_1} \le \tau_{i_2} \le \cdots \le \tau_{i_k}$, where k is the number of elements in the set $i \in C_j$. From formulas (6) and (7) we have:

$$F_{j0}(t) = \begin{cases} e^{-\lambda_{i_1}} \dots e^{-\lambda_{i_s}} \; where \; t < \tau_{i_1}, \\ e^{-\lambda_{i_2}} \dots e^{-\lambda_{i_s}} \; where \; \tau_{i_1} < t < \tau_{i_2}, \\ \dots \\ e^{-\lambda_{i_q}} \dots e^{-\lambda_{i_s}} \; where \; \tau_{i_{q-1}} < t < \tau_{i_q}, \\ \dots \\ 1 \; where \; t \ge \tau_{i_s}. \end{cases}$$

From the last equality, we find the estimates for the average $M\tau_{j0}$ and the variance $D\tau_{j0}$ random variable τ_{j0}:

$$M\tau_{j0} = \sum_{q=1}^{s} \left(1 - e^{-\lambda_{i_q}}\right) e^{-\lambda_{i_{q+1}}} \dots e^{-\lambda_{i_s}} \tau_{i_q}. \quad (8)$$

$$M\tau_{j0}^2 = \sum_{q=1}^{s} \left(1 - e^{-\lambda_{i_q}}\right) e^{-\lambda_{i_{q+1}}} \dots e^{-\lambda_{i_s}} \tau_{i_q}^2,$$

$$D\tau_{j0} = M\tau_{j0}^2 - (M\tau_{j0})^2 \quad (9)$$

3. Conclusion

Analysis of factors affecting the organizational and technological reliability of monolithic construction operations leads to a conclusion that the level of such reliability can be assessed by the indicator of duration of construction and installation works [10, 13-16]. This requires the use of a mathematical model for assessing the duration of construction and installation works taking into account the time losses of elimination of violations. To begin with, the duration of separate works should be determined on the basis of the calendar plan followed by an assessment of the probability of extending the term of construction as required for elimination of violations prescribed by construction supervision authorities. In this connection it is necessary to develop methods of increasing the organizational and technological reliability of monolithic construction projects by means of timely adjustment of the construction calendar plan.

References

[1] Bolotova A S and Ginzburg A V 2016 Analysis of organizational and technological reliability of monolithic construction operations *Economics and Entrepreneurship* **10** 647–651

[2] Bolotova A S 2016 Methods of increasing the organizational and technological reliability of monolithic construction operations *Science Review* **18** 186–190

[3] Bolotova A S and Sviridov V N 2016 Main challenges of organizing and exercising control of quality indicators in monolithic construction operations *Science Review* **24** 25–29

[4] Rubtsov I V, Treskina G E and Bolotova A S 2015 Classification of monolithic reinforced-concrete structure defects and their impact on quality *Scientific review* **18** 58–63

[5] Bolotova A S 2014 Process safety and quality control of monolithic construction, in *Construction - formation of living environment* (Moscow: MSUCE) pp. 501–504

[6] Abdullayeva G I, Velichkin V Z and Soldatenko T N 2013 Improving the organizational and technological reliability of construction of continually extended structures by the failure forecasting method *Civil Engineering Journal* **3** 43-50

[7] Feller W 1967 *Introduction to probability theory and its applications, vol. 1* (Moscow: Mir)
[8] Ginzburg A V 2016 Information model of the life cycle of a construction project *Industrial and Civil Construction* **9** 61-65
[9] Lapidus A A 2014 Efficiency potential of organizational and technological solutions of a construction project *Herald of the Moscow State University of Civil Engineering* **1** 175–180
[10] Lapidus A and Abramov I 2018 Formation of production structural units within a construction company using the systemic integrated method when implementing high-rise development projects *E3S Web of Conferences* **33** 03066
[11] Topchiy D, Skakalov V and Yurgaitis A 2018 Integrated verification construction supervision as a tool to reduce the developer's risks when implementing new and redevelopment projects *International Journal of Civil Engineering & Technology* 1 85–993
[12] Abramov A, Poznakhirko T and Sergeev A 2016 The analysis of the functionality of modern systems, methods and scheduling tools *MATECWebConf* **86**
[13] Topchiy D and Kochurina E 2018 Estimation of the degree of influence of environmental factors on construction in a dense urban environment *System technologies* **1** 107–111
[14] Topchiy D and Skakalov V 2017 Structural and functional modeling of multi-level and multi-criteria links of organizational and technological, managerial structures and information support when implementing construction control over the course of re-development of industrial facilities Prospects of science **10** 44–50
[15] Topchiy D and Kochurina E 2018 Environmental situation in construction, reconstruction and re-profiling of facilities in high-density urban development *MATEC Web Conf* **193**
[16] Topchiy D and Tokarskiy A 2018 Formation of the organizational-managerial model of renovation of urban territories *MATEC Web Conf* **196**

IOP Conf. Series: Materials Science and Engineering **786** (2020) 012060 doi:10.1088/1757-899X/786/1/012060

Use of program ANSYS in the analysis of interaction between a rigid foundation and reinforced soil

V M Antonov[1] and I A Al-Naqdi[2]

[1] Department of buildings and structures Construction, Tambov state technical University, Tambov, Russia
[2] University of Technology, Baghdad, Iraq

E-mail: antonov280562@mail.ru, iabdulmonem@gmail.com

Abstract. Work is devoted to model and process the results of interaction between a rigid metal plate and a reinforced sand base using the software "ANSYS".
One of the effective ways to increase the strength is the inclusion of various materials (reinforcement) in the soil below the structure. It increases the resistance of the soil to stress and shear, limits lateral deformation. Organic, synthetic, metal and stone materials are used as reinforcing elements. The studies showed that the effectiveness of the reinforcement and the bearing capacity of the reinforced ground depend on the position, length, roughness and fixity condition of the reinforcement [27].

1. Introduction

In the laboratory of "soil Mechanics" TSTU more than 20 years, conducted testing of reinforced bases. This article presents the results of the tank tests of a two-layers base, the upper layer of which (in the first series of the experiments was 30cm, in the second 10cm) was represented by distributiona cushion of fine homogeneous sand, the lower (in the first series was 30cm, in the second 50cm) imitated a weak base. Sawdust was used for the lower layer.

In order to strengthen the base, smooth and corrugated steel rods were used, which were placed both under the sole of the model and outside the sole. Cement-sand injections with different angles of inclination and volumes were also used. Rods and injections were supposed to limit lateral deformations and change the stress state in the soil [1-5, 8, 10], adding reinforcing fibers basically can increase soil tensile strength [26]. Reinforcing also increases bearing capacity of soil [14-22] and reducing settlement [23-25]. During experimental tests, were determined the current displacement(s),where s – is the displacement of the center of the model, and displacement prior to ultimateload (s_u). Deformations were measured by indicators of the hour type ICH-10, fixed on the reference frame. The repeatability of the experiments was 2..3 times [6, 7, 9].

To assess the effect of reinforcing soil on the development of deformation in the base, a numerical solution by FEM for the scheme "Foundation – base" was performed. The results were processed using finite element method by ANSYS / Multiphysics [11-13].

To simulate soil-matrices, a three-dimensional model / Drucker-Prager with dimensions of 550 mm x 650 mm x 750 mm was studied, which consisted of 6256 elements, each of which had a length of 40 mm, a width of 40 mm and a depth of 40 mm as in Fig.1.

Injections were considered as a rod diameter = 6 mm with modulus of elasticity E = 6000 MPa, density ρ = 1800 kg/m^3 and Poisson's ratio ν = 0.17. InjectionwassimulatedusingelementLink180.

IOP Conf. Series: Materials Science and Engineering **786** (2020) 012060 doi:10.1088/1757-899X/786/1/012060

The development of displacement was limited in the x and y directions, i.e. the displacements could develop only in the vertical direction z. The plate was modeled using a linear isotropic model by Solid 45 element type, the model consisted of 32 elements. An element is defined by eight nodes that have three degrees of freedom at each node.

Plate diameter 100 mm, modulus of elasticity E = 203940 MPa, Poisson's ratio ν = 0.3; density ρ = 7700 kg/m^3.

a) b)

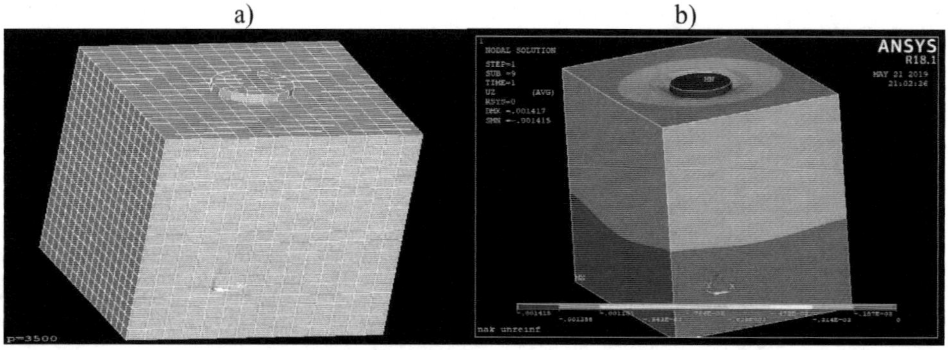

Figure 1. Model by ANSYS (a) and Stress-strain state of unreinforced sand.

Injections of cement mortar into a well with a diameter of 6mm to a depth of ¼ D and ½ D (D-diameter of the Foundation model) on two or three sides were carried out. With angle of injection 30^0,45^0 and 60^0. At the end of the hardening of the mortar, the vertical load on the base was transferred by means of a lever through a plate D=100mm. The volume of cement mortar, angle of inclination and depth of injections were changed. The projection scheme is shown in Fig.2.

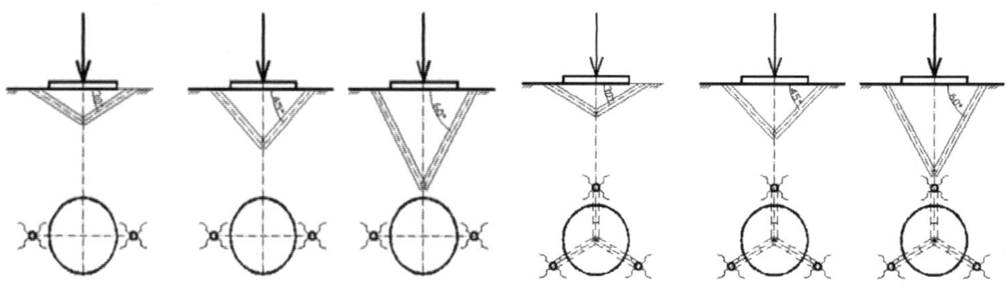

Figure 2. Scheme for the arrangement of injections.

Characteristics of the layer of fine sand - General modulus of deformation E = 48000 kPa, Poisson's ratio ν = 0.3; density ρ = 1510 kg/m^3, cohesion c = 6 kPa, angle of internal friction φ = 38°. For sawdust layer - ρ = 0,5 g/sm^3, ω = 50%,modulus of deformation E = 220 kPa, cohesion c = 2 kPa; the angle of internal friction φ = 30°, Poisson's ratio ν = 0,016.

The effectiveness of reinforcement was evaluated by values:

a) coefficient of relative strain at a fixed load,

$$k_s = s/s_{0,} \quad (1)$$

(where s is the settlement-in case of reinforced, s_0 is settlement, for unreinforced base at same load);

b) the reinforcement ratio,

$$k_\mu = F_{us}/\mu_s, \quad (2)$$

IOP Conf. Series: Materials Science and Engineering **786** (2020) 012060 doi:10.1088/1757-899X/786/1/012060

(whereμ_S is the percentage of reinforcement of base,$\mu_S = V_S/V, V_S$ is the volume of reinforcement,V is the volume of the base used in the work, conventionally defined as V = B. L. H, L is the length, B is width, and H is the height; all of the base);

c) coefficient of relative bearing capacity of the base,

$$k_F = F_{us}/\ F_u, \tag{3}$$

(where F_{us} -bearing capacity of reinforced base -ultimate load), F_u - the same but for non-reinforced.

The results of comparison of numerical simulations and experimental values of settlement for the angle of inclination of 45° injections, are shown in Fig.3 (a-e). The development of the main stresses under the foundation in the software package "ANSYS" is shown in Figure 3 (a,b).

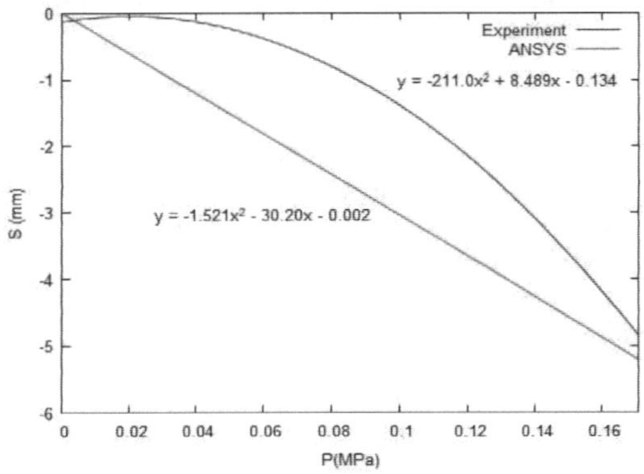

(a)

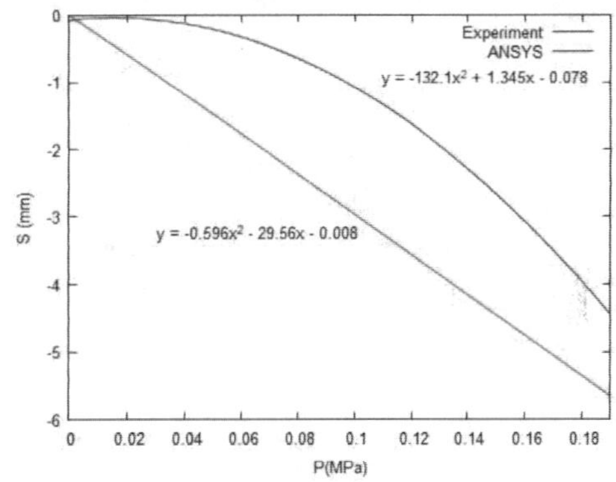

(b)

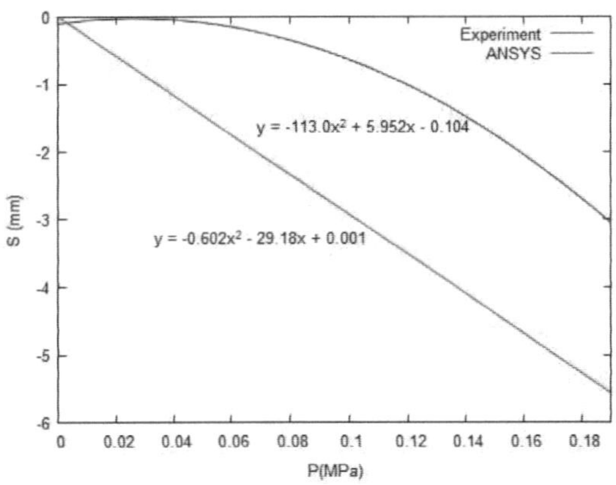

(c)

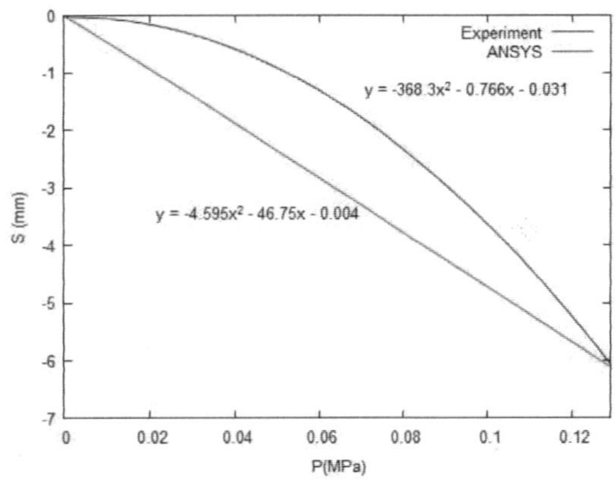

(d)

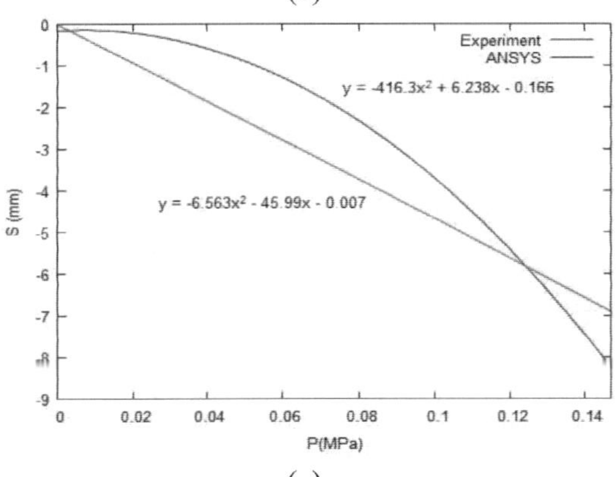

(e)

Figure 3. Load-settlement curve in case, angle of inclination equals 45^0 for: a- unreinforced base-30 cm sand, b- reinforced by 2 injections- 30 cm sand, c- reinforced by 3 injections- 30 cm sand, d- reinforced by 2 injections- 10 cm sand, e-reinforced by 3 injections- 10 cm sand.

IOP Conf. Series: Materials Science and Engineering **786** (2020) 012060 doi:10.1088/1757-899X/786/1/012060

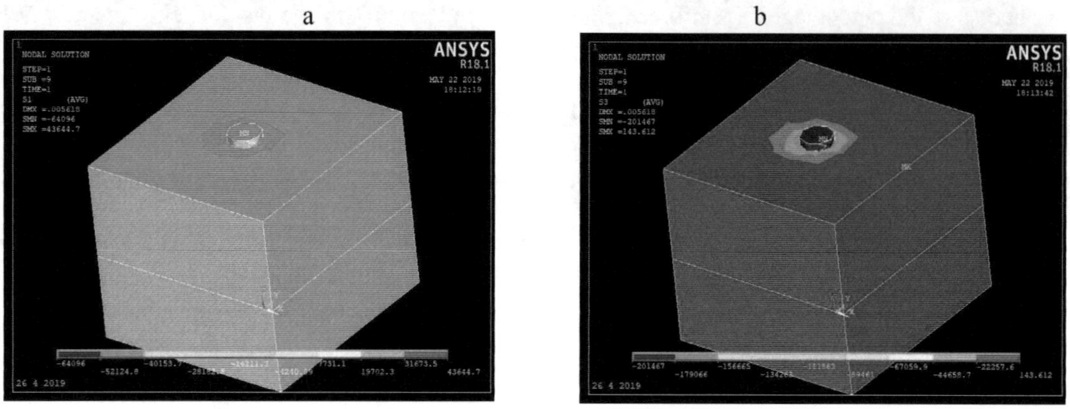

Figure 4. The main stresses $\sigma_1(a)$ *and* $\sigma_3(b)$, in the base of the model at a pressure under the sole P=190 kPa during two injections at ½ D, the angle of injection 45^0, and a layer of sand 30 cm.

The results of laboratory experiments are shown in table.1 (settlement values were determined at P = 0.17 MPa.

Table 1. Experimental results of the base, reinforced by cement-sand injection

n_p	$\bar{l}$	β°,	F_{us}[kN]	S [mm]	μ,%	k_μ	k_f	k_μ
$\bar{h} = \frac{30}{30} = 1,\ s_0 = 5.05\ mm, F_{us,0} = 1.35\ kN$								
2	0.25	30	1.65	2.47	0.90	0.49	1.22	1.83
		45	1.50	2.79	1.12	0.55	1.11	1.35
		60	1.50	3.46	1.57	0.68	1.11	0.95
	0.5	30	1.80	2.69	1.36	0.53	1.33	1.32
		45	1.65	3.58	1.66	0.71	1.22	0.99
		60	1.50	3.52	2.35	0.70	1.11	0.64
3	0.25	30	1.50	4.01	1.36	0.79	1.11	1.10
		45	1.35	4.43	1.65	0.85	1.00	0.82
		60	1.35	4.60	2.35	0.91	1.00	0.57
	0.5	30	1.80	2.05	2.04	0.40	1.33	0.88
		45	1.50	2.23	2.50	0.44	1.11	0.60
		60	1.65	2.38	3.53	0.47	1.22	0.47
n_p	$\bar{l}$	β°,	F_{us}[kN]	S [mm]	μ,%	k_μ	k_f	k_μ
$\bar{h} = \frac{10}{50} = 2, s_0 = 5.95\ [mm], F_{us,0} = 0.675\ [kN]$								
2	0.25	30	0.75	5.29	0.90	0.89	1.11	1.10
		45	0.75	5.08	1.12	0.85	1.11	0.82
		60	0.75	4.42	1.57	0.74	1.11	0.57
	0.5	30	1	2.77	1.36	0.46	1.48	0.88
		45	1	2.46	1.66	0.41	1.48	0.60
		60	0.85	3.71	2.35	0.62	1.26	0.47
3	0.25	30	0.75	3.75	1.36	0.63	1.11	0.55
		45	0.82	2.34	1.65	0.39	1.22	0.50
		60	0.75	4.01	2.35	0.67	1.11	0.32
	0.5	30	1.27	2.59	2.04	0.43	1.89	0.62
		45	1.12	2.48	2.50	0.41	1.67	0.45
		60	1.05	3.01	3.53	0.50	1.56	0.30

2. Summary

The settlement of the reinforced base is less than the unreinforced, and the carrying capacity is higher. A significant influence on the development of deformation, having the angle of injection (the values of K_s when injection angle 30^0,45^0 and 60^0 to a depth of 1/4D, respectively, was equal to k_s=0.49;0,55;0,68 by p= 0,17 MPa, μ=0,903, 1.07, 1.57), and the volume of cement mortar(0,53/0,41;0,71/0,45;0,7/0,47).

It is shown that increasing the depth of the reinforcement zone from ¼ D to ½D reduces the rate of settlement developed by 2 times regardless of the thickness of the weak layer. Deformation at a lower thickness of the weak layer is lower, since in this case the weak layer is not within the compressible zone.

The bearing capacity of the reinforced base was increased up to 1.5 times depending on the angle of inclination and volume of the injected mortar.

References

[1] Eksperimental'nye issledovaniya peschanyh osnovanij, armirovannyh vertikal'nymi sterzhnyami / V.M. Antonov, S.N. Vybornov // Sb. trudov V1 nauchnoj konferencii TGTU. TGTU, Tambov - 2001, C. 202-203.

[2] Hryanina, O.V. Izmeneniya napryazhyonnogo sostoyaniya gruntov osnovaniya vvedeniem v nego armiruyushchih elementov / O.V. Hryanina, G.G. Boldyrev // Sovremennye problemy fundamentostroeniya: Sb. tr. Mezhdunar. nauch.-tekhn. konf. – Volgograd, 2001. – C. 96 – 98.

[3] Antonov, V.M. Eksperimental'noe issledovanie armirovannyh peschanyh osnovanij / V.M. Antonov // Sovremennye problemy fundamentostroeniya: Cb. tr. Mezhdunar. nauch.-tekhn. konf. – Volgograd, 2001. – C. 9–10.

[4] L utilization de la barre d amature pour l augmentation de la capacity portante de la base. (Vliyanie sterzhnevogo armirovaniya na nesushchuyu sposobnost' osnovaniya.) / Antonov V.M.,Vyazovov V.B., Vybornov S.N., Diavara S. //Vestnik TGTU. Tom 10 - № 2 - 2004/ - C. 567-578.

[5] Antonov, V.M. Vliyanie armirovaniya na prochnostnye i deformacionnye harakteristiki peschanogo osnovaniya / V.M. Antonov, V.V. Ledenev, D.A. Prokin // Nadyozhnost' i dolgovechnost' stroitel'nyh materialov konstrukcij i osnovanij fundamentov: Cb. tr. Mezhdunar. nauch.-tekhn. konf. – Volgograd: VolgGASA, 2005. – C. 9 – 13.

[6] Antonov, V.M., Ledenev V.V.Eksperimental'nye issledovaniya armirovannyh osnovanij ./ Vestnik central'nogo regional'nogo otdeleniya RAASN., Vyp.6, Voronezh-Tver',2007, C.3-11

[7] Antonov, V.M.,Ledenev V.V., Vliyanie urovnya nagruzheniya na deformacii polzuchesti peskov Vestnik otdeleniya stroitel'nyh nauk RAASN., Vyp.13, Tom 1, Moskva-Orel, 2009, C. 23-27

[8] Antonov, V.M. Eksperimental'nye issledovaniya armirovannyh osnovanij: monografiya / V.M. Antonov. - Tambov: Izd-vo GOU VPO TGTU – 2011.-78s

[9] Antonov, V.M., Kozlova YU.A., Kozlov A.A.Vliyanie armirovaniya dispersnogo osnovaniya na NDS P-obraznoj ramy.//LAP LAMBERT AcademicPublishing, 2016/ C.88-91

[10] Eksperimental'nye issledovaniya diskretno armirovannyh osnovanij / V.M. Antonov, I.A. Al'-Nakdi, M.A. Popov // Ustojchivoe razvitie regiona: arhitektura, stroitel'stvo, transport: Materialy 5-j mezhdunarodnoj nauchno-prakticheskoj konferenciya Instituta arhitektury, stroitel'stva i transporta FGBOU VO "Tambovskij gosudarstvennyj tekhnicheskij universitet".– Tambov: Izd-vo Pershina R.V., 2018. – S. 148-156.

[11] Ravishankar, Pallavi, NeelimaSatyam. "Numerical Modelling To Study Soil Structure Interaction For Tall Asymetrical Building." In *International Conference on Earthquake Geotechnical Engineering Istanbul, Turkey*. 2013. 5p.

[12] Hashiguchi K., Saitoh K., Okayasu T. Tsutsumi S. Evaluation of typical conventional and unconventional plasticity models for prediction of softening behaviour of soils.*Geotechnique*

52, no. 8, 2002. pp. 561-578.

[13] Drucker D. C., Prager, W. Soil mechanics and plastic analysis for limit design. Quarterly of Applied Mathematics, vol. 10, no. 2, 1952. pp. 157-165.

[14] Raghavendra, Hadadi B. "Analysis of soil-reinforcement interaction in reinforced soil beds." *Proceedings of the Institution of Civil Engineers-Ground Improvement* 161, no. 1, 2008. pp. 9-15.

[15] Sharma, Radhey, Qiming Chen, Murad Abu-Farsakh, and Sungmin Yoon. "Analytical modeling of geogrid reinforced soil foundation." *Geotextiles and Geomembranes* 27, no. 1 (2009): 63-72.

[16] Chakraborty, Manash, and Jyant Kumar. "Bearing capacity of circular foundations reinforced with geogrid sheets." *Soils and Foundations* 54, no. 4 (2014): 820-832.

[17] Chen, Qiming, and Murad Abu-Farsakh. "Ultimate bearing capacity analysis of strip footings on reinforced soil foundation." *Soils and Foundations* 55, no. 1 (2015): 74-85.

[18] W.R. Azzam, A.M. Nasr. "Bearing capacity of shell strip footing on reinforced sand" Journal of Advanced Research, Volume 6, Issue 5, 2015, pp. 727-737

[19] Cicek, Elif, ErolGuler, and TemelYetimoglu. "Effect of reinforcement length for different geosynthetic reinforcements on strip footing on sand soil."*Soils and Foundations* 55, no. 4 (2015): 661-677.

[20] Khing, K. H., B. M. Das, V. K. Puri, E. E. Cook, and S. C. Yen. "The bearing-capacity of a strip foundation on geogrid-reinforced sand."*Geotextiles and Geomembranes* 12, no. 4 (1993): 351-361.

[21] Basudhar, P. K., SantanuSaha, and Kousik Deb. "Circular footings resting on geotextile-reinforced sand bed." *Geotextiles and Geomembranes* 25, no. 6 (2007): 377-384.

[22] Latha, G. Madhavi, and Amit Somwanshi. "Bearing capacity of square footings on geosynthetic reinforced sand." *Geotextiles and Geomembranes* 27, no. 4 (2009): 281-294.

[23] Madhav, Madhira R., and Hormoz B. Poorooshasb. "A new model for geosynthetic reinforced soil." *Computers and Geotechnics* 6, no. 4 (1988): 277-290.

[24] Abu-Farsakh, Murad, Qiming Chen, and Radhey Sharma. "An experimental evaluation of the behavior of footings on geosynthetic-reinforced sand."*Soils and Foundations* 53, no. 2 (2013): 335-348.

[25] Alawaji, H. A. "Settlement and bearing capacity of geogrid-reinforced sand over collapsible soil." *Geotextiles and Geomembranes* 19, no. 2 (2001): 75-88.

[26] Li, Jian, Chaosheng Tang, Deying Wang, Xiangjun Pei, and Bin Shi. "Effect of discrete fibre reinforcement on soil tensile strength."*Journal of Rock Mechanics and Geotechnical Engineering* 6, no. 2 (2014): 133-137.

[27] Shahin, Hossain Md, TeruoNakai, Yukihiro Morikawa, Saki Masuda, and Susumu Mio. "Effective use of geosynthetics to increase bearing capacity of shallow foundations."*Canadian Geotechnical Journal* 54, no. 12 (2017): 1647-1658.

IOP Conf. Series: Materials Science and Engineering **786** (2020) 012061 doi:10.1088/1757-899X/786/1/012061

Development and research of the tower for hanging supports on hop-gardens

P A Smirnov[1], A E Makushev[2], A O Vasiliev[1], R V Andreev[1], A G Lozhkin[3] and V I Medvedev[1]

[1] Department of technical service, Chuvash State Agricultural Academy, 29 K/Marks Street, Cheboksary 428000, Russian Federation
[2] Rector of Chuvash State Agriculture Academy, Chuvash State Agricultural Academy, 29 K/Marks Street, Cheboksary 428000, Russian Federation
[3] Department of technical service, Chuvash State Agricultural Academy, 29 K/Marks Street, Cheboksary 428000, Russian Federation

E-mail: 3777222@bk.ru

Abstract. In recent years, in the conditions of economic sanctions and food embargo in Russia, the issue of import substitution has become acute. These phenomena are directly related to agricultural products and, in particular, to the production of hops. Currently, about 90% of commercial hops are imported to the Russian market from abroad. Current trends in the improvement of modern hop-growing technology require the introduction of high-tech innovations into the cultivation process. Its productivity and product quality depend on timely and proper technological operations in the production of hops. Given the increasing prices of energy, it is necessary to introduce improved low-cost technologies in the production of hops. The most important aspect at the present stage is the mechanization of such a process as hanging supports using mobile towers. The use of these towers can increase productivity in this operation in 5-6 times compared to manual labor. The tower must contain a horizontal frame with a trailed device, a working platform, hydraulic cylinders for its lifting, a system of guide rings for support beams. A small ladder should be provided for lifting workers to the site. As a result of the design, the optimal parameters of the scissor-type perspective tower were determined, providing the required stability, both static and dynamic.

1. Introduction

Hops belong to agricultural crops with a small volume of production – the world area of its plantations is more than 51 thousand hectares [1, 2]. Hop farming in Russia was, and in a number of European countries, a large and important branch of agricultural production. In the second half of the last century, in our country and abroad there was a tendency of specialization in hop growing, enlarging the area under the main crops. In recent years, in the conditions of economic sanctions and food embargo in Russia, the issue of import substitution has become acute. These processes are directly related to agricultural products, and, in particular, to hops. Currently, about 90% of commercial hops are imported to the Russian market from abroad.

hop growing as a traditional branch of the Chuvash Republic [3 - 4].

Its productivity and product quality depend on timely and proper technological operations in the production of hops. Given the increasing prices of energy, it is necessary to introduce improved low-cost technologies in the production of hops.

The purpose of the research is to expand the areas of increasing the efficiency of the hop industry, the use of resource-saving low-cost technology of cultivation of hops, which involves the use of a set of technological measures. Among them, the most important at the present stage is the mechanization of such a process as hanging the supports with the help of mobile towers. The development of resource-saving technologies of cultivation, processing, and use of hops can reduce the cost of cultivation of hops by 30-35% [11 - 15].

2. Materials and methods

BH-4 towers and PPH-1 platforms are used for partially mechanized mounting of hops supports. But the towers for the garter of twine for trellis, designed and produced earlier, do not correspond to the dimensions established by the requirements of the Traffic Regulations of the Russian Federation in width. In practice, they do not correspond to the width and height of the entrance gates and bridges. This aspect severely limits its operation not only in a limited area of the farm but also during transportation.

The use of towers for hanging supports of hops allows increasing productivity in this operation in 2-3 times. The technological process of the tower is as follows. At the edge of the hop garden, workers fill tufts of supports into the guide rings of the tower. After that, a tractor with a tower enters the hop- garden and, using a hydraulic cylinder, raises the site with the workers to the desired height. The workers pull out the ends of the supports from the beam and tie them to the upper longitudinal wire through 0.8 - 1 m. When the support unit moves, it is pulled out of the beam and sags on the trellis. Workers behind the tower, tie the lower ends of the supports to the string, stretched along the row.

Productivity per shift of a nine-person crew is 2-3 hectares, the number of longitudinal wires to which support is tied up for one pass of the unit is 4, and working cassettes for supports are 4. Height of raising of the working platform to 3.5 m. The tower is aggregated with tractors of 0.9-1.4 classes.

In this regard, the projected tower for hanging hops supports is assumed to be with adjustable gauge width: in the working position - up to 3.3 m, in the transport position - 2.5 m. The offset point of the trailer also has a significant impact on the stability of the movement of the tower [5].

Thus, when designing, it is necessary to fulfill the established regulatory requirements for carrying out calculations of the stability of mobile towers for hanging hop supports, for which correct consideration of dynamic loads is of great importance [6, 7].

The data to justify the parameters of the tower for hanging hops supports are determined by the size of the hop landing (Figure 1).

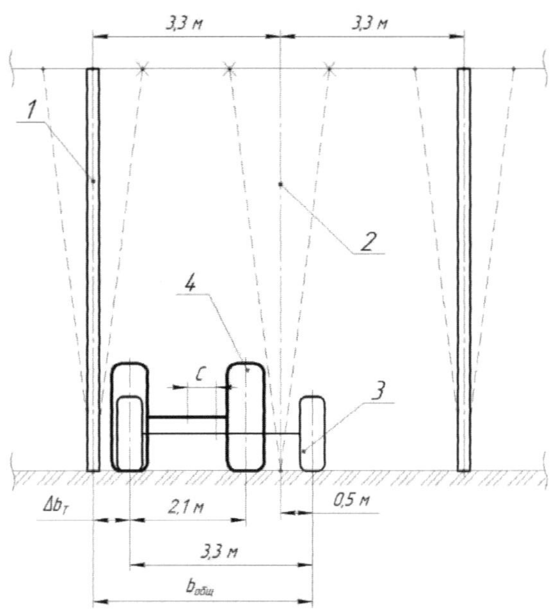

Figure 1. The layout of the elements of the tower in the rows of hops
× - tying when passing; 1 - pillar row; 2 - rowless column; 3 - tower chassis; 4 - tractor chassis.

The choice of the tower scheme is carried out according to four main criteria.

1. Ensuring the lateral stability of the tower for the comfortable and safe operation of personnel [8, 9].
2. The absence of hitting the support wheels of the tractor and the tower on the rows of hops.
3. Modernization, implementation of automated binding of the twine knot at a given step.
4. Calculation of the displacement of the trailer point of the tower to ensure maximum longitudinal stability [10].

In most cases, the tractor moves strictly in the center of the aisle, therefore, the technological gap is defined as

$$b_u = b_m + 2\Delta b, \quad (1)$$

where, b_u – row width of hops, m;

b_m – tractor track gauge, m;

Δb – the distance from the hop row to the nearest track of the tractor wheel, m.

Hence,

$$\Delta b = \frac{b_u - b_m}{2} \quad (2)$$

The technological gauge of the tower for hanging hops should be as close as possible to the width of the aisle, i.e. $b_B = b_m$.

Calculation of the trailer point displacement relative to the center of the row spacing or the axis of symmetry of the tractor.

When designing a tower for hanging hops supports, we consider that the simplest and at the same time the ergonomic design is of scissor type.

Preliminary baseline data for design:

1) a 2-lever scheme has been adopted for reliability and stability, taking into account the specifics of work in the field;
2) the required maximum height of the scissor lift - $H_n = 3.8$ m;
3) the height of the platform (chassis) of the lift is selected taking into account the size of the support wheels, on which the dynamic stability of the structure $h_n = 1.0$ m depends;
4) the required angle of elevation of the levers $\alpha = 45 \ldots 55$ ° from the analysis of the required factors and features of the work of the tower.

Calculation of the inertial forces arising during acceleration of the lift and during braking Acceleration (deceleration) is determined by the general formula

$$a = \frac{v_p - v_0}{t_p}, \quad (3)$$

where, v_0 - the initial speed of movement, m / s;

v_p - working speed, m / s;

t_p - acceleration time, s.

In the general case, the force acting on one person (average weight 75 kg) is determined according to 2nd laws of dynamics:

$$F_u = m \cdot a \quad (4)$$

Calculation of the length of the scissor lift arms.

For the initial data for the calculation of the geometrical parameters of the lift, we take the angle of the solution of the levers $\alpha = 45 \ldots 55$ °, as well as the required height of the tower lift $H_n =$ 3.5 m

The geometric parameters of the levers are related to each other by the following relations (Figure 2):

$$\sin\alpha = \frac{H_H/2}{L} = \frac{H_H}{2L} \quad (5)$$

IOP Conf. Series: Materials Science and Engineering **786** (2020) 012061 doi:10.1088/1757-899X/786/1/012061

$$\cos\alpha = \frac{L_H}{L} \tag{6}$$

$$\tan\alpha = \frac{H_H/2}{L_H} = \frac{H_H}{2L_H} \tag{7}$$

Calculation of forces on the supports and levers of the scissor mechanism. Source data: personnel mass $m_p = 225$ kg;

the mass of the cradle (basket) $m_l = 100$ kg; weight of the scissor mechanism $m_n = 100$ kg.

The total mass of the lifting mechanism, platform (basket) and personnel $m_\Sigma = m_p + m_l + m_n$

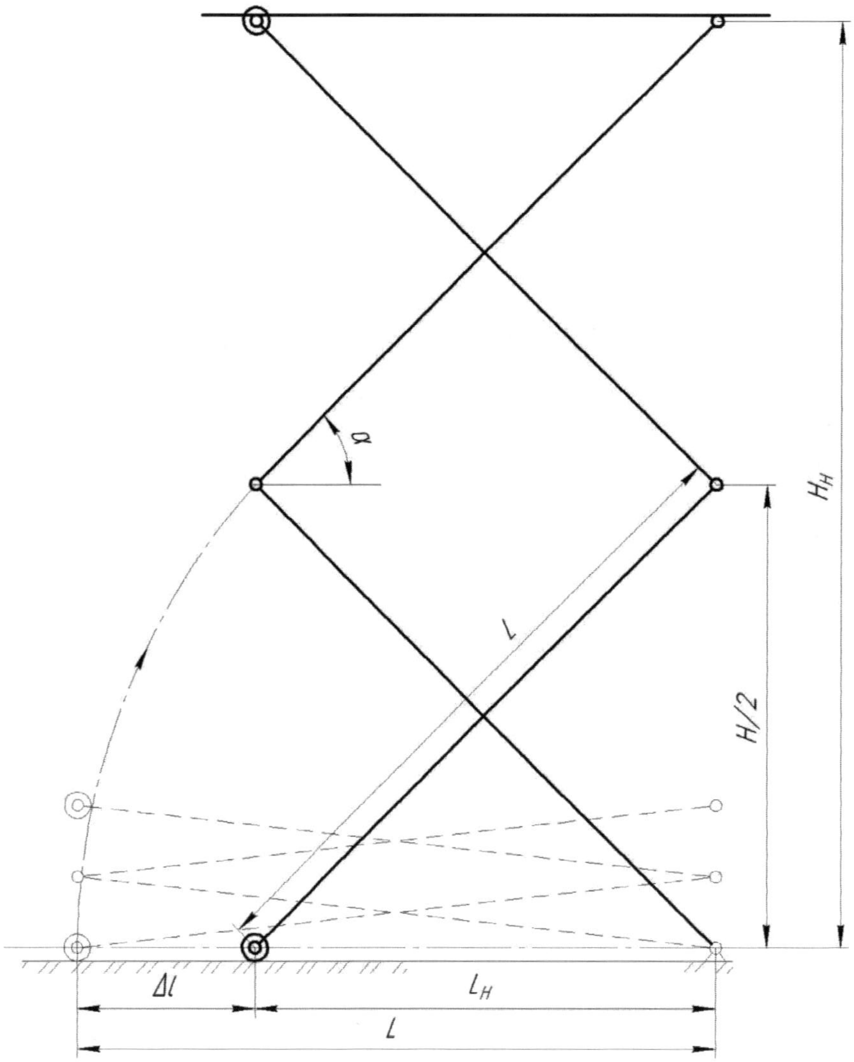

Figure 2. Scheme for calculating the length of the scissor lift arms.

The gravity falling on one lower support:

$$F_z^H = \frac{G_\Sigma}{4} = \frac{m_\Sigma \cdot g}{4}. \tag{8}$$

The gravity of the upper supports:

$$F_z^B = \frac{G_B}{4} = \frac{(m_n + m_л)\cdot g}{4}. \tag{9}$$

IOP Conf. Series: Materials Science and Engineering **786** (2020) 012061 doi:10.1088/1757-899X/786/1/012061

Since during acceleration and deceleration, inertial forces arise that act on the lower and upper fixed supports, as well as creating an overturning moment, it is important to obtain their value at the design stage of the tower geometry.

3. Results and discussion

In the final calculations, according to (1) and (2), when the initial data is substituted, the distance from the hop row to the nearest tractor wheel track is Δb = 0.65 m.

The total width of b_{total} = 4.0 m, therefore, the center of displacement of the tower is located relative to the middle of the aisle. The center of the tractor is located at a distance of 1.61 m from the pole row. Consequently, the asymmetry of the tower relative to the axis of the tractor should be C = 0.6 m when using inter-row spacings on hop gardens 3.3 m.

To use the tower on the hop gardens with a row spacing of 2.5 m, no trailer point offset is required.

To take the initial data when determining the inertial forces when starting off, a series of measurements of the starting time of the tractor MTZ-82 was carried out, with which it was planned to aggregate the designed tower for hanging hop supports. For this, the total mass of the lifting mechanism, platform (basket) and personnel, which amounted to m_Σ = 425 kg, was approximately calculated by computer simulation.

Since it is planned to work on the 3rd gear with a reduction gear, the tractor speed is 7.4 m / s. In order to maximally approach the working conditions on the hop gardens with a tower, a trailer with a total weight of 450 kg was installed on the tractor.

The results of measuring the starting time of a tractor with a trailer are shown in Table 1.

Table 1. Timing of tractor starting

No of repetition	Starting time, s
1	3.0
2	2.7
3	2.1
4	2.9
5	1.0
6	2.6
7	3.2
8	2.5
9	3.2
10	1.5
Average	2.47

At the same time, during the tests, no correlation was observed between the starting time and the field gradient.

Taking into account the starting from the state of rest when v_0 = 0 m / s and with the acceleration time value: t_p = 2.47 s. We obtain, using (3), the acceleration at the start of a = 0.4 m / s^2.

However, as it was revealed, with a sharp starting or braking time can reach t_p = 1.0 s. In this case, we obtain the acceleration value up to a = 0.4 m / s^2.

In the average statistical case, the inertia force acting on one person (average weight 75 kg) according to (4) is approximately equal to F_u = 30 N, and with a sudden start, its values increase to F_u = 75 N.

As a result of calculations of the geometry of the designed tower for hanging hops, the optimal combination of the solution angle α and the lever length L was found according to (5), (6) and (7). The result is shown in figure 3.

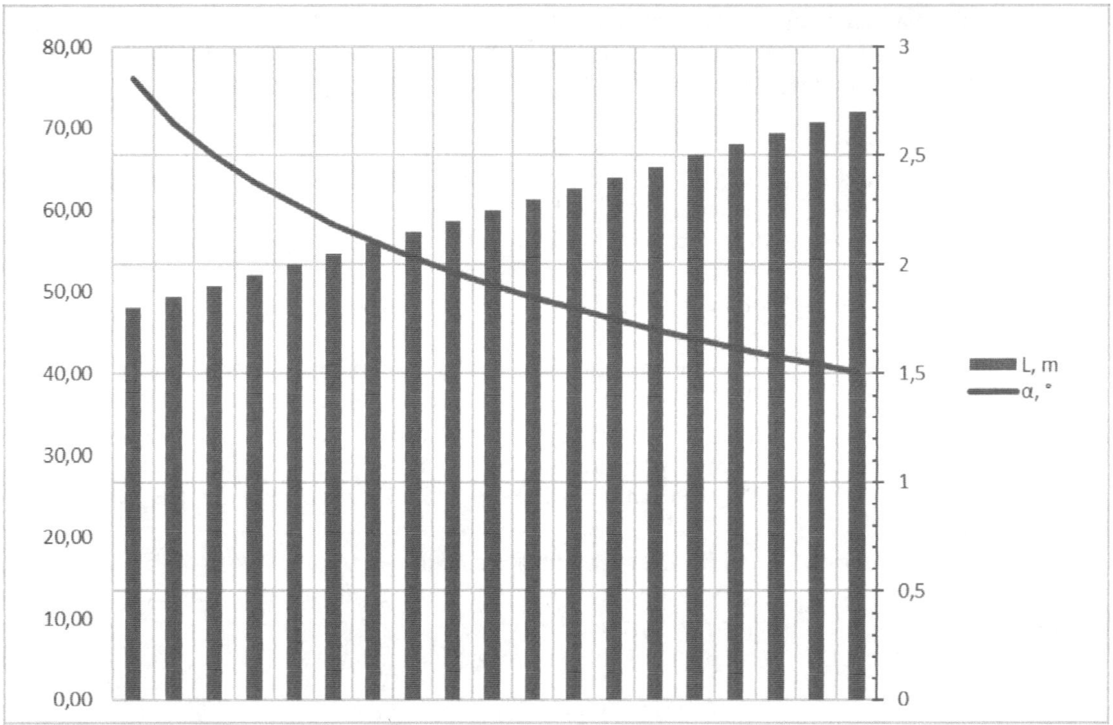

Figure 3. Diagram of changes in the angle of solution α depending on the length of the lever L.

It can be seen from the diagram that the optimal geometry starts from the intersection of the solution angle curve and the lever length, which corresponds to $\alpha = 52\,°$ and $L = 2.25$ m.

Thus, the longitudinal length of the grip in the raised (working) state with the maximum raised position is $L_n = 1.47$ m.

Choose a length of 2.25 m along the axis of the roller lever. Then, taking into account the substitution, we obtain the horizontal shortening value $\Delta L = 0.81$ m (Figure 2).

Taking into account the location of the axles of the roller axle, as well as the mode of operation of hydraulic cylinders, the length of the guide length cannot exceed 1 m, which is provided with a small margin.

With further substitution, using (8) and (9), we obtain the force of gravity per one lower support $F_z^H = 1042.3$ N, as well as the force of gravity per upper support $F_z^B = 797$ N.

These values should be considered when designing the height of the handrails and the width of the basket for staff.

4. Conclusion

The value of the overturning momentum rapidly increases with a multiple of the angle of inclination. In order to give the tower the best stability in the designed tower, wheels are provided with the ability to move them along the axis of rotation. In the transport position, the distance between them is determined by the maximum permissible dimensions of vehicles for movement on public roads, and in the working position, this distance increases. Taking into account the fact that the required maximum height of the scissor lift must reach $H_n = 3.8$ m with the length of the horizontal arm $L = 2.25$ m. The height of the platform (chassis) of the lift $H_n = 1.0$ m was chosen taking into account the size of the supporting wheels, on which the dynamic stability of the structure depends. The gauge size in the transport position on the middle line is 2.50 m, and in the working position - 3.30 m. Such chassis parameters provide the required stability, both static and dynamic.

References

[1] Zakharov A., Makushev A., Evgrafov O., Zakharov D. and Tsaregorodtsev E. (2017). Formation of the hop-growing cluster in the Chuvash Republic. *Proceedings of the 30th International Business Information Management Association Conference, IBIMA 2017 - Vision 2020.* pp. 4802- 4810.

[2] Vasiliev A.O., Andreev R.V., Pushkarenko N.N., Smirnov P.A. and Ivanov E.A. (2017) Modern ways of improving the mechanization of hop cultivation. *Proceedings of the 30th International Business Information Management Association Conference, IBIMA 2017 - Vision 2020.* pp. 5289- 5294.

[3] Lozhkin A.G., Makushev A.E., Pushkarenko N.N., Kornilova L.M., Stepanova V.V. and Fedorova O.N. (2017) Revival of hop-production in the Chuvash Republic: problems, challenges, and opportunities. *Proceedings of the 30th International Business Information Management Association Conference, IBIMA 2017 - Vision 2020.* C. 5295-5299.

[4] Schnaitter M., Wimmer A., Kollmannsberger H., Gastl M., and Becker T., (2016), Influence of hop harvest date of the 'Mandarina Bavaria' hop variety on the sensory evaluation of dry-hopped top- fermented beer. *Journal of the Institute of Brewing*, **122(4)**, pp. 661-666.

[5] Vasiliev A. O., (2013), Improving the quality indicators of the cabbage harvesting machine by improving its support-trailer system, PhD dissertation, Chuvash State Agricultural Academy, Cheboksary. (In Russian)

[6] Gonzales, H.B., Casada, M.E., Hagen, L.J., Maghirang, R.G., Barden, C.J. (2018) Porosity and drag determination of a single-row vegetative barrier (Maclura pomifera) *Transactions of the ASABE* **61(2)**, pp. 641-652.

[7] Okamoto, Y., Nakamura, S. (2015) Static and seismic analysis of multi-span cable-stayed bridges with steel and CFT towers. *IABSE Conference, Nara: Elegance in Structures – Report* pp. 358-359.

[8] Sakhapov R.L., Nikolaeva R.V., Gatiyatullin M.H., Makhmutov M.M. (2016) Modeling the dynamics of the chassis of construction machines. *Journal of Physics: Conference Series.* **Vol. 738. № 1**. 012119.

[9] Sakhapov R.L., Nikolaeva R.V., Gatiyatullin M.H., Makhmutov M.M. (2016) Risk management model in road transport systems. *Journal of Physics: Conference Series*. **Vol. 738. № 1**. 012008.

[10] Alatyrev, S.S., Andreev, R.V., Vasiliev, A.O., Grigoryev, A.O., Tikhonov, S.V. (2018) Influence of the parameters of support wheels on the amount of the device of the trailer from the tracking course *IOP Conference Series: Materials Science and Engineering* **457(1)**, 012002.

[11] Smirnov, P.A., Makushev, A.E., Kazakov, Y.F., Vasilyev, A.O., Andreev, R.V. (2019) Influence of types of tractor running gears on the value of hop garden row spacing compaction *INMATEH - Agricultural Engineering* **57(1)**, 19-28.

[12] Kornilova, L.M., Makushev, A.Y., Tolstova, M.L., Pushkarenko, N.N., Gubanova, N.G. (2018) Challenges to implement strategies of import substitution in the field of hop supplying to breweries and factors to intensify hop production *Proceedings of the 32nd International Business Information Management Association Conference, IBIMA 2018 - Vision 2020: Sustainable Economic Development and Application of Innovation Management from Regional expansion to Global Growth* 1855-1858.

[13] Kornilova, L.M., Makushev, A.Y., Tolstova, M.L., Pushkarenko, N.N., Stepanova, V.V. (2018) Economic efficiency of hop-producing organizations *Proceedings of the 32nd International Business Information Management Association Conference, IBIMA 2018 - Vision 2020: Sustainable Economic Development and Application of Innovation Management from Regional expansion to Global Growth* 1852-1854.

[14] Andreev, R.V. (2014), Improving the drive of the cutting apparatus of the cabbage harvesting machine, PhD dissertation, Chuvash State Agricultural Academy, Cheboksary. (In Russian)

[15] Korotkov, A.V. (2015) The effect of liming aftereffect, fertilizers and green manure crops on the parameters of fertility of the gray-forest heavy loamy soil and hop yield *Bulletin of the Mari State University. Series: Agricultural sciences. Economics* **4 (4)**, 29-33. (In Russian)

IOP Conf. Series: Materials Science and Engineering **786** (2020) 012062 doi:10.1088/1757-899X/786/1/012062

Improving the efficiency of excavators at low temperatures

S A Zenkov[1] and D Lkhanag[2]

[1] Bratsk State University, 40, Makarenko St., Bratsk, 665709, Russia
[2] Mongolian University of Science and Technology, 8th khoroo, Baga toiruu 34, Sukhbaatar district,Ulaanbaatar, Mongolia

E-mail: mf@brstu.ru

Abstract. In the paper the analysis is carried out for using polymer antiadhesive sheets in order to reduce adhesion of wet cohesive soil to working bodies of excavation machines when operating under sub-zero temperatures. When extracting thawed mine rock under subzero air temperature, intensive freezing of cohesive ground to contacting bucket surfaces, pickup apparatus elements, conveyor belts, treating facilities and links takes place. Freezing intensity depends on humidity, grain size composition, physical properties and temperature of mined rock. Material properties of working bodies which contact mined ground also have significant impact on adhesion of freezing of mined rock. Clay rocks are the most prone to freezing on. An experiment was held using PPL-EI (ППЛ-ЭИ) and PPL-UI (ППЛ-УИ) polymer coatings on adhesion of cohesive overburden rocks to mine excavator buckets under subzero temperatures. After mathematic analysis of experimental results regression equations were obtained, which were applied to engineering solution of excavator bucket.

1. Introduction

The expertise of operating excavation machines shows that while mining cohesive soils with environment temperature below zero freezing and adhesion of soil to working bodies dramatically reduces performance of machines which leads to longer time required to eliminate consequences of freezing [1-10].

Performance falls because of lower useful capacity of a bucket due to incomplete unloading, increased frontal resistance to cutting (digging) caused by soil adhesion to the working body, resistance of soil loading into the bucket and longer downtime needed to clean working bodies.

The most widely used machines on “Erdenet” open-pit ore mine are machines with bucket working bodies (EKG-4,6, EKG-8I, EKG-10 excavators), BelAZ-548 mining trucks and DET-250 bulldozers. Performance of these machines is defined by design and engineering factors, bucket configuration and type of the ground in operation (figure 1).

One of the approaches to reduce the adhesion is creating an intermediate layer out of liner polymer coatings on the contact boundary between soil and working body (preventive treatment) [2,3].

Liner coatings manufacturing is based on ultra-high molecular polymers. The expertise of operating various engineering equipment in fields of industry dealing with mining and processing mineral resources proves that the most efficient way to reduce adhesion and freezing of wet mined rock and mineral materials to working equipment units contacting them is protecting (lining) with polymer coatings that have hydrophobic and antiadhesive properties [2,3]. These are polymer antiadhesive liner plates (PALP) manufactured by LLC “As-Tik KP” and polymer antiadhesive sheets (PAS) recommended by “Technotex” group of companies.

Figure 1. "Erdenet" open-pit mine.

Polymer material graphite-reinforced plastic, designed in Research institute of efficiency and safety of mining operations (НИИОГР), has been successfully implemented on assets of RF Ministry of Energy as preventive solution. However, further implementation of this material is limited by insufficient shock resistance and lack of preventive effect under sub-zero temperatures. After the research it was concluded that the most suitable material according to technical specifications is modified polyurethane, fluoroplastic F-3M (Ф-3М) [2].

Depending on the working mode, the most efficient polymer antiadhesive sheets are PAS liner sheets of different grades. Polymer antiadhesive sheets (PAS) grades are as following: PPL-EI: abrasion resistance – 400 units; shock resistance – 9 u; operating temperature range - -45 to +90°C; application – contacting materials with f coefficient up to 5 u [2]; PPL-UI (high-modulus): abrasion resistance – 200 units; shock resistance – 25 u; operating temperature range - -80 to +100°C; application – contacting materials with f coefficient up to 8 u. Application of ultra-high molecular polymers allows to improve performance due to the following properties: low wear, low friction coefficient, high impact resistance, chemical resistance to acids, alkalies, salts and other aggressive environments, wide operating temperature range from -50 to +90°C, high durability within this range, sound absorption, crack resistance and hydrophobic properties that prevent wet materials from sticking to and freezing over the surface of polymer coating.

2. Formulation of the problem and the method of solution

In order to examine accuracy of represented data an experiment was conducted on a special shift bench [3].

Methods of multi-factor planning were applied for chosen objectives of experimental research, which enabled one to obtain maximum volume of valuable data about the processes under research with minimal amount of experiments. The effect was evaluated based on the magnitude ratio of specific adfreezing factor, which is considered to be shearing stress for start of ground sample shifting along the operating surface. This parameter is defined by the following formula:

$\tau = \frac{P_s}{S}$, where P_s – force required to shift ground sample along various surfaces, N; S – applied part are of frozen ground sample, m^2.

In order to conduct active experiment and build a mathematical model, rotatable central composite design is chosen according to recommendations [3] and specifics of experimental research being carried out. Levels of factors and variability intervals when conducting the experiment are enlisted in table 1.

Table 1. Levels of factors and variability intervals for the experiment

Factors	Levels of factors					Variability intervals
	-2	-1	0	1	2	
X_1 –soil dispersity, D_e , mm	$9\cdot10^{-3}$	$7\cdot10^{-3}$	$5\cdot10^{-3}$	$3\cdot10^{-3}$	$1\cdot10^{-3}$	$2\cdot10^{-3}$
X_2 – normal pressure on the ground P, kPa	0	10	20	30	40	10
X_3 – gravimetric moisture of soil, W, %	7.5	12.5	17.5	22.5	27.5	5.0
X_4 – environment temperature, T,℃	5	-5	-15	-25	-35	10
X_5 – soil to metal contact time, t, min	0.5	10.5	20.5	30.5	40.5	10

In order to define necessary amount of replicated experiments in the research a series of trial experiments was set. It included experiments on shifting all ground types used in the research. Experiments were conducted with and without preventive treatment with environment temperature T = -15 °C, soil to metal contact time t_c = 20.5 min., gravimetric moisture of soil W = 17.5 %, normal pressure on the ground P = 20 kPa.

The experiments were being carried out on a shift bench (figure 2), with two polymer antiadhesive sheets being used. Polymer antiadhesive sheets (PAS) are squeezed with clamps which prevents them from moving along the carriage. PAS types used in the experiment were the two most suitable in terms of parameters and recommended by "Technotex" group of companies: PPL-UI (10 mm) and PPL-EI (6 mm).

Experimental data processing resulted into the following regression equations that define quantitative and qualitative character of response function alteration depending on the factors in the field of research when implementing the matrix of experiment.

Soil shearing stress without antiadhesive sheets being applied:

$$R_{w.e.} = 20.28 + 0.74x_1 + 1.69x_2 - 1.91x_3 - 1.24x_4 - 1.01x_5 - 0.88x_1^2 - 1.49x_2^2 - 0.31x_3^2 -$$
$$0.163x_4^2 - 0.51x_5^2 + 1.42x_1x_2 - 0.283x_1x_3 - 0.046x_1x_4 - 1.058x_1x_5 - 0.092x_2x_3 + 1.47x_2x_4 -$$
$$4.167x_2x_5 + 0.02x_3x_4 + 2.183x_3x_5 - 3.254x_4x_5$$

Soil shearing stress for PPL-EI:

$$R_{EI} = 2.596 + 0.06x_1 + 0.225x_2 - 0.627x_3 - 0.152x_4 + 0.007x_5 - 0.158x_1^2 + 0.203x_2^2 + 0.053x_3^2 -$$
$$0.171x_4^2 - 0.103x_5^2 + 0.025x_1x_2 - 0.679x_1x_3 + 0.058x_1x_4 - 0.225x_1x_5 - 0.236x_2x_3 - 0.09x_2x_4 -$$
$$0.333x_2x_5 - 0.679x_3x_4 + 0.08x_3x_5 - 0.225x_4x_5$$

IOP Conf. Series: Materials Science and Engineering **786** (2020) 012062 doi:10.1088/1757-899X/786/1/012062

Soil shearing stress for PPL-UI:

$$R_{UI} = 3.028 + 0.506x_1 + 0.127x_2 - 0.247x_3 - 0.522x_4 + 0.289x_5 - 0.103x_1^2 + 0.308x_2^2 + 0.058x_3^2 - 0.203x_4^2 - 0.241x_5^2 + 0.334x_1x_2 - 0.203x_1x_3 + 0.178x_1x_4 - 0.290x_1x_5 - 0.159x_2x_3 - 0.090x_2x_4 - 0.771x_2x_5 - 0.128x_3x_4 + 0.359x_3x_5 - 0.334x_4x_5$$

Figure 2. Shift bench.

The regression equations mentioned above were used to build floor-quasifactorial shearing stress dependencies on external factors with the rest of variable factors being fixed.

3. Results and discussion

Analysis of obtained results proves that the surfaces under research have different stress values. Shearing stress variations for PPL-EI and for 3-grade steel are similar to each other, but differ 17 kPa on average in thrust, while the highest stress being observed with 17.5 % humidity.

Shearing stress reduces when temperature drops for all types of coating. 10 degrees temperature fall results into 1 kPa average increase of shearing stress for all types of coating. Average difference in shearing stress between polymer coatings and steel equals 18 kPa. The difference between PPL-EI and PPL-UI equals 0.5 kPa which is insignificant. When temperature reaches -25 degrees, shearing stress for PPL-EI surface stops increasing and stays almost constant.

Shearing stress on contact time dependency varies for all types of coating. For steel shearing stress increases with contact time increase up to 20.5 minutes and stays almost constant with the further increase in contact time. In case of polymer coatings shearing stress does not vary significantly within all contact time interval.

4. Conclusion

Polymer antiadhesive sheets differ in chemical composition and structure, therefore, performance for each of them has been tested by experiment. The most suitable type of polymer coating in terms of shearing characteristics is PPL-EI. Analysis of obtained data proves that the difference in shearing stress between polymer antiadhesive sheets and steel is evident and equals 17-18 kPa, which was proven by experiment.

The engineering solution chosen is fixing polymer antiadhesive sheets (PPL-EI) to EKG-8I excavator bucket by bolt fixture. Model of a bucket with polymer sheets coating was built in KOMPAS 3D graphic editor (figure 3). Using PPL-EI polymer antiadhesive sheets has led to reduction of cycle duration for excavation of wet adhesive soil due to 17% less adhesion and freezing over the working body of the excavator; to lower power consumption during operation due to better cleaning of the bucket from soil.

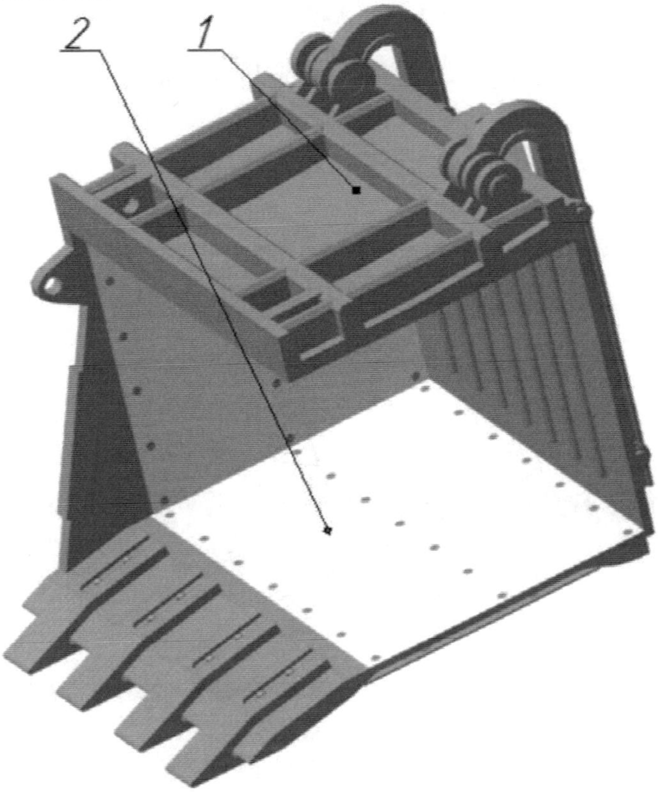

Figure 3. 3D model of EKG-8I excavator bucket with antiadhesive sheets: 1 – the bucket, 2 – PPL-EI.

References

[1] Tong J, Ren L, Yan J, Ma Y and Chen B 1999 *Int. Agricultural Eng. J.* **8** 1-22

[2] Zadneprovsky R P 1992 *Working bodies earthwork and reclamation machines and equipment for soil and high humidity materials* (Moscow: Mashinostroenie) 176

[3] Zenkov S A, Kurmashev E.V., Krasavin O.Y. 2009 Analysis of increasing the productivity of excavators at the use of piezoceramic transducers *Systems Methods Technologies* **4** 38-41

[4] Wang X L, Ito N, Kito K and Garcia P P 1998 *J. of Terramech.***35** 87-101

[5] Azadegan B and Massah J 2012 Effect of temperature on adgesion of clay soil to steel *Cercetări Agronomice în Moldova* XLV **2(150)** 21–27

[6] Rajaram G and Erbach D C 1999 Effect of wetting and drying on soil physical properties *Journal of Terramechanics* **36** 39–49

[7] Chen B, Liu D, Ning S and Cong Q 1995 Research on the reducing adhesion and scouring of soil of lugs by using unsmoothed surface electro-osmosis method *Transactions of the Chinese Society of Agricultural Engineering* **11(3)** 29–33

[8] Ignatyev K A, Filonov A S, Lkhanag D and Battseren I 2013 Definitions of time from the surface soil breakout body work in a high impact *Scientific transactions* (Ulaanbaatar: MUST) **3/139** 144–146

[9] Ignatyev K A, Filonov A S and Zarubin D A 2012 Application of piezoceramic radiators for combating adhesion or soils to excavating part of an earthmoving machine *Science and Education: materials of the II Int. Research and Practice Conf.* I (Munich: Vela Verlag Waldkraiburg publishing office) pp. 251–256

[10] Sharma V K, Drew L O and Nelson L 1977 *Transactions of the ASAE* **20** 46-51

IOP Conf. Series: Materials Science and Engineering **786** (2020) 012063 doi:10.1088/1757-899X/786/1/012063

Vibroprotective mechanism with the quasizero stiffness for seats of operators of construction and road machines

S V Tomleeva [1,*], A G Saveliev[2], G D Moiseev[2] and P G Kolesnikov[3]

[1]Bryansk State Engineering-Technology University, 3, Dimitrova prospect, Bryansk, Russia

[2]Moscow Automobile and Road Construction State Technical University (MADI), 64, Leningradsky prospect, Moscow, Russia

[3]Siberian State Technological University, 82 Mira prospect, Krasnoyarsk, Russia

[*]E-mail: tomkuzsv@mail.ru

Abstract. The paper considers vibration isolation device with a quasi-zero stiffness, which nonlinear element is made in the form of a closed elastic rod, an elastic line described by the circle. The authors list the features of stiffness and strength characteristics of the device, and select the parameters that ensure the full protection on a portion of the suspension.

1. Introduction

The effect of vibration on the human body has been studied to some extent [1], which resulted in requirements for the permissible level of vibration at the workplace, which should be taken into account when equipping the workplace of the operator of construction and road vehicles.

Approaches to solving this problem are following: passive vibration protection, where elements with unregulated (constant) and adjustable stiffness are used and shock absorber resistance coefficients are selected accordingly; active vibration protection based on the implementation of instantaneous counteraction to the input disturbance; pulsed vibration protection based on the use of systems with an abrupt change in the stiffness of the suspension elements. There is also a new developing direction of vibration protection based on the use of devices of forming a nonlinear stiffness characteristic having quasi-zero values in the position of static equilibrium of the protected object in elements of vibration protection. These devices function as passive, having the properties of active vibration protection devices. Elements that provide quasi-zero stiffness are hopping elements that have a different structural solution (Mises truss, popping membrane, and others).

2. Setting the problem

One of the main tasks in calculating a vibration isolator is to ensure a low frequency of its free vibrations, since the smaller it is, the wider the frequency range of the driving force, which ensures optimal vibration protection. Achieving a low frequency of free oscillations is possible only when using a "soft suspension" with elastic elements with a low stiffness coefficient. The presence of a segment with quasi-zero stiffness on the graph of the stiffness of the hopping elements gives reason to consider them as elastic elements of vibration protection systems.

The disadvantages of such systems are: a relatively small segment of the power characteristic, which provides optimal vibration protection; ensuring the position of the static equilibrium of the system when changing the weight of the protected object (weight of the driver). However, recently, these problems have been successfully solved [2]. When pneumatic springs and controlled dry friction are included in the vibration-protective system, it is possible to achieve an increase in the required length of the power characteristic up to 10 times and return the system to an unstable equilibrium position automatically.

In this paper, we consider the vibration-proof device of the driver's seat of a construction vehicle driver, which includes a non-linear suspension of quasi-zero stiffness.

3. Analytical Model

The scheme of the device is presented on the figure 1 [3].

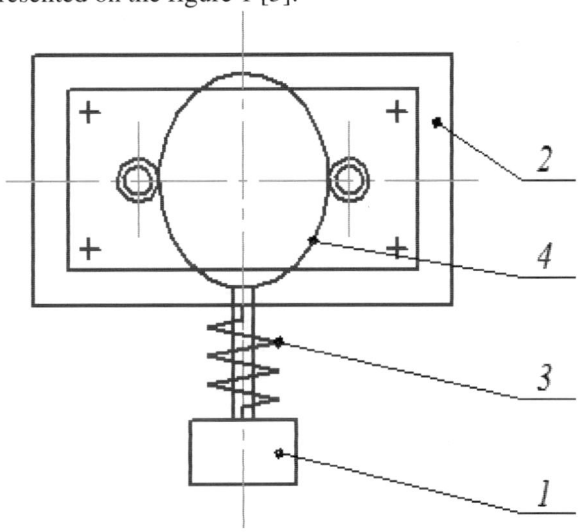

Figure 1. General view of the vibration protection device.

The element forming the nonlinear stiffness function of the device is made in the form of a closed elastic rod. When mechanical vibrations are transmitted through the body 1 to the protected object 2 in a spring of constant stiffness 3, a dynamic elastic force arises due to displacement from the position of static equilibrium. In parallel with the spring, the hopping element 4, which is an elastic closed rod of rectangular cross-section, is also shifted in the vertical direction, slipping between two rollers pivotally attached to the protected object 2. The distance between the centers of the rollers is fixed and always less than the diameter of the elastic ring, therefore, when moving the rod between the rollers at the contact points, some forces arise in the vertical direction, which are opposite to the action of the spring elastic force 3. Thus, the external action forms two oppositely directed parallel forces that compensate each other friend and do not allow excitation to be transmitted to the protected object 2.

To analyze the vibration-protective qualities of the proposed device, it is necessary to obtain the power characteristic of the device, to make sure that it has segments characteristic of non-linear suspensions of quasi-zero stiffness.

Let's consider one of the parts symmetrically to the vertical (Figure 2). The action of the rollers on the rod is replaced by the action of a concentrated force P applied at the point of contact of the roller with the rod. To determine the stiffness function of this element, we will use the technique described in

[4], according to which it is necessary to find the vertical displacement of the point of application of force P and divide the force into this displacement:

$$C = \frac{P}{f}. \tag{1}$$

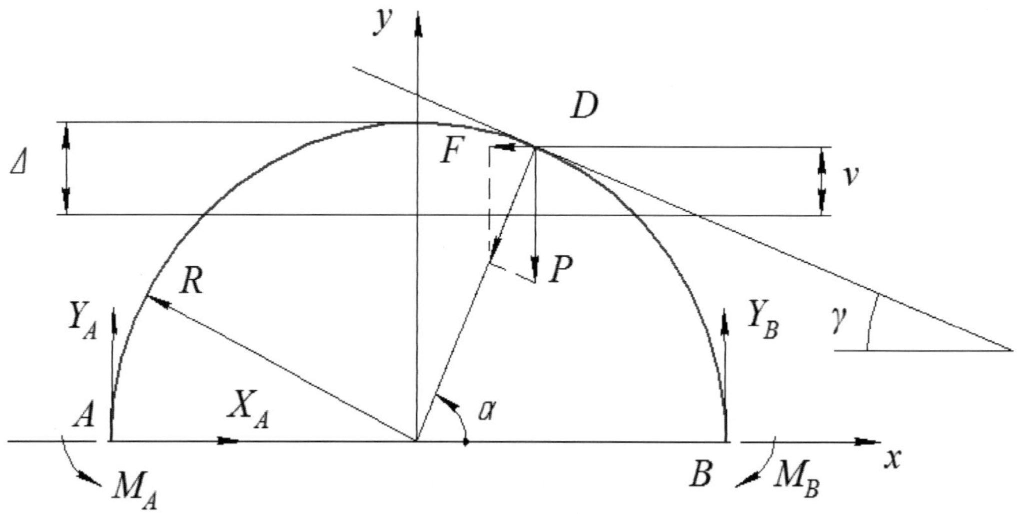

Figure 2. Calculation scheme of the core element.

We restrict ourselves to considering a rod of small curvature, since in our case the condition $h/R \ll 10$ is fulfilled, where R is the radius of curvature of the elastic line of the rod, h is the height of the section of the rod.

To calculate the displacement of the rod, we use the Mohr method. For rods of small curvature, the displacement of the point of application of force is determined by the formula:

$$f = \int_0^l \frac{Mds}{EJ}\frac{\partial M}{\partial P}, \tag{2}$$

Where E – elastic modulus of the rod material, J – the moment of inertia of the cross section of the rod, M – the bending moment acting in the section of the rod, P - the acting force. The expression of bending moment includes Y_B, M_B. To determine them, we use the condition for fixing the end section B of the rod:

- the vertical movement of point B is zero ($f_B = 0$);
- the angle of rotation of the cross section of the rod at point B is zero ($\theta_B = 0$).

These conditions are expressed by the formulas:

$$\begin{cases} f_B = \dfrac{1}{EJ}\displaystyle\int_s M \frac{\partial M}{\partial Y_B} ds = 0 \\ \theta_B = \dfrac{1}{EJ}\displaystyle\int_s M \frac{\partial M}{\partial M_B} ds = 0 \end{cases} \tag{3}$$

Having solved the resulting system of equations, we find Y_B, M_B. Substituting them into the expression of the bending moment M, we find f by the formula (2):

$$f = \frac{R^3}{EI}\Big[P\big(\frac{\pi}{2}-\frac{\alpha}{2}+\frac{3\sin 2\alpha}{4}+(\pi-\alpha)(\cos\alpha)^2\big)+Y_B\big(-\sin\alpha-\frac{\pi}{2}+\frac{\alpha}{2}-\frac{\sin 2\alpha}{4}-\cos\alpha(\pi-\alpha)\big)+M_B(\sin\alpha+\cos\alpha(\pi-\alpha)/R\Big], \quad (4)$$

where $Y_B = \frac{P}{\pi}(\pi-\alpha+0.5\sin 2\alpha)$;

$$M_B = \frac{PR}{\pi}(\pi-\alpha+0.5\sin 2\alpha-\sin\alpha-\cos\alpha\,(\pi-\alpha)).$$

Knowing the stiffness function of the elastic rod defined by formula (1), we obtain the force characteristic F of the elastic element in the x-axis direction:

$$F = P\cos\gamma\cos\alpha,$$

where $P = C\cdot v$;

$$v = \Delta - \mathrm{R}(1-\sin\alpha);$$

Δ – specified preload of the center point at which $x = 0, \alpha = \frac{\pi}{2}$.

Since $= 90^o - \alpha$, when eventually

$$F = \frac{1}{2}P\sin 2\alpha \quad (5)$$

The graph of the power characteristic *F(x)* is shown in Figure 3 by a dashed line.

4. Results

The calculations were carried out with the following numerical values: h = 0.002m – section height of a flexible rod; b = 0.012 m – section width; R = 0.20m – radius of curvature of the elastic line of the rod; $E = 2.1\cdot 10^5\ MPa$.

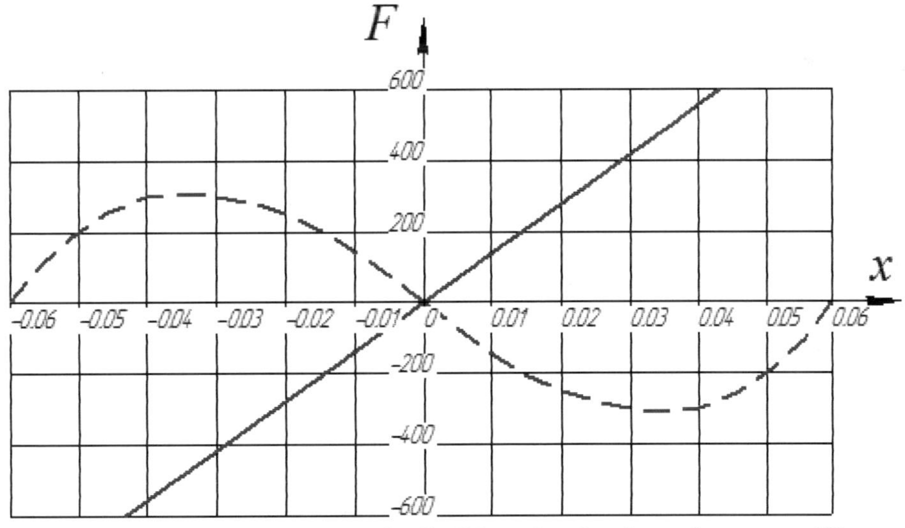

Figure 3. Power characteristics of a flexible rod and springs of constant stiffness.

The figure shows graph of a constant stiffness spring. When these elements work together, a general characteristic of the vibration protection device formed, as shown in Figure 4. In the vicinity of the origin, is we clearly distinguish the segment where F = 0. This is the section where the condition of complete vibration protection of the object is fulfilled. To ensure this condition, it is necessary to appropriately select the device parameters and spring stiffness so that the angle of inclination of the tangent to the power characteristic of the flexible rod and the spring stiffness coefficient are equal in modulus. For this, it is necessary to take the derivative with respect to x of the function described by formula (5) and equate it at x = 0 to C_1, where C_1 is the stiffness of spring 3 shown in Figure 1.

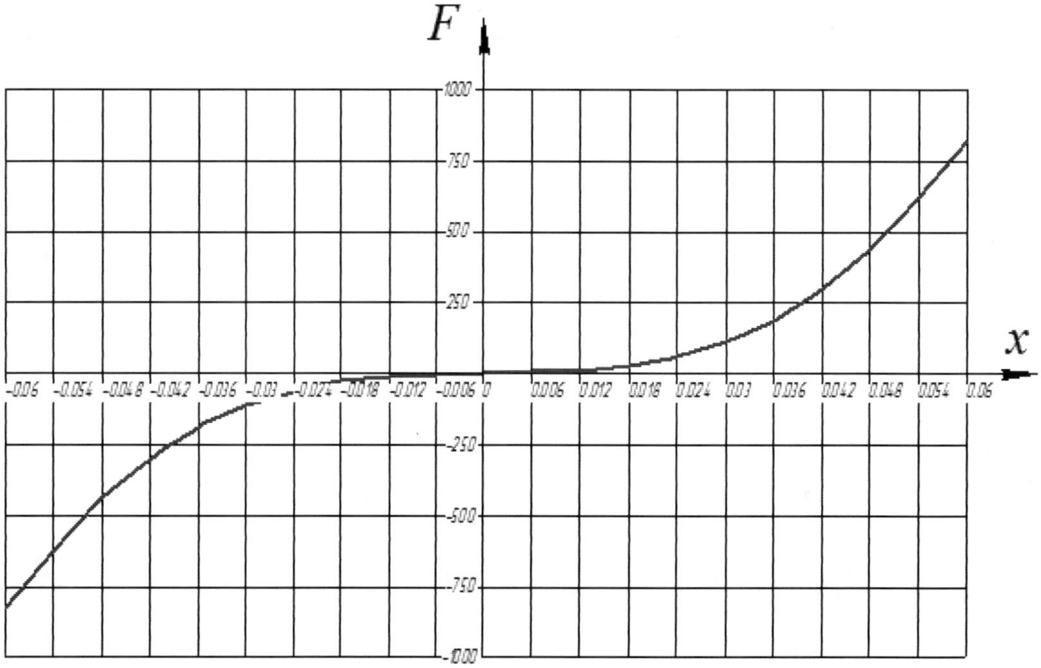

Figure 4. Graph of the power characteristics of the vibration protection device.

As can be seen, in the vicinity of zero, the stiffness of the device is close to zero. Thus, this vibration protection device refers to the type of devices with quasi-zero stiffness and has all their properties.

5. Conclusion

As a result of the calculations, the power and stiffness characteristics of the vibration protection device with the hopping element in the form of a closed rod are obtained. The nature of the graphs shows that the calculated device refers to the type of devices with quasi-zero stiffness and, with a certain selection of parameters, can provide optimal vibration protection over a wide range of input disturbance frequencies.

References

[1] Kosarev V V and Babanov S A 2011 Professionalnye bolezni: uchebnoe posobie / M : INFRA – M, 252

[2] Zotov A N 2007 Vibroizolyatory s kvazinulevoy zhestkostyu // Izvestiya vysshikh uchebnykh zavedeniy «Gornyy zhurnal» **2** 147-51

[3] Ustroystvo dlya umensheniya mekhanicheskikh kolebaniy transportnogo sredstva: a s 494299 SSSR **1980740** 45 2

[4] Tomleeva S V Metodika realizatsii raschetnoy bazy nelineynogo gibkogo vibrozashchitnogo elementa pereskoka / *VINITI* **761-V96**, 7 3 96

IOP Conf. Series: Materials Science and Engineering **786** (2020) 012064 doi:10.1088/1757-899X/786/1/012064

Study of the influence of the driver's reaction time on the road traffic situation at a controlled intersection using simulation modeling

R R Zagidullin

Kazan State University of Architecture and Engineering, 420043, Kazan, Zelenaya str., 1, Russia

E-mail: r.r.zagidullin@mail.ru

Abstract. The role of the human factor in ensuring safe intensification of road traffic is great. This can be achieved with good traffic management and a sharp reduction in motor vehicles collisions (MVC). Nowadays, in countries with a high level of motorization, MVCs lead to great amount of deaths and material losses. This happens because, no matter how reliable the vehicles, roads, and traffic rules, the outcome of the MVC in the vast majority of cases depends on the actions of the driver. The continuing rapid growth of the fleet leads to an increase in the intensity of traffic flows, which imposes ever-increasing demands on drivers. Therefore, improving the reliability of drivers becomes an urgent task, without the resolution of which it is impossible to achieve a significant reduction in MVC and improve traffic safety. The information obtained in the process of studying the patterns of perception by a driver of a road stop will help to solve this problem; therefore, this work is devoted to the urgent topic of the analysis of the influence of the driver's reaction time on traffic safety and traffic capacity of the road network and is performed using the simulation method.

1. Introduction

Safe traffic on roads is determined by many factors: compliance with traffic rules, mutual respect of drivers, pedestrian behavior when crossing highways. One of the main conditions for trouble-free transport movement is the response time of the driver.

The following factors affect the response of the driver and his perception of a dangerous situation:

1. Gender - men who drive a vehicle respond more quickly to the appearance of a danger signal, the response time for them is 1.8 seconds, and for women - 2.8 seconds. However, they perceive a simple situation almost the same way.

2. Age - for car owners whose age does not exceed 30 years, the perception of a dangerous situation occurs quicker than for drivers from 40 years and above. But older people make the right decisions faster, and their reaction time is stable. To solve a simple situation, a young motorist needs 0.17 sec., for a complex one - 1.54 sec. By the age of 60, the indicators change: for a simple situation - 0.26 sec., for a complex - 2.05 sec.

3. Experience - in the event of an emergency on the road you can always see an experienced driver. He does not panic and does not fuss, his actions are quick and accurate.

4. Physical training - sports aimed at developing reaction and endurance help drivers to perceive a dangerous situation faster and choose the right strategic actions.

5. Workplace - a combination of little things that can distract a motorist (an uncomfortable seat, stuffiness in the cab, loose doors, improperly installed cargo in the trunk, noisy passengers) increase the reaction time.

6. Time of day - a person's biological clock is set so that at night there is a decrease in concentration, he often wants to sleep. At night, the period of perception increases by 20-25%. The predawn hours and time before sunset are also difficult for the driver. Therefore, he reacts longer even to a simple traffic situation, and this can lead to MVC.

7. Weather conditions - rain, snow, fog, icing of the road complicate driving, increase the reaction speed of the driver.

8. Medications - there is a large list of medicines that can not be taken if you plan to drive. These may be the most common medications that relieve pain symptoms that help with colds.

9. Alcohol - it's no secret that alcohol and car driving are incompatible concepts. The responsible driver will not allow himself to drink alcohol on the eve of the trip. Most of the MVC occurs because alcohol reduces the concentration of attention, narrows the vision, inhibits motor reflexes. The time to prevent MVC is increased several times.

10. Working conditions - oddly enough, it is easier for the driver to respond to a danger signal within the city than on suburban routes. The monotonous road relaxes and reduces the level of attentiveness, as a result, the motorist incorrectly assesses the situation [1].

11. We should also note the use of gadgets, which despite the prohibition of their use while driving, still take a lot of driver's attention.

12. Most often, it is the speed of decision that prevents MVCs on the road. An important role is played by the time period during which the driver manages to take the necessary actions. In the paper, we consider traffic situations in details taking into account the reaction time of drivers that occurs in the zone of influence of traffic control.

2. The main part

Given the complexity of determining traffic conditions, due to the large number of influencing factors, the study used the method of simulation of traffic flows, allowing:

- to take into account any combination of road conditions, the availability of means of traffic management, as well as the whole variety of situations that arise when traffic flows;
- significantly reduce the duration of the study and the preparation of practical measures to improve traffic conditions;
- establish the main characteristics of traffic flows and give them a quantitative and qualitative assessment, as well as clarify the formulation of analytical problems and verify the reliability of analytical dependencies [2].

AIMSUN specialized program, a microscopic model for simulating the movement of vehicles, was used to simulate the conditions of automobile traffic flows taking into account the reaction time of the driver and the subsequent assessment of the traffic situation. Models of this class allows to describe in detail the behavior of each of the road users.

The object of the study is the intersection of six-lane main streets of citywide significance with allocated lanes for public transport and dividing lanes with traffic lights (Figure 1).

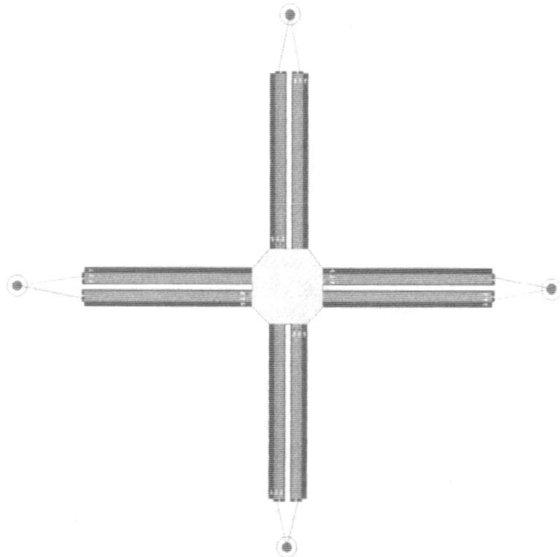

Figure 1. Intersection of main streets of citywide significance with allocated lanes for public transport.

The work of traffic lights at an intersection with a cycle time of 90 seconds is shown in Figure 2. First and second phases are equal to 39 seconds with an intermediate cycle of 6 seconds of which 3 seconds are a yellow signal.

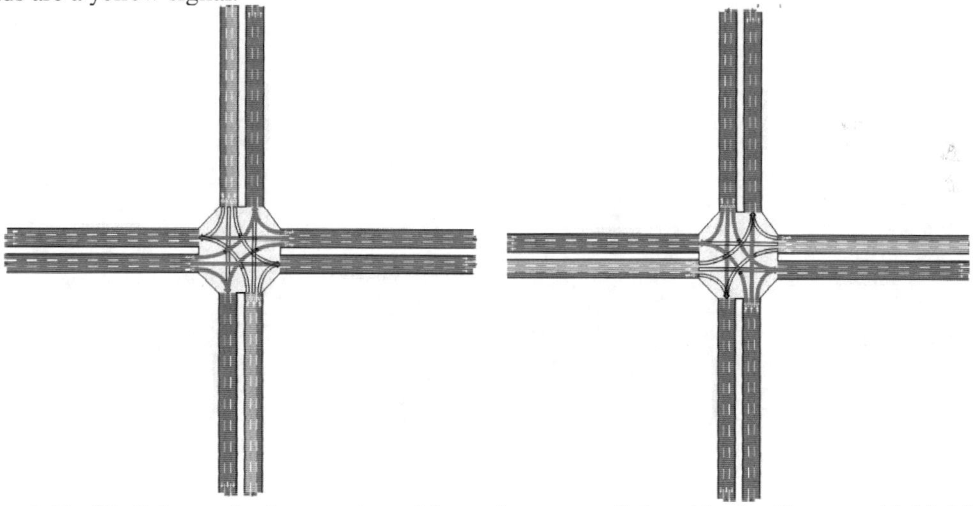

Figure 2. Traffic lights at the intersection of the main streets of citywide significance with highlighted lanes.

IOP Conf. Series: Materials Science and Engineering **786** (2020) 012064 doi:10.1088/1757-899X/786/1/012064

Table 1. Intersection traffic intensity matrix.

	East	West	North	South	Total
East		900	200	100	1200
West	900		100	200	1200
North	100	200		900	1200
South	200	100	900		1200
Total	1200	1200	1200	1200	4800

To conduct an experiment to determine traffic conditions, 6 values of the reaction time of drivers (t_{pv}) were allocated: 0.5; 1.0; 1.5; 2.0; 2.5; 3.0 sec. The values of motion parameters obtained during simulation are presented in table 2.

Table 2. The results of simulation.

Movement parameters \ t_{pv}	0,5	1,0	1,5	2,0	2,5	3,0
Travel time, sec/km	206,62	205,34	307,22	381,29	461,64	476,19
Delay time, sec/km	136,83	146,39	248,15	322,15	401,98	416,26
Stop time, sec	116,55	122,2	215,82	286,95	364,05	376,92
Total travel time, h	67,03	66,63	98,5	117,13	130,75	124,88
Total traveled distance, km	1165,96	1166,07	1152,73	1103,46	1019,94	943,2
Number of stops, #/vehicles/km	3,95	4,04	5,62	6,33	7,37	7,8
Density, vehicles/km	18,39	18,27	27,24	33,22	36,31	34,95
Capacity, vehicles/h	4876	4877	4820	4614	4265	3945
Speed, km/h	25,02	26,8	18,77	13,36	10,1	9,32

Indicators of delay time, density, speed and traffic capacity with increasing driver response time from 0.5 to 3.0 seconds are shown in Figure 3.

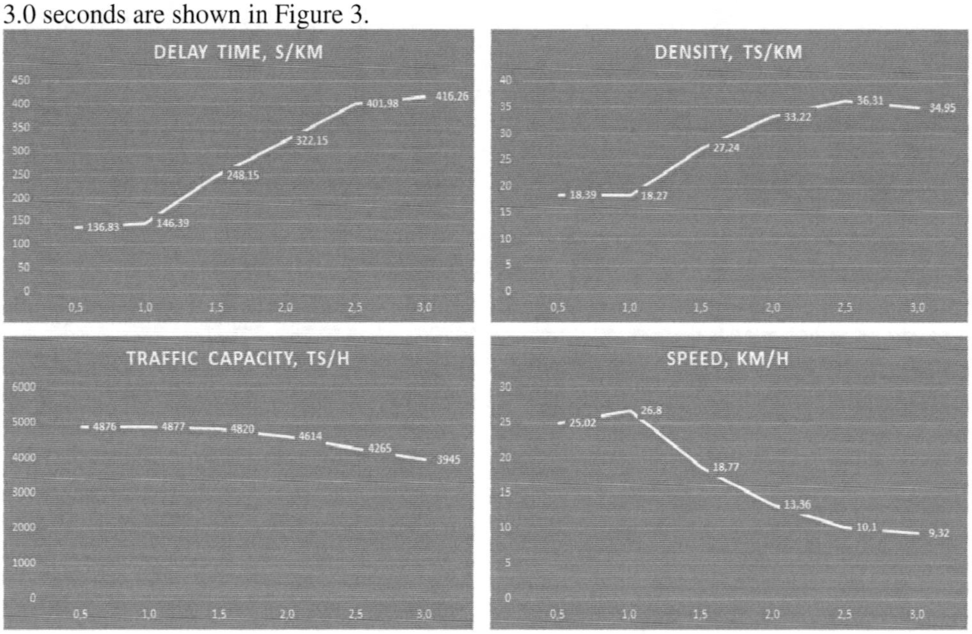

Figure 3. Charts of delay time, density, capacity, speed.

The delay time increased by 204%, the density increased by 90%, the speed decreased by 63%, the capacity decreased by 19%.

3. Conclusion

According to the results of the comparative analysis, it is clear that the reaction time of the driver has a significant impact on the road traffic situation at the regulated intersection of main streets. A study of the perception of the driver of the road situation shows that there are quite large reserves in solving the problem of ensuring road safety. This is, first of all, an increase in knowledge in the norms and methods of road design due to relying not only on the stability of the car against skidding and rollover, but also on the psychophysiological capabilities of the driver.

Rationing the parameters of roads is a technical and economic task and the success of its solution depends on the perfection of the methods of economic calculations and the availability of indicators that take into account both direct costs during the construction of the road and the costs associated with the subsequent operation of the road, accident rate on it and the convenience of drivers and passenger movements. The road is the place of work of thousands of people, and the working conditions on it should not only be safe, but also convenient. The information obtained in the process of studying the patterns of perception by the driver of the road situation will help solve this problem.

There are two ways of use of the listed facts. The first is the development of methods to assess, taking into account the created road conditions, the reliability and convenience of drivers and, if necessary, make adjustments to design decisions. The second way is to improve the norms and methods of road design, the use of which eliminates the creation of visually obscure roads that are uncomfortable or dangerous for traffic. Both of these paths do not exclude, but complement each other.

Further development of science and improvement of the practice of road design and traffic management require the expansion of knowledge about the driver, his reactions, his capabilities as an operator of a complex system. To ensure the reliability of drivers, a high level of professional skills and good psychophysiological training are required. An individual approach to training can be of great help in the formation of such qualities, taking into account the psychological characteristics of the trainees. The introduction of psychological selection and the improvement of medical control over the condition of drivers before the trip and on the road can contribute to increasing the reliability of drivers.

References

[1] Information on https://autopravilo ru/sovety/vremya-reakcii-voditelya-chto-oznachaet-i-ot-kakix-faktorov-zavisit html
[2] Zavadskiy YU V 1977 Reshenie zadach avtomobilnogo transporta metodom imitatsionnogo modelirovaniya / YU V Zavadskiy – M : Transport, – 72
[3] Babkov V V 1982 Dorozhnye usloviya i bezopasnost dvizheniya – M : Transport, – 288
[4] Vaysman A I 1993 Zdorove voditeley i bezopasnost dvizheniya – M : Transport, – 136
[5] Dryu D R 1972 Teoriya transportnykh potokov i upravlenie imi – M : Transport, – 424
[6] Klinkovshteyn G I and Afanasev M E 1992 Organizatsiya dorozhnogo dvizheniya – M : Transport, – 208
[7] Konoplyanko V I 1991 Organizatsiya i bezopasnost gorodskogo dvizheniya – M : Transport, – 182
[8] Lobanov E M 1980 Proektirovanie dorog i organizatsiya dvizheniya s uchetom psikhofiziologii voditelya – M : Transport, – 311
[9] Lobanov E M Vizgalov V M i dr 1996 Proektirovanie i izyskanie peresecheniy avtomobilnykh dorog – M : Transport, – 232
[10] Lobanov E M Silyanov V V Sitnikov YU M 1993 Propusknaya sposobnost avtomobilnykh dorog – M : Transport, – 146
[11] Silyanov V V 1977 Teoriya transportnykh potokov v proektirovanii dorog i organizatsii dvizheniya – M : Transport, – 350

Modification of cement-bound mixtures with sodium formate additives for the construction of pavement bases at low air temperatures

E A Vdovin* and V F Stroganov

Kazan State University of Architecture and Engineering, 420043, Kazan, Zelenaya str., 1, Russia

*E-mail: vdovin007@mail.ru

Abstract. The paper considers the influence of an antifrosty additive of sodium formate (SF) on the kinetics of hardening of cement-bound mixes (CBM) in the bases of road pavements at air temperatures from 0^0C to (-15) ^{0}C with a cement content of 7.0% and 10.0%. The authors establish that the highest intensity of CBM hardening was noted in the first 7 days of hardening, at which the strength was 50-60% of the strength of the material at the age of 28 days. The SF modifier provides an increase in compressive strength (Rcs) of the material depending on the amount of additive and air temperature. The study makes it possible to establish the optimal consumption of SF for different temperature regimes of hardening of mixtures and at the indicated cement contents. The use of SF provides a 30% set of strength at a temperature not lower than (-5) °C. Modification of CBM with additives of SF provides the opportunity to increase the volume of construction of pavement foundations by performing work during periods of the year with lower air temperatures.

1. Introduction

The development of the state's transport system is determined by many factors, including volumes of road construction [1-4]. The increase in volumes is possible with the use of materials ensuring work in adverse conditions, especially at low air temperatures (below 0^0C) [5-10]. Another direction in the development of the country's transport strategy is the use of stone materials and soils modified with functional additives, reinforced with cement [11-33]. The use of anti-frost additives represented by a wide range of chemical compounds: chlorides, nitrates, nitrites, carbonates, sulfates and formates, etc., is known as functional additives for work at low air temperatures [34–42]. Taking into account chemical and environmental safety, compounds based on low molecular weight, low basic carboxylic acids have become very popular [42–45]. The most popular compound is sodium formate (SF) - the sodium salt of formic acid. Its widespread use as an antifrosty additive to cement concrete is well-known [46–47]. This substance has a relatively low freezing temperature, up to (-17) ^{0}C, which attracted the attention of specialists in the field of winter concreting. However, data on the use of SF in the road industry in literary sources are very scarce [48, 51].

Basing on the properties of SF and the relevance of its application to increase the volume of road construction, the following goal of the work is formulated: to study the dependences of the influence of antifrosty SF additives on the kinetics of CBM hardening in road surfaces at air temperatures from 0^0C to (-15) ^{0}C with a cement content of 7.0% , 10.0% and determination of the feasibility of using a functional additive of SF when arranging CBM pavement layers at low temperatures.

2. Materials and methods

Studies were carried out on crushed stone mixtures of limestone and dolomite rocks, including substandard crushed stone of stone quarries of fractions (0-40) mm. The particle size distribution and physico-mechanical properties of the mixtures are shown in Tables 1 and 2.

Table 1. Granulometric composition of crushed stone mixture.

Maximum grain size, (mm)	The total residue on the sieve size of the holes, (mm)								
	40	20	10	5	2,5	1,25	0,63	0,315	0,14
40	5	37	56	70	80	85	86	88,5	91

Table 2. Physico-mechanical properties of crushed stone mixtures.

Fraction Size (mm)	Bulk density (kg / m3)	Water absorption (%)	Abrasion Grade	Strength grade	Frost resistance indicator
0-40	1600	9,1	IV	200	10

To strengthen the mixtures we used the most recommended for road construction Portland cement brand CEM I 42.5N. The mineral and chemical compositions of cement clinker are shown in Tables 3 and 4.

Table 3. Mineral composition of cement clinker.

Name	C_3S	C_2S	C_3A	C_4AF
Content, (%)	58	17	8	13

Table 4. The chemical composition of cement clinker.

Name	SiO_2	Al_2O_3	Fe_2O_3 *a)*	*CaO b)*	*MgO*	SO_3	R_2O (Na_2O+ +0,65 K_2O)	CaO_{cb}
Content, (%)	21,1	5,85	4,2	65,4	1,13	1,03	1,07	0,16

As an antifrosty additive we used sodium formate acid (HCOONa). When selecting CBM formulations, the cement content was adopted taking into account the brand strength established for road pavement. The tests were carried out on mixtures with 7% and 10% cement by weight of the crushed stone mixture.

The temperature regime of CBM hardening was changed from 0 ° C to (-15) ° C. As a criterion for evaluating the effectiveness of antifrosty additives, we took an indicator of the set of compressive strength during 28-day hardening at a negative temperature. In accordance with the requirements of consumers, at least 30% of the brand strength should be provided at a temperature of (-15) °C (± 5) °C at 28 days of age. To compare the test results, the process of hardening of control mixtures was studied under normal conditions (+ 20 ° C).

To manufacture the samples with dimensions of 15 * 15 * 15cm, we used metal molds lubricated with a release agent. The cement-bound mixture was compacted on a laboratory vibratory platform of type S-I35-A, which provided vertical vibrations of the filled form with a frequency of 2900 ± 100 vibrations per minute and an amplitude of 0.5 ± 0.05 mm. After preliminary aging for 2-2.5 hours, the samples were placed in chambers with different temperature and humidity conditions: in a normal storage chamber (temperature 200^{0}C, with a relative humidity of 95%), in refrigeration chambers with constant temperatures 0^{0}C, (-5) ^{0}C , (-10) ^{0}C, (-15) ^{0}C.

The compressive strength of CBM samples was determined on an MS-500 press. The kinetic dependences of the compressive strength are constructed from the averaged values of the index in points for various amounts of the SF additive (Figure 1-2).

3. Results and discussion

Similarly to cement concrete, the highest rate of curing during compression of CBM was determined in the first 7 days of hardening (Figure 1-2). We found that mixtures, regardless of the content of the physical density and temperature, have a strength in the range of 50-60% of the strength of the material at the age of 28 days.

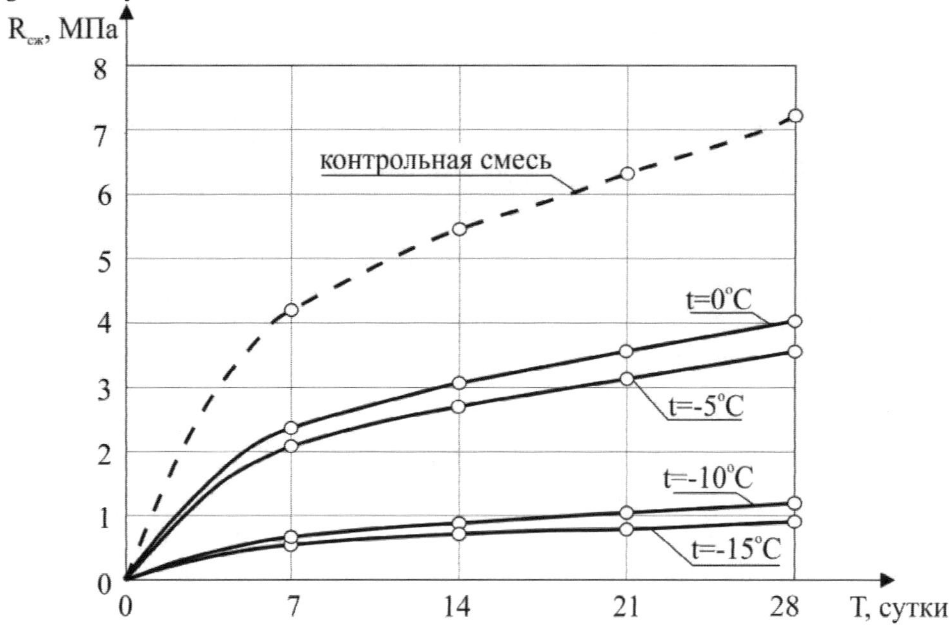

Figure 1. Hardening kinetics of CBM with the addition of SF and cement content of 7% at different temperatures.

The research revealed that the SF contributes to an increase in the strength of the material depending on the amount of added additive and the temperature of the outside air (Figure 3.4). In these figures, in addition to the curves of the dependences of the strength of the mixtures, lines of the levels of vintage strengths and the strength of the control mixtures are plotted, and the dotted lines indicate the minimum strengths at the age of 28 days, which should be provided with the introduction of an anti-frost additive.

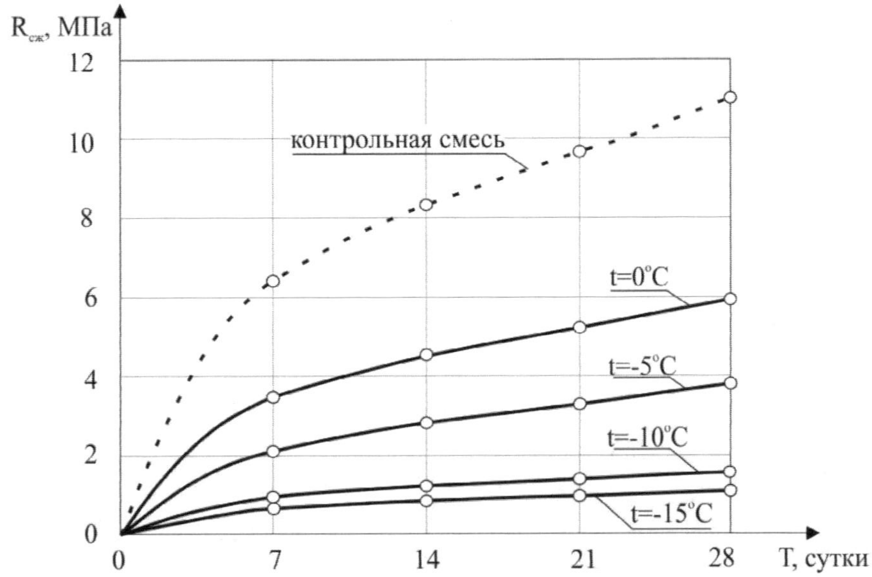

Figure 2. Hardening kinetics of CBM with the addition of SF and cement content of 10% at different temperatures.

We determined that the use of SF provides a 30% set of strength at a temperature not lower than -5 ° C (Figure 3.4). With a 7% cement content, the level of material strength is not lower than 40 MPa (grade M40) with SF 2.5-3.0% by weight of cement. For mixtures with a high cement consumption (10%), a lower addition of SF is required: 2-2.5% to achieve the M40 grade.

Figure 3. The dependence of the compressive strength of CBM on the amount of SF additives and temperature of hardening with a cement content in the mixture of 7%.

A further increase in SF of 3.5-4.0% in the mixture provides brand strength above 60 MPa (grade M60). Modification of CBM with additives of SF provides the ability to set the required strength for layers of pavement and increase the volume of road construction by performing work during periods of the year with lower air temperatures. Investigations (Figs. 3, 4) made it possible to establish the optimal SF consumption for temperature hardening mixtures from 0 ° C to (-15) ° C with a cement content of 7% and 10%.

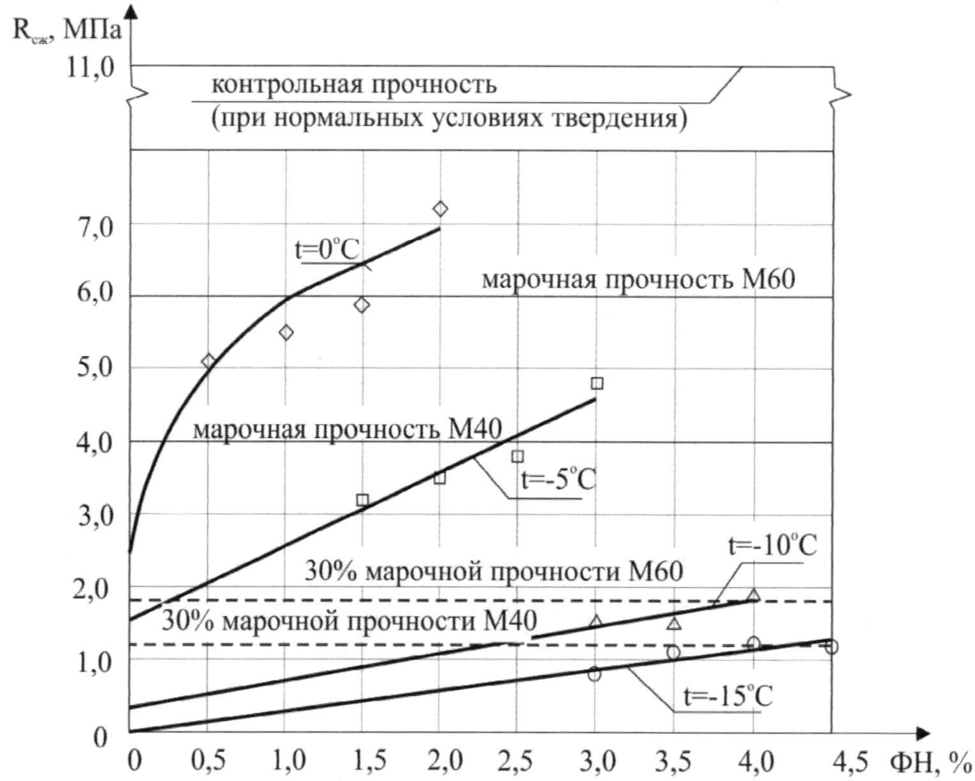

Figure 4. The dependence of the compressive strength of CBM on the amount of SF additives and temperature of hardening with a cement content in the mixture of 10%.

Analyzing the data of kinetic and strong dependences (Figure 1-4), we should note that the addition of SF provides a greater increase in strength compared to cement, although the amount of functional modifier is much smaller in absolute value. An increase in cement content by almost 1.5 times (from 7% to 10%) practically does not change the level of material strength at low temperatures, while the strength of control samples CBM (hardening at 20 ° C for 28 days) increased by 1, 5 times.

Lowering the hardening temperature from 0 ° C to (-15) ° C provides a multiple (8-10 times) decrease in the strength of the material by 28 days. In the interval (-10) - (- 15) °C, the used amount of SF 4.0-4.5% does not provide a noticeable increase in strength. The question of studying the effect of increased amounts of SF (more than 4.5%) in CBM compositions for pavement requires further study. Moreover, the modification of mixtures with the introduction of additional components of additives that can increase the hardening efficiency at temperatures below (-10) - (- 15) 15° C and provide the specified construction and technical properties of pavement materials is of a great interest. The subsequent hardening of CBM at positive temperatures will provide further hardening, which is known from the theory and practice of winter concreting [36–38, 42], but requires research for CBM.

4. Conclusion

1. In accordance with the purpose of the work, the dependences of the effect of the antifreeze additive SF on the kinetics of hardening CBM in pavement bases at air temperatures from 0^0C to (-15) ^{0}C with a cement content of 7.0%, 10.0% were studied. The highest rate of hardening of CBM is

determined in the first 7 days. It was found that mixtures, regardless of the content of the physical density and temperature, have a strength in the range of 50-60% of the strength of the material at the age of 28 days. It is determined that the use of SF provides a 30% set of strength at a temperature not lower than (-5) ^{0}C. At a 7% cement content, the level of material strength is not lower than 40 MPa (grade M40) with a pressure factor of 2.5-3.0% by weight of cement. For mixtures with a high cement consumption (10%), a lower addition is required: 2-2.5% to achieve the M40 grade. A further increase in the SF of 3.5-4.0% in the mixture provides brand strength above 60 MPa (grade M60). Investigations made it possible to establish the optimal consumption of SF for temperature hardening mixtures from 0 ° C to (-15) ° C and with a cement content of 7% and 10%.

2. We should note that it is advisable to use a functional SF additive when arranging CBM pavement layers at low temperatures, which under these conditions provides the opportunity to set the required strength for pavement layers and increase the volume of road construction by performing work during periods of the year with adverse conditions.

3. Attention should be paid to the prospects of the development of research on the use of antifreeze additives based on SF in CBM compositions for road pavements in the direction of study:

- modification of the material with an increased amount of antifrosty additives;
- the possibility of increasing the hardening efficiency at temperatures below (-10) - (- 15) ^{0}C by adding additional components to the modifier;
- the processes of subsequent hardening of the material in pavements at positive temperatures.

References

[1] D Ludlow V and Sakhrani C Wu 2019 Managing State Transportation Research Programs, TRB, Wachington, D C, 164p

[2] A Kim, N Sadatsafavi, S D Anderson and P Bishop 2017 Preparing for the Future of TransportationConstruction: Strategies for State Transportation Agencies, *Journal of Management in Engineering* **33(3)**

[3] A Kim, N Sadatsafavi, S D Anderson and P Bishop 2017 Using Scenario Planning for Identifying Major Future Trends and Their Implications for State Transportation Agencies, International Conference on Sustainable Infrastructure 237-49

[4] S Kermanshachi, S D Anderson, P Goodrum, T R B Taylor 2017 Project Scoping Process Model Development to Achieve On-Time and On-Budget Delivery of Highway Projects, Transportation Research Record: *Journal of the Transportation Research Board*, **2630** 147-55

[5] F M Sayfullina, L R Mustafina and D N Semenov 2018 Innovatsionnye tekhnologii kak osnova ustoychivogo razvitiya dorozhnogo stroitelstva Voprosy innovatsionnoy ekonomiki, **8(4)** 705-14

[6] M V Karmaza and R V Motylev 2018 Technologies for the use of antifrosty additives in winter concreting, International Conference Technical and Natural Sciences, *Journal of Technical and Natural* **9**, , OEAPS Inc , Berlin, Germany, 8-14

[7] R Myrdal 2007 Accelerating admixtures for concrete, SINTEF REPORT Building and Infrastructure Concrete and Innovation Center, Project, **3** 1-35

[8] N R Panga and J A Naqash 2019 Development of high early strength in concrete incorporating alccofine and non-chloride accelerator, SN Applied Sciences

[9] Sh Gholami, J Hu, Y Kim and M Mamirov 2019 Performance of Portland Cement-Based Rapid-Patching Materials with Different Cement and Accelerator Types, and Cement Contents, Transportation Research Record: *Journal of the Transportation Research Board*

[10] N T Todd, P Suraneni and W Weiss 2017 Hydration of Cement Pastes Containing Accelerator at Various Temperatures: Application to High Early Strength Pavement Patching, Advances in Civil Engineering Materials

[11] S K Saxena, M Kumarb and N B Singh 2018 Effect of Alccofine powder on the properties of Pond fly ash based Geopolymer mortar under different conditions, *Environmental Technology and Innovation* **9** 232-42

[12] B V Kavyateja, J G Jawahar and C Sashidhar 2019 Investigation on ternary blended self compacting concrete using fly ash and alccofine, International *Journal of Recent Technology and Engineering* **7** 447-51

[13] L Hu, J Hao and L Wang 2014 Laboratory evaluation of cement treated aggregate containing crushed clay brick, *Journal of Traffic and Transportation Engineering (English Edition)* **1(5)** 371-82

[14] A T Visser 2017 Potential of South African road technology for application in China, *Journal of Traffic and Transportation Engineering (English Edition)* **4(2)** 113-7

[15] M Iwański, P Buczyński and G Mazurek 2016 Optimization of the road binder used in the base layer in the road construction, Construction and Building Materials, **125** 1044-54

[16] H Ö Öz 2018 Properties of pervious concretes partially incorporating acidic pumice as coarse aggregate, Construction and Building Materials, **166** 601-9

[17] V Mymrin, E K Aibuldinov, K Alekseev, V Petukhov, M A Avanci, A Kholodov, A Taskin, R E Catai and A Iarozinski 2019 Efficient road base material from Kazakhstan's natural loam strengthened by ground cooled ferrous slag activated by lime production waste, *Journal of Cleaner Production* **231** 1428-36

[18] A Rana, P Kalla, H K Verma and J K Mohnot 2016 Recycling of dimensional stone waste in concrete: A review, *Journal of Cleaner Production* **135** 312-31

[19] L D Poulikakos, C Papadaskalopoulou, B Hofko, F Gschösser, A C Falchetto, M Bueno, M Arraigada, J Sousa, R Ruiz, C Petit, M Loizidou and M N Part 2017 Harvesting the unexplored potential of European waste materials for road construction, Resources, *Conservation and Recycling* **116** 32-44

[20] V Vishwakarma and D Ramachandran 2018 Green Concrete mix using solid waste and nanoparticles as alternatives – A review, *Construction and Building Materials* **162** 96-103

[21] Absalyamova S et al 2016 "The impact of corruption on the sustainable development of human capital " *Journal of Physics: Conference Series* **738(1)**

[22] F Autelitano and F Giuliani 2016, Electric arc furnace slags in cement-treated materials for road construction: Mechanical and durability properties, Construction and Building Materials, **113** 280-9

[23] Em Mengue, H Mroueh, L Lancelot and R Medjo Eko 2017 Physicochemical and consolidation properties of compacted lateritic soil treated with cement, *Soils and Foundations* **57(1)** 60-79

[24] Es Melese, H Baaj, S Tighe, S Zupko, T Smith 2019 Characterisation of full-depth reclaimed pavement materials treated with hydraulic road binders, *Construction and Building Materials* **226** 778-92

[25] Op E Oluwatuyi, B O Adeola, El A Alhassan, Em S Nnochiri, Ab E Modupe, Ol O Elemile, T Obayanjua and G Akerele 2018 Ameliorating effect of milled eggshell on cement stabilized lateritic soil for highway construction, *Case Studies in Construction Materials*, **9** e00191

[26] B Dołżycki and P Jaskuła 2019 Review and evaluation of cold recycling with bitumen emulsion and cement for rehabilitation of old pavements, *Journal of Traffic and Transportation Engineering* (English Edition), **6(4)** 311-23

[27] D Ribeiro, R Néri and R Cardoso 2016 Influence of Water Content in the UCS of Soil-Cement Mixtures for Different Cement Dosages, Procedia Engineering, **143** 59-66

[28] H Mola-Abasi and I Shooshpasha 2016 Influence of zeolite and cement additions on mechanical behavior of sandy soil, Journal of Rock Mechanics and Geotechnical Engineering, **8(5)**, 746-52

[29] H F Hassan, R Taha, A Al Rawas, B Al Shandoudi, Kh Al Gheithi, A M Al Barami 2005 Potential uses of petroleum-contaminated soil in highway construction, Construction and Building Materials, **19(8)** 646-52

[30] V Anggraini, Af Asadi, Ag Syamsir and B B K Huat 2017 Three point bending flexural strength of cement treated tropical marine soil reinforced by lime treated natural fiber, *Measurement* **111** 158-66

[31] L Nguyen, B Fatahi and H Khabbaz 2016 Predicting the Behaviour of Fibre Reinforced Cement Treated Clay, Procedia Engineering, **143** 153-60

[32] M Rezaeian, P M Vaz Ferreira and Ab Ekinci 2019 Mechanical behaviour of a compacted well-graded granular material with and without cement, *Soils and Foundations*, **59(3)** 687-98

[33] Th Lenoir, M Preteseille and S Ricordel 2016 Contribution of the fiber reinforcement on the fatigue behavior of two cement-modified soils, *International Journal of Fatigue*, **93(1)** 71-81

[34] O N Ilina and N V Konovalov 2013 Dorozhno-stroitelnyy material na osnov mestnykh mineralnykh materialov, obrabotannykh kompleksnym vyazhushchim, Izvstiya KGASU, **2(24)** s 295-300

[35] F Karagöl, R Demirboğa, M A Kaygusuz, M M Yadollahi and R Polat 2013 The influence of calcium nitrate as antifreeze admixture on the compressive strength of concrete exposed to low temperatures, *Cold Regions Science and Technology*, **89** 30-5

[36] F Karagöl, R Demirboğa and H Kh Waleed 2015 Behavior of fresh and hardened concretes with antifreeze admixtures in deep-freeze low temperatures and exterior winter conditions, *Construction and Building Materials* **76** 388-95

[37] P C Aïtcin and A B Eberhardt 2016 Historical background of the development of concrete admixtures, Science and Technology of Concrete Admixtures,

[38] R Polat 2016 The effect of antifreeze additives on fresh concrete subjected to freezing and thawing cycles, *Cold Regions Science and Technology* **127** 10-7

[39] M Çulluan and M Arslan 2013 The effects of antifreeze use on physical and mechanical properties of concrete produced in cold weather, *Composites Part B: Engineering* **50** 202-9

[40] R Demirboğa, F Karagöl, R Polat and M A Kaygusuz 2014 The effects of urea on strength gaining of fresh concrete under the cold weather conditions, *Construction and Building Materials*, **64** 114-20

[41] M Çullu and M Arslan 2014 The effects of chemical attacks on physical and mechanical properties of concrete produced under cold weather conditions, Construction and Building Materials, **57** 53-60

[42] N Xie, Yu Dang and X Shi 2019 New insights into how MgCl2 deteriorates Portland cement concrete, *Cement and Concrete Research* **120** 244-55

[43] L Kivekäs 2019 Research and use of antifreeze admixtures in Finland

[44] J Suh, W S Yumb, Y Jeong, H Park, J E Oh 2019 The cation-dependent effects of formate salt additives on the strength and microstructure of CaO-activated fly ash binders, *Construction and Building Materials*, **194** 92-101

[45] H Schreiberová, P Bílý, J Fládr, K Šeps, R Chylík and T Trtík 2019 Impact of the self-healing agent composition on material characteristics of bio-based self-healing concrete, *Case Studies in Construction Materials*, **11** e00250

[46] A Hartmann, M Khakhutov and J -Ch Buhl 2014 Hydrothermal synthesis of CSH-phases (tobermorite) under influence of Ca-formate, *Materials Research Bulletin* **51** 389-96

[47] O Tene, Smes veshchestv, primenimaya, v chastnosti, v kachestve dobavki dlya betonnoy smesi: pat 2433096 Ros Federatsiya 2009134179/03

[48] O V Milovidova, V M Sychev, E V Rozhkov, S YU Pliner and S F SHmotev 2018 Sostav dlya stabilizatsii prirodnykh i tekhnogennykh gruntov: pat 2670468 Ros Federatsiya 0002670468

[49] V V CHelnokov 2016 Muravi, gryzushchie led, KHimiya i zhizn

IOP Conf. Series: Materials Science and Engineering **786** (2020) 012065 doi:10.1088/1757-899X/786/1/012065

[50] S V Gushchin and A A Drozd 2014 Vliyanie formiata natriya na tverdenie betona pri otritsatelnykh temperaturakh, *Problemy sovremennogo betona i zhelezobetona*, **6**
[51] S V Lanovetskiy 2015 Issledovanie ustoychivosti peresyshchennykh rastvorov formiata natriya, *Inzhenernyy vestnik Dona*, **2(2)**
[52] A V Kochetkov 2011 Otsenka sootvetstviya primeneniya formiata natriya v kachestve protivogololednogo materiala Zakonu «O tekhnicheskom regulirovanii», *Stroitelnye materialy*, **7** 44-6

IOP Conf. Series: Materials Science and Engineering **786** (2020) 012066 doi:10.1088/1757-899X/786/1/012066

Assessment of the impact of capital construction on the adjacent street-road network with traffic at unregulated intersections

A V Zedgenizov

Irkutsk National Research Technical University, Russia, 664074, Irkutsk, Lermontov st, 83, room I-213

E-mail: azedgen@gmail.com

Abstract. The article discusses the problems of assessing the influence of centers of gravity on the adjacent street-road networks in the process of their functioning, expansion or reprofiling. The paper shows the main models for estimating the throughput of unregulated intersections proposed in the domestic and foreign literature. The author gives sequence of calculating the transport demand transforming into the intensity of the traffic flow on the road network. Moreover, the study proposes recommendations on assessing the level of traffic service and load factor based on transport demand and throughput, adjacent street-road network.

1. Introduction

The throughput of the street-road network (SRN) is the main criterion for assessing its potential capabilities, i.e. the size of the transport supply, formed both in its individual sections, and on the scale of the city (city district). The problems of assessing the throughput of roads and its elements are widely studied and presented in the domestic and foreign literature [1-7, etc.].

Along with the transport supply, one of the main criteria for assessing the effectiveness of the functioning of the SRN is the ratio of transport demand, transforming into the intensity of individual transport on the SRN and the transport supply, expressed in SRN capacity. This ratio is called the SRN load level and is expressed by the load factor. For normal driving conditions, such a system should meet the condition of balance i.e. transport demand should not exceed transport supply. Thus, it should be concluded that the effectiveness of the functioning of the SRN as a whole will mainly depend on the traffic organization at its individual nodes (intersections).

2. The main part

According to federal law No. 433 of December 29, 2017, “On the Organization of Road Traffic in the Russian Federation and on Amending Certain Legislative Acts of the Russian Federation”, the following provisions apply in the Russian Federation today:

- the need to develop a plan during the construction or reconstruction of capital construction objects (CCO) in relation to the network of roads and (or) their sections adjacent to the indicated CCO;

- the need to prepare engineering survey materials, research results of existing and predicted traffic parameters, and statistical information;
- requirements for ensuring the effectiveness of traffic organization, which must be taken into account when placing a CCO in accordance with the legislation on urban development. CCO placement in violation of traffic organization performance requirements is not allowed;
- the need to calculate the capacity (number of parking spaces) of public parking lots is determined in accordance with the standards of urban planning;
- the need to develop a plan when arranging vehicle entry into and out of public parking, and vehicle movement inside a public parking lot.

The level of traffic service is a concept that has been firmly entrenched in domestic regulatory documents relatively recently, and this is primarily due to the development of motorization, an increase in the number of types of vehicles and, accordingly, the need for traffic organization with a very wide range of factors and traffic conditions. This concept is a derivative of the basic characteristics of the functioning of SRN elements and, above all, intersections. These characteristics include the load coefficient (level) $z = N/P$, where N is the traffic intensity, vehicle/h, P is the actual (practical) throughput, vehicle/h.

The model for estimating the throughput of unregulated intersections is represented by the expression:

$$P_{\mathrm{p}} = N_{\mathrm{gl}}(\mathrm{A}\frac{e^{-\beta_{q1}\cdot\gamma\cdot\Delta t_{\mathrm{gr}}}}{1-e^{-\beta_1\cdot\gamma\cdot\delta t}} + B\frac{e^{-\beta_{q2}\cdot\gamma\cdot\Delta t_{\mathrm{gr}}}}{1-e^{-\beta_2\cdot\gamma\cdot\delta t}} + C\frac{e^{-\beta_{q3}\cdot\gamma\cdot\Delta t_{\mathrm{gr}}}}{1-e^{-\beta_3\cdot\gamma\cdot\delta t}}) \quad (1)$$

where N_{gl} – traffic on the main road, vehicle/h; $\gamma = N_{gl}/3600$; A – coefficient characterizing freely moving cars; B – coefficient characterizing a partially connected part of the flow of cars; C - coefficient characterizing the connected part of the flow of cars; $A = \epsilon_m - \epsilon_p$ – for uphill sections; ϵ_m - coefficient taking into account the number of slowly moving cars in the flow; ϵ_p - coefficient taking into account the steepness of the slope and the length of the elevation; Δt_{gr} – boundary interval accepted by the driver; δt - interval between exits of cars from the queue on the secondary road, sec; $\beta_{q_1}, \beta_{q_2}, \beta_{q_3}$ – coefficients characterizing traffic flow density; A+B+C = 1.

Thus, given the transport demand, transforming into the intensity of individual transport, which can be calculated by the expression:

$$N_{ИТ/час} = E_{цмт} \cdot \frac{d_{ИТ}}{P_{ИТ}} \cdot k_{СН} \quad (2)$$

where d_{IT} – the proportion of visitors with individual cars per hour; P_{IT}– average number of passengers, people.; k_{cn} – daily unevenness coefficient for the considered hour; E_{CHMT} – transport demand for local nod, corr/day.

The load factor z for unregulated intersections will look like this:

$$z = \frac{E_{\mathrm{CHMT}}\cdot\frac{d_{\mathrm{it}}}{P_{\mathrm{it}}}\cdot k_{\mathrm{cn}}}{N_{\mathrm{gl}}(\mathrm{A}\frac{e^{-\beta_{q1}\cdot\gamma\cdot\Delta t_{\mathrm{gr}}}}{1-e^{-\beta_1\cdot\gamma\cdot\delta t}}+B\frac{e^{-\beta_{q2}\cdot\gamma\cdot\Delta t_{\mathrm{gr}}}}{1-e^{-\beta_2\cdot\gamma\cdot\delta t}}+C\frac{e^{-\beta_{q3}\cdot\gamma\cdot\Delta t_{\mathrm{gr}}}}{1-e^{-\beta_3\cdot\gamma\cdot\delta t}})} \quad (3)$$

In general, we should note that the 2.32 model is highly saturated with factors that take into account the presence of connected parts of the traffic flow "bundles", slopes, flow density, etc., which, according to the author, is a reflection of the intervals between vehicles in the main stream. Depending on these intervals, the speed of the main flow, as well as the time intervals between vehicles in the secondary flow, the throughput of the secondary direction at an unregulated intersection will differ. Thus, a technique that takes into account mainly the listed factors is presented in [6, 7, 8].

The driver decides to enter an unregulated intersection while approaching the intersection and moving in a secondary direction, which depends on the size of the time interval between two consecutive vehicles in the main direction. If the driver considers one of these intervals acceptable, he starts moving, but if the interval seems too small to him, he will wait for an acceptable interval. If the driver waits too long, then it may turn out that in the end, for an acceptable one, he will count an interval that will be less than one of those that he has already missed.

The time interval that the driver needs to start moving in a secondary direction is called the boundary interval. At this stage, when calculating, it should be taken into account that at all approaches to the intersection, each group of traffic (left, straight and right) carries out movement along a designated lane, so far without taking into account the real organization of traffic and the number of lanes at the approaches (Figure 1). The principle of movement at an unregulated intersection in accordance with the traffic rules and rankings is that the lower the rank value, the higher the priority for the passage crossing the traffic flow.

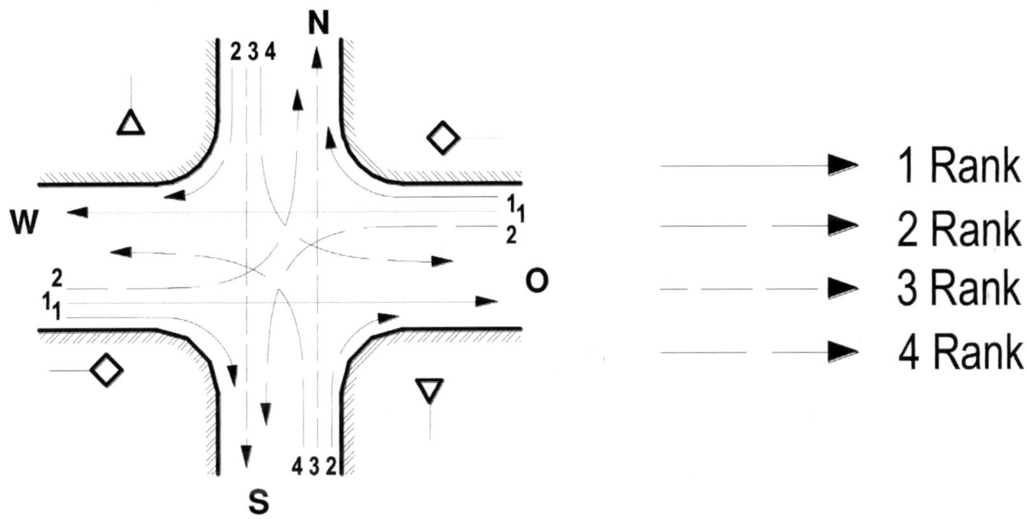

Figure 1. Scheme of movement at an unregulated intersection.

For example, flows of the first rank are not inferior to anyone, as a rule, they move along the road indicated by the sign 2.1 "Main road" in the forward or right-handed direction. The second rank is forced to give way to the first (from the main direction to the left and from the secondary to the right). The third rank implies passage directly from the secondary direction and is forced to give way to the first and second rank. The fourth rank (to the left from the secondary direction) gives way to everyone (1st, 2nd and 3rd rank), its throughput is the lowest.

The throughput of first-rank flows under ideal conditions (the absence of left and right turns, parked vehicles on the far right lane, longitudinal curves and curves in the plan, etc.) can be equated to an ideal saturation stream (approximately 1900 vehicles/h). When calculating the capacity of flows of the first rank (from the main direction to the right, in addition to the listed coefficients, we use coefficients that take into account the presence of pedestrian traffic, public transport stops, parking on the far right lane).

The throughput of flows of the second rank is estimated by the basic throughput, which is calculated for all ranks other than the first (2, 3 and 4 ranks). It should be noted that the basic throughput takes into account the influence of flows of the first rank only and for this reason the bandwidth of the flows of the second rank is equal to their basic throughput. The calculation is carried out by the expression:

$$G_i = \frac{3600}{t_f} e^{-\frac{q_p}{3600}\left(t_g - \frac{t_f}{2}\right)}, \quad (4)$$

where G_i – basic throughput of the secondary flow i, vehicle/h; q_p – the intensity of the priority direction of the first rank, which conflicts with the considered one, vehicle/h; t_g – the average boundary interval, sec; t_f – average interval, sec.

With increasing traffic density, the average boundary interval decreases. The average boundary interval is taken on the basis of field studies of the most common cases at unregulated intersections.

To determine the throughput of a flow of 3rd rank, it is necessary to determine the likelihood that the flow of 2nd rank will move without jams:

$$p = 1 - \frac{N}{c} \tag{5}$$

where p – the probability of unhindered movement of flows of the 2nd and 3rd ranks; N – the intensity of the movement of the flow, units/h;; c – throughput of the considered flow, vehicle/h.

The probability of an unhindered movement of a flow of rank 2 is calculated for each approach. The capacity of the 3rd rank flow is determined based on the formula:

$$c_3 = \prod_{i=1}^{n} p_{02i} G_3 , \tag{6}$$

where G_3 – 3rd rank basic flow throughput, units / h;

The capacity of the 4th rank flow is determined based on the formula:

$$c_4 = \prod_{i=1}^{n} p_{02i} \prod_{j=1}^{m} p_{03j} G_4 , \tag{7}$$

where $\prod_{i=1}^{n} p_{02i} \prod_{j=1}^{m} p_{03j}$ – the product of the probabilities of the unhindered movement of all flows of the 2nd and 3rd rank, which are inferior to the considered flow of the 4th rank.

In the methodology for assessing the throughput of unregulated intersections, it is important to remember that the throughput of each direction is calculated taking into account the fact that each direction (traffic flow) moves along a dedicated lane. It does not take into account mutual interference (delays) resulting from combining flows of different ranks into one or more physical bands on the way. As a rule, secondary approaches have no more than one lane on the approach, and the main no more than two (Figure 2). A particular case of identifying the effectiveness of an unregulated intersection is to determine the throughput of a mixed lane (Figure 3).

IOP Conf. Series: Materials Science and Engineering **786** (2020) 012066 doi:10.1088/1757-899X/786/1/012066

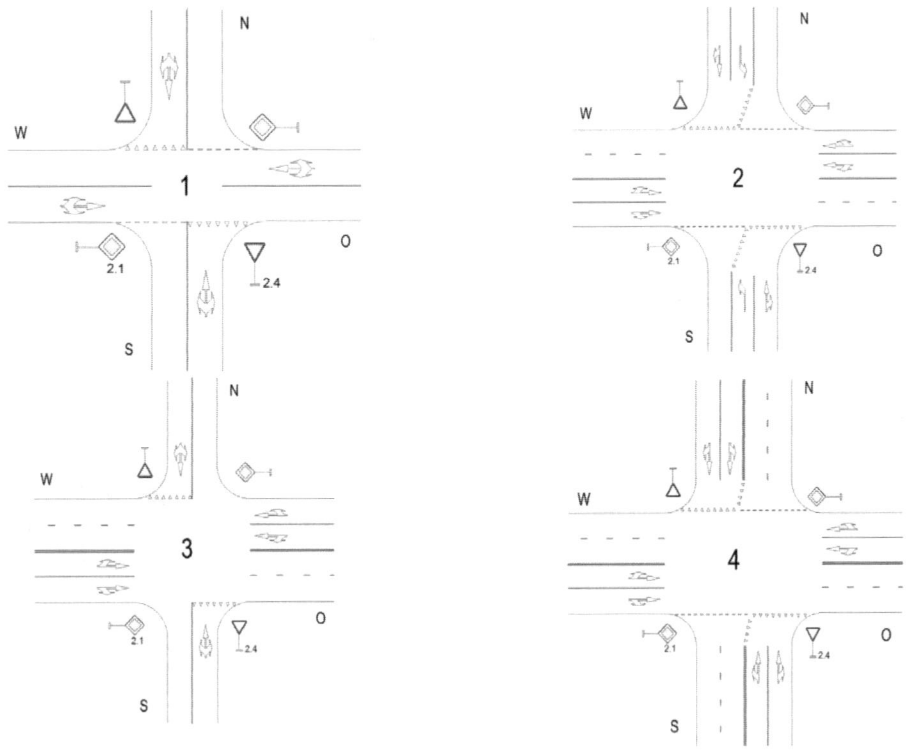

Figure 2. Possible options for organizing traffic at an unregulated intersection.

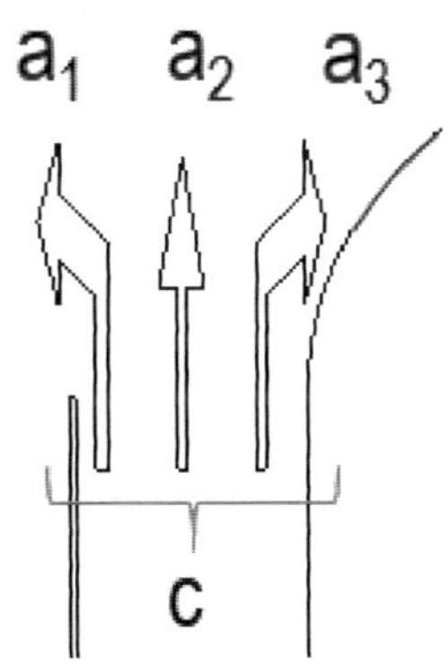

Figure 3. Mixed lane at unregulated intersection.

The mixed bandwidth is determined by the following formula:

$$c = \frac{1}{\sum_{i=1}^{n} \frac{a_i}{c_i}} \quad (8)$$

where c – throughput of the mixed lane, units/h;; a_i – is the fraction of the i-th flow in the sum of all n flows in the considered lane; c_i – hroughput of the i-th direction, units / h.

Thus, given (4) - (8), the load factor of the i-th flow can be represented by the expression:

$$z = \sum_{i=1}^{n} \frac{E_{цмт} \cdot \frac{d_{um}}{P_{um}} \cdot k_{сн}^{уб} \cdot k_d}{\left(\frac{3600}{t_f} e^{-\frac{q_p}{3600}\left(t_g - \frac{t_f}{2}\right)} \right) \cdot \left(\prod_{j=1}^{m} 1 - \frac{N_{02j}}{c_{02j}} \right) \cdot \left(\prod_{j=1}^{m} 1 - \frac{N_{03j}}{c_{03j}} \right)} \tag{9}$$

where N_i – the intensity of the i-th direction of the traffic flow in the movement group, vehicle./h; q_p – the traffic intensity of rank 1, which conflicts with the direction of movement, vehicles / h; k_d - coefficient taking into account the share of traffic flow distributed between intersections; $k_{сн}^{уб}$ - coefficient of daily unevenness in descending order; n – number of traffic flows in a traffic group; t_g – is the average boundary interval, sec; t_f – is the average interval of succession, sec; N_{02j} – is the traffic intensity of the second rank, which conflicts with the direction of movement, vehicle/h (in the absence it is excluded); N_{03j} – is the traffic intensity of the third rank, which conflicts with the direction of movement, vehicle/h (in the absence it is excluded); c_{02j} and c_{03j} - throughput of the traffic flow of the second and third ranks, respectively, conflicting with the direction of movement, vehicle/h (in the absence it is excluded); m – the number of traffic flows of the second and third ranks, respectively, conflicting with the direction of movement.

The value of the impact of the CCO on the adjacent SRN in the case of re-profiling of existing or commissioning of new ones should be estimated on the basis of mathematical models relying, on the one hand, on the methodological recommendations of the industry document for assessing the capacity of roads, and on the other, on transport demand arising to this CCO.

References

[1] Gorev A E 2010 Osnovy teorii transportnykh sistem: ucheb posobie / A E Gorev – SPb : SPbGASU, – 214
[2] Efremov I S 1980 Teoriya gorodskikh passazhirskikh perevozok: ucheb posobie dlya vuzov– M : Vyssh shk– 535
[3] Organizatsiya dorozhnogo dvizheniya v gorodakh: Metodicheskoe posobie pod obsh red YU D SHelkova: Nauchno-issledovatelskiy tsentr GAI MVD Rossii - M : 1995 — 143
[4] Otraslevoy dorozhnyy metodicheskiy dokument Metodicheskie rekomendatsii po otsenke propusknoy sposobnosti avtomobilnykh dorog ODM 218 2 020-2012
[5] Tekhnicheskie sredstva organizatsii dorozhnogo dvizheniya: Ucheb dlya vuzov – M : Transport, 1990 – 255
[6] Tekhnicheskie sredstva organizatsii dorozhnogo dvizheniya: metod ukazaniya k laboratornym rabotam i kursovomu proektu / sost : A V Zedgenizov, A G Levashev Irkutsk: Izd-vo IrGTU, 2010 36
[7] Greenshields B D, Shapiro D and Ericksen E L 1947 Traffic Performance at Urban Street Intersections Technical Report1, Yale Bureau of Highway Traffic
[8] Raff M S and Hart J W A volume Warrant For Urban Stop Signs Eno Foundation for Highway
[9] Zedgenizov A 2018 The methodology of estimation of transport demand by the parameters of location of the mass gravitation centers in the urbanized territories *Paper presented at the MATEC Web of Conferences*, **212**, doi:10 1051/matecconf/201821205002

IOP Conf. Series: Materials Science and Engineering **786** (2020) 012067 doi:10.1088/1757-899X/786/1/012067

Development of transport infrastructure of urban mobility based on cable metro technology

A V Panfilov[1,*], A A Korotkiy[1], E A Panfilova[1] and I A Lagerev[2]

[1]Don State Technical University, Gagarin square 1,Rostov-on-Don, 344000, Russia

[2]Bryansk State University, Bezhitskaya 14, Bryansk, Bryansk, Russia

[*]E-mail: panfilov@ikcmysl.ru

Abstract. The article presents materials on the use of cable metro technology in a highly urbanized environment, as an improvement in urban mobility in the format of a smart city. The presented transport system is significantly different from the traditional cablecar, which is described in the comparative characteristics of each system. The article describes the principle of operation of the cable metro, based on the use of mechatronic motion modules.

1. Introduction

Social and economic transformations, together with achievements in the field of transport technologies, social networks, wireless services and cloud spaces, contribute to the growth of human mobility in the urban environment. More and more people use Web applications, choosing the most convenient option for moving, optimizing it for cost, travel time, waiting time, number of transfers, etc.

According to the results of a social survey [1], 83% of respondents said that innovative transport is a key feature of a smart city, and almost half (45%) of the of respondents believe that the city can never reach a world class level without good digital communications.

Currently, smart cities are characterized by six key indicators - smart economy, smart mobility, smart environment, smart people, smart living, smart government.

Each indicator is determined by factors. There are special indicators to assess the level of development of a factor in a particular city. Of the six key characteristics, let us focus on smart mobility [2]. To assess the level of smart mobility in practice for a particular city, we use the following factors:

1) the ability to easily get to all areas and significant places of the city;
2) openness and accessibility of the city at the national and international levels;
3) the availability of information infrastructure (Internet, GLONASS, GPS, etc.);
4) the availability of a sustainable, innovative and safe transport system.

Mobility is a key component of a "smart city", which must be adapted to the information infrastructure of the city. Public transport is the most effective way of moving a person in megacities.

With the development of transport technologies, a new type of passengers appears who are configured to receive information about the route in real time and no longer want to travel without continuously receiving data throughout their movement.

Transport logistics are becoming a "multimodal multiplier" for smart cities, increasing the efficiency of public transport [3,4]. Digitization of the urban transport system through real-time information transfer, the introduction of mobile applications and other technologies improves the quality of passenger service and allows them to receive more thoughtful and effective solutions along the travel route. By receiving and using real-time data using predictive analytics, transporters can significantly change their work in a positive direction.

Public transport, cities and society are on the verge of rapid change, as advances in mobility and technology are increasingly used in our lives. The growth of urbanization in our cities will continue to influence the development of innovations in the transport sector [5,6].

2. Analysis of the transport infrastructure of urban mobility

Today, engineering can offer a person different options to move in an urbanized environment: railways, road, water, air transport, high-speed railways, magnetic suspension trains, monorail, trolleybus, high-speed tram, rail bus, string transport, cable cars. 13 structurally different modes of transport in total. Together with various constructive options, there about 200 ideas. Each has its own advantages and disadvantages.

Unfortunately, no type of transport meets the current requirements for the creation of efficient means of transportation for the conditions of large cities, megacities and urban agglomerations, which must combine environmental friendliness, safety and speed of movement of passengers, low economic characteristics of transporting passengers and cargo. Constant traffic jams, gas contamination of streets and other inconveniences of urban movement is an objective initiator of the development of new innovative types of urban transport (Figure 1).

Figure 1. Movement in an urbanized environment.

The logical direction of the fundamental overcoming of the problems associated with the use of ground public transport is the transition to another high-altitude level. The transfer of traffic flows to urban overground space still remains a promising, but little developed direction for improving the transport infrastructure of modern cities. Technically, this can be realized using cable transport technologies. One of the innovative modes of transport, which began to be introduced in the city of Moscow, is the cable cars used as urban transport.

3. Cable car technology

The cable metro is an innovative type of cable transport [5,6] (a type of passenger cable car), and is designed to carry passengers in an urbanized environment, between settlements and directly in megacities located on rough terrain, including those with the presence of water barriers (river, lake, swamp, bay, etc.), as well as trunk lines, transport and waterways, high-voltage power lines, oil, gas, water pipes, high-rise buildings and structures, taking into account safe dimensions of approximation.

The main difference between the cable metro and traditional passenger cable cars is the presence of intermediate stations designed not only for boarding and disembarking passengers in rolling stock, but also, which is a fundamental difference, for transferring passengers to other directions and forwarding the rolling stock to other routes, and also almost infinite length of transportation of rolling stock by one traction rope. A mechatronic distributed drive ("active roller battery"), installed discretely at intermediate stations or linear supports, acts as the mover of the traction rope. The design of overhead cable cars with respect to the possibility of using it as public passenger transport in large metropolitan cities, urban agglomerations and within natural territories was protected by a number of patents [3,7].

Actually, the term "cable metro" was introduced into use in scientific works [3,4,7] in the process of solving the following engineering problem: the creation of a universal passenger public transport system for an urbanized environment based on the mechatronic modular design and construction principle with the possibility of unlimited completion without limiting the shape of the profile of transport lines in plan and height, without limiting the length of the path in space, without limiting the number and configuration of special nodal stations of transport lines.

The transport system, which general structural diagram is shown in Figure 2 includes a path supported by intermediate supports, and the vehicle in the form of passenger cars is a mechatronic system. It contains mechatronic motion modules interconnected by a mechanical link in the form of a path consisting of one or more strands of an annular traction rope, and a control system based on a hierarchical decentralized network with a single protocol for the exchange of information between mechatronic modules.

IOP Conf. Series: Materials Science and Engineering **786** (2020) 012067 doi:10.1088/1757-899X/786/1/012067

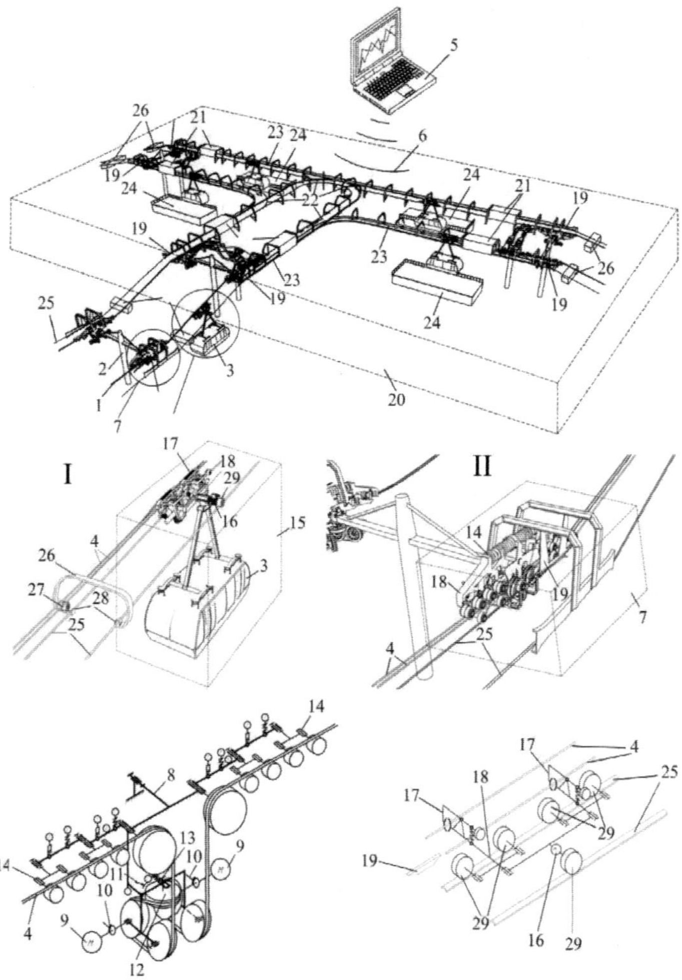

Figure 2. The general structural diagram of the cable metro.

Despite the fact that the cable metro looks like a traditional suspended cable car, it is based on either new or fundamentally different technical solutions (see Table 1) [12].

Table 1. Comparative characteristics of the cable metro and passenger cableway.

Passenger cableway	Cable metro
Disadvantage: the use of obsolete equipment.	Advantage: the use of mechatronic technologies to create a controlled multi-rope transport system in an urbanized environment.
Disadvantage: limited length of the transportation route. Reason: the presence of one concentrated drive traction rope movement. Negative consequences: 1) with an increase in the length of the ropeway, the power of its drive increases, as well as the dimensions of the equipment and supporting metal structures and, as a result, the cost indicators of the project; 2) ropes of great length require ropes of great aggregate strength, which leads to an increase in the mass and cost of the rope, drive and metal structures; 3) therefore, on the long routes, several cable cars are installed one after another; 4) it is energy-consuming, expensive and time-consuming during construction, especially in urban environments.	Advantage: unlimited length of the route, the ability to continue the metro line with the expansion of urban development. Reason: mechatronic motion modules implement the idea of a drive linearly distributed along the route for driving passenger cabins. Positive effects: 1) low power of one mechatronic module, since it provides the movement of the traction rope in a limited section of the route; 2) lower total power and energy consumption of mechatronic modules in comparison with traditional single drive on equal length paths (up to 30%); 3) small overall dimensions of equipment and supporting metal structures; 4) the possibility of using traction and load-bearing ropes of lower aggregate strength and, accordingly, a smaller diameter (3-5 times), dead weight (up to 7 times) and cost; 5) the possibility of increasing the distance between adjacent supporting metal structures along the route up to 3000 m; 6) improvement of dynamic characteristics by reducing the mass of moving elements.

Table 1.

Disadvantage: predominantly linear trajectory for passenger cabs. Reason: the reciprocating movement of the traction rope between the two end points of the route. Negative consequences: 1) the lack of the ability to switch from one line of the cable car to another without a transfer; 2) the impossibility of forming an arbitrary or circular transport route.	Advantage: the ability to create any routes within the entire system of cable lines, including ring routes (i.e., the logistic capabilities of the cable metro are equivalent to traditional modes of transport). Reason: the presence of devices to reorient the direction of movement of passenger cabins. Positive effects: 1) the ability to transfer passenger cabins to another line of the cable metro at junction stations without transferring passengers; 2) the possibility of creating sludge points for passenger cabins during the period of reduced passenger flow on the line; 3) the possibility of rapid transfer of cabins on the line of increased passenger flow, at rush hour, etc.
Disadvantage: a complete stop of the movement of all passenger cabs on the line in the event of a drive failure of the traction rope movement. Reason: the presence of one concentrated drive traction rope. Negative consequences: 1) the need for urgent one-time emergency rescue and repair work along the length of the route; 2) a long duration of emergency rescue operations, disruption of passenger plans, delay, etc .; 3) a rather difficult psychological situation for passengers; 4) difficulties in organizing the system of maintenance and repair of the cableway, because carrying out these works requires stopping cabs.	Advantage: almost zero probability of cabs stopping in case of failure of the traction rope drive. Reason: the presence of a cab motion drive linearly distributed along the route based on a plurality of mechatronic motion modules. Positive effects: 1) there is no need for urgent repair of the failed mechatronic motion module and emergency rescue operations; 2) the possibility of organizing a high-quality system of maintenance and repair of mechatronic motion modules, because its implementation is possible during the operation of the cable metro without stopping it; 3) the absence of psychological consequences and inconvenience for passengers in case of failure of the mechatronic motion module.

Table 1.

Disadvantage: the inability to stop the cab at the station for boarding and unloading passengers. Reason: one-piece connection of the passenger cabin to a constantly moving traction rope. Negative consequences: 1) unsafe when boarding and disembarking passengers on the go; 2) it is impossible to arrange the landing of disabled people and persons with disabilities, children, passengers with prams and bulky luggage, etc .; 3) when landing several people, it is necessary to move after the cabin; 4) it is impossible to use cabins of profitable capacity (30 ... 40 people).	Advantage: the ability to stop the cab at the station for boarding and unloading passengers. Reason: the presence of an automated control system for the movement of cabs that tracks their location and automatically performs the connection-disconnection of the cabin from a constantly moving rope. Positive effects: 1) the complete safety of boarding and unloading passengers in a stopped cabin; 2) safe landing of disabled people and persons with disabilities, children, passengers with prams and bulky luggage, etc .; 3) it is possible to use passenger cabins of profitable capacity (30-40 people each).
Disadvantage: lack of an automated cab movement control system. Reason: the "rigid" scheme of the cableway, not implying the possibility of significant regulation of traffic parameters. Negative consequences: 1) the inability to ensure the required level of comfort and safety of passengers; 2) the inability to control the movement parameters of passenger cabins, including their stopping time, speed, etc.	Advantage: the presence of an automated control system for the movement of cabs. Reason: the use of information technology based on RFID tags (transponders) integrated into the traction rope and forming a solid track structure. Positive effects: 1) increased safety of movement; 2) the ability to track the movement and control the movement of each passenger cabin, including its location, speed, the need to stop; 3) identification of cases of disturbance in public order in the cabs, video detection and detention of violators.

4. Design features of the cable metro

The mechatronic module of the cable car is equipped with an acceleration and deceleration engine and automatic grips (Figure 3). They are mounted on the carriage, and in their number and size correspond to the threads of the annular traction rope with vehicles suspended in the form of passenger cars. Automatic grippers interact with a gripper actuation device mounted on intermediate supports.

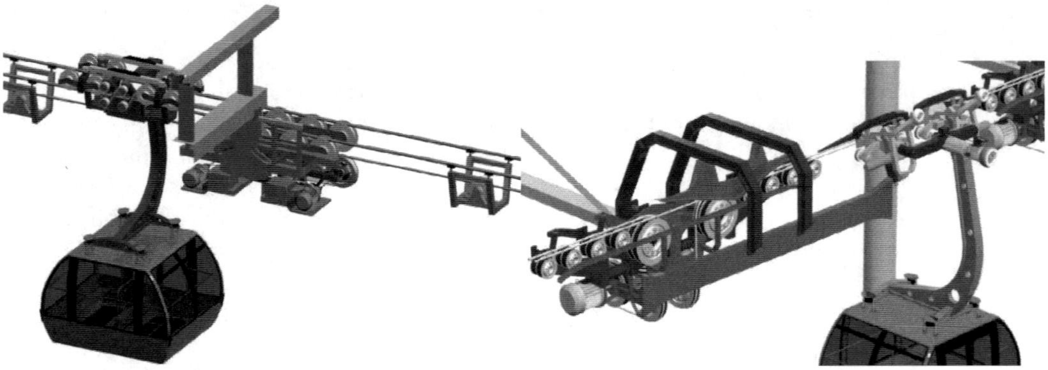

Figure 3. Three-dimensional model of the mechatronic module of the vehicle.

The considered transport system of the cable metro works as follows. The paths for the passenger cab vehicles are made of strands of an annular traction rope passing over a land-based urbanized infrastructure, taking into account the permissible dimensions for approaching various types of infrastructural facilities, including those that can be based on them. The strands of the annular traction rope are connected to the vehicle by means of a carriage and automatic grippers installed on it. Pulling force or braking force on the threads of the annular traction rope is created by mechatronic modules of the annular traction rope located on the intermediate supports.

The control system uses a decentralized network with a single information exchange protocol (for example, PROFIBUS). It has an economic advantage in terms of reducing prices by reducing cable connection and reducing footprint, as well as a technological advantage in terms of reducing debugging and commissioning times, flexibility to connect additional devices to the control system, and the possibility of developing system intelligence when realizing automation tasks outside CPU unit using multifunctional controllers.

The presence of the mechatronic module of the station allows us to change the route of movement of passenger cars to other paths due to automatic "arrows" of entry-exit, paired with one of the hard tracks of the corresponding route. This function allows passengers to minimize travel time between the start and end points of the route.

Currently, several technical solutions are known aimed at the development and improvement of certain aspects of cable cars for urban environments, which should be considered as directly related to the concept of "cable metro".

A further development of the ideology of creating urban cableways for passenger transport in an urbanized environment, based on innovative technology of the cable metro, is a cableway with mechatronic propulsion devices [8,10]. The proposed technical solution is aimed at improving the reliability, safety and efficiency of controlling the mechanisms of movement of the cableway and ensuring uniform load distribution between the drives, eliminating excessive sagging and tension of the traction rope.

5. Examples of urban mobility transport infrastructure projects based on cable metro technology

To confirm the operability and verify the effectiveness of the Cable Metro transport system, the authors develop schemes for its implementation in a number of Russian cities, including Bryansk, Rostov-on-Don, St. Petersburg, Stavropol, Moscow, Sevastopol, Sochi, Yekaterinburg, Krasnoyarsk, Novocherkassk, Skolkovo, taking into account their natural and geographical features, as well as the terrain, the specifics of urban development and the location of communal infrastructure.

The concept of innovative urban transport for the development of the cable metro network in Rostov-on-Don was developed at the Don State Technical University [9].

For the developed routes we propose typical architectural and planning solutions of cable metro stations (Figure 4), integrated into the existing street and transport infrastructure of Rostov-on-Don. At cable metro stations, it is possible to place commercial and socially-oriented infrastructure for passengers in the form of cafes, shops, pharmacies, payment centers, bank branches, toilets, etc. (Figure 5).

a) b)

Figure 4. Three-dimensional model of intermediate supports and rolling stock cable cars integrated in the street network of Rostov-on-Don: a – Krasnoarmeiskaya str..; b -. Nagibina str.

Figure 5. Three-dimensional model of the internal view of the cable metro stations.

Bryansk is an ideal city for the construction of the cable metro, since in relation to its development and relief, the conditions that are necessary for the most efficient functioning of the cable metro are well suited [11]. At the same time, these same features of the city's development and topography are a barrier to the development of its transport infrastructure based on traditional types of automobile and electric transport. The existing transport infrastructure of Bryansk is complicated by the fact that the territory of the city has significant area, and four districts are separated from each other by the Desna, Bolva and Snezhet rivers with wide floodplains and an extensive network of railways with bridges and overpasses. Due to the imperfection of the planning structure, the backbone network of the center of Bryansk is heavily congested with transit traffic. During peak hours, the speed of vehicles does not exceed 20 km / h.

Under the conditions of dense urban development in the city of Bryansk, inter-city transportation based on traditional types of ground transport has practically exhausted its capabilities for the development of passenger traffic. The growth in road transport is accompanied by an increase in the number of cars on the road. In relation to 2009, in 2018 there was an increase in the number of cars by 34%, trucks - by 38%, buses - by 20%. At the same time, the total number of passengers transported by

IOP Conf. Series: Materials Science and Engineering **786** (2020) 012067 doi:10.1088/1757-899X/786/1/012067

road and trolleybus public transport decreased from 78 million to 65.5 million, which could obviously be associated with the development of personal motor transport of city residents, leading to an additional increase in the load on the transport infrastructure of the city [11].

Taking into account the passenger flows in Bryansk, Bryansk State University named after academician I.G. Petrovsky made a preliminary assessment of the number and possible locations of the cable metro lines (Figure 6).

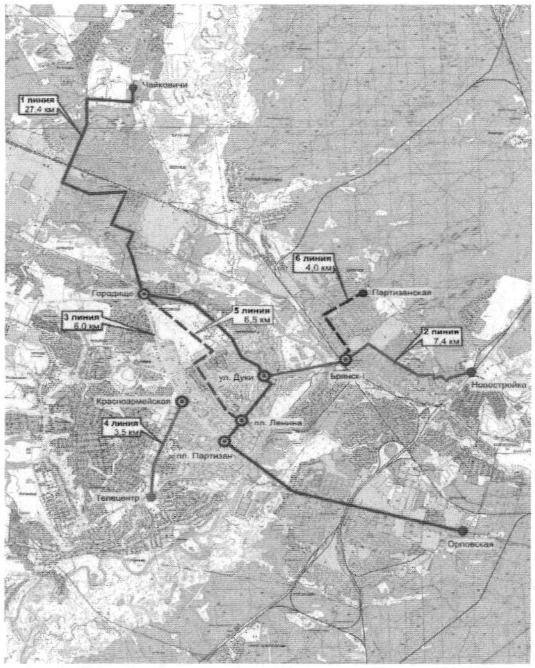

Figure 6. The location of cable metro lines in Bryansk.

The project for the construction of a passenger cable transport system based on cable metro technology for the city of Yekaterinburg was developed at the Novocherkassk ICC "Mysl" and LLC "Skado" in conjunction with CJSC "Ural Expert Center" (Figure 7)

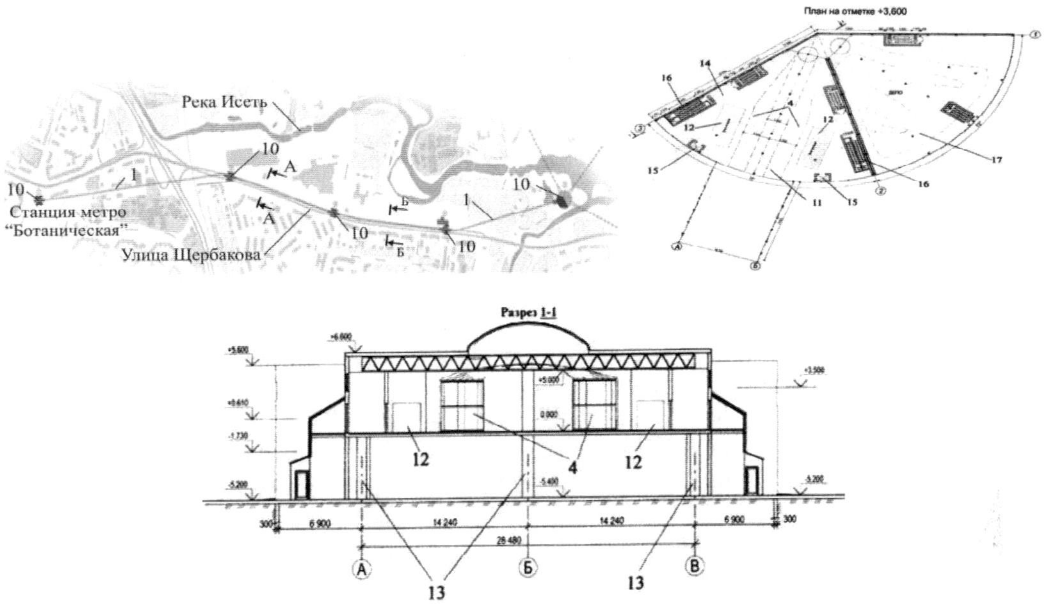

Figure 7. Cable car line in Yekaterinburg along the river.

In relation to the conditions of the city of Yekaterinburg, a passenger cable car transport suspension system based on the cable metro operates as follows. Passengers come by underground metro to the Botanicheskaya station in Yekaterinburg, either by commuter trains, or by personal cars and place them in multi-storey car parks, or come to shopping centers. All of the above facilities are connected by the transport system module integrated into the passenger transport logistics of the city of Yekaterinburg. Station modules have different geometric shapes in plan and an unlimited number of levels in height for their integration into the prevailing architectural appearance of an urbanized environment, forming multimodal complexes, which allows passengers to comfortably make transfers and minimize routes when changing the mode of transport.

The project for the construction of a passenger cable transport system based on cable metro technology for the city of Yekaterinburg is integrated in the logistics of passenger transport. It suggests that one of the multimodal complexes of the cableway metro transport system should be combined with the Botanicheskaya metro station, and the path should go along Shcherbakova Street to Zimnyaya Street, with a subsequent continuation to Himmash and the Utkus skiing complex. Within the limits of the city territory, a cable metro line crosses the water barrier - the floodplain of the Iset River.

6. Conclusion

The innovative transport system "cable metro" is an important element in the development of transport infrastructure of urban mobility and can be used in the technology of "smart city", as it has several advantages over other modes of transport.

Technical and economic calculations of the expected values of the characteristics of the passenger transport system "Cable Metro" showed that it has the following characteristics [3,5,7]:

- maximum passenger traffic - 3.0 thousand passengers / hour;
- construction cost of 1 km of track - 0.3 billion rubles;
- maximum speed - 40 km / h;
- average speed - 34 km / h;
- energy efficiency coefficient - 0.42;
- specific energy consumption - 25 W • h / t • km;

- the cost of 1 pass. / km - 2.0 rubles;
- the present value of the organization of movement at the maximum passenger flow is 1.8 thousand rubles / pass .;
- present value of rolling stock - 0.4 thousand rubles / pass. / year;
- specific energy consumption - 0.3 ... 0.5 L/100 pass. / Km;
- area occupied by 1 passengers on the street - 0 m^2.

A comparison of the cable metro and traditional forms of urban transport for a number of basic technical and economic indicators allows us to formulate the following advantages of the cable metro:

- social:
  - the fare is at the level of traditional types of urban public transport;
  - speed (up to 50 ... 60 km / h) and clear predictability of travel time due to the absence of undetermined downtime in traffic jams, at traffic lights, etc .;
  - absence of loss of time for forced downtime and waiting in traffic jams;
  - convenience of embarkation and disembarkation, including for people with disabilities;
  - comfort and safety of movement, as the rolling stock has a smooth ride and there is no likelihood of a collision with another vehicle;
  - the ability to carry hand luggage and bulky goods;
- technical:
  - movement above the ground without intersection with the routes of other types of urban public, freight and special transport;
  - lack of traffic accidents;
  - creation of direct routes of any configuration;
  - relative energy consumption for moving (up to 40 km / h) is 5 ... 10 times lower than that of a modern car;
  - practically insensitive to terrain, ground and underground objects and communications;
  - the existing road-street network and engineering infrastructure are not violated;
  - fits well with the architectural look of the city;
  - construction of bridges, overpasses, embankments and tunnels is not required;
  - volumetric excavation, violating the landscape and requiring subsequent restoration, is not required,;
  - resistance to the effects of adverse climatic factors and natural disasters (earthquakes, floods, landslides, hurricane winds, low or high ambient temperatures, etc.);
  - the use of electric traction, which does not adversely affect the environment, low noise and vibration, the absence of harmful emissions;
  - minimal need for building materials and structures, relatively low consumption of structural steels, non-ferrous metals, semi-finished products and building materials;
- economic:
  - relatively low cost of design and construction (up to 30 million rubles / 1 km of track);
  - the cost of constructing a highway with infrastructure 2 ... 5 times cheaper than modern railways and roads;
  - lowest costs for the organization of movement and the purchase of rolling stock;
  - the lowest specific consumption and the highest energy utilization coefficient for the transportation of 1 passenger;
  - savings in travel costs through the introduction of energy-efficient technologies and independence from petroleum products;
  - minimal land allocation (on average 0.1 hectares of land per 1 km of the route);
  - no redemption or alienation of expensive land plots, demolition of existing buildings and structures for laying a ground road is required;

○ a significant part of the areas of landing stations can be leased and used for commercial or other purposes.

An analysis of the conditions of the most appropriate use of the cable metro as an innovative transport system for regular passenger traffic, carried out in a number of works, showed that the cable metro works most effectively in cases where other types of urban transport are ineffective, in particular:

- in the presence of dense urban development with the presence of relatively narrow streets (2 ... 4 lanes for traffic);
- in the conditions of spatial dispersion of urban areas in the absence or insufficiency of alternative highways between them;
- in case of difficulty or inability to lay new highways within the urban area or between remote urban areas or satellite cities within the urban area;
- if there are wide rivers or river floodplains within the city limits;
- with a ravine or hilly terrain;
- to effectively solve the problem of ensuring environmental protection from the harmful effects of transport infrastructure on human living conditions and the environment.

References

[1] V P Kupriya vskiy, A V Akimov, O N Pokusaev, V V Alenkov, D E Namiot and S A Sinya 2017Intellektualnaya mobil st i mobil st kak usluga v Umnykh Gorodakh //*International Journal of Open Information Tech logies ISSN*: 2307-8162 **5(12)**

[2] D P Romanov, A M Suvorova and V P Kupriyavskiy 2019 Umnaya mobil st // *International Journal of Open Information Tech* logies ISSN: 2307-8162 **7(3)**

[3] A A Korotkiy, A V Lagerev, B CH Meskhi, V M Prikhodko, G V Kustarev, V B Maslov, D A Korotkiy, M V Kirsa v, A V Panfilov and I A Lagerev Pat 120617 Rossiyskaya Federatsiya, MPK B61B 7/00 Transportnaya sistema («Kanat e metro») / – 2012121284/11; 27

[4] A A Korotkiy, A V Panfilov, D SH Kvebliani, B D Popiashvili 2013 Pat 2471661 Rossiyskaya Federatsiya, MPK B61B 1/00, B61B 15/00, E01C 1/04 Transportnyy multimodalnyy kompleks /– 2011138208/11; zayavl 16 09 2011; 1

[5] Korotkiy A A 2012 Vliyanie passazhiropotokov na kharakter funktsionirovaniya passazhirskogo multimodal go terminala gorodskoy kanat y dorogi// *Izvestiya vysshikh uchebnykh zavedeniy Severo-Kavkazskiy region Ser : Tekhnicheskie nauki* — **3** 81–4

[6] Korotkiy A A and Panfilov A V 2014 Kanatnye dorogi vogo pokoleniya kak element passazhirskoy transport y infrastruktury gor go klastera Sochi-2014// *Bezopas st truda v promyshlen sti* – **6** 38–41

[7] A A Korotkiy, L V Stotskaya, N A Rozhkov, A V Kurlykov, G V Kustarev, A A Tumasov, D A Korotkiy and A V Panfilov 2011 Pat 2471662 Rossiyskaya Federatsiya, MPK B61B 7/00 Transportnaya sistema («Kanat e metro») /– 2011138129; 1

[8] A A Korotkiy, N A Glebov, V A Tytskiy Pat 2507095 Rossiyskaya Federatsiya, MPK B61B 7/00 Podvesnaya kanatnaya doroga s mekhatronnymi dvizhitelyami /– 2012126463/11; zayavl 25 06 2012; opubl 20 02 2014 – Byul **5**

[9] Meskhi, B CH and Korotkiy A A 2011 Kontseptsiya razvitiya seti kanat go metro v Rostove-na-Donu // *Vestn Donsk gos tekhn un-ta* –**11(8)** 1348–55

[10] Panfilov, A V , PrikhodkoV M and Kustarev G V 2013 In vatsionnyy passazhirskiy kanatnyy transport dlya urbanizirovan y sredy s mekhatronnymi modulyami dvizheniya // *In vatsii i investitsii* **6** 187–91

[11] Lagerev A V , Lagerev I A and Korotkiy A A 2012 Kontseptsiya in vatsion y sistemy gorodskogo transporta «Kanat e metro goroda Bryanska» // *Vestnik Bryanskogo gosudarstven go tekhnicheskogo universiteta* **3** 12–5

[12] Lagerev, A V , Lagerev I A and MiltoA A 2015 Preliminary Dynamics and Stress Analysis of Articulating n-Telescoping Boom Cranes using Finite Element Method // *International Review on Modelling and Simulations* **8(2)** 223–6

Intersections at different levels - an alternative solution to the development of the road network

O A Loginova* and R V Nikolaeva

Kazan State University of Architecture and Engineering, 420043, Kazan, Zelenaya str., 1, Russia

*E-mail: loginova@kgasu.ru

Abstract. Modern road network should provide conditions for safe and comfortable movement, both for drivers and for pedestrians. Transport industry development specialists face such transport problem as the low capacity of the road network. Low capacity causes congestion in the road network, disruptions in the organization of normal traffic flow, traffic jams, an increase in the number of traffic accidents. One way to solve this transportation problem is to use alternative solutions when designing. These solutions include constructing the intersections at different levels. The aim of the study is to compile data on alternative diamond-shaped intersections at different levels with a change in the side of movement. The main results of the study are that at the moment there are three schemes of a diamond-shaped intersection with a change in the side of the movement. These types of intersections have a smaller number of conflict points, compared with the intersection of the diamond type. The experience in the operation of road sections with diamond intersections showed a decrease in accident rates. Cyclists and pedestrians can use the junctions as well. The results of the study give an idea of the level of design of alternative types of intersections abroad. Considered interchanges can be used not only when at crossings of high-category roads, but also in cities, in the conditions of the existing development.

1. Introduction

Every year there is a tendency of increasing number of cars, which makes it necessary to carry out measures for the reconstruction of the existing road network and the construction of new roads. The existing road network does not correspond to socio-economic needs, which is primarily due to the high growth rate of motorization. Imbalances in the development of the road network strongly affect the uneven loading of roads, which in turn stimulates the growth of traffic accidents, especially in the areas with low traffic capacity and high traffic intensity. One of the most emergency sections of the road network is the intersection of highways. Accident statistics show that, on average, 60% of all traffic accidents occur exactly at the one level intersections of roads.

With a high intensity of traffic, intersections of one level do not satisfy the requirements of movement, since they have insufficient capacity. In this case, there are congestions and traffic jams, a high density of movement, and the need for maneuvering creates unforeseen and dangerous situations. This leads to an increase in the number of traffic accidents.

In order to improve the conditions of traffic flow and reduce the number of conflict points at intersections, transport interchanges are built with different levels. There are various intersection patterns at different levels. The advantage of these intersections is the absence of interception points of traffic flows, there are only merging and branching points.

2. Influence of planning and schemes of intersections at different levels on the accident rate

Intersections at different levels can have different configurations and traffic patterns. The effect on traffic safety of various geometric elements of intersections at different levels was studied in detail by American (Lundy, 1967; Cirillo, 1968, 1970; Yates, 1973) and Norwegian (Wold, 1995) researchers. These studies examined the safety of different types of intersections at different levels, the influence of the values of the turning radius at the ramps, the significance of the secondary road pass location (above or below the main road), and the length of the acceleration and deceleration lanes. The results of these studies are presented in Table 1.

Table 1. Impact of different types of intersections at different levels on the number of accidents.

Intersection types at different levels	The percentage change in the number of accidents		
	Impact on the types of accidents	Best result	Limits of results variation
Diamond intersection at different levels instead of pipe type intersection	All types of accidents in the intersection zone	-25	(-50; -23)
Diamond intersection at different levels instead intersection with straight ramps	All types of accidents in the intersection zone	-25	(-59; +40)
Diamond-shaped intersection at different levels instead of the clover leaf intersection	All types of accidents in the intersection zone	-6	(-17; +6)
Diamond intersection at different levels instead intersection at 2 levels with 2 straight ramps	All types of accidents in the intersection zone	-9	(-25;+10)
Increase of the turning radius at the ramps	All types of accidents in the intersection zone	-23	(-28; -17)
The minor road passes over the main road instead of the main road above the minor	All types of accidents in the intersection zone	-4	(-17; +10)
Lengthening the acceleration lane for 30 meters	All types of accidents on the acceleration lane	-11	(-17; -5)
Lengthening the deceleration lane for 30 meters	All types of accidents on the deceleration lane	-7	(-13; +0)

The results of the study showed that diamond-shaped intersections at two levels are the safest intersection types. First, the diamond-shaped intersection is simple in layout and provides good visibility for drivers, which greatly reduces the likelihood that the driver can choose the wrong direction to enter or exit. Second, the ramps of the diamond-shaped intersection at different levels have a direct form.

Curvy ramps have a higher degree of risk than direct ones. Third, in most cases at the diamond-shaped intersections a minor road passes over the main road.

3. Modifications of diamond-type intersections

The era of designing and developing of traffic interchanges began in 1916, when the American Arthur Hale received a patent for a clover leaf interchange [2]. Since then, the number and modification of intersections in different levels has significantly grown. Many intersections and junctions of highways were developed with two levels, which were based on the "clover leaf". One type of such intersections is a "diamond" type interchange, which is based on an incomplete "clover leaf" [3, 4]. In this type of intersection, there are no left-turning exits that are present in the "clover leaf".

In 2003, Gilbert Chlewicki proposed a modification of the diamond-shaped intersection [5]. This is a diamond-shaped interchange with a change in the sides of the movement. In the proposed solution, traffic flow on the minor road moves from the right side to the left side and vice versa. The change of traffic flows happens before and after the overpasses.

There are several variations of this interchange. The first option - a minor road runs along the overpass over the main road. The change in the sides of the movement takes place before and after the overpass, i.e. above the main road (Figure 1).

The second option - a minor road passes under the overpass, the main road passes over it. A change in the direction of movement takes place before and after the overpass, i.e. under the main road (Figure 2) [5].

The third option is a modification of the second, when the diamond and the nodes of the movement change are shifted away from the intersection point of the axes of the main and minor roads [6].

Figure 1. Changing the sides of the movement before and after the overpass.

Figure 2. Changing the sides of the movement under the overpass.

4. Investigation of the features of diamond-shaped intersections with changing traffic movement sides

4.1 Elements of diamond intersections with changing traffic movement sides

Diamond-shaped intersections with changing sides of the movement have five unique working areas [5]:

1. Approach zone - refers to the street segment between the upper intersection and the first intersection of the roundabout.
2. Intersection zone - refers to the intersection itself and the entrances and exits from this intersection.
3. Exit zone - refers to the area between the diverging area of the freeway and the cross street.
4. Entry zone - refers to the area between the second intersection and the junction of the freeway.
5. Leaving area - refers to the area between the second intersection and the next adjacent signal at the intersection.

IOP Conf. Series: Materials Science and Engineering **786** (2020) 012068 doi:10.1088/1757-899X/786/1/012068

The working zones of diamond-shaped interchanges with a change in the sides of the movement are presented in Figure 3.

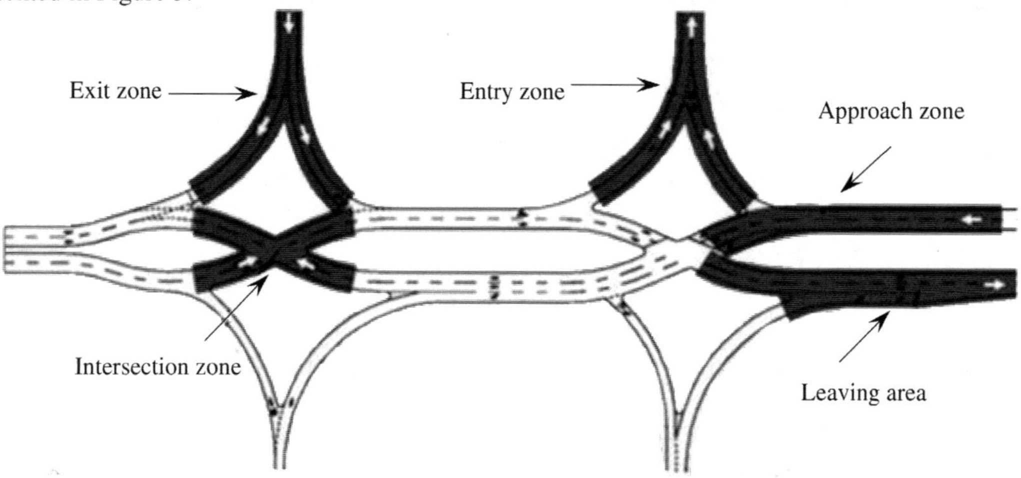

Figure 3. Working areas of diamond-shaped intersections with changing sides.

One of the advantages of the diamond-shaped intersection is the solution of the left-turn exits. The proposed option is much more convenient compared to left-turning exits at the intersections of the clover leaf type, where in order to make a left turn, the driver has to turn right at the exit. The left-turning exits of the "clover leaf" are made in the form of a ring of small radius in the layout. In the considered variants, after passing the traffic light at the first node of the change of direction, the driver moves to the left side of the road and sees a turn to the left, i.e. there is no loss of driver orientation. The left exit is smoother, it is not necessary to do a full circle, like at the intersection of the "clover leaf" type. We can say that this is the same left turn exit as in the "clover leaf", but pulled along the main road.

This solution of the diamond-shaped intersection has a positive effect on the emotional state of the driver and on his perception of the road situation [6, 7]. Since the change of direction occurs before the crossing of the overpass, the driver has time to think about the maneuver while the car was standing in front of the traffic light. There is no loss of orientation, in opposite to the "clover leaf", where driver has to turn right for the left-side exit. Thus, the emotional state of the driver remains stable and does not interfere with the movement at the diamond intersection.

It should also be noted that the diamond-shaped intersection with a change in the direction can be used in urban conditions, because it is quite narrow. All exits are extended along the main road. This allows us to enter the intersection in a dense urban area without affecting existing buildings.

4.2 Safety of diamond intersections with changing sides.

The number of conflict points present at the intersection determines the degree of danger of the intersection. Conflicting points arise at the intersection of traffic, pedestrian and bicycle flows. Conflicts between vehicles can be described as merging, branching or crossing of the flows, where conflicts pose the greatest risk for accidents. Although traffic control devices can reduce the number of crashes by eliminating conflict points or controlling them with traffic signs, traffic lights or road markings, traffic control and intersection geometry cannot completely eliminate the mistakes made by road users.

The design of the diamond-shaped intersection with a change in the sides of the movement showed that it improves the movement of turning vehicles on the road, which contributes to a significant reduction in the number of conflict points between vehicles compared to conventional intersections of the rhombus type. A diamond-shaped intersection with a change in the sides of the movement reduces the severity of conflicts, since conflicts between left-side movements and opposing through movements are eliminated.

The decrease in the number of conflict points at the diamond-shape intersection with a change in the sides of the movement as compared with the intersection of the rhombus type is associated with a change in the direction of traffic flows and the placement of traffic lights in front of these nodes [8, 9].

The interchange developed by Chlewicki has 14 conflict points: 2 intersection points, with traffic lights in front of them, 6 merging points and 6 branching points of traffic flows (Figure 4). The presence of traffic light regulation forces drivers to be more attentive, to move at a slower speed when moving from one side of the road to another. We should note that the diamond-shaped interchange does not have a long section of flow maneuver, like a "clover leaf" or "rum" type, where there is an intersection of cars that entering the left exit and leaving it t [10, 11, 12, 13, 14, 15]. Diamond shape avoids this due to changes in the movement sides and the presence of traffic lights that limit the speed of traffic flow. It is worth noting that the diamond-shape intersection has 26 conflict points: 10 intersection points, 8 merge points and 8 branch points of the traffic flows.

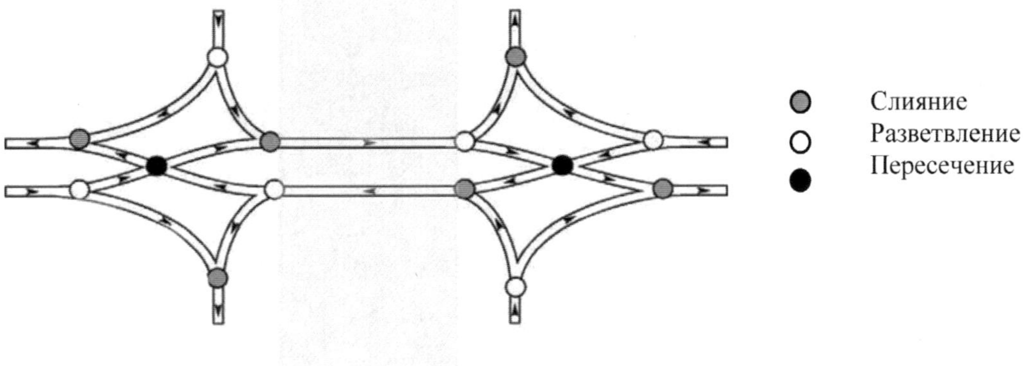

Figure 4. The number of conflict points at the diamond-shaped intersection with a change in the sides of the movement.

At the development of geometrical parameters of a diamond-shaped intersection with changing sides of the movement, we should take into account that this type of intersection will be effective for road crossings of I and II technical categories or for crossings of streets of urban significance with continuous traffic and traffic light regulation. On the roads of I and II categories we can project 4 or more lanes. The presence of two lanes in the same direction is the minimum at which this interchange will function. One lane is for traffic in the forward direction. Another lane is for driving to the left, but it will have a merging of the flows that make a left turn and left u-turn. Therefore, in order to avoid an increase in the number of conflict points, we should design two lanes for moving in the forward direction, one lane for making a left turn, and one more lane for a left u-turn (Figure 1). Thus, the overpass should be designed with four lanes in each direction.

At the development of diamond-shaped intersection in urban areas, it is imperative to consider the presence of sidewalks for pedestrians. If the side change diamond is below the main road, then pedestrian crossings and sidewalks should be built along the minor road (Figure 2). Sidewalks can be located on the overpass if the main road passes underneath. The literature provides an example of the organization of pedestrian traffic [14]. In particular, when the diamond is located above the main road, the sidewalk can be located between the traffic change flows, i.e. along the overpass axis. At the traffic lights the walkways are divided into right and left sides of the junction and pedestrians should cross the road using "zebra" (Figure 1).

4.3 The formation of a speed mode on a diamond-shaped intersection with a change in the sides of the movement

At the roundabout, traffic flows merge in the right-hand lane of the main direction (with the exception of junctions with straight left turns) and to assess the throughput of the confluence zone, it is necessary to determine the distribution of the intensity of the direct direction along the lanes of the carriageway. This distribution depends on the intensity and composition of the movement (Figure 5). On city highways with higher traffic, the distribution of traffic along the lanes of the carriageway is more stable.

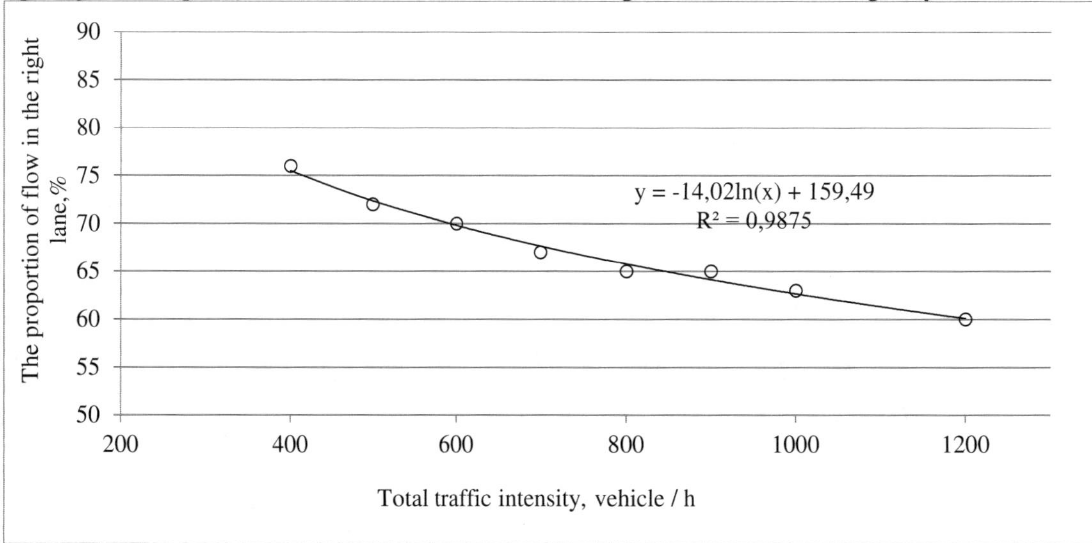

Figure 5. The distribution of traffic intensity on the lanes of roads.

The intensity of the main direction has the same effect on the boundary time interval in the confluence zone as at any conflict point: at average densities of the main flow, an increase in traffic intensity in the right lane causes a decrease in the boundary time interval in the confluence zone of the traffic intersection, and at low and high density the time interval in the confluence zone of the traffic intersection remains almost constant (Figure 6).

The boundary interval can be reduced by reducing the difference in the speeds of the main and inflowing flows. The most unfavorable case, corresponding to the smallest throughput of the confluence zone, is a preliminary stop of the car before leaving the exit. This situation can be eliminated by the installation of transition-speed lanes. However, the difficulty of exiting can be completely eliminated in the absence of traffic on the transition-speed lane. Otherwise, the exit conditions from the congress will be determined precisely by this traffic intensity.

Free flow speeds are limited by the diamond-shaped geometry with changing sides of the movement. Studies of diamond-shaped interchanges with changing directions of movement throughout the United States showed that the free flow between cars is lower than the established speed limit. Free flow rates for left turn and right turn movements are also limited by intersection geometry.

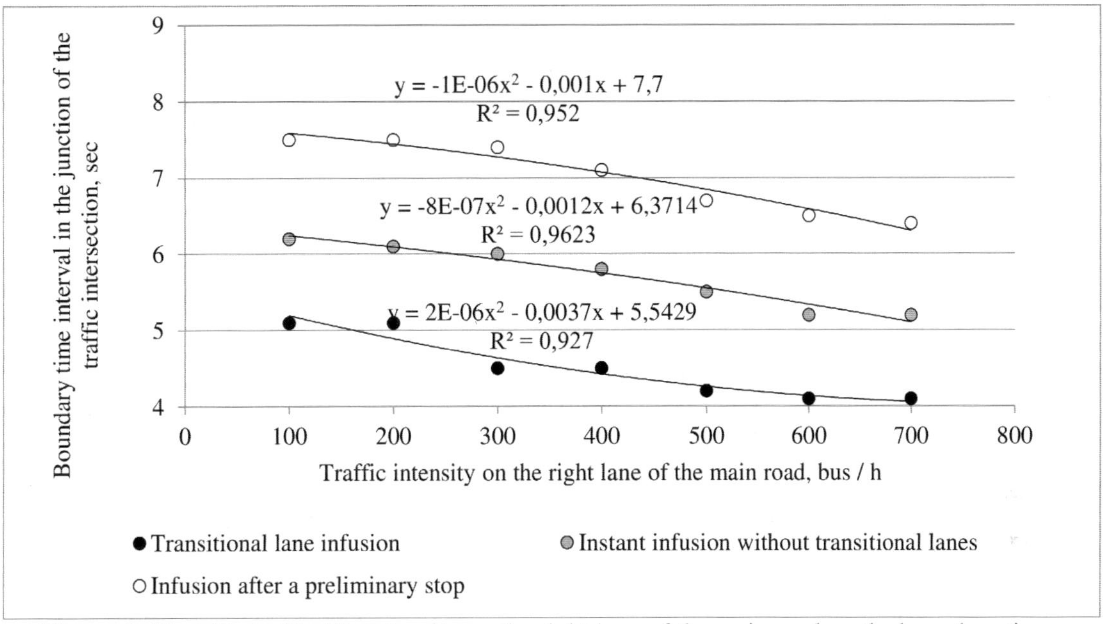

Figure 6. The effect of traffic intensity on the right lane of the main road on the boundary time interval in the confluence zone of the traffic intersection (85% coverage flow).

Studies of the free flow velocity at 6 diamond-shaped interchanges with changing sides of the movement are given in Table 2 [5].

Table 2. Parameters of the speed of the car on the diamond interchanges with a change in the sides.

№	Intersection	Speed limit (km/h)	Vehicle speed (km/h)	Speet of left turn off the highway (km/h)	Speed between vehicles (km/h)
1	MO 13, Sprinfield, MO	64,36	38,62	24,14	40,23
2	National st., Springfield, MO	64,36	40,23	33,80	46,67
3	Bessemer st., Alkoa, TN	56,31	41,84	24,94	51,50
4	Dorsett Rd, Maryland, MO	48,27	41,84	37,82	49,89
5	Front st., Kansas-city, MO	56,31	38,79	32,19	43,13
6	Winton Rd, Rochester, NY	72,4	46,51	29,93	50,05

Table 2 shows that vehicle speeds range from 38.62 to 41.83 km / h regardless of speed limits. Similarly, speeds above or below the bridge between cars range from 40.22 to 49.88 km / h. Therefore, the geometric design of the interchange controls the vehicle's free-flow speeds more than the set speed limit. This can reduce bandwidth, but also offers benefits to calm traffic.

The distance between traffic lights at the junction has a direct effect on bandwidth. A shorter distance between the two signals will tend to provide higher throughput, because the signals can be better coordinated and less time is required to move between cars. A greater distance increases the likelihood of starvation of demand on a downward cross signal for through traffic. Figure 7 presents a theoretical example of the percentage of unused time passing vehicles due to the travel time between cars [5].

IOP Conf. Series: Materials Science and Engineering **786** (2020) 012068 doi:10.1088/1757-899X/786/1/012068

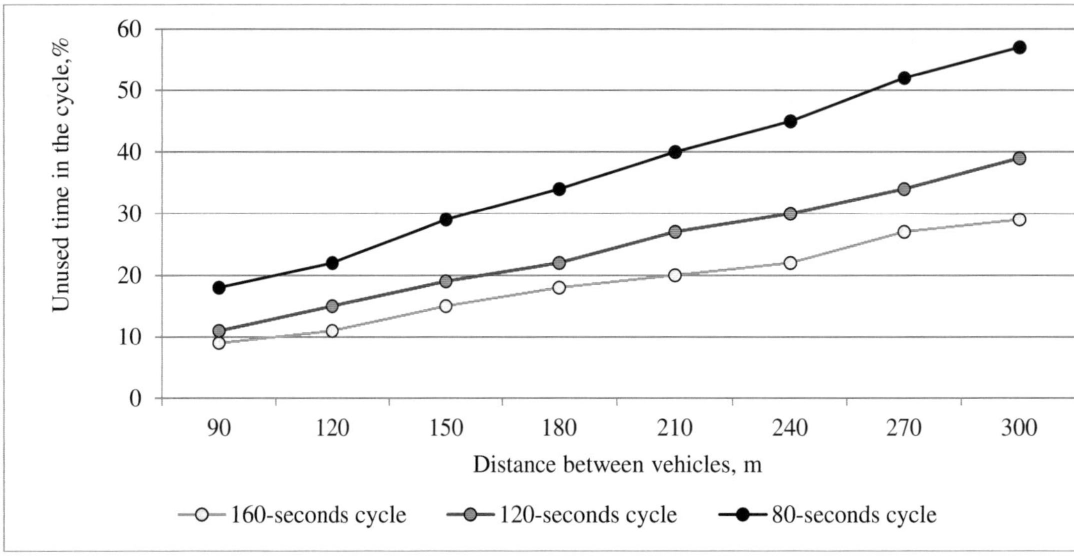

Figure 7. An example of a percentage starvation of demand for a diamond-shaped intersection with a change in the sides of the movement (assumed speed = 50 km / h).

The graph is based on an estimated vehicle speed of 50 km / h and shows three different cycle lengths (80, 120 and 160 seconds). As the distance increases, the travel time between the cars takes up a larger part of the cycle, assuming that both signals turn green at the same time. In practice, part of this unused time can be captured by setting the signal offsets according to the distance of travel, but the opportunity for unused time still increases with a large distance.

5. Conclusion

The increase in the number of vehicles leads to a gradual exhaustion of the resources of the road network. In this regard, the resulting traffic congestion becomes a serious problem, due to which the capacity of the road network decreases and the number of traffic accidents increases. To solve the problem of increasing the capacity of the road network, we can highlight the basic approach: introducing constructive changes to the road network - building traffic intersections at different levels.

Studies have shown that the safest intersection types are diamond-shaped intersections at two levels. The main advantage of this design is the absence of left-handed turning flows, which would cross the opposite direction. This is achieved by a temporary change in the side of movement within the junction. From the point of view of road safety, a diamond-shaped intersection with a change in sides of the movement is also safer than the classic diamond-shaped intersection. This is influenced by two factors. The first is just two conflicting points of intersection of flows. The second is a reduction in the speed of movement on the minor road due to the geometrical characteristics. The cons of this type of intersection are rather large occupied area (relative to other types of diamond-shaped junctions). This is due to the characteristics of the geometric elements of the minor road. There is also possible some kind of confusion among drivers in the place where the movement changes the side - keeping to the right, they can enter the oncoming traffic. Therefore, these places should be very well signposted and guide drivers.

The diamond-shaped intersection with a change in the side of the movement is a relatively new interchange, which has not yet been applied in Russia. The installation of these intersections at different levels will entail additional costs for informing and training drivers, cyclists and pedestrians on the rules of travel and passage along this type of interchange.

Such non-traditional types of intersections are based on variant design. That allows us to offer its own set of solutions for the organization of traffic, pedestrians and cyclists, for specific conditions. The use of a diamond-shaped intersection with changing sides of the movement allows to increase traffic safety, reduce delays, and is available for cyclists and pedestrians, compared to traditional types of intersections. The presented solutions are not always familiar to domestic designers and may cause difficulties in the design of this type of interchanges.

References

[1] Elvik R, Mysen A B and Vaa T 1997 Trafikksikkerhetshandbok: Oversikt over virkninger, kostnader og offentlige ansvarsforhold for 124 trafikksikkerhetstiltak, Transportokonomisk Institutt, Oslo, 754

[2] Fedotov V A and Buyanov E S 2011 Problemy funktsionirovaniya transportnykh razvyazok tipa «klevernyy list» i ikh resheniya // *Dorogi i mosty* 25 pp 99–125

[3] Pospelov P I, SHevyakov A P and SHCHit B A 2013 Planirovka rombovidnykh peresecheniy avtomobilnykh dorog v raznykh urovnyakh // *Nauka i tekhnika v dorozhnoy otrasli* **3** pp 11–3

[4] Variants of transport interchange : pat 2468138 Rus Federation **2011130936** 33 28

[5] Bastian Schroeder; Chris Cunningham; Brian Ray, Andy Daleiden, Pete Jenior and Julia Knudsen 2014 Kittelson & Associates, Inc Diverging Diamond Interchange Informational Guide U S Department of Transportation Federal Highway Administration Office of Safety, 226

[6] Joe G Bared, Tom Granda and Abdul Zineddin 2017 Drivers Evaluation of the Diverging Diamond Interchange TECHBRIEF U S Department of Transportation Federal Highway Administration Office of Safety 11

[7] Nikolaeva R V 2016 Predlozheniya po proektirovaniyu avtomobilnykh dorog s uchetom vospriyatiya voditelem dorozhnykh usloviy // *Izvestiya KGASU* **2(36)** 252-8

[8] Nikolaeva R V and Loginova O A 2015 Optimizatsiya dorozhnoy obstanovki pri proektirovanii avtomobilnykh dorog s uchetom psikhofiziologii voditelya // *Izvestiya KGASU* **4(34)** 357-62

[9] Mustafin M M and Loginova O A 2017 Proektirovanie levopovorotnykh otvetvleniy transportnykh razvyazok, s primeneniem perekhodnykh krivykh peremennoy skorosti VGV_kurve : Elektronnyy sbornik statey po materialam XV studencheskoy mezhdunarodnoy nauchno-prakticheskoy konferentsii SibAK Novosibirsk, 7–11

[10] James McCarthy, Joe Bared, Wei Zhang and Mark Doctor *Design at the crossroads Publication Number:* FHWA-HRT-13-005 **77(1)**

[11] V Chilukuri, S Siromaskul, M Trueblood and T Ryan 2011Diverging Diamond Interchange Performance Evaluation Missouri Department of Transportation 62

[12] Chlewicki G 2011 Should the Diverging Diamond Interchange Always Be Considered Transportation Research Record Washington, D C, **2223** 88

[13] Tahmina Khan and Michael Anderson 2016 Drivers' Evaluation of the Diverging Diamond Interchange // International Journal of Traffic and Transportation Engineering **6** 38-50

[14] Gubaidullina T, Ivanova N, Absalyamova S and Yerina T 2018 Analysis of national strategies for sustainable development with regard to fundamental conceptual premise *In Journal of Physics: Conference Series* **1141(1)** 012018IOP Publishing

[15] J A Martinez and R L Cheu 2012 Double Crossover Versus Conventional Diamond Interchanges Both with Frontage Roads // *Journal of Transportation of the Institute of Transportation Engineers* **4(1)** 1-16

IOP Conf. Series: Materials Science and Engineering **786** (2020) 012069 doi:10.1088/1757-899X/786/1/012069

The role of the transport infrastructure of the Republic of Tatarstan in the development of regional entrepreneurship

S Absalyamova*, T Absalyamov and C Mukhametgalieva

Kazan Federal (Volga region) University, Institute of Management, Economics and Finance, 4 Butlerova street, 420012 Kazan, Russia

*E-mail: s.absalyamova@yandex.ru

Abstract. The article considers the impact of transport infrastructure on the development of regional entrepreneurship. On the example of hosting sport mega-events and the increased interest in the cultural heritage, the article shows mutual influence of tourism in the region and the development of the transport infrastructure of the republic. The authors study the main components of the transport and communication framework, which allow to increase the tourist flow, passenger traffic and cargo flow.

1. Introduction

Regional transport infrastructure has distinctive features: its sectors operate nationwide as a single system, and at the same time its activities are characterized by a distinct regional aspect; There is a close interconnection and interdependence of transport infrastructure with other components of the regional system.

Strengthening the regional functions of the transport infrastructure is consistent with the goals of the Transport Strategy of the Russian Federation for the period until 2030. However, the goals of regional development determine the need not only to implement transport infrastructure projects in the territory of a region, but to coordinate regional and transport development strategies taking into account the positive and negative factors of transport infrastructure impact on the region

Studies of the impact of the state of transport infrastructure on the economies of other countries have shown that its development reduces the costs of domestic and foreign trade, but there are practically no studies of the complex impact of transport infrastructure on regional development as a whole. The main problem here is the difficulty of establishing causal relationships of this influence.

In this regard, an important aspect of the research task is the development of theoretical and methodological provisions for assessing the place and role of transport infrastructure in the socio-economic development of the region.

2. The main part

According to the World Economic Forum, Russia is at 64th place out of 144 possible quality transport infrastructure. And in terms of road quality - at 123rd place. The transport system of Russia has become one of the constraints on the development of Russian regions. Nevertheless, a qualitative transformation of the Russian transport system in the near future is quite possible. Its development can contribute to economic growth at a level of up to + 1% to GDP growth rates annually. For this, the model must meet modern social, environmental and technological challenges and maximize the development of related industries. The transport system can and should become a driver of economic growth until 2035.

The contribution of the transport industry to the gross regional product structure can be estimated by considering the share of the transport component in the output of goods and services of all sectors of the national economy of the Republic of Tatarstan. The contribution of the transport component to the gross regional product of the Republic of Tatarstan in 2017 amounted to 6.4%.

Table 1. Contribution of the transport industry to the GRP of the Republic of Tatarstan, billion rubles.

Indicator	2015	2016	2017
Gross Regional Product (GRP)	1883	1940	2072
Gross Value Added (GVA) "Transport"	116	122,4	131,6
GVA "Transport" physical volume index, in% to the previous year (in comparable prices)	100,5	101,9	103,9
The share of GVA "Transport" in GRP,%	6,3	6,3	6,4

Let's consider the impact of transport infrastructure on the development of individual sectors of the regional economy using tourism as an example.

The Republic of Tatarstan has a unique tangible and intangible heritage, reflecting the history of the formation and development of many peoples of the Volga region since ancient times and representing value not only on the republican and all-Russian, but also on a global scale [2].

Today, over 7000 objects of historical and cultural value are registered with the state, of which 1540 are under protection of federal, regional (republican) and local (municipal) significance. More than half of historical and cultural monuments are located in 13 historical settlements (Kazan, Bolgar (Spassk, Kuybyshev), Bilyarsk (Bilyar), Elabuga, Bugulma, Buinsk, Mamadysh, Mendeleevsk, Menzelinsk, Tetyushi, Chistopol, Laishevo and Sviyazhsk).

Cultural heritage objects are important for the Republic of Tatarstan due to the fact that they are:

- a mandatory attribute of the Republic of Tatarstan as a state with a rich historical and cultural past;
- a condition for a comfortable environment for life and relaxation of residents of the republic and guests of the capital;
- visible symbols of historicism and traditions of the Republic of Tatarstan;
- a confirmation of Russia's integration into the pan-European cultural context;
- one of the drivers for the development of the regional economy.

The influence of cultural heritage objects on the development of regional infrastructure can be compared with the influence of such cultural and sports mega-events as the celebration of the 1000th anniversary of Kazan, the World Student Universiade 2013, the World Aquatics Championship 2015, the Confederations Cup 2017, and the FIFA World Cup - 2018 , etc.

Jobs are also being created in areas that are indirectly related to the main activities of tourist facilities (in restoration and construction, in the service sector, in transport and in agriculture). The number of such additional jobs is many times greater than the number of full-time employees of museums, and this also creates seasonal jobs for schoolchildren, senior citizens and for part-time workers [1].

The influence of the activities of cultural heritage objects on stimulating the development of regional infrastructure (road, hotel, catering, consumer services, etc.) is of a great significance.

An important stage in inter-regional cooperation in 2016 was the creation of the Great Volga Way along the Volga Federal District. The inter-regional project "The Great Volga Way" includes 22 tourist routes that have a territorial, cultural and historical community.

For the purposes of sustainable spatial development of the Republic of Tatarstan, including sustainable tourism, large transport and communication systems, such as the intersection of highways, large railway stations, airports, logistics centers, river ports, are of particular importance [4]. Since Soviet times, the republic has a transport framework that is based on the system of federal and regional highways.

The development of the transport framework of the Republic of Tajikistan is associated with the development and activation of existing and the formation of new transport networks in the directions of west-east, north-south, north-west and southeast [7].

The development of the transport and communication framework of the Republic of Tatarstan is carried out due to [6]:

- development of the northern corridor (part of the Transsiberian (TS) international corridor)
- reconstruction of the federal highway M-7 with the expansion of the carriageway, construction of the ring road to "intercept" transit vehicles further from the capital.
- active formation of high-speed railways through the reconstruction of existing and construction of new ones;
- the formation of a new southern transport corridor - the construction of a highway of regional significance in the direction of Staroe Drozhzhanoye - Buinsk - Tetyushi - Bolgar - Bazarnye Mataki - Aksubaevo – Cheremshan – Almetyevsk – Muslyumovo – Aktanysh based on existing roads;
- the development of the western meridional transport framework due to the reconstruction and modernization of existing roads and railways and the formation of high-speed passenger traffic;
- construction of a new section of the Kazan-Arsk-Baltasi highway with a continuation to the north to Kirov;
- development of the eastern meridional corridor due to reconstruction, modernization and new construction of high-speed roads and railways;
- continuing the development of the transport corridor in the direction north-west - south-east. In this direction, as part of the formation of the automobile route in the direction of St. Petersburg - Vologda - Kostroma - Kirov - Kazan - Orenburg - the border of the Republic of Kazakhstan, a project for the construction of the Shali high-speed highway is already being implemented in the Republic of Tatarstan - Magpies Mountains - Alekseevskoye - Almetyevsk - M5 (Bavly);
- the formation of the middle meridional interregional transport corridor in the direction of Baltasi - Bogatye Saba - Tyulyachi - Chistopol - Nurlat. Within the estimated term of the territorial planning scheme, the southern part of the corridor should be formed along the directions of Chistopol - Nurlat and Alekseevsk - Bazarnye Mataki with continuation to Samara;
- development of infrastructure of airports of international and regional significance. Completion of the reconstruction of the Kazan International Airport, reconstruction of the Begishevo and Bugulma airports [3].

The main measure in the development of water transport, which is part of the system of transport corridors on the territory of the Republic of Tatarstan, is the provision of regular high-speed shipping lanes in the two directions of Tetyushi - Kamskoye Ustye - Zelenodolsk and Menzelinsk - Kamskoye Ustye, which continue southward towards Volgograd, Rostov- on-Don, Astrakhan, the Caspian, Azov and Black Seas [8]. In the western direction there is regular communication with Nizhny Novgorod, Yaroslavl, Moscow and the northern regions of Russia. In the east - with Ufa and Perm [3].

The implementation of the Great Water Ring of Europe (the Volga River - the Don River - the Sea of Azov - the Black Sea - the Danube River - the Main River - the Rhine River - the Baltic Sea - the Volga River) will have unique value not only for freight international transport, but for the development of international tourism. Without a doubt, this will affect the realization of the tourism potential of the Republic of Tatarstan, and cultural heritage sites are a moving part of this sector.

Thus, the historical, cultural and natural heritage in their totality is a specific and important economic resource of the region, it can become not only a factor in the development of cultural life, but also the basis of a special branch of specialization, one of the promising directions of stimulating socio-economic policy and the development of transport infrastructure of the republic.

To conclude, let's have a brief look at the graphs of relation between the annual number of tourists in Russian Federation and number of small businesses (Figure 1), and paved road density and number of small businesses (Figure 2).

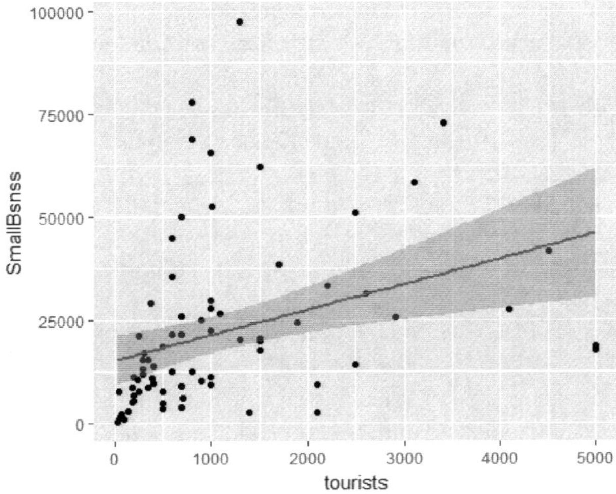

Figure 1. Annual number of tourists in Russian regions vs. number of Small Businesses, 2017.

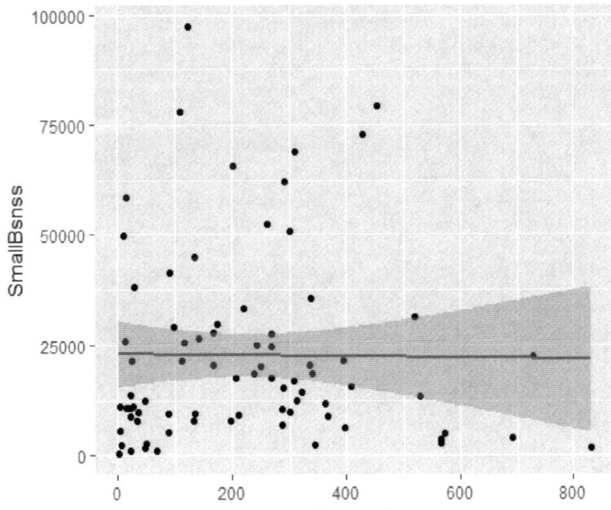

Figure 2. Density of paved motorways in Russian regions vs. number of Small Businesses, 2017.

3. Conclusion

The analysis allows us to conclude that transport, as a multifunctional system, has a positive synergistic effect on the economic development of the region and its various industries. In order to ensure the socio-economic development of the region, it is necessary to determine priority areas for the development of transport infrastructure in the framework of regional policy. Regional policy should include the development of regional transport infrastructure as a factor in their development. Regional policy should also stimulate public-private partnerships for the development of transport infrastructure.

References

[1] Absalyamova S , Absalyamov T , Khusnullova A and Mukhametgalieva C 2016 The impact of corruption on the sustainable development of human capital/ *Journal of Physics: Conference Series* **738(1)**, 012009

[2] Absalyamov T 2015 The influence of cultural and sport mega-events on sustainable development of the city //*Procedia-Social and Behavioral Sciences* **188** 197-201

[3] Dolgosrochnaya Strategiya razvitiya transportnogo kompleksa Respubliki Tatarstan (s pozitsij ustojchivogo razvitiya) URL:http://mindortrans tatarstan ru/dolgosrochnaya-strategiya-razvitiya-transportnogo

[4] Gubaidullina T , Ivanova N , Absalyamova S and Yerina T 2018 Analysis of national strategies for sustainable development with regard to fundamental conceptual premise / *Journal of Physics: Conference Series* **1141(1),** 012018

[5] Doklad Itogi deyatelnosti Gosudarstvennogo komiteta Respubliki Tatarstan po turizmu za 2016 god – URL: http://tourism tatarstan ru/rus/documents htm

[6] Strategiya sotsialno-ekonomicheskogo razvitiya Respubliki Tatarstan na period do 2030 URL:http://tatarstan2030 ru/

[7] Kiselev V A 2018 i dr Optimizatsiya transportnoj infrastruktury gorodov //Transportnoe delo Rossii **5** 138-40

[8] Radchenko D and Ponomarev Y 2019 Prostranstvennoe razvitie transportnoj infrastruktury i stepen ee vliyaniya na sovokupnuyu faktornuyu proizvoditelnost v Rossii (Spatial Development of Transport Infrastructure and the Degree of Its Influence on Aggregate Factor Productivity in Russia) //Available at SSRN 3361023

Road safety as a factor in the socio-economic development of the country

R V Nikolaeva

Kazan State University of Architecture and Engineering, 420043, Kazan, Zelenaya str., 1, Russia

E-mail: nikolaeva1@bk.ru

Abstract. In the conditions of modern society, the prospective development of the country's economy cannot be considered without a well-established transport system. The discrepancy between the level of development of the transport network and the level of demand for road transport leads to inefficient and huge costs. The main costs are associated with the elimination of the consequences of road accidents. The article considers road safety as a particularly significant socio-economic problem of the country. The relationship between economic development and the level of road safety was studied using the example of the Republic of Tatarstan (a subject of the Russian Federation). As main indicators we selected population, level of motorization, length of transport networks, gross regional product, accident rates and damage as a result of deadly accidents. Studies have shown that there is an interdependence of road safety and the economy of the republic. Moreover, it is important to assess the direct and indirect socio-economic costs of the consequences of road accidents. This allows us to measure the burden that a traffic accident causes to society, and emphasizes the return on investment in road safety.

1. Introduction

The development of the transport system of any country takes place in a socio-economic context. Transport is an important element of the economy and is one of the general development tools, which means that a large burden is placed on the transport system. The economic development of the country is increasingly linked to the mobility of people, goods and information. Therefore, it can be said that the connection between the quantity and quality of transport networks and the level of economic development is obvious.

The level of development of the transport system of the country is associated with socio-economic indicators as well. The transport system, represented by a highly developed transport network, is efficient and provides great social and economic opportunities. The positive effect will be expressed by better access to production, employment and additional investments for further development. Transport systems that do not have sufficient capacity or reliability contribute to the economic costs that reduce the quality of life. The main costs in this case are associated with the elimination of the consequences of motor vehicle collisions (MVC).

MVC costs undermine the economic well-being of society. It is enough to say that the average age of dead in MVC is 20-40 years, i.e. victims are citizens of the most productive age. Economic losses associated only with injuries as a result of MVC, on average, are estimated at 2% of the gross domestic product (GDP) in Europe [1, 2]. According to the Russian Union of Motor Insurers, only the material damage from MVC in Russia is 2.5% of GDP.

Thus, it is possible to designate the mutual influence and interconnection between the country's socio-economic development and road safety.

2. Literature review

The study of the relationship between the socio-economic situation of the country and MVCs is being studied quite actively. The main studies in this area, conducted in various countries, are presented below.

In France, accidents were assessed by three groups of workers: managers, skilled workers and unskilled workers (Lenguerrand et al, 2006) [3].

British scientists compared injuries per 10,000 inhabitants in 15% of the poorest areas and 15% of the richest areas of the Lothian region in Scotland (Abdalla, Barker and Raeside, 1997) [4]. The number of accidents with pedestrians was studied by (Graham, Glaister and Anderson, 2005) [5], which resulted in the development of a negative binomial regression model to explain the number of MVCs.

In Norway, Denstadli and Hjorthol (2002) compared data on the average daily length of travel for people, taking into account their incomes [6].

Studies in Sweden have researched the relationship between socioeconomic variables and the frequency of road traffic injuries (Laflamme and Engstrom, 2002) [7].

We should mention two studies in the United States on the relationship between social status and MVCs. The first study showed that as per capita income increases, death rates among drivers and passengers of motor vehicles decrease sharply. In poor areas, the rate is about 2.5 times higher than in the richest areas. (Baker et al, 1992) [8].

The second American study compared death rates among motor vehicle drivers per million kilometers depending on race and socio-economic status. One of the most interesting findings of the study is that the differences in the level of mortality are partly due to differences in the behavior of drivers. Thus, 42% of men with the highest level of education wore seat belts, compared with only 19% of men with the lowest level of education (Braver, 2003) [9].

Studies of the relationship between various indicators of social status and participation in traffic accidents led to different conclusions. A French study found that people with high social status are more likely to participate in MVC than people with low social status. All other studies have shown that low social status - regardless of how it is measured - is associated with a higher level of participation in traffic accidents. This has been consistently found in studies conducted in the UK, Sweden, Canada, and the United States of America. Thus, the prevalence of evidence suggests that people with low social status are associated with a higher risk of becoming a participant of MVC than people with high social status.

3. The relationship between the transport system and the economy

The economic importance of the transport industry can be assessed from a macroeconomic and microeconomic point of view.

At the macroeconomic level (the importance of transport for the whole economy), the transport and mobility that it provides are linked to the level of production, employment and income within the national economy. In many developed countries, transport accounts for between 6 and 12% of gross domestic product (GDP).

At the microeconomic level (the importance of transportation for certain parts of the economy), transportation is associated with the costs of producer, consumer and production. Thus, the importance of specific transport activities and infrastructure can be assessed for each sector of the economy. Usually, higher income levels are associated with a greater share of transport in consumption expenditures. Transport accounts for an average of 10% to 15% of the costs of all sectors of the country, while it accounts for about 4% of the cost of each unit of production in production, but this figure varies considerably depending on the industry.

The relationship between the transport system and the country's economy can be formulated as follows [10]:

1) the transport system has a complex effect on the economic development of the region both at the macro level, setting the vector of its development, and in relation to individual firms that determine the use of infrastructure facilities, which is expressed as:

- providing jobs, generating income for people employed in the country's economy;
- ensuring the transport availability of resources, production capacity;
- the spatial development of the country, the involvement and development of new territories, the creation of new jobs, and the development of business in the adjacent territories;
- formation of budget revenues;
- reducing the accident rate, the number of road accidents, road congestion, vehicle wear.

2) The country's economy in turn affects the transport system through the following channels:

- expansion of labor, material and technical resources for the construction of transport infrastructure facilities;
- management and regulation of traffic flows;
- budget investments in transport infrastructure;
- formation of demand for transport services.

4. The impact of the development of the Republic of Tatarstan on road safety

To empirically confirm the hypothesis of the impact of the country's economic development on the level of road safety, we conducted a study. The main objective of this study is to assess the impact of the transport system on road safety. The study was based on data for the period 1998-2018 for the subject of the Russian Federation - the Republic of Tatarstan.

Indicators of the level of road safety are functions of many variables, and we selected the following as the main ones:

1) socio-economic indicators: population, gross regional product;
2) transport indicators: the length of transport networks, the level of motorization;
3) Accident rates: MVCs, number of injured and dead.

Time series of the presented indicators for the Republic of Tatarstan are shown in Figure 1.

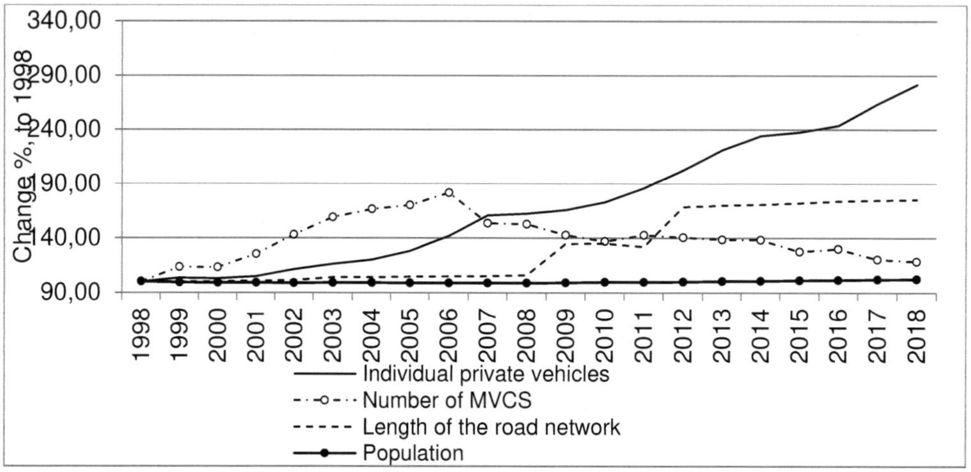

Figure 1. Dynamics of development of road transport, road network and changes in accident rates in the Republic of Tatarstan.

The analysis shows that since 2006, the Republic of Tatarstan demonstrates a growing mismatch between high rates of growth in the number of vehicles and a slow increase in the length of the paved roads network, which accounts for the most of freight and passenger transportation. Peak jumps in the growth of the transport network of the republic are associated with the holding of international events, for which large investments were made in the development of the transport system [11].

One of the most informative indicators of accidents is the social risk (public health risk), taking into account the number of deaths in MVC per 100,000 inhabitants. This indicator can give an objective assessment of the level of traffic safety [12]. Figure 2 shows the relationship between the level of motorization and social risk of the Republic of Tatarstan for the period 1998-2018.

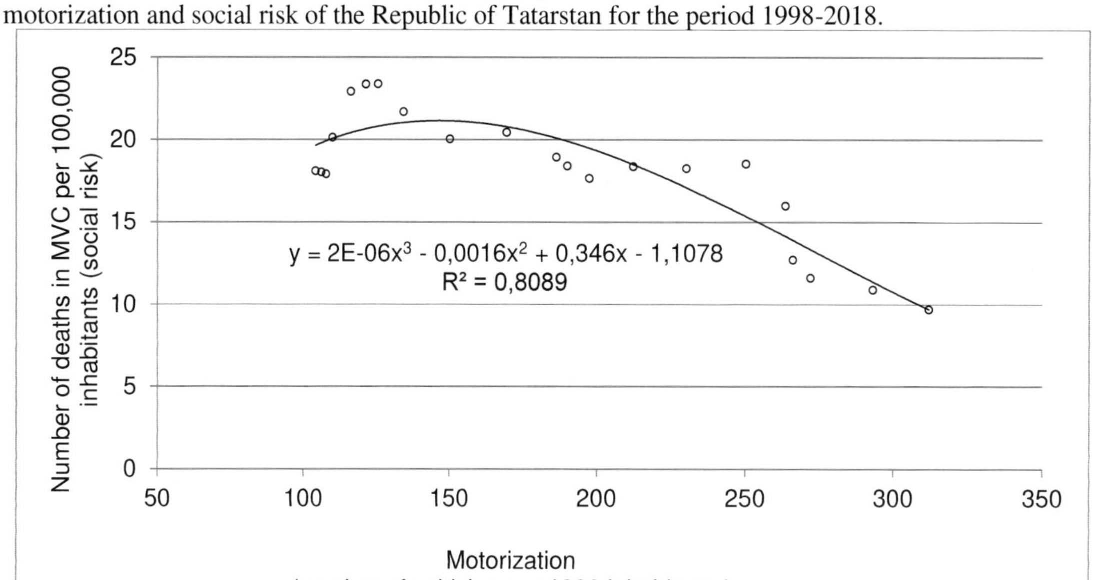

Figure 2. The relationship of the level of motorization and road traffic risk in the Republic of Tatarstan (1998-2018).

Analysis of the relationship between the level of motorization and the risk of road traffic in the Republic of Tatarstan showed that with an increase in the level of motorization, there is a tendency towards a decrease in the number of people died in MVCs. Figure 2 highlights certain critical points characterizing the qualitative changes in the studied relationship between the level of motorization and social risk. Studies have shown that with the growth of motorization, social risk decreases. Moreover, there are transition zones in the range of motorization varying from 250–260 vehicles per 1,000 inhabitants, when the relative death amount gradually moves into a stabilization zone.

A huge role in the formation of accidents is played by a level of development of the transport system. Imbalances in the development of the road network strongly affect the uneven loading of roads, which in turn stimulates the growth of MVC, especially in areas with low traffic capacity and high traffic intensity [13].

The existing shortage of roads has a negative impact on the level of road safety. The risk level of MVCs increases. Figure 3 shows the relationship between the density of roads and social risk of the Republic of Tatarstan for 1998-2018.

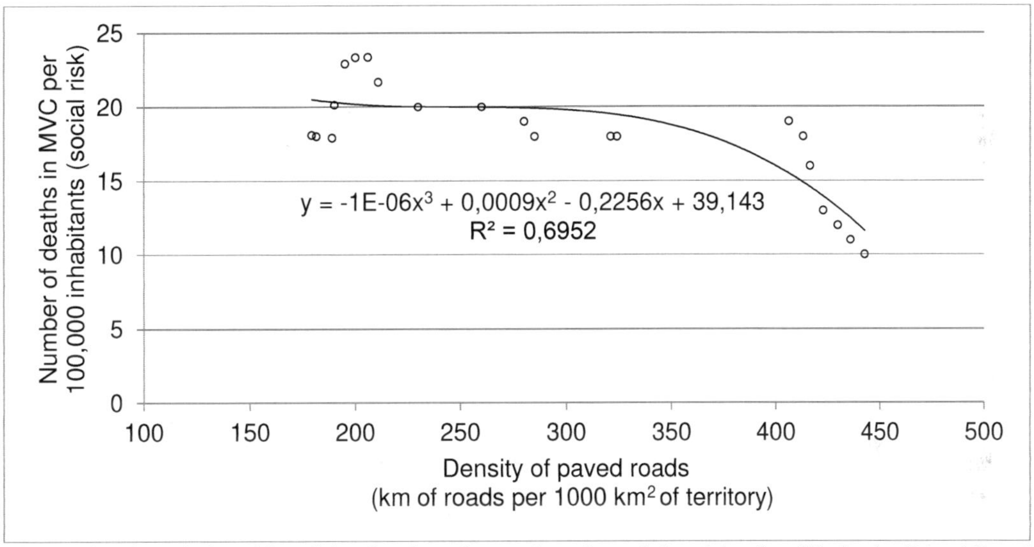

Figure 3. The relationship of the density of paved roads and the risk of traffic in the Republic of Tatarstan (1998-2018).

The analysis of the presented dependence shows that with an increase in the density of the transport network in the Republic of Tatarstan, there is a gradual decrease in the relative risk of involvement of people in MVCs. With a highway density of 410 km per 1000 km^2 of territory, we note a decrease in the number of fatalities, which is associated with attracting investment in the transport system in 2011-2012 for the XXVII World Summer Universiade 2013 in Kazan [14, 15]. All this indicates the need for a significant increase in the length of roads in order to reduce the number of MVCs.

The solution to the problem of the shortage of the road network is becoming increasingly important, if we keep in mind the rapid growth in the number of vehicles that is faster than the development of the roads. If we proceed from the number of vehicle fleets, we can conclude that accidents should include not only a deficit in the total length of roads, but also a low technical level of the existing road network and, above all, a shortage of high-grade roads – motorways and high-speed roads with huge capacity [16].

In this regard, in order to develop the road network of the Republic of Tatarstan, and to increase the level of traffic safety as well, it is necessary to increase the length of public roads with hard surface.

The development of the road network is associated with the attraction of huge capital expenditures. Together with the development of the road network, it is also necessary to provide a system of measures to eliminate dangerous sections and increase the capacity of highways. Among the most important tasks - are the measures aimed at reducing the severity of MVCs (installation of barrier fences separating off-road traffic, construction of elevated and underground pedestrian crossings, road lighting, improving traffic management and other measures that generally reduce accident rates).

Nowadays, road safety management decisions need to be made on the basis of an economic assessment of damage from MVCs. The assessment of damage from MVCs is based on the principles of a planned economy and the laws governing the growth of national income and depends on the size of the gross regional product (GRP) of a country and the number of people employed in the economy [17]. Figure 4 shows the results of the calculation of socio-economic damage from the deaths in the MVCs in the Republic of Tatarstan.

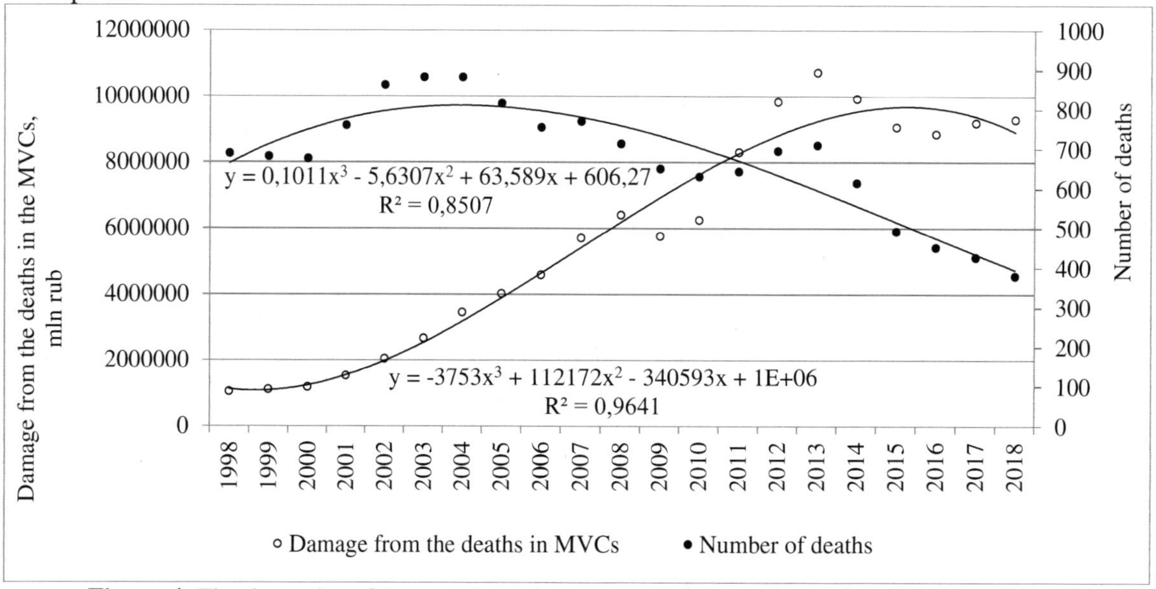

Figure 4. The dynamics of damage from deaths in MVCs in the Republic of Tatarstan.

Based on the analysis of the relationship between the level of motorization and the risk of road traffic (Figure 2) and the socio-economic damage from MVCs in the Republic of Tatarstan, we can assume that with a further increase in vehicles, the number of people affected in MVCs will decrease and the damage will continue to increase.

With the growth of vehicles, other negative consequences will increase. First of all, there will be an increase in the imbalance between the development of the transport network and the level of motorization, and there will also be an increase in the damage caused by the underdevelopment of the transport system. Figure 5 shows the relationship between the density of roads and the damage caused by the deaths in the MVCs in the Republic of Tatarstan for 1998-2018.

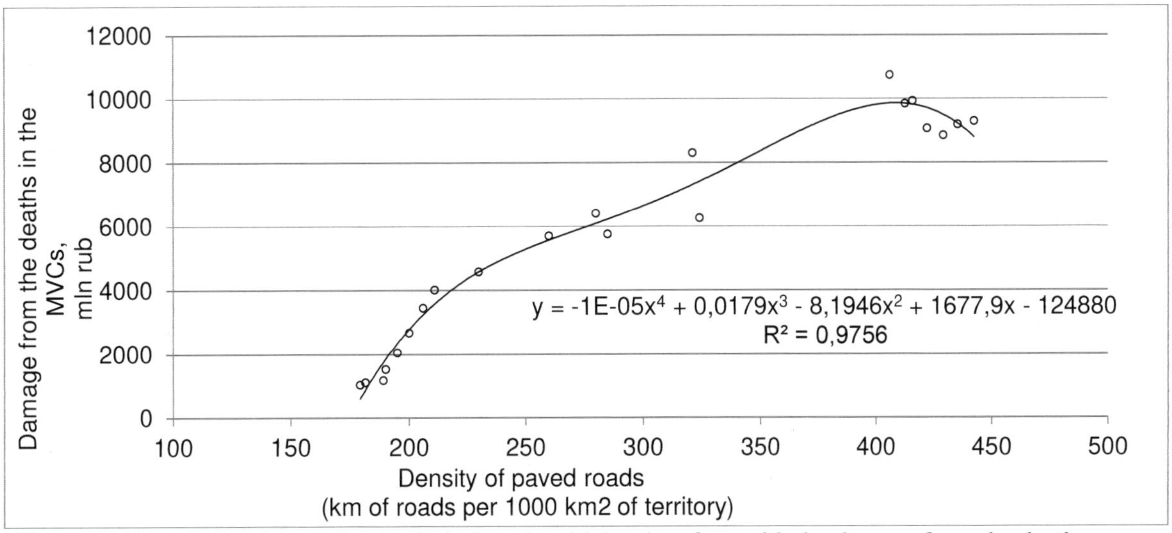

Figure 5. The relationship of the density of roads with hard surface with the damage from the deaths in the MVCs of the Republic of Tatarstan (1998-2018).

The analysis shows that having reached the peak density of highways at 410 km per 1000 km^2 of territory, the damage from loss of life in MVCs decreases. This can be associated with the development of the transport network (which is associated with attracting investments in the transport system in 2011-2012, as mentioned above). Thus, we can say that the development of the transport network of the Republic of Tatarstan contributes to the improvement of road safety. However, it is also necessary to take into account that an increase in the length of the road network in areas with an insufficiently developed network may contribute to an increase in the number of MVCs on them, since in a number of cases they are beginning to assume unusual transportation functions, which contributes to an additional increase in traffic load. In some cases, along with the general improvement of traffic conditions on such roads, there is a known effect of "migration" of MVCs to neighboring areas due to the remaining heterogeneity of traffic conditions, if the reconstruction of road sections was local in nature. In this regard, to ensure the planned level of traffic safety while increasing the length of the road network, the development features and traffic load of the road network should be taken into account, as well as the creation of uniform traffic conditions on each of them in order to uniformly reduce the overall level of observed accidents.

5. Socio-economic efficiency of transport infrastructure development

To improve road safety, it is necessary to reduce the disproportion between the growth of vehicles and the length of the transport networks of the Republic of Tatarstan. Currently, part of the roads in the republic are in a dangerous condition, and considerable funds are required to bring them to proper condition. The relevance of this issue is due to the need to improve road safety, primarily through ensuring the effectiveness of budget expenditures allocated for these purposes.

The socio-economic efficiency of improving road conditions is determined by the impact of the modernization and development of the road network on the costs of road transport and on the demand for road transport. The effects associated with an increase in the speed of vehicles and an increase in the reliability of the year-round traffic on the road network are also taken into account.

Assessment of the socio-economic efficiency of the implementation of measures to improve road conditions can be performed in accordance with the methodology for the phased improvement of transport and operational qualities of roads, which is based on the principle of calculating the transport effect.

According to this methodology, reduction of the cost of road transport is created by lowering the tariffs for the transport of goods and passengers, which in turn affects the cost of goods, works and services. Reducing the cost of production contributes to lower producer prices, which leads to an increase in demand for manufactured goods and services. These factors will lead to an increase in the GRP of the Republic of Tatarstan. A reduction in the transport margin on the goods and services produced leads to an additional (in addition to a decrease in producer prices) decrease in the acquisition prices of goods and services (buyers prices), which further increases the demand for them and the republic's GRP.

The marginal analysis of the profitability of the transport complex reveals the connection between the growth rates of capital investments in the industry and profits. This analysis shows that with the growth of the industry's capitalization, its profits increase at an accelerated pace, i.e. each additional ruble of investments brings more and more return.

Studies have shown that by 1 rub. invested in the construction of roads, the effect in 5 years will be 3 rub., and with investments in the repair and reconstruction of roads 6 rub. The most effective are investments in measures to ensure road safety: installation of road signs - 27 rubles, lighting of road sections - 12 rubles, installation of barrier fences - 11 rubles etc.

Thus, investments in highways have a significant positive effect on the macroeconomic indicators and the growth rate of the country's economy.

6. Conclusion

Improving road safety is one of the priorities of the country's economy. This situation is caused, first of all, by economic necessity, since movement by means of transport is designed to minimize the time costs, which is a criterion of efficiency from the standpoint of modern human. The improvement and development of transport networks contributes to the mobility of labor, goods and resources, the expansion of trade relations and increased competitiveness and accessibility of economic regions. All this together determines the need to modernize the transport network to improve road safety.

Studies show that there is a relationship between a country's economic development and road safety, as measured by the number of MVCs. Economic changes in the development of the country have a direct impact on the level of motorization, transport risk, the number of MVCs. In this situation, it is necessary to clearly formulate goals for road safety and, in particular, to set quantitative goals.

References

[1] Khegay Y A 2014 Bezopasnost dorozhnogo dvizheniya - vazhneyshaya chast sotsialno-ekonomicheskogo razvitiya strany // *Teoriya i praktika obshchestvennogo razvitiya* **9** 111-5

[2] Meloyan V G 2010 Otsenka ushcherba ot dorozhno-transportnykh proisshestviy v kontekste ekonomicheskoy bezopasnosti Rossii // Novye tekhnologii **4** 105-08

[3] Lenguerrand E , Martin J L , Chiron M and Laumon B 2006 Road crash involvement and professional status: a prospective study of the French GAZEL cohort Paper submitted to Accident Analysis and Prevention

[4] Abdalla I M , Barker D and Raeside R 1997 Road accident characteristics and socio-economic deprivation Traffic Engineering and Control, December, 672-676

[5] Graham D , Glaister S and Anderson R 2005 The effects of area deprivation on the incidence of child and adult pedestrian casualties in England Accident Analysis and Prevention, **37** 125-35

[6] Denstadli, J M and Hjorthol, R 2002 Den nasjonale reisevaneundersøkelsen Nøkkelrapport Rapport 588 Oslo, Transportøkonomisk institutt

[7] Laflamme L and Engström K 2002 Socioeconomic differences in Swedish children and adolescents injured in road traffic accidents: cross sectional survey *British Medical Journal*, **324** 396-7

[8] Baker S P , O'Neill B , Ginsburg M J and Li G 1992 The injury fact book Second edition New York, Oxford University Press

[9] Braver E R 2003 Rice, Hispanic origin, and socioeconomic status in relation to motor vehicle occupant death rates and risk factors among adults *Accident Analysis and Prevention*, **35** 295-309

[10] Kataeva YU 2013 V Integralnaya otsenka urovnya razvitiya transportnoy infrastruktury regiona // *Vestnik PGU* Seriya: Ekonomika **4(19)** 66-73

[11] Nikolaeva R V 2017 Vzaimosvyaz mezhdu ekonomikoy strany i bezopasnostyu dorozhnogo dvizheniya//Tekhnika i tekhnologiya transporta **3(4)** 15

[12] Sakhapov R L , Nikolaeva R V , Gatiyatullin M H and Makhmutov M M 2016 Risk management model in road transport systems/ *Journal of Physics: Conference Series* **738(1)** 012008

[13] Avtomobilnye dorogi i mosty Vliyanie razvitiya i sostoyaniya dorozhnoy seti na uroven bezopasnosti dvizheniya na dorogakh Rossii Obzornaya informatsiya Vypusk 4 Moskva 2003 URL: http://base1 gostedu ru/56/56230/ (data obrashcheniya 5 05 2015)

[14] Gubaidullina T, Ivanova N , Absalyamova S and Yerina T 2018 Analysis of national strategies for sustainable development with regard to fundamental conceptual premise *In Journal of Physics: Conference Series* **1141(1)** 012018) IOP Publishing

[15] Absalyamov T 2015 The influence of cultural and sport mega-events on sustainable development of the city *Procedia-Social and Behavioral Sciences* **188** 197-201

[16] Anokhin B B 2007 Osnovnye tendentsii avtomobilizatsii naseleniya i ikh uchet v, programmakh razvitiya avtomobilnykh dorog // *Transport Rossiyskoy Federatsii* **9** 42 - 6

[17] Sakhapov R and Nikolaeva R 2017 Economic aspects of traffic safety administration Transportation Research Procedia 578-83

IOP Conf. Series: Materials Science and Engineering **786** (2020) 012071 doi:10.1088/1757-899X/786/1/012071

Improving the methods of training specialists in the sphere of construction and utilities (on the example of the Volgograd region

E M Maznitsa*, M K Belyaev, O V Maksimchuk

Volgograd State Technical University Institute of Architecture and Construction Volgograd, Russia

E-mail: Dilemaz@mail.ru

Abstract. The existing methods of training engineering personnel in the sphere of road, power, utility machines and other special equipment do not take into account the changed realities of the national economy. For maintenance engineers it is now not enough to know the basics of the operation of machines and mechanisms, it is necessary to have management skills. The operator should study not only the technical characteristics of the equipment, but also the principles of pricing for utility services, the basics of tariff regulation, the methodology for calculating tariffs and the cost of utility services. Therefore, the development of practical management skills in the housing and utilities sector, the study of the fundamentals of accounting and tax reporting, estimated rationing and legal training are becoming a modern addition to the classical training of specialists - operators of the construction and utilities sectors.

In recent years, there has been a significant decline in private investment in the communal infrastructure. This is especially noticeable in the water, heat and power supply sectors. There are a lot of reasons including the high level of risks and the low solvency of household consumers in a crisis economy. The solution of these problems is determined by the influence of various factors [1]. For example, one of the main limiting factors for the effective implementation of housing and communal services reform is the lack of experienced qualified personnel in the management and operation of utilities, which is associated with a sharp change in the functioning of the housing sector in modern Russia [2]. More recently, the state was responsible for the work of the housing sector, for heat in houses, water supply and sewage, electricity, cleaning, etc. and now - in one decade everything changed dramatically, the responsibility is blurred, it is completely incomprehensible for the population how the housing sector is working now [3].

Discussing today the reasons for the decline in GDP growth and changes in economic policy that are necessary for the revival of "constructive" production, it can be concluded that there are signs of recession in the Russian economy. The current movement towards recession is a consequence of the deep deindustrialization of the domestic economy [4]. Deindustrialization always leads to economic stagnation, and Russia's experience confirms this. The industrial sector survived at least two hard blows.

The first - in the 90s, when everything collapsed, including industrial communications. In the early 2000s, oil prices began to rise, and golden rain fell on the country. The industry was revived. But then there was a change of authority, and with it the economic model of development changed: in fact, instead of the economy of supply, the economy of demand for the export-raw material base was developed. Such ill-conceived economic policy dealt a second blow to the industrial sector [5].

In the context of the crisis and the possible post-crisis development of the Russian economy, entrepreneurship education should become the most important factor in the modernization of education and growth of economy. It is possible that the current crisis will lead to an even greater segmentation of the regional business education market. Of course, it is important to single out the positive moment of the crisis, as the most suitable time for improving the content of educational programs and improving the quality of the educational process, but the current situation has a downside. It seems that the need for a full-fledged, serious business education will always remain, as well as in managerial trainings, seminars and short programs. In our opinion for Russia and our region, one of the main areas of business education should be innovative management and the development of new areas of entrepreneurship [6].

The peculiarity of the current situation of the innovative development of the Volgograd region is the presence of a significant research and production base, combined with the lack of specific innovations related to the lack of enterprises' own, borrowed and attracted funds, low effective demand for scientific and technical products. Available examples of high-quality goods in general do not solve the problem of low competitiveness of the main products of enterprises [7]. Managers of enterprises in modern conditions are not ready to invest in especially risky scientific and technical projects at the stage of scientific research and experienced constructive works. Investor interest arises, as a rule, if a prototype or an experimental batch of a new material or product has been obtained as a result of research work. In the scientific and technical sphere, effective innovative entrepreneurship is impossible without managers in the commercialization of technologies and in the management of innovative projects. Now there is a situation where the rapidly developing Russian economy is facing a shortage of specialists in introducing new technologies, most specialists want to work in quiet places without much risk and responsibility. For example, we need specialists who will solve the engineering task of transferring the heating system of residential buildings from hydrocarbon fuel sources (fuel oil, diesel fuel, etc.) to alternative and renewable types of energy (solar, wind, wave, as well as hydroelectric and thermal energy ocean) [8].

It seems that here the Higher School needs to move from training theorists to practitioners, for example, everywhere there is not enough competent specialists - real estate operators. How to prepare them? Why do not talented advanced young professionals come in such a promising in innovation and financial terms, the industry as housing and utilities? Should there be any specifics in training such specialists at the regional level? And is it necessary to take into account the peculiarities of regional economic development?

In our opinion, in the formation of educational programs in various areas of education, including in the housing and utilities sphere, regional specificity, the level of modernization of the economy of a particular region should be taken into account [9]. As for the Volgograd region, the degree of lag in its economic development, not even from the advanced, but from the ordinary and medium-sized regions of Russia, is increasing every year. The low dynamics of the economy puts pressure on social indicators, which, in turn, affect demographics. In order to reverse the negative trends, serious efforts are needed and, above all, to improve the quality of government in the region. It seems that separate, very successful and fairly large-scale projects will not be able to provide a breakthrough of the regional economy as a whole, it is too big and inertial. They cannot replace the daily hard work in creating a favorable environment for the development of the region. Entrepreneurs and the population are key economic players, and the state cannot and should not replace them in a market economy. But they respond to incentives. Consequently, the task of an effective economic policy of regional authorities is to create a strong motivation to develop business right here and now [10].

Since the beginning of the 90s of the twentieth century, the region has developed a steady shortage of capacity for heat, electricity, water supply and sanitation. Over the past few years, funds from budgets of all levels have been allocated in insufficient quantities. This led to serious losses:

Table 1. Indicators of utility system functioning of Volgograd [11, p. 215].

Indicators	unit	power supply	heat supply	water supply	water disposal
Total network length	km	3 464	898	1 897	1 058
The length of the networks needing replacement	km	866	415	498	485
line 2 : line 1 x 100 %	*%*	*25.0*	*46.2*	*26.3*	*45.8*
The amount of leave in the network		1 973 949, thous. kWh	5 413, thous. Gcal	185 581, thous. cub. m	158 333, thous. cub. m
Total sales of goods and services		1 372 251, thous. kWh	4 417, thous. Gcal	131 361, thous. cub. m	104 149, thous.cub. m
The volume of losses	%	601 698, thous. kWh	996, thous. Gcal	54 220, thous. cub. m	
(line3 – line 4) : line 3 x 100 %		*32,2*	*18,4*	*29,2*	
Normative volume of losses		335 966, thous. kWh	708, thous. Gcal	33 405, thous. cub. m	
The actual specific energy consumption per unit of resource			106 614 kWh Gcal	652 kWh thous. cub. m	567 kWh thous. cub. m
Depreciation of utility infrastructure systems (network)	%	82	74	72	77
Number of accidents, t*otal * for a period of more than 8 hours*		66 *	136 *	4 614	7 121
System failure *(the ratio of the number of accidents to the length of networks)*		0.02 *	0.16 *	2.43	6.73
Loss ratio *(the ratio of the number of loss to network ratio)*		173.700, kWh km	1.109, Gcal km	28.582, cub. m km	

Source: according to municipal unitary enterprise "Volgograd Interdistrict Electric Networks", municipal unitary enterprise "Volgograd Utilities" and municipal unitary enterprise "City Water Channel of Volgograd"

One of the areas of training of modern engineering personnel can be an educational program for managers of housing and communal services. The Institute of Architecture and Construction of the Volgograd State Technical University has long developed such a program. To improve the methods of training engineering personnel for operation, engineers need serious economic training, especially in terms of tax and tariff accounting. The simplest calculation of thermal energy consumption in

IOP Conf. Series: Materials Science and Engineering **786** (2020) 012071 doi:10.1088/1757-899X/786/1/012071

accordance with the method of Methodical documents in construction 41-4.2000 (approved by order of the State Construction Committee of Russia of 06/05/2000 No. 15), Set of Rules and Regulations No. 23-01-99 of 01/01/2000, requires the specialist to know these standards requirements:

Consumption is calculated by the formula:

$$Q = q * \frac{t_{\text{вн}} - t_{\text{cpo}}}{t_{\text{вн}} - t_{\text{pa}}} * 24 * n \quad (1)$$

where

q – quantity of heat

t vn – internal temperature

t cp – average temperature

t pa – counting temperature

n – the number of days of the month counted in the calculation.

In our opinion, the formation of such direction of education as "Training of Specialists for Management Companies in the Municipal Sphere" and "Management of Housing and Communal Services" should be an important step in developing a regional development strategy for the executive authorities and local governments. The training of specialists in introducing innovations, including the housing and utilities sector, can be carried out at the expense of regional or local budgets, or with the use of subsidies from these budgets. The support of the authorities may also include financial, credit, property, technical, consulting, informational direction. The necessary methodological support includes training of specialists in the technical sphere, who:

- - can execute various agreements with all interested individuals and legal entities (owners, district administrations, land committees, etc.);
- - know the basics of accounting and management accounting and reporting;
- - can properly exploit the common property of the participants in the homeowners' partnership;
- - know the basics of team management.

The approximate list of blocks of the educational program is as follows:

Table 2. The approximate composition of the educational program "Management of housing and communal services".

The name of the topic, the name of the issues studied in lectures and seminars	Lectures, hour	Practical lesson, hour	Forms of control, hour
Module "Technical and operational training"			
Basic concepts and operating conditions for residential real estate			
Functional features of the operation of foundations, basements, attics, roofs, etc.			
Pipeline operation (heating, ventilation, water supply and drainage, sewage)			
Operation of power equipment in residential buildings			
Repair and reconstruction in residential buildings			
Module "Financial and tariff preparation"			
Regulatory base for calculating utility tariffs			
Calculations. Wages, charges, taxes, etc.			
The module "Legal training"			
Normative base of housing and utilities system			
Types and forms of contracts of various civil and legal orientation			
Module "Management Training"			
Analysis of specific business situations			

References

[1] Maznitsa E 2011 Opredelenie investitsionnogo i finansovogo potentsiala pri sozdanii malogo predpriyatiya [Definition of investment and financial potential when creating a small enterprise] «Narodnoe khozyaistvo» [National economy, Section "Proceedings of the IV International Scientific and Practical Conference" Modern Problems of the National Economic Complex "April 26-28, 2011": *MII Nauka Publishing House*, Moscow, **1** 211- 5

[2] Prilozhenie k Strategii razvitiya zhilishchno-kommunal'nogo khozyaistva v Rossiiskoi Federatsii na period do 2020 goda [Appendix to the Strategy for the Development of Housing and Communal Services in the Russian Federation for the Period up to 2020] Rasporyazhenie Pravitel'stva RF ot 26 yanvarya 2016 g N 80-r «Ob utverzhdenii Strategii razvitiya zhilishchno-kommunal'nogo khozyaistva v RF na period do 2020g» [Order of the Government of the Russian Federation dated January 26, 2016 N 80-p "On approval of the Strategy for the development of housing and communal services in the Russian Federation for the period up to 2020"] GARANT system: http://base garant ru/71313950/#ixzz5g5g9hQ6J (access date 02/19/2019)

[3] Maznitsa E and Brishten S 2013 Formirovanie sistemy pokazatelei i primenenie metoda analiza ierarkhii dlya opredeleniya effektivnosti innovatsii [Formation of a system of indicators and application of the hierarchy analysis method for determining the effectiveness of innovations] *Ekonomicheskii analiz – teoriya i praktika* [Economic analysis - theory and practice **13(316)** 24 - 33

[4] Maznitsa E , Arcov S and Fyedorovyh E 2004 Public-private partnership as basis of modernization of the housing sphere of russia / *Globalization and its socio-economic consequences, 15th International Scientific Conferenc*e , Proceedings, (Part III) - Rajecke Teplice, Slovak Republic, 7th – 8th October 1338-47 (Web of Science) Access mode: http://ke uniza sk/sites/default/files/content_files/proceedings_part_iii pdf

[5] Maznitsa E 2011 Perspektivy innovatsi-onnogo razvitiya regional'nogo ZhKKh na osnove tochechnogo (lokal'nogo) reformirovaniya: monografiya [Prospects for the innovative development of regional utilities based on point (local) reform: monograph] Volgograd: VolgGASU, 107

[6] Maznitsa E 2014 Razvitie svyazei vysshei shkoly i biznesa kak sposob sokhraneniya investitsionnoi aktivnosti v regionakh Rossii v usloviyakh finansovogo krizisa Aktual'nye problemy razvitiya vertikalnoi integratsiisistemy obrazovaniya, nauki i biznesa: ekonomicheskie, pravovy I sotsialnye aspekty [Development of relations between higher education and business as a way to preserve investment activity in the regions of Russia in the context of the financial crisis Actual problems of the development of vertical integration in the education, science and business systems: economic, legal and social aspects]:materials of the II International Scientific Practical ConferenceOctober 23-24, 2014 - T 2 / ed S L Igolkina – Voronezh *Voronezh scientific and technical information center* **243** 35 -43

[7] Pridachuk M P , Cyzho L N and Boldyreva M S 2019 The development of the system of organizational and environmental indicators for the development of the housing sector in Russia // International Scientific Conference "Competitive, Sustainable and Safe Development of the Regional Economy" The materials will be posted on the website of the publishing house Atlantis Press and indexed in Conference Proceedings

[8] Alkemade F , Heimeriks G , Schoen A , Villard L and Laurens P 2015 Tracking the internationalization of multinational corporate inventive activity: national and sectoral characteristics *Research policy,* ***44*****(9)**, 1763-72 doi: 10 1016/j respol 2015 01 007

[9] Bannò, M , Piscitello, L , &Varum, C (2013) Determinants of the Internationalization of Regions: The Role and Effectiveness of Public Policy Measures *Regional Studies,* ***49*****(7)** 1208-22 doi:10 1080/00343404 2013 821570

[10] Barzi F, Cortelezzi F, Marseguerr G and Zoia M G 2015 Determinants of the Internationalization of Regions: The Role and Effectiveness of Public Policy Measures INNOVATION-

MANAGEMENT POLICY & PRACTICE, Cortelezzi, Flavia/0000-0001-7481-1541, 10 1080/14479338 2015 1036541, 281-307[4] Begg, I (1999) *Cities and Competitiveness CURS Urban Stud Urban Studies* **36(5)** 795-809 doi:10 1080/0042098993222

[11] Maznitsa E, O V Maksimchuk N I and Borisova L N 2017 Chizho Modernizatsiya zhilishchno-kommunalnogo khozyaistva goroda na osnove realizatsii innovatsionnykh reshenii [Modernization of the Housing and Public Utilities of the City through the Implementation of Innovative Solutions] Volgograd State Technical University – Volgograd: VolgGTU

[12] Belyaeva L 2009 The level and quality of life *Challenges in measuring and interpreting Sociological Studies*, **1** 33-42

[13] Cardenas Davila, Cesar F ; Vargas Hernandez and Jose G 2015 Strategic-spatial analysis of the implementation of business opening politics of Mexico FAEDPYME INTERNATIONAL REVIEW-FIR JUL-DEC 8-25

[14] de Dominicis, Laura; Florax, Raymond J G M ; de Groot, Henri L F 2013 Regional clusters of innovative activity in Europe: are social capital and geographical proximity key determinants? *APPLIED ECONOMICS* 10 1080/00036846 **663474** 2325-35

[15] Kennedy A (2016) Smart Specialization, Regional Growth and Applications to European Union Cohesion Policy *REVIEW OF INTERNATIONAL POLITICAL ECONOMY*, **23(1)** 65-92 10 1080/09692290 1105845

IOP Conf. Series: Materials Science and Engineering **786** (2020) 012072 doi:10.1088/1757-899X/786/1/012072

Ways to improve safety and environmental friendliness of the city's transport system

I V Makarova*, L M Gabsalikhova, G R Sadygova and K A Magdin

Kazan Federal University, Russian Federation

*E-mail: kamIVM@mail.ru

Abstract. The transport sector is one of the largest environmental polluters and hydrocarbons consumers. The oil prices instability and the need to reduce the negative impact on the environment, cause countries to introduce alternative technologies. In the article, measures to reduce harmful substances emissions with vehicles exhaust gases are divided into three groups: constructive, organizational-technological, and organizational-managerial. The environmental friendliness of urban transport systems is ensured by an increase in the “green” vehicles share in the fleet. This electric transport: trams, trolley buses, electric buses, as well as vehicles with natural gas' engine. Reducing air pollution, noise and vibration make electric vehicles more attractive and competitive, but we must bear in mind that the silent movement of electric vehicles and electric buses may pose a threat to the pedestrian’s safety. Road traffic accidents (RTA) worsen the situation, as they aggravate the difficult traffic situation, causing congestion, traffic jams, and, consequently, increase the negative impact. The study authors selected the intersection Avtozavodsky Avenue - Syuyumbike Avenue (Naberezhnye Chelny), as a RTA concentration place in 2015-2019. We investigated emissions and noise pollution by road transport. In the article, we proposed measures aimed at reducing emissions and noise.

1. Introduction

Road transport is a source of chemical and physical pollution. Chemical pollution is caused by vehicles engines emissions. Physical pollution is caused by acoustic and electromagnetic fields arising during operation of vehicle's systems and aggregates, as well as from the interaction of its especial elements (tires, body) with the environment (road, atmosphere) when driving.

One of the transport situation degradation's factors in cities is the existing disproportion between the road network development and the vehicles fleet growth rate, which leads to traffic conditions degradation, congestion, increased delays, increased fuel consumption. This situation becomes characteristic of many large and medium-sized cities [1]. Along with the progressing atmospheric air quality reduction, the environmental problems associated with the motor vehicles noise impact on the environment are also increased. To solving the approaching environmental crisis humanity places great hopes on alternative energy sources and environmentally friendly transport [2]. The cities' transport complex is becoming more environmentally friendly by an increase in the “green” vehicles share in the fleet. This electric transport: trams, trolley buses, electric buses, as well as vehicles with natural gas' engine. Reducing air pollution, noise and vibration make electric vehicles more attractive and competitive transport mode. The environmental reasons for the use of vehicles on natural gas also are major [3]. To reduce the motor transport's negative impact on the environment, it is necessary to update the urban fleet by transiting public transport and municipal vehicles to environmentally friendly fuels (compressed and liquefied natural gas, liquefied petroleum gases), as well as created infrastructure for refueling these vehicles.

2. Existing Solutions in the Sphere of Increasing Safety and Environmental Friendliness of the City's Transport System

2.1 Methods to Reduce Emissions of Harmful Substances with Vehicle's Exhaust Gases

Measures to reduce harmful substances emissions with vehicles exhaust gases can be divided into three groups.

The first activities group is related to improving the vehicles and engines design, improving the fuels quality and includes:

1) changes in fuel supply processes to engine cylinders, including the use of electronic fuel injection;
2) provision of exhaust gas recirculation and mounting of microprocessor-based engine control systems;
3) application of gasoline and diesel fuels of high ecological purity, characterized by optimal fractional composition;
4) using of vehicles on alternative energy sources [4, 5].

The second group includes organizational-technical measures:

1) ensuring regular monitoring of the exhaust gases toxicity and smokiness;
2) limitations introduction to the vehicles movement on certain lanes at certain day hours;
3) route drivers' orientation;
4) fleet formation in accordance with the not exceeding emissions' conditions (forecast number, fleet structure by age and type of fuel; adjustment of the vehicles fleet numerical composition);
5) equipping the truck fleet with exhaust gas neutralization systems [6].

The third group – organizational-management activities - includes:

1) selection of optimal urban planning solutions, including the transport interchanges construction , underground crossings, the circular motion using;
2) traffic and its structure redevelopment for a long period and separate day hours [7, 8];
3) vehicles-free zones creation, primarily in the Central cities part;
4) introduction of paid travel during peak hours in urban areas with the highest traffic intensity;
5) location of public transport stops optimization;
6) using to traffic organization the automated traffic management systems in combination with the air pollution control system [9];
7) improving the traffic lights efficiency;
8) introduction of one-way traffic schemes.

2.2 Methods which Used to Protect from Transport Noise

Road transport is the noise impact main source on the environment, which intensity increases in dense urban construction conditions. Many researchers note that in the overwhelming majority of cases, it is traffic noise that determines the city's noise background.

Transport noise has a negative impact on the person's basic organs and systems: the hearing aid, the central nervous system, the cardiovascular system, the vision organs, the immune system. The constant impact of noise high levels, adversely affecting human health, reducing his work productivity, creative activity and the rest effectiveness, causes significant damage to society and the state [10]. In the works

[11, 12] it is indicated that the constant impact traffic noise increases the risk of developing high pressure, hypertension.

By the main methods allowing to reduce the traffic flows noise impact are: architectural-planning solutions; design-technological and organizational-administrative methods.

Architectural-planning solutions include: the buildings location at the maximum possible distance from highways, urban railways, no closer than 100 m to them; the use of new soundproofing building materials, triple windows glazing or double-glazed windows, windows sealing; landscaped areas. An effective measure is the use of noise screens, which are sound-reflecting, sound-absorbing and combined.

Design and technological measures include the improvement of vehicle design and infrastructure solutions.

Organizational-administrative measures provide for the traffic flow differentiation according to place, time and its qualitative composition. Reserve of decrease in traffic noise levels, especially on high-traffic lines, is the flow separation of trucks and cars, with adequate segregated lanes provision.

2.3 Prospects for Market Development and Ways to Solve the Problems With Electric Buses Safety Operation in Russia

Buses are vehicles that create a noise high level in urban areas. The article authors [13] are investigating the possible noise extent reduction in urban areas through electric buses. Taking into account the traffic peculiarities on different bus routes, the authors found that the noise from electric buses is practically not reduced on roads with heavy traffic.

The paper authors [14] analyzed the societal benefits derived from noise reduction through the electric buses using in the Gothenburg city. The results show that the largest benefits from electrical buses are obtained during acceleration, for example, at bus stops, and for maximum levels indoors.

Electric transport can actively develop not in all segments. In freight traffic, where there are no standard routes, the development of charging infrastructure for electric transport will be difficult for a long time. From this point of view, this direction is more promising in the city buses segment.

Electric buses sales will depend primarily not on private customers, but on federal, regional, municipal financing. Thus, the article authors [15] studied the key factors hindering the electric buses use as passenger transport, such as their cost, lack of charging infrastructure. The findings also indicate that the electric buses introduction is more dependent on state support.

The papers authors [16, 17] believe that the main attention in the electric bus market development should be given to a key requirement for the urban automobile transport electrification: the charging infrastructure development for refueling the city buses network.

Traffic noise in urban areas is a key factor for informing road users. The condition of the road surface, the type of vehicle and the age of the pedestrian affect the reactions timeliness to avoid collisions [18].

A high percentage of accidents are caused by electric and hybrid vehicles [19]. In the authors' study [20, 21] the main problem reported by all drivers was related to the electric buses noiselessness for pedestrians.

For preventing impact crashes, the electric vehicles with pedestrians it is necessary to install noise generators [22].

3. Results and discussion

Naberezhnye Chelny city with over half a million people population is largest industrial center of Zakamsky region Tatarstan Republic. The city is characterized by a high level of road transport networks development. During the problem area study, an electronic map of the RTA in the Naberezhnye Chelny city (based on the results of 2015-2019) was studied. As a RTA result, concentration points were identified, as a which result traffic along one or several lanes was blocked for a long time, as led to traffic difficulties and jams on this road network part. Thus, according to the analysis results, the

IOP Conf. Series: Materials Science and Engineering **786** (2020) 012072 doi:10.1088/1757-899X/786/1/012072

intersection Avtozavodsky Avenue - Syuyumbike Avenue (24 RTA) was chosen as the object under study. Field measurements were carried out by visual observation. As these observations result, a time was detected, during which at the selected intersection there is the greatest vehicles accumulation. Peak time were divided into: morning high hours - from 6:30 to 7:30; evening high hours - from 16:30 to 18:00.

3.1 Calculation of Motor Transport Emissions

For the road (or its section) in the presence of an adjustable intersection, the total pollutants emission, M, g / km, is calculated by the formula:

$$M = \sum_1^n(M_{P1} + M_{P2}) + \sum_1^{n_1}(M_{L3} + M_{L4}) + \sum_1^m(M_{P3} + M_{P4}) + + \sum_1^{n_1}(M_{L1} + M_{L2}) \quad (1)$$

where, M_{P1}, M_{P2}, M_{P3}, M_{P4} - is the pollutants emission into the atmosphere by vehicles in the intersection area when the traffic light signal prohibits vehicles from passing, g / km;

M_{L1}, M_{L2}, M_{L3}, M_{L4} - pollutants emission into the atmosphere by vehicles moving along this road during the period under consideration, g / km.

Here indices 1 and 2 correspond to each of the two movement directions on a highway with greater traffic intensity, 3 and 4 - for a highway with a lower traffic intensity.

n, m - is the vehicles flow stops number before the intersection on the roads forming it for a 20-minute time period.

n_1, m_1- - the periods number of vehicles flow movement in the intersection area at allowing traffic signal for a 20-minute period of time.

The i-th pollutant emission by particular direction vehicles movement in the intersection area with traffic signals prohibiting traffic for a 20-minute additional inspection period M_{Pi}^3, g / km, is calculated by the formula (2):

$$M_{Pi}^3 = \frac{P_c}{60}\sum_1^{N_c}\sum_1^k(M'_{\Pi i,k} * G_k) \quad (2)$$

where P_c - the duration of the traffic light prohibitory signal (including yellow) within 20 minutes, s;

N_c- the action cycles number of traffic light signal prohibiting the movement within 20-minute period of time, s;

M'_{ik}- i-th pollutant's specific emission by k-th group's vehicles, which are in the queue at the traffic light prohibiting signal, as show in Table 1, g / min;

G_k- the k-th group vehicles amount queuing around the intersection at the each cycle end of the traffic light signal action prohibiting the movement.

Table 1. Specific pollutant emissions values for vehicles in the crossroad area.

Car group name	Emissions of pollutants, g / min						
	CO	NO2	CH	Soot	SO2	Formaldehyde	benzapilene
Cars	0,5	0,015	0,1	0,015	0,5×10-2	0,4×10-3	0,15×10-6
Trucks and buses total weight up to 3,5 t	2,0	0,04	0,3	0,08	0,9×10-2	1,4×10-3	0,4×10-6
Trucks from 3,5 to 12 t	2,5	0,12	0,66	0,9	1,7×10-2	7,2×10-3	1,1×10-6
Trucks over 12 t	2,7	0,14	0,83	1,1	2,4×10-2	9,5×10-3	1,3×10-6
Buses over 3,5t	1,9	0,1	0,57	0,67	1,5×10-2	4,8×10-3	0,9×10-6

To reduce the negative impact on motor vehicles emissions in the Naberezhnye Chelny city, the following measures are recommended:

1) The optimal town-planning solutions choice which provide for the underpasses and traffic interchanges construction.

2) Control's strengthening on the toxicity and smoke content of vehicles exhaust gases on highways, herewith, special attention should be paid to outdated vehicles models, especially trucks

3) Routing drivers' orientation, transit transport permits optimization.

4) Improving the traffic lights efficiency.

5) Upgrading the urban fleet by transiting public and municipal transport to environmentally friendly fuels, as well as creating infrastructure for refueling these transport types.

3.2 Calculation of Motor Transport Noise Impact in Cities

To measure the sound pressure at the intersection, a ShI-01 type noise meter-vibrometer was used. The noise levels at the intersection were measured under the following climatic conditions: precipitation lack in the form of rain, snow; atmospheric pressure - 1.013 Pa (760 mmHg), permissible variation ± 5%; ambient air temperature - not lower than -5 and not higher than +30 ° C; wind speed - no more than 5 m/s.

The traffic flows noise characteristics at the intersection were measurements in four points. Table 2. shows the noise measurements results in the morning and evening pick hours at the Vakhitov - Suyumbike intersection.

Table 2. Noise measurement results.

Time	Vehicles number	Minimal noise	Average noise	Maximum noise
6:30-6:50	1540	69,2	79,5	86,4
6:50-7:10	1535	69,8	78,1	88,3
7:10-7:30	1508	70,0	79,9	88,6
7:30-7:50	1473	70,3	78,1	86,4
16:30-16:50	1734	71,6	80,1	85,2
16:50-17:10	1773	72,2	79,8	88,5
17:10-17:30	1759	71,4	79,6	88,1
17:30-17:50	1719	71,7	79,7	87,6

Were also measured noise from a single car. The noise was recorded at different its movement speeds. The experiment results are shown in Table 3, from which it follows that the noise varies in proportional to the vehicle speed.

Table 3. Experiment results.

Speed, in km per hour	20	40	60	80
Noise, dB-A	64,5	73,8	84,2	93,1

To reduce noise pollution from motor vehicles in the Naberezhnye Chelny city, we can offer the following ways:

1) Introduction of new technologies to road building and pavements repair.

2) Priority public transport development in order to reduce the road network load, negative effect on the environment and improve the citizen's life quality. Public transport popularization among residents. Urban public transport modernization.

3) Popularization and development of cycling, bicycle lines building, the rental bicycles organization.

4) Installation of acoustic screens along most busy highways.

4. Conclusion

The growth in the vehicles fleet together with a relatively unaltered transfer capacity city streets led to a decrease in the average traffic speed, which resulted in an increase in fuel consumption and time spent on transportation, growth in the RTA likelihood and number, an increase in environmental pollution from vehicles exhaust gases, and a degradation in emotionally-psychological population condition and other negative consequences. Existing methods to reduce vehicle emissions and noise exposure were analyzed.

The use of alternative environmentally friendly motor fuels types is part of the design-technological measures to reduce emissions from road transport. Most automotive companies in the world see the future in the energy-saving technologies application and the vehicles transfer to electrical transmission using environmentally friendly energy systems. However, some studies confirm the increase in the accidents number involving pedestrians due to the low noise level of vehicles on electric propulsion, therefore additional protective measures are necessary.

The electric buses role in the public transport system is important as a counter-measure to climate change and to negative environmental impacts when using non-renewable natural resources as motor fuel. Gas-powered buses and trucks are also a promising transport mode, since they have a high return on investment due to low fuel prices and lower operating costs, while gas is the most environmentally friendly fuel type. Although electric buses and electric cars are an environmentally friendly transport type, it cannot be considered as an alternative to gas because it is about 3 times more expensive, besides, it is necessary to create charging infrastructure for electric transport, while a gas stations network already exists in most Russian regions. Electric buses can only afford large cities.

As a research result conducted in the Naberezhnye Chelny city, proposals were made for organizing events aimed at reducing the road transport negative impact on the environment. Calculated values of pollutants specific emissions for vehicles in the intersection zone. To measure the sound pressure at the intersection, a ShI-01 type noise meter-vibrometer was used. Moreover, we measured noise from a single car. The noise was recorded at different vehicle speeds. The results show that the noise varies in directly proportional to the vehicle speed.

References

[1] Makarova I, Khabibullin R, Belyaev A and Belyaev E 2013 Urban transport system management in the context of region sustainable development strategy *Transport Problems* **8(2)** 107-11

[2] Khan M 2017 Identifying and addressing barriers for the sustainable development of natural gas as automotive fuel Int. *J. of hydrogen energy* **42** 25453– 73

[3] Engerer H and Horn M 2010 Natural gas vehicles: *An option for Europe Energy Policy* **38** 1017– 29

[4] O van Vliet, A S Brouwer, T Kuramochi, Ma van den Broek and A Faai Energy use, cost and CO2 emissions of electric cars // *Journal of Power Sources* **196(4)**, 2298-310

[5] Thomas C 2009 Fuel cell and battery electric vehicles compared Int. J. *of Hydrogen Energy* **34** 6005–20

[6] Alkemade U and Schumann B 2006 Engines and exhaust after treatment systems for future automotive applications *Solid State Ionics* **177** 2291– 96

[7] Makarova I, Khabibullin R, Belyaev E and Mavrin V 2016 Increase of City Transport System Management Efficiency with Application of Modeling Methods and Data Intellectual Analysis. In: Sładkowski A., Pamuła W. (eds) Intelligent Transportation Systems – Problems and Perspectives. Studies in Systems, Decision and Control, **32** 37-80

[8] Makarova I, Shubenkova K, Mavrin V and Magdin K 2019 Environmental safety of city transport systems: Problems and influence of infrastructure solutions *Lecture Notes in Networks and Systems* **68** 24-34

[9] Pernebekov S, Sadykov Z, Meiirbekov A and Tortbayeva D 2014 Modeling of traffic flows with due regard to ecological criteria *Life Science Journal*, **11(5)** 300-2

[10] Vienneau D, Schindler C, Perez L, Probst-Hensch N and Röösli M 2015 The relationship between transportation noise exposure and ischemic heart disease: *A meta-analysis Environmental Research* **138** 372– 80

[11] Babisch W, Wolke G, Heinrich J and Straff W 2014 Road traffic noise and hypertension – *Accounting for the location of rooms Environmental Research* **133** 380– 7

[12] Barregard L 2011 *Traffic noise and hypertension Environmental Research* **111** 186– 7

[13] Laib F, Braun A, Rid W 2019 Modelling noisė reductions using electric buses in urban traffic. A case study from Stuttgart *Transportation Research Procedia* **37** 377– 84
[14] Larsson K and Holmes M 2017 Social benefit analysis of reduced noise from electrical city buses in Gothenburg The *J. of the Acoustical Society of America* **141** 3733
[15] Mohamed M, Ferguson M and Kanaroglou P 2017 What hinders adoption of the electric bus in Canadian transit? Perspectives of transit providers Transportation Research D
[16] Xylia M, Leduc S, Patrizio P, Kraxner F and Silveira S 2017 Locating charging infrastructure for electric buses in *Stockholm Transportation Research* **78** 183– 200
[17] Wang Y, Huang Y, Xu J and Barclay N 2017 Optimal recharging scheduling for urban electric buses: A case study in Davis *Transportation Research* E **100** 115– 32
[18] Mendonc C, Freitas E, Ferreirac J, Raimundoc I and Santos J 2013 Noise abatement and traffic safety: The trade-off of quieter engines and pavements on vehicle detection *Accident Analysis and Prevention* **51** 11– 7
[19] Garay-Vega L, Hastings A, Pollard J, Zuschlag M and Stearns M 2010 Quieter Cars and The Safety of Blind Pedestrians: Phase I Natl. Highw. Transp. Saf. Agency 1–151
[20] Woodcock A, Topolavic S and Osmlond J 2013 The multi-faceted public transport problems revealed by a small scale transport study Advances in Human Aspects of Road and Rail Transportation 321– 31
[21] Cocron P and Krems J 2013 Driver perceptions of the safety implications of quiet electric vehicles *Accident Analysis & Prevention* **58** 122– 31
[22] Gabsalikhova L, Sadygova G and Almetova Z 2018 Activities to convert the public transport fleet to electric buses *Transportation Research Procedia* **36** 669– 75

IOP Conf. Series: Materials Science and Engineering **786** (2020) 012073 doi:10.1088/1757-899X/786/1/012073

Development of transport infrastructure in smart cities

I Makarova*, A Boyko, I Giniyatullin and A Ahmadeeva

Department of Transport Systems Service, Kazan Federal University, Russia, Kazan, Kremlevskaya street, 18

* E-mail: kamivm@mail.ru

Abstract. Smart Mobility is one of the main directions of development of the transport system in intelligent cities. In this case, along with management intellectualization, the issue of transition to "green", safe and sustainable modes of transport, such as bicycles, should be resolved. Democracy and availability, cheapness and environmental friendliness, promotion of a healthy lifestyle are the reason for the growing popularity of this type of transport throughout the world. This document proposes one of the possible ways to develop bicycle transport in a smart city, which allows you to expand the number of users by reducing the physical requirements for cyclists. The proposed development is the concept of Smart Bike, which controls the condition of the cyclist and the environment and turns on the electric motor in critical situations. This reduces the cyclist's physical activity, as well as the consumption of the battery, which has a positive effect on the ecology of Smart Cities.

1. Introduction

The world is undergoing the largest wave of urban growth in history. More than half of the world's population now lives in towns and cities, and by 2030 this number will swell to about 5 billion. Much of this urbanization will unfold in Africa and Asia, bringing huge social, economic and environmental transformations.

UN projections show the world's rural population has already stopped growing, but the world can expect to add close to 1.5 billion urbanites in the next 15 years, and 3 billion by 2050. How the world meets the challenge of sustainable development will be intimately tied to this process.

In 1950 about 2/3 of the population worldwide lived in rural settlements and 1/3 in urban settlements. By 2050, we will observe roughly the reverse distribution, with more than 6 billion people living in the messy, burgeoning athmosphere of urbanized areas.

According to the Sustainable Urbanization Policy Brief, urban centres currently occupy less than 5% of the world's landmass. Nevertheless, they account for around 70% of both global energy consumption and greenhouse gas emission. Innovation in urban infrastructure and technology is essential when addressing this issue. For instance, greenhouse gas emissions could be reduced by up to 1.5 billion CO_2e annually by 2030, primarily through transformative change in transport systems in the world's 724 largest cities.

Urban planning decisions and strategic design thinking in the context of rapid urbanization account for social equity, mobility patterns, global competitiveness and energy-efficiency.

Strain on traditional service delivery from increasing populations poses a significant challenge to the sustainable growth of cities. Managing physical resources, such as energy or water as well as managing healthcare, traffic and city logistics, among others, is a major challenge that cities need to address to ensure sustainable growth in the future.

Smart City is an urban environment, which combines a variety of technologies to reduce the negative impact on the environment in general and, thereby, to provide a more comfortable living conditions. The main goal of the Smart City idea is to create a sustainable model of urban development and to preserve the quality of life for urban citizens. One of the main challenges to be solved for the successful operation of the urban ecosystem is the problem of citizens' mobility. At the same time, the transport system should cause minimal negative impact on the environment.

2. General trends in the development of smart cities

Cities are hubs for ideas, commerce, culture, science, productivity, social development and much more. At their best, cities have enabled people to advance socially and economically. But, many challenges exist to maintaining cities in a way that continues to create jobs and prosperity while not straining land and resources. Urban challenges include congestion, lack of funds to provide basic services, a shortage of adequate housing and declining infrastructure. The challenges cities face can be overcome in ways that allow them to continue to thrive and grow, while improving resource use and reducing pollution and poverty.

In November 2008, IBM's Chairman, CEO and President Sam Palmisano, during a speech at the Council on Foreign Relations, outlined a new agenda for building a "Smarter Planet". The speech emphasized how the world's systems and industries are becoming more instrumented, interconnected and intelligent, and that leaders and citizens can take advantage of this state of affairs to improve these systems and industries.

In January 2010 Sam Palmisano gave a follow-up speech to the Chatham House called the "Decade of Smart". He highlighted dozens of initiatives in which leaders created smarter systems to solve the planet's most pressing problems. The speech aimed to inspire others to follow the leads of these innovators by helping to create a smarter planet.

The first question is what is meant by a 'smart city'. The answer is, there is no universally accepted definition of a smart city. It means different things to different people. The conceptualisation of Smart City, therefore, varies from city to city and country to country, depending on the level of development, willingness to change and reform, resources and aspirations of the city residents. To provide for the aspirations and needs of the citizens, urban planners ideally aim at developing the entire urban eco-system, which is represented by the four pillars of comprehensive development-institutional, physical, social and economic infrastructure. This can be a long term goal and cities can work towards developing such comprehensive infrastructure incrementally, adding on layers of 'smartness'.

Some typical features of comprehensive development in Smart Cities are described below.

1) Promoting mixed land use in area based developments in order to make land use more efficient;

2) Housing and inclusiveness - expand housing opportunities for all;

3) Creating walkable localities –reduce congestion, air pollution and resource depletion, boost local economy, promote interactions and ensure security. The road network is created or refurbished not only for vehicles and public transport, but also for pedestrians and cyclists, and necessary administrative services are offered within walking or cycling distance;

4) Preserving and developing open spaces - parks, playgrounds, and recreational spaces in order to enhance the quality of life of citizens, reduce the urban heat effects in Areas and generally promote eco-balance;

5) Promoting a variety of transport options: Transit Oriented Development (TOD), public transport and last mile para-transport connectivity;

6) Making governance citizen-friendly and cost effective - increasingly rely on online services to bring about accountability and transparency, especially using mobiles to reduce cost of services and providing services without having to go to municipal offices. Forming e-groups to listen to people and obtain feedback and use online monitoring of programs and activities with the aid of cyber tour of worksites;

7) Applying Smart Solutions to infrastructure and services in area-based development in order to make them better. For example, making Areas less vulnerable to disasters, using fewer resources, and providing cheaper services.

ITS covers a number of types of transport including rail, freight and car and the associated technologies are a plenty. Today we start building our future vision of an ITS to not only run efficient transport networks, but also have access to predictive analytics to make better decisions with more accuracy, faster. This can benefit future services planning based on population trends and land zone changes, as well as having access to behavioural analytics from ticketing systems to aid flow and manage demand, provide specialized services to segmented users' groups and even realignment of services based on customer insights.

3. Smart city and smart mobility

A "Smart City" is an urban region that is highly advanced in terms of overall infrastructure, sustainable real estate, communications and market viability. It is a city where information technology is the principal infrastructure and the basis for providing essential services to residents.

Transport system is one of the major intellectual systems in the Smart City. To ensure its sustainability and safety, the work is being done in three ways: smart infrastructure, smart vehicles and smart users. Solutions concern the creation of an efficient and integrated mobility system that allows for organizing and monitoring seamless transport across different modes, increasing the use of environmentally friendly alternative fuels and creating new opportunities for collective mobility.

Addressing the mobility challenge calls for a paradigm shift in urban planning, encouraging compact cities as a way to increase accessibility and to reduce the need for transportation altogether. Because the purpose of mobility is to gain access to destinations, activities, services and goods, urban planning should therefore be resident-centred, so that functional endpoints – the reasons for travel – are as close as possible to each other, in effect reducing distances and transportation needs. Thus, city residents should be able to address their needs using as little travel as possible.

Bicycle infrastructure planning should include the creation of bike parking's, bike sheds and bikeways as well as it should be taken into account the terrain and the structure of population, who want to use the bike to get around the city. Despite a fast growing literature on the bike lanes design the problem of terrain identification and topographic conditions modeling is still actual. The most common method of bicycle wayfinding is the shortest path method. As far as bicycle routing is not always possible to avoid hilly terrain, bike-lifts and electric drives creation can solve the problem of overcoming steep climbs.

Effective segregation of road traffic is central to most campaigners' demands for greater safety, and SkyCycle is one of the most extreme methods of separating bicycles from cars on the agenda. By building a new network of cycle paths high above existing roads and railways, the scheme would create a series of cycle superhighways across London, with specific entry and exit points throughout the city. "Any long distance bicycle commute would be less likely to encounter large heavy goods vehicles if we are able to build a few of these elevated cycleways", says Sam Martin, founder of Exterior Architecture, the firm behind the designs [2].

Brian Macdowall, of the Alliance for British Drivers, thinks the project could work in principle. The problem with inserting cycle lanes at the roadside usually means narrowing the road itself, which can

IOP Conf. Series: Materials Science and Engineering **786** (2020) 012073 doi:10.1088/1757-899X/786/1/012073

aggravate congestion. "If the cycle lanes could run along major routes without loss of road space to motorists, then it would help". He is concerned that the cost would prevent the plans becoming a reality, however. Martin estimates the initial cost at £200m for the first 6.5km artery [3]. To solve transport problem in large cities Dominican architect Richard Morita Castillo offers to build dedicated lanes for cyclists . Innovation lies in placing these lanes over city highway. Such overpasses for cyclists Castillo calls ecobahns (like autobahn) or Cycling underground. A similar solutions are proposed for Moscow and Kazan. And for Russia it is not only about the division of space and reduction the number of conflict points, but it is also necessary protection for cyclists in difficult weather conditions.

4. Development of DSS for assessing the effectiveness and safety of infrastructure projects

The development of cycling in cities will lead to an increase in the number of cyclists and the size of their transport work. Therefore, safety requirements for bicycle paths will constantly increase. In this regard, it is important, based on interdisciplinary research, to establish and evaluate the values of safety indicators at different stages of automobile, cycling, and pedestrian traffic from beginning to end the route, i.e. when potentially there is a risk of accidents. The authors of the article believe that the safety of cycling in an urbanized area can be assessed by the magnitude of the hazard level σ, which reflects the likelihood of accidents involving cycling. When $\sigma = 1$, an accident is unavoidable; if $\sigma = 0$, the probability of an accident is zero. The degree of danger is affected by: the presence of intersections, the frequency and speed of movement of vehicles, the type of bike lane, as well as its width.

The safety level of cyclists depends on a variety of statistical and dynamic factors:

1) The presence of isolated bike paths in the city;
2) The need and ability to use the bike in conjunction with public transport;
3) Availability of bicycles (personal and rented);
4) Individual characteristics of bicycles (average speed of overcoming a route, travel time, average speed of movement, distance to the onset of fatigue and the need for repair, time for rest and repair);
5) Transport behavior (the presence of a bicycle helmet, compliance with traffic rules, conflict with other road users - car drivers, cyclists, pedestrians).

When drawing up routes of urban movements, it is necessary to choose the safest possible route, taking into account the peculiarities of the traffic participant. Categorizing a route can be performed by the presence of various factors complicating traffic conditions. Depending on the type of travel, the routes in the city can be road, pedestrian, bicycle and combined. The combined route, as a rule, includes the areas where the movement is carried out by public transport, and the movement between the starting (end) point and the stop of public transport is on foot or by bicycle. These sites are assessed using the multi-criteria route safety assessment methodology developed by us. The enlarged diagram of the assessment algorithm is shown in Fig.1.

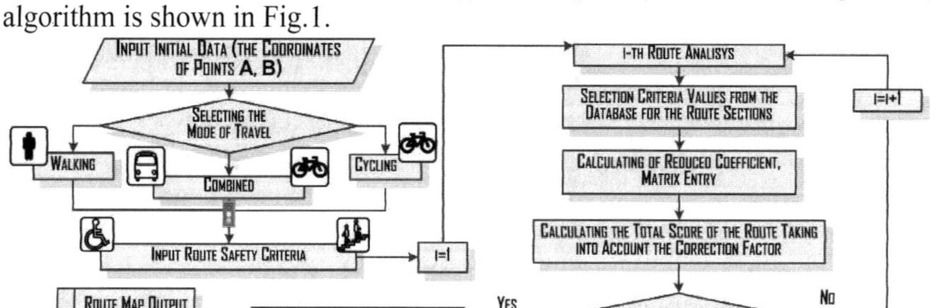

Figure 1. Assessment algorithm.

At the first stage, factors that determine the category of the route by complexity for pedestrians and cyclists were identified. Since the significance of a factor for different categories of road users is determined by their characteristics, to determine the weight of factors, categories of pedestrians and

IOP Conf. Series: Materials Science and Engineering **786** (2020) 012073 doi:10.1088/1757-899X/786/1/012073

cyclists were identified by age and health status. The route can be estimated using a complex indicator, which is calculated using the weighted average formula: (1)

$$K = \sum_{i=1}^{n} \frac{K_i \cdot \alpha_i}{\alpha_i} \quad (1)$$

Here Ki is the value of the i-th parameter, i is the weight of the indicator. In order to summarize dissimilar indices in the formula their reduced values are calculated.

Factors that determine the safety of the route can be both objective (eg, terrain, presence of unregulated intersections, etc.), and subjective due to the features and physical condition of the road user (age, health condition, etc.). Adequacy of the assessment will depend on the correctness of the selected factors and their combined inclusion. For example, the same route can be safer in the daylight than in the dark, in the summer than in the winter, etc.

Since a significant part of moving around the city is made up of mixed routes, the transfers number on the route should be estimated, because it is the interchange points safety on such routes that can be decisive. In articles of researchers studying ways to improve the safety of cyclists, existing bike routes are evaluated. We propose to predict the effectiveness of projects, choosing from several options that which will initially be most appropriate in each case considered. For these purposes, it is proposed to develop a control decision support system, the intellectual core of which will be imitative models. A conceptual model of a decision support system is shown in Figure 2.

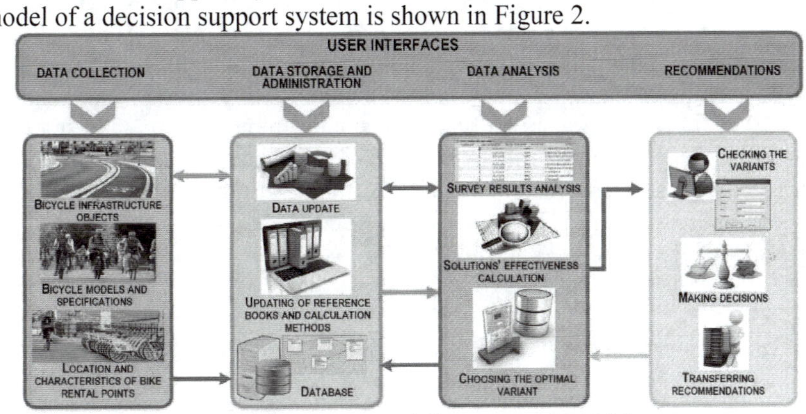

Figure 2. Conceptual diagram of the DSS.

The correctness of the work of the DSS is largely determined by the quality of the source information and the adequacy of the method of its processing. This is ensured by the presence of:

1) a module for collecting, storing and administering data on the quantitative and qualitative properties of bicycle infrastructure objects, parameters of the transport system, as well as monitoring data on the parameters of cycling and pedestrian flows;

2) module for analysis of statistical and dynamic data;

3) a module for decision making with the purpose of analyzing and evaluating data and the subsequent development of recommendations (calculation of the track width based on passenger traffic and throughput of the existing bicycle infrastructure; justification of the need for other infrastructure facilities);

4) module for development of simulation models before / after changes in accordance with the received recommendations).

The developed module (figure 3) for choosing the optimal route will be included in the composition of this system. With it help, it will be possible to assess the security of both existing and projected infrastructures.

IOP Conf. Series: Materials Science and Engineering **786** (2020) 012073 doi:10.1088/1757-899X/786/1/012073

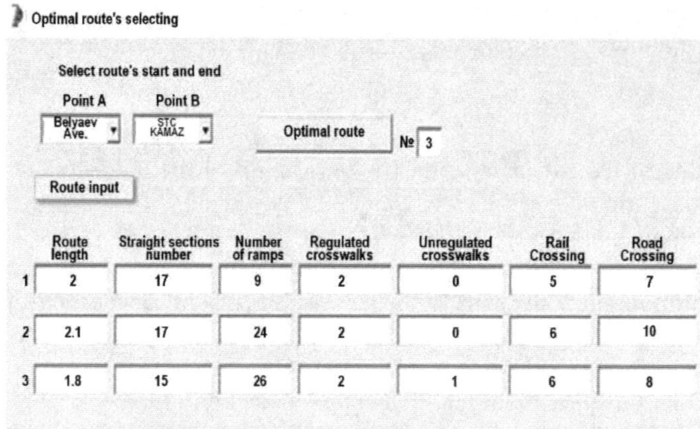

Figure 3. View of the developed module.

5. Conclusion

Transition to eco-friendly and non-motorized modes of transport is one of the key elements in ensuring sustainable urban mobility. It also contributes to improving the environment, reducing emissions of CO2, NOX, CH, PM. There will be a decrease in the load on the transport system of the smart city. In addition, the experience of developed countries shows that environmentally friendly and efficient mode of transport also contributes to the development of the economy. In addition, health and longevity also benefit from cycling. However, there are still many obstacles to the widespread use of bicycles as an alternative mode of movement. Some of these problems can be solved by integrating the bicycle with public transport. This is possible by creating bicycle paths and parking lots. Successful development of Smart City needs complex solutions. This will create a comfortable urban environment for citizens, as well as get a synergistic effect that will contribute to the sustainable development of Smart City.

References

[1] Un environment, summary of the sixth global environment outlook geo – 6, regional assessments: key findings and policy messages, unep/ea 2/inf/17, available: http://web unep org/geo/assessments/regional
[2] Daily mail reporter, "skycycle the elevated london bike network", london media, available: http://www constructionchat co uk/articles/skycycle-london-bike-network/, 4 september 2012
[3] Sam judah, 8 radical solutions to protect cyclists, bbc news magazine, available: http://www bbc com/news/magazine-24998730, 20 november 2013
[4] F R whitt and D G Wilson 2004 bicycling science, 2nd ed cambridge, ma: the mit press, 364
[5] D G wilson and J Papadopoulos 2004 bicycling science, 3d ed cambridge, ma: the mit press, 476
[6] J Parkin and J Rotheram 2010 "design speeds and acceleration characteristics of bicycle traffic for use in planning, design and appraisal", *transport policy*, **17(5)** 335-41
[7] J larsen et al 2013 "build it but where? the use of geographic information systems in identifying locations for new cycling infrastructure", *international journal of sustainable transportation*, **7** 299-317
[8] A Forsyth and K Krizek 2011 "urban design: is there a distinctive view from the bicycle?", *journal of urban design*, **16**, 531-49
[9] G Rybarczyk, "simulating bicycle wayfinding mechanisms in an urban environment", urban, planning and transport research: *an open access journal*, **2** 89-104
[10] "Bici-metro": an alternative project to sky cycles is available for global implementation! available: https://05dimensions wordpress com/2012/10/22/bici-metro-an-alternative-project-to-sky-cycles-is-available-for-global-implementation/

[11] Varlamov 2014 “bicycles’ roads”, live journal, available: http://varlamov ru/979334 html
[12] Aytbagin 2014 “kazan architects offered the alternative bicycle lanes: velopoliten”, inkazan journal, available: http://inkazan ru/news/society/21-01-2014/alternativu-velodorozhkam-predlozhili-kazanskie-arhitektory-velopoliten?slug_for_redirect=2014%2f01%2f21%2 falternativu-velodorozhkam-predlozhili-kazanskie-arhitektoryi-velopoliten
[13] Trofimenko Y and Shashina E 2017 methodology and results of assessing safety of bicycle infrastructure in russian cities transportation research procedia, 19-20
[14] Makarova I , Pashkevich, A and Shubenkova K 2018 safe routes as one of the ways to re-duce the number of road accidents victims lecture notes in networks and systems 21(73-84)

IOP Conf. Series: Materials Science and Engineering **786** (2020) 012074 doi:10.1088/1757-899X/786/1/012074

Alternative power resources in the motor vehicles supply

I V Makarova[1,*], L A Gubacheva[2], S Andreev[3] and A Andreev[2]

[1]Kazan Federal University, Naberezhnye Chelny, Russia

[2]Vladimir Dahl Luhansk National University, Lugansk, LPR

[3]Department of Public Administration and Management, State Telecommunication University, Kyiv, Ukraine

E-mail: kamivm@mail.ru

Abstract. The rapid development of engineering, technology and mass motorization during the transition to the fourth industrial revolution is accompanied by negative processes in ecosystems and leads to a change in the economic paradigm - to the circular economy. Reducing resource consumption while decreasing the growth of waste production and consumption will reduce the negative load on the environment. The search for global solutions to ensure comfortable living conditions on the planet is determined by minimizing the adverse environmental impact of solid industrial and household waste (MSW). The article presents a new technology of energy- and resource saving in automobile transport using MSW as an alternative fuel and exhaust gas cleaning filters for internal combustion engines, that will allow to reduce pollution by the slowly decomposing polyethylene of the world ocean, which is now becoming a global scale catastrophe.

1. Introduction

Industry 4.0 assumes the rational use of natural and technical resources. Energy from waste in the programs Industry 4.0 is an important factor in improving the ecological purity of the environment and the quality of life of the population of cities [1-4].

The fourth industrial revolution in Europe is taking place under conditions of accelerated growth in the vehicles fleet with internal combustion engines, especially diesel ones, which causes a significant deterioration in air quality in urban agglomerations. The most negative impact on the state of the air environment have emissions of nitrogen mono- and dioxide in the exhaust gases of internal combustion engines, including those using natural gas as a fuel. Nitrogen oxides, possessing high toxicity, are able to form even more toxic complexes and play a significant role in the formation of photochemical red fogs (smog), which is observed in many cities with developed industry (especially in summer with low airflow through streets and avenues) [5-10]. Hazardous chemicals and other pollutants (for example, plastic waste) are still produced in large volumes. They are universally present in the human body and the environment and accumulate in stocks of raw materials and products. According to the World Health

Organization, in 2016, the burden of disease due to exposure to individual chemicals was estimated at 1.6 million lives (this estimate is likely to be underestimated).

Research of United Nations Environment Assembly and partner's estimations showed that in 2014 about 280 million tons of plastic (polyethylene (PE), polypropylene (PP) and polyvinyl chloride (PVC)) were produced, and only a small percentage of its waste reused. Instead, plastic, which has lost its consumer properties, found its last refuge in the world ocean, thus causing damage to marine ecosystems, equal to several billion dollars annually, since they are not biodegradable in the marine environment. It was found that under marine conditions, fragmentation occurs rather slowly and can take up to 5 years, during which the ocean continues to be fouled with plastic garbage (Figure 1). In recent years, plastic micro-particles (0.1–5 mm in diameter) are becoming more and more disturbing, since they often become food for marine organisms, including birds, fish, mussels, worms, and zooplankton [11-12].

The fourth session (Nairobi, 11–15 March 2019) of the United Nations Environment Assembly made disappointing key findings on the Global Chemicals Outlook II. The global goal to minimize the adverse effects of chemicals and waste will not be achieved by 2020. This task can be accomplished, but this requires urgent and more ambitious actions by all interested parties around the world. Thus, the creation of industries for the processing of solid industrial waste, for example, polyethylene, with the production of additional marketable products in the form of power generator gas is relevant task.

Figure 1. Clogging the ocean with plastic garbage.

2. The purpose of the work

Reducing the temperature of the mixture in the combustion zone. It is the most important measure for the disposal of industrial waste, aimed at reducing the concentration of harmful substances in any equipment for burning solid, liquid and gaseous fuels and in internal combustion engines.

3. Methods of achieving the goal

The use of synthetic combustible gas on the basis of generator gas, obtained from solid industrial and household wood waste, including with the addition of polyethylene waste, both in mobile and stationary gas-generating installations, can be taken as a promising direction.

The use of alternative sources of motor fuel has its own history. Thus, the Second World War, the "war of motors," as it was later called, required a tremendous strain on all sectors of the national economy, including oil-producing ones. At the same time, the lion's share of liquid fuel (in all the belligerent countries, and not only in the USSR) was directed to the needs of the front, to provide armoured formations, aircraft, and the navy. But at the same time, economical gas-generating trucks, tractors, and truck tractors [13–14] were perfectly suited for the needs of the rear. Currently, the use of gas generators in the CIS countries is often presented as evidence of "poverty" and industrial "backwardness". Meanwhile, such power plants were used much more in Europe in general, and in Germany in particular, especially after the start of the Second World War [15].

In 1936 in Western Europe there were about 10 thousand gas-generating motor vehicles. With the outbreak of hostilities, their number increased by more than 50 times. Only in Germany there were almost 300 thousand of such trucks, and they were used in the Wehrmacht as logistic support vehicles and in military transport connections. Figure 2 presents the Volkswagen Type 82 car with a box engine and a gas generator operating on wood (1944) [16].

Figure 2. Volkswagen Type 82 (Kübelwagen).

According to the technical data, Volkswagen KdF-11 ("Volkswagen Käfer"), produced since 1933, had a four-cylinder four-stroke air-cooled engine with an opposite arrangement of cylinders in the rear part (engine capacity is 1131 cm^3, power - 25 hp, shaft speed - 3300 rev / min, maximum torque - 53 NM. The consumption of gasoline with an octane number of 74 with a compression ratio of 5.8: 1 was 7.5 litres per 100 km of run, and the content of carbon monoxide CO in the exhaust gases during track control was 3.0%. It should be noted that the efficiency was estimated by the ratio of the maximum torque to the working volume in litres and was 4.7 NM / L.

The authors carried out the research and development work to improve the environmental level and fuel efficiency of the Ukrainian people's car ZAZ 968M-005 with a V-shaped four-stroke four-cylinder air-cooled engine located at the rear of the body and having a working volume of 887 cm3. The results of the experimental trips (with a mileage of 20 thousand km) showed that the MeMZ-966G engine had power28 hp and it is achieved at a shaft rotational speed of 4300 rpm, and a maximum torque of 53 NM at a shaft rotational speed of 2700 rpm. With a compression ratio of 6.5: 1, operation on gasoline with an octane number of 72 is differed by an average consumption of 9.2 litres per 100 km of run. The content of carbon monoxide CO in the exhaust gases during track control at traffic police stations was recorded at 3.0%. The serviceability of the conventional MeMZ-966G engine, determined by the ratio of the maximum torque to the working volume of the engine, was 61 NM / L [17].

After upgrading the MeMZ-966G engine in 1997, using its own know-how technology, the authors managed to increase the maximum torque to 83 NM at the same working volume and compression ratio. With a frequency rotation of 2900 rpm, serviceability increased 1.77 times (94 Nm / l) and approached the level of the 1997 PORSHE 911 with a six-cylinder boxer engine at the rear section. As the long-term operation has shown, due to the engine forcing, the average fuel consumption decreased to 6.5 litres per 100 km, and the carbon monoxide CO content in the exhaust gases during track control decreased to 0.3-1.5% under other equal conditions. Maximum power 39.5 hp was achieved at a shaft rotational speed of 4300 rpm, i.e. increased 1.41 times [18]. Thus, while reducing the fuel consumption, and hence thermal pollution of the environment by 1.5 times, the level of CO emissions also decreased by 2-10 times with different operating modes of the MeMZ-966G engine.

At the same time the conducted pilot experiments on the recirculation of energy supply showed the prospects of using both liquid and gaseous motor fuels at the automobile transport unit. It will ensure a reduction in consumption of liquid motor fuels by 3-4 times while reducing emissions of toxic substances with exhaust gases. A significant disadvantage of using conventional technology for converting automobiles with gasoline engines to generator gas derived from wood waste is the inevitable drop in torque on the output shaft and power by 50% or more, under all other equal conditions. It leads to a loss of speed and actual load capacity transportation of goods and passengers.

The later calculations for similarly designed engine structures ZMZ-53, MeMZ-966G, Volkswagen Type 82, D240 showed the possibility of increasing the torques after the transition to the low calorific value generating gas produced on the modernized gas generator [19].

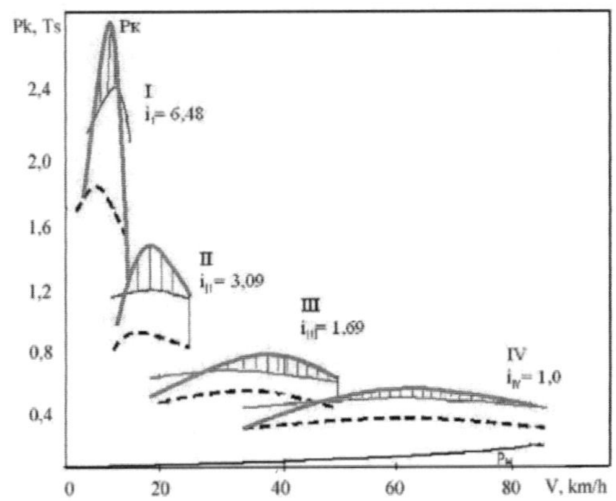

Figure 3. Traction characteristics of the GAZ-53 car when working on gasoline and generator gas: —ordinary gasoline, ▬▬ordinary gas-generating gas, ▬ the proposed gas-generating gas.

Figure 3 shows the actual traction characteristic (the dependence of the thrust force Pk on the speed V) of a GAZ-53 car with a ZMZ-53 engine running on gasoline.

For comparison, the calculated traction characteristics are presented for the same vehicle with the ZMZ-53 engine converted to a conventional gas generator on wood and after upgrading using know-how technology and operating on a gas generator on a mixture of solid wood and polyethylene waste. The analysis showed that when the GAZ-53 car is converted to a conventional generating gas, its dynamics fall by 33%, while after the proposed modernization of the engine and gas generator it can grow by 17%, all other equal conditions.

4. The research results and discussion

The following outlines the basics of a simplified justification of ways to improve gas-generating processes with using different types of solid fuels based on the principles of heat engineering [20].

The heat of combustion of a wide range of organic substances depends on the elemental composition in% and is well determined by the D.I. Mendeleev formula:

$$Q_i = 0{,}34\ C + 1{,}03\ H - 0{,}11(O - S) - 0{,}025\ W,\ \text{MJ/kg}. \quad (1)$$

An approximate stoichiometric equation for the combustion of methane in a mixture with air in an engine with an excess coefficient $\alpha = 1.0$ is:

$$CH_4 + 2\cdot(O_2 + 4N_2) = CO_2 + 2H_2O + 8N_2 + Q_1, \quad (2)$$

It is known from heat engineering that when “burning” 1 m3 of air and burning organic fuel, 3.8 MJ of heat is released, and when methane is burned, according to the formula presented, heat will be released in a volume of 860 MJ, which is equivalent to the heating value of natural gas at 38 MJ / m^3.

For comparative calculations of the change in the process of converting an automobile engine to mixed types of fuel, an approximate formula of gasoline of C_8H_{18} on isooctane type with a specific density of 692 kg / m^3 and calorific value of 44 MJ / kg was taken.

The approximate stoichiometric equation for the combustion of C_8H_{18} gasoline mixed with air in an engine with an excess coefficient $\alpha = 1.0$ is:

$$C_8H_{18} + 12{,}5(O_2 + 4\ N_2) = 8\ CO_2 + 9\ H_2O + 50\ N_2 + Q_2, \quad (3)$$

where the calculated value of the released heat will be Q_2 = 5310 MJ per 114 kg of gasoline when heating the total mass of 1914 kg of burned gases at about 2770^{0}C with an average heat capacity of the process of 1 MJ / kg deg. According to the theory of internal combustion engines, about half of the released heat is consumed on the engine shaft and the temperature of the gases in the cylinders before opening the exhaust valves drops to 1385^{0}C and less, taking into account the heat losses in the walls of the cooled cylinders, piston and combustion chamber. After the exhaust process ending, the exhaust gas temperature drops to 700-900^{0}C when gas enters the silencer system. This was observed in actual operating conditions of the MeMZ-966G forced engine. Thus, non-production losses of the fuel power carrier exceed 50% or more with a very large level of emission of toxic and greenhouse gases and substances, and thermal pollution of the surrounding air environment due to harmful afterburning of excess hydrocarbons.

An exemplary process for producing generator gas from wood waste with a moisture content of 10% consists in feeding an ordinary gas generator to the reaction zone. This can be represented by a formula adapted to the partial component volume parts of the incoming gases:

$$0.054(C_6H_{10}O_5 + H_2O) + 0,097(O_2 + 4N_2) = 0,161\,H_2 + 0,209\,CO + 0,092CO_2 + 0,023CH_4 + 0,388N_2 + 0,009\,O_2 + 0,117H_2O + Q_3. \quad (4)$$

An approximate stoichiometric equation for combustion of a generator gas mixed with air in an engine with an excess coefficient $\alpha = 1.0$ is:

$$0,161\,H_2 + 0,209\,CO + ,092\,CO_2 + 0,023\,CH_4 + 0,388\,N_2 + 0,009\,O_2 + 0,117\,H_2O + 0,222\,(O_2 + 4N_2) = 0,324\,(H_2O + CO_2) + 1,276\,N_2 + Q_4, \quad (5)$$

where the calculated value of the released heat is Q_4 = 95 MJ per 9.7 kg of wood waste under the total mass of the burnt gases is 55.7 kg heated to approximately 1590^{0}C with an average heat capacity of the process of 1 MJ / kg deg.

According to the theory of internal combustion engines, about half of the released heat is consumed on the realization of the useful torque on the engine shaft, and the temperature of the gases in the cylinders before opening the exhaust valves drops to 795^{0}C and lower taking into account the heat losses in the walls of the cooled cylinders, piston and combustion chamber. After the exhaust process ending, the temperature of the exhaust gases drops to 350-450^{0}C when gas enters the silencer and neutralizer system. The estimated calorific value of wood waste in the processing into generator gas in a conventional gas generator will be: Q_{ggn} = 9.85 MJ / kg.

An exemplary process for producing generator gas from a mixture of 90% solid household wood waste with a moisture content of 10% and 10% polyethylene waste consists in feeding an air to the reaction zone of a conventional gas generator and can be represented by a formula that is adapted to the partial component volume parts of the incoming gases and ethylene:

$$0.054(C_6H_{10}O_5 + H_2O) + 0,097(O_2 + 4N_2) + 0,076(CH_2) = 0,161\,H_2 + 0,209\,CO + 0,092CO_2 + + 0,023CH_4 + 0,388N_2 + 0,009\,O_2 + 0,117H_2O + 0,038\,C_2H_4 + Q_5. \quad (6)$$

An approximate stoichiometric equation for the combustion of a new generating gas mixed with air in an engine with an excess coefficient $\alpha = 1.0$ is:

$$0,161\,H_2 + 0,209\,CO + ,092\,CO_2 + 0,023\,CH_4 + 0,388\,N_2 + 0,009\,O_2 + 0,038\,C_2H_4 + 0,117\,H_2O + 0,374\,(O_2 + 4N_2) = 0,440\,(H_2O + CO_2) + 1,884\,N_2 + Q_6, \quad (7)$$

where the calculated value of the released heat is Q_6 = 160 MJ per 10.77 kg of a mixture of 90% wood waste and 10% polyethylene waste and under heating a total mass of 85 kg of burnt gases at about 1880^{0}C with an average heat capacity of the process of 1 MJ / kg deg.

According to the theory of internal combustion engines, about half of the released heat is consumed on the engine shaft and the temperature of the gases in the cylinders before opening the exhaust valves drops to 940°C and lower, taking into account the heat losses in the walls of the cooled cylinders, piston and combustion chamber. After the exhaust process ending, the exhaust gas temperature drops to 380-490^{0}C when gas enters the silencer and neutralizer system. The calculated calorific value of a mixture consisting of 90% wood waste and 10% polyethylene waste during processing into generator gas in a conventional gas generator will be: Q ggn = 14.7 MJ / kg. (M. tak: To mixture having 90% wood waste and 10% polyethylene waste during processing into generator gas in a conventional gas generator calculated calorific will be)

The special interest is the changes assessment in the technical and economic indicators of the vehicles conversion from liquid motor fuel to the mixed power consumption of liquid fuel and generator gas. As a baseline, we can take the option of using 114 kg of gasoline for the vehicles operation, in which 5310 MJ is generated.

For the first option (a vehicle with a mixed fuel consumption - 28.3 kg of gasoline and generator gas produced from wood waste), the required mass of wood waste can be determined from the condition for obtaining heat in a volume equal to the difference between the total heat and the heat generated by the gasoline part: 5310-1335 = 4005 MJ. Then the estimated required mass of wood waste will be: $M_w = Q_w / Q_{ggn} = 4005 / 9.85 = 420$ kg.

For the second option (using generating gas from a mixture of 90% wood waste and 10% polyethylene waste) the total mass of waste is required: $M_w = 4005 / 14.7 = 278$ kg, of which 251 kg are wood and 27 kg are polyethylene waste.

If for the base case with the fuel price of 1 litre 50 roubles, the fuel cost will be 8250 roubles, then for the first option of mixed consumption of 28.3 kg of gasoline worth 2062 roubles and wood waste worth 1260 roubles. (at a price of 3 roubles / kg), total expenses will amount to 3322 roubles, i.e., 2.48 times less.

And for the second option, it will take 890 roubles on a mixture of wood and polyethylene waste. This in total amounts to 2952 roubles, i.e., 2.78 times less than gasoline.

5. Conclusion

1. The use of gas generators consuming solid waste of wood- and polyethylene-containing origin, as a regenerative source of power cattier when recirculation of exhaust gases, will reduce the temperature of combustion of the fuel-air mixture in the cylinders of the internal combustion engine and reduce emissions of nitrogen oxide and carbon in the exhaust gases.
2. The use of technology forcing engines such as ZMZ-53, MeMZ-966G, Volkswagen Type82, D240, when using the equipment of gas generators, operating on solid waste will allow to compensate the loss of their dynamical properties and preserve the characteristics of motor vehicles and tractors.
3. Modernized transport units with gas generators expand the scope of application and technical and economic indicators of autonomous energy systems operating in remote areas and consuming local types of fuel and solid waste of local production and livelihoods.
4. Modernization of engines can be performed in the sites of operation of transport and technological machines in the conditions of workshops and service stations.
5. The use of gas generator equipment will reduce the consumption of liquid motor fuels by 3-4 times due to the use of local solid fuels and solid waste.
6. The fuel cost for vehicles in the presence of gas generating equipment of recirculation type can be reduced by 2-3 times.

References

[1] Digital Transformation Monitor, Germany: Industrie 4 0, January 2017 URL: https://ec europa eu/growth/ tools-databases/dem/
[2] Securing the future of German manufacturing industry, Recommendations for implementing the strategic initiative INDUSTRIE 4 0, Final report of the Industrie 4 0 Working Group, April 2013
[3] Munts V A et al 2012 Burning of volatiles in fluidized bed combustion of solid fuels Translated from Élektricheskie Stantsii *Power Technology and Engineering* **46(3)** 16 – 22
[4] Diesel - declared carcinogen IARC concludes diesel exhaust fumes are carcinogenic – URL: http://www ohsrep org au/hazards/chemicals/diesel-declared-carcinogen
[5] Why Are Vehicles' Exhaust Fumes Harmful To Humans? – URL: https://www scienceabc com/humans/ why-are-vehicles-exhaust-fumes-harmful-to-humans html
[6] Markov V A , Bashirov R M and Gabitov I I 2002 Toxicity of exhaust gases of diesel engines 2nd ed , Rev and add - M : Publishing House of N E Bauman Moscow State Technical University , - 376
[7] Gritsuk I V et al 2011 Reducing the toxicity of exhaust gases of a converted gas-diesel engine for use in a stationary power plant Bulletin of scientific works of DonIZT **25**

IOP Conf. Series: Materials Science and Engineering **786** (2020) 012074 doi:10.1088/1757-899X/786/1/012074

[8] Smirnova T et al A new fuel for urban transport URL: engine aviaport ru/issues/02/page42 html

[9] Balasingham K 2015 Industry 4 0: Securing the Future for German Manufacturing Companies [Master Thesis]: University of Twente School of Management and Governance Business Administration (M Sc) Academic

[10] Scala F and Salatino P 2002 Modelling fluidized bed combustion of high-volatile solid fuels *Chemical Engineering Science* **57(7)** 1175-96

[11] Stafford R and Jones Peter J S 2019 Viewpoint – Ocean plastic pollution: A convenient but distracting truth? *Marine Policy* **103** 187-91

[12] Nicola J 2019 Beaumont et al Global ecological, social and economic impacts of marine plastic /*Marine Pollution Bulletin* **142** 189-95

[13] Tokarev G G 1955 Gas-generating cars Mashgiz 205

[14] Yudushkin N G 1955 Generator tractors State scientific and technical edition of engineering literature 242

[15] Kipriyanov F A 2016 Study of the gas generator Modern scientific research and innovation, **1(57)** 146-52

[16] Leichter personen-kraftwagen type82 beschreibung bedienungsanweisung instandsetzungsanleitung/ Vokswagenwerk G m b H Kundendienst - 1943 23

[17] Andreev A and Andreev S 2004 Volkswagen Kafer "VW 82" power units (Dr Ing H P F PORSCHE) and Zaporozhets ZAZ-968M-005: features and prospects *Vestnik* V Dahl SNU-Lugansk,**7-2(77)** 53-6

[18] Patrakhaltsev N N and Savastenko A A 2005 Forcing of internal combustion engines by supercharged: - M : Legion-Avtodata, 176

[19] Andreev A A and Andreev S A 2013 On improving the serviceability of gas generator cars /*Vestnik V Dahl SNU- Lugansk*, **9(198)** 17-23

[20] Baskakov A P and Vitt O K et al 1991 Heat Engineering: Textbook for Universities General ed Baskakov A P 2nd ed Rev and add Energoatomizdat 224

IOP Conf. Series: Materials Science and Engineering **786** (2020) 012075 doi:10.1088/1757-899X/786/1/012075

Improving the quality of the investigation of a road traffic accident

S I Tikhomirov*, Y Y Komarov and V M Volchkov

Volgograd State Technical University, Volgograd, Russia

E-mail: 1704tihii87@mail.ru

Abstract. The article deals with the issue of improving the quality of auto expertise and as a result, with the issue of investigating road accidents. The main parameter used in the auto expertise is vehicle deceleration. But today we consider the characteristics of the roadway and the load of the car when decelerating a car. As for temperature factor it is not only ignored, but the possibility to take it into account doesn't even exist. The impact of the ambient temperature on vehicle deceleration was studied. There was simulation of vehicle braking which physically justified the impact of ambient temperature. On the basis of the practical tests conducted, the dependence of vehicle deceleration on the ambient temperature was determined. This dependence will allow auto experts make calculations more objectively which will positively affect not only the course of road accident investigation but also the identification of factors conducive to development and emergence of the accident.

1. Introduction

Addressing road transport accidents involves a set of preventive measures a special place among which belongs to the study of the road accident causes. Auto expertise allows justice system to establish these reasons in full since it takes into account not only the existing legal but also technical aspects. The technical ability of the participants in the accident to prevent the accident is determined. In conducting technical calculations vehicle deceleration is one of the main parameters used to influence all the calculations. The expert chooses this parameter individually based on the available characteristics of a vehicle. Thus, the validity and accuracy of the deceleration value selected by an expert may affect not only the results of the auto expertise conclusions but also the lives of the accident participants [11].

2. Relevance and scientific value of the article with a brief review of the literature

At present, if it is not possible to experimentally establish the deceleration value by the expert divisions of the Ministry of Justice and the Ministry of Internal Affairs of Russia it is selected from the corresponding tables (Table 1). Moreover, the selected parameter will depend on the following factors: type of road surface, vehicle category and load. All the other factors are not possible to take into account for choosing the deceleration value [10].

Table 1. The values of the established deceleration of M1vehicle on the roadway part with adhesion coefficient of φ= 0,7.

Vehicle category	The values of the established vehicle deceleration, m/s²		
	Without load	50% load	100% load
M1	6,8	6,6	6,3

The value of the established deceleration is influenced by various external factors, including the model of the car, the type of tires rubber, the type and condition of the road surface, the ambient temperature, etc. Let us estimate one of them namely the ambient temperature and its impact on the value of the established vehicle deceleration.

The factors that characterize the road surface, its wear and condition have the greatest effect on the deceleration. Temperature is the next most influential factor. This statement that the ambient temperature, all things being equal, influences the vehicle deceleration is reflected in the works of A. A. Kochetkov, E.V. Balakina, S.A. Evtyukov and other authors [2].

In addition, the systematic study of vehicle braking process shows quite a noticeable influence of ambient temperature on the deceleration, especially at low temperatures.

3. Statement of the problem

Vehicle deceleration during braking determined mainly by the adhesion coefficient of the tire to the road surface depends on a large number of factors (the condition of the tire tread and the road surface). It might be therefore asked to what extent the influence of the ambient temperature on vehicle deceleration value (and accordingly, the stopping distance) is physically justified.

In this regard, we perform a simulation of vehicle braking in order to determine the temperature in the area where the tire contacts the road as well as the possible influence of the ambient temperature on it.

We will carry out practical tests by which we will establish the dependence of vehicle deceleration on the ambient temperature.

4. Theoretical part

First of all, it should be noted that the temperature in the area where the tire contacts the road surface during vehicle movement is not equal to the ambient temperature but is determined by two points:

First, the tire of the vehicle wheel is warmed up in the process of movement (without braking) due to its deformation, which is accompanied by irreversible loss of energy;

A lot of papers are devoted to the problem of tire heating during car rolling [15, 19].

The loss of energy Q spent on heating the material is usually measured as:

$$Q = Q_0 \varepsilon^2 \tag{1}$$

where ε - amplitude of the variable strain; Q_0 - proportionality coefficient.

Q_0 value has a very complex structure and it is determined not only by friction inside the rubber-like material but also due to internal friction between the various components of the complicated composite tire structure (micro-friction between dissimilar layers, between the matrix and the fibers of the reinforced parts of the composite).

Let us choose a heat source Q which provides a qualitative match with experimental data.

The wheel consists of a hollow tire heated by deformation. The inner part of the wheel is a metallic disc. The heat released inside the tire is discharged into the air that fills the tire as well as into the metal forming the basis of the wheel. Then it is transferred to the surrounding air both from the outer surface of the tire and from the surface of the metal part of the wheel. The scheme of the heat flows is presented on Figure 1 [14].

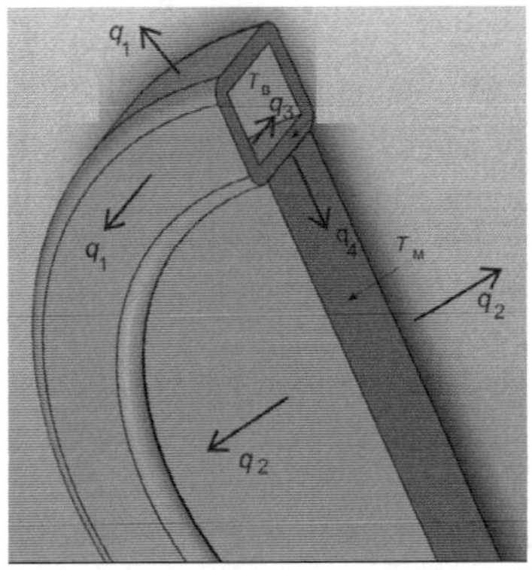

Figure 1. Scheme of the wheel heat flows.

T_{outer} - is the ambient temperature; T_{air} - is the air temperature inside the tire; T_m - is the temperature of the metal disc of the wheel; T_t - is the tire temperature. Heat flows are denoted as: q_1 - is the flow from the tire into the environment; q_2 - is the flow from the metal disc into the environment; q_3 - is the flow from the tire to the air filling the tire; q_4 - is the flow from the tire to the metal disc of the wheel.

The temperature distribution in the tire is described by the equation of transient thermal conductivity:

$$\rho_t C_t \frac{\partial T_ш}{\partial t} + \nabla(\lambda_t \nabla T_t) = Q \quad (2)$$

- temperature; ρ, C, λ - is density, heat capacity and coefficient of thermal conductivity, respectively; Q - is heat generation inside the tire due to deformation; t - is time.

Equation (2) should be solved under the following initial and boundary conditions:

$T(0) = T_{outer}$ (initial temperature = ambient temperature);

$$-\lambda_t \frac{\partial T_ш}{\partial n} = h_{t,outer}\left(T_{outer} - T_t\right) \quad (3)$$

$$-\lambda_t \frac{\partial T_ш}{\partial n} = h_{t,air}\left(T_{air} - T_t\right) \quad (4)$$

$$-\lambda_t \frac{\partial T_t}{\partial n} = h_{t,m}\left(T_t - T_m\right) \quad (5)$$

Relations (3) - (5) determine the heat transfer from the tire to the environment, from the tire to the air inside the tire and from the tire to the metal disc, respectively. The heat transfer coefficients $h_{t,outer}$, $h_{t,\,air}$, and $h_{t,m}$ determine the heat transfer from the tire into the surrounding air, the heat transfer from the tire to the air inside the tire and the heat transfer from the tire to the metal disc, respectively [16].

Similarly, the temperature distribution in a metal disc is described by the following equation:

$$\rho_m C_m \frac{\partial T_m}{\partial t} + \nabla(\lambda_m \nabla T_m) = 0 \quad (6)$$

Unlike equation (3), there is no term defining the internal heat source here while the other symbols correspond with those in equation (3). The boundary conditions for equation (7) are:

$$-\lambda_m \frac{\partial T_m}{\partial n} = h_{outer}, {}_{M}\left(T_{outer} - T_m\right) \quad (7)$$

$$-\lambda_m \frac{\partial T_m}{\partial n} = h_{Mt}\left(T_m - T_t\right) \quad (8)$$

Condition (7) determines the rejection of heat from a metal disc into the surrounding air with the corresponding heat transfer coefficient $h_{outer,m}$.

Condition (8) determines the heat transfer from the tire to the metal disc of the wheel.

We shall notice that since the heat conduction problem for solids (tires and metal disc) is further solved with a joint boundary, conditions (5) and (8) become redundant because the equality of heat flows on the adjacent boundary between solids is performed automatically. Naturally, the tire and wheel disc are characterized by their thermal properties (density, heat capacity and thermal conductivity) [5].

We consider the air inside the tire as a kind of abstract body that can be heated by heat transfer from the tire to this body. Therefore, the heating equation of this abstract body with the thermophysical air properties should be added to the heat conduction equation (8):

$$\rho_{\mathrm{air}} C_{\mathrm{air}} V_{\mathrm{air}} \frac{dT_{\mathrm{air}}}{dt} = q_{\mathrm{tair}} = h_{\mathrm{air}} \left(T_{\mathrm{t}} - T_{\mathrm{air}}\right) S_1 \tag{9}$$

where q_{air}, C_{air}, V_{air} - is density, heat capacity and air volume inside the tire;

S_1 - is the area of the inner surface of the tire.

The term on the right represents a heat source which is characterized by the boundary condition (4).

We will solve the problem of heating the metal rim as a whole since the transfer of heat from the tire to the metal will be limiting due to the high thermal diffusivity of the metal:

$$\rho_{\mathrm{m}} C_{\mathrm{m}} V_{\mathrm{m}} \frac{dT_{\mathrm{m}}}{dt} = q_{\mathrm{tm}} = h_{in\_\mathrm{met}} \left(T_{\mathrm{t}} - T_{\mathrm{m}}\right) S_2 - h_{\mathrm{outer,m}} \left(T_{\mathrm{m}} - T_{\mathrm{ambient}}\right) S_3 \tag{10}$$

Thus, equations (2), (9) and (10) are to be numerically solved with respect to three unknown variables T_t, T_m and T_{air}. For an equation with partial derivatives the boundary conditions are determined by relations (3), (4) and (5). For ordinary differential equations boundary conditions are not required. As the initial conditions for T, T_m and T_{air} we simply choose the equality of these temperatures to the ambient temperature at the beginning.

To complete the problem statement it is required to set the heat transfer coefficients $h_{air,t}$, h_{air} and $h_{outer,m}$.

The conditions of convective heat transfer are set by the boundary conditions (3), (7) and the convention coefficient h_c. Convection coefficients are determined by thermophysical properties (thermal conductivity, density and viscosity) and are expressed in terms of the Nusselt number, Nu [6]:

$$h_c = \frac{Nu \cdot \lambda}{d_e} \tag{11}$$

where d_e -is the typical size of the body streamlined by the flow;

The Nusselt number in its turn also depends on the parameters of the blowing air flow.

This dependence is determined by the Reynolds number, Re:

$$Nu = \zeta \cdot 0.018 \cdot \mathrm{Re}^{0.8} \tag{12}$$

where ζ - is the correction for the channel curvature which is determined by the ratio of the channel radii and its external curvature

$$\varsigma = 1 + 1.8\left(\frac{2d_{\mathrm{ch}}}{d_{\mathrm{cur}}}\right) \tag{13}$$

where d_{ch} is the diameter of the channel (pipe), d_{cur} is the diameter of curvature.

The Reynolds number which is the ratio of inertia forces to viscous forces in a flow is calculated in the usual way:

$$\mathrm{Re} = \frac{V \cdot d_e}{\mu} \tag{14}$$

where V - is the typical speed; d_e - is the typical size, μ - is the coefficient of dynamic viscosity.

The energy of the heat source Q is determined by the deformation of the tire which is concentrated mainly in the area where the tire contacts the road. The heat released in the contact area is transferred along with the wheel rotation. Therefore, the heat source for each point of the tire should be represented as a stage periodic function, and the convective term of heat transfer from the contact area should be added to the heat conduction equation (8). Since wheel rotation is rather quick the typical heat distribution in the tire due to its thermal conductivity takes much less time than the transfer due to the wheel rotation. That's why value Q might be considered as a constant equal to the heat average generated in the tire per time unit [3].

We used *Comsol Multiphysics* pack to solve the problem concerning the temperature T in the tire and the temperature T_t [4].

After selecting the value of the heat source Q the calculations for different values of the ambient temperature were made (-10°C, 0°C, 10°C, 20°C, 30°C) that are shown on Figure 2.

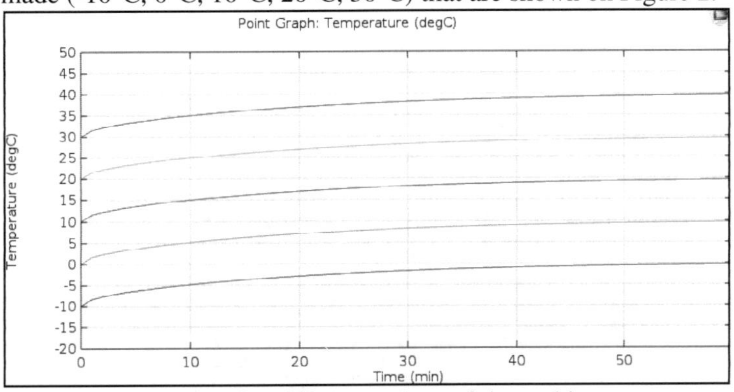

Figure 2. Tire heating for different values of the ambient temperature.

It is clear that these data correspond to a certain tire with its thermophysical and mechanical properties and to a certain road surface which was tested. These results may differ to some extent for other tires and road conditions.

Thus we can see that the ambient temperature significantly determines the temperature of the tire after a long distance travelled by the vehicle.

In the course of rolling the tire is approximately heated by 10 – 30°C which means its temperature differs from the ambient temperature but finally, all things being equal, it will be higher than the ambient temperature.

The process of tire heating when sliding down the road is determined by the equation of heat conduction that is similar to equation (9). The term of the heat source Q is absent here owing to heat release in its cyclic deformation during rolling:

$$\rho C\frac{\partial T}{\partial t}+\nabla(\lambda\nabla T)=0 \tag{15}$$

Instead, a surface heat source Q_n is added which might be estimated as the kinetic energy of the vehicle $mv^2/2$ at the beginning of its full braking and it should turn into thermal energy until the vehicle stops.

The tire is supposed to be stationary in the model while the road and the surrounding air are moving towards (in the direction opposite to the real vehicle movement) at speed V, which varies from the initial speed V_0 to zero.

There is a surface heat source in the contact area which is characterized by the kinetic energy of the vehicle. Heat from the surface source is distributed into the tire volume from which it is released into the environment due to convection, transferred to the air inside the tire and to the metal disc corresponding to the wheel rim. A convective term providing heat removal from the contact area shall be added to the heat conduction equation for the road:

$$\rho C\left(\frac{\partial T}{\partial t}+V(t)\cdot\nabla T\right)+\nabla\cdot(\lambda\nabla T)=0 \tag{16}$$

The initial temperature of the tire T_t - is equal to the temperature of its heating as a result of rolling and the initial temperature of the road T_r equals to the ambient temperature $T_{ambient}$:

$$T_t(0)=T_h \quad T_r(0)=T_{ambient} \tag{17}$$

When we solve the problem of the heat flows equality at the boundary between two bodies (tire and road) the boundary conditions are determined automatically in the contact area. The condition of heat transfer from the tire to the metal wheel is automatically performed, as well. The remaining boundary conditions correspond to the heat exchange with the environment:

$$-\lambda\frac{\partial T}{\partial n}=h(T-T_{\text{ambient}}) \tag{18}$$

where the heat transfer coefficient h is different for the tire and the road.

The stated problem was solved with the help of Comsol Multiphysics pack which is shown on Figure 3.

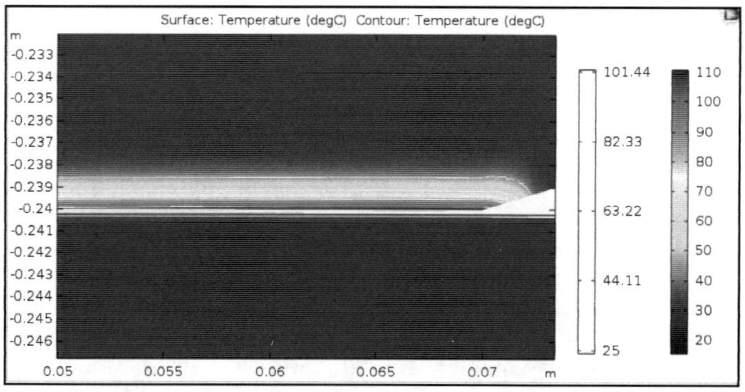

Figure 3. Temperature distribution in the area the tire contacts the road at initial vehicle speed of 60 km/h.

The tire temperature in the contact area at various ambient temperatures is shown on Figure 4.

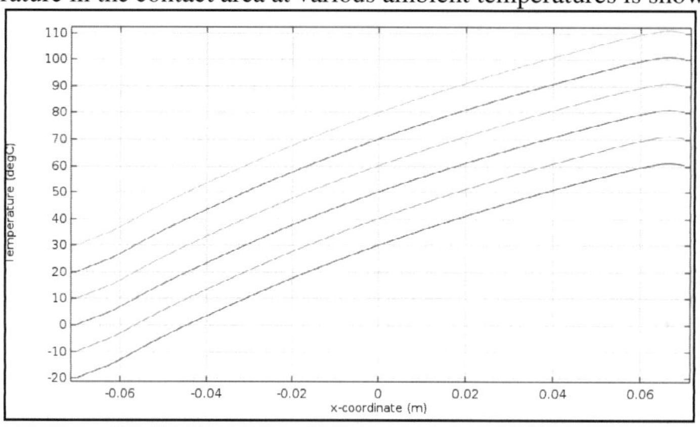

Figure 4. Temperature distribution along the contact area at different ambient temperatures.

Thus, it is physically justified that the ambient temperature (all things being equal) will noticeably determine the final temperature in the contact area. This means that the specialists should take into account the impact of the ambient temperature while conducting auto expertise.

Practical relevance, suggestions, results of implementations and experimental studies 7560 practical tests were carried out in order to establish the dependence of the vehicle deceleration on the ambient temperature [9]. Based on those tests, it has been determined that the vehicle deceleration decreases when we use summer tires what can be described by the piecewise linear dependence on Figure 5.

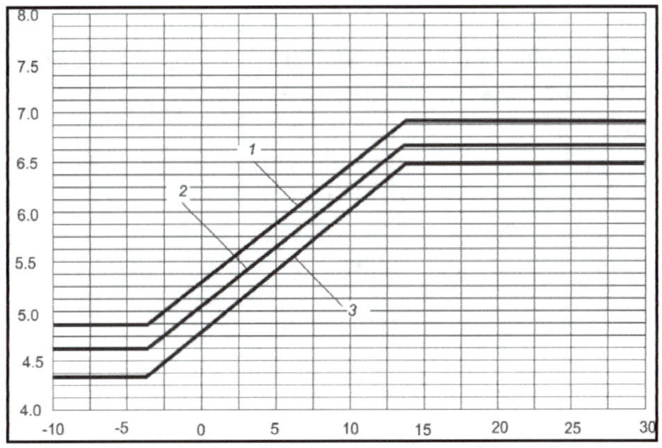

Figure 5. Deceleration values of a vehicle with summer tires without load – "1", with 50% load - "2" and with 100% load - "3".

It follows that all subsequent data characterizing the mechanism of the incident, in the calculation of those there is a parameter of the established vehicle deceleration will also change depending on the ambient temperature.

At minus temperatures below -5 °C if the vehicle was equipped with summer tires the established vehicle deceleration value of all the investigated vehicles would remain unchanged and equal to j_{-5}^{0}. With an increase of the ambient temperature from -5 °C to 20 °C, deceleration had been rising by linear principle until j_{20}^{0} was reached. A further increase of temperature from 20 to 30 °C had practically no effect on the established vehicle deceleration value. Such a course of the studied process suggests that the law of the change of the established deceleration depending on the ambient temperature can be described by a piecewise linear dependence

$$j = \begin{cases} j_{-5} \text{ at } t < -5\,°\text{C} \\ j_{-5} + at \text{ at } -5\,°\text{C} \le t \le +20\,°\text{C} \\ j_{20} \text{ at } t > +20\,°\text{C} \end{cases} \tag{19}$$

The use of regression analysis allowed us to estimate the value of *a* coefficient (slope ratio). It turned out to be equal $a = 0.07$ m/s^2 for almost all types of vehicles. The values of j_{-5}^{o} varied over a larger range and they obviously depend on the physical and mechanical properties of the road surface and the wheels rubber of the vehicle. With an increase of the ambient temperature from -5 °C to +20 °C, the established deceleration rises. Consequently, when we choose the established deceleration value without adjustment to the ambient temperature our calculations can be incorrect distorting the real picture of events, and giving a wrong assessment to the actions of the road accident participants [1].

The studied dependence will not only help improve the objectivity of auto expertise, but also the quality of the road accidents investigation which will allow a truly innocent person to avoid punishment, and vice versa, those whose actions caused the incident to bear responsibility for their actions. The absence of more precise parameters objectively reflecting the influence of one or another factor, in this case the temperature factor, makes the process of investigating the true causes of the accident not only meaningless but also dangerous [9]. In the conduct of auto expertise not only the question of the conformity of the actions of accident participants is examined but all circumstances of the accident as a whole. The necessary and sufficient conditions for the emergence and further development of the accident are established, too. In case false conditions contributing to the emergence and further development of the accident are identified, the measures aimed at eliminating and reducing the risk of a similar accident repetition at the same site may not only have any practical impact, but also have the opposite effect.

5. Findings (conclusion)

We performed the vehicle braking simulation during which it was determined that the ambient temperature affects the final temperature in the contact area and as a result the vehicle deceleration.

Practical tests have been carried out based on which we have established the dependence of the vehicle deceleration on the ambient temperature with summer tires and with various vehicle loads.

The obtained results allow the experts to qualitatively improve the conclusions of the auto expertise. This will help to avoid misunderstandings in the investigation of road accidents and will eliminate the factors that really affect not only the emergence but also the development of those accidents.

References

[1] About acceptance of technical regulations of the customs Union "about safety of wheeled vehicles" [electronic resource]: 09 12 2011 N 877 (Red - [Moscow, 2015] - Access from help - legal system "ConsultantPlus"

[2] Balakina E V 1990 Coefficient of tire adhesion to the road surface : monograph - Tires : the machine, job, operation, repair Moscow : Transport, 271

[3] C C Clark 2006 (Chairman of the editorial Board) and others pneumatic tire/ / national administration of road safety USA Dep Transport, 708

[4] Volchkov M , Styazhin V N , Kalinin I V , Tarasova I A and Shvedov E G 2013 Computer simulation of mathematical physics problems (comsol multiphisics) : studies In allowance ; VSTU - Volgograd, 68

[5] G M Bartenev and V V Lavrentev 1972 Friction and wear of polymers Publishing house "Chemistry", 240

[6] *Handbook of chemistry*, **I-VII** 1960-70

[7] Ryabov I M , Pozdeev a V , Gudkov D V , Malinin N N , Hasanov M M and Murtuzov M M 2014 Features of automobile tire as elastic element// *Tyre plus (Ukraine)* **2** 8-10

[8] Knoroz V I 1978 the Work tires - Moscow: Transport, 238

[9] Komarov, Y Y 2015 Factors that increase the objectivity of automotive expertise / Y Y Komarov, S I Tikhomirov // Transport of the Russian Federation 32-4

[10] Komarov Y Y 2012 Examination of road accidents in examples and tasks: studies textbook for universities - M : Hot line - Telecom, 290

[11] Kuzmin N A 2001 Processes and patterns of changes in the technical condition of cars in operation : studies manual 216

[12] Larin A A 2011 Study of the formation of the thermal state of pneumatic tires in the process of stationary rolling Bulletin of the national center Techn University "KHPI" : SB nauch Tr The themed slots issue : Dynamics and strength of machines - *Kharkov : NTU "KHPI"* **63** 64-78

[13] Gudkov V A , Tarnovsky V N , Gudkov D V and Tretyakov O B 2008 *Main operational characteristics of tires Bus plus* **3** 8-11

[14] Borzunova T L , Volchkov V M , Stein V N , Goats ar A and Volggtu 2005 Methods of mathematical physics for engineers (computer packages): studies manual Volgograd: RPK "Polytechnic", 76

[15] Mikheev M A 1977 fundamentals of heat transfer / Mikheev M A , Mikheeva I M - M : Energy, - 344

[16] Panovko Y G 1960 Internal friction at vibrations of elastic systems / Y G Panovko - Moscow: Fizmatgiz, 193

[17] Gudkov V A , Ryabov M I , Gudkov D V , Malinin N N , Mamakunem M M and R M Ostrov 2011 *Peculiarities of heating of the tires during vehicle operation Bus plus* (Ukraine) **4** 12-4

[18] Wong H 1979 Basic formulas and heat exchange data for engineers - Moscow: Atomizdat, 89

[19] Yudaev B N 1973 Heat Transfer - Moscow: High school, 208

IOP Conf. Series: Materials Science and Engineering **786** (2020) 012076 doi:10.1088/1757-899X/786/1/012076

Improving the methods of training specialists in the sphere of construction and utilities (on the example of the Volgograd region)

E M Maznitsa[*], M K Belyaev and O V Maksimchuk

Volgograd State Technical University Institute of Architecture and Construction, 400074, Academicheskaya st., 1, Volgograd, Russia

E-mail: Dilemaz@mail.ru

Abstract. The existing methods of training engineering personnel in the sphere of road, power, utility machines and other special equipment do not take into account the changed realities of the national economy. For maintenance engineers it is now not enough to know the basics of the operation of machines and mechanisms, it is necessary to have management skills. The operator should study not only the technical characteristics of the equipment, but also the principles of pricing for utility services, the basics of tariff regulation, the methodology for calculating tariffs and the cost of utility services. Therefore, the development of practical management skills in the housing and utilities sector, the study of the fundamentals of accounting and tax reporting, estimated rationing and legal training are becoming a modern addition to the classical training of specialists - operators of the construction and utilities sectors.

In recent years, there has been a significant decline in private investment in the communal infrastructure. This is especially noticeable in the water, heat and power supply sectors. There are a lot of reasons including the high level of risks and the low solvency of household consumers in a crisis economy. The solution of these problems is determined by the influence of various factors [1]. For example, one of the main limiting factors for the effective implementation of housing and communal services reform is the lack of experienced qualified personnel in the management and operation of utilities, which is associated with a sharp change in the functioning of the housing sector in modern Russia [2]. More recently, the state was responsible for the work of the housing sector, for heat in houses, water supply and sewage, electricity, cleaning, etc. and now - in one decade everything changed dramatically, the responsibility is blurred, it is completely incomprehensible for the population how the housing sector is working now [3].

Discussing today the reasons for the decline in GDP growth and changes in economic policy that are necessary for the revival of “constructive” production, it can be concluded that there are signs of recession in the Russian economy. The current movement towards recession is a consequence of the deep deindustrialization of the domestic economy [4]. Deindustrialization always leads to economic stagnation, and Russia's experience confirms this. The industrial sector survived at least two hard blows.

The first - in the 90s, when everything collapsed, including industrial communications. In the early 2000s, oil prices began to rise, and golden rain fell on the country. The industry was revived. But then there was a change of authority, and with it the economic model of development changed: in fact, instead of the economy of supply, the economy of demand for the export-raw material base was developed. Such ill-conceived economic policy dealt a second blow to the industrial sector [5].

In the context of the crisis and the possible post-crisis development of the Russian economy, entrepreneurship education should become the most important factor in the modernization of education and growth of economy. It is possible that the current crisis will lead to an even greater segmentation of the regional business education market. Of course, it is important to single out the positive moment of the crisis, as the most suitable time for improving the content of educational programs and improving the quality of the educational process, but the current situation has a downside. It seems that the need for a full-fledged, serious business education will always remain, as well as in managerial trainings, seminars and short programs. In our opinion for Russia and our region, one of the main areas of business education should be innovative management and the development of new areas of entrepreneurship [6].

The peculiarity of the current situation of the innovative development of the Volgograd region is the presence of a significant research and production base, combined with the lack of specific innovations related to the lack of enterprises' own, borrowed and attracted funds, low effective demand for scientific and technical products. Available examples of high-quality goods in general do not solve the problem of low competitiveness of the main products of enterprises [7]. Managers of enterprises in modern conditions are not ready to invest in especially risky scientific and technical projects at the stage of scientific research and experienced constructive works. Investor interest arises, as a rule, if a prototype or an experimental batch of a new material or product has been obtained as a result of research work. In the scientific and technical sphere, effective innovative entrepreneurship is impossible without managers in the commercialization of technologies and in the management of innovative projects. Now there is a situation where the rapidly developing Russian economy is facing a shortage of specialists in introducing new technologies, most specialists want to work in quiet places without much risk and responsibility. For example, we need specialists who will solve the engineering task of transferring the heating system of residential buildings from hydrocarbon fuel sources (fuel oil, diesel fuel, etc.) to alternative and renewable types of energy (solar, wind, wave, as well as hydroelectric and thermal energy ocean) [8].

It seems that here the Higher School needs to move from training theorists to practitioners, for example, everywhere there is not enough competent specialists - real estate operators. How to prepare them? Why do not talented advanced young professionals come in such a promising in innovation and financial terms, the industry as housing and utilities? Should there be any specifics in training such specialists at the regional level? And is it necessary to take into account the peculiarities of regional economic development?

In our opinion, in the formation of educational programs in various areas of education, including in the housing and utilities sphere, regional specificity, the level of modernization of the economy of a particular region should be taken into account [9]. As for the Volgograd region, the degree of lag in its economic development, not even from the advanced, but from the ordinary and medium-sized regions of Russia, is increasing every year. The low dynamics of the economy puts pressure on social indicators, which, in turn, affect demographics. In order to reverse the negative trends, serious efforts are needed and, above all, to improve the quality of government in the region. It seems that separate, very successful and fairly large-scale projects will not be able to provide a breakthrough of the regional economy as a whole, it is too big and inertial. They cannot replace the daily hard work in creating a favorable environment for the development of the region. Entrepreneurs and the population are key economic players, and the state cannot and should not replace them in a market economy. But they respond to incentives. Consequently, the task of an effective economic policy of regional authorities is to create a strong motivation to develop business right here and now [10].

Since the beginning of the 90s of the twentieth century, the region has developed a steady shortage of capacity for heat, electricity, water supply and sanitation. Over the past few years, funds from budgets of all levels have been allocated in insufficient quantities. This led to serious losses:

Table 1. Indicators of utility system functioning of Volgograd [11, p. 215].

Indicators	unit	power supply	heat supply	water supply	water disposal
Total network length	km	3 464	898	1 897	1 058
The length of the networks needing replacement	km	866	415	498	485
line 2 : line 1 x 100 %	%	25.0	46.2	26.3	45.8
The amount of leave in the network		1 973 949, thous. kWh	5 413, thous. Gcal	185 581, thous. cub. m	158 333, thous. cub. m
Total sales of goods and services		1 372 251, thous. kWh	4 417, thous. Gcal	131 361, thous. cub. m	104 149, thous.cub. m
The volume of losses	%	601 698, thous. kWh	996, thous. Gcal	54 220, thous. cub. m	
(line3 – line 4) : line 3 x 100 %		32,2	18,4	29,2	
Normative volume of losses		335 966, thous. kWh	708, thous. Gcal	33 405, thous. cub. m	
The actual specific energy consumption per unit of resource			106 614 kWh Gcal	652 kWh thous. cub. m	567 kWh thous. cub. m
Depreciation of utility infrastructure systems (network)	%	82	74	72	77
Number of accidents, total * for a period of more than 8 hours		66 *	136 *	4 614	7 121
System failure (the ratio of the number of accidents to the length of networks)		0.02 *	0.16 *	2.43	6.73
Loss ratio (the ratio of the number of loss to network ratio)		173.700, kWh km	1.109, Gcal km	28.582, cub. m km	

Source: according to municipal unitary enterprise "Volgograd Interdistrict Electric Networks", municipal unitary enterprise "Volgograd Utilities" and municipal unitary enterprise "City Water Channel of Volgograd"

One of the areas of training of modern engineering personnel can be an educational program for managers of housing and communal services. The Institute of Architecture and Construction of the Volgograd State Technical University has long developed such a program. To improve the methods of training engineering personnel for operation, engineers need serious economic training, especially in terms of tax and tariff accounting. The simplest calculation of thermal energy consumption in accordance with the method of Methodical documents in construction 41-4.2000 (approved by order of the State Construction Committee of Russia of 06/05/2000 No. 15), Set of Rules and Regulations No. 23-01-99 of 01/01/2000, requires the specialist to know these standards requirements:

Consumption is calculated by the formula:

$$Q = q * \frac{t_{вн} - t_{сро}}{t_{вн} - t_{ра}} * 24 * n \quad \text{, where} \qquad (1)$$

q – quantity of heat
t vn – internal temperature
t cp – average temperature
t pa – counting temperature
n – the number of days of the month counted in the calculation.

1. Summary

In our opinion, the formation of such direction of education as "Training of Specialists for Management Companies in the Municipal Sphere" and "Management of Housing and Communal Services" should be an important step in developing a regional development strategy for the executive authorities and local governments. The training of specialists in introducing innovations, including the housing and utilities sector, can be carried out at the expense of regional or local budgets, or with the use of subsidies from these budgets. The support of the authorities may also include financial, credit, property, technical, consulting, informational direction. The necessary methodological support includes training of specialists in the technical sphere, who:

- can execute various agreements with all interested individuals and legal entities (owners, district administrations, land committees, etc.);
- know the basics of accounting and management accounting and reporting;
- can properly exploit the common property of the participants in the homeowners' partnership;
- know the basics of team management.

The approximate list of blocks of the educational program is as follows:

Table 2. The approximate composition of the educational program “Management of housing and communal services”.

The name of the topic, the name of the issues studied in lectures and seminars	Lectures, hour	Practical lesson, hour	Forms of control, hour
Module "Technical and operational training"			
Basic concepts and operating conditions for residential real estate			
Functional features of the operation of foundations, basements, attics, roofs, etc.			
Pipeline operation (heating, ventilation, water supply and drainage, sewage)			
Operation of power equipment in residential buildings			
Repair and reconstruction in residential buildings			
Module "Financial and tariff preparation"			
Regulatory base for calculating utility tariffs			
Calculations. Wages, charges, taxes, etc.			
The module "Legal training"			
Normative base of housing and utilities system			
Types and forms of contracts of various civil and legal orientation			
Module "Management Training"			
Analysis of specific business situations			

References

[1] Maznitsa E 2011 Opredelenie investitsionnogo i finansovogo potentsiala pri sozdanii malogo predpriyatiya [Definition of investment and financial potential when creating a small enterprise] «Narodnoe khozyaistvo» [National economy, Section "Proceedings of the IV International Scientific and Practical Conference" Modern Problems of the National Economic Complex: MII *Nauka Publishing House*, Moscow, **1** 211- 5 -

[2] Prilozhenie k Strategii razvitiya zhilishchno-kommunal'nogo khozyaistva v Rossiiskoi Federatsii na period do 2020 goda [Appendix to the Strategy for the Development of Housing and Communal Services in the Russian Federation for the Period up to 2020] Rasporyazhenie Pravitel'stva RF ot 26 yanvarya 2016 g N 80-r «Ob utverzhdenii Strategii razvitiya zhilishchno-kommunal'nogo khozyaistva v RF na period do 2020g» [Order of the Government of the Russian Federation dated January 26, **80-p**

[3] Maznitsa E and Brishten S 2013 Formirovanie sistemy pokazatelei i primenenie metoda analiza ierarkhii dlya opredeleniya effektivnosti innovatsii [Formation of a system of indicators and application of the hierarchy analysis method for determining the effectiveness of innovations] *Ekonomicheskii analiz – teoriya i praktika* **13(316)** 24 -33

[4] Maznitsa E , Arcov S and Fyedorovyh E 2016 Public-private partnership as basis of modernization of the housing sphere of russia /Globalization and its socio-economic consequences, 15th International Scientific Conference , Proceedings, (Part III) - Rajecke Teplice, Slovak Republic, 1338-47 (Web of Science) Access mode: http://ke uniza sk/sites/default/files/content_files/proceedings_part_iii pdf

[5] Maznitsa E 2011 Perspektivy innovatsi-onnogo razvitiya regional'nogo ZhKKh na osnove tochechnogo (lokal'nogo) reformirovaniya: monografiya [Prospects for the innovative development of regional utilities based on point (local) reform: monograph] Volgograd: VolgGASU, 107

[6] Maznitsa E 2014 Razvitie svyazei vysshei shkoly i biznesa kak sposob sokhraneniya investitsionnoi aktivnosti v regionakh Rossii v usloviyakh finansovogo krizisa Aktual'nye problemy razvitiya vertikalnoi integratsiisistemy obrazovaniya, nauki i biznesa: ekonomicheskie, pravovy I sotsialnye aspekty [Development of relations between higher education and business as a way to preserve investment activity in the regions of Russia in the context of the financial crisis Actual problems of the development of vertical integration in the education, science and business systems: economic, legal and social aspects]:materials of the II International Scientific Practical ConferenceOctober 23-24, 2014 - T 2 / ed S L Igolkina – Voronezh *Voronezh scientific and technical information center* **243** 35 -43

[7] Pridachuk M P , Cyzho L N and Boldyreva M S 2019 The development of the system of organizational and environmental indicators for the development of the housing sector in Russia // International Scientific Conference "Competitive, Sustainable and Safe Development of the Regional Economy" The materials will be posted on the website of the publishing house Atlantis Press and indexed in Conference Proceedings

[8] Alkemade F , Heimeriks G , Schoen A , Villard L and Laurens P 2015 Tracking the internationalization of multinational corporate inventive activity: national and sectoral characteristics *RESEARCH POLICY*, **44(9)** 1763-72 doi: 10 1016/j respol 2015 01 007

[9] Bannò M , Piscitello L and Varum, C 2013 Determinants of the Internationalization of Regions: The Role and Effectiveness of Public Policy Measures *Regional Studies*, **49(7)** 1208-22 doi:10 1080/00343404 2013 821570

[10] Barzi, Federica; Cortelezzi, Flavia; Marseguerra, Giovanni; Zoia and Maria Grazia 2015 Determinants of the Internationalization of Regions: The Role and Effectiveness of Public Policy Measures INNOVATION-MANAGEMENT POLICY & PRACTICE, Cortelezzi, Flavia/0000-0001-7481-1541, 10 1080/14479338 1036541, 281-307 Begg, I (1999) Cities and Competitiveness CURS *Urban Stud Urban Studies*, **36(5)**, 795-809 doi:10 1080/0042098993222

[11] Maznitsa E, O V Maksimchuk N I and Borisova L N 2017 Chizho Modernizatsiya zhilishchno-kommunalnogo khozyaistva goroda na osnove realizatsii innovatsionnykh reshenii [Modernization of the Housing and Public Utilities of the City through the Implementation of Innovative Solutions] Volgograd State Technical University – Volgograd: VolgGTU

[12] Belyaeva L 2009 The level and quality of life Challenges in measuring and interpreting *Sociological Studies*, **1** 33-42

[13] Cardenas Davila, Cesar F, Vargas Hernandez and Jose G 2015 Strategic-spatial analysis of the implementation of business opening politics of Mexico FAEDPYME INTERNATIONAL REVIEW-FIR JUL-DEC 8-25

[14] de Dominicis, Laura; Florax, Raymond J G M ; de Groot and Henri L F 2013 Regional clusters of innovative activity in Europe: are social capital and geographical proximity key determinants? *APPLIED ECONOMICS* 10 1080/00036846 **663474** 2325-35

[15] Kennedy A (2016) Smart Specialization, Regional Growth and Applications to European Union Cohesion Policy *REVIEW OF INTERNATIONAL POLITICAL ECONOMY*, **23:1** 65-92, 10 1080/09692290 2015 1105845, JAN 2 2016

IOP Conf. Series: Materials Science and Engineering **786** (2020) 012077 doi:10.1088/1757-899X/786/1/012077

Development of a model of the single information space of the enterprise supporting the replacement of layers

B M Shifrin[1], I V Eliseev[1,*], V A Sokolova[1] and V A Kalyashov[2]

[1]Saint-Petersburg State Forest Technical University

[2]Saint Petersburg State University of Architecture and Civil Engineering

*E-mail: yeliseyef@yandex.ru

Abstract. The main purpose of this article is to substantiate and develop the methodology of replacement of generations of layers in an attempt to minimize destruction of links between adjacent layers of the single information space of the enterprise, caused by extreme instability of states of the single information space.

1. Introduction

Nowadays, one of the most popular development trends of information technologies is an attempt to unify all information resources and systems, handling the data in a remote multiuser mode, this is especially necessary for the development and support of automated systems in construction and engineering, requiring constant updating. Such an integration of the data and the handling software within the company constitutes the single information space (SIS) of the enterprise.

In theory, the mission of the SIS of the enterprise is to avoid information's doubling thanks to automatic synchronization of databases of various information systems (IS). This will enable to reduce organisational barriers between various departments, owing to quick exchange of electronic data, as well as to cut down time and labour costs falling on the paperwork. Thus, the controllability of a business can be increased through transparency of operations with data and online monitoring of SIS users' activity.

It is noted in the modern literature [1-4] that, besides technological aspects of the creation of the SIS of the enterprise, there are points of the similar importance: the management, the organizational and the financial. According to management points, the leaders of the enterprise should not be simply interested in deployment of the SIS but rather actively favoring it. Within organizational aspect, the enterprise's personnel supposed to be the immediate users of the single information space, is to comprehend purposes and objectives of its creation. Furthermore, as every solid campaign, the development, acquisition and launching of the firmware of the SIS requires an appropriate budget, for its financial costs.

2. The body of the text

An universal SIS of the enterprise can be represented by a hierarchy of a set of n layers $\{S\}=\{S_1,\ldots, S_i, \ldots, S_n\}$, which possess the accumulated information (the level of knowledge) $\{I\}=\{I_1,\ldots, I_i, \ldots, I_n\}$ and influence each other (Figure 1).

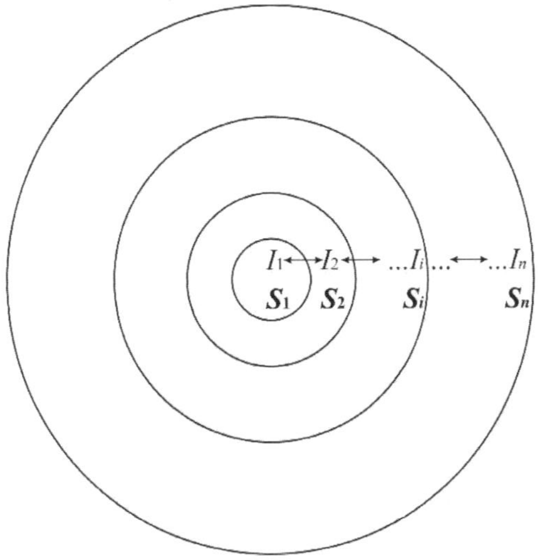

Figure 1. The universal SIS of the enterprise.

Consequently, the dependence of the knowledge of a higher level on the knowledge of lower level can be formalized by means of conversion function, as follows:

$$I_i = I_{i-1,i}(\{F_{i-1,i}\}, I_{i-1}),\ i = (2 \div n), \tag{1}$$

where $\{F\}$ is the set of various factors affecting the conversions (for instance, the topology and the capacity of data networks, the technical architecture of the hardware etc.), whilst the dependence the knowledge of the lower level on the knowledge of the higher level can be formalized in the following way:

$$I_i = I_{i+1,i}(\{F_{i+1,i}\}, I_{i+1}\},\ i = (1 \div n-1). \tag{2}$$

In a simplified form, the lifecycle of the IS related to the layer S_i can be presented as follows (Figure 2).

IOP Conf. Series: Materials Science and Engineering **786** (2020) 012077 doi:10.1088/1757-899X/786/1/012077

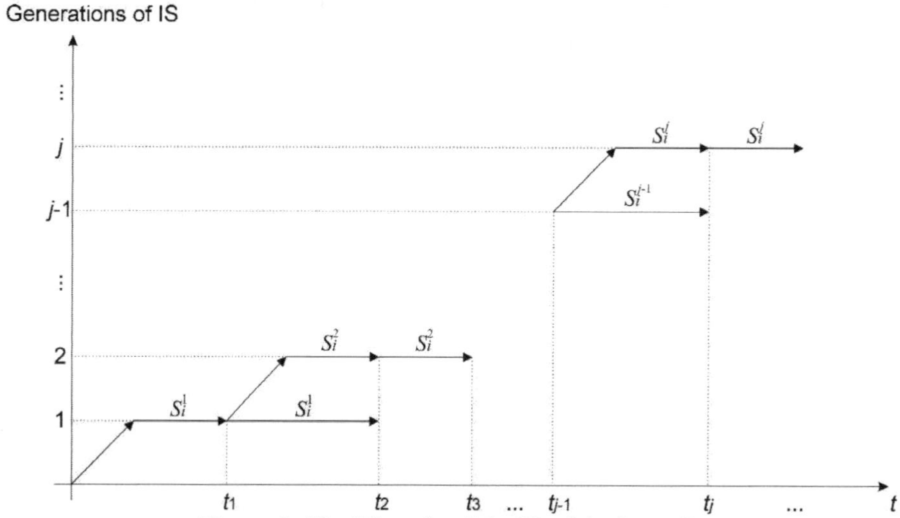

Figure 2. The lifecycle of the IS of the layer S_i.

To the time t_1 the system will stop its modernization, but the knowledge accumulated during its running time is relevant, that is to the time t_2 there are two simultaneous systems belonging to various generations the first and the second (S_i^1, S_i^2). The phase between t_2 and t_3 represents the situation where the preceding generation is to be completely replaced by the new generation of the IS, etc.

Let us, by using the automatic approach [5-10], compose the transition graph for the layer S_i. (Figure 3).

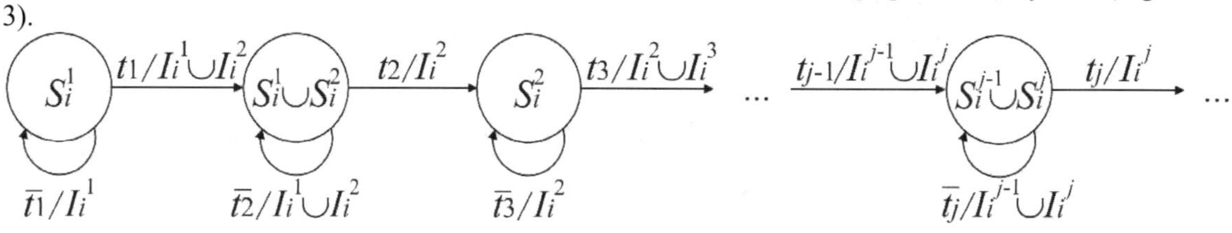

Figure 3. The transition graph for the layer S_i, by generations.

The graph nodes (the positions) are generations of the lifecycle of the IS of the layer Si (S_i^1, S_i^1, S_i^2, S_i^2,..., S_i^{j-1}, S_i^j,...).

The graph arcs show the conditions under which the position will change and those under which those will not (so called loops). Near arc, it is accepted to show under which input actions the transition is about to be, and which output signals are to be generated.

In this way, the Figure 3 shows that until time t1 ($\overline{t_1}$) is reached, the IS of the first generation with the knowledge I_i^1 level is being used; after reaching time t1, there will be two systems with accumulated knowledge I_i^1, I_i^2, belonging to various generations (S_i^1, S_i^2) will be in use, etc.

Now let us formulate the global criteria of minimization the difference between transitions to other layers, in changing the generation.

Using the expression (1), for the set of generations 1, 2, …, j, ... we obtain the following dependences of the knowledge of the higher level on the knowledge of the lower level:

$$I_i^1 = \varphi_{i-1,i}^1(\{F_{i-1,i}^1\}, I_{i-1}^1),\ I_i^2 = \varphi_{i-1,i}^2(\{F_{i-1,i}^2\}, I_{i-1}^2),\ \ldots,\ I_i^j = \varphi_{i-1,i}^j(\{F_{i-1,i}^j\}, I_{i-1}^j),\ \ldots, i = (2 \div n).$$

Then, the set of goal functions minimizing the losses caused by destruction of dependencies of the knowledge of the higher level on the knowledge of the lower level for n layers can be formulated in the following way:

$$d^{1,2}_{i-1,i}(\varphi^{1}_{i-1,i}, \varphi^{2}_{i-1,i}) \to \min ,$$
$$d^{2,3}_{i-1,i}(\varphi^{2}_{i-1,i}, \varphi^{3}_{i-1,i}) \to \min ,$$
$$\ldots,$$
$$d^{j-1,j}_{i-1,i}(\varphi^{j-1}_{i-1,i}, \varphi^{j}_{i-1,i}) \to \min ,$$
$$\ldots$$

Likewise, by using the expression (2), for the set of generations 1, 2, …, j, ... the following dependencies the knowledge of the higher level on the knowledge of the lower level are obtained:

$$I^{1}_{i} = \varphi^{1}_{i+1,i}(\{F^{1}_{i+1,i}\}, I^{1}_{i+1}),$$
$$I^{2}_{i} = \varphi^{2}_{i+1,i}(\{F^{2}_{i+1,i}\}, I^{2}_{i+1}),$$
$$\ldots,$$
$$I^{j}_{i} = \varphi^{j}_{i+1,i}(\{F^{j}_{i+1,i}\}, I^{j}_{i+1}),\ i = (1 \div n\text{-}1).$$

Then, the set of goal functions, which minimize the losses cause by destruction the dependencies of the knowledge of the higher level on the knowledge of the lower level for layers n, are formalised as follows:

$$d^{1,2}_{i+1,i}(\varphi^{1}_{i+1,i}, \varphi^{2}_{i+1,i}) \to \min ,$$
$$d^{2,3}_{i+1,i}(\varphi^{2}_{i+1,i}, \varphi^{3}_{i+1,i}) \to \min ,$$
$$\ldots,$$
$$d^{j-1,j}_{i+1,i}(\varphi^{j-1}_{i+1,i}, \varphi^{j}_{i+1,i}) \to \min ,\ldots$$

In the Figure 1, the universal SIS of the enterprise is displayed as a hierarchy of layers. For many modern companies engaged in IT activity, the SIS can be represented in the following way (Figure 4).

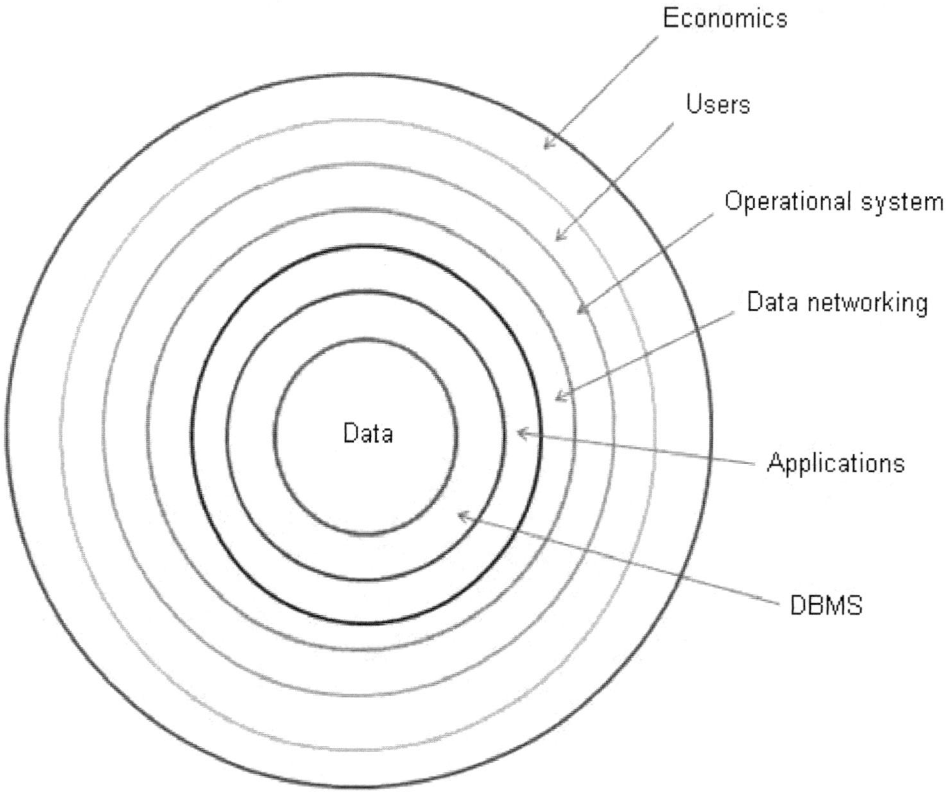

Figure 4. The SIS of an IT-enterprise.

Information technologies should easily adopt changes of the outside environment and requirements of the business. This is in fact the goal of improvement of mechanisms of software re-usage and increasing their compatibility and transferability. Processes of migration and modernization, known as collective name "re-engineering", are of great demand [11].

Everything in this world is subject to change, business requirements to information systems included. As a result, such systems are subject to be enlarged, widened and integrated. Their number and function grow. This leads to the qualitative transformation of the automation concept itself, required re-organization of information infrastructure covering items automated. Sooner or later, the re-engineering is noted. This is not only typical for "inherited" systems but for relatively new developments as well.

A ground to rebuild the information system can now be of a politic or psychology nature: the lead programmer has moved away; the relations to the platform developer have been decayed; the level of technical support or contractor's financial level has been changed. With no doubt, the key factor is the total cost of ownership of the information system.

Consequently, due to various circumstances – and those are not mandatory the technical issues –one popular database management system is often today changed one for another. Moreover, projects of transition to another DBMG appear, and this presumes organization of the support of more than one platform.

Many replicated solutions are developed for the enterprises of various scales and for the wide society of clients, everyone of each has its own preferences. Even having developed the concept in accordance with requirements of a company, one can face the requirement to apply different servers. Say, a central office of a geographically distributed company can be operated on Oracle platform, and their affiliates (handling lower data volumes and having little personnel) are run by MS SQL Server, whilst mobile users are supported through MS SQL Desktop or Sybase.

In accordance with our hierarchy, let us consider layer S_2, where the objects stored in the bases, the libraries etc. are used in capacity of accumulated information I_2.

The global problem, solved in present days to some extent, is to keep data available (layer S_1), where ERP data (the financial, accounting and operative information etc.) are used in capacity of I_1 , in changing the DBFS.

Having used the ratio (1) and accepted the current level of the generation of DBMS as j, we come to the following information dependences φ_{12}^{j} , φ_{12}^{j+1} between layers S_1 and S_2:

$$I_2^j = \varphi_{12}^j(\{F_{12}^j\}, I_1^j);\ I_2^{j+1} = \varphi_{12}^{j+1}(\{F_{12}^{j+1}\}, I_1^{j+1}),$$

where the set of main affecting factors $\{F_{12}^{j}\}$, $\{F_{12}^{j+1}\}$ includes:

1. the extreme values of the information being stored and handled by the IS;
2. the condition of the IT fleet;
3. the natural obsolescence of development tools;
4. the increase of payload of the IS die to increase of computational processes and/or their users;
5. the development of integration mechanisms with adjacent IS based on other platforms;
6. the alteration of business-processes of the enterprise (the classification of company's profile);
7. the alteration of license policy of software's suppliers.

Then, the set of goal functions D, which minimize the loss in transiting of the DBFS layer from S_2^j to S_2^{j+1} , can be presented in the form of degrees of difference of information dependences φ_{12}^{j} , φ_{12}^{j+1} :

$$D(\varphi_{12}^j, \varphi_{12}^{j+1}) \min.$$

Two main factors affecting in a negative way in changing the generation of the database:

1. the degree of risk to loose the functionality of the system for users, thanks to difference between information components (let us name it d_1);
2. the degree of loose of the availability level of historical data and the persistence of the registration (let us name it d_2).

Then, the following is obtained:

$$d_1(\varphi_{12}^j, \varphi_{12}^{j+1}) \min;\ d_2(\varphi_{12}^j, \varphi_{12}^{j+1}) \min.$$

Presently, two main classic approaches to the migration between DBMS are used: the approach I (building of the system «from ground zero») and the approach II (transition of data with the stand steel of the current system).

The approach I will in any event dissatisfy the developed model by the goal function d_2, and the approach II is efficient if the time of transferring data is limited – normally the transition time is not more than 60 hours. Otherwise, such a transition will not satisfy the developed model by goal function d_1.

A conclusion can be drawn, that in a climate of running of big enterprises' corporative information systems which as a matter of fact consists of various databases (more 10) with sound volume of information (several terabytes per each DB) and the requirement to implement persistent analytical functions, the existing approaches are not efficient in usage, and a principally new methodology of step-by-step migration between DBMS of every DB without loss of overall functionality of the system, is to be developed.

3. Conclusions

The grave problem of the development of the single information space of the enterprise lies in the fact that the position of the SIS is very unstable and is subject to many factors: the digenesis of hardware,

IOP Conf. Series: Materials Science and Engineering **786** (2020) 012077 doi:10.1088/1757-899X/786/1/012077

the digenesis of developers, the alteration of business requirements, the alteration of country's legislation etc.

Most likely, it may be affirmed that the instability of the SIS leads to intersection in functionality of various IS within the enterprise, causing significant losses. In this connection, the idea to develop the methodology of replacement of generations of layers in an attempt to minimize destruction of links between adjacent layers, is seen actual.

Reference

[1] Sadykov A M 2017 Problemy sozdaniya i avtomatizacii upravleniya edinogo informacionnogo prostranstva predpriyatiya V sbornike:Innovacii v tekhnologiyah i obrazovanii Sbornik statej uchastnikov IX Mezhdunarodnoj nauchno-prakticheskoj konferencii S 162-4

[2] Katasonova G R and Sotnikov A D 2017 Ispol'zovanie metodov formalizacii pri organizacii edinogo informacionnogo prostranstva predpriyatij V sbornike:Komp'yuternaya integraciya proizvodstva i IPI-tekhnologii Materialy VIII Vserossijskoj nauchno-prakticheskoj konferencii Orenburgskij gosudarstvennyj universitet S 555-8

[3] Kuharenko A M and Anohina M E 2018 Rol' edinogo informacionnogo prostranstva predpriyatiya v povyshenii effektivnosti biznesa V sbornike:Tekhnologii informacionnogo obshchestva Sbornik trudov XII Mezhdunarodnoj otraslevoj nauchno-tekhnicheskoj konferencii S 339-40

[4] Lapina T I , Dimov E M , Petrik E A and Lapin D V 2018 Upravlenie dostupom k informacionnym resursam v informacionnyh sistemah *Modelirovanie, optimizaciya i informacionnye tekhnologii* 6 **4(23)** 523-34

[5] Eliseev I V 2007 Osnovnye polozheniya avtomatnogo proektirovaniya sistem logicheskogo upravleniya ob"ektami lesopromyshlennogo kompleksa Izvestiya Sankt-Peterburgskogo gosudarstvennogo agrarnogo universiteta **178** 92

[6] Eliseev I V 2007 Osnovnye polozheniya avtomatnogo proektirovaniya sistem logicheskogo upravleniya ob"ektami lesopromyshlennogo kompleks *Izvestiya Sankt-Peterburgskoj lesotekhnicheskoj akademii* **178** 92-5

[7] Shifrin B M 2008 Primenenie SWITCH-tekhnologii dlya razrabotki logicheskoj modeli upravleniya processom sushki shpona Izvestiya Sankt-Peterburgskoj lesotekhnicheskoj akademii **185** 186-194

[8] Shifrin B M 2009 Primenenie SWITCH-tekhnologii dlya razrabotki logicheskoj modeli upravleniya processom zagruzki shpona Izvestiya Sankt-Peterburgskoj lesotekhnicheskoj akademii **186** S 151-5

[9] Shifrin B M , Eliseev I V and Pendrikov E S 2018 Sintez diskretnyh avtomatov: uchebno-metodicheskoe posobie SPb: SPbGLTU, 36

[10] Eliseev I V , Shifrin B M and Sokolova V A 2018 Razrabotka logicheskih sistem upravleniya tekhnologicheskimi processami lesopileniya pilomaterialov *Sistemy Metody Tekhnologii* **4(40)** 157-62

[11] Shapot M and Popov E V 2004 Reinzhiniring baz dannyh *Otkrytye sistemy* SUBD **4**

Organizational capacity of cdio syllabus in actualization of the objectives of engineering education from regional perspective

G V Bukalova[1], A N Dorofeev[2,3] and A N Novikov[1]

[1]Orel State University, Orel, Russia

[2]National Research University Higher School of Economics, Moscow, Russia

[3]Financial University under the Government of the Russian Federation, Moscow, Russia

E-mail: 57orleya@gmail.com

Abstract. Engineering and technical progress is transforming the automotive industry rapidly. Conformance of professional competencies of college graduates to the requirements of the regional industry is seen as the main competitive advantage at the regional labor market. Both the business and engineering academic communities recognize the importance and stand ready for continual actualization of educational objectives. An important (from both theoretical and practical standpoints) question is the form of presentation of training results declared by a higher education institution for the purpose of assessing its relevance by the business community.Using the CDIO Syllabus architecture as the basis of a structured system for polling representatives of the industry and the academic community of the region, i.e. the two main stakeholders, certain information was obtained for updating educational objectives of automotive engineering undergraduate programs. The study conducted made it possible to distinguish and detail value and target-based opinions of respondents, as well as to evaluate, in absolute and relative terms, the level of relevancy of a graduate's competencies declared by a college.

Presently, a strategic program to develop the national system of competencies and proficiencies within the Russian industry is being implemented [Natalia Shmatko, 2016, pp. 340-354] , [Gurtov V. A., Garifullina N. Yu., Sigova S. V., 2016, pp. 68–75]. The purpose of this reform is to transform today's labor market into a market of professional competencies. The production sector of the automotive industry also develops a new set of requirements to labor resources, which views personnel competencies as main productive values. This major restructuring of the Russian national qualification system has created an intense impetus of changes within the system of engineering training. Updating training programs with the aim of "improvement of quality of professional training, as well as ensuring its flexibility and adaptability with respect to labor market challenges" is deemed of vital importance.

Professional training of engineering and technical staff for the automotive industry is closely related to the modern trends of development of the said industry's regional sector. Conformance of professional competencies of college graduates to the requirements of the regional industry is seen as a major asset and the main competitive advantage at the regional labor market. It is common knowledge that "when facing a highly competitive professional labor market college graduates that have inadequate competency are bound to accept positions requiring a lower level of training, which is lower than the training level of such graduates".

The main source of engineer and technical staff of automotive service enterprises in the Oryol Region is the Polytechnic College named after N. N. Polikarpov, a division of the Oryol State University. A study of relevance of the College graduates' competences included into automotive engineering training curriculum was carried out on the basis of Vehicle Service and Repair Department. During the study, the degree of conformance of the professional competency of a graduate who has completed the undergraduate program "Operation of Vehicles and Production Machinery and Complexes" declared by the College with the requirements of the regional industry to qualitative characteristics of engineering and technical staff of a vehicle service business was determined.

Empirical data for the study were obtained using feedback of respondents, managers and employees of the most competitive car servicing enterprises of the Oryol Region. The respondent selection also included representatives of academic staff and students of the University.

The competency-based approach was selected at the methodological framework of studies. The applied method of study used a formalized expert poll having the form of a questionnaire survey, which used modular questionnaires to allow for a comparative analysis of respondent feedback. Actual significance of graduate competencies declared by the College was adopted as the subject of evaluation. An important issue, which arose during the preparation of the survey, was the problem of selecting the form of representing training objectives to the professional community in view of the difference of terminology used to describe competencies by the industry from that used in training standards.

CDIO Syllabus, the architecture of educational objective representation was used as the form of presentation of educational objectives to the professional community with the purpose of analyzing their relevance. This methodological approach was applied in order to focus respondents' attention on major structural elements of a college graduate's professional competency, and not just on their constituent individual competencies that are frequently perceived by the academic circles and the industry differently. The analyzed set of graduate competencies was comprised of four blocks: 1) fundamental knowledge, 2) professional competencies, 3) communicative skills, and 4) system engineering. A 100-point grading scale was used.

Survey results revealed major differences in the perception of relevant significance of education standards (competencies) by respondents. The results of evaluations given by employers, graduates, teachers and students expressed in relevant significance points are shown in Figure 1.

Consolidation of response analysis results reveals that there is a substantial gap between employer's expectations relating to professional training level of college graduates and the regulation-based competency level declared by the college. Interestingly, these competencies are not part of professional or operational sphere, which could be expected, they lie in managerial, ethical and moral sphere instead.

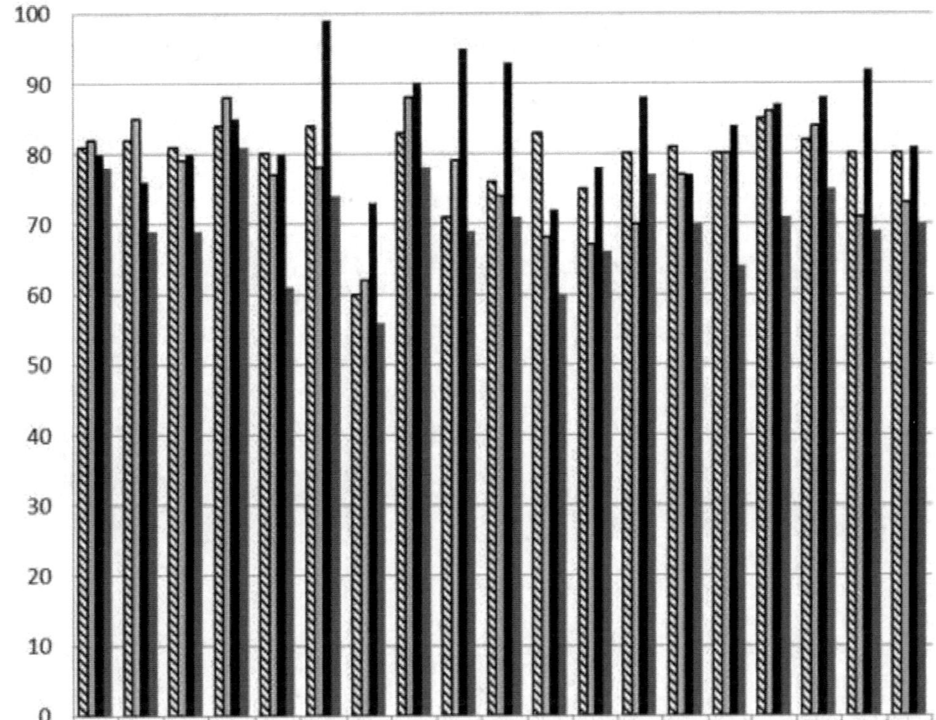

	1.1	1.2	1.3	2.1	2.2	2.3	2.4	2.5	3.1	3.2	3.3	4.1	4.2	4.3	4.4	4.5	4.6	4.7	4.8
Student	81	82	81	84	80	84	60	83	71	76	83	75	80	81	80	85	82	80	80
Teacher	82	85	79	88	77	78	62	88	79	74	68	67	70	77	80	86	84	71	73
Employer	80	76	80	85	80	99	73	90	95	93	72	78	88	77	84	87	88	92	81
Graduate	78	69	69	81	61	74	56	78	69	71	60	66	77	70	64	71	75	69	70

Figure 1. Respondent Evaluation of Relevance of Regulation-Based Training Objectives (College Graduate Competences).

1.1 Knowledge of math and natural sciences; 1.2 Basic engineering knowledge; 1.3 Knowledge of engineering techniques and methods; 2.1 Ability to solve production-related issues in terms of their technical and economic expediency; 2.1 Ability to solve production-related issues in terms of their technical and economic expediency; 2.2 Ability to experiment and research and ability to gain professional knowledge individually; 2.3 Competencies reflecting consistency of professional reasoning; 2.4 Ability to assume a meaningful social stance in terms of professional activity effectiveness; 2.5 Competencies reflecting the ability to take professional responsibility in terms of ethical standards; 3.1 Ability to take part in group production activities; 3.2 Ability to maintain business communication; 3.3 Communication skills in foreign languages; 4.1 Social and environmental competencies; 4.2 Ability to identify production processes from the standpoint of economy; 4.3 Ability to plan production and managerial activities; 4.4 Ability to design; 4.5 Ability to take part in production; 4.6 Ability to apply engineering solutions in practice; 4.7 Ability to manage engineering activities; 4.8 Entrepreneurship as part of engineering activities.

The most significant inconsistencies between positions of employers and college teachers are the significance of competencies within the group "Professional accountability in terms of ethical standards". On the average, college teachers evaluated competencies of this block 20% lower than employers. The significance of the group "Leadership as part of engineering activities" was evaluated by college teachers 23% lower than by employers. Some material divergences of positions of employers and college teachers were revealed in relation to the significance of competencies within the group "Consistency of professional reasoning".

The level of significance of competencies within the Consistency of professional reasoning group was 21% higher for employers than for college teachers. It should be noted that competencies within this group comprise an essential element of systems engineering implementation, the modern widely acknowledged methodology used in production organization.

The lowest grades of the analyzed competencies were given by graduates employed by the industry. The results of the poll gave evidence that this group of respondents was the most pessimistic in their estimates of relevance of education objectives subjected to analysis.

The results of the study were summarized based on the values of the relative competency block relevance indicator (see the Table 1).

Table 1. Relative Competency Block Relevance Indicator.

Block of Analyzed Competencies	Respondent response (points)			
	Employers	Teachers	Students	Employed Graduates
1. Basic knowledge	75	82	81	73
2. Professional competencies	85.4	78.6	78.6	78.2
3. Communication skills	87	74	72.6	67
4. System engineering	84	76	80	70

The results of the study with regard to the *Basic knowledge* block demonstrated a clear focus of teachers and students on gaining theoretical knowledge within future professional sphere.

Relative relevance indicator of the *Professional competencies block* revealed a minor (8%) difference in its evaluation by employers and teachers. Assessment of employed graduates was much different from that of the other groups of respondents: the difference amounted to 18%.

A major difference in the evaluation of competencies within the *Communication skills block* by employers and teachers was also revealed. This points to a conclusion that teachers need to focus more on improving the significance of training objectives described by competencies within the Interpersonal skills group.

Relative relevance indicator of the *System engineering block* allowed to reveal a high level of conformity of competency relevance evaluation within this block between employers and students. This denotes a fairly developed motivation of students in the acquisition of competencies of the block. Nevertheless, numeric values of the indicator reveal that the relevance of this block's competencies to the teachers is much (by 10%) lower than that expected by employers. This reinforces the already mentioned necessity to change the attitude of teachers to the development of the Systems engineering block competencies.

Conclusion

A comparative analysis of relevant significance of the discussed competencies revealed a definitive exceedance of employer values over the perception of these competencies by college teachers. In 80% of cases, competency relevance estimates given by employers exceeded those of college teachers. This clearly indicates the need for a meaningful alteration of value and conceptual position of college teachers implementing the teaching process of the undergraduate program "Operation of Vehicles and Production Machinery and Complexes".

One peculiarity of the expert poll results for graduates employed within the industry was that their competency significance evaluations were lower (by 3% to 23%) than those of employers throughout the assessed competency groups. Responses of employer respondents lay within the strategic planning outlook of "how things should look in the future". Responses of employed graduates classified as specialists reflect the situation "as is".

Acquisition of competencies by college students is dependent upon the perception of these competencies as being significant to achieve training objectives by academic staff. Revealed considerable inconsistencies between perception of significance of training objectives by college teachers and expectations of major regional employers pose a real threat to quality level of professional training of the undergraduate program "Operation of Vehicles and Production Machinery and Complexes".

References

[1] Shmatko N 2016 "Researchers' competencies in the coming decade: attitudes towards and expectations of the Russian innovation system", Foresight, Vol 18 No 3, pp 340-354
[2] Gurtov V A , Garifullina N Yu , Sigova S V 2016 Forecasting Recruitment Needs of the Russian Economy: Qualitative Aspects Studies on Russian Economic Development, Vol 27, No 1, pp 68–75
[3] Formirovanie Nacional'noj sistemy kvalifikacij: sbornik normativnyh dokumentov (2014) M : Pero
[4] Meissner D , Shmatko N 2019 Integrating professional and academic knowledge: the link between researchers skills and innovation culture J Technol Transf, 44:1273–1289
[5] Varshavskaya, E YA , Kotyrlo, E S 2019 Vypuskniki inzhenerno-tekhnicheskih i ekonomicheskih special'nostej: mezhdu sprosom i predlozheniem Voprosy obrazovaniya /Educational Studies Moscow 2 98–128
[6] Metodicheskie rekomendacii po aktualizacii dejstvuyushchih federal'nyh gosudarstvennyh obrazovatel'nyh standartov vysshego obrazovaniya s uchetom prinimaemyh professional'nyh standartov Utverzhdeno ministrom obrazovaniya i nauki RF (2015) №DL – 02/05vn URL: http: // fgosvo ru/support/43/5/8/
[7] Chiocchia G, Gola M and Guglieri G 2017 National Qualification Frameworks for Aerospace Engineering Education in Western Europe and in the Russian Federation: the AIRQUAL Project 2009-11 Aerotecnica Missili & Spazio 96(2), 110–6
[8] Pukharenko Yu V , Norina N V and Norin V A 2017 Russian engineering education in the era of change European Journal of Engineering Education, 42(2) 171-18
[9] Bukalova G V and Novikov A N 2016 Obrazovatel'nye normativy kak sredstvo dekompozicii professional'nyh standartov 5-aya Mezhdunarodnaya nauchno-prakticheskaya konferenciya "Aktual'nye voprosy innovacionnogo razvitiya transportnogo kompleksa" sbornik nauch trudov Orel FGBOU VO «OGU imeni I S Turgeneva» - S 349-357
[10] Terentyeva I , Kirillova O , Kirillova T , Pugacheva N , Lunev Ar, Chemerilova I and Luchinina A 2018 Arrangement of cooperation between labour market and regional vocational education system International Journal of Educational Management 8 1041-155
[11] Kliucharev G A , Latov Iu V 2016 Continuing Education, Sociological Research, 55(4) 225-44
[12] Lunev A, Petrova I , Zaripova V 2013 Competency-based models of learning for engineers: a comparison, European Journal of Engineering Education, 38(5) 543-55
[13] Bukalova, G V , Novikov, A N 2018 Aktualizaciya obrazovatel'nyh normativov – osnova opredeleniya izmenenij v soderzhanii inzhenernogo obrazovaniya «Mir transporta i tekhnologicheskih mashin» 3(62) 129-60
[14] Chuchalin A (2018): Evolution of the CDIO approach: BEng, MSc, and PhD level, European Journal of Engineering Education
[15] Chuchalin A, Malmqvist J , Tayurskaya M 2016 Professional development of Russian HEIs' management and faculty in CDIO standards application, European Journal of Engineering Education, 41(4) 426-37

[16] Kosyakov A , Svit U , Sejmur S , Bimer S 2014 Sistemnaya inzheneriya Principy i praktika M : DMK- Press

[17] Perfil'eva, O V 2014 Universitety i regional'noe razvitie: teoreticheskij analiz i metodologiya issledovaniya Izvestiya Saratovskogo un-ta Ser Ekonomika Upravlenie Pravo 14(3) 479– 88

[18] Postalyuk, N YU 2014 Mekhanizmy translyacii kvalifikacionnyh trebovanij rynka truda v programmy professional'nogo obrazovaniya i obucheniya Professional'noe obrazovanie v Rossii i za rubezhom 3(15) 37-40

IOP Conf. Series: Materials Science and Engineering **786** (2020) 012079 doi:10.1088/1757-899X/786/1/012079

New Way To Determine The Technical Condition Of Ball Joints

A N Novikov[1], S V Eremin[1] and A G Shevtsova[2,*]

[1]Orel State University named I.S. Turgenev «OSU named I.S. Turgenev» Orel, Russia

[2]Belgorod state technological University named V.G. Shukhov «BSTU named V.G. Shukhov» Belgorod, Russia

E-mail: shevtsova-anastasiya@mail.ru

Abstract. Among the multitude of vehicle suspension assemblies, the ball joint is one of the most important bearing assemblies, the operability of which directly depends on the operating conditions. Ball joints are used to connect the suspension and steering components and are kinematic interfaces that have three angular degrees of freedom when moving the working elements (levers). The reliability of the hinges has a significant impact on the safety of the car. In most cases, the ball joint is an indiscriminate node, which makes it difficult to determine the technical condition in a car service center. One of the methods of diagnosing are laboratory bench examinations of ball joints. To obtain reliable information about the wear of ball joints and the development of a technique for determining their technical condition in the conditions of an auto service company, a laboratory bench research area is needed. The completed research aimed at developing methods, tools and algorithms for determining the technical state of ball joints in the conditions of an auto service company. As a result of the research, the author's team developed a vibration method for determining the technical state of ball joints and a mathematical model for changing the technical condition of the ball joint of the front suspension of the MacPherson type of a car, in bench tests.

1. Analysis of test methods for ball joints

The efficiency of the hinges is determined by the gap in their interfaces. The excess of the permissible norms of these gaps is determined by the driver subjectively to the appearance of vibrations on the steering wheel, knocking and extraneous sounds from the suspension and running gear of the car, loss of controllability of the car while driving, but these signs also appear in the case of malfunction of other chassis and steering elements -blocks, bearings, steering rods and their tips, shock absorbers and hinges of their fastening). The above facts lead to a decrease in the durability of the suspension, increased wear of tires and wheel bearings. There is a decrease in active safety of the car [1].

One of the objectives of this study is to develop a method for diagnosing ball joints with varying degrees of wear, providing the most reliable information about their technical state [2].

Tests of ball joints, like any other devices, mechanisms and elements are divided into several types:

1. reliability, life, durability (cyclic resource tests);
2. for breaking, breaking, failure, resistance to maximum load (failure test);
3. on the definition of hidden gaps, magnitudes of defects, processes, composition (diagnosis).

For forecasting, as well as determining the installed service life of ball joints, as well as for determining their technical condition, automobile enterprises carry out various accelerated road, laboratory and bench tests and studies of suspensions [3,4].

Sometimes obtaining reliable information when carrying out road research on durability and reliability seems to be the most rational, since the approach of the loading regimes is the greatest when tested to operational conditions. However, the above presented method, for all its simplicity of the testability test plan, and comparability of results has drawbacks, such as high cost, complexity and duration of implementation [5].

The most effective way of investigating the ball joints is to reproduce it in the laboratory. At the same time, the accuracy of the reproduction of the process determines the completeness and reliability of the information.

Tests of the front suspension assembly simplify the testing of ball joints and can significantly reduce the additional difficulties that arise when creating the necessary test conditions, close to the real operating conditions. This allows you to obtain results with a high degree of certainty. The lack of such tests is the impossibility of obtaining reliable information about the changes in the parameters of individual suspension elements such as ball joints, etc. Negative moments are also unjustified costs (energy, time, material, etc.) for testing the suspension as a whole in conditions where it is necessary to obtain reliable information on the nature of the change in the parameters of a particular suspension element [6].

To obtain the necessary information on wear and the technical condition of the ball joints, bench laboratory tests are necessary. The test bench used to test the ball joints must fully reproduce the movements and loads that are hinged under real conditions.

In addition, to be able to fully trust the results of bench tests, it is necessary to verify them with data obtained in real conditions [7].

2. Determination of the basic parameters of the operation of the ball joints of the front suspension type "McPherson"

For ball joints of the front suspension of the McPherson type of a passenger car, the parameter characterizing its technical condition is the axial clearance h. Limit is the state of the ball joint, in which the axial and radial clearances reach 0.7 mm, when the axial and radial loads ± 981 H are applied to the finger. These values are valid for automotive ball pins with a diameter of incomplete sphere d = 25-35 mm.

The dependence of the axial clearance h can be represented as a function:

$$h = f(F_L, F_{LAT}, F_V, N, \mu) \tag{1}$$

where F_L – axial loads directed along the longitudinal axis of the car, H; F_{LAT} – lateral loads inside the ball joint, H; F_V - vertical loads inside the ball joint, H; N – the operating time of the unit, is expressed in kilometers of car run; μ – coefficient of friction.

The loads and their direction are determined in the spherical hinges of the front suspension of the "McPherson" type on the example of a small class "Daewoo Nexia" passenger car under various conditions.

The diagnostic feature characterizing the technical condition of the ball joint is the movement of the finger relative to the hinge body in the vertical plane in view of the presence of a gap between the polymer insert and the ball pin. This movement during the operation of the suspension leads to the development of a gap h, which characterizes the appearance of vibrations in the hinge.

The mathematical model of changing the technical condition of the ball joint of the front suspension of the "McPherson" type of a passenger car under the conditions of bench research can be presented in the form of a differential equation:

$$m_1\ddot{x} + b_1\dot{x} + c_1x + (\dot{x} - \dot{y})b_2 + (x - y)c_2 = 0 \tag{2}$$

where x – vertical mass transfer m_1, m; $\dot{x}$ – mass speed m_1, m/s; $\ddot{x}$ – acceleration of mass m_1, m/s^2; y – kinematic effect of mass m_2 (forced displacement of the "free-running part of the installation"), m; $\dot{y}$ – the speed of the "free-running part of the installation", m/s; b_1 - viscoelastic resistance in the interface "lever-frame stand", m^2/s; b_2 - viscous resistance in conjugation "finger head - polymer insert", m^2/s; c_1 – coefficient of stiffness of interface "lever - frame of the stand", kg/s^2; c_2 - coefficient of stiffness of interface "head of the finger - polymer insert", kg/s^2.

In the course of the study of the equation obtained, its mathematical transformation was carried out and it was reduced to a dimensionless form:

$$\ddot{\xi} + \frac{\varepsilon 1}{\eta}\dot{\xi} + \frac{g1}{\eta^2}\xi = \frac{\varepsilon 2}{\eta}cos\tau + \frac{g2}{\eta^2}sin\tau \tag{3}$$

Modeling the change in the technical state of the ball joint of the McPherson suspension of a passenger car in the Mathcad software environment made it possible to establish that the main parameters affecting the magnitude of the acceleration are: axial dimensionless gap Δ, ε_1 and ε_2 are the damping factors in the lever and hinge coupling, respectively, g_1 and g_2 are the stiffness coefficients in the arm and hinge coupling, respectively, md (m) is the function, white noise simulation, η is the relative frequency.

3. Justification of the vibration method for determining the axial clearance in the ball joints of the suspension of cars

From the point of view of accuracy of measurement, informativeness, simplicity and speed of diagnosis of the ball joint, the most promising ones are diagnostic methods based on the use of capacitive vibration sensors installed on the investigated couplings of suspension and running gear. Their movement leads to the appearance of vibration in them, which leads to the formation of an electrical signal that can be converted into a gap size. Such methods of diagnosing in the conditions of service stations are most effectively used together with electro-hydraulic detectors [8].

The parameter studied, characterizing the technical condition of the ball joint, is the axial clearance h. As the axial clearance increases to h = 0.7 mm or more, the hinge is unsuitable for subsequent operation [9,10].

The aim of the experimental studies is to obtain the dependences between the value of the axial clearance h in the ball joints and the readings of the vibration sensors installed on the working elements of the stand. On the basis of the results obtained, we develop a method for determining the axial clearance h in the ball joints installed on vehicles with vibration sensors, without disassembling and removing the hinges from the vehicle.

The presence of vibration is due to the presence of gaps between the spherical part of the ball joint and the polymer insert. Vibrations appear in the ball joint when the vehicle is moving, with the opposite loads appearing in the unit under consideration when the wheel is shifted, when it moves over the unevenness of the roadway, acceleration, deceleration and car movement during the turn [11].

The vibration displacement (vibration displacement) is measured with a low-frequency vibration with an upper boundary of the frequency components of 100-200 Hz. Vibration speed, the main parameter of vibration, it characterizes the vibrational energy. The amplitude of the components of the vibration velocity in a sufficiently wide frequency band (10 - 1000 Hz) is uniform, which simplifies the measurement and increases the reliability. Vibration acceleration characterizes the dynamic force action of the elements on each other inside the mechanism, which leads to the considered vibration. In our case, this is the interaction of the spherical part of the finger and the polymer liner in the ball joint. The application of vibration acceleration is justified, since it does not need to be specially converted, it is measured in m /s^2.

4. Development of an experimental stand for diagnostic tests

The basis of the stand is one side of the modernized front suspension type "McPherson" car "Daewoo Nexia". During the work of the stand kinematics of motion and all loads in the ball joint remain, similar to the loads in the hinge of the car during its operation. As a source of reciprocating motion, a vertical double-acting hydraulic cylinder is used - 40x250-11-8811 (Figure 1a, 1b).

a) b)

Figure1. General view of the stand for diagnosing ball joints: a) mechanical part, b) hydraulic part: 1 - vertical (main) hydraulic cylinder, 2 - vertical hydraulic cylinder lever, 3 - rotary cam, 4 - stand frame, 5 - lever, 6 - Emergency shutdown button, 7 - horizontal hydraulic cylinder, 8 - horizontal hydraulic cylinder lever, 9 - controller, 10 - the hydraulic tank for driving the main hydraulic cylinder, 11 - the hydraulic distributor of the main hydraulic cylinder, 12 - the manometer, 13 - the drive for the main hydraulic cylinder, 14 - the hydraulic cylinder for the drive of the additional hydraulic cylinder, 15 - the additional hydraulic cylinder drive pump, 16 - hydraulic cylinder, 17 - the drive of the main hydraulic cylinder.

A safety valve with a built-in pressure regulator allows the vertical load to be varied from 150 to 4063 H (Table 1). The frequency of the stand is determined experimentally. When the pressure in the main hydrolysis line is 1.8 MPa, it is 2 Hz, which corresponds to the movement of the car on a road with a ground covering at a speed of 5-15 km/h.

Table 1. Maximum forces acting on the ball joint in the vehicle suspension and in the diagnostic stand.

Maximum strength, [H]	Car «Daewoo Nexia»	Diagnostic stand
Longitudinal, F_L	149	12-355
Lateral, F_{LAT}	1512	1591
Vertical, F_V	179	150-4063

The research method includes the installation of a hinge on the bench, the control measurement of the axial gap, the warming up of the stand, the installation of the test stand in the working position, the start and adjustment of the software, the launch of the stand, the reading of the vibration sensors. Then the stand stops, the data is checked and saved using the LabVIEW SignalExpress software.

The experimental substantiation of the method for diagnosing the technical condition of the ball joints of the vehicle suspension has been made, which will improve the safety of traffic [12-15]. The main purpose of the experimental studies was to obtain the dependences between the value of the axial clearance h in the hinges and the readings of the vibration probes. These data are necessary for the development of a technology that allows the determination of the axial clearance h in ball joints that are in real operation using vibration sensors, without disassembling and removing the ball joints from the vehicle.

5. Experimental research

For the experiment under laboratory bench tests, thirty ball joints were used, which have a gap in the range 0.1-0.9 mm with an interval of 0.13-0.2 mm. As a standard, the ball joint No. 1 was used, a new one, without using the car. The axial clearance h of which was measured at the stand with the aid of a specialized device was 0.01 mm. The remaining hinges were dismantled from the suspension of the car at different mileage values.

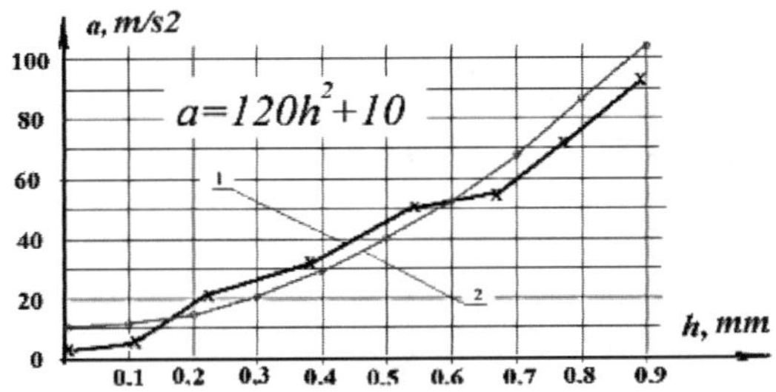

Figure 2. Graph of the dependence of the vibration acceleration on the value of the axial clearance of the hinges:
1 - graph obtained experimentally, 2 - reduced graph.

The resulting equation has the form:

$$h = \sqrt{\frac{a-10}{120}} \tag{4}$$

where a - dimensional acceleration, m/s^2;

As shown by the analysis of the joint graphs, the dependence is manifested when the gap is in the range from 0.01 mm to 0.9 mm, that is, for all hinges presented in the study.

The results obtained can serve as a basis for developing a method for diagnosing the clearance in the ball joints. The dependences between the values of the maximum value of the amplitude and the value of the axial clearance in the hinges are determined.

The error in measuring the value of the axial clearance by the vibration method in comparison with the measurement of the gap by the control device within the gap value of 0.1-0.9 mm does not exceed 10%.

The values of the maximum and average amplitudes of the vibration acceleration are an order of magnitude lower than the same characteristics obtained during bench laboratory studies. This is due to the fact that the vibrations that arise in the suspension of the car and its ball joints, in studies in the conditions of the car service are largely absorbed by the tires of the car. The obtained results can serve as a basis for the technique of diagnosing the axial clearance in ball joints.

6. Conclusions

The analysis of the front suspension of passenger cars was carried out, it showed that 80% of cars have a front suspension of the McPherson type and an analysis of the existing methods of determining the technical condition of the ball joints, which revealed their shortcomings - the complexity of the design, low measurement accuracy, and material resources for diagnosis. This confirms the relevance of the developed vibration method for determining the technical state of ball joints.

Scientifically-methodical approaches and a mathematical model of the technical condition of the ball joint of the front suspension of the McPherson type of a passenger car were developed, with bench research.

Experimental studies were carried out to determine the value of the axial clearance in 30 samples of ball joints using a vibratory method, numerical values of vibration acceleration and their correspondence to the value of the axial clearance:

- for laboratory bench tests from 1.4 m/s^2, which corresponds to the value of the axial clearance 0.01 mm, up to 95 m/s^2 - corresponds to an axial gap of 0.9 mm;
- for autoservice plant conditions from 1.0 m/s^2, which corresponds to the value of the axial clearance 0.01 mm to 2.8 m/s^2 - corresponds to an axial clearance of 0.9 mm.

On the basis of the analysis of the graphs of the dependence of the maximum amplitude of vibration acceleration a on the time of the experiment, in test samples with an axial clearance value of 0.01-0.9 mm, empirical relationships were obtained to determine the actual value of the axial clearance h.

Comparison of the graphs of the dependence of vibration acceleration on the magnitude of the axial clearance in mathematical modeling and in the experimental study showed that the greatest discrepancy between them does not exceed 10%. This confirms the adequacy of the developed mathematical model of the technical condition of the ball joint and its compliance with the bench research conditions.

Reference

[1] S Iana and S Fischer 2000 Numerical analysis of displacements in spatial mechanisms with ball joints *Mechanism and Machine Theory* **35** 1623-40

[2] E Ossa, C Palacio and M Paniagua 2011 Failure analysis of a car suspension system ball joint *Engineering Failure Analysis* **18** 1388–94

[3] R Sakhapov, R Nikolaeva, M Gatiyatullin and M Makhmutov 2016 Modeling the dynamics of the chassis of construction machines *Journal of Physics: Conference Series* **738(1)** 012119

[4] R Sakhapov, R Nikolaeva, M Gatiyatullin and M Makhmutov 2016 Risk management model in road transport systems *Journal of Physics: Conference Series* **738(1)** 012008

[5] J Kang and K Kim 2010 Squeak noise in lead screw systems: Self-excited vibration of continuous model *Journal of Sound and Vibration* **329(17)** 3587-95

[6] V Berdichevsky and W Yu 2014 An Equivalent Plate Modeling of Corrugated Structures *International Journal of Solids and Structures* **51** 2073-83

[7] V Smirnov The fundamentals of vibration measurement [Electronic resource] Vibrodiagnostics for beginners and specialists "INCOTES" Diagnosis of machines and mechanisms - Access mode: http://www vibration ru/osn_vibracii shtml, free

[8] R Ahmadimov 2009 A new approach in the design of the car suspension ball joint *Modern high technology* **4** 23 – 5

[9] I Mikhailovsky 2007 The development of the manufacturing and testing of ball studs front suspension of cars made of steel Mat 65th scientific-technical Conf Sat doc T 1 - Magnitogorsk MSTU 23-5

[10] A Novikov, A Katunin, M Tebekin and I Novikov 2015 Vibration diagnostics as a way of defining car suspension bracket ball elements technical condition *International Journal of Applied Engineering Research* **10(24)** 44884-8

[11] A Shevtsova, I Novikov and A Borovskoy 2015 Research of influence of time of reaction of driver on the calculation of the capacity of the highway *Transport Problems* **3** 53-9

[12] I Novikov, A Shevtsova and A Katunin 2016 Determination of the adaptation of traffic light control to the changing structure of traffic flow *International Journal of Pharmacy and Technology* **4** 26635-43

[13] V Sarbayev, A Novikov, I Novikov and A Shevtsova 2016 Definition of requirements to the accounting of rotary movement in the analysis of vehicles collisions International Journal of *Pharmacy and Technology* **8** 22588-96

[14] A Novikov, A Katunin, I Novikov and A Shevtsova 2017 Research of influence of dynamic characteristics for options controlled intersection *Procedia Engineering* **187** 664 – 71

[15] A Novikov, Novikov and A Shevtsova 2018 Study of the impact of type and condition of the road surface on parameters of signalized intersection *Transportation Research Procedia* **36** 548–55

IOP Conf. Series: Materials Science and Engineering **786** (2020) 012080 doi:10.1088/1757-899X/786/1/012080

Using experience of the dynamic models of piston internal combustion engines

I E Agureev*, M Y Elagin, R N Khmelev, K Y Platonov and V A Pyshnyi

Department of "Cars and automotive industry", Tula State University, Tula, Russia

E-mail: aiah@yandex.ru

Abstract. This article describes using experience of the hierarchical system of dynamic models for researching the operation and calculation of piston engines. The results of practical using and the possibilities of dynamic models developed by the authors are shown.

1. Introduction

The modern stage of theoretical investigations of reciprocating internal combustion engines (ICE) is characterized by the great variety of the applied mathematical models which complexity constantly increases. Mathematical simulation as the research instrument in the modern theory of reciprocating engines occupies an important place and promotes their progress development. Evolution of ICE simulation is developing now in a direction of more and more detailed account of assemblage of various factors, raises of an exactitude and giving to models of natural properties.

The complete analysis of papers has shown that it is paid more and more attention to the problems of the engine construction «as a whole system» because of computing machinery development and considerable advances in the field of ICE simulation. The whole system engine models as a whole system are the mathematical models reflecting process of engine functioning in time as a uniform system.

This paper sets the classification of models and the comparative analysis of the mathematical models applied to researches of functioning and calculation of ICE as a whole system. Dynamic models as the most perspective are observed more detailed. The results of ICE calculations which were got by the dynamical models developed by authors are shown. The results of the research project are published with the financial support of Tula State University within the framework of the scientific project № 2017-22.

2. Types of the mathematical models applied to ice mathematical description «as a whole»

Currently according to the character features it is possible to merge the models of ICE «as a whole», which applied to ICE functioning researches and calculation, in the following four types: thermodynamic, mechanical, informational, dynamical.

The thermodynamic models [1-3] are the most wide-spread and represent the mathematical models of ICE working processes. The basic set of equations is the equations (1) of working fluid mass and energy balances added with relationships for calculation of a heat capacity of working fluid, heat exchange, gas exchange, fuel delivery, mixing, combustion and emission of harmful substances in this case are used:

$$\begin{cases} \frac{dm}{dt} = \sum_{j=1}^{k} G_{nj} - \sum_{j=1}^{k} G_{pj}; \\ \frac{dU}{dt} = \sum_{j=1}^{k} G_{nj} i_{nj} - \sum_{j=1}^{k} G_{pj} i_{pj} + \dot{Q} - p\frac{dW}{dt}, \end{cases} \quad (1)$$

where G_{nj} and G_{pj} – input and output mass flux density through a j valve bore respectively; $\dot{Q}$ – heat release (or removal); i – specific enthalpy of working fluid.

The observed models, as a rule, describe only an intracyclic exchanges of engine working processes parameters with high confidence. Mode of ICE behavior in this case is set by values of crankshaft speed and cyclic fuel delivery (excess air ratio).

Mechanical models [4-6] are founded on the system dynamic equation «the engine – a user»:

$$\frac{d\omega}{dt} = \frac{M_{\partial} - M_c}{J_{np}} \quad (2)$$

where $M_{\partial} = M_i - M_n$; M_n – the moment of mechanical losses; M_i – the indicated engine torque.

The indicated mean torque of the engine is defined in observed models by means of functional dependences of a view:

$$M_{\partial} = f(\omega, p_k, h_p) \text{ [5]; } M_i = f(\eta_i) \text{ [4]; } M_i = f(p_i) \text{ [6]; } M_n = f(\omega),$$

Where p_k – pressure of charging air; h_p – a position of controls fuel delivery; p_i – an indicated mean pressure; η_i – indicated efficiency, $\eta_i = f(\alpha, \omega)$.

The mechanical models allow to gain relationship at various laws of a loading, however they demand considerable quantity of experimental data on the engine, so it restricts their application.

The information models [7, 8] are founded on the automatic control theory. In this case the mathematical identification of ICE elements by the standard links used in the theory of automatic control takes place. Every element is observed as a controlled process having input and output magnitudes link between them is presented by differential equations of a view [8]:

$$T_{2A_i}^2 \frac{d^2 A_i}{dt^2} + T_{1A_i} \frac{dA_i}{dt} + A_{i0} = K_i \Delta M_c, \quad (3)$$

where A_i – observable ICE parameter; A_{i0} – initial value of observable ICE parameter; T_{ij} – factors of differential equations; K_i – coefficients of amplification; ΔM_c – the law of a resist torque variation operating on a engine crankshaft. Behavior of the engine in various modes of regimes is analyzed by means of linear differential equations (3) and a superposition principle.

The observed models are applied to perfecting of engines automatic control systems. As a deficiency of information models it is important to note necessity of experimental definition of factors of differential equations.

The dynamic ICE models are divided into two types: quasidynamic and actually dynamic models.

The feature of quasidynamic models [9, 10] is joint application for the describing of the engine operation of the set of equations engaging the equations (1) and (2). Thus the average engine torque is defined from an indicator diagram gained by calculations. In this case engine operation is observed as the process consisting of a consecutive row of work cycles different from each other, identical to corresponding cycles of steady-state behavior. Application of the quasidynamic models is linked with assumption about existence of a resistant to steady-state behavior of operation of the engine during separately taken cycle without detection of this fact that distorts dynamic properties of the engine. Thus calculation of angular velocity variation is carried on a large grid with a step equivalent to cycle time.

Thermodynamic, mechanical, information and quasi-dynamic models according to the classification presented in the work [11], are usually «mean value models».

Actually the dynamic models or «event based models» [12-17] are deprived noted deficiencies as describe engine operation as taking into account intracyclic, and an intercyclic variation of the engine work parameters. In the fundamentals of observed models differential equations (1) and (2) are assumed. Thus the torque instantaneous value in the equation (2), is defined as follows (for the single-cylinder engine):

$$M_{\partial} = a_4\left[(p - p_0)f_n - F_{mp} - a_3 m_n \omega^2 - a_2 m_n \frac{d\omega}{dt}\right],$$

where p_0 – gas pressure under the piston; $F_{\text{òð}}$ – a frictional forces in mechanism; $m_{\text{ï}}$ – a reduced mass of parts of the engine reciprocating;

$$a_2 = r_\kappa \cdot \left(\sin\varphi + \frac{\lambda}{2}\sin 2\varphi\right);\ a_3 = r_\kappa \cdot (\cos\varphi + \lambda\cos 2\varphi);$$

$$a_4 = r_\kappa \sin\varphi\left[1 + \frac{\lambda\cos\varphi}{1 - \lambda/4\cdot(1 - \cos 2\varphi)}\right];\ \lambda = \frac{r_{\kappa p}}{l_{ш}},$$

where $r_{\kappa p}$ – crank radius; $l_{ш}$ – length of a connecting rod.

3. ICE Dynamic models

The most perspective models among introduced above are dynamic models for the following reasons:

A. These models at higher level then other models present ICE transient condition as a whole systems because of full account intracyclic and intercyclic parameter changes of the engine operation.

In particular, the calculation result [14] difference of ICE transient condition for dynamic and quasi-dynamic models in feature determination of fuel-handling system reach 23 %, parameters the inlet and outlet of the systems – 2 %, effective indexes of engine operation – 12 % (figure. 1).

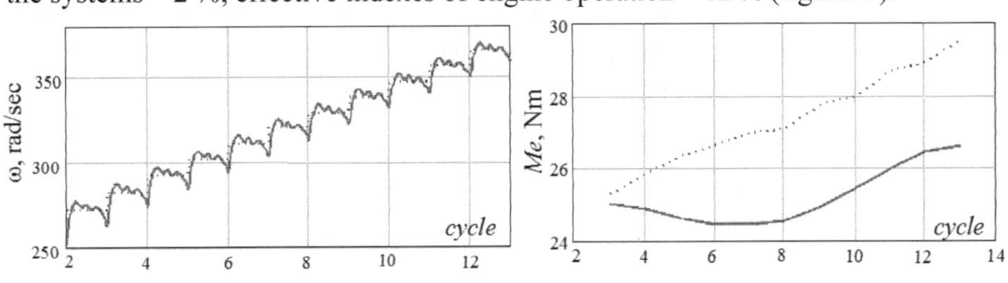

Figure 1. Calculation results of angular velocity and torque in transient regime, calculated by dynamic and quasidynamic models.

B. Multicyclic calculation of ICE transient condition is made by these models, the steady state is considered as a special case of the transition. Thus, the models allow in fully to perform calculations both static (figure. 2), and dynamic characteristics (figure. 3) of ICE [14].

C. The models are opened so process models in ICE cylinders and systems may be connected with them [13-15]. According to the results of computational experiments on dynamic models developed

IOP Conf. Series: Materials Science and Engineering **786** (2020) 012080 doi:10.1088/1757-899X/786/1/012080

recommendations [15] to improve the efficiency of the diesel TMZ-450D in the composition of the electric unit, in particular:

- rational choice of parameters of the air-gas path and gas exchange system allowed to increase the torque and power of the diesel engine by 7.5 %;
- rational choice of parameters of the automatic speed control system allowed to reduce the slope of the regulatory branch from 5.07 % to = 4.80 %, and the transition time – from τ = 1.1 s to τ = 0.44 s

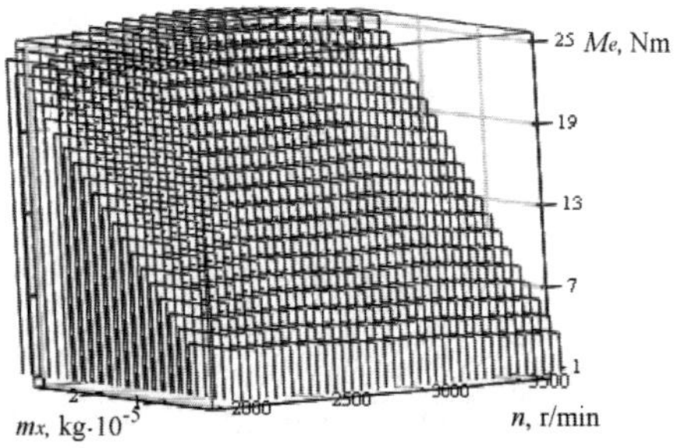

Figure 2. ICE static characteristics.

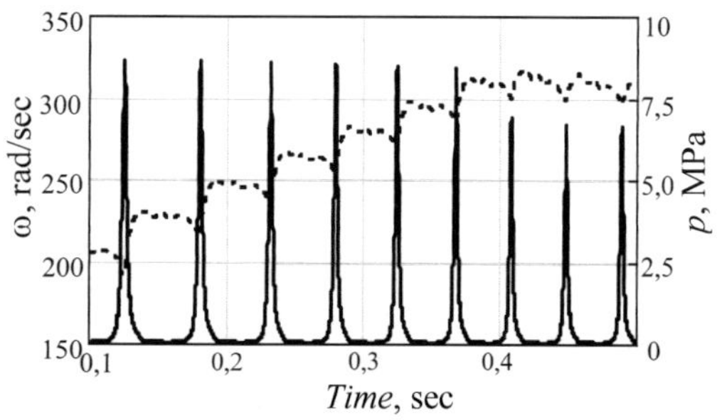

Figure 3. Modification of indexes operating characteristic (indexes) of diesel engine TMZ-450D in transient operating conditions.

For the VAZ-2111 engine, experimentally practical recommendations for the modernization of the intake system were developed and confirmed. At the same time, the inter-cylinder power unevenness and the average indicator pressure at the frequency of 2000 rpm were reduced by 21 %, at the frequency of 5000 rpm by 16 %, the power characteristics of the engine improved by 6-8 %.

The developed dynamic models can be successfully used in the design calculations of the ICE, which is illustrated in [14, 15].

D. The use of dynamic models allows to observe ICE as nonlinear dissipative thermomechanical system [12]. This system is applicable to apparat of nonlinear dynamics. Nonlinear dynamics [18-20] becomes the theoretical base for research of the complete effects inherent ICE, for example:

1) It is possible to exist several ways simultaneously in development of transient processes. Such possibility is caused by the choice of initial and/or boundary conditions; accepted values of model parameters. In the first case a transition to any regime among a large variety of the possible regimes having their own attractive areas is observed. In the second case reconstruction of a large variety of possible regimes and appropriate model solutions can be observed.

The boundary between steady modes of behavior, i.e., equilibrium position ($\omega = 0$) and stable limit cycle was constructed using dynamic model. This boundary was simulated for the same sequence changes regulators (figure. 4). The angular velocity is selected as the initial condition.

2) Bifurcational dependence and dynamic (phase) variables from design parameters of a construction, properties of working mediums, etc. The bifurcation diagram is shown in figure. 5 which divides the range of inertia moment of gyrating ICE details into parts, which determine availability of stable limit cycle and stable equilibrium state ($\omega = 0$). The shown boundary lines are conditions of bifurcation transition: loss of stable limit cycle. This bifurcation is widely observed in the ICE theory and practice in the problem of starting qualities and the problem of the engine cutoff (stability loss) during excess of allowable value by loading moment. The stable ICE operation is arranged above the graph in figure. 5.

3) Existence of complicated stationary regimes which are presented not by unique, repeating from cycle to cycle indicator diagram. These regimes are irregular (chaotic) regimes. At the deeper level bifurcation ICE properties are observed in a known problem of intercyclic nonidentity [21, 22]. It is

shown that intercyclic nonidentity can be observed as outcome of transition to an irregular (chaotic) regime. In particular, the data cited in operations [21, 22] shows dependence of ICE work process type from mixture composition. It should be noted that the ICE dynamic model can show the ability to bifurcations to the irregular (chaotic) regime even in the 0-dimensional form. On figure. 6 the sequence of indicator diagrammes is shown in figure. 6 which correspond to the cycles of period 1, 2, 4, 8, 3, and also the chaotic regime that can be considered as reflecting intercyclic nonidentity.

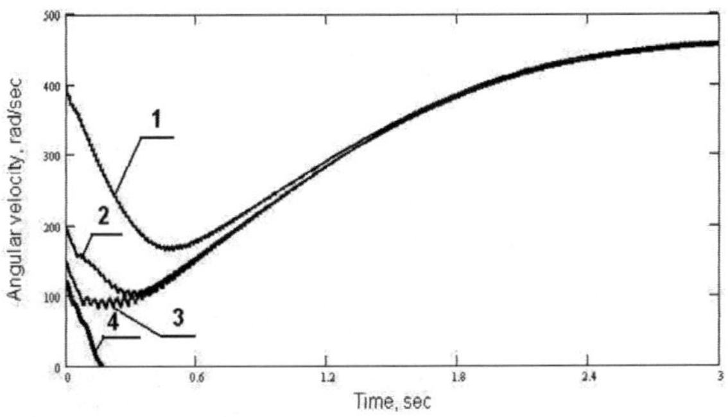

Figure 4. The dependence of angular velocity from initial conditions.

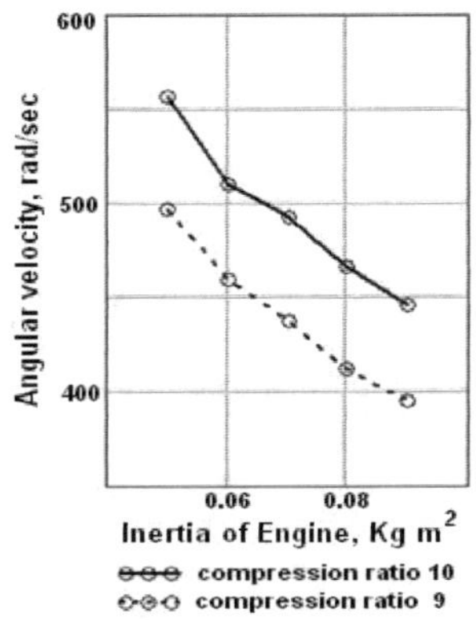

Figure 5. The bifurcation diagram (boundary lines of stability loss of limit cycles) for monocylindrical petrol ICE.

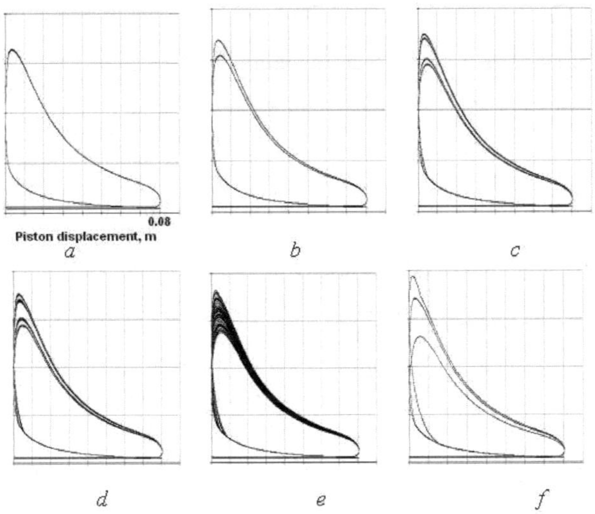

Figure 6. Projections of phase portraits (indicator diagrams) of various attractors of ICE model depending on exterior loading parameter:
a) – a cycle of period 1;
b) – a cycle of period 2;
c) – a cycle of period 4;
d) – a cycle of period 8;
e) – a chaotic regime;
f) – a cycle of period 3.

4) Ability to generation of the ordered (dissipative) structures at the most various levels of the exposition. The observed dynamic models can successfully be applied at projecting calculations of ICE that is illustrated in research work [12].

4. Conclusion

In this paper the perspective application and the further development of ICE dynamic models is shown by the analysis. These models allow to describe engine operation as the whole dynamic system taking into account complex interconnections of its elements. The piston engine as a complex system has special system properties not inherent to the individual elements. These properties are the result of combining the functions in time and space. Therefore, the property to execute the set objective function is realized only by system as a whole, instead of its separate elements.

Thus, the dynamic models are the effective instrument of optimizing calculations of piston engines on the steady-state and transient conditions.

References

[1] Kuleshov A S et al 2010 Diesel-RK program: Simulation and optimization of internal combustion engines working processes // Proc of Internat Conf "Engine 2010" - Bauman MSTU, Moscow, Russia 287-91
[2] Gavrilov A A, Ignatov M S and Efros V V 2003 Calculation of reciprocating engine cycles // Vladimir: VlSU 124
[3] Vyrubov D N et al 1983 Internal combustion engines: Theory of piston and combined engines Edited by Orlin A S, Kruglov M G // Moscow, Engineering industry 372
[4] Kuznetsov A G 2010 Development of methods and means of diesel work effectiveness increase for dynamic regime // Moscow 32
[5] Markov V A et al 2007 Dependence of toxicity of diesel exhaust gases in transient processes on external speed characteristic // Proc of Internat Conf "Engine 2007" Bauman MSTU, Moscow, Russia 329-35
[6] Epshtein A S 1968 Transient processes calculation of combined engine D100 Development problems of combined internal combustion engines // Moscow: Engineering industry 205-24
[7] Krutov V I 1989 Automatic regulation and steering of explosive motors // Moscow Engineering industry 416
[8] Khairtdinov I N 2003 Methods development of dynamic mathematical model for diesels researching by transient load // Kazan, KSAU 20
[9] Regner G, Loibner E, Krammer J, Walter L and Truemner R Analysis of transient drive cycles with the use of the collaborative modeling, http://www aps-c ru/publications/boost_cruise_rus pdf
[10] Malozemov A A 2011 Effectiveness increase of multipurpose energotechnological complexes by engine - generating sets perfecting // Chelyabinsk 34
[11] Guzzella L and Onder C Introduction to Modeling and Control of Internal Combustion Engine Systems 2010 Springer-Verlag Berlin Heidelberg 354
[12] Agureev I E 2001 Nonlinear dynamic models of reciprocating internal combustion engines: Synergetic approach to construction and the analysis // Tula: TSU publishing 224
[13] Maliovanov M V 1998 Dynamic theory of internal combustion engines (creation utility and development stages) // Tula: TSU publishing **2** 189-96
[14] Khmelev R N 2011 Theoretical bases development of identification parameters of piston engine as the whole system for their higher efficiency functioning // Tula 229
[15] Maliovanov M V, Khmelev R N 2012 Development of methods of design calculations reciprocating internal combustion engines // proceedings of international scientific-practical conference "Modernization and research in the TRANS-tailor complex", Volume 1 Perm, PNIPU 290-3
[16] Falcone P, Gennaro M, Fiengo G, Santini S, Glielmo L and Langthaler P 2003 "Torque generation model for Diesel Engine" // 42nd IEEE Conference on Decision and Control 2003, Hawaii, USA
[17] Kao M and Moskwa J J 1995 Turbocharged Diesel Engine Modeling for Nonlinear Engine Control and State Estimation // ASME Transactions, *Journal of Dynamic Systems Measurement and Control*, **117(1)** 20-30
[18] Magnitskiy N A and Sidorov S V 2004 New methods of chaotic dynamics // Moscow: Editorial URSS 320
[19] Malinetskiy G G, Potapov A B and Podlazov A B 2006 Non-linear dynamics: Methods of approaching, results, expectancies // Moscow, ComBook 280

[20] Kuznetsov S P 2006 Dynamic chaos (course of lectures) // Moscow: physico-mathematical literature publishing house p 356
[21] Daw C, Finney C E A and Green J Bet al 1996 A simple model for cyclic variations in a spark-ignition engine // SAE Technical Paper Series, Paper 962086
[22] Daw C, Kennel M B and Finney C E A et al 1998 //Observing and modeling nonlinear dynamics in an internal combustion engine // *Physical Review E* **57(3)** 2811–9

IOP Conf. Series: Materials Science and Engineering **786** (2020) 012081 doi:10.1088/1757-899X/786/1/012081

The basis of an automated design system of the packed absorber using neural networks

O S Kharitonova[1,*], I V Vyatkina[2], V V Bronskaya[3,4], L E Khairullina[4], R S Shaikhetdinova[5] and N V Kotova[1]

[1]Department of Chemical Technology of Petroleum and Gas Processing, Kazan National Research Technological University, 68 Karl Marx Street, Kazan 420015, Russia

[2]Department of Bilingual Education, Kazan National Research Technological University, 68 Karl Marx Street, Kazan 420015, Russia

[3]Department of Chemical Process Engineering, Kazan National Research Technological University, 68 Karl Marx Street, Kazan 420015, Russia

[4]Department of Information Systems, Kazan Federal University, 35 Kremlyovskaya Street, Kazan 420008, Russia

[5]Department of Technology of Engineering Materials, Kazan National Research Technological University, 68 Karl Marx Street, Kazan 420015, Russia

*E-mail: olga.220499@mail.ru

Abstract. An optimal artificial neural network has been developed for an application to determine the height and diameter of the packed absorber apparatus using neural networks. The obtained results can be used for modeling a wide class of objects of chemical technology with the possibility of formalization of calculation procedures.

1. Introduction

Nowadays there is a growing interest in artificial neural networks. Almost all scientific and technical disciplines conduct research related to the introduction of neural networks to perform applied tasks for simplifying processes or enhancing speed and productivity of these processes

Due to the massive introduction of artificial neural networks, they are of great interest for various spheres of human life activity, including mechanical engineering.

The growing significance of neural network modeling is due to the growing level of complexity of modern chemical-technological systems and the need of detailed mathematical description [1-3].

2. The main part

Packed columns are widely used in industry devices with continuous phase contact, in which the interphase area is created by bodies of a certain size and a special shape. The packing material that is in the column is placed on the support grilles with holes for liquid and gas. The fluid evenly refluxes the packing material using a distributor and flows in the form of a thin film down the packing material surface. The fluid equilibrium distribution throughout the packing material is usually not achieved due to the wall effect (high loading density of the packing material in the central part of the device). Therefore, the fluid tends to spread out to the column walls from the center. To avoid it, the sections for the packing material which height is not more then 3-4 meters are created in the columns, installing fluid redistributors. As a result, the fluid is redirected from the periphery of the column to its axis.

The technological calculation of the packed absorber is reduced to the determination of the height and the diameter of device, which determine the capital costs in the design of technological equipment. The standard calculation contains the definition of mass transfer coefficients, which generally have a significant deviation from the experimental values due to the inaccuracy of measurements, the presence of impurities, the imperfection of the standard calculation algorithm etc [4-8].

First of all, mol per cent are converted into relative mass concentrations:

$$\overline{Y} = \frac{yM_A}{(100-y)M_B} \tag{1}$$

where M_A is the absorber molar mass, M_B the inert gas molar mass.

Then we find the consumption of the target component transferred from one phase to another:

$$\dot{M} = G(\overline{Y}_b - \overline{Y}_f) \tag{2}$$

using the inert gas flow formula

$$G = V\rho \tag{3}$$

After it is necessary to find the phase-equilibrium constant

$$m = \frac{EM_A}{PM_B} \tag{4}$$

where E is Henry coefficient. Henry coefficient is represented as table values in the literature. We approximate the available values and get the formula $E = 24289*t^2 + 3*10^6*t + 7*10^7$.

Knowing the value of the phase-equilibrium constant you can find the minimum absorbent consumption:

$$L_{\min} = \frac{Mm}{y_b}, \tag{5}$$

after which the actual consumption is found:

$$L = 1.5L_{\min} \tag{6}$$

Then you need to find the gas velocity. The gas velocityω is found in the following way. At first we calculate the fictitious gas velocity ω_3 at the flooding point according to the equation:

$$\lg\left(\frac{\omega_3^2\sigma\rho_2\mu_l^{0.16}}{gV_{fr}^3\rho_l}\right) = A - 1.75\left(\frac{L}{G}\right)^{0.25}\left(\frac{\rho_g}{\rho_l}\right)^{0.125} \tag{7}$$

where σ — specific surface of the nozzle, m^2/m^3, g — gravitational acceleration, m/s^2; V_{fr} — free space of nozzle, m^3/m^3; ρ_l and ρ_g — gas and liquid density, kg/m^3, μ_l — dynamic viscosity coefficient of liquid, $mPa*s$; A — coefficient depending on the type of nozzle (in this case -0.073).

IOP Conf. Series: Materials Science and Engineering **786** (2020) 012081 doi:10.1088/1757-899X/786/1/012081

Then we determine the operating the gas velocity ω:

$$\omega = 0.8\omega_3 \tag{8}$$

After it is required to calculate the diffusion coefficient for the gas phase:

$$D_g = \frac{4.3*10^{-7}T^{3/2}}{P(\upsilon_A^{1/3}+\upsilon_B^{1/3})^2}\sqrt{\frac{1}{M_A}+\frac{1}{M_B}} \tag{9}$$

where T — temperature, K, M_A and M_B — molal masses of gases A and B, υ_A and υ_B — molal volumes of gases A and B defined as the sum of atomic volumes of the elements that go to make up the gas.

It's also required to calculate the diffusion coefficient for the gas phase:

$$D_l = \frac{1*10^{-6}}{AB\sqrt{\mu}\left(\upsilon_A^{1/3}+\upsilon_B^{1/3}\right)^2}\sqrt{\frac{1}{M_A}+\frac{1}{M_B}} \tag{10}$$

where μ — dynamic viscosity coefficient of liquid, $mPa*s$, M_A and M_B — molal masses of dissolved matter and solvent; υ_A and υ_B — molal volumes of dissolved matter and solvent; A and B — coefficients depending on of the properties of the dissolved matter and solvent (in this case A = 1, B = 4.7).

Mass transfer coefficient in the gas phase β_g is found according to the equation

$$Nu_g^{'} = 0.407\,\mathrm{Re}_g^{0.655}\left(\mathrm{Pr}_g^{'}\right)^{0.33} \tag{11}$$

Here $Nu_g^{'} = \frac{\beta_g d_e}{D_g}$ is the Nusselt number for the gas phase. Therefore β_g is equal to:

$$\beta_g = 0.407\left(D_g/d_e\right)\mathrm{Re}_g^{0.655}\left(\mathrm{Pr}_g^{'}\right)^{0.33} \tag{12}$$

where $\mathrm{Re}_g = \frac{4\omega\rho_g}{\sigma\mu_g}$ — Reynolds number for the gas phase in the nozzle; $\mathrm{Pr}_g^{'} = \frac{\mu_g}{\rho_g D_g}$ — Prandtle number for the gas phase; μ_g — gas viscosity, $Pa*s$; d_e — equivalent diameter of the nozzle, m.

Mass transfer coefficient in the gas phase β_l is found according to the equation

$$Nu_l^{'} == 0.0021\,\mathrm{Re}_l^{0.75}\left(\mathrm{Pr}_l^{'}\right)^{0.5} \tag{13}$$

where $Nu_l^{'} = \frac{\beta_l\delta_{red}}{D_l}$ is the Nusselt numder for the liquid phase. Consequently,

$$\beta_l = 0.0021\left(D_l/\delta_{red}\right)\mathrm{Re}_l^{0.75}\left(\mathrm{Pr}_l^{'}\right)^{0.5} \tag{14}$$

where $\mathrm{Re}_l = \frac{4L}{S\sigma\varphi\mu_l}$ — Reynolds number for liquid phase; $\mathrm{Pr}_g^{'} = \frac{\mu_l}{\rho_l D_l}$ — Prandtle number for liquid; $\delta_{red} = \left(\frac{\mu_l^2}{\rho_l^2 g}\right)^{1/3}$ — reduced thickness of liquid film, m; $S = \frac{\pi D^2}{4}$ — cross section of column, m^2; $D = \sqrt{\frac{4V}{\pi\omega}}$ — column diameter, m; φ — the moisness coefficient of the nozzle.

The technological calculation of the packed absorber is reduced to the determination of the height and diameter of the device. The standard calculation contains the definition of mass transfer coefficients, which in most cases have a significant deviation from the experimental values due to the inaccuracy of measurements, the presence impurities, the imperfections of the standard calculation algorithm, etc. In connection with it there is a question of determining the mass transfer coefficients using artificial intelligence technologies.

Work on artificial neural networks or just neural networks was initially dictated by the fact that the human brain calculates quite differently than an ordinary computer. In general, a neural network is a machine that is designed to simulate how the brain completes a specific task or function; a network is implemented using electronic components or created a model as software on your computer.

Today sigmoidal function is the most common form of activation function used in the construction of artificial neural networks. It is defined as a strictly increasing function. An example of an activation function of this type is a function called logistic:

$$\varphi(\upsilon) = \frac{1}{1+\exp(-a\upsilon)} \quad (15)$$

where a is the slope parameter of the sigmoid function. You can get sigmoidal functions of different slopes changing a (Figure 1). At the limit when the slope parameter approaches infinity, the sigmoidal function becomes just a threshold. While the threshold function takes values of 0 or 1, the sigmoidal function takes a continuous range of values from 0 to 1. Sigmoidal function is differentiable.

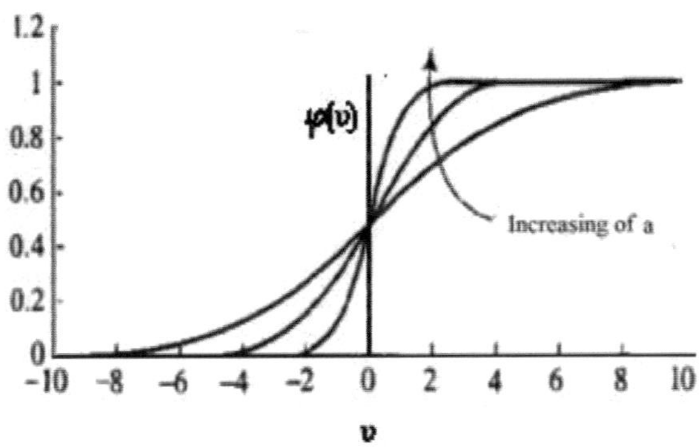

Figure 1. The graph of sigmoidal activation function.

The artificial neural network is a massive parallel distributed processor consisting of simple blocks that has a natural tendency to store empirical knowledge and makes it available for use. The processor reminds the brain because the network receives information from the environment through the educational process and the network has interneurons bonds (synaptic weights) used to store acquired knowledge.[9-12]

3. Results

AForge.NET framework [13] was chosen to implement the artificial neural network of technological calculation of the absorption process. The main applications of the framework are image processing and filters use, artificial neural networks of different architectures, genetic algorithms what is called fuzzy logic, robotics and so on. The framework consists of a set of libraries and samples of applications designed for specific areas of artificial intelligence and computer vision: image and video processing libraries, computer vision libraries, fuzzy computing libraries, neural networks and machine learning libraries.

A multilayer network of direct distribution, which is the most suitable topology for empirical modeling and engineering applications, are used in this paper. A teacher learning method was selected from all network learning methods. The training procedure requires a set of process inputs and outputs. During learning, weights and offsets are iteratively corrected to minimize the target function. The learning algorithm - backpropagation - moves the network parameters in the direction of the negative gradient. The activation function is sigmoidal one with a factor of 2.

The program part was written having formed a general idea of the necessary neural network structure. An artificial neural network in C# language was implemented in Visual Studio 2017 development environment using the AForge.Neuro library. During the work a full-fledged application Windows Forms, which implemented the functions of loading a training selection from a file, setting the necessary training accuracy and entering a vector for obtaining results of neural network operation was created. The application interface is shown in Figure 2. The keyboard input check was implemented — the user has the ability to enter only numbers and separating characters, and full points are automatically replaced with binary points for convenience to record data in the program.

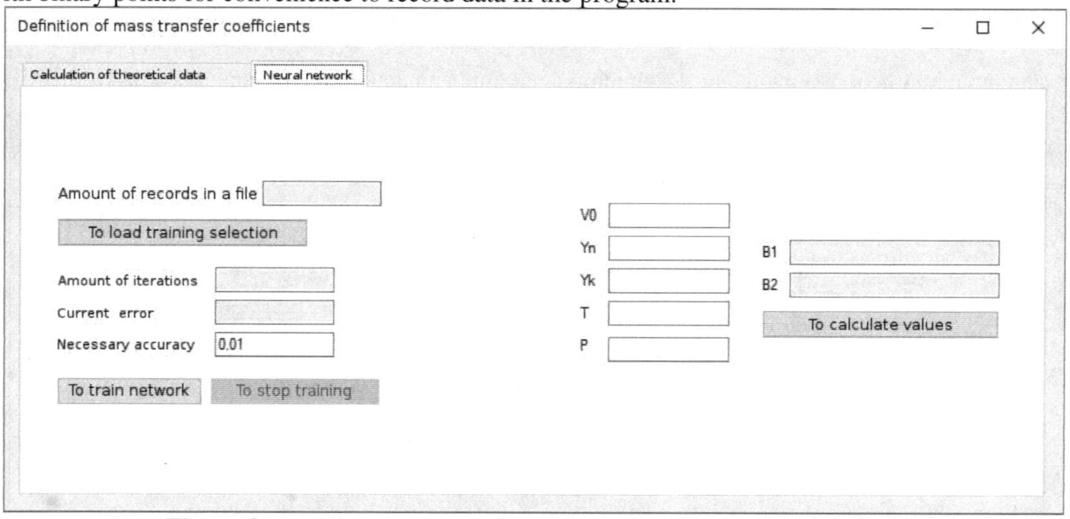

Figure 2. The application interface for working with neural network.

Loading input data from a file and saving data to a file in both parts of the application is implemented. The CSV format was chosen for the files for easy recording and reading. The input vectors are written as "V_0; y_n; y_k; t; P". In order to select the structure, errors for each mass transfer coefficients were calculated. The results obtained during the operation of the neural network and obtained by the experiment were compared.

4. Conslusion

Errors of each mass transfer coefficient were calculated to select the architecture. The results obtained during the operation of the neural network and obtained by the experience were compared. The best architecture is the 5-10-2 one based on table 1.

Table 1. Input and output parameters of the neural network.

Architecture	Error β_g	Error β_l
5-5-2	4,74752*10^{-5}	8,11155*10^{-7}
5-6-2	5,74983*10^{-5}	6,36088*10^{-7}
5-7-2	4,70527*10^{-5}	7,4511*10^{-7}
5-8-2	6,50883*10^{-5}	8,27173*10^{-7}
5-9-2	6,68836*10^{-5}	9,02349*10^{-7}
5-10-2	4,59501*10^{-5}	7,14545*10^{-7}
5-11-2	5,35965*10^{-5}	8,31043*10^{-7}

The results of the artificial neural network work showed that the accuracy of determining the height and diameter of the absorber in comparison with theoretical calculations increased at average by 15-20%. The result of the work is an application to determine the height and diameter of the packed absorber device using neural networks.

References

[1] Aminova G A, Manuiko G V, Bronskaya V V, Ignashina T V, Dyakonov G S, Bashkirov D V and Demidova E V 2008 *Journal of Engineering Physics and Thermophysics* **81** 1247-51
[2] Manuiko G V, Aminova G A, Dyakonov G S, Ahmetov I G and Bronskaya V V 2015 Theoretical *Foundations of Chemical Engineering* **49** 246-51
[3] Manuiko G V, Aminova G A, Bronskaya V V, Ignashina T V, Dyakonov G S and Bashkirov D V 2008 *Theoretical Foundations of Chemical Engineering* **42** 336-9
[4] Jess A and Wasserscheid P 2013 Chemical Technology (Wiley)
[5] Comyns E A 2014 Encyclopedic dictionary of named processes in Chemical technology (CRC Press, Taylor & Francis Group)
[6] Eduardo Eyng, Fl'vio Vasconcelos da Silva, Fernando Palz and Ana Maria Frattini Fileti 2009 Neural Network Based Control of an Absorption Column in the Process of Bioethanol Production Brazilian archives of biology and technology
[7] Rój E 2004 Simulation of the Sieve Plate Absorption Column for nitric oxide absorption process using neural networks (Pulawy: Instytut Nawozow Sztuznych Press)
[8] Shahsavand A, Derakhshan Fard F and Sotoudeh F 2011 Application of artificial neural networks for simulation of experimental CO2 absorption data in a packed column Journal of Natural Gas Science and Engineering 3
[9] Suleymanov Y, Gafarov F and Khusnutdinov N 2013 Journal of Integrative Neuroscience 12 103-16
[10] Moudar H Zgoul 2012 Use of artificial neural networks for modelling rate dependent behaviour of adhesive materials Int *J Adhes and Adhes* **36** 1-7
[11] Charhyia Nadir Nizar Ali and Dudul Sanjay Vasant 2008 *Sensor for classification of material type and its surface properties using radial basis networks IEEE Sens J* **8** 1981-91

[12] Styrcz A, Mrozek J and Mazur G 2011 A neural-network controlled dynamic evolutionary scheme for global molecular geometry optimization Int J Appl Math and Comput Sci 21 559-66
[13] Documentation C# language https://docs microsoft com/ru-ru/dotnet/csharp/
[14] Documentation C# language https://www tutorialspoint com/csharp/
[15] Documentation AForge NET [Electronic resource] http://www aforgenet com/framework/documentation html

Possible scope of filler beam bridges on the Russian Railways

L K Diachenko*, N A Labutin and A V Lang

Emperor Alexander I St. Petersburg State Transport University

*E-mail: leonid_dyachenko@mail.ru

Abstract. One of the most stable trends is the use of multi-component solutions in the practice of bridge construction, one of which is the use of rigid reinforcement elements in bridge girders and decks. The purpose of this work is to assess the possibility of using filler bridge girders on the Russian railway network and determine the scope of their effective application. In the course of the study, the analysis of design solutions was carried out taking into account foreign experience in design and operation and the study of relevant regulatory documents. Also in the course of the work, a comparison of the cost indicators of rigid-reinforced girders and the existing standard projects of the reinforced concrete and steel simple beam bridges was performed. The results of the study are presented by the formed list of the main advantages and disadvantages of girders with rigid reinforcement and the definition of the area of their most effective application. The practical significance of the work is because steel-concrete girders with rigid reinforcement are the most effective solution for the short span bridges up to 15 meters length from both technical, operational and economic sides. The proposed solution allows reducing the resource intensity of the construction and labor costs for its construction and maintenance.

1. Introduction

Filler beam are a structure of closely spaced steel rolling profiles combined with concrete [1]. The main advantages of this solution are high load capacity, rigidity, aesthetics, ease of maintenance and construction. A general view of the cross section of the structure is shown in Figure1.

The first bridge structures with the use of rolling profiles as reinforcing elements appeared in France at the end of the XIX century [2, 3]. However, the availability of constructions of this type have become available only in the 1970s. At the same time, design rules and regulations were developed and the effectiveness of the solution was confirmed by the European States and the International Union of Railways (UIC) [3, 4, 5]. Subsequently, part of the developed norms was included in the European standards for the design of Eurocode [6].

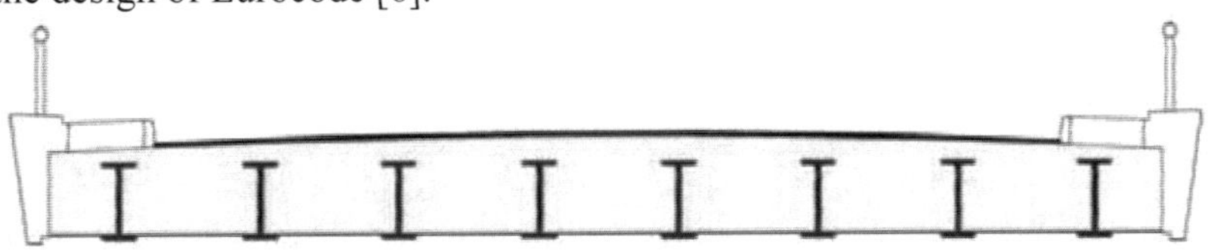

Figure1. Scheme of the cross-section of the filler beam.

IOP Conf. Series: Materials Science and Engineering **786** (2020) 012082 doi:10.1088/1757-899X/786/1/012082

Leading countries in the number of bridge structures with rigid reinforcement [1] is Germany (26%), Italy (24%) and France (20%). The total percentage of distribution on the European railway network is between 20 and 25%.

2. Design description

As reinforcement elements, the most widely I-beam assortment EU 53-62 steel S235-355. As a rule, transverse reinforcement is performed only in the stretched zone by reinforcing bars passed through the holes in the wall of the rolling profiles.

The thickness of the concrete layer above the steel elements according to Eurocode is determined from the following conditions:

1. the minimum thickness of 7 cm (protective layer, the distribution of efforts on the reinforcement elements, preventing buckling of beams reinforcement);
2. the thickness is not more than 1/3 of the total height of the section.

To improve the efficiency of the material in the section, it is possible to use closed elements [7] and T-beams with an arrangement in the stretched section zone. Thus, from all possible variants of designs, the most widespread structure with rigid reinforcement in the form of a wide-beam I-beam, located in a compressed and stretched zones and the distribution of concrete over the entire cross-sectional area (Figure2.). The graph of distribution of internal forces in the compressed zone of concrete, compressed and stretched zones of the I-beam is presented in General in Figure3.

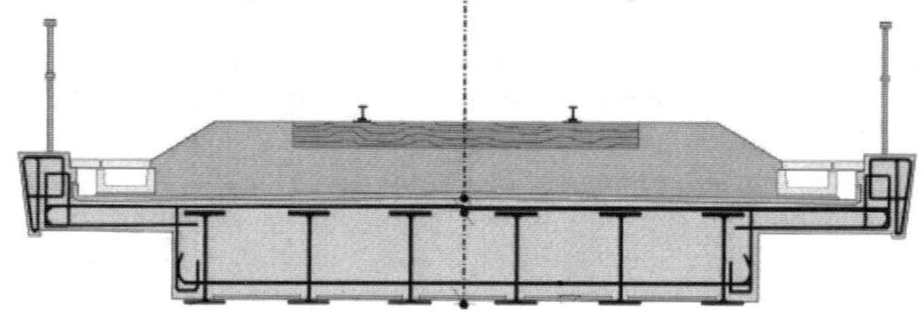

Figure2. Cross-section of the bridge with rigid reinforcement of German Railways.

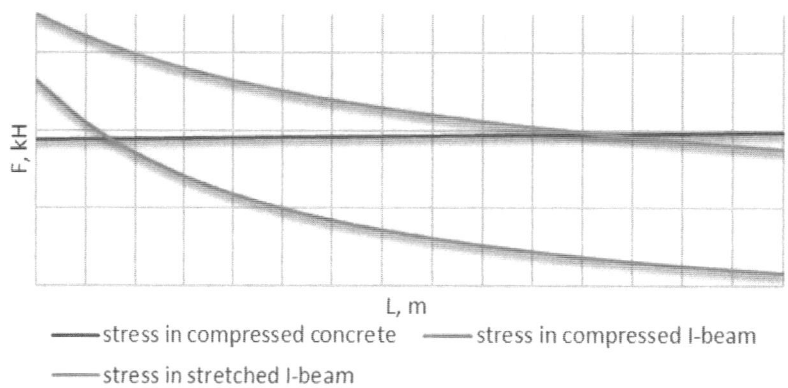

Figure3. A graph of the efforts of the elements of the cross section depending on the span length.

It should be noted the possibility of producing prestressed structures with rigid reinforcement. The presence of prestressing provides minimal deflections from the time load, which allows the use of these superstructures on high-speed Railways [8].

3. Advantages of filler beam

The analysis of existing structures allows to reveal their following advantages:

1. small construction height;

2. high rigidity of the girders;
3. high structural endurance;
4. variety of construction methods;
5. possibility of effective utilization.

These advantages determine the possibility of using filler beams on high speed railways [9,10].

4. Determination of the scope of possible application

The experience of such companies as Deutsche Bahn [10,11] and Czech Railways [4] shows that span structures with filler beam are most effective for covering spans from 6 to 20 m [6, 7]. This area of application relates to small bridges – one of the most common types of bridges on Railways (from 100 to 250 structures per 1000 km) [12]. Span structures of small bridges in Russia are usually built according to standard projects.

To compare the technical and economic indicators and determine the scope of possible application, a preliminary calculation of giders was made according to the method given in [3, 6]. The calculation was carried out on the effects of temporary loads CK (K=14) with the use of assortment GOST 26020-83. Materials submitted B30 class concrete and steel 15HSND design resistance of 295 MPa.

According to the requirements [3, 6], the number of I-beams in the section is determined based on the approximate width of the span structure of 4.9 m. the distance between the axes of the beams is determined:

$$b_s = h / 3 + 0,6$$

where h is the height of the I-beam

The number of I-beams n and the span width B'_{pr} are determined by formulas respectively:

$$n = \frac{B_{pr}}{b_{st} + b_m}$$

$$B'_{pr} = n \cdot b_{pr} + (n-1) \cdot b_m + 2 \cdot 0,8$$

Where B_{pr} – approximate width of the superstructure equal to 4.9 m;
b_{st} – width of I-beam shelf;
b_m – distance between I-beams;
n – number of I-beams;
0,8 – minimum distance from the axis of the extreme I-beam to the outer face of the cross section.

The design scheme of the cross section is shown in Figure4. The moment of internal forces M_{RD} in the cross sections of this type of structures is determined by the expression:

$$M_{RD} = F_{sc} X_{FSC} + F_{sc} X_{FBC} + F_{sc} X_{FST}$$

Where F_{BC}, F_{SC}, F_{ST} – the stress in the compressed concrete, compressed and stretched zones of rebar;
X_{FBC}, X_{FSC}, X_{FST} – shoulders internal efforts

The calculation was carried out for spans with the value of a span from 3 to 18 m in increments of 0.5 m. For each span length was made to select the most economical combination of the height of the I-beam and the thickness of the covering layer of concrete.

IOP Conf. Series: Materials Science and Engineering **786** (2020) 012082 doi:10.1088/1757-899X/786/1/012082

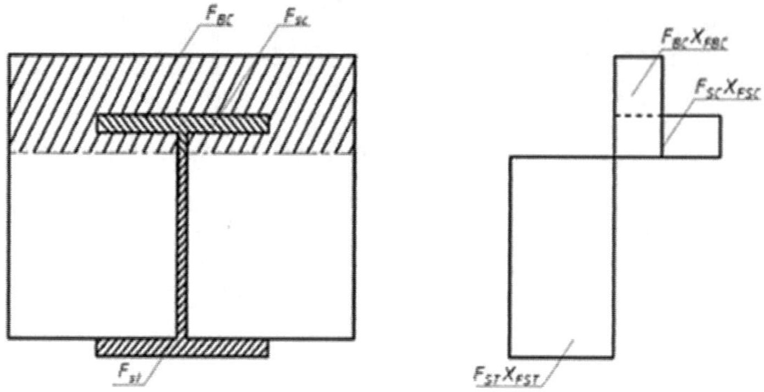

Figure4. Design scheme of cross section.

Comparison and analysis of the results and indicators of existing typical reinforced concrete girders with filler beam of similar length confirmed that this solution is the most economical and effective for covering spans up to 15 m (Figure 5).

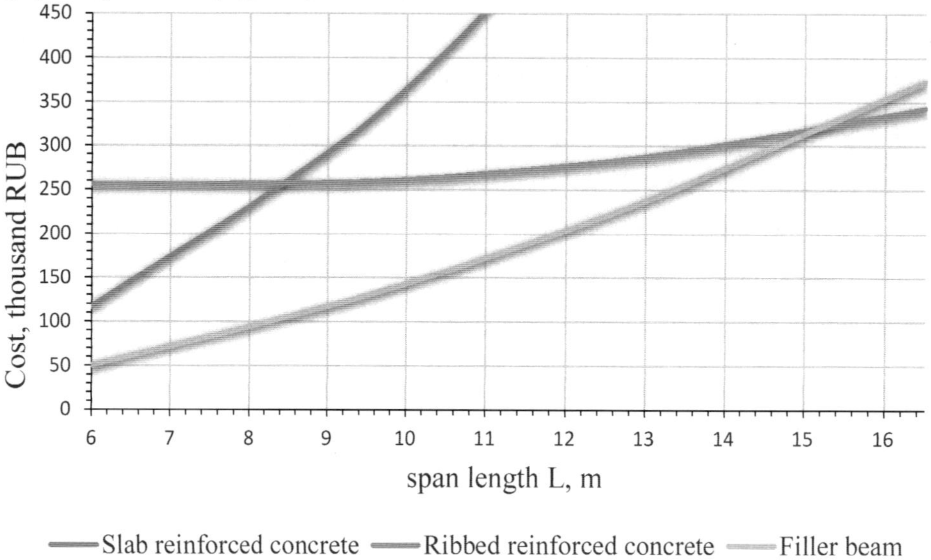

Figure5. Cost of superstructures depending on the size of the span.

5. Determination of the main characteristics of the section

The preliminary calculation of cross sections made by the method [3, 6] for girders with a length of 3 to 18 meters allowed obtaining dependences that allow determining the size of the main elements of the section in the first approximation.

The obtained calculated values of I-beam height depending on the span length and their subsequent approximation (Figure6) allowed concluding about the possibility of assigning the height of the I-beam expression.

$$h = L / 20$$

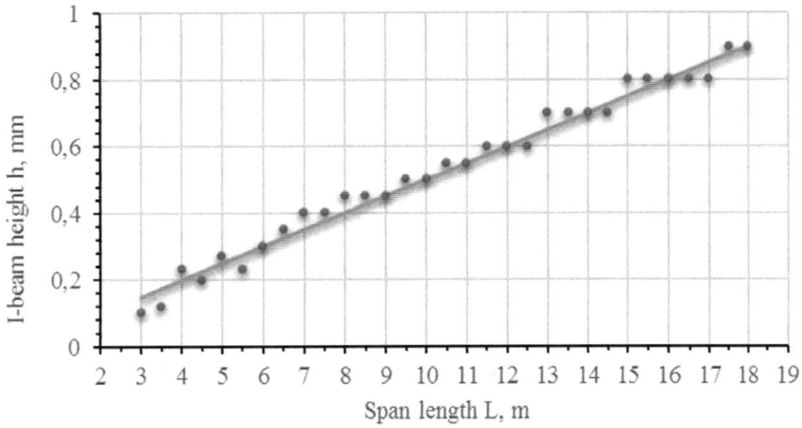

Figure 6. Calculated values of the height of the I-beam, depending on the span length and the approximating function.

Similarly, the dependence of the thickness coefficient of the concrete layer to the height of the I-beam c/h depending on the span length was obtained (Figure 7.). The approximating function is:

$$c/h = 13,2L^{-1.7}$$

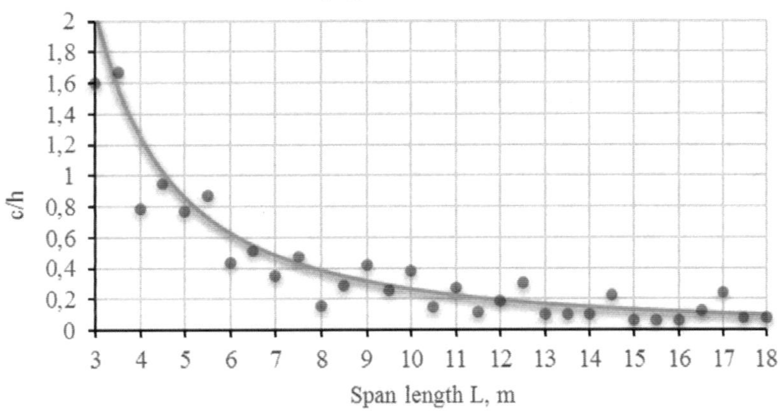

Figure 7. Calculated values of the ratio of the concrete layer thickness and I-beam height depending on the span length and the approximating function.

Thus, the dependences were obtained, allowing determining in the first approximation in the sketch calculation the basic necessary dimensions of the cross section elements.

6. Conclusion

Despite the fact that the use of girder bridges with filler beam requires the development of appropriate technical conditions and methods of calculation, they have great potential for implementation on the network of Russian Railways. The analysis of the design solution and its technical and economic indicators shows that the superstructures with filler beam are the most effective and economical solution for the overlap of spans up to 15 m.

Expressions were also obtained to determine the basic dimensions of the cross section in the first approximation in the sketch calculation.

References

[1] Obdrobinak J , Bujnak J and Zilka J 2012 Study on short span deck bridges with encased steel beams University of Zilina, Faculty of Civil Engineerig, Slovakia
[2] Rademacher D and Martin P 2016 Fast and easy dimensioning of filler beam bridges, 19th IABSE Congress, Stockholm, Sweden
[3] International Union of Railways UIC Design tables for filler beam railway bridges № ISBN 2-919969-37-4, Luxembourg, 1999
[4] ČD MVL 511 2005 Bridge guideline – Superstructures of railway bridges with encased steel beams Prague: SŽDC Czech Republic
[5] UIC, SNF and ARBED 1997 Design tables for filler beam railway bridges Recommendations for design of joist-in-concrete railway bridges Luxembourg
[6] EN 1994-2 2004 : Eurocode 4 : Design of composite steel and concrete structures P 2 : General rules and rules for bridges
[7] Kvocak V , Dubecky D and Spisak M 2013 The composite section analysis of encased beams with closed shape Bulletin of Lviv *Polytechnic National University* **756** 135-9
[8] Staquet S , Espion B and Toutlemonde F 2010 Innovation for Railway Bridge Decks: A Prebent Steel-VHPC *Beam Structural Engineering International Journal*, **2** 134-7
[9] L Dyachenko, A Benin and A Dyachenko 2018 Research of interaction of the "train – bridge" system with bridge deck resonant vibrations, MATEC Web of *Conferences Siberian Transport Forum - TransSiberia* DOI: 10 1051/matecconf/201823905002
[10] L Dyachenko, A Benin and A Dyachenko 2018 Design of Dynamic Parameters for Simple Beam Bridges on High-Speed Railways, IOP Conf Series: *Materials Science and Engineering* **463** 022048 DOI: 10 1088/1757-899X/463/2/022048
[11] DB Netz 2016 AGBemessungstabellen für Eisenbahnbrücken aus Walzträgern in Beton - 2 Auflage Luxembourg
[12] DB Netz AG: Richtlinie 804 2013 – Eisenbahnbrücken und sonstige Bauwerke planen, bauen und instand halten Ausgabe
[13] Bogdanov G I 2013 Istoria mostostoeniya : ucheb posobie Saint-Petersburg: Nestor-Istoriya, – 168

IOP Conf. Series: Materials Science and Engineering **786** (2020) 012083 doi:10.1088/1757-899X/786/1/012083

Comparison of various calculation models for the bridge dynamic analysis

A V Lang*, L K Diachenko and N A Labutin

Emperor Alexander I St. Petersburg State Transport University

*E-mail: langandrew@mail.ru

Abstract. The purpose of this study is to determine the rational use areas of various computational models types while dynamic analysis. This topic is relevant because of the need to carry out precise dynamic calculations for structures at the high-speed rail, in order to ensure the design of a reliable structures.The object of the study is a single-span girder for the high-speed rail. Research methods are numerical analysis using various models; modal calculation of the oscillation shapes and frequencies; comparative analysis. The result of this study are recommendations on the areas of rational use of various types of models.

1. Introduction

Dynamic analysis of structures requires the use of modern software systems [1-4]. However, for each specific type of calculation, a rational choice of the most appropriate calculation model is necessary: it should allow obtaining reliable results at minimal amount of time.

In this article three different types of models of single-span girder for high-speed railroad were compared, a detailed comparative analysis was made, and recommendations on the areas of rational use of each type were given [5-7].

2. Description of the study object and models

The study was based on the project of reinforced concrete single-span girder for high-speed railroad. It has a box cross section with inclined walls. The total length is 34.2 m; the calculational length is 33.1 m. Near the support zones (within 1.5 m on each side), the span has a thickening of the slabs and walls. The material of the girder is B35 concrete [8-9].

During the study, several finite element models of this girder were developed:

- model based on beam finite elements
- model based on shell finite elements
- model based on three-dimensional (volume) finite elements

For modeling and calculation were implemented FE-analysis software packages [10-14]. A model based on beam finite elements was made using two cross sections - above the support and in the middle of the span – with taking into account the transition part between them.

IOP Conf. Series: Materials Science and Engineering **786** (2020) 012083 doi:10.1088/1757-899X/786/1/012083

Shell model was made using rectangular shell elements that simulate the walls and slabs of the girder. All oblique elements of the girder (haunches, transition sections) are made by stepwise approximation to rectangular plate elements.

A model based on three-dimensional finite elements was created using an automatic 3D-model meshing. This model is closest to the real structure geometry.

The total masses of all models were slightly different due to the various modeling specifics. The masses errors were equalized by adjusting the mass of the material.

In the research process a modal analysis of the considered models was made.

3. Frequency analysis of various girder models

The first (bending) oscillation mode by the beam, shell and volume models shows a high correlation in the oscillations type and in the frequencies values (Figure 1). Deviations of the frequency of beam and shell models from the frequency of the volume model are just 3.76% and 2.15% respectively.

a) beam model 5,58 Hz | b) shell model 5,37 Hz | c) volume model 5,46 Hz

Figure 1. 1st bending oscillation mode.

By the 1st torsional frequency, three models show a high correlation (Figure 2). At the same time, the beam model has the highest frequency among all. The higher torsional frequency of the beam model is explained by the fact that this model has a higher transverse stiffness, because all the cross sections in it are absolutely rigid according to the plane-sections hypothesis. The shell model is divided into separate elements not only in the longitudinal but also in the transverse direction; therefore, it has a lower transverse rigidity and, consequently, a lower torsional frequency. The model based on volume finite elements has lower transverse rigidity as well.

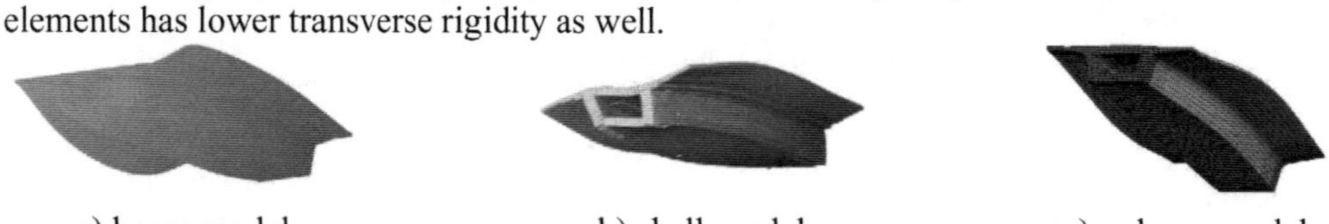

a) beam model 15,78 Hz | b) shell model 14,58 Hz | c) volume model 15,65 Hz

Figure 2. 1st torsional oscillation mode.

The 2nd flexural oscillation form, as well as the 1st one, shows a fairly broad correlation for the beam, shell and volume models: 16.96, 14.67 Hz and 15.51 Hz, respectively. Deviations of the beam and shell models frequencies from the frequency of the volume model are 7.6% and 0.82%. The difference in frequencies is explained again by the different stiffness of the cross sections. By the type of oscillations of the shell and volume models, it can be noted that the cantilever part of the sections oscillates synchronously with the beam part, but has a different oscillations amplitude (Figure 3).

a) beam model 16,96 Hz | b) shell model 14,67 Hz | c) volume model 15,51 Hz

Figure 3. 2nd flexural oscillations mode.

The 3rd bending oscillation mode shows similar correlation.

4. Investigation of intermediate oscillation modes

In general, a big amount of intermediate oscillation modes characterize the shell and volume models.

In the shell and volume models exists a mode of horizontal oscillations caused by horizontal deformations of the structure (Figure 4). The beam model does not show this oscillation mode. In this case, the shell and volume models show a high correlation (the deviation is 3.11%).

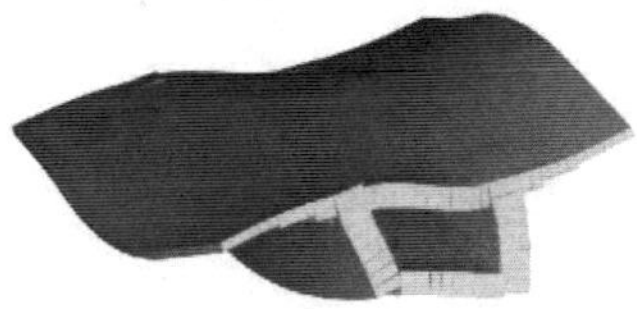

a) shell model
8,62 Hz

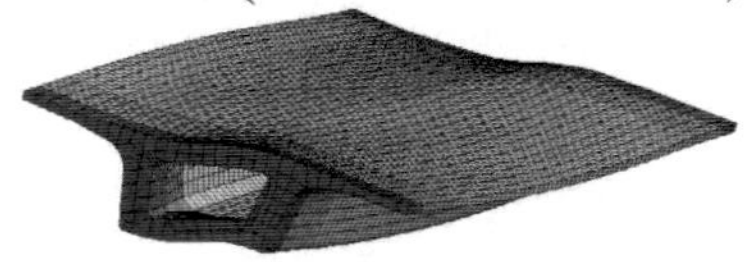

b) volume model
8,36 Hz

Figure 4. Horizontal oscillation modes of the girder.

In this part some forms of local oscillation model based on three-dimensional elements are considering. Due to the high dimensionality of this model type, it becomes possible to take into account any oscillation modes of each structural element.

This is well reflected using an example of girder upper slab oscillations (Figure 5).

a) 22,31 Hz

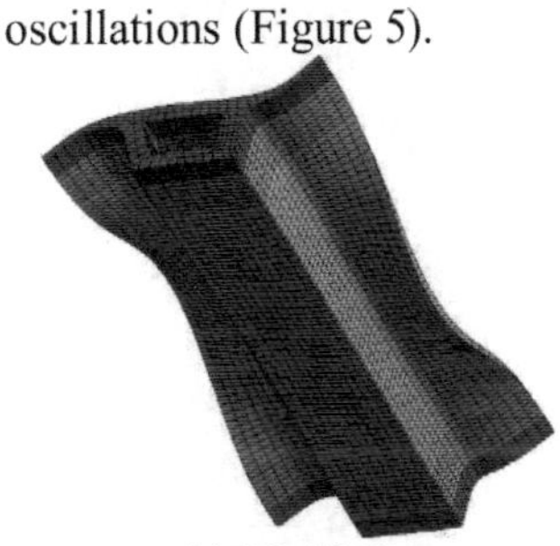

b) 22,70 Hz

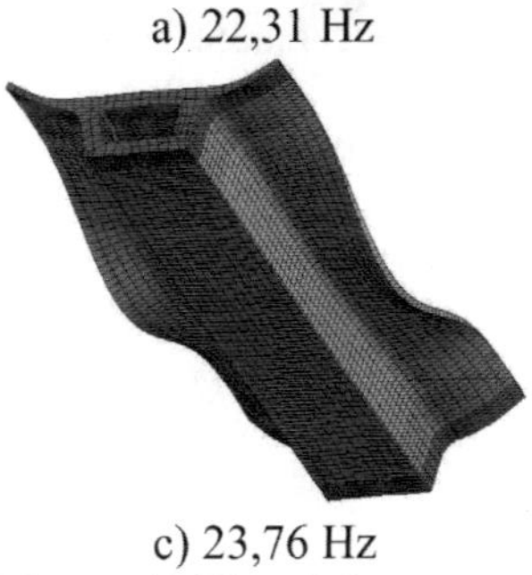

c) 23,76 Hz

d) 24,83 Hz

Figure 5. The girder upper slab oscillation modes (volume elements model).

5. Comparative analysis

According to the models calculation and comparison results, the following conclusions can be drawn:

The beam model shows the main oscillation modes of the structure. The oscillations of the slab part and the beam part in each section are identical in phase and amplitude, because during calculation, the hypothesis of plane sections is used. Thus, using this model, the basic, most high-amplitude oscillations of the entire structure can be estimated. However, this model does not allow tracing the individual structural elements oscillations, first of all the upper slab, which can also negatively affect the superstructure reliability and its operation safety. The beam model advantage regarding to dynamic calculations is a lower dimension, and, consequently, a higher calculation speed.

The shell model allows to calculate a much larger amount of oscillation modes and to track local modes and frequencies. In the calculation by this model it is possible to take into account the oscillations of the girder slabs - both the cantilevers of the upper slab and the sections of the upper and lower slabs between the walls.

However, the shell model has a significant drawback, namely, the high modeling complexity at its low precision: transitional zones are modeled by averaged rectangles; many difficulties arise during the connecting different structural parts into a single model.

The model based on three-dimensional (volume) elements has all the advantages of the shell model, also it has a much higher accuracy and provides the biggest amount of different oscillation shapes and frequencies.

6. Conclusion

Taking into account the above factors, we can conclude that the model based on beam elements is suitable for the calculation of the most common oscillation modes, specific to the whole structure. In particular, this type is well suited for approximate calculations during the variant designing of structures: it does not require much time and high computational power.

The model type based on three-dimensional finite elements is the most applicable for deeper dynamic analysis, especially for the calculation of individual structural elements.

Thus, the volume model allows to completely calculate the spectrum of the structure possible oscillations, which makes it possible to prevent the resonance effects occurrence of the whole structure or its individual elements.

References

[1] B Lena 2004 Dynamic Analysis of a Railway Bridge subjected to High Speed Master Thesis, Department of Civil and Architectural Structural Engineering, Royal Institute of Technology Stockholm

[2] L Fry A rough assessment of railway bridges for high speed trains, Engineering Structures, 548–56

[3] RW Clough, JP Penzien, Dynamics of Structures McGraw-Hill ISBN 0-07-011392-0

[4] S Alampalli 2000 Effects of Testing, Analysis, Damage, and Environment on Modal Parameters, *Mechanical Systems and Signal Processing* **14(1)** 63–74

[5] R Brincker, C E Ventura and P Andersen 2003 Why Output Only Modal Testing is a Desirable Tool for a Wide Range of Practical Applications, in Proc of the 21st International Modal Analysis Conference on Structural Dynamics, Kissimmee, Florida, USA, 141–5

[6] D J Ewins 2000 Modal Testing: Theory, Practice and Application Baldock, Hertfordshire: Research Studies Press Ltd 400

[7] T Lidström 1993 Finite Element Modelling Supported by Object Oriented Methods Licentiate Thesis,

[8] L Dyachenko, A Benin and A Dyachenko 2018 Research of interaction of the "train – bridge" system with bridge deck resonant vibrations, MATEC *Web of Conferences Siberian Transport Forum - TransSiberia* DOI: 10 1051/matecconf/201823905002

[9] L Dyachenko, A Benin and A Dyachenko 2018 Design of Dynamic Parameters for Simple Beam Bridges on High-Speed Railways, IOP Conf Series: *Materials Science and Engineering* **463** 022048 DOI: 10 1088/1757-899X/463/2/022048

[10] Sergio G M Neves, A Azevedo and Rui Calçada 2008 Development of an efficient finite element model for the dynamic analysis of the train-bridge interaction Conference: IABMAS – 4th *International Conference on Bridge Maintenance*, Safety and Management At: Seoul, Korea DOI: 10 1201/9781439828434 ch79

[11] R Delgado and SM Santos Modelling of railway bridge-vehicle interaction on high speed tracks, Computers & Structures; T 63 P 511-523 (1997) DOI:10 1016/S0045-7949(96)00360-4

[12] N Zhang, H Xia and W Guo Vehicle 2008 –bridge interaction analysis under high-speed trains Journal of Sound and Vibration, **309(3-5)** 407–25

IOP Conf. Series: Materials Science and Engineering **786** (2020) 012083 doi:10.1088/1757-899X/786/1/012083

[13] Y B Yang, J D Yau and Y S Wu 2004 Vehicle-Bridge Interaction Dynamics: With Applications To High-Speed Railways Singapore: World Scientific Publishing Company
[14] Sergio G M Neves, Pedro Aires Montenegro, A F M Azevedo and Rui Calçada 2014 A Direct Method for Analysing the Nonlinear Vehicle-Structure Interaction in High-Speed Railway Lines, Conference: The Second International Conference on Railway Technology: Research, *Development and Maintenance* At: Ajaccio, Corsica DOI: 10 4203/ccp 104 81

IOP Conf. Series: Materials Science and Engineering **786** (2020) 012084 doi:10.1088/1757-899X/786/1/012084

Comparative research of orderliness dinamics of road safety systems in the volga federal district and the Russian Federation

A I Petrov[1,*], S A Evtyukov[2], V I Kolesov[1] and D A Petrova[3]

[1]Tyumen Industrial University

[2]Saint Petersburg State University of Architecture and Civil Engineering

[3]Ural Federal University named after the first President of Russia B. N. Yeltsin

*E-mail: ArtIgPetrov@yandex.ru

Abstract. This article considers the issues of dynamics of road safety provision systems in three multiscaled systems (Russia, Volga Federal District or VFD and separate subjects of VFD). The indicator of relative information entropy Hn of road safety provision process was taken as the characteristic of system orderliness. The ideology, methodics and results of quantitative assessment of road safety provision systems information entropy H were considered. Time series of modification of relative information entropy Hn of road safety provision systems in 2004…2018 in Russia, Volga Federal District and individual subjects of VFD were presented. Special attention was paid to consideration of the researched process in the Orenburg region and Samara region. This regions is characterized by non-standard, different from general trend dynamics of information entropy H of road safety provision system. The general conclusion of the article is that during last 15 years information entropy of road safety provision systems significantly decreased, i.e. orderliness of formation process of road traffic participants safety greatly increased. However, results achieved in this sphere by VFD regions are still far from best world results.

1. Introduction

Management – transfer of manageable system from current state to desirable (i.e. complex of works on practical implementation of image of the future). During management activities structural and technological orderliness of system and its functioning process increases [1]. In other words, any management activity is intended to fight with entropy – process of growth of uncontrolled chaos in system. Quality of management can be accurately measured by evaluation of system entropy and dynamics of its decline.

State of any transport-technological system is described by enormous set of different characteristics that can be classified by various features (efficiency, quality of functioning, development speed, etc.) [2, 3]. While managing the system, it is necessary to control dozens of traffic parameters and adjust production process in time. It is a complex task. That is the reason of strict regulation of road traffic participants behavior (with the aim of formation of optimal low-entropy system state). It is supposed that such regulation will decrease the degree of chaos in transport-technological system and increase its reliability. For that purpose, system of restrictions imposed on participants of transport process - road traffic rules - has been widely used worldwide for a long time. Increasing of road safety is expressed in development of complex traffic light regulations and initiation of freedom limitation of road traffic participants.

Specifics of transport-technological systems consist in their non-stationarity, continuous variability and not full controllability. In this connection transport management is one of the most difficult types of management activities, entropy in transport systems is quite large. One of the most demonstrative examples of transport systems' entropy is traffic accident rate that is presented only in negative way despite laws of dialectics. According to the estimations of World Health Organization [4] annually in the result of road accidents nearly 1,35 mln. people die, more than 30 mln. people get injured, economic loss because of road accident rate makes $ 550 bln.

2. Formulation of the problem

Management by objectives is used for control of large systems. The distinctive feature of this method is clear understanding of cause-effect relationships, formulation of goals and development of constrained environment for management. Particularly in Russia such approach was formulated in the Target Federal Programme on road safety (for a period of 2013…2020) [5] and moved to the Strategy of road safety [6].

But these documents only declare general directions of road safety development and don't formalize peculiar methods of solving designated problems. There is a variety of instruments that can provide road safety, they can be both engineering and juridical. Efficiency of their usage can be different in various regions of the Russian Federation. It can be explained by national and social-economical specifics of regions. In this connection state of road safety differently forms in each region. And still there is no tool for measurement the degree of road safety system orderliness.

Taken analysis of this issue [1] showed that one of the most efficient instrument of evaluation of the system state from the position of orderliness or forwardness to opposition to negative impact of chaos is assessment of entropy.

This article presents ideology, method of quantitative assessment of information entropy in the road safety provision sphere and results of assessment of dynamics of relative entropy of road safety provision systems in regions of Volga Federal District.

3. Main concepts and definitions

Fundamental concepts of this article are Orderliness and Information entropy. It is necessary to formulate their definitions.

Orderliness – system property, identifying the result of implementation of set of rules and forbiddances that structure system and limit its modifications [7]. Orderliness is an antonym to chaos. Applied to road safety provision processes, orderliness decreases freedom of actions of road traffic participants and at the same time probability of violation or road traffic rules. That automatically leads to decline of probability of road accidents formation and enhancement of final road accidents statistics.

Information entropy – measure of uncertainty (disorganization) of system state [8]. The concept of information entropy, introduced by C.E. Shannon in his paper «A Mathematical Theory of Communication» [9], is a probability of some event. The less the probability of event, the more information it includes [1]. Applied to the assessment of orderliness of road safety provision processes, it means that if information entropy is low, then organizers of transport and technological process can guarantee safe functioning of the system. In the conditions of high uncertainty of system state, we can get set of possible outcomes, including undesirable (road accidents and their consequences).

IOP Conf. Series: Materials Science and Engineering **786** (2020) 012084 doi:10.1088/1757-899X/786/1/012084

4. Method of quantitative assessment of information entropy in road safety provision sphere

The main principles of this method were developed by authors of this article and had already been presented in earlier published works [7, 8, 10]. Moreover, results were showed at scientifically-practical conferences in Saint-Petersburg in 2016 and 2018, in Oryol in 2017...2019, in Moscow in 2018, in Orenburg in 2017, in Tyumen in 2016...2019. Thereby main ideas of this method will be just mentioned in this article.

1. The process of road accident rate formation should be considered from the position of process of quantitative information transmission through the chain «Population – The number of vehicles – The number of road accidents – The number of victims – The number of deaths» (Figure 1). The scale of consideration or road accident formation process can differ: large (World, Continent), medium (Country, individual regions of country), small (city, district of city).

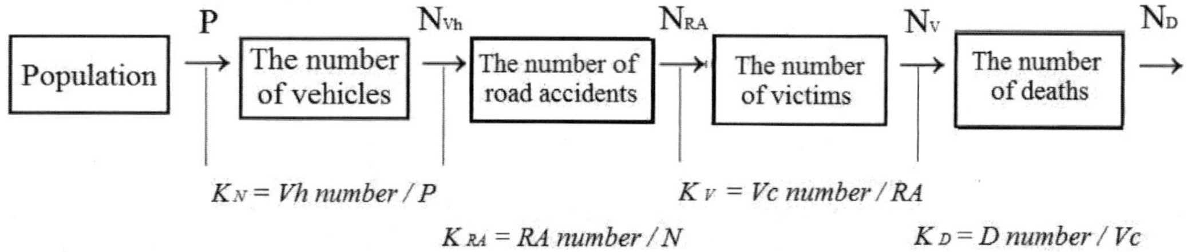

Figure 1. The cause-effect chain of road accident rate formation [10].

2. Calculation of transition coefficients K_i between blocks of the cause-effect chain of road accident rate formation (Figure 1). We will describe this process as 4 sub-processes that have specific coefficients K_N (transformation of number of population into the number of transport vehicles in transport fleet), K_{RA} (transformation of vehicle fleet into the number of road accidents), K_V (transformation of the number of road accidents into the number of road accidents victims), K_D (transformation of the number of road accidents victims into the number of lost in road accidents people).

3. Identifying of the positive of the contribution Q relatively to weights of appropriate elements of examined transformational process within the chain «Population – <...> – The number of deaths in road accident» (1):

$$Q = Q_N + Q_{RA} + Q_V + Q_D = \ln(1/K_N) + \ln(1/K_{RA}) + \ln(K_V) + \ln(1/K_D) \tag{1}$$

The physical meaning of the positive of the contribution Q of different elements of the chain «Population – <...> – The number of deaths in road accident» into the final result of road accident rate is the measure of information amount or derivative of examined process entropy.

4. Identifying the structure of weight coefficients w_i for assessing the positive of the contribution Q of different elements of the chain «Population – <...> – The number of deaths in road accident». Availability of calculated values w_N , w_{RA} , w_V w_D of positive allows to solve the main problem of entropic analysis – assess the impact of different elements of the chain «Population – <...> – The number of deaths in road accident» on formation of final road accident rate. Above-stated researches were held for each year from period of 2004...2018.

5. Calculation of entropy H in road safety provision systems of Russian Federation and far eastern regions by classic C.E. Shannon's [9] formula (2):

$$H = -\sum_{i=1}^{n} w_i \cdot \ln w_i \tag{2}$$

where n system elements count (in our case n = 4);

w_i weight coefficients, satisfying the normalization condition, $\sum_{i=1}^{n} w_i = 1$.

6. Calculation of relative information entropy (3) of the road safety provision systems of the Russian Federation and Volga Federal District regions:

$$H_n = H/H_{\max} = H/\ln(n) \tag{3}$$

Values of *Hn* varies in range from 0 to 1, where 1 means total disorganization or process, while 0 indicates total orderliness and absence of the chaos in system. In reality entirely ordered or disordered systems and processes don't exist. Actual range of *Hn* values is 0.6 (some European countries)...0.9 (some African countries) for the road safety provision sphere.

5. Initial data and results of calculation of relative entropy *Hn* of road safety provision systems in Russia and Volga Federal District

As mentioned above specific methods of assessment of human-technical systems orderliness were previously considered in [7, 8]. This article is presenting only results of estimation of relative entropy *Hn* of road safety provision systems in the Russian Federation, VFD and two regions with opposite rates of entropy change (during 2004...2018).

Tables 1...4 show initial data [11] and results of calculation of relative entropy *Hn* of road safety provision systems in the Russian Federation, Volga Federal District, Orenburg region and Samara region accordingly.

Table 1. Initial data [11] that was used for analysis of road safety orderliness dynamics in the Russian Federation and results of calculation of relative entropy *Hn*.

Year	Population, thousands of people	Vehicles fleet, thousands of units.	The number of road accidents in year, units	The number of victims, people	The number of deaths in road accidents, people	Value of relative entropy *Hn*
2004	144591.0	31439225	207686	284784	34425	0.782
2005	143763.0	33355337	222475	307667	33858	0.783
2006	143041.0	34465080	228309	317011	32640	0.783
2007	142430.0	35446415	233052	324531	33238	0.781
2008	142007.0	38194754	217557	299885	29840	0.769
2009	141904.0	40688499	202967	282305	27603	0.762
2010	141956.0	41648965	199083	276762	26544	0.759
2011	142912.6	43325312	199868	279801	27953	0.755
2012	143030.1	45471096	203597	286609	27991	0.751
2013	143347.1	47881812	204068	285462	27025	0.745
2014	146301.9	52175879	199720	278748	26963	0.734
2015	146307.7	56469971	184000	254311	23114	0.720
2016	146832.3	58025620	173694	241448	20308	0.718
2017	146899.0	59790545	169432	234462	19088	0.713
2018	146828.2	60578772	168099	232907	18214	0.712

Table 2. Initial data [11] that was used for analysis of road safety orderliness dynamics in the Volga Federal District and results of calculation of relative entropy *Hn*.

Year	Population, thousands of people	Vehicles fleet, thousands of units.	The number of road accidents in year, units	The number of victims, people	The number of deaths in road accidents, people	Value of relative entropy *Hn*
2004	31036.0	6839880	39445	53647	6786	0.779
2005	30710.0	6944388	43614	59673	6799	0.779
2006	30511.0	7094131	45928	63362	6465	0.783
2007	30346.0	7360952	45775	63647	6636	0.780
2008	30241.6	7730508	43749	60301	6117	0.772
2009	30157.8	8176113	41041	57139	5751	0.765
2010	30109.0	8176198	40414	56306	5581	0.765
2011	29880.4	8511217	40315	56722	5630	0.762
2012	29808.7	8952948	41710	58740	5755	0.757
2013	29772.3	9373412	44302	62085	5700	0.755
2014	29715.4	10331296	42983	60512	5382	0.743
2015	29715.4	11289174	39101	54271	4476	0.727
2016	29636.6	11496727	38574	53801	4035	0.727
2017	29542.7	11805704	37246	51530	3859	0.721
2018	29397.3	12226022	37585	51943	3646	0.717

Table 3. Initial data [11] that was used for analysis of road safety orderliness dynamics in the Orenburg region and results of calculation of relative entropy *Hn*.

Year	Population, thousands of people	Vehicles fleet, thousands of units.	The number of road accidents in year, units	The number of victims, people	The number of deaths in road accidents, people	Value of relative entropy *Hn*
2004	2163.0	268395	1635	2193	344	0.790
2005	2150.0	514319	3137	4573	503	0.787
2006	2137.9	517806	3587	5116	463	0.792
2007	2126.0	594805	3597	5266	475	0.781
2008	2119.1	600816	3345	5108	460	0.784
2009	2111.5	611236	3110	4474	414	0.769
2010	2113.0	610746	2901	4079	356	0.764
2011	2031.3	649366	2676	3887	366	0.754
2012	2023.8	674703	2449	3589	369	0.745
2013	2016.1	664229	2734	4006	355	0.754
2014	2001.1	793543	2679	3929	419	0.724
2015	2001.1	922857	2362	3351	353	0.691
2016	1989.6	961980	2319	3368	306	0.693
2017	1977.7	996067	2312	3223	295	0.678
2018	1963.0	1000365	2226	3117	273	0.678

Table 4. Initial data [11] that was used for analysis of road safety orderliness dynamics in the Samara region and results of calculation of relative entropy *Hn*.

Year	Population, thousands of people	Vehicles fleet, thousands of units.	The number of road accidents in year, units	The number of victims, people	The number of deaths in road accidents, people	Value of relative entropy *Hn*
2004	3218.1	813833	3642	4817	725	0.742
2005	3201.0	853428	4496	5948	748	0.752
2006	3188.9	884473	5241	7154	752	0.766
2007	3178.0	868799	5189	7259	713	0.773
2008	3172.8	921971	4679	6411	586	0.760
2009	3171.4	967046	4384	6106	584	0.754
2010	3170.0	971609	4267	5997	532	0.756
2011	3215.4	976211	4210	5923	491	0.757
2012	3213.4	1029569	4600	6649	591	0.758
2013	3213.3	1078536	4757	6810	578	0.753
2014	3212.7	1163565	4373	6309	534	0.741
2015	3212.7	1248594	3883	5563	430	0.728
2016	3203.7	1241582	3853	5498	431	0.727
2017	3193.5	1244407	3724	5236	374	0.725
2018	3183.0	1300244	4021	5638	344	0.724

6. Estimation of dynamics of relative entropy *Hn* of road safety provision systems in the Russian Federation and Volga Federal District (VFD)

Figure 2 shows the trend of change of relative entropy for road safety provision systems in Russia and Volga Federal District in 2004…2018.

Value of *Hn* decreases in both cases, therefore the degree of chaos declines in compared road safety provision systems and their orderliness increases. Processes of regulation of road safety provision system functioning in Russia and VFD proceeds in parallel (Figure 2). But in such big system as Russia the proceeding is smoother than in VFD where the process of entropy decline is characterized by greater oscillation amplitude.

Figure 3 shows the trend of change of relative entropy for road safety provision systems in the VFD regions in 2004…2018.

Analysis of Figure 3 allows to understand that, despite regional specifics, practically in all regions level of road safety provision system orderliness increases, because relative entropy *Hn* of road safety formation processes persistently decreases.

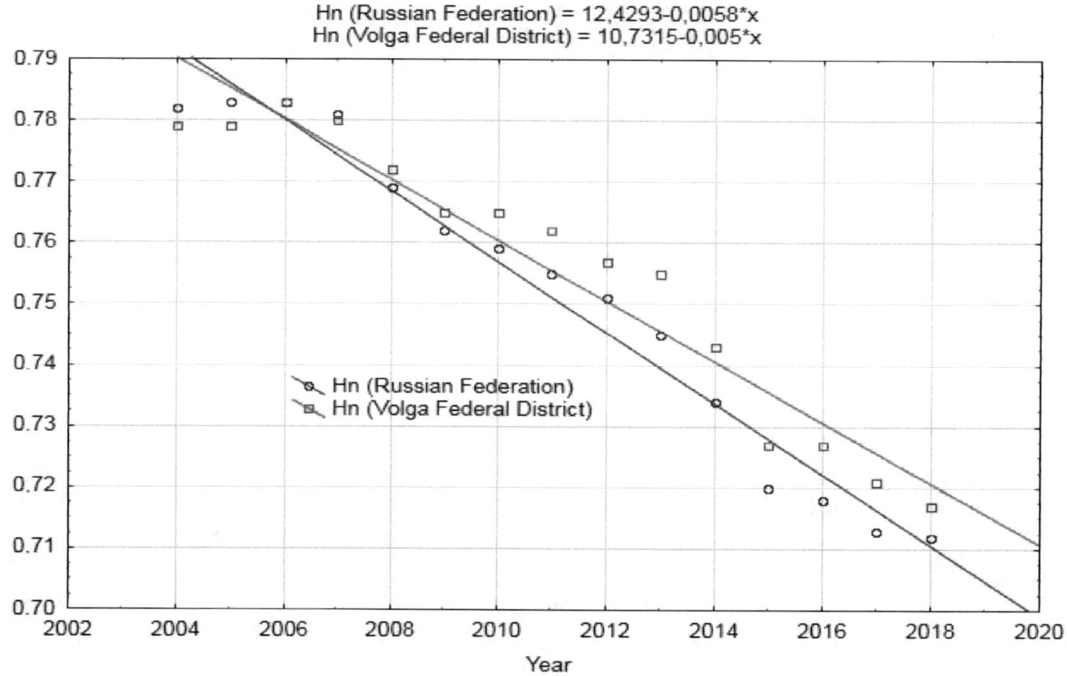

Figure 2. Time series of modification of values of relative entropy *Hn* of road safety provision systems in the Russian Federation and Volga Federal District.

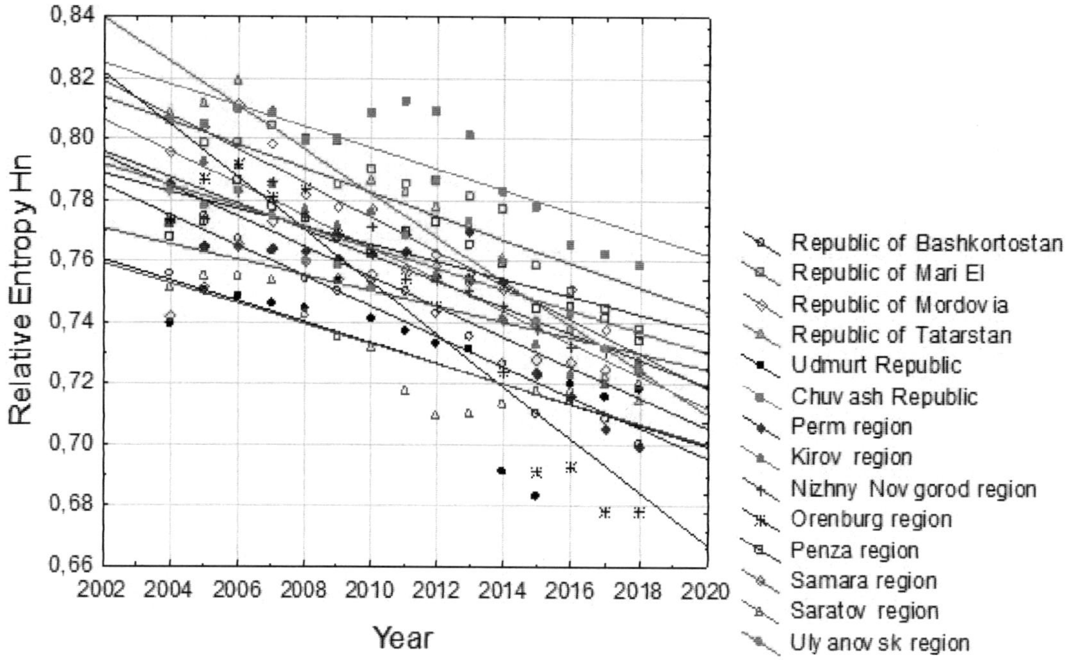

Figure 3. Time series of modification of values of relative entropy Hn of road safety provision systems in the VFD regions.

Table 5 shows the trend of change of relative entropy *Hn* for road safety provision systems in the VFD regions in 2004…2018.

IOP Conf. Series: Materials Science and Engineering **786** (2020) 012084 doi:10.1088/1757-899X/786/1/012084

Table 5. Results of calculation of relative entropy *Hn* in the subjects of the Volga Federal District (2004…2010).

Subjects of the Volga Federal District	Values of relative entropy *Hn* of road safety provision systems in the Volga Federal District subjects						
	2004	2005	2006	2007	2008	2009	2010
Republic of Bashkortostan	0.756	0.775	0.767	0.763	0.754	0.750	0.753
Republic of Mari El	0.772	0.798	0.798	0.804	0.800	0.799	0.790
Republic of Mordovia	0.796	0.804	0.811	0.798	0.782	0.778	0.777
Republic of Tatarstan	0.808	0.811	0.819	0.809	0.799	0.785	0.786
Udmurt Republic	0.739	0.750	0.748	0.746	0.744	0.753	0.741
Chuvash Republic	0.806	0.805	0.810	0.808	0.799	0.800	0.808
Perm region	0.785	0.765	0.765	0.764	0.763	0.761	0.762
Kirov region	0.773	0.793	0.791	0.785	0.777	0.771	0.776
Nizhny Novgorod region	0.773	0.774	0.783	0.786	0.775	0.768	0.771
Orenburg region	0.774	0.787	0.792	0.781	0.784	0.769	0.764
Penza region	0.768	0.773	0.786	0.778	0.774	0.768	0.762
Samara region	0.742	0.752	0.766	0.773	0.760	0.754	0.756
Saratov region	0.751	0.755	0.755	0.753	0.743	0.735	0.731
Ulyanovsk region	0.783	0.778	0.783	0.775	0.760	0.758	0.751

Republic of Bashkortostan	0.750	0.743	0.735	0.726	0.710	0.714	0.708	0.700
Republic of Mari El	0.785	0.786	0.781	0.777	0.758	0.750	0.744	0.738
Republic of Mordovia	0.769	0.762	0.754	0.751	0.724	0.751	0.738	0.736
Republic of Tatarstan	0.783	0.778	0.771	0.761	0.739	0.738	0.722	0.714
Udmurt Republic	0.737	0.733	0.731	0.691	0.683	0.720	0.716	0.718
Chuvash Republic	0.812	0.809	0.801	0.783	0.778	0.766	0.762	0.758
Perm region	0.763	0.755	0.770	0.753	0.723	0.716	0.705	0.699
Kirov region	0.769	0.755	0.754	0.741	0.733	0.723	0.720	0.727
Nizhny Novgorod region	0.762	0.753	0.750	0.745	0.738	0.732	0.730	0.724
Orenburg region	0.754	0.745	0.754	0.724	0.691	0.693	0.678	0.678
Penza region	0.770	0.773	0.766	0.759	0.744	0.745	0.741	0.734
Samara region	0.757	0.758	0.753	0.741	0.728	0.727	0.725	0.724
Saratov region	0.717	0.709	0.710	0.713	0.717	0.717	0.720	0.720
Ulyanovsk region	0.768	0.786	0.773	0.760	0.740	0.743	0.731	0.725

7. Evaluation of dynamics of relative entropy *Hn* of road safety provision systems in the Volga Federal District (VFD) regions

By approximation of time series of modification of value of relative entropy *Hn* of road safety provision systems in the VFD regions during 2004…2018 by linear function $Hn = f(Years)$ we got values of parameter b of model $Hn = a - b(Years)$. Values of this parameter are presented in table 6.

Table 6. Quantitative values of parameter *b* of model $Hn = a - b(Years)$ for regions of the Volga Federal District.

Region of Volga Federal District	Quantitative values of parameter *b* of model $Hn = a - b(Years)$	Coefficient of model determination R^2
Republic of Bashkortostan	-0,0050	0,898
Republic of Mari El	-0,0039	0,645
Republic of Mordovia	-0,0056	0,889
Republic of Tatarstan	-0,0072	0,931
Udmurt Republic	-0,0034	0,530
Chuvash Republic	-0,0035	0,677
Perm region	-0,0050	0,733
Kirov region	-0,0053	0,879
Nizhny Novgorod region	-0,0043	0,894
Orenburg region	-0,0086	0,873
Penza region	-0,0029	0,728
Samara region	-0,0026	0,537
Saratov region	-0,0033	0,716
Ulyanovsk region	-0,0034	0,620

Analysis of quantitative values of parameter *b* of model $Hn = a - b(Years)$ specific for different VFD regions allows to make conclusion about specifics of dynamics of growth of orderliness of regional road safety provision systems.

8. Discussion of results

According to the data from table 4 during the last 15 years (2004…2018):

1. The highest pace of growth of road safety provision system orderliness was in Orenburg region (b = -0.0086 or 0.86 % in a year) and Republic of Tatarstan (b = -0.0072 or 0.72 % in a year);
2. Relatively small growth of road safety provision systems orderliness was registered in Samara region (b = -0.0026 or 0.26 % in a year) and Penza region (b = -0.0029 or 0.29 % in a year);
3. In other VFD regions changes of road safety provision systems orderliness equals to the range [-0.0033; -0.0056] or [-0.33 % in a year; -0.56 % in a year];
4. Considering established trends of dynamics of decreasing *Hn* (≈ 0.5 % in a year), nearly 15…22 years required in order to achieve the level of the best world practices of the road safety provision systems orderliness (*Hn* = 0.55…0.60).

9. Conclusion

In the Road safety strategy in the Russian Federation for 2018-2024 [6] the target for Human Risk for 2024 equals to 4 people/100 thousand people or in other words 6000 dead people in road accidents per year in the country with 150 million population. Is it an achievable goal? Taking into account the decline of *Hn* by 0.5 % per year during 2019…2024, it is expectable that *Hn* will decrease by 3 % relatively to $Hn_{RF\ 2018} = 0.712$, i.e up to $Hn_{RF\ 2024} = 0.691$. This level of dynamics of relative information entropy in road safety provision system of Russia will predict the value of the mortality rate in road accidents as approximately 12.5 thousand people per year. For achievement of ambitious plans, formulated in the Road safety strategy in the Russian Federation up to 2024, *Hn* should have been decreased more significantly than 0.5 % per year. It is hardly possible.

IOP Conf. Series: Materials Science and Engineering **786** (2020) 012084 doi:10.1088/1757-899X/786/1/012084

References

[1] Schrödinger E 1944 What is Life? The Physical Aspect of the Living Cell Cambridge at the University Press

[2] Koryagin M and Dementiev A and Sokolov V 2018 Management system of urban transport market *Transportation Research Procedia* **36** 334–40

[3] Vuchic VR 1999 Transportation for livable cities N J: Rutgers University, Brunswick)

[4] World Health Organization Homepage Available at: https://www who int/news-room/detail/07-12-2018-new-who-report-highlights-insufficient-progress-to-tackle-lack-of-safety-on-the-world's-roads

[5] Russian Federation Target Federal Programme on road safety (for a period of 2013…2020) Available at: http://www fcp-pbdd ru/about_program/

[6] Road safety strategy in the Russian Federation for 2018-2024 Available at: http://static government ru/media/files/g6BXGgDI4fCEiD4xDdJU wlxudPATBC12 pdf

[7] Petrov A and Kolesov V 2018 Entropic analysis of dynamics of road safety system organization in the largest Russian cities IOP Conf Series: *Earth and Environmental Science* **177** 012015

[8] Kolesov V and Petrov A 2018 System dynamics of process organization in the sphere of traffic safety assurance *Transportation Research Procedia* **36** 286-94

[9] Shannon CE A 1948 Mathematical Theory of Communication *Bell System Technical Journal* **27(379-423)** 623-56

[10] Kolesov V and Petrov A 2017 Cybernetic Modeling in Tasks of Traffic Safety Management *Transportation Research Procedia* **20** 305-10

[11] State Inspection for Road Traffic Safety of the Ministry of Internal Affairs of Russia Homepage Available at: https://www gibdd ru/stat

International Scientific Conference Interstroymeh - 2019 (ISM - 2019) IOP Publishing
IOP Conf. Series: Materials Science and Engineering **786** (2020) 012085 doi:10.1088/1757-899X/786/1/012085

Assessment of sustainability of improving road safety process in the volga federal district

A I Petrov[1,*] and D A Petrova**Ошибка! Закладка не определена.**[2]

[1]Federal State Budget Educational Institution of Higher Education «Industrial University of Tyumen»

[2]Federal State Autonomous Educational Institution of Higher Education «Ural Federal University named after the first President of Russia B.N. Yeltsin»

*E-mail: ArtIgPetrov@yandex.ru

Abstract. The article examines the issues of assessing the sustainability of traffic safety improvement process in the cities of the Volga Federal District of the Russian Federation. In 2015…2018 a sharp decrease in the overall level of road traffic accidents in the Russian Federation was recorded. However, in different regions and cities of the country this positive process runs extremely heterogeneously, with various speeds and different levels of qualitative changes in the field of road safety. The T-Wilcoxon criterion is an instrument, used in analyzing accident rate statistics, which can help in the argumentation of the opinion on the stability of this process or, vice versa, on the chaotic state and weak expression. On the example of accident rate statistics in subjects of the Volga Federal District of Russia, the article proves that improvement of road safety can be characterized as sustainable.

1. Introduction

A significant decrease [1] in the number of road accidents, dead and injured people in road accidents in the last four years (2018/2015) in Russia is the cause for the formulation of general conclusion about the success of road traffic safety Federal Program realization. How stable is this positive trend? Whether everywhere the process of road safety improvement can be considered as qualitative? To answer these questions, we will use the T-Wilcoxon criterion.

2. Methods of assessment of researched process stability

T-Wilcoxon criterion is designed to compare two dependent samples between themselves regarding the attribute expression [2, 3, 4, 5, 6, 7, 8, 9, 10, 11, 12]. Particularly, with its help, it is possible to determine the degree of changes intensity in dependent selections at different time periods. T-criterion is based on the ranking of the absolute values of the difference between two sets of sample values in the first and second experiments. In our case, the data sets for 2018 and 2015 will be compared by two most important characteristics of road traffic accidents (Human Risk *HR* and Road Traffic Accident Severity Coefficient $C_{sev.\ RTA}$) in subjects of the Volga Federal District of Russia (VFD) of the Russian Federation.

Zero hypothesis of the research H0 usually [3, 5] formulated as «Intensity of shifts in the typical direction doesn't exceed the intensity of shifts in the atypical direction». Applied to our case, it consists in the absence of statistical differences between the time distributions defined for the same selections of 2015 and 2018. According to the zero hypothesis, differences between selections values of different years of research are not enough to accept that non-random factors cause this difference.

Hypothesis H1 classically [3, 5] formulated as «Intensity of shifts in the typical direction exceeds the intensity of shifts in the atypical direction», i.e. the change in the data distribution on the characteristics of the accident rate of 2018, relative to similar data distribution of 2015, did not occur accidentally - some specific factors had an impact on this data shift, e.g. organizational and managerial efforts that were invested in the solution of increasing road safety problem.

The essence of the research is to prove hypothesis H1 and refute zero hypothesis H0. Two results are possible. In the case of the hypothesis H1 evidence, it is necessary to conclude the non-randomness of the shift and the effectiveness of some managerial influence, aimed at changing the actual situation in the sphere of road safety. If hypothesis H1 is not proved, i.e. the zero hypothesis H0 will be confirmed, then it can be claimed that the quality of the road safety management leaves much to be desired.

When using the methods of mathematical statistics, level of statistical significance p-level, in other words, «probability that we found differences as significant, but they are actually accidental», plays an important role [8].

In statistics, three levels of statistical significance are distinguished: the lowest (5% or $p = 0.05$); sufficient (1% or $p = 0.01$) and the highest (0.1% or $p = 0.001$). If the probability of randomness of received results is more than 5% ($p = 0.05$), then the lowest level of statistical significance is not reached, and hypothesis H1 is not proved.

3. Used road safety statistics

To study the quality of road safety improvement process comparisons between two sets of data about road accidents with victims (2015 and 2018) were made in subjects of the Volga Federal District of Russia of the Russian Federation (totally 14 regions). The initial information was received on the website of the State Inspection for Road Traffic Safety of the Ministry of Internal Affairs of Russia [1].

Tables 1 and 2 shows the initial (for subjects of the VFD of Russia) information, necessary for calculation values of such significant characteristics of accident rate [6, 7] as Human Risk *HR* and Road Traffic Accident Severity Coefficient $C_{sev.\ RTA.}$.

Table 1. Characteristics of road accident rate in subjects of the Volga Federal District of Russia (2015).

Subjects of the Volga Federal District	Numerical values (2015)			
	Road accidents, units [1]	Road fatalities, people [1]	*Human Risk HR*, deaths/ 100 K people	*Road Traffic Accident Severity Coefficient* $C_{sev.\ RTA.}$
Republic of Bashkortostan	4563	604	14.83	9.53
Republic of Mari El	1035	101	14.69	6.77
Republic of Mordovia	937	153	18.91	11.66
Republic of Tatarstan	4969	492	12.76	7.45
Udmurt Republic	1379	209	13.77	11.90
Chuvash Republic	1613	208	16.80	8.78
Perm Krai	4161	421	15.96	7.31
Kirov region	1689	192	14.72	8.04
Nizhny Novgorod Region	5342	522	15.96	7.11
Orenburg region	2362	353	17.64	10.53
Penza region	2097	233	17.19	7.73
Samara region	3883	430	13.38	7.73
Saratov region	3401	362	14.52	7.74
Ulyanovsk region	1670	196	15.52	8.50

Table 2. Characteristics of road accident rate in subjects of the Volga Federal District of Russia (2018).

Subjects of the Volga Federal District	Numerical values (2018)			
	Road accidents, units [1]	Road fatalities, people [1]	*Human Risk HR*, deaths/ 100 K people	*Road Traffic Accident Severity Coefficient* $C_{sev.\ RTA.}$
Republic of Bashkortostan	4406	550	13.58	9.06
Republic of Mari El	787	83	12.20	7.31
Republic of Mordovia	1096	148	18.60	8.85
Republic of Tatarstan	4612	379	9.72	6.24
Udmurt Republic	2003	157	10.42	5.96
Chuvash Republic	1250	162	13.24	8.93
Perm Krai	3444	293	11.22	6.24
Kirov region	1756	174	13.68	7.21
Nizhny Novgorod Region	5528	386	12.01	5.21
Orenburg region	2226	273	13.91	8.76
Penza region	1913	212	16.08	7.65
Samara region	4021	344	10.81	6.10
Saratov region	3213	321	13.15	6.99
Ulyanovsk region	1330	164	13.24	8.60

4. Results of calculation of Wilcoxon T-critrion for researched data sets

Tables 3 and 4 show results of T-Wilcoxon criterion assessment [10] relatively changes of Human Risk *HR* and Road Traffic Accident Severity Coefficient $C_{sev.\ RTA}$. values in cities of VFD of Russia during 2015-2018. «Increasing of value» is taken as an atypical shift.

Table 3. Calculations of atypical shifts ranks sum $T_{emp.\ (2018/2015)}$ relative to characteristic of Human Risk *HR* in subjects of the Volga Federal District of Russia of Russia.

Subjects of the Volga Federal District	Values of HR, deaths/100 K people		Shift of HR		Ranked number of shift
	Before (2015)	After (2018)	Factual	Absolute	
Republic of Bashkortostan	14.83	13.58	-1.25	1.25	4
Republic of Mari El	14.69	12.20	-2.49	2.49	7
Republic of Mordovia	18.91	18.60	-0.31	0.31	1
Republic of Tatarstan	12.76	9.72	-3.04	3.04	9
Udmurt Republic	13.77	10.42	-3.35	3.35	10
Chuvash Republic	16.80	13.24	-3.56	3.56	11
Perm Krai	15.96	11.22	-4.74	4.74	14
Kirov region	14.72	13.68	-1.04	1.04	2
Nizhny Novgorod Region	15.96	12.01	-3.95	3.95	13
Orenburg region	17.64	13.91	-3.73	3.73	12
Penza region	17.19	16.08	-1.11	1.11	3
Samara region	13.38	10.81	-2.57	2.57	8
Saratov region	14.52	13.15	-1.37	1.37	5
Ulyanovsk region	15.52	13.24	-2.28	2.28	6
Atypical shifts ranks sum Temp.					0

Table 4. Calculations of atypical shifts ranks sum $T_{emp.\ (2018/2015)}$ relative to characteristic of Road Traffic Accident Severity Coefficient $C_{sev.\ RTA}$. in subjects of the Volga Federal District of Russia.

Subjects of the Volga Federal District	Values of Csev. RTA.		Shift of Csev. RTA.		Ranked number of shift
	Before (2015)	After (2018)	Factual	Absolute	
Republic of Bashkortostan	9.53	9.06	-0.47	0.47	4
Republic of Mari El	6.77	7.31	0.54	0.54	5
Republic of Mordovia	11.66	8.85	-2.81	2.81	13
Republic of Tatarstan	7.45	6.24	-1.21	1.21	9
Udmurt Republic	11.90	5.96	-5.94	5.94	14
Chuvash Republic	8.78	8.93	0.15	0.15	3
Perm Krai	7.31	6.24	-1.07	1.07	8
Kirov region	8.04	7.21	-0.83	0.83	7
Nizhny Novgorod Region	7.11	5.21	-1.90	1.90	12
Orenburg region	10.53	8.76	-1,77	1.77	11
Penza region	7.73	7.65	-0,08	0.08	1
Samara region	7.73	6.10	-1,63	1.63	10
Saratov region	7.74	6.99	-0,75	0.75	6
Ulyanovsk region	8.50	8.60	0,10	0.10	2
Atypical shifts ranks sum Temp					10

Atypical shifts ranks sum (in tables 3 and 4 positive shifts are atypical) is formed during the process of summation of ranked numbers specific for positive shifts cases.

The procedure of making decisions about detection of shifts or statistically significant differences between selections of 2015 and 2018 consists in the comparison of atypical shifts ranks sum values $T_{emp.}$ with tabular values $T_{cr.}$

Table 5 shows the fragment of the table of T-Wilcoxon criterion critical values ($T_{cr.}$) for two levels of statistical significance. Reasoning can be based on the results of comparison $T_{emp.}$ with $T_{cr.}$ (p= 0.01) [2, 3, 4, 5, 8, 9, 12].

Table 5. The fragment of table of T-Wilcoxon criterion critical values ($T_{cr.}$) [8].

n	Level of statistical significance		n	Level of statistical significance	
	p = 0.05	p = 0.01		p = 0.05	p = 0.01
5	0	-	13	21	12
6	2	-	14	25	15
7	3	0	15	30	19
8	5	1	16	35	23
9	8	3	17	41	27
10	10	5	18	47	32
11	13	7	19	53	37
12	17	9	20	60	43

Rule of acceptance the hypothesis H_1: if the empirical value of criterion $T_{emp.} \leq T_{cr.}$, appropriate to the level of statistical significance p = 0.01, then promoted statistical hypothesis is considered to be proved [2, 3, 4, 5, 8, 9].

5. Results of research

Comparing the value of atypical shifts ranks sum Temp. = 0 (for the case of HR) and Temp. = 10 (for the case of Csev. RTA.) with tabular (Tcr.) values of T-Wilcoxon criterion (Tcr. = 15 for the case p = 0.01, N = 14) it can be concluded that Temp. < Tcr., i.e. Temp. is in the zone of significance and changes of values of HR and Csev. RTA. are not accidental and hypothesis **H1** is proved.

6. Explanation of the results

Studies [6, 13] show that sharp improvement of level of traffic safety has become possible in the Russian regions because of increase in attention of the state to questions of people's life quality in recent years. The analysis of the process of transformation of living conditions of the people in the safety performance of traffic dedicated to the article [14]. Article [15] is devoted to assessment of spatial features of road and transport accident rate in regions of Russia.

7. Conclusion

On the example of subjects of the Volga Federal District we draw a conclusion that process of road safety improvement is quite sustainable in cities of Russian Federation and its quality can be assessed as positive.

References

[1] State Inspection for Road Traffic Safety of the Ministry of Internal Affairs of Russia Official website Indicators of security status of traffic Available at: https://www gibdd ru/stat/
[2] Donnelly R A 2007 (jr) Statistics Moscow: Astrel-AST
[3] Yeliseyeva I I and Yuzbashev M M 2004 General theory of statistics Moscow: Finance and statistic
[4] Morozov E A A 2014 step-by-step algorithm of actions when using methods of mathematical statistics in psychology and pedagogical researches *Scientific dialogue* **3(27)** 29–45
[5] Orlov A I 2004 Application-oriented statistics Moscow: Examinatio
[6] Petrov A I 2015 Peculiarities of the formation of the motor transport accident rate in space and time Tyumen: TSOGU

[7] Petrov A I 2013 City Transport External environment Stability of public transport in the conditions of adverse influence of the external environment Tyumen: TSOGU
[8] Sidorenko E V 2002 Methods of mathematical processing in psychology: practical guidance St Petersburg: Speech
[9] G N Uilkinkson Steady statistical methods of assessment of data / Under the editorship of RL Loner, 1984 Moscow: Mechanical engineering
[10] Wilcoxon F 1945 Individual comparisons by ranking methods *Biometrics Bulletin* **1(6)** 80–3
[11] Siegel S 1956 Non-parametric statistics for the behavioral sciences New York: McGraw-Hill
[12] Berestneva O G, Marukhina O V and Shevelev G E 2012 Application of mathematical statistics Tomsk: TPU
[13] Petrov A I 2018 Road Traffic Accident Rate as an Indicator of the Quality of Life Economic and Social Changes-Facts Trends Forecast **3(45)** 154-72
[14] Kolesov V and Petrov A 2017 Cybernetic modeling in tasks of traffic safety management *Transportation Research Procedia* **20** 305-10
[15] Petrov A and Petrova D 2018 Assessment of Spatial Unevenness of Road Accidents Severity as Instrument of Preventive Protection from Emergency Situations in Road Complex IOP Conference Series: *Materials Science and Engineering* **142** 012116

Imitation and analytical approaches to assessment of condition and modeling of city transport system nodes

E A Mamaev, N A Kovaleva, A I Khashev* and O V Mulenko

Rostov State Transport University, Rostov-on-Don, Russia

*E-mail: mikasya@yandex.ru

Abstract. The issues of the city transport system and transport nodes assessment concerning the possibilities of express assessment by analytic methods and accurate forecast of its development by transport (imitation) models are discussed. Localization of the assessments for the transport services consumers and transport companies enables to implement the functions of the city transport regulation by the municipality. The social and economic factors of the city transport development for increasing the transport quality and public access are considered.

1. Introduction

The organization of rational city transport system is based on the use of the analytical and imitational approaches allowing to assess and forecast basic parameters of the transport operation, consider the interest of the transport process members as well as assess the influence of the transport system infrastructure development level on the implementation of social and economic purposes of the municipality and city population. Recently, in the Russian Federation Traffic Management Complex Schemes (TMCS) are developed for settlements, urban agglomerations, and regions based on the transport flows modeling. For the assessment of the taken measures results, the different levels models shall be developed, including express assessment models for the transport system operation which enables to quickly determine the efficiency and performance as well as to calculate basic traffic parameters and provide fast assessment and correction of the changes results.

The issues of the city transport processes modeling are addressed at many works, including the discussion of the particular aspects of the city transport system stability improvement by intellectualization of public transport and provision of its priority at the road (Makarova et al., 2016), use of alternative means of transport for the public transportation (Gorbachev, 2017), increase of the traffic safety (Batishcheva et al., 2017, Safiullin et al., 2018), improvement of the throughput functionality due to crossroads optimization (Chubukov et al., 2017), and issues of an ecological burden caused by transport flows (Yakimov, 2017, Yakimov et al., 2018).

The issue of the absence of any accurate assessment criteria for the transport system management (Sharov et al., 2014) shall be solved at the state level by classifying the cities and regions by the number of population, economic links development level (Konovalova et al., 2017), the transport flows laid across the settlement considering modern requirements to delivery chains organization (Mikhaylyk et al., 2019).

Use of the imitation modeling is predetermined by the necessity to select rational traffic schemes and change of the road net configuration (Makarova et al., 2014, Zulfadly et al., 2014).

2. Methods

The development of the assessment system and measures for the development of the city transport shall take into consideration the trends and values which significantly affect at the strategic management level such as:

– grow of the automobilization level (both individual and commercial cargo transport), transit cargo traffic, transport traffic intensiveness and necessity of the car parking places;
– increase of street and road network load;
– decrease of the road safety level;
– provision of the transport availability to the population;
– decrease of the average travel time and etc.

Significant infrastructure projects are specific in their methodological approaches to the consistency analysis and efficiency assessment.

Firstly, the transport which covers the demand for «now and here» transportation services requires adequate forecasting of the future market. Errors in the forecasts to greater or smaller limit influence the final price of products and thus reduce their compatibility. In addition, large transport companies and industry have no methods of direct influence on the transportation services market.

Secondly, the transport systems are widely-distributed and their elements are significant under the condition of the balanced power and resources distribution for effective transportation organization. The narrow places in the transport network can influence on global throughput capacity and barrier its development.

Thirdly, transport system development issues delegated to state (regional, municipal) management level include financial and business elements presented by large transport companies, the «holders» of the public transport infrastructure. In addition, the single transport infrastructure (single transport network) is one of dominants of stable development of interindustry an interregional connections.

These and other factors, connected with the methods of decision taking determine the peculiarities of strategic decisions taking and assessment of their efficiency at the different stages of implementation, see table 1.

The algorithm of the strategic decision taking concerning city transport includes large transport companies and holdings forming basic cargo flows, federal and regional administrations with their financial, economic and social responsibility zones as well as the system of territory planning and town-planning policy determining the structure and scheme of passenger traffic flow.

Table 1. First stages of infrastructure projects implementation for the city transport.

Stage of decision taking	Basic tasks	Stage implementation participants	Used methods and tools
1. Design basis memorandum	Determination of the project and pre-analysis of the change of technical and operational values of the transport system	Initiative group, organization.	Foresight-methods, expert analysis, SWOT-analysis and etc.
2. Technical assignment for project validation	Technical evaluation of the project feasibility and cost effectiveness indicator	Customers-initiators of the project and consulting organizations	Technical requirements to the project basing on GOSTs, OSTs, SNiPs, TU and other normative acts for designing.
3. Calculations of project validation	Qualitative assessment of the technical properties and feasibility of the project	Design and scientific and consulting institutions	Analytic and statistical methods, imitational modeling
4. Investments justification	Assessment and justification	Consulting group representing the interests of the project agents and market	Economic and finance analysis, projects management methods
5. Economic assessment of investments efficiency	Assessment of economic efficiency of investments	Companies and organizations participating in the project	Methods of investments efficiency assessment (industry and interindustry)
6. Development of investments project business plan	Presentation of the project with its attractiveness assessment for all agents (business, capital, state structures)	Consulting group or organization which develop business-plan on contractual basis	Methods of investments project business-plan development

Main feature of the first stage of taking the decision concerning the city transport system development is associated with an uncertainty of future condition and the necessity to achieve adequate forecast. Methodological apparatus of this stage includes expert and cause-consequence analysis, statistical methods, foresight-methods with a following common feature: assessment basing on benchmarking and comparative analysis of trends, schematically implemented in the diagrams (Ishikawa, Gantt, Pareto diagram), matrixes (SWOT, BCG, Shell/DPM. Ansoff, McKincey, Arthur D. Little), graphs (tasks pyramids, strategic plans) and etc.

Nowadays, the growth of city population and development of residential construction continuously lead to the expansion of cities borders. This is the reason for a large number of «narrow places» – conflicting routes, insufficiency of the throughput capacity of the neighboring roads, great number of pedestrian and motor roads crossings, traffic speed limits and stops at the traffic lights, increase of the negative impact on the environment. Due to the grow of the railway traffic intensity and speed and absence of grade-separated crossings the costs of motor transport and accidents risk also increase (Mamaev E.A. et al., 2014, Mamaev E.A. et al., 2018).

Basic loss associated with the complications of street and road network shall be determined considering the use of the analytical imitational mathematic models.

Analytical models are the dependences and equations allowing to take decision on the parameters characteristics of the studied model for the determination of the rational operation of the studied facility (for maximum and average loads).

The main difference between the imitational and the analytical models is the description of the system behavior by a set of the elements behavior algorithms (probable implemented in different environments, but linked together) which describes the system processes.

The use of the imitational models is considered reasonable for the assessment of the condition and determination of vehicles movement speed within a railway crossing.

Analytical method of assessment of time loss by the vehicles at a single level crossing shall be presented as following. Average value of time loss by commercial vehicles V at «narrow places» zone consists of two elements:

$$V = (V_{speed} + V_{delay}) \cdot 265, \qquad (1)$$

where V_{speed} is time loss by vehicles due to traffic speed reduction, hours per day;

V_{delay} is time loss by vehicles due to the delay caused by waiting an opportunity for passing, hours per day;

265 – number of working days.

During the pass through the «narrow places», the transport traffic speed reduces resulting in the delay of the whole traffic flow.

$$T_{\text{скор}} = N_{\text{AT}} \cdot \left(\frac{S}{V_0} - \frac{S}{V_1} \right), \qquad (2)$$

where N_{AT} – vehicle traffic intensity at the zone under consideration, units per day;

S – average length of the «narrow place», including slowing-down and acceleration zones, km;

V_0 – vehicles speed in the «narrow place», km/h;

V_1 – normative vehicle speed, km/h.

Time loss by vehicles due to the delay caused by waiting an opportunity for passing is directly related to the traffic intensity, speed, road crossing type, traffic participants. Average delay of vehicles:

$$T_{\text{зад}} = N_{\text{перес}} \cdot \left(\frac{t_{\text{ст}}^2 \cdot N_{\text{AT}}}{24 \cdot 120} + \frac{t_{\text{ст}}}{2} \right), \qquad (3)$$

where N_{perec} – average number of stops due to flows crossings and stop-overs;

T_{CT} – average stop time while waiting for the traffic pass, hours.

Combination of the imitational and analytical methods is the basics of the assessment methods for the transport infrastructure projects and methods used in TMCS, based on the use of the results of imitation experiment in analytical calculations of the city transportation support and specific regions and nodes.

3. Main part

The abovementioned efficiency assessment model for the traffic organization on the example of the railway crossing can be implemented with the use of imitational transport modeling (Chislov et al., 2016). One of the most popular tools of imitational transport modeling is PTV Vissim, which enables to simplify the required researches by avoiding the necessity to intrude into road traffic.

Let's consider the application of the abovementioned approach to the multilevel railway and motor road junction at the street and road network in Rostov-on-Don at one of the crossings, Figure 1. To increase the road safety and vehicles time loss at this road section, the possibility of the construction of the multilevel railway and motor road junction according to the scheme on Figure1 is considered.

The imitation model of this Rostov-on-Don transport network is created.

During the rush-hour, the traffic intensity at this crossroad is 1500–1700 vehicles/hour, $N_{\text{AT}} = 7000$ units/day., $S = 0.2$ *km*.

IOP Conf. Series: Materials Science and Engineering **786** (2020) 012086 doi:10.1088/1757-899X/786/1/012086

Based on the statistical surveys and implementation of imitational model of the transport network node the following values were determined: $V_0 = 11.92$ km/h and $V_0 = 40.21$ km/h, Figure 1.

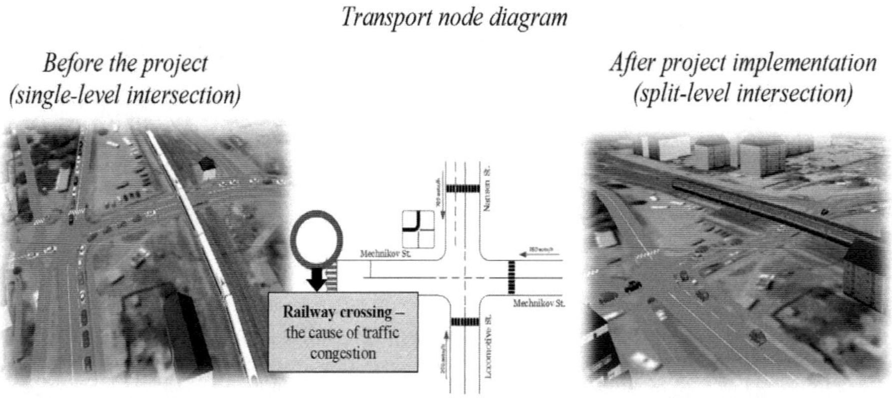

Figure 1. PTV Vissim model of multilevel railway and motor road junction

The traffic characteristics in the railway crossing at different crossing types are analysed.

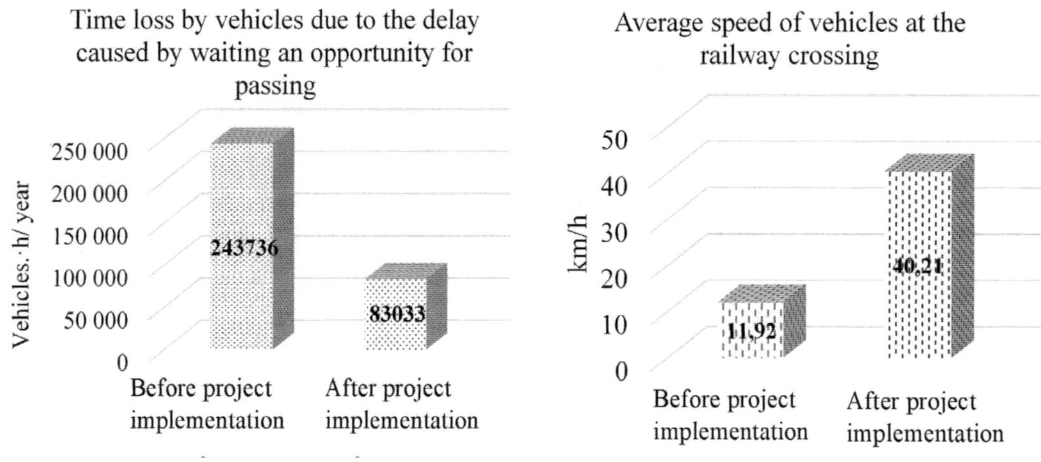

Figure 2. Analysis of immitation modelling results.

Figure 2 shows the average traffic speed (km/h) in railroad crossing before and after the project implementation.

The values are calculated at t_{ct}=4 min, N_{perec}=60.

Using the vehicles speed obtained by imitational modelling the time loss by the vehicles at the railway crossing with single and multi level crossing is calculated.

As seen from Figure 2, the loss before the project implementation are 160 thou. vehicles h/year.

4. Conclusions

Consequently, the combination of the analytical and imitational methods during the analysis of city transport networks is an efficient and adequate tool for strategic decisions taking.

The changes of city transport network shall be made for the consecutive development of the throughput capacity and optimal scheme of traffic organization as well as for the city and agglomeration development.

References

[1] Makarova I, A Pashkevichb and K Shubenkova 2017 Ensuring Sustainability of Public Transport System through Rational Management, In: Proceedings of the 16th International Scientific Conference Reliability and Statistics in Transportation and Communication, RelStat 2016, October 2016, Riga Procedia Engineering, 137-146

[2] Gorbachev R 2017 Architecting Automated Control System of Above Ground Transport System of H-BAHN Type, In: Proceedings of the IVth International Conference on Engineering and Telecommunication (EnT) 124-6

[3] Batishcheva O and A Ganichev 2017 Creative approaches to traffic safety organization issues *Transportation Research Procedia* **20** 36-40

[4] Safiullin, R, M Kerimov, A Afanasyev and A Marusin 2018 A model for justification of the number of traffic enforcement facilities in the region *Transportation Research Procedia* **36** 493-9

[5] Yakimov M 2017 Optimal models used to provide urban transport systems efficiency and safety *Transportation Research Procedia* **20** 702-8

[6] Yakimov M and Y Trofimenko 2018 Developing an urban public passenger transport route network with account for natural resource limitations *Transportation Research Procedia* **36** 801–9

[7] Chubukov A, V Kapitanov, O Monina, V Silyanov and U Brannolte 2017 Calculation of traffic capacity of signaled intersections *Transportation Research Procedia* **20** 125-31

[8] Sharov M and A Mikhailov 2014 Urban Transport System Reliability Indicators In: Proceedings of the 12th International Conference "Organization and Traffic Safety Management in Large Cities", SPbOTSIC-2016, St Petersburg Transportation research Procedia 591-95

[9] Konovalova T and L Zarovnaya 2017 The assessment model for economic efficiency of traffic safety improvements *Transportation Research Procedia* **20** 311-5

[10] Mikhaylyk M, V Baginova and E Mamaev 2019 Empirical markers in the concept of digital logistics of multichannel supply chains Topical Problems of Architecture, Civil Engineering and Environmental Economics (TPACEE 2018) E3S *Web Conf* **91**, 2019 URL: https://doi org/10 1051/e3sconf/20199108056

[11] Makarova, I, Belyaev E, Mavrin V and Suleimanov I 2014 City transport system improvement through the use of simulation modeling system *International Journal of Applied Engineering Research*, **22(9)** 15649–55

[12] Zulfadly A, Syahriah B and Mariana M 2014 Improving the quality of public transportation system: application of simulation model for passenger movement Procedia – *Social and Behavioral Sciences* **153** 542-52

[13] Mamaev E A 2018 Object model of city transport system in optimal management *Inzhenerhiy vestnik Dona* **2(49)** – Access mode: http://www ivdon ru/ru/magazine/archive/n2y2018/4899

[14] Mamaev E A 2014 On assessment of economy loss caused by inefficient traffic organization in transport network *Vestnik RGUPS* **2** 64-9

[15] Chislov O, E Mamaev, A Guda, V Zubkov and V Finochenko 2016 Algorythmic and software support of efficient design of railway transport technological systems *International Journal of Applied Engineering Resear*ch **11(23)** 11428-38

AUTHOR INDEX